PERFORMANCE ANALYSIS AND MODELING OF DIGITAL TRANSMISSION SYSTEMS

Coded Modulation Systems
John B. Anderson and Arne Svensson

Communication System Design Using DSP Algorithms: With Laboratory Experiments for the TMS320C6701 and TMS320C6711
Steven A. Tretter

A First Course in Information Theory
Raymond W. Yeung

Interference Avoidance Methods for Wireless Systems
Dimitrie C. Popescu and Christopher Rose

Nonuniform Sampling: Theory and Practice
Edited by Farokh Marvasti

Performance Analysis and Modeling of Digital Transmission Systems
William Turin

Stochastic Image Processing
Chee Sun Won and Robert M. Gray

PERFORMANCE ANALYSIS AND MODELING OF DIGITAL TRANSMISSION SYSTEMS

William Turin
AT&T Labs—Research
Florham Park, New Jersey

Kluwer Academic / Plenum Publishers
NEW YORK, BOSTON, DORDRECHT, LONDON, MOSCOW

Library of Congress Cataloging-in-Publication Data

Turin, William.
Performance analysis and modeling of digital transmission systems/William Turin.
p. cm. — (Information technology: transmission, processing, and storage)
Rev. ed. of: Digital transmission systems, c1999.
Includes bibliographical references and index.
ISBN 0-306-48191-X
1. Digital communications—Evaluation. 2. Digital communications—Mathematical models. I. Turin, William. Digital transmission systems. II. Title. III. Series.

TK5103.7.T8675 2004
621.382—dc22

2004041417

The files are organized into the following directories: Chapter 1 through Chapter 7, and Appendix 5 through Appendix 7. For details on using the CD-ROM, please refer to the book Introduction.

For further information, please write to the publisher at abozicevic@wkap.com to obtain the author's current contact information.

ISBN: 0-306-48191-X

233 Spring Street, New York, New York 10013

http://www.kluweronline.com

10 9 8 7 6 5 4 3 2 1

A C.I.P. record for this book is available from the Library of Congress

Printed in the United States of America

To my wife Galina

CONTENTS

PREFACE

This book is an expanded third edition of the book Performance Analysis of Digital Transmission Systems, originally published in 1990. Second edition of the book titled Digital Transmission Systems: Performance Analysis and Modeling was published in 1998. The book is intended for those who design communication systems and networks. A computer network designer is interested in selecting communication channels, error protection schemes, and link control protocols. To do this efficiently, one needs a mathematical model that accurately predicts system behavior.

Two basic problems arise in mathematical modeling: the problem of identifying a system and the problem of applying a model to the system analysis. System identification consists of selecting a class of mathematical objects to describe fundamental properties of the system behavior. We use a specific class of *hidden Markov models* (HMMs) to model communication systems. This model was introduced by C. E. Shannon more than 50 years ago as a Noisy Discrete Channel with a finite number of states. The model is described by a finite number of matrices whose elements are estimated on the basis of experimental data. We develop several methods of model identification and show their relationship to other methods of data analysis, such as spectral methods, autoregressive moving average (ARMA) approximations, and rational transfer function approximations.

The model proved to be general enough to approximate large varieties of stochastic processes, which explains its popularity in various applications, such as automatic control, queueing systems, telecommunications, economics, psychology, finance, and so on. The model popularity is also illustrated by the variety of its name in different fields (hidden Markov model, stochastic sequential machine, correlated Markov processes, probabilistic automaton, partially observed Markov process, state space process, linear stochastic process, generalized Gilbert model to name a few). The HMM based technology is now a preeminent method of speech recognition, handwriting recognition, image recognition, document analysis, and artificial intelligence. Since their invention, the HMMs were successfully applied to modeling error sources in communication channels. Recent interest in HMMs can be explained by their popularity in modeling wireless communication channels.

Because of the HMM's popularity there are many publications devoted to its theory. The majority of publications are are dealing with estimating the model parameters ("model training") and the model-based recognition. In applications to telecommunications, however, the other problem is very important: the model-based evaluation of the system performance parameters. In this book, we develop methods for calculating the system performance parameters on the basis of HMMs. We derive closed-form expressions for various system performance characteristics (e.g., efficiency of forward error detection and error correction schemes and communication protocols). The model-related calculations are based on methods of matrix and operator probability theory which are developed in this book. The proposed methods are illustrated by numerous examples. The results analytical methods are compared with experimental data and computer simulations.

This book is intended for specialists in telecommunications and deals with practical aspects of model construction and application. It is based on the author's more than 40 years experience in processing experimental data and analyzing real-world communication systems. Most of the methods considered have been used in practice.

Several theoretical problems, such as various types of presentation and simplification of the model, which are directly related to creating the model from experimental data, are discussed in this book. Because some mathematical background is essential to understanding the proposed methods, to make the book self-contained I present a brief mathematical introduction related to matrix algebra, Markov chains, statistical inference, and error correcting codes.

We consider also the continuous time discrete state HMM as a limiting case of its discrete counterpart. This allowed us to show the avenue for extending the results developed for the discrete time HMM to the continuous time models instead of rederiving them essentially without significant changes.

In this third edition we added Chapter 7 and Appendix 7 which present the continuous state HMM and Markov processes, respectively. In contrast with the traditional approach, we present this theory as a generalization of the theory of the discrete state HMM developed in the previous chapters. We derive the major facts and algorithms of the theory on the basis of computing certain probability density functions while the traditional approach is more "statistical" and is based on the estimation theory. The probabilistic approach allowed us not only to simplify the presentation, but also to explain the relationship between the discrete state and continuous state HMM theories. Appendix 7 presents an abbreviated theory of the continuous state Markov processes Chapter 3 has been completely rewritten to emphasize new developments in HMM theory and its applications to wireless communications. Special attention was paid to the theory and applications of the expectation-maximization (EM) algorithm and its generalizations. We present new directions in application of the EM algorithm not only to estimating parameters of probability distributions but also to curve fitting and function maximization.

In the previous editions, we considered only the worst case scenario for the two-way communications when transmitters and receivers have always a message to transmit. This concept completely ignores the message arrival process. However, in many systems (especially in wireless systems) the message arrival process plays a significant role. In Chapter 5 of this edition we describe methods for analyzing performance of the two-way systems assuming that the message arrivals are described by the batch Markov arrival process (BMAP).

We added new material to Appendix 5 related to differentiating with respect to a matrix which is often used (sometimes incorrectly) in technical literature. We not only give a clear definition of the derivative, but also explain how to use it correctly by considering examples and applications. We have also added some material related to quadratic forms and block matrix inversion used in the new Chapter 7. In Chapter 4, new material dealing with the increasingly popular maximum a posteriori (MAP) symbol and sequence decoding in channels with memory has been added.

The MATLAB® code of the majority of algorithms presented in this book and Simulink® HMM models are given in the accompanying CDROM. Since most of the equations of this book are written in the matrix form, the MATLAB programming language and environment represent the perfect choice for their implementation and, most importantly, the *matrix way of thinking* about the problems and and their solutions which is, unfortunately, not very popular in technical publications. For example, the vast majority of publications on the system performance analysis in channels with memory are using the two-state HMM while the matrix notations allow us to find the solutions for the general channel HMM.

I am grateful to AT&T Laboratories for supporting my work on the material used in this book. I am also immensely grateful to late Dan Leed, without whose support and encouragement this book never would have appeared. I would like to thank my colleagues Saeed Ghassemzadeh, Carol Martin, Rittwik Jana, and Jack Winters with whom I worked on the material presented in Sec. 7.7 of Chapter 7. I am thankful to Courtney Esposito of The MathWorks Book Program for the support in preparing the M-files presented in the accompanying CDROM. Special thank goes to Professor Jack Wolf for his kind support and encouragement and Ana Bozicevich for the assistance in preparing this edition. Helpful comments on the previous editions of the book made by Sam Boodaghians, Charlie Canada, Hector Corrales, Adam Irgon, Dan Jeske, Robert van Nobelen, Mahmood Noorchashm, Jay Padgett, Vasant Prabhu, Nambirajan Seshadri, and Ward Whitt are gratefully acknowledged.

William Turin

NOTATION

J	the set of all integers
N	the set of all positive integers
$\sum\limits_{i_1,\ldots,i_n}$	sum over all values of $i_1,\ldots,i_n$
$\sum\limits_{i\in F}$	sum over all values of i that belong to a set F
$\mathbf{A} = [a_{ij}]_{m,n}$	a rectangular matrix with elements a_{ij} consisting of m rows and n columns
$\mathbf{A}+\mathbf{B}$	matrix sum: $\mathbf{A}+\mathbf{B} = [a_{ij}]_{m,n} + [b_{ij}]_{m,n} = [a_{ij}+b_{ij}]_{m,n}$
$\mathbf{AB}$	matrix product: $\mathbf{AB} = [a_{ij}]_{m,n} \cdot [b_{ij}]_{n,p} = [\sum\limits_{k=1}^{n} a_{ik}b_{kj}]_{m,p}$
$\lambda\mathbf{A}$	$= \lambda[a_{ij}]_{m,n} = [\lambda a_{ij}]_{m,n}$ a product of a number and a matrix
$\mathbf{A}' = [a'_{ij}]_{n,m}$	the transpose of the matrix $\mathbf{A} = [a_{ij}]_{m,n}$: $a_{ij} = a'_{ji}$
$diag\{a_i\}_n$	$= \begin{bmatrix} a_1 & 0 & \cdots & 0 \\ 0 & a_2 & \cdots & 0 \\ \cdots & \cdots & \cdots & \cdots \\ 0 & 0 & \cdots & a_n \end{bmatrix}$ a diagonal matrix
$\mathbf{I} = diag\{1\}$	a unit matrix
det **A**	matrix **A** determinant
$\mathbf{A}^{-1}$	matrix **A** inverse ($\mathbf{AA}^{-1} = \mathbf{A}^{-1}\mathbf{A} = \mathbf{I}$)
$\mathbf{A}^{\dagger}$	matrix **A** pseudoinverse (see Appendix 5, Sec. 5.1.10)

$\mathbf{a} = row\{a_i\}_n$ $= [a_1, a_2, \ldots, a_n]$ a row matrix

$\mathbf{a} = col\{a_i\}_n$ $= [a_1, a_2, \ldots, a_n]'$ a column matrix

$\mathbf{1}$ $= col\{1\}$ a column matrix consisting of ones

$\mathbf{A} = [\mathbf{A}_{ij}]_{m,n}$ a block matrix consisting of m rows and n columns of sub-matrices (matrix blocks) $\mathbf{A}_{ij}$

$block\ row\{\mathbf{A}_i\}_n = [\mathbf{A}_i]_{1,n} = [\mathbf{A}_1, \mathbf{A}_2, \ldots, \mathbf{A}_n]$ a block row matrix

$block\ col\{\mathbf{A}_i\}_n = [\mathbf{A}_i]_{n,1} = [\mathbf{A}_1, \mathbf{A}_2, \ldots, \mathbf{A}_n]'$ a block column matrix

$$block\ diag\{\mathbf{A}_i\}_n = \begin{bmatrix} \mathbf{A}_1 & 0 & \cdots & 0 \\ 0 & \mathbf{A}_2 & \cdots & 0 \\ \cdots & \cdots & \cdots & \cdots \\ 0 & 0 & \cdots & \mathbf{A}_n \end{bmatrix} \quad \text{a block diagonal matrix}$$

$rank\ \mathbf{A}$ matrix $\mathbf{A}$ rank: the highest order of nonzero determinants which are composed of the elements of rows and columns of the matrix

$\mathbf{A} \otimes \mathbf{B}$ $= [a_{ij}]_{m,n} \otimes \mathbf{B} = [a_{ij}\mathbf{B}]_{m,n}$ Kronecker product of the matrices

$\mathbf{A} > \mathbf{B}$ denotes that all the elements of the matrix $\mathbf{A}$ are greater than the corresponding elements of the matrix $\mathbf{B}$: $a_{ij} > b_{ij}$ for all i and j

$\mathbf{A} \leq \mathbf{B}$ denotes that all the elements of the matrix $\mathbf{A}$ are not greater than the corresponding elements of the matrix $\mathbf{B}$: $a_{ij} \leq b_{ij}$ for all i and j

$\mathbf{e}^m$ denotes a sequence of m identical symbols $\mathbf{e}$: $\mathbf{e}^m = \mathbf{e}, \mathbf{e}, \ldots, \mathbf{e}$

x_i^j denotes a sequence $x_i, x_{i+1}, \ldots, x_j$

i. i. d. independent identically distributed (variables)

$Pr\{A\}$ probability of A

$Pr\{A \mid B\}$ conditional probability of A

$\mathbf{Pr}\{A\}$ matrix probability of A

$\mathbf{x} \in D$ $\mathbf{x}$ belongs to D

$\mathbf{x} \notin D$ $\mathbf{x}$ does not belong to D

$\bar{x} = \mathbf{E}\{x\}$ an expected value (mean) of x

$i \bmod p$ i modulo p: a remainder of division of i by p

$erf(x)$ error function $\dfrac{2}{\sqrt{\pi}} \displaystyle\int_0^x e^{-x^2} dx$

$erfc(x)$ co-error function $\dfrac{2}{\sqrt{\pi}} \displaystyle\int_x^\infty e^{-x^2} dx = 1 - erf(x)$

$\mathsf{N}(\boldsymbol{x},\boldsymbol{\Sigma})$	$(2\pi)^{-m/2}\lvert\boldsymbol{\Sigma}\rvert^{-1/2}\exp(-0.5\boldsymbol{x}'\boldsymbol{\Sigma}^{-1}\boldsymbol{x})$
$\mathsf{G}(\boldsymbol{x},\mathbf{P})$	$\exp(-0.5\mathbf{x}'\mathbf{C}\mathbf{x})$
$REM(m,n)$	remainder of m/n
ACK	positive acknowledgement
APP	a posteriori probability
ARQ	Automatic Repeat-reQuest (system)
AWGN	additive white Gaussian noise
BCH	Bose-Chaudhuri-Hocquenghem (code)
BSC	binary symmetric channel
BWA	Baum-Welch algorithm
CSOC	convolutional self-orthogonal code
DFT	discrete Fourier transform
EFS	error-free second
EM	expectation-maximization
FBA	forward-backward algorithm
FEC	forward error correction
FFT	fast Fourier transform
GF(p)	size p Galois field
GBN	Go-Back-N (protocol)
HGMM	hidden Gauss-Markov model
HMM	hidden Markov model
LS	least squares (method)
MAP	maximum a posteriori
MDS	maximum-distance-separable (code)
ML	maximum likelihood
NAK	negative acknowledgement
PDF	probability density function
PH	phase-type (distribution)
RS	Reed-Solomon (code)
UWB	Ultra-Wide Bandwidth (channel)

INTRODUCTION

During the past three decades, digital transmission networks have become ubiquitous and complex. Many analytical tools and software packages have been created to analyze and evaluate the behavior of these systems. However, these tools often use simplistic models of the system elements' behavior such as exponential distribution of the intervals between failures, Poisson distribution of packets arriving at a node, and binomial distribution of errors in a block. These models are selected to simplify the analysis, but they often differ from the actual networks sufficiently to make the computed results of doubtful value.

Computer simulation, on the other hand, can usually model real systems quite closely. But performance studies take a significant amount of computer time, so that in many cases it is not feasible to carry out complete system optimization when test cases are run in ordinary computer centers. Therefore, it is useful to develop mathematical tools and models that perform analytical calculations of the performance of real networks without incurring inaccuracies due to unrealistic assumptions. This book develops general methods for modeling digital transmission systems and applies these methods to practical performance analysis.

In Chapter 1 we discuss the rationale underlying the choice of models for error source in digital transmission channels. In order to be able to analyze and predict a system's performance, it is necessary to have some simplified theoretical description of the system's operation—*the system model.* Usually, a communication system has several channels that may be statistically dependent. Therefore, the channel model should be able to describe the transmission distortion in a set of dependent channels. For example, in a full duplex system one station sends messages over the direct channel and receives acknowledgements over the return channel. If these channels are statistically dependent, we need a model that describes errors in the set of two channels. Another example represents the multiple access channels where we need to describe the dependence between the channels (especially when evaluating the performance of the space-time codes [32]). In general, we have a set of T transmitters that send symbols $\mathbf{a} = (a^{(1)}, \ldots, a^{(T)})$ over the communication channel and a set of R recevers that receive symbols $\mathbf{b} = (b^{(1)}, \ldots, b^{(R)})$. We call this set of channels a *vector channel.*

Channel impairments distort the transmitted symbols, and therefore the received sequence contains errors whose values are defined by some measure of difference between transmitted and received symbols. This distortion is usually characterized by the conditional probabilities $Pr(\mathbf{b}_i^{i+m} \mid \mathbf{a}_i^{i+m})$ of receiving the sequence of symbols $\mathbf{b}_i^{i+m} = (\mathbf{b}_i, \mathbf{b}_{i+1}, \ldots, \mathbf{b}_{i+m})$ when the sequence $\mathbf{a}_i^{i+m} = (\mathbf{a}_i, \mathbf{a}_{i+1}, \ldots, \mathbf{a}_{i+m})$ was sent. Here i is an integer ($i \in \mathbf{J}$) and m is a nonnegative integer ($m \in \mathbf{N}$). The channel may also be described by the conditional probabilities of errors $Pr(\mathbf{e}_i^m \mid \mathbf{a}_i^m)$. If, in addition, the error sequence does not depend on the transmitted sequence: $Pr(\mathbf{e}_i^{i+m} \mid \mathbf{a}_i^{i+m}) = Pr(\mathbf{e}_i^{i+m})$ for all $i \in \mathbf{J}$ and $m \in \mathbf{N}$, then the channel is said to be *symmetric*. The channel is called *stationary* if these probabilities do not depend on i.

We assume in the sequel that the channel is stationary and symmetric. The complete description of this channel can be achieved by the multidimensional distributions of errors $Pr(\mathbf{e}_0^m)$. This description assumes that the channels are ideally synchronized, that is, the number of received symbols is equal to the number of transmitted symbols. Small changes in this description account for the synchronization loss. Even though the multidimensional distributions give us a complete error source description, their practical applications are quite limited because of their computational complexity. This complexity is one of the reasons for using an approximate channel error sequence description based on an *error source model.*

The most important requirement for an error source model is its agreement with the experimental data. Other important goals are model simplicity and convenience of use. Yet another feature is also desirable: system model generality. The model should be a particular case of a broad family of models with wide applicability. The menu should be rich enough to enable an investigator to choose a model that will agree with experimental data, so that model prediction will be credible.

The family of models that we use for the system description is the family of *hidden Markov models* (HMMs). [22] These models were introduced by Shannon in 1948 [21] as the finite state channels (FSC). (However, they were studied in mathematical literature since 1912. [27]) They are also known as *stochastic sequential machines* (SSMs) [23] *partially observable Markov processes,* and *probabilistic automata* (PA). [24] We do not provide a historical perspective related to HMMs. (The reader interested in the history of HMM developments and many references on the subject is recommended to read the paper [28] by Y. Ephraim and N. Merhav.) As is shown in Chapter 1, this family is general enough to approximate any finite-order multidimensional distribution. The model is capable of describing both the correlation between errors in a channel and the correlation between errors in different channels. Many error source models described in technical journals belong to this family. [1—20]

As a specific example we consider physical causes of error bursts in satellite and fading radio channels and explain why these channels can be modeled by HMMs. In addition, we analyze the relationship between the HMM and some stochastic processes frequently used for dynamic system modeling, speech processing, image processing and queueing theory (autoregressive processes, renewal processes, Markov functions, semi-Markov processes, and matrix processes).

Because of the model's generality, the number of parameters that define an HMM may be large; we consider methods for minimizing this number. The problem of simplifying the model description is closely related to the mathematical problem of lumping the states of a class of semi-Markov processes. We derive an algorithm to minimize the number of model parameters.

The next step is to develop mathematical tools to use with the HMM for system performance evaluation. Chapter 2 describes the needed tools, which are based on the notion of matrix probability. These tools include extension of basic probability theorems to matrix probabilities, closed-form expressions of matrix generating functions, recursive equations, signal-flow-graph methods, and factorization methods. These methods are then applied to the calculation of some basic performance measures (bit error probability, error-free-second probability, average error-burst length, error-free run distribution, number of errors in a block distribution, etc.).

Statistical methods of estimating model parameters are developed in Chapter 3. We introduce a deterministic version of the *expectation maximization* (EM) algorithm and develop numerous versions of the algorithm for approximating a stochastic process with an HMM.

In this edition, we consider new applications of the algorithm and, in particular, its application to finding a maximum of a function of several variables. After presenting a general algorithm we develop its version for solving the problem of fitting a matrix-geometric distribution to a general one-dimensional distribution. We give a detailed description of the algorithm which should help to understand a more general case of fitting various HMMs to other stochastic processes by the *Baum-Welch algorithm (BWA)*.

The EM algorithm and its special case, the BWA, are iterative and their convergence depends on the initial values of the model parameters. We introduce many heuristical methods for finding good initial parameter values. The memory requirement of the BWA grows proportionally to the experimental data size which is usually huge in telecommunications applications. We develop an unidirectional and parallel BWA whose memory requirements are independent on the size of the data.

We also consider the problem of the matrix process parameter estimation and show that it can be reduced to the simpler problem of estimating the coefficients of a finite difference equation. This problem is closely related to some modern techniques of linear system identification, to order and parameter estimation of autoregressive moving average (ARMA) models, and to Padé approximation theory. The proposed methods are of greater interest to statistical inference relating to semi-Markov processes. This alternative approach allows us to find a model with the reduced parameter space.

In Chapter 4, we give a short introduction to the theory of error correction coding. Special attention is paid to the modern statistical techniques which include *maximum a posteriori* (MAP) decoding and, in particular, the so-called turbo decoding. [25] We show that these techniques are closely related to the hidden Markov modeling and fitting. The presentation is simple and does not assume previous knowledge of the coding theory.

After the brief introduction, we use the matrix probability theory developed in Chapter 2 to calculate performance parameters of error-detection and error-correction codes if errors are modeled by an HMM. There were many attempts to solve this problem approximately. We give exact formulae for calculating basic performance parameters that characterize the effectiveness of error-correction codes. We determine matrix probabilities of errors and of erasures in a received message for several commonly used block codes. The error matrix probability is expressed as a function of the parity check matrix through the matrix generating function for the general linear block code. The Fourier transform is applied to the probability evaluation. For the channel with independent errors these equations express the probability of incorrect decoding through the code-weight generator polynomial. We also analyze the performance of some convolutional codes, in particular convolutional self-orthogonal codes popular in satellite communications.

In Chapter 5, we compare several communication protocols and automatic repeat request (ARQ) schemes such as stop-and-wait and go-back-N. We determine message error probability, message erasure probability, message delay distribution, the throughput performance of the protocol, message duplication probability, message loss probability, synchronization loss probability, and the average time of synchronization recovery. We describe a communication protocol as an HMM and then analyze its performance using matrix probability methods developed in previous chapters. Message arrival process plays an important role in many communication systems. We develop methods for analyzing the performance of such systems assuming that the message arrivals are described by the batch Markov arrival process (BMAP).

In Chapter 6, we consider continuous-time HMMs which are very popular in queueing theory. We develop their theory as a limiting case of the theory of their discrete counterparts. There are several reasons for this approach. We find the conditions for the existence of the limit. These conditions are very important in practice, because in many applications discrete processes are approximated with the continuous processes (for example, packet arrivals in digital communication networks are often modeled by continuous-time processes). The other reason is that we can use the results developed in the book for the discrete-time HMMs and transform them for the continuous-time models.

In Chapter 7 we consider HMMs whose state-space is not necessarily discrete. We call them a continuous state HMM for short, but they are actually the generalization of the discrete HMMs and may contain a mixture of the discrete and continuous state HMMs. Our approach to studying continuous state HMMs is more "probabilistic" than the traditional "statistical" approach in that that we analyze various probability density functions (PDFs) associated with these processes rather than estimating their parameters and states on the basis of experimental data.

The traditional approach is concentrated on the special case — *hidden Gauss-Markov models* (HGMM) — while we study first the properties of the general HMMs and then specialize them to the HGMMs. This approach allows us to consider continuous state HMMs as a generalization of their discrete state counterparts, but also to simplify the derivation of their basic properties. For example, we demonstrate that the forward algorithm represents the Kalman filter for the HGMM. We derive the Rauch-Tung-Striebel (RTS) smoother much simpler than it is done traditionally and we also generalize it to non-Gaussian HMMs.

We introduce *operator probabilities* as a generalization of the matrix probabilities and show that, formally, the methods developed for the matrix probabilities in the previous chapters can be extended to operator probabilities. This gives the reader a clear view on the forward-backward algorithm as applied to the continuous state HMM. In general, the algorithms applied to the continuous state HMMs are more complex than their discrete state versions, because integration of PDFs is more complex than multiplication of matrices. We emphasize that, for the HGMMs, the integration can be replaced by the corresponding operations with the mean vectors and covariance matrices which is the major reason for their popularity. For example the forward-backward algorithm can be performed using the Gaussian elimination from the mean-covariance matrix of the entries corresponding to the observations. We also illustrate this approach by developing the BWA for the HGMM. Finally, we show the application of the HGMM to modeling wireless channel in the time domain and the ultra-wide band channel in the frequency domain.

To avoid impeding the development of concepts in the text, complex derivations are relegated to the appendices. They include the proofs of the conditions of Markov and semi-Markov lumpability, methods of matrix series summation, asymptotic approximation of matrix probabilities, block graph applications, derivation of the union bound on Viterbi algorithm performance, and derivation of the Reed-Solomon code weight generator.

To make the book self contained, we presented several tutorials. Appendix 3 contains an introduction to statistical methods and statistical inference related to the Markov chains. Appendix 5 provides a review of matrix algebra relevant to the material discussed in this book. It contains many topics that are not covered in the standard courses such as Kronecker products, general spectral decomposition, functions of matrices, partial fraction expansion for the matrices, matrix inversion lemma, matrix series and polynomials, functions of matrices, quadratic forms, and derivatives with respect to a matrix. There is a lot of controversy in applying the derivatives with respect to a matrix in technical literature because they represent a combination of partial derivatives and the matrix structure of the result. Therefore, it is not clear how to handle the case of dependent variables. For example, if the matrix is symmetrical, how to handle the derivatives with respect to the equal elements. Some authors define these derivatives assuming that the elements are independent and replacing them with the equal values in the result; some other authors derive formulae for two cases corresponding to nonsymmetric and symmetric matrices separately. But then the question arizes what to do in the infinite number of other cases of dependent variables. In Appendix 5, we give a solution to this problem and present a clear method for deriving properties of the derivatives. Appendix 6 provides a brief tutorial on Markov chains. Appendix 7 presents a brief theory of continuous state Markov processes and Gaussian distribution. The main reason for this appendix is to "lighten" Chapter 7 which covers the continuous state HMM. This separation allows the reader understand that, for example, the AR process is a Markov process while ARMA process is an HMM. We present the results as a generalization of the corresponding results for the Markov chains developed in Appendix 6 by replacing matrices with the corresponding operators. This approach allows the reader to understand why the Riccati equation is used to find the stationary PDF and why there is a fast algorithm for finding the power of an operator.

The tutorials cover all the the material that is needed to understand the book. However, they are organized in such a way that allows a reader to learn complex concepts starting with simple ones. Most of the chapters have problem exercises. For additional exercises, we recommend Ref. 26.

The book contains numerous algorithms for solving various problems related to statistical analysis of communication systems. Most of the algorithms are presented by the pseudo-code in the text and implemented in MATLAB® M-files containing the MATLAB code on the included CDROM. We use the term MATLAB function (or program) for the code written by The MathWorks, Inc., which is a part of the MATLAB a high-performance language for technical computing. The name MATLAB stands for *matrix laboratory*. It was originally written to provide easy access to state-of-the-art software for matrix computations. Since all the problems in this book are presented using matrix algebra, MATLAB programming language is a natural choice for solving them; the other reasons are its broad availability and popularity. We assume the reader's familiarity with MATLAB programming language and do not provide a tutorial because excellent books and guides are available. [26] However, the M-files for the examples of the book and its appendices can serve as a tutorial (or a demo) in both matrix algebra and MATLAB.

The Communications Toolbox [27] (reqirinig Signal Processing Toolbox) provides many functions for simulating realistic systems and coding applications, (including the Galois field computations) considered in Chapter 4. The Simulink® is a software package for modeling and simulating dynamical systems [28] considered in Chapters 2,4,and 7. The Statistics Toolbox provides many tools for solving the problems considered in Chapter 3. In this book, we call the MATLAB functions all the functions from MATLAB and the previously listed toolboxes.

MATLAB is an interactive language which means that its commands are programs which need to be loaded and executed. Because of it, the "for-loops" that are common in non-interactive languages are undesirable. Instead we should try (if possible) to use the MATLAB powerful index facility. In other words, think of any problem using vector-matrix and index-set terms. (This approach is often called an object-oriented programming.)

Simulink together with the Communications Blockset and DSP Blockset allow us to perform simulation of complex systems by using block models of their parts available in these blocksets. On the accompanying CDROM, we present block models for simulating HMM channels (with discrete and Gaussian state observation distributions) and illustrate their applications using Simulink and the blocksets.

To improve performance of the MATLAB code, it can be translated into the C-code and compiled to create a non-interactive program. To speed up the execution, MATLAB provides flexible and powerful utilities for linking with programs written in C or Fortran enabling a user to customize and optimize critical parts of the program.

References

1. E. N. Gilbert, "Capacity of a burst-noise channel," *Bell Syst. Tech. J.*, **39**, 1253-1266, Sept. (1960).
2. A. A. Alexander, R. M. Gryb, and D. W. Nast, "Capabilities of the telephone network for data transmission," *Bell Syst. Tech. J.*, **39**(3), 471-476, May (1963).
3. A. G. Usol'tsev and W. Ya. Turin, "Investigation of the error distribution laws in tone telegraph channels with frequency modulation," *Telecommunications,* no. 7, (1963).
4. O. V. Popov and W. Ya. Turin, "The probability distribution law of different numbers of errors in a combination," *Telecommunications and Radioengineering,* No. 5, (1967).
5. L. N. Kanal and A. R. K. Sastry, "Models for channels with memory and their applications to error control," *Proc. of the IEEE,* **66**(7), 724-744, July (1978).
6. E. O. Elliott, "Estimates of error rates for codes on burst-noise channels," *Bell Syst. Tech. J.,* **42**, 1977-1997, Sept. (1963).
7. R. H. McCullough, "The binary regenerative channel," *Bell Syst. Tech. J.,* **47**, 1713-1735, Oct. (1968).
8. E. L. Cohen and S. Berkovits, "Exponential distributions in Markov chain models for communication channels," *Inform. Control,* **13**, 134-139, (1968).
9. B. D. Fritchman, "A binary channel characterization using partitioned Markov chains," *IEEE Trans. Inform. Theory,* **IT-13**, 221-227, Apr. (1967).
10. S. Tsai, "Markov characterization of the HF channel," *IEEE Trans. Commun. Technol.,* **COM-17**, pp. 24-32, Feb. (1969).
11. P. J. Trafton, H. A. Blank, and N. F. McAllister, "Data transmission network computer-to-computer study," in *Proc. ACM/IEEE 2nd Symp.* on *Problems in the Optimization of Data Communication Systems* (Palo Alto, CA), 183-191, Oct. (1971).
12. R. T. Chien, A. H. Haddad, B. Goldberg, and E. Moyes, "An analytic error model for real channels," in *Conf. Rec., IEEE Int. Conf. Communications (ICC)*, 15-7-15-12, June (1972).
13. E. L. Bloch, O. V. Popov and W. Ya. Turin, "Error number distribution for the stationary symmetric channel with memory,"*The Second International Symposium on Information Theory,* Acad. Kiado, Budapest, (1972).

14. V. Ya. Turin, "Probability distribution laws for the number of errors in several combinations," *Telecommunications and Radioengineering*, **27**(12), (1973).

15. V. Ya. Turin, "Estimation of the probability of the delay exceeding in decision feedback systems,"*Problems of Information Transmission,* **10**(3), (1974).

16. W. Turin, and M. M. Sondhi, "Modeling error sources in digital channels," *IEEE Journ. Sel. Areas in Commun.,* **11**(3), 340-347, (1993).

17. F. Swarts and H.C. Ferreira, "Markov characterization of channels with soft decision outputs," *IEEE Trans. Commun.,* **41**, 678-682, May (1993).

18. M. Sajadieh, F.R. Kschischang, and A. Leon-Garcia, "A block memory model for correlated Rayleigh fading channels," *Proc. IEEE Int. Conf. Commun.,* 282-286, June (1996).

19. S. Sivaprakasam and K.S. Shanmugan, "An equivalent Markov model for burst errors in digital channels." *IEEE Trans. Commun.,* **43**, 1347-1355, April (1995).

20. M. Zorzi and R.R. Rao, "On the statistics of block errors in bursty channels," *IEEE Trans. Commun.,* **45**, 660-667, Jun. (1997).

21. C. E. Shannon, "A Mathematical Theory of Communication," *Bell Syst. Tech. J.*, **27**, 379-423, 623-656, July, October, (1948). (Also in *Claude Elwood Shannon Collected Papers*, New Jersey, A. Sloan and A. D. Wyner Eds, IEEE Press, Piscataway, New Jersey, 1993.)

22. Rabiner, L. and Juang, B.-H. *Fundamentals of Speech Recognition*, (Prentice Hall, Englewood Cliffs, New Jersey, 1993).

23. J. W. Carlyle, "Reduced forms for stochastic sequential machines," *J. Math. Anal. and Appl.*, **7**(2), 167-175, (1963).

24. A. Paz, *Introduction to Probabilistic Automata*, (Academic Press, New York, 1971).

25. C. Berrou and A. Glavieux, "Near optimum error correcting coding and decoding: turbo-codes," *IEEE Trans. Commun.,* **COM-44**(10), 1261-1271, Oct. (1996).

26. M. F. Neuts, *Algorithmic Probability: A Collection of Problems.* (Chapman & Hall, London, 1995).

27. A. A. Markov, "On trials associated into a chain by unobserved events," Russian Acad. Sci. News (IAN), **6**(6), 551-572, (1912).

28. Y. Ephraim and N. Merhav, "Hidden Markov processes ," *IEEE Trans. Inf. Theory* **48**(6), 1518 -1569, 2002.

29. *MATLAB The Language of Technical Computing: Using MATLAB* (The MathWorks, Inc., South Natick, MA, 1984-2003).

30. *Communications Toolbox For Use with MATLAB: User's Guide* (The MathWorks, Inc., South Natick, MA, 1984-2003).

31. *Simulink Model-Based and System-Based Design: Using Simulink* (The MathWorks, Inc., South Natick, MA, 1990-2003).

32. V. Tarokh, N. Seshadri, and A. R. Calderbank, "Space-time codes for high data rate wireless communication: performance analysis and code construction," *IEEE Trans. Inf. Theory,* **44**(2), 744-765, (1998).

1

ERROR SOURCE MODELS

1.1 DESCRIPTION OF ERROR SOURCES BY HIDDEN MARKOV MODELS

1.1.1 Finite State Channel

A discrete finite state channel (FSC) model was introduced by Shannon.[38] According to this model, the channel can be in any state from the state space $S=\{\alpha_1,\alpha_2,\ldots,\alpha_u\}$. Usually, states are enumerated by integers: $S=\{1,2,\ldots,u\}$. If the channel is in state $s_{t-1}\in S$ and the input to the channel is $\mathbf{a}_t\in A$, the channel outputs symbol $\mathbf{b}_t\in B$ and transfers to state $s_t\in S$ with probability $Pr(\mathbf{b}_t,s_t \mid \mathbf{a}_t,s_{t-1})$. The probability of the final state s_t and receiving a sequence $\mathbf{b}_1^t = (\mathbf{b}_1,\mathbf{b}_2,\ldots,\mathbf{b}_t)$ conditional on the initial state s_0 and transmitted sequence $\mathbf{a}_1^t = (\mathbf{a}_1,\mathbf{a}_2,\ldots,\mathbf{a}_t)$ has the form

$$Pr(\mathbf{b}_1^t,s_t \mid \mathbf{a}_1^t,s_0) = \sum_{s_1^{t-1}} \prod_{i=1}^{t} Pr(\mathbf{b}_i,s_i \mid \mathbf{a}_i s_{i-1}) \tag{1.1.1}$$

It is convenient to write this equation in the matrix form. Define $\mathbf{P}(\mathbf{b}_t \mid \mathbf{a}_t)$ a conditional *matrix probability* (MP) of the output symbol $\mathbf{b}_t$ given the input symbol $\mathbf{a}_t$ as a matrix whose ij-th element is $Pr(\mathbf{b}_t,s_t=j \mid \mathbf{a}_t,s_{t-1}=i)$:

$$\mathbf{P}(\mathbf{b}_t \mid \mathbf{a}_t) = \left[Pr(\mathbf{b}_t,s_t \mid \mathbf{a}_t,s_{t-1})\right]_{u,u}$$

It is easy to see that Eq. (1.1.1) can be written in the following matrix form (see Appendix 5 and Appendix 6.1)

$$\mathbf{P}(\mathbf{b}_1^t \mid \mathbf{a}_1^t) = \mathbf{P}(\mathbf{b}_1 \mid \mathbf{a}_1)\mathbf{P}(\mathbf{b}_2 \mid \mathbf{a}_2)\cdots\mathbf{P}(\mathbf{b}_t \mid \mathbf{a}_t) = \prod_{i=1}^{t}\mathbf{P}(\mathbf{b}_i \mid \mathbf{a}_i) \tag{1.1.2}$$

where $\mathbf{P}(\mathbf{b}_1^t \mid \mathbf{a}_1^t)$ is the matrix whose ij-th element is $Pr(\mathbf{b}_1^t,s_t=j \mid \mathbf{a}_1^t,s_0=i)$. We call this matrix a conditional MP of the output sequence $\mathbf{b}_1^t$ given the input sequence $\mathbf{a}_1^t$.

The conditional probability $Pr(\mathbf{b}_1^t \mid \mathbf{a}_1^t,s_0)$ of receiving $\mathbf{b}_1^t$ conditional on the channel initial state s_0 and the input sequence $\mathbf{a}_1^t$ is the sum over s_t of $Pr(\mathbf{b}_1^t,s_t \mid \mathbf{a}_1^t,s_0)$. The summation is equivalent to multiplying $\mathbf{P}(\mathbf{b}_1^t \mid \mathbf{a}_1^t)$ by the matrix column $\mathbf{1}$ all of whose elements are ones:

$$\mathbf{p}(\mathbf{b}_1^t \mid \mathbf{a}_1^t) = \mathbf{P}(\mathbf{b}_1^t \mid \mathbf{a}_1^t)\mathbf{1}, \quad \mathbf{1} = [1\ 1\ \cdots\ 1]'$$

Here, $\mathbf{p}(\mathbf{b}_1^t \mid \mathbf{a}_1^t)$ is the matrix column whose elements are the probabilities $Pr(\mathbf{b}_1^t \mid \mathbf{a}_1^t, s_0)$:

$$\mathbf{p}(\mathbf{b}_1^t \mid \mathbf{a}_1^t) = \left[Pr(\mathbf{b}_1^t \mid \mathbf{a}_1^t, s_0=1)\ \ Pr(\mathbf{b}_1^t \mid \mathbf{a}_1^t, s_0=2)\ \ \cdots\ Pr(\mathbf{b}_1^t \mid \mathbf{a}_1^t, s_0=u)\right]'$$

The conditional probability $Pr(\mathbf{b}_1^t \mid \mathbf{a}_1^t)$ of receiving $\mathbf{b}_1^t$ conditional on the channel input sequence $\mathbf{a}_1^t$ is given by

$$Pr(\mathbf{b}_1^t \mid \mathbf{a}_1^t) = \sum_{s_0=1}^{u} Pr(s_0) Pr(\mathbf{b}_1^t \mid \mathbf{a}_1^t, s_0)$$

which can be written as

$$Pr(\mathbf{b}_1^t \mid \mathbf{a}_1^t) = \mathbf{p}_0 \mathbf{p}(\mathbf{b}_1^t \mid \mathbf{a}_1^t) = \mathbf{p}_0 \mathbf{P}(\mathbf{b}_1^t \mid \mathbf{a}_1^t)\mathbf{1} = \mathbf{p}_0 \prod_{i=1}^{t} \mathbf{P}(\mathbf{b}_i \mid \mathbf{a}_i)\mathbf{1} \qquad (1.1.3)$$

where $\mathbf{p}_0 = [Pr(s_0=1)\ Pr(s_0=2)\ \cdots\ Pr(s_0=u)]$ is the matrix row of the channel initial state probabilities.

In summary, an FSC can be described as $\{S, A, B, \mathbf{p}_0, \mathbf{P}(\mathbf{b} \mid \mathbf{a})\}$ where S is the channel state space, A is its input alphabet, B is its output alphabet, $\mathbf{p}_0$ is the initial state probability vector, and $\mathbf{P}(\mathbf{b} \mid \mathbf{a})$ is the conditional MP of the output symbol $\mathbf{b} \in B$ given the input symbol $\mathbf{a} \in A$. If the channel is not discrete, the matrix probabilities are replaced with the corresponding *matrix probability density functions* (MPDF).

The FSC is often called a channel with memory because of the output symbols dependecies. Equation (1.1.3) plays a central role in evaluating performance of digital communication systems in channels with memory. It allows us to develop the matrix form equations for various probability distributions which are valid for any FSC and to use powerful methods of the matrix theory for analyzing the system performance. The equation is similar to the corresponding equation for the channels without memory. We can expect to explore this similarity to develop simple equations and algorithms for computing the performance parameters. However, since matrix products are noncommutative, sometimes it is not simple to accomplish. The major part of this book is devoted to developing equations and algorithms for calculating the system performance parameters on the basis of Eq. (1.1.3)

For each sequence of the channel input symbols, the sequence of the channel states is a (nonhomogeneous) Markov chain † with the state transition probability matrix

$$\mathbf{P}_{\mathbf{a}_t} = \sum_{\mathbf{b}} \mathbf{P}(\mathbf{b} \mid \mathbf{a}_t)$$

Note, that the channel states are "hidden" (not directly observable).

The FSC models are popular in many applications. In system theory they are called *stochastic sequential machines* [7] or *probabilistic automata.* [6,29]

In many applications, an output of one channel serves as an input to the other. It is not difficult to prove that if both channels are FSCs, then the combined channel is also an FSC.

† Readers needing an introduction to Markov chains are advised to read Appendix 6 of this book.

Indeed, let $\{S_1, A, B, \mathbf{p}_{10}, \mathbf{P}_1(\mathbf{b} \mid \mathbf{a})\}$ and $\{S_2, B, C, \mathbf{p}_{20}, \mathbf{P}_2(\mathbf{b} \mid \mathbf{a})\}$ be two FSCs. Define a new set of states as a combination of the first and second channel states (its state space is a Cartesian product of state spaces $S_1 S_2$). If the first channel is in state $s_t^{(1)}$ and the second channel is in state $s_t^{(2)}$, then the combined channel is in state $s_t = (s_t^{(1)}, s_t^{(2)})$. Given an input symbol $\mathbf{a} \in A$ to the first channel, the combined channel transfers to state $s_{t+1} = (s_{t+1}^{(1)}, s_{t+1}^{(2)})$, first channel outputs $\mathbf{b} \in B$, and second channel outputs $\mathbf{c} \in C$ with probability

$$Pr(\mathbf{b},\mathbf{c},s_{t+1} \mid \mathbf{a},s_t) = Pr(\mathbf{b},s_{t+1}^{(1)} \mid \mathbf{a},s_t^{(1)})Pr(\mathbf{c},s_{t+1}^{(2)} \mid \mathbf{b},s_t^{(2)})$$

independently of the previous states and input and output symbols. Thus, the combined channel with the input alphabet A and the output alphabet BC is an FSC. According to the previous equation, the conditional MP of its output symbol $(\mathbf{b},\mathbf{c})$, given an input symbol $\mathbf{a}$, can be written in the following matrix form

$$\mathbf{P}(\mathbf{b},\mathbf{c} \mid \mathbf{a}) = \mathbf{P}_1(\mathbf{b} \mid \mathbf{a}) \otimes \mathbf{P}_2(\mathbf{c} \mid \mathbf{b}) \tag{1.1.4}$$

where $\otimes$ denotes the Kronecker product of matrices (see Appendix 5). The combined channel state initial probability vector can be also presented by the Kronecker product of the components initial probability vectors: $\mathbf{p}_0 = \mathbf{p}_{10} \otimes \mathbf{p}_{20}$. Thus,

$$\{S_1 S_2, A, BC, \mathbf{p}_{10} \otimes \mathbf{p}_{20}, \mathbf{P}_1(\mathbf{b} \mid \mathbf{a}) \otimes \mathbf{P}_2(\mathbf{c} \mid \mathbf{b})\}$$

describes the combined FSC.

If we are interested only in the second channel output, the output conditional MP is

$$\mathbf{P}(\mathbf{c} \mid \mathbf{a}) = \sum_{\mathbf{b}} \mathbf{P}(\mathbf{b},\mathbf{c} \mid \mathbf{a}) = \sum_{\mathbf{b}} \mathbf{P}_1(\mathbf{b} \mid \mathbf{a}) \otimes \mathbf{P}_2(\mathbf{c} \mid \mathbf{b})$$

Thus, this FSC can be described as $\{S_1 S_2, A, C, \mathbf{p}_{10} \otimes \mathbf{p}_{20}, \sum_{\mathbf{b}} \mathbf{P}_1(\mathbf{b} \mid \mathbf{a}) \otimes \mathbf{P}_2(\mathbf{c} \mid \mathbf{b})\}$. If the first channel output is not discrete, the sum of the MPs in the previous equation is replaced with an integral of the corresponding MPDFs.

We can prove similarly that a network of FSCs is also an FSC. It would be nice if the channel input could be described by a similar model. This model is introduced in the next section.

1.1.2 Hidden Markov Models

Let us now address the problem of the channel input (source) modeling. We assume that the source can be modeled by an *autonomous* FSC (whose output does not depend on its input). Thus, the source is described by its alphabet A, state space S_s, matrix row $\mathbf{p}_s$ of the state initial probabilities, and the MP $\mathbf{P}_s(\mathbf{a})$ whose elements are the probabilities $Pr(\mathbf{a},j \mid i)$ of transferring from the source state i to state j and producing the symbol $\mathbf{a}$. This source model is sometimes called a *finite state generator* (FSG) while an FSC is called a *transducer.* The probability of producing a sequence $\mathbf{a}_1^t$ by the FSG is given by Eq. (1.1.3), which takes the form

$$Pr(\mathbf{a}_1^t) = \mathbf{p}_s \prod_{i=1}^{t} \mathbf{P}_s(\mathbf{a}_i)\mathbf{1}$$

Elements of the symbol MP $\mathbf{P}_s(\mathbf{a})$ can be written as

$$Pr(\mathbf{a},j \mid i) = Pr(j \mid i)Pr(\mathbf{a} \mid i,j)$$

which allows us to give an alternative description of the model: The model states constitute a Markov chain with the transition probability matrix

$$\mathbf{P} = \left[Pr(j \mid i)\right] = \sum_{\mathbf{a}} \mathbf{P}_s(\mathbf{a})$$

The model states are not directly observable (hidden), the (observed) symbol sequence is a probabilistic function of the hidden states and the symbol conditional probability, given a sequence of states, depends only on the previous and the current states (or, in other words, on the state transition $i \rightarrow j$). If this probability depends only on the current state $Pr(\mathbf{a}_t \mid i,j) = Pr(\mathbf{a}_t \mid j)$, the model is called a discrete *hidden Markov model* (HMM). [32]

In general, an HMM is described as $\{S, A, \mathbf{p}_0, \mathbf{P}, \mathbf{F}(\mathbf{a})\}$, where S is the set of its (hidden) states, A is the set of its observations, $\mathbf{p}_0$ is the state initial probability vector, $\mathbf{P}$ is the state transitional probability matrix, and $\mathbf{F}(\mathbf{a}) = diag\{Pr(\mathbf{a} \mid j)\}$ is a diagonal matrix of the state observation probabilities (or probability density functions).

For the HMM, the observation transitional probability has the form $Pr(\mathbf{a},j \mid i) = Pr(j \mid i)Pr(\mathbf{a} \mid j)$, which means that the MP $\mathbf{P}(\mathbf{a})$ can be written as

$$\mathbf{P}(\mathbf{a}) = \mathbf{PF}(\mathbf{a}) \tag{1.1.5}$$

Thus, we can say that an HMM is a special case of the FSG whose observation MP has the form of Eq. (1.1.5). However, if we denote a pair of subsequent states as a new state (i,j), we can treat a general FSG as an HMM. For this reason, any autonomous FSC will be called an HMM in the sequel.

The state initial probability vector $\mathbf{p}_0$ is one of the model-independent parameters. In the majority of applications, it is assumed that the underlying Markov chain is stationary (see Appendix 6). If the matrix $\mathbf{P}$ is *regular* (see Appendix 6), the state initial probability vector can be found as a solution of the system

$$\boldsymbol{\pi}\mathbf{P} = \boldsymbol{\pi} \qquad \boldsymbol{\pi}\mathbf{1} = 1 \tag{1.1.6}$$

Therefore, a stationary HMM can be defined by the collection of MPs $\mathbf{P}(\mathbf{a})$ only, since the initial state probability vector can be determined as the (unique) solution of Eq. (1.1.6).

If the FSC input is modeled by an HMM $\{S_s, A, \mathbf{p}_s, \mathbf{P}_s(\mathbf{a})\}$ and the channel is an FSC $\{S, A, B, \mathbf{p}_0, \mathbf{P}(\mathbf{b} \mid \mathbf{a})\}$, then the input-output combination $(\mathbf{a},\mathbf{b})$ is an autonomous FSC or, in other words, an HMM. We call this HMM an *input-output* HMM (IOHMM). Its observation $(\mathbf{a},\mathbf{b})$ matrix probability $\mathbf{P}_c(\mathbf{a},\mathbf{b})$, according to Eq. (1.1.4) is given by

$$\mathbf{P}_c(\mathbf{a},\mathbf{b}) = \mathbf{P}_s(\mathbf{a}) \otimes \mathbf{P}(\mathbf{b} \mid \mathbf{a})$$

The combined state initial probability vector is $\mathbf{p}_c = \mathbf{p}_s \otimes \mathbf{p}_0$.

1.1.3 Discrete-Time Finite State Systems

A discrete-time system (DTS) is defined by the so-called state-space equations:

$$\begin{aligned} s_t &= S(s_{t-1}, \mathbf{a}_t, \xi_t) \\ \mathbf{b}_t &= O(s_t, \mathbf{a}_t, \eta_t) \end{aligned} \tag{1.1.7}$$

where $\mathbf{a}_t$ is the system input (control) symbol, s_t is the system state, $\mathbf{b}_t$ is the system output (reaction), and ξ_t and η_t are independent identically distributed (i.i.d.) variables (noises). Since we are considering only finite state spaces, we assume that $S(s_{t-1},\mathbf{a}_t,\xi_t)$ has a finite range. (In practice, if the state space is not finite, the finite space is created by the proper state quantization.)

It is not difficult to see that the DTS is a special case of the FSC. Indeed, it follows from the first equation of (1.1.7) that for any fixed sequence of input symbols $\mathbf{a}_1^t$ the sequence of states is a Markov chain with the state transition probability matrix

$$\mathbf{P}_t = \left[Pr(s_t=j \mid s_{t-1}=i,\mathbf{a}_t)\right] = \left[Pr(j = S(i,\mathbf{a}_t,\xi))\right]$$

According to second equation of (1.1.7), the output $\mathbf{b}_t$ are conditionally independent, given the state and input sequences, and have the cumulative probability distribution function

$$F_i(\mathbf{b}) = Pr(O(i,\mathbf{a}_t,\eta)<\mathbf{b})$$

Thus, Eq. (1.1.7) define an FSC whose conditional MP (or MPDF) has the form

$$\mathbf{P}(\mathbf{b}_t \mid \mathbf{a}_t) = \left[Pr(s_t = S(s_{t-1},\mathbf{a}_t,\xi_t))Pr(\mathbf{b}_t = O(s_t,\mathbf{a}_t,\eta_t))\right] \quad (1.1.8)$$

Example 1.1.1. Consider the following DTS

$$s_t = sign(s_{t-1}+a_t+\xi_t), \qquad b_t = s_t+\eta_t$$

where $sign(x) = 1$ for $x \geq 0$ and $sign(x) = -1$ for $x < 0$. It follows from the first equation that the system state space is binary ($S=\{-1,1\}$). Suppose that the input alphabet is binary ($A=\{-1,1\}$), ξ_t is a binary variable with probabilities $Pr(\xi_t=-1) = p$, $Pr(\xi_t=1) = q$; η_t has a zero-mean normalized Gaussian distribution with the PDF $(2\pi)^{-0.5}\exp(-0.5\eta^2)$.

Let us describe this system as an FSC. Suppose that $a_t = 1$, then the transition probability $Pr(s_t=1 \mid s_{t-1}) = Pr(s_{t-1}+1+\xi_t\geq 0)$. Thus, $Pr(s_t=1 \mid s_{t-1}=-1) = Pr(\xi_t\geq 0) = q$ and $Pr(s_t=1 \mid s_{t-1}=1) = Pr(2+\xi_t\geq 0) = 1$. The state transition probability matrix has the form

$$\mathbf{P}_{a_t} = \begin{bmatrix} Pr(s_t=-1 \mid s_{t-1}=-1) & Pr(s_t=1 \mid s_{t-1}=-1) \\ Pr(s_t=-1 \mid s_{t-1}= 1) & Pr(s_t=1 \mid s_{t-1}= 1) \end{bmatrix} = \begin{bmatrix} p & q \\ 0 & 1 \end{bmatrix} \quad \text{for } a_t = 1$$

Since $b_t = s_t+\eta_t$, the state output PDF has the form $(2\pi)^{-0.5}\exp[-0.5(b_t-s_t)^2]$. Therefore, the conditional MPDF can be written as

$$\mathbf{p}(b_t \mid a_t=1) = \frac{1}{\sqrt{2\pi}}\begin{bmatrix} p\exp(-0.5(b_t+1)^2) & q\exp(-0.5(b_t-1)^2) \\ 0 & \exp(-0.5(b_t-1)^2) \end{bmatrix}$$

If $a_t=-1$, we obtain similarly

$$\mathbf{P}(b_t \mid a_t=-1) = \frac{1}{\sqrt{2\pi}}\begin{bmatrix} \exp(-0.5(b_t+1)^2) & 0 \\ p\exp(-0.5(b_t+1)^2) & q\exp(-0.5(b_t-1)^2) \end{bmatrix}$$

As we have seen before, if the FSC input is modeled by an HMM, then the channel output is also an HMM. Therefore, if a finite state DTS input is an HMM, then its output is an HMM.

Thus, we have shown that any DTS can be presented as an FSC. The converse statement is also valid. Indeed, suppose that we have an FSC with the output conditional MP (or MPDF) $\mathbf{P}(\mathbf{b}_t \mid \mathbf{a}_t)$. Denote $\mathbf{F}(\mathbf{b} \mid \mathbf{a}_t) = \mathbf{P}(\mathbf{b}_t<\mathbf{b} \mid \mathbf{a}_t)$ the corresponding cumulative MP distribution. Its elements are $Pr(\mathbf{b}_t<\mathbf{b},s_t \mid \mathbf{a}_t,s_{t-1})$, which can be written as

$$Pr(\mathbf{b}_t<\mathbf{b},s_t \mid \mathbf{a}_t,s_{t-1}) = Pr(s_t \mid \mathbf{a}_t,s_{t-1})Pr(\mathbf{b}_t<\mathbf{b} \mid \mathbf{a}_t,s_{t-1},s_t)$$

Given an input sequence, let us now try to simulate the state sequence according to $Pr(s_t \mid \mathbf{a}_t,s_{t-1})$. It is well known that a random variable x with the cumulative probability distribution function $F(x)$ can be generated from a uniformly distributed over the interval (0,1) variable u as $x = F^{-1}(u)$, where $F^{-1}(u)$ is the inverse function: [34]

$$F^{-1}(u) = inf\{x:F(x)\geq u\}$$

Thus, if we denote

$$G(s;\mathbf{a}_t,s_{t-1}) = \sum_{s_t<s} Pr(s_t \mid \mathbf{a}_t,s_{t-1})$$

the state cumulative probability distribution function, the next state can be generated as

$$s_t = G^{-1}(u_t;\mathbf{a}_t,s_{t-1})$$

where $\{u_t\}$ is a sequence of independent (0,1) uniformly distributed random variables. This equation has the form of the first equation in (1.1.7).

After obtaining s_t, we can similarly generate the system output

$$\mathbf{b}_t = H(\mathbf{u}_t;\mathbf{a}_t,s_{t-1},s_t)$$

according to its conditional probability $Pr(\mathbf{b}_t<\mathbf{b} \mid \mathbf{a}_t,s_{t-1},s_t)$. This equation is slightly different from the second equation in (1.1.7). But if we define a new state as (s_{t-1},s_t), we obtain equations of the form of (1.1.7).

Example 1.1.2. Describe as a DTS the Gilbert-Elliott model [13,19]

$$\mathbf{P} = \begin{bmatrix} p_{00} & p_{10} \\ p_{01} & p_{11} \end{bmatrix} \quad \mathbf{F}(0) = \begin{bmatrix} h_1 & 0 \\ 0 & h_2 \end{bmatrix} \quad \mathbf{F}(1) = \begin{bmatrix} 1-h_1 & 0 \\ 0 & 1-h_2 \end{bmatrix}$$

If the previous state was s_{t-1}, the next state s_t is selected with the probability p_{s_{t-1},s_t}. Therefore, if ξ_t is (0,1) uniformly distributed random variable, then it is easy to see that

$$s_t = 0.5sign(\xi_t - p_{s_{t-1},0}) + 0.5$$

The state output b_t can be obtained similarly:

$$b_t = 0.5sign(\eta_t - h_{s_t}) + 0.5$$

1.1.4 Error Source HMM

In evaluating a communication system performance parameters, it is convenient to consider sequences of errors $\mathbf{e}_1^t$ instead of the sequences of the channel outputs $\mathbf{b}_1^t$. Formally, this means just a variable substitution in all the previous equations because the error sequences are deterministic functions of the channel input and output. Thus, Eq. (1.1.3) can be written as

$$Pr(\mathbf{e}_1^t \mid \mathbf{a}_1^t) = \mathbf{p}_0 \prod_{k=1}^{t} \mathbf{P}(\mathbf{e}_k \mid \mathbf{a}_k)\mathbf{1}$$

If the matrices $\mathbf{P}(\mathbf{e} \mid \mathbf{a}) = \mathbf{P}(\mathbf{e})$ do not depend on $\mathbf{a}$, then the channel is called *symmetric* and formula (1.1.4) can be written as

$$Pr(\mathbf{e}_1^t) = Pr(\mathbf{e}_1^t \mid \mathbf{a}_1^t) = \mathbf{p}_0 \prod_{k=1}^{t} \mathbf{P}(\mathbf{e}_k)\mathbf{1} \tag{1.1.9}$$

which means that error sequences in a symmetric FSC are modeled by an HMM $\{S, E, \mathbf{p}_0, \mathbf{P}(\mathbf{e})\}$, where E is the set of all possible errors. In the following discussion we consider only symmetric channels, and therefore formula (1.1.9) plays the main role. The stationary error source is described by the error set and the error matrix probabilities only because the state initial probabilities are found from Eq. (1.1.6). For the stationary HMM, the symbol error probability has the form

$$Pr(\mathbf{e}) = \boldsymbol{\pi}\mathbf{P}(\mathbf{e})\mathbf{1}$$

Consider now the case in which the FSC consists of h binary channels. Let $\mathbf{e} = (e^{(1)},...,e^{(h)})$ be the channel error vector ($e^{(i)} = 0$ if i-th channel bit was received correctly and $e^{(i)} = 1$ otherwise). The error source state transition is described as a Markov chain with the matrix $\mathbf{P}$. If the source is in state j, then the conditional probability of an error in the i-th channel is $Pr(e^{(i)} = 1 \mid j) = \varepsilon_j^{(i)}$ and $Pr(e^{(i)} = 0 \mid j) = 1 - \varepsilon_j^{(i)}$. A diagonal matrix $\mathbf{F}(\mathbf{e}) = diag\{Pr(\mathbf{e} \mid j)\}$ of state error probabilities can be written as

$$\mathbf{F}(\mathbf{e}) = \prod_{i=1}^{h} (\mathbf{I} - \mathbf{E}_i)^{1-e^{(i)}} \mathbf{E}_i^{e^{(i)}}$$

where $\mathbf{E}_i = diag\{\varepsilon_j^{(i)}\}$ and $\mathbf{I} = diag\{1\}$ is the unit (or identity) matrix. According to Eq. (1.1.5) the error matrix probability has the form

$$\mathbf{P}(\mathbf{e}) = \mathbf{PF}(\mathbf{e}) \tag{1.1.10}$$

For this model, the symbol error probability is given by

$$p(\mathbf{e}) = \boldsymbol{\pi}\mathbf{PF}(\mathbf{e})\mathbf{1} = \boldsymbol{\pi}\mathbf{F}(\mathbf{e})\mathbf{1} = \sum_{j=1}^{u} \pi_j Pr(\mathbf{e} \mid j) \tag{1.1.11}$$

We illustrate this model in the next section.

1.1.5 Gilbert's Model

Gilbert [19] initiated the study of the Markov chain error source models. According to his model, the source of errors has two states: G (for good) and B (for bad or burst), which constitute a Markov chain with the matrix

$$\mathbf{P} = \begin{bmatrix} Q & P \\ p & q \end{bmatrix} = \begin{bmatrix} 0.997 & 0.003 \\ 0.034 & 0.966 \end{bmatrix}$$

The model state transition diagram is shown in Fig. 1.1. In state G the probability of error is equal to zero ($\varepsilon_1^{(1)} = 0$); in state B the probability of error is equal to $\varepsilon_2^{(1)} = 1 - h = 0.16$. Thus,

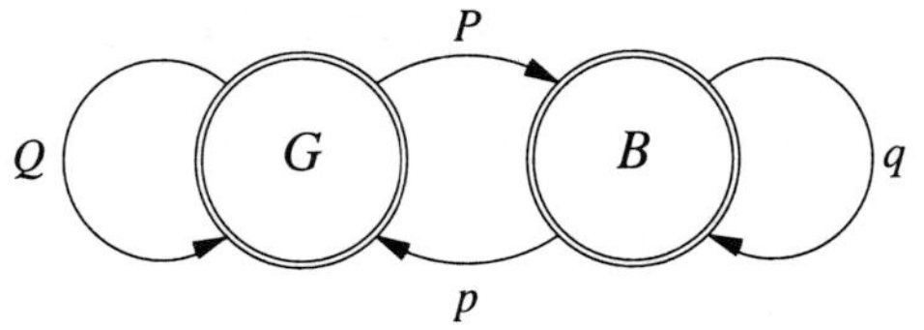

Figure 1.1. Gilbert's model state diagram.

$$\mathbf{F}(0) = \begin{bmatrix} 1 & 0 \\ 0 & h \end{bmatrix} = \begin{bmatrix} 1 & 0 \\ 0 & 0.84 \end{bmatrix} \qquad \mathbf{F}(1) = \begin{bmatrix} 0 & 0 \\ 0 & 1-h \end{bmatrix} = \begin{bmatrix} 0 & 0 \\ 0 & 0.16 \end{bmatrix}$$

According to formula (1.1.10), we obtain

$$\mathbf{P}(0) = \mathbf{PF}(0) = \begin{bmatrix} Q & Ph \\ p & qh \end{bmatrix} = \begin{bmatrix} 0.997 & 0.00252 \\ 0.034 & 0.81144 \end{bmatrix}$$

$$\mathbf{P}(1) = \mathbf{PF}(1) = \begin{bmatrix} 0 & P(1-h) \\ 0 & q(1-h) \end{bmatrix} = \begin{bmatrix} 0 & 0.00048 \\ 0 & 0.15456 \end{bmatrix}$$

To complete the model description, we need to find the state stationary distribution from Eq. (1.1.6), which, in our case, have the form

$$\pi_1 Q + \pi_2 p = \pi_1$$
$$\pi_1 P + \pi_2 q = \pi_2$$
$$\pi_1 + \pi_2 = 1$$

Solving this system we obtain $\pi_1 = p/(P+p) \qquad \pi_2 = P/(P+p)$ so that

$$\boldsymbol{\pi} = \begin{bmatrix} \frac{p}{P+p} & \frac{P}{P+p} \end{bmatrix} = \frac{1}{P+p}[\,p \quad P\,] = [0.91892 \quad 0.08108]$$

Thus, Gilbert's model can be described as $\{S=(0,1),\ A=(0,1),\ \boldsymbol{\pi},\ \mathbf{P}(0),\ \mathbf{P}(1)\}$. We will use this model for illustration throughout the book. The model parameter numerical values are used in the cases in which analytical expressions are too complex and numerical expressions are easier to understand and verify.

The bit-error probability can be found using Eq. (1.1.11):

$$p_{\text{err}} = \boldsymbol{\pi}\mathbf{P}(1)\mathbf{1} = \pi_2(1-h) = \frac{(1-h)P}{P+p} = 0.01297$$

To calculate the probability of the 01 error combination we use formula (1.1.9):

$$Pr(01) = \boldsymbol{\pi}\mathbf{P}(0)\mathbf{P}(1)\mathbf{1}$$

We have

$$\boldsymbol{\pi}\mathbf{P}(0) = \frac{1}{P+p}[\,p \quad P\,]\begin{bmatrix} Q & Ph \\ p & qh \end{bmatrix} = \frac{1}{P+p}[\,p \quad Ph\,]$$

$$\boldsymbol{\pi}\mathbf{P}(0)\mathbf{P}(1) = \frac{1}{P+p}[p \quad Ph]\begin{bmatrix} 0 & P(1-h) \\ 0 & q(1-h) \end{bmatrix} = \frac{1}{P+p}[0 \quad P(p+qh)(1-h)]$$

And, finally,

$$Pr(01) = \boldsymbol{\pi}\mathbf{P}(0)\mathbf{P}(1)\mathbf{1} = \frac{1}{P+p}[0 \quad P(p+qh)(1-h)]\begin{bmatrix} 1 \\ 1 \end{bmatrix} = \frac{P(p+qh)(1-h)}{P+p}$$

1.1.6 Equivalent Models

In practical applications, it is convenient to simplify the HMM description (reduce the number of states, simplify its matrix structure, etc). For this purpose we need to define a class of equivalent models and select a simpler HMM from the class.

We say that models λ and μ are equivalent if the processes of occurrence of errors described by these models are equivalent in distribution. [18] In other words, the models are equivalent if all their corresponding multidimensional probability distributions of errors are identical: $Pr(\mathbf{e}_1,\mathbf{e}_2,...,\mathbf{e}_n \mid \lambda) = Pr(\mathbf{e}_1,\mathbf{e}_2,...,\mathbf{e}_n \mid \mu)$ for all $n = 1,2,...$ and all $\mathbf{e}_k$, $k = 1,2,...,n$. According to this definition, two models with matrices $\mathbf{P}_\lambda(\mathbf{e})$ and $\mathbf{P}_\mu(\mathbf{e})$ are equivalent if

$$\boldsymbol{\pi}_\lambda \mathbf{P}_\lambda(\mathbf{e}_1)\mathbf{P}_\lambda(\mathbf{e}_2)\cdots\mathbf{P}_\lambda(\mathbf{e}_n)\mathbf{1} = \boldsymbol{\pi}_\mu \mathbf{P}_\mu(\mathbf{e}_1)\mathbf{P}_\mu(\mathbf{e}_2)\cdots\mathbf{P}_\mu(\mathbf{e}_n)\mathbf{1} \qquad (1.1.12)$$

for all $n = 1,2,...$ and all $\mathbf{e}_k$, $k = 1,2,...,n$.

It is convenient to use the following sufficient condition of equivalency. [7] If there exists a matrix $\mathbf{T}$ such that

$$\mathbf{T}\,\mathbf{P}_\mu(\mathbf{e}) = \mathbf{P}_\lambda(\mathbf{e})\,\mathbf{T} \qquad \mathbf{T}\,\mathbf{1} = \mathbf{1} \qquad (1.1.13)$$

for all $\mathbf{e}$, then the models with matrices $\mathbf{P}_\lambda(\mathbf{e})$ and $\mathbf{P}_\mu(\mathbf{e})$ are equivalent.

Indeed, let $\boldsymbol{\pi}_\lambda$ be the matrix of the stationary probabilities of the Markov chain with the matrix

$$\mathbf{P}_\lambda = \sum \mathbf{P}_\lambda(\mathbf{e})$$

Then $\boldsymbol{\pi}_\mu = \boldsymbol{\pi}_\lambda \mathbf{T}$ is the stationary vector of the Markov chain with the matrix

$$\mathbf{P}_\mu = \sum \mathbf{P}_\mu(\mathbf{e})$$

as is seen from the following equations:

$$\boldsymbol{\pi}_\mu\,\mathbf{P}_\mu = \boldsymbol{\pi}_\lambda\,\mathbf{T}\,\mathbf{P}_\mu = \boldsymbol{\pi}_\lambda\,\mathbf{P}_\lambda\,\mathbf{T} = \boldsymbol{\pi}_\lambda\,\mathbf{T} = \boldsymbol{\pi}_\mu$$

$$\boldsymbol{\pi}_\mu\,\mathbf{1} = \boldsymbol{\pi}_\lambda\,\mathbf{T}\,\mathbf{1} = \boldsymbol{\pi}_\lambda\,\mathbf{1} = 1$$

Since

$$\boldsymbol{\pi}_\mu\,\mathbf{P}_\mu(\mathbf{e}_1)\,\mathbf{P}_\mu(\mathbf{e}_2)\cdots\mathbf{P}_\mu(\mathbf{e}_n)\,\mathbf{1} = \boldsymbol{\pi}_\lambda\,\mathbf{T}\,\mathbf{P}_\mu(\mathbf{e}_1)\,\mathbf{P}_\mu(\mathbf{e}_2)\cdots\mathbf{P}_\mu(\mathbf{e}_n)\,\mathbf{1} =$$

$$\boldsymbol{\pi}_\lambda\,\mathbf{P}_\lambda(\mathbf{e}_1)\,\mathbf{P}_\lambda(\mathbf{e}_2)\cdots\mathbf{P}_\lambda(\mathbf{e}_n)\,\mathbf{T}\,\mathbf{1} = \boldsymbol{\pi}_\lambda\,\mathbf{P}_\lambda(\mathbf{e}_1)\,\mathbf{P}_\lambda(\mathbf{e}_2)\cdots\mathbf{P}_\lambda(\mathbf{e}_n)\,\mathbf{1}$$

equations (1.1.12) are satisfied and therefore the models are equivalent.

In the particular case when the matrix $\mathbf{T}$ is nonsingular, equation (1.1.13) takes on the form

$$\mathbf{P}_{\mu}(\mathbf{e}) = \mathbf{T}^{-1}\ \mathbf{P}_{\lambda}(\mathbf{e})\ \mathbf{T} \qquad \mathbf{T1} = \mathbf{1}$$

Thus, the matrices $\mathbf{P}_{\mu}(\mathbf{e})$ and $\mathbf{P}_{\lambda}(\mathbf{e})$ are *similar* (see Appendix 5.1.5). They may be reduced to their simplest (normal) form.

If conditions of equivalency are satisfied approximately $[Pr(\mathbf{e}_1^n \mid \lambda) \approx Pr(\mathbf{e}_1^n \mid \mu)]$, we say that the model μ approximates the model λ in distribution. We address the questions of approximating error source models with HMMs in Sec. 1.5.4. In Chapter 3, we develop methods of such approximations.

1.1.7 The HMM Generality

We justify the HMM generality by showing that it permits us to model various types of error dependence in different channels. To demonstrate this, let us describe an HMM in which errors in different channels occur independently. Suppose that the model of the source of errors in the $k-$th channel is described by the matrices $\mathbf{P}_k(e^{(k)}) = [Pr(e^{(k)}, i_k \mid j_k)]$. The sources of errors in different subchannels are statistically independent if

$$Pr(\mathbf{e}, j \mid i) = \prod_{m=1}^{h} Pr(e^{(m)}, j_m \mid i_m) \tag{1.1.14}$$

where $\mathbf{e} = (e^{(1)}, e^{(2)}, ..., e^{(h)})$, $i = (i_1, i_2, ..., i_h)$, and $j = (j_1, j_2, ..., j_h)$. This equation can be written in the following matrix form (see Appendix 5)

$$\mathbf{P}(\mathbf{e}) = \mathbf{P}_1(e^{(1)}) \otimes \mathbf{P}_2(e^{(2)}) \otimes \cdots \otimes \mathbf{P}_h(e^{(h)}) \tag{1.1.15}$$

Thus, the MP of the composite vector channel errors whose component subchannels are statistically independent can be expressed as a Kronecker product of the subchannel error MPs.

In many practical cases it is possible to justify the subchannels independence if they are physically isolated. In these cases, we use formula (1.1.15) to describe the error source model. We will use it in Chapter 5 to model the error source in the two-way systems.

The other extreme is the channel deterministic dependence. We can describe a vector channel in which errors occur simultaneously in all its subchannels. Indeed, if the conditional error probability $\varepsilon_j^{(i)}$ equals to 0 for all the channels ($i = 1, 2, ..., h$) and some subset of states (say $j = 1, 2, ..., m$) and equals 1 otherwise, then

$$\mathbf{E}_i = \begin{bmatrix} 0 & 0 \\ 0 & \mathbf{I} \end{bmatrix} \qquad \mathbf{I} - \mathbf{E}_i = \begin{bmatrix} \mathbf{I} & 0 \\ 0 & 0 \end{bmatrix}$$

According to representation (1.1.10), $\mathbf{F}(\mathbf{1}) = \mathbf{E}_1$, $\mathbf{F}(\mathbf{0}) = \mathbf{I} - \mathbf{E}_1$, and $\mathbf{F}(\mathbf{e}) = 0$ otherwise. Therefore, in this case errors occur (or do not occur) simultaneously in all channels. Thus, the HMM can describe different degrees of error dependence in the channels.

This model can also reflect different degrees of dependence between errors in the same channel. Suppose that error sequence is an order k Markov chain (see Appendix 6). Then for any $\mathbf{e}_0^n = \mathbf{e}_0, \mathbf{e}_1, ..., \mathbf{e}_n$ and $n \geq k$, the conditional probability of an error at the moment n

$$Pr(\mathbf{e}_n \mid \mathbf{e}_0^{n-1}) = Pr(\mathbf{e}_n \mid \mathbf{e}_{n-k}^{n-1}) \tag{1.1.16}$$

depends only on k previous errors $\mathbf{e}_{n-k}^{n-1}$. This Markov chain can be described as an HMM.

Indeed, consider a process whose states are defined as k-tupples $S_n = (\mathbf{e}_{n-k+1}^{n})$. According to equation (1.1.16), the sequence of states S_n is a simple Markov chain. It is also obvious that state error probability $p_{S_n}(\mathbf{e}) = 1$ if $\mathbf{e} = \mathbf{e}_n$ and is equal to 0 otherwise. Therefore, the order k Markov chain is an HMM.

We can construct the order k Markov chain whose multidimensional probability distributions are identical to the order $k+1$ distributions $Pr(\mathbf{e}_0^k)$ of any error source by using the following equation $Pr(\mathbf{e}_n \mid \mathbf{e}_{n-k}^{n-1}) = Pr(\mathbf{e}_0^k) / Pr(\mathbf{e}_0^{k-1})$. Since we always have only finite sets of experimental data, we can construct an HMM for practically any statistical data. However, the size of the matrix, which defines the HMM, grows exponentially with k. This growth is usually called the *curse of dimensionality*.

Example 1.1.3. Let us illustrate the discussed technique of model building in the case of the second-order Markov chain ($k = 2$) and one binary symmetric channel ($h = 1$). In this case, the Markov chain S_n has four states: (0,0), (1,0), (0,1), and (1,1). The chain transition probability matrix is

$$\mathbf{P} = \begin{bmatrix} Pr(0 \mid 0,0) & 0 & Pr(1 \mid 0,0) & 0 \\ Pr(0 \mid 1,0) & 0 & Pr(1 \mid 1,0) & 0 \\ 0 & Pr(0 \mid 0,1) & 0 & Pr(1 \mid 0,1) \\ 0 & Pr(0 \mid 1,1) & 0 & Pr(1 \mid 1,1) \end{bmatrix}$$

The state conditional probabilities of errors are equal to 1 when the chain is in the state (0,1) or (1,1) and equal to 0 in other states. Therefore,

$$\mathbf{F}(0) = \begin{bmatrix} 1 & 0 & 0 & 0 \\ 0 & 1 & 0 & 0 \\ 0 & 0 & 0 & 0 \\ 0 & 0 & 0 & 0 \end{bmatrix} \qquad \mathbf{F}(1) = \begin{bmatrix} 0 & 0 & 0 & 0 \\ 0 & 0 & 0 & 0 \\ 0 & 0 & 1 & 0 \\ 0 & 0 & 0 & 1 \end{bmatrix}$$

According to (1.1.10),

$$\mathbf{P}(0) = \mathbf{PF}(0) = \begin{bmatrix} Pr(0 \mid 0,0) & 0 & 0 & 0 \\ Pr(0 \mid 1,0) & 0 & 0 & 0 \\ 0 & Pr(0 \mid 0,1) & 0 & 0 \\ 0 & Pr(0 \mid 1,1) & 0 & 0 \end{bmatrix}$$

$$\mathbf{P}(1) = \mathbf{P}\,\mathbf{F}(1) = \begin{bmatrix} 0 & 0 & Pr(1 \mid 0,0) & 0 \\ 0 & 0 & Pr(1 \mid 1,0) & 0 \\ 0 & 0 & 0 & Pr(1 \mid 0,1) \\ 0 & 0 & 0 & Pr(1 \mid 1,1) \end{bmatrix}$$

It is usually said that impulse noise is the major contributor to the channel error bursts. However, other sources also produce error bursts. We analyze some of these sources in the following sections.

1.1.8 Satellite Channel Model

The major results of the classical information transmission theory are based on the assumption that channel noise is *additive white Gaussian noise* (AWGN) and channel errors are *independent and identically distributed* (i. i. d.). A satellite channel is considered an example for which this theory is good enough.†

However, the real satellite channel error statistics showed that these assumptions are far from reality. Even if we assume that the space noise is the AWGN, the channel output errors are not i. i. d. It may be explained by the various types of channel impairments (such as intersymbol interference, nonlinearities, adjacent channel interference, modem defects, etc.), and different digital signal transformations (such as differential encoding, scrambling, error-detection and error-correction coding, etc.). Consider, for example, a typical satellite communication system as shown in Fig. 1.2.

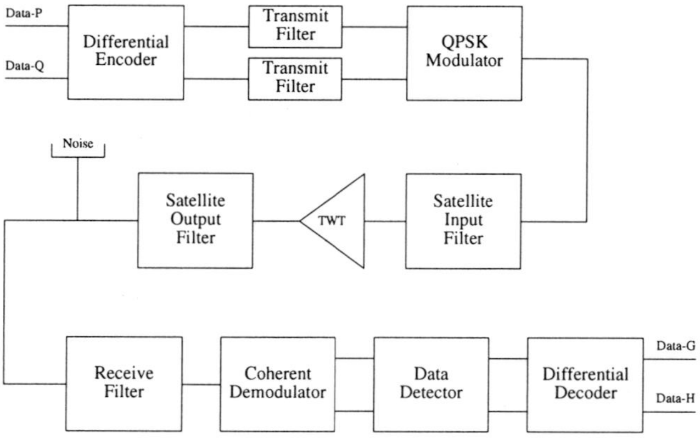

Figure 1.2. Functional block diagram of satellite channel.

The transmitter model consists of a differential encoder, transmit shaping filters, and quadrature phase-shift keying (QPSK) modulator. The transponder model consists of a channel input filter, traveling wave tube (TWT) amplifier, and the transponder output filter. The receiver model includes a receive filter, symbol timing and carrier recovery circuitry, coherent demodulator, data detector, and differential decoder.

Two independent data streams (Data-P and Data-Q in Fig. 1.2) appear on the differential encoder input and two data streams (Data-G and Data-H) appear on the differential decoder output, thereby forming the set of two binary channels. We will show that, in the presence AWGN, the error sequence on the differential decoder output can be described as an HMM.

It is convenient to use complex number notations to describe signal transformations in QPSK-modulated systems. We, therefore, denote $z_n = p_n + \mathrm{j}\, q_n$ ($p_n = \pm 1$, $q_n = \pm 1$, $-\infty < n < \infty$) as the differential encoder input data and $w_n = c_n + \mathrm{j}\, d_n$ ($c_n = \pm 1$, $d_n = \pm 1$,

† The major part of this section is reprinted from the IEEE *Journal on Selected Areas in Communications*, **6**(1), January 1988, with the permission of the IEEE.

$-\infty < n < \infty$) as its output data. It is easy to verify that the differential encoder operation [39] may be defined by

$$w_n = -0.5(1+\mathrm{j})\, w_{n-1} z_n^* \tag{1.1.17}$$

where z_n^* denotes the complex conjugate of z_n.

The input to the satellite nonlinear amplifier is equal to

$$s(t) = \mathrm{Re}\; u(t)\, e^{\mathrm{j}\,\omega_c t} \tag{1.1.18}$$

where ω_c is the carrier frequency,

$$u(t) = \sum_{n=-\infty}^{\infty} w_n p(t-nT) \tag{1.1.19}$$

$p(t)$ is a pulse shape which is the convolution of the original rectangular pulse with the transmit filter impulse response and the satellite input filter low-pass equivalent, [36] and T is the rectangular pulse duration.

The satellite TWT amplifier operation could be described by the equation [39]

$$s_1(t) = \mathrm{Re}\; u_1(t)\, e^{\mathrm{j}\,\omega_c t} \tag{1.1.20}$$

where

$$u_1(t) = v[R(t)]\, u(t) \tag{1.1.21}$$

with $v(R) = (f(R)/R)\, e^{\mathrm{j}\, g(R)}$. Functions $f(R)$ and $g(R)$ characterize the TWT AM/AM and AM/PM conversion and $R(t) = |u(t)|$ is the TWT input signal amplitude.

Assume that the satellite output signal is distorted by the AWGN. Uplink noise may be compounded with that of the downlink by using the method described in Ref. 40. Assume also that the receiver carrier phase tracking is ideal and that the variances of the recovered carrier and of the symbol clock produce insignificant performance loss, given the transmission of purely random data bits. Then, the receiver data sampler output is

$$u_3(nT) = u_2(nT) + N(nT) \tag{1.1.22}$$

where $u_2(t)$ and $N(t)$ are the low-pass representations of the signal and noise [36] on the output of the receiver filter. The complex signal $u_2(t)$ is a convolution of the TWT output signal $u_1(t)$ with the low-pass equivalents of the satellite output filter and the receiver input filter. The inphase and quadrature components of the received signal represent the real and imaginary parts of $u_3(nT) = s_{I,n} + \mathrm{j}\, s_{Q,n}$.

The data detector makes a decision about the transmitted bit, depending on the difference between the sampled level and the decision threshold. The data detector outputs the sequence

$$\omega_n = e_n + \mathrm{j}\, f_n = sign(s_{I,n}) + \mathrm{j}\; sign(s_{Q,n}) \tag{1.1.23}$$

Finally, the differential decoder performs the operation which is inverse to that of the differential encoder:

$$\zeta_n = -0.5(1+\mathrm{j})\, \omega_n^* \omega_{n-1} \tag{1.1.24}$$

Equations (1.1.17) through (1.1.24) may be treated as a satellite channel physical model. Let us show that we can approximate this model with an HMM. Consider first the channel without the differential encoding. The signal sample $s(nT)$ on the input to the TWT suffers from intersymbol interference: it depends not only on w_n but also on several neighboring values w_i for $i=n-l,\ldots,n+l$ due to the signal filtering. Usually, this signal degradation is insignificant because the filters are designed to minimize the interference (equalized channel). However, the TWT significantly increases the intersymbol interference and, according to (1.1.20) and (1.1.21), adds the cross-coupling between the inphase and quadrature components. This means that we cannot neglect the intersymbol interference in the signal $u_3(nT)$ on the data detector input.

Define the channel state as a vector $S_n=(w_{n-l},\ldots,w_{n+l})$. The vector size (the number of samples that need to be considered) depends on the TWT parameters and also on the offset of its operating point from saturation. If the source bit stream is purely random, the sequence of states S_n is a Markov chain. Since $u_2(nT)$ is a deterministic function of states S_n, then, according to Eq. (1.1.22), the receiver input sequence $u_3(nT)$ is an HMM with the observation $u_3(nT)$ PDF of the form

$$p(u_3(nT) \mid S_n) = (2\pi)^{-1}\sigma^{-2}\exp(-0.5|u_3(nT)-u_2(nT)|^2/\sigma^2)$$

According to Eq. (1.1.23), the state conditional error probabilities can be written as

$$p_{S_n}(\mathbf{e}) = p_{S_n}(e_I,e_Q)=p_{S_n}(e_I)p_{S_n}(e_Q) \tag{1.1.25}$$

where

$$p_{S_n}(e_I)=0.5\,erfc(|\mathrm{Re}\,u_2(nT)|/\sigma\sqrt{2}) \qquad p_{S_n}(e_Q)=0.5\,erfc(|\mathrm{Im}\,u_2(nT)|/\sigma\sqrt{2}) \tag{1.1.26}$$

and σ^2 is the power of the noise on the output of the receiver filter and

$$erfc(x) = \frac{2}{\sqrt{\pi}}\int_x^\infty e^{-x^2}dx$$

The average error probabilities for the inphase and quadrature channels are equal to

$$p(e_I) = \sum_{S_n} p(S_n)p_{S_n}(e_I) \quad \text{and} \quad p(e_Q) = \sum_{S_n} p(S_n)p_{S_n}(e_Q)$$

respectively. In these equations, $p(S_n)$ denotes stationary probabilities of the HMM states. In the particular case when purely random data bits are transmitted, all these stationary probabilities are equal to 4^{-2l-1}.

We have shown that, in the absence of the differential coding, the error sequence may be described as an HMM. The differential coding increases the model's complexity, but the model may still be described as an HMM. According to Eq. (1.1.24), the differential encoder output depends on two consecutive samples on the data detector output. If we add ω_{n-1} to the state definition by introducing states $S_{d,n} = (S_n,\omega_{n-1})$, then error sequences on the differential decoder output will also be modeled by an HMM.

1.1.9 Fading Wireless Channels

Radio networks are becoming increasingly popular because they provide seamless communications for mobile users. Digital communication over wireless channels allows the users to transfer data and provides many additional capabilities to voice and image transmission (such as privacy, encryption, compression, error correction, etc). However, the wireless channel signal distortions (due to fading, intersymbol interference, adjacent channel and co-channel interference, specular reflections and shadowing, nonlinearities, and noise) lead to a bursty nature of errors with a high error rate which significantly reduces the communication reliability. Because of the error burstiness, the channel modeling with HMM is very popular. [35,37,41,46,47] In this section, we consider a simplified wireless channel whose block diagram is shown in Fig. 1.3.

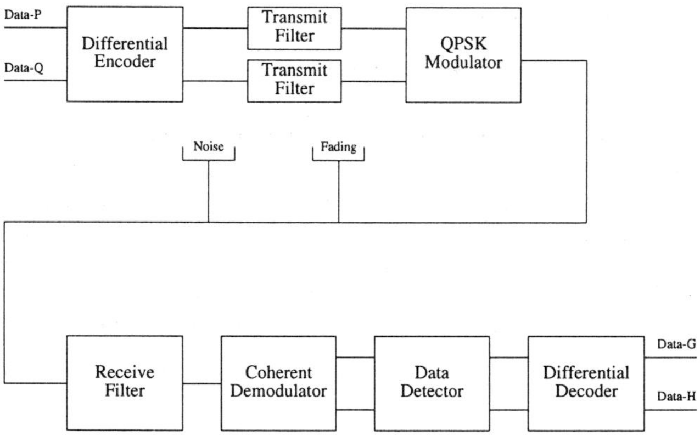

Figure 1.3. Functional block diagram of wireless channel.

This figure is similar to Fig. 1.2 (the satellite block is replaced with fading). Because of this similarity, all the equations of the previous section, except for Eq. (1.1.21), can be used to describe the wireless channel. Equation (1.1.21) which models the satellite TWT amplifier must be replaced by the equation which models flat fading [Ref. 31, p. 716]. $u_1(t) = a(t)u(t)$, where $a(t)$ is a complex random process whose magnitude is less than one. For the $\pi/4$-DQPSK modulation used in wireless communications, [25] Eq. (1.1.17) and (1.1.24) also change. Equation (1.1.17) is replaced by

$$w_n = 2^{-0.5}\mu(z_n)w_{n-1}z_n \qquad \mu(z) = \begin{cases} 1 & \text{if } \operatorname{Re}(z)\operatorname{Im}(z)<0 \\ -1 & \text{if } \operatorname{Re}(z)\operatorname{Im}(z)>0 \end{cases}$$

and Eq. (1.1.24) is replaced by

$$\zeta_n = 2^{0.5}\mu(\omega_n\omega_{n-1}^*)\omega_n\omega_{n-1}^*$$

Even though these equations are different, the effect of the two differential decoders is the same: the decoded symbol ζ_n depends only on $\omega_n\omega_{n-1}^*$. It is usually assumed that the fading is slow so that the fade amplitude is constant during a symbol transmission and the phase can be perfectly recovered from the received signal. In this case, repeating the derivations of the previous section, we obtain

$$p_{\alpha_n}(\mathbf{e}) = p_{\alpha_n}(e_I, e_Q) = p_{\alpha_n}(e_I)p_{\alpha_n}(e_Q) \tag{1.1.27}$$

where

$$p_{\alpha_n}(e_I) = 0.5erfc(|\text{Re}u_2(nT)|/\sigma\sqrt{2}) \quad p_{\alpha_n}(e_Q) = 0.5erfc(|\text{Im}u_2(nT)|/\sigma\sqrt{2}) \tag{1.1.28}$$

$\alpha_n = |a(nT)|$ is the fading envelope. These equations can be simplified: [31, p.716]

$$p_{\alpha_n}(e_I) = p_{\alpha_n}(e_Q) = 0.5erfc(\alpha_n\sqrt{E_b/N_0}) \tag{1.1.29}$$

where E_b is the energy per bit and $N_0 = 2\sigma^2$.

If the fade envelope is described as an HMM with the MPDF $\mathbf{p}(\alpha)$, then the symbol error sequence is also an HMM with the MP

$$\mathbf{P}(\mathbf{e}) = \int_{\alpha} \mathbf{p}(\alpha)p_{\alpha}(\mathbf{e})d\alpha \tag{1.1.30}$$

We can use this equation in theoretical studies if we have an analytical expression for the fading MPDF $\mathbf{p}(\alpha)$. In practice, measured samples are quantized. Let $y_0 = 1 > y_1 > y_2 > ... > y_m = 0$ be the quantization levels. If the envelope sample α belongs to the i-th quantization interval $[y_{i-1} \le \alpha < y_i]$, we replace the sample with some value $\hat{\alpha}(i)$ from the interval. Usually, the interval's centroid is selected. However, if we need to estimate the error performance upper bound, the quantization interval lower bound should be used. For the quantized fading amplitude HMM, the symbol error MP has the form

$$\mathbf{P}(\mathbf{e}) = \sum_{i=1}^{m} \mathbf{P}(\hat{\alpha}(i))p_{\hat{\alpha}(i)}(\mathbf{e}) \tag{1.1.31}$$

where $\mathbf{P}(\hat{\alpha}(i))$ is the quantized fading MP

$$\mathbf{P}(\hat{\alpha}(i)) = \int_{y_{i-1}}^{y_i} \mathbf{p}(\alpha)d\alpha$$

Usually, the quantized fading is modeled as a Markov chain whose states are the fading levels; [35,41,43,46] the fading modeling with HMMs is discussed in Sec. 1.5.4.

We have demonstrated that the system without the differential coding can be modeled as an HMM. The sequence ζ_n on the differential decoder output can also be presented as an FSC whose input symbols are (w_n, w_{n-1}). We leave the proof to the reader (see Problem 7).

1.1.10 Concatenated Channel Error Sources

Suppose that the link between receiving and transmitting stations is composed of two independent FSCs. As we have seen in Sec. 1.1.1, the combined channel is also an FSC. Suppose that both channels are symmetrical. Let $\mathbf{P}_1(\mathbf{e})$ and $\mathbf{P}_2(\mathbf{e})$ be the error MP for the first and second channels, respectively. If we assume that an error in the composite channel is a function of the component errors only, $\mathbf{e} = \phi(\mathbf{e}_1, \mathbf{e}_2)$, then the composite channel is an HMM whose error MP has the form:

$$\mathbf{P}(\mathbf{e}) = \sum_{\mathbf{e}_1} \mathbf{P}_1(\mathbf{e}_1) \otimes \mathbf{P}_2(\phi^{-1}(\mathbf{e}_1, \mathbf{e}))$$

where $\phi^{-1}(\mathbf{e}_1, \mathbf{e})$ is the inverse function to $\phi(\mathbf{e}_1, \mathbf{e}_2)$ with respect to $\mathbf{e}_2$ [a solution for $\mathbf{e}_2$ of the equation $\mathbf{e} = \phi(\mathbf{e}_1, \mathbf{e}_2)$]. In particular, if errors are additive ($\mathbf{e} = \mathbf{e}_1 + \mathbf{e}_2$), this equation becomes

$$\mathbf{P}(\mathbf{e}) = \sum_{\mathbf{e}_1} \mathbf{P}_1(\mathbf{e}_1) \otimes \mathbf{P}_2(\mathbf{e}-\mathbf{e}_1)$$

If we assume that the models are described by the matrices of the form (1.1.5) $\mathbf{P}_1(\mathbf{e}) = \mathbf{P}_1\mathbf{F}_1(\mathbf{e})$, $\mathbf{P}_2(\mathbf{e}) = \mathbf{P}_2\mathbf{F}_2(\mathbf{e})$ for the first and second channels in the link, then the composite channel states can be described by the pairs of the component channel states with the transition probability matrix $\mathbf{P} = \mathbf{P}_1 \otimes \mathbf{P}_2$. and the conditional MP of errors

$$\mathbf{F}(\mathbf{e}) = \sum_{\mathbf{e}_1} \mathbf{F}_1(\mathbf{e}_1) \otimes \mathbf{F}_2(\mathbf{e}-\mathbf{e}_1)$$

Example 1.1.4. Suppose that a channel is composed of a terrestrial channel whose error source is described by the Gilbert's model (see Sec. 1.1.3) and a satellite channel with independent errors that occur with the probability p_0.

The correct bit is received ($e=0$) if both channels did not corrupt it ($e_1=0$ and $e_2=0$) or both channels distorted it ($e_1=1$ and $e_2=1$). This means that $\phi(0,0) = \phi(1,1) = 0$. If only one channel distorted the bit, it is received incorrectly, that is, $\phi(1,0) = \phi(0,1) = 1$. Thus defined function is called an *exclusive-or* and is denoted as $e = e_1 \oplus e_2$. The composite channel is described by the HMM with the MPs

$$\mathbf{P}(0) = \mathbf{P}_1(0) \otimes \mathbf{P}_2(0) + \mathbf{P}_1(1) \otimes \mathbf{P}_2(1)$$
$$\mathbf{P}(1) = \mathbf{P}_1(1) \otimes \mathbf{P}_2(0) + \mathbf{P}_1(0) \otimes \mathbf{P}_2(1)$$

where

$$\mathbf{P}_1(0) = \begin{bmatrix} Q & Ph \\ p & qh \end{bmatrix} \quad \mathbf{P}_1(1) = \begin{bmatrix} 0 & P(1-h) \\ 0 & q(1-h) \end{bmatrix}$$
$$\mathbf{P}_2(0) = q_0 = 1-p_0 \quad \mathbf{P}_2(1) = p_0$$

Performing the matrix multiplication, we obtain the matrices describing the composite channel

$$\mathbf{P}(0) = \begin{bmatrix} Qq_0 & P(p_0+hq_0-hp_0) \\ pq_0 & q(p_0+hq_0-hp_0) \end{bmatrix} \quad \mathbf{P}(1) = \begin{bmatrix} Qp_0 & P(q_0+hp_0-hq_0) \\ pp_0 & q(q_0+hp_0-hq_0) \end{bmatrix}$$

1.2 BINARY SYMMETRIC STATIONARY CHANNEL

1.2.1 Model Description

The single channel case ($h=1$) plays the most important role in applications. In this case, the probability $Pr(e_1^m)$ of occurrence of a sequence of errors $e_1^m = e_1, e_2, \ldots, e_m$ can be computed by the formula (1.1.9):

$$Pr(e_1^m) = \boldsymbol{\pi} \prod_{i=1}^{m} \mathbf{P}(e_i)\mathbf{1} \tag{1.2.1}$$

Multidimensional error-free run distributions have the form

$$f_\lambda(\lambda_1^m) = Pr(0^{\lambda_1}10^{\lambda_2}\cdots 0^{\lambda_m}1 \mid 1)$$
$$= \boldsymbol{\pi}\mathbf{P}(1)\prod_{i=1}^{m}\{\mathbf{P}^{\lambda_i}(0)\mathbf{P}(1)\}\mathbf{1}/\boldsymbol{\pi}\mathbf{P}(1)\mathbf{1} \tag{1.2.2}$$

where the symbol $0^\lambda = 00...0$ denotes a sequence of λ consecutive zeroes. The distributions of the intervals between the correct symbols are given by

$$f_l(l_1^m) = Pr(1^{l_1}01^{l_2}\cdots 1^{l_m}0 \mid 0)$$
$$= \boldsymbol{\pi}\mathbf{P}(0)\prod_{i=1}^{m}\{\mathbf{P}^{l_i}(1)\mathbf{P}(0)\}\mathbf{1}/\boldsymbol{\pi}\mathbf{P}(0)\mathbf{1} \tag{1.2.3}$$

It is possible that in these equations $\lambda_i = 0$ and $l_i = 0$ so that $f_\lambda(0, 0,..., 0) = Pr(11\cdots 1 \mid 1) = Pr(1^m \mid 1)$ and $f_l(m) = Pr(1^m 0 \mid 0)$ are dependent (see Problem 2). This dependency creates an ambiguity in experimental data processing. To eliminate the ambiguity we consider the distributions that do not permit $\lambda_i = 0$ or $l_i = 0$:

$$p_\lambda(\lambda_1^m) = Pr(0^{\lambda_1}10^{\lambda_2}\cdots 0^{\lambda_m}1 \mid 1, \lambda_1>0,\ \lambda_2>0,...,\lambda_m>0)$$
$$= \boldsymbol{\pi}\mathbf{P}(1)\prod_{i=1}^{m}\{\mathbf{P}^{\lambda_i}(0)\mathbf{P}(1)\}\mathbf{1}/\boldsymbol{\pi}\mathbf{P}(1)[\mathbf{P}(0)(\mathbf{I}-\mathbf{P}(0))^{-1}\mathbf{P}(1)]^m\mathbf{1} \tag{1.2.4}$$

$$p_l(l_1^m) = Pr(1^{l_1}01^{l_2}\cdots 1^{l_m}0 \mid 0,\ l_1>0,\ l_2>0,...,l_m>0)$$
$$= \boldsymbol{\pi}\mathbf{P}(0)\prod_{i=1}^{m}\{\mathbf{P}^{l_i}(1)\mathbf{P}(0)\}\mathbf{1}/\boldsymbol{\pi}\mathbf{P}(0)[\mathbf{P}(1)(\mathbf{I}-\mathbf{P}(1))^{-1}\mathbf{P}(0)]^m\mathbf{1} \tag{1.2.5}$$

These distributions are connected with the distributions (1.2.2) and (1.2.3):

$$p_\lambda(\lambda_1^m) = f_\lambda(\lambda_1^m)/c_\lambda$$
$$c_\lambda = 1 - \sum_{\lambda_1=0}^{1}\sum_{\lambda_2=0}^{1}\cdots\sum_{\lambda_m=0}^{1} f_\lambda(\lambda_1,\lambda_2,...,\lambda_m) + f_\lambda(1,1,...,1)$$
$$p_l(l_1^m) = f_l(l_1^m)/c_l$$
$$c_l = 1 - \sum_{l_1=0}^{1}\sum_{l_2=0}^{1}\cdots\sum_{l_m=0}^{1} f_l(l_1,l_2,...,l_m) + f_l(1,1,...,1)$$

In the case of a single variable, these equations become

$$p_\lambda(\lambda) = f_\lambda(\lambda)/[1-f_\lambda(0)] \tag{1.2.6}$$
$$p_l(l) = f_l(l)/[1-f_l(0)] \tag{1.2.7}$$

where

$$f_\lambda(\lambda) = \boldsymbol{\pi}\mathbf{P}(1)\mathbf{P}^\lambda(0)\mathbf{P}(1)\mathbf{1}/\boldsymbol{\pi}\mathbf{P}(1)\mathbf{1} \qquad \lambda=0,1,...$$
$$f_l(l) = \boldsymbol{\pi}\mathbf{P}(0)\mathbf{P}^l(1)\mathbf{P}(0)\mathbf{1}/\boldsymbol{\pi}\mathbf{P}(0)\mathbf{1} \qquad l=0,1,...$$

Example 1.2.1. Let us illustrate the distributions calculation for Gilbert's model, described in Sec. 1.1.5. The error-free run distribution (1.2.2)

$$f_\lambda(\lambda) = \boldsymbol{\pi}\mathbf{P}(1)\mathbf{P}^\lambda(0)\mathbf{P}(1)\mathbf{1}\,/\,\boldsymbol{\pi}\mathbf{P}(1)\mathbf{1} \tag{1.2.8}$$

can be calculated using spectral representation of $\mathbf{P}(0)$ (see Eq. (A.5.21) of Appendix 5)

$$\mathbf{P}(0) = \mathbf{T}^{-1}\begin{bmatrix} g_1 & 0 \\ 0 & g_2 \end{bmatrix}\mathbf{T} \tag{1.2.9}$$

assuming that g_1 and g_2 are *different* eigenvalues of the matrix $\mathbf{P}(0)$, which are found from the characteristic equation

$$\det(\mathbf{P}(0) - g\mathbf{I}) = \det\begin{bmatrix} Q-g & Ph \\ p & qh-g \end{bmatrix}$$
$$= g^2 - (Q+qh)g + h(Q-p) = 0$$

The roots of this equation are given by $g_{1,2} = b \pm \sqrt{b^2 + h(p-Q)}$, where $b = (Q+hq)/2$. The transform matrix $\mathbf{T}$ is composed from the matrix $\mathbf{P}(0)$ left eigenvectors

$$\mathbf{T} = \begin{bmatrix} p & g_1 - Q \\ p & g_2 - Q \end{bmatrix} \qquad \mathbf{T}^{-1} = [\,p(g_2 - g_1)]^{-1}\begin{bmatrix} g_2 - Q & -g_1 + Q \\ -p & p \end{bmatrix} \tag{1.2.10}$$

Using the spectral representation (1.2.9), we obtain [see Eq. (A.5.25) of Appendix 5]

$$\mathbf{P}^\lambda(0) = \mathbf{T}^{-1}\begin{bmatrix} g_1^\lambda & 0 \\ 0 & g_2^\lambda \end{bmatrix}\mathbf{T}$$

so that Eq. (1.2.8) becomes $f_\lambda(\lambda) = a_1(1-g_1)g_1^\lambda + a_2(1-g_2)g_2^\lambda$ where

$$a_1 = \frac{(1-h)(p-Q+qg_1)}{(1-g_1)(g_1-g_2)} \qquad a_2 = 1 - a_1$$

For the numerical values of parameters given in Sec. 1.1.5 we have (see M-file `polygeom`)

$$f_\lambda(\lambda) = 0.000469 \cdot 0.997461^\lambda + 0.154091 \cdot 0.810979^\lambda$$

Quite analogously, we find the error series distribution

$$f_l(l) = \boldsymbol{\pi}\mathbf{P}(0)\mathbf{P}^l(1)\mathbf{P}(0)\mathbf{1}\,/\,\boldsymbol{\pi}\mathbf{P}(0)\mathbf{1}$$

which, after some algebra, can be reduced to

$$f_l(l) = \begin{cases} 1 - a & \text{for } l = 0 \\ a(p+qh)(1-h)^{l-1}q^{l-1} & \text{for } l > 0 \end{cases}$$

where $a = P(1-h)/(p+P)$ is the bit-error probability (see Sec. 1.1.5). Typically, this distribution is not geometric, but the distribution $p_l(l)$, which is defined by Eq. (1.2.7), is geometric:

$$p_l(l) = (p+qh)(1-h)^{l-1}q^{l-1} \qquad l = 1,2,\ldots$$

The distribution $p_\lambda(\lambda)$ is bigeometric: $p_\lambda(\lambda) = b_1(1-g_1)g_1^\lambda + b_2(1-g_2)g_2^\lambda$, where $b_i = a_i g_i/(p+qh)$. Thus, the model may be characterized by the geometric distribution of the series of errors and the bigeometric distribution of the error-free runs.

As we mentioned in Sec. 1.1.6, it is possible to make certain transformations of the matrices $\mathbf{P}(0)$ and $\mathbf{P}(1)$ to obtain an HMM that is equivalent to the original one. In particular, it was shown that the similarity transformations (1.1.13)

$$\mathbf{P}_1(0) = \mathbf{T}^{-1}\mathbf{P}(0)\mathbf{T} \qquad \mathbf{P}_1(1) = \mathbf{T}^{-1}\mathbf{P}(1)\mathbf{T} \tag{1.2.11}$$

where the matrix $\mathbf{T}$ satisfies the conditions

$$\mathbf{T}^{-1}\mathbf{P}(0)\mathbf{T} \geq 0 \qquad \mathbf{T}^{-1}\mathbf{P}(1)\mathbf{T} \geq 0 \qquad \mathbf{T}\mathbf{1} = \mathbf{1} \tag{1.2.12}$$

yield an equivalent model with the matrices $\mathbf{P}_1(e)$. Using this fact, we can select the matrix $\mathbf{T}$ so as to simplify the description of the model. Let us consider several useful descriptions.

1.2.2 Diagonalization of the Matrices

Assume that the roots $g_1, g_2, \ldots, g_n$ of the characteristic polynomial det $(\mathbf{P}(0) - g\mathbf{I})$ are different. Then there is a matrix $\mathbf{T}$ (see Appendix 5.1.5) such that

$$\mathbf{P}_1(0) = \mathbf{T}^{-1}\mathbf{P}(0)\mathbf{T} = diag\{g_i\} \tag{1.2.13}$$

If all $g_i \geq 0$, the first of the conditions in Eq. (1.2.12) is satisfied. If the last condition in Eq. (1.2.12) does not hold, then it is possible to satisfy it by an appropriate normalization. To do so we choose the matrix $\mathbf{T}_1 = \mathbf{TB}$ with $\mathbf{B} = diag\{b_i\}$, $\mathbf{T}^{-1}\mathbf{1} = col\{b_i\}$, $b_1 \neq 0, \ldots, b_n \neq 0$ as a transforming matrix: $\mathbf{T}_1\mathbf{P}(0)\mathbf{T} = \mathbf{P}_1(0)$ and $\mathbf{T}_1\mathbf{1} = \mathbf{T}^{-1}\mathbf{T1} = \mathbf{1}$. If, at the same time, $\mathbf{P}_1(1) = \mathbf{T}_1^{-1}\mathbf{P}(1)\mathbf{T}_1 \geq 0$, then we have a model with a diagonal matrix $\mathbf{P}_1(0)$ which is equivalent to the original one. Similar transformations can be performed with the matrix $\mathbf{P}(1)$. The convenience of using a diagonal matrix in the calculations is obvious. For example, according to Eq. (1.2.4)

$$p_\lambda(\lambda) = \sum_{i=1}^{n} A_i g_i^\lambda \tag{1.2.14}$$

In this case the distribution of lengths of the intervals between the errors is polygeometric, and the parameters of the progressions are the eigenvalues of the matrix $\mathbf{P}(0)$.

Example 1.2.2. Let us illustrate these transformations for Gilbert's model. We cannot use the matrix $\mathbf{T}$ of Example 1.2.1, because it does not satisfy the normalization condition $\mathbf{T1} = \mathbf{1}$ of Eq. (1.2.12). To obtain an equivalent representation, we need to normalize these matrices (divide each row by the sum of its elements). For the model parameter numerical values given in Sec. 1.1.5, we have $g_1 = 0.99746$, $g_2 = 0.81098$, and the normalized matrices become

$$\mathbf{T} = \begin{bmatrix} 0.9866342 & 0.01336582 \\ -0.2236539 & 1.22365391 \end{bmatrix} \qquad \mathbf{T}^{-1} = \begin{bmatrix} 1.0110435 & -0.0110435 \\ 0.1847939 & 0.8152061 \end{bmatrix}$$

Thus,

$$\mathbf{P}_1(0) = \mathbf{TP}(0)\mathbf{T}^{-1} = \begin{bmatrix} 0.99746 & 0 \\ 0 & 0.81098 \end{bmatrix} \qquad \mathbf{P}_1(1) = \mathbf{TP}(1)\mathbf{T}^{-1} = \begin{bmatrix} 0.00047 & 0.00207 \\ 0.03493 & 0.15409 \end{bmatrix}$$

1.2.3 Models with Two Sets of States

In certain applications of the HMM with the matrices $\mathbf{P}(0) = \mathbf{P}(\mathbf{I} - \mathbf{E})$ and $\mathbf{P}(1) = \mathbf{PE}$, it is convenient to have two sets of states: "good" states where error probability is small and "bad" states where this probability is large. It is possible to find an equivalent model in which the probabilities of errors for some states take the values $\varepsilon_g = \min \varepsilon_i$ and for other states $\varepsilon_b = \max \varepsilon_i$. To achieve it we decompose a state for which $\varepsilon_g < \varepsilon_j < \varepsilon_b$ into two substates j_1 and j_2. Let ε_g be the error probability in the first substate and ε_b in the second substate. Define the transition probabilities

$$p_{i_1 j_1} = p_{i_2 j_1} = \alpha p_{ij} \qquad p_{i_1 j_2} = p_{i_2 j_2} = (1 - \alpha) p_{ij} \tag{1.2.15}$$

We choose α in such a way that the error probability in the combined state is

$$\varepsilon_j = \alpha\varepsilon_g + (1 - \alpha)\varepsilon_b \tag{1.2.16}$$

Thus, $\alpha = (\varepsilon_b - \varepsilon_j)/(\varepsilon_b - \varepsilon_g)$.

Let the states of the initial chain be ordered in such a way that the first are those in which the conditional probabilities of errors are ε_g; the next are those that correspond to ε_j, which is neither equal to ε_g nor ε_b; and the last states are those in which the probabilities of errors are ε_b. Then the matrices **P** and **E** can be written in the block form

$$\mathbf{P} = \begin{bmatrix} \mathbf{P}_{11} & \mathbf{P}_{12} & \mathbf{P}_{13} \\ \mathbf{P}_{21} & \mathbf{P}_{22} & \mathbf{P}_{23} \\ \mathbf{P}_{31} & \mathbf{P}_{32} & \mathbf{P}_{33} \end{bmatrix} \qquad \mathbf{E} = \begin{bmatrix} \mathbf{E}_g & 0 & 0 \\ 0 & \mathbf{E}_i & 0 \\ 0 & 0 & \mathbf{E}_b \end{bmatrix} \tag{1.2.17}$$

where $\mathbf{E}_g = \varepsilon_g \mathbf{I}_g$, $\mathbf{E}_b = \varepsilon_b \mathbf{I}_b$, $\mathbf{E}_i = diag\{\varepsilon_j\}$ for $\varepsilon_g < \varepsilon_j < \varepsilon_b$.

The matrices $\mathbf{P}_1$ and $\mathbf{E}_1$ of the chain transformed according to Eq. (1.2.15) and (1.2.16) have the form

$$\mathbf{P}_1 = \begin{bmatrix} \mathbf{P}_{11} & \mathbf{P}_{12}\mathbf{E}_p & \mathbf{P}_{12}\mathbf{E}_q & \mathbf{P}_{13} \\ \mathbf{P}_{21} & \mathbf{P}_{22}\mathbf{E}_p & \mathbf{P}_{22}\mathbf{E}_q & \mathbf{P}_{23} \\ \mathbf{P}_{21} & \mathbf{P}_{22}\mathbf{E}_p & \mathbf{P}_{22}\mathbf{E}_q & \mathbf{P}_{23} \\ \mathbf{P}_{31} & \mathbf{P}_{32}\mathbf{E}_p & \mathbf{P}_{32}\mathbf{E}_q & \mathbf{P}_{33} \end{bmatrix} \qquad \mathbf{E}_1 = \begin{bmatrix} \mathbf{E}_g & 0 & 0 & 0 \\ 0 & \mathbf{E}_g^{(1)} & 0 & 0 \\ 0 & 0 & \mathbf{E}_b^{(1)} & 0 \\ 0 & 0 & 0 & \mathbf{E}_b \end{bmatrix} \tag{1.2.18}$$

where $\mathbf{E}_p = diag\{(\varepsilon_b - \varepsilon_j)/(\varepsilon_b - \varepsilon_g)\}$, $\mathbf{E}_q = diag\{(\varepsilon_j - \varepsilon_g)/(\varepsilon_b - \varepsilon_g)\}$, $\mathbf{E}_g^{(1)} = \varepsilon_g \mathbf{I}_i$, and $\mathbf{E}_b^{(1)} = \varepsilon_b \mathbf{I}_i$.

It is easy to verify that if $\boldsymbol{\pi} = [\ \boldsymbol{\pi}_1,\ \boldsymbol{\pi}_2,\ \boldsymbol{\pi}_3\ ]$ is the stationary state probability probability vector of the original chain, the stationary vector of the transformed chain is $[\ \boldsymbol{\pi}_1,\ \boldsymbol{\pi}_2\mathbf{E}_p,\ \boldsymbol{\pi}_2\mathbf{E}_q,\ \boldsymbol{\pi}_3\ ]$.

Of course, all the previous arguments are heuristic. Let us give a formal proof of equivalency of the representations (1.2.17) and (1.2.18). The matrix $\mathbf{P}_1$ can be rewritten as

$$\mathbf{P}_1 = \mathbf{TP}\begin{bmatrix} \mathbf{I}_g & 0 & 0 & 0 \\ 0 & \mathbf{E}_p & \mathbf{E}_q & 0 \\ 0 & 0 & 0 & \mathbf{I}_b \end{bmatrix} \quad \text{where} \quad \mathbf{T} = \begin{bmatrix} \mathbf{I}_g & 0 & 0 \\ 0 & \mathbf{I}_i & 0 \\ 0 & \mathbf{I}_i & 0 \\ 0 & 0 & \mathbf{I}_b \end{bmatrix} \tag{1.2.19}$$

But then $\mathbf{P}_1\mathbf{E}_1\mathbf{T} = \mathbf{TPE}$, $\mathbf{P}_1(\mathbf{I} - \mathbf{E}_1)\mathbf{T} = \mathbf{TP}(\mathbf{I} - \mathbf{E})$, so that (1.1.13) is satisfied, thereby proving the equivalency of Eq. (1.2.17) and (1.2.18).

It is easy to verify that all the proofs are also valid in the case where ε_g is replaced by any number ε_x such that $0 \le \varepsilon_x \le \varepsilon_g$, and ε_b is replaced by ε_y such that $\varepsilon_b \le \varepsilon_y \le 1$, and the decomposition is performed only for the states in a certain subset. In particular, every binary HMM is equivalent to the *canonic binary HMM* with $\varepsilon_x = 0$ and $\varepsilon_y = 1$.

Example 1.2.3. Let us find a canonic model (with $\varepsilon_x = 0$ and $\varepsilon_y = 1$) that is equivalent to Gilbert's model with the matrices

$$\mathbf{P} = \begin{bmatrix} Q & P \\ p & q \end{bmatrix} \qquad \mathbf{E} = \begin{bmatrix} 0 & 0 \\ 0 & 1-h \end{bmatrix}$$

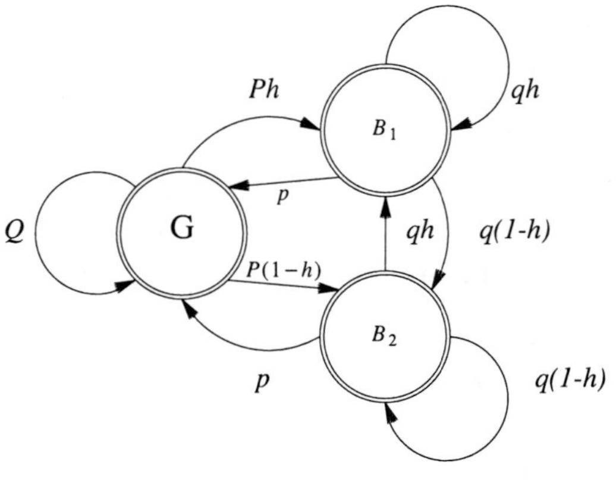

Figure 1.4. Gilbert's model expanded state diagram.

In this case $\varepsilon_g = 0$ and there are no states with the probability of errors $\varepsilon_x = 1$ if $h \neq 0$. Therefore $\mathbf{E}_p = h$, $\mathbf{E}_q = 1-h$, and from (1.2.19) we obtain

$$\mathbf{P}_1 = \begin{bmatrix} 1 & 0 \\ 0 & 1 \\ 0 & 1 \end{bmatrix} \begin{bmatrix} Q & P \\ p & q \end{bmatrix} \begin{bmatrix} 1 & 0 & 0 \\ 0 & h & 1-h \end{bmatrix} = \begin{bmatrix} Q & Ph & P(1-h) \\ p & qh & q(1-h) \\ p & qh & q(1-h) \end{bmatrix}$$

The matrix $\mathbf{E}_1$ is found from (1.2.18) and takes the form

$$\mathbf{E}_1 = \begin{bmatrix} 0 & 0 & 0 \\ 0 & 0 & 0 \\ 0 & 0 & 1 \end{bmatrix}$$

Therefore, the two-state model in which errors occur only in the second state, with the probability $1-h$, is equivalent to the three-state model in which errors never occur in the first two states and always happen in the third state. Flow graphs of the model are depicted in Figs. 1.1 and 1.4.

1.2.4 Block Matrix Representation

In the HMM applications it is convenient to represent the matrices in the block form because this form simplifies some calculations.

Consider a Markov chain with the matrix **P**, and partition the set of its states into two subsets, A_0 and A_1. For simplicity, we attribute all the states with the initial numbers of $i = 1, 2, \ldots, R$ to the subset A_0, and the remaining states with the numbers of $i = R+1, \ldots, n$ we attribute to the subset A_1. In correspondence with this partition, we rewrite the matrix **P** in the form of blocks

$$\mathbf{P} = \begin{bmatrix} \mathbf{P}_{00} & \mathbf{P}_{01} \\ \mathbf{P}_{10} & \mathbf{P}_{11} \end{bmatrix} \tag{1.2.20}$$

We assume that an error occurs, when the Markov chain is in the states of the set A_1, and that no error occurs in the states of the set A_0. It is clear that this model is a particular case of the HMM when

$$\mathbf{P}(0) = \mathbf{P}(\mathbf{I} - \mathbf{E}) = \begin{bmatrix} \mathbf{P}_{00} & 0 \\ \mathbf{P}_{10} & 0 \end{bmatrix} \quad \mathbf{P}(1) = \mathbf{PE} = \begin{bmatrix} 0 & \mathbf{P}_{01} \\ 0 & \mathbf{P}_{11} \end{bmatrix} \tag{1.2.21}$$

where

$$\mathbf{E} = \begin{bmatrix} 0 & 0 \\ 0 & \mathbf{I} \end{bmatrix}$$

The advantage of representation (1.2.20) is that the occurrence of errors is deterministically connected to the state of the Markov chain: $e_i = 0$ if the chain state $s_i \le R$ and $e_i = 1$ otherwise. An HMM in which observations are deterministic functions of its states is called a *Markov function.*

In calculations involving block matrices, the formulas are not usually as compact as in calculations using the matrices $\mathbf{P}(e)$, but the orders of the matrices are smaller which leads to faster computations. Probability distribution (1.2.1) takes the form

$$Pr(e_1^m) = \boldsymbol{\pi}_{e_1} \mathbf{P}_{e_1 e_2} \cdots \mathbf{P}_{e_{m-1} e_m} \mathbf{1} = \boldsymbol{\pi}_{e_1} \prod_{i=1}^{m-1} \mathbf{P}_{e_i e_{i+1}} \mathbf{1} \tag{1.2.22}$$

Distribution (1.2.2) of lengths of intervals between the errors has the form

$$\begin{aligned} f_\lambda(\lambda_1^m) &= Pr(0^{\lambda_1} 10^{\lambda_2} \cdots 10^{\lambda_m} 1 \mid 1) \\ &= \boldsymbol{\pi}_1 \prod_{i=1}^{m} \{\mathbf{P}_{10} \mathbf{P}_{00}^{\lambda_i - 1} \mathbf{P}_{01}\} \mathbf{1} / \boldsymbol{\pi}_1 \mathbf{1} \end{aligned} \tag{1.2.23}$$

and distribution (1.2.3) of the lengths of intervals between correct symbols can be written as

$$\begin{aligned} f_l(l_1^m) &= Pr(1^{l_1} 01^{l_2} \cdots 01^{l_m} 0 \mid 0) \\ &= \boldsymbol{\pi}_0 \prod_{i=1}^{m} \{\mathbf{P}_{01} \mathbf{P}_{11}^{l_i - 1} \mathbf{P}_{10}\} \mathbf{1} / \boldsymbol{\pi}_0 \mathbf{1} \end{aligned} \tag{1.2.24}$$

In these equations, $\boldsymbol{\pi}_0$ and $\boldsymbol{\pi}_1$ are subblocks of the stationary state probability vector $\boldsymbol{\pi} = [\boldsymbol{\pi}_0 \;\; \boldsymbol{\pi}_1]$ which is the unique solution of Eq. (1.1.6). This equation can be written in the following block form

$$\boldsymbol{\pi}_0 \mathbf{P}_{00} + \boldsymbol{\pi}_1 \mathbf{P}_{10} = \boldsymbol{\pi}_0 \quad \boldsymbol{\pi}_0 \mathbf{P}_{01} + \boldsymbol{\pi}_1 \mathbf{P}_{11} = \boldsymbol{\pi}_1 \quad \boldsymbol{\pi}_0 \mathbf{1} + \boldsymbol{\pi}_1 \mathbf{1} = 1 \tag{1.2.25}$$

The similarity transformation (1.2.11) with the matrix

$$\mathbf{T} = block\ diag\{\mathbf{T}_e\} = \begin{bmatrix} \mathbf{T}_0 & 0 \\ 0 & \mathbf{T}_1 \end{bmatrix} \tag{1.2.26}$$

leads to

$$\mathbf{R}_{ij} = \mathbf{T}_i^{-1} \mathbf{P}_{ij} \mathbf{T}_j \qquad (i,j = 0,1) \tag{1.2.27}$$

If the conditions

$$\mathbf{T}^{-1} \mathbf{PT} \ge 0 \qquad \mathbf{T1} = \mathbf{1} \tag{1.2.28}$$

hold, then the transformed matrix is stochastic and the model with the matrix

$$\mathbf{P}_1 = \begin{bmatrix} \mathbf{R}_{00} & \mathbf{R}_{01} \\ \mathbf{R}_{10} & \mathbf{R}_{11} \end{bmatrix}$$

is equivalent to the original model with the matrix **P**.

If all the eigenvalues of the matrices $\mathbf{P}_{00}$ and $\mathbf{P}_{11}$ are positive and different, then there exists a matrix **T** of the form (1.2.26), which diagonalizes the matrices $\mathbf{P}_{00}$ and $\mathbf{P}_{11}$. If conditions (1.2.28) hold as well, then the matrix **P** can be transformed to the form

$$\mathbf{P}_1 = \begin{bmatrix} diag\{\gamma_i\} & \mathbf{R}_{01} \\ \mathbf{R}_{10} & diag\{g_i\} \end{bmatrix} \qquad (1.2.29)$$

For this model, the interval distributions (1.2.23) and (1.2.24) are simplified because it is easy to compute powers of diagonal matrices: $[diag\{\gamma_i\}]^m = [diag\{\gamma_i^m\}]$ (see Appendix 5).

We showed previously that the transformation

$$\mathbf{TP}_1 = \mathbf{PT} \qquad \mathbf{T1} = \mathbf{1} \qquad (1.2.30)$$

with the rectangular matrix **T** as in Eq. (1.2.26) is more general. In this case, the blocks of the matrices $\mathbf{P}_1$ and **P** are related by

$$\mathbf{T}_{i0}\mathbf{R}_{0j} + \mathbf{T}_{i1}\mathbf{R}_{1j} = \mathbf{P}_{i0}\mathbf{T}_{0j} + \mathbf{P}_{i1}\mathbf{T}_{1j} \qquad (i,j = 0,1) \qquad (1.2.31)$$

This type of transformation may not only simplify the matrix structure but also may decrease the matrix size.

Example 1.2.4. In Example 1.2.3 we showed that Gilbert's model can be described by the Markov function with the matrix

$$\begin{bmatrix} Q & Ph & P(1-h) \\ p & qh & q(1-h) \\ p & qh & q(1-h) \end{bmatrix} = \begin{bmatrix} 0.997 & 0.00252 & 0.00048 \\ 0.034 & 0.81144 & 0.15456 \\ 0.034 & 0.81144 & 0.15456 \end{bmatrix}$$

Let us apply the transformation (1.2.27) to simplify the model description.

Choosing $\mathbf{T}_0$ as the matrix **T** of example 1.2.2

$$\mathbf{T}_0 = \begin{bmatrix} 0.9866342 & 0.01336582 \\ -0.2236539 & 1.22365391 \end{bmatrix} \qquad \mathbf{T}_1 = 1$$

we obtain

$$\mathbf{T} = \begin{bmatrix} 0.9866342 & 0.01336582 & 0 \\ -0.2236539 & 1.22365391 & 0 \\ 0 & 0 & 1 \end{bmatrix} \qquad \mathbf{T}^{-1} = \begin{bmatrix} 1.0110435 & -0.0110435 & 0 \\ 0.1847939 & 0.8152061 & 0 \\ 0 & 0 & 1 \end{bmatrix}$$

Thus,

$$\mathbf{P}_1 = \mathbf{TPT}^{-1} = \begin{bmatrix} 0.99746 & 0 & 0.00253 \\ 0 & 0.81098 & 0.18902 \\ 0.18432 & 0.66112 & 0.15456 \end{bmatrix} = \begin{bmatrix} g_1 & 0 & 1-g_1 \\ 0 & g_2 & 1-g_2 \\ a_1 & a_2 & p_{33} \end{bmatrix}$$

where $p_{33} = q(1-h)$. Several modifications of Gilbert's model use this matrix structure.[17,43,42] We will consider these models in Sec. 1.5.1.

One can restore the original Gilbert model parameters using the relations of Example 1.2.1.

$$h = \frac{g_1 g_2}{g_1 + g_2 + p_{33} - 1} \qquad q = \frac{p_{33}}{1-h} \qquad Q = g_1 + g_2 - qh$$

1.3 ERROR SOURCE DESCRIPTION BY MATRIX PROCESSES

1.3.1 Matrix Process Definition

The requirement of nonnegativity of the elements of matrices describing a Markov function limits the class of transformations that simplify these matrices. It is obvious, however, that if the matrices $\mathbf{P}(e)$ undergo an arbitrary transformation of similarity $\mathbf{T}^{-1}\mathbf{P}(e)\mathbf{T}$ with a matrix $\mathbf{T}$ whose elements are complex numbers (see Example 1.2.2), then the probability $Pr(e_1^n)$ of a sequence of errors can be found by a formula which is quite analogous to Eq. (1.2.1):

$$Pr(e_1^n) = \mathbf{y} \prod_{i=1}^{n} \mathbf{M}(e_i)\mathbf{z}$$

where

$$\mathbf{M}(e_i) = \mathbf{T}^{-1}\mathbf{P}(e_i)\mathbf{T} \qquad \mathbf{y} = \boldsymbol{\pi}\mathbf{T} \qquad \mathbf{z} = \mathbf{T}^{-1}\mathbf{1}$$

which can easily be verified:

$$Pr(e_1^n) = \boldsymbol{\pi} \prod_{i=1}^{n} \mathbf{P}(e_i)\mathbf{1}$$

$$= \mathbf{y}\mathbf{T}^{-1} \prod_{i=1}^{n} \mathbf{P}(e_i)\mathbf{1} = \mathbf{y} \prod_{i=1}^{n} \mathbf{M}(e_i)\mathbf{T}^{-1}\mathbf{1} = \mathbf{y} \prod_{i=1}^{n} \mathbf{M}(e_i)\mathbf{z}$$

The matrices $\mathbf{M}(e)$, $\mathbf{y}$, and $\mathbf{z}$ can have complex elements.

Processes with distributions defined by previous relations represent a particular case of matrix or pseudo-Markov processes. According to Ref. 15, we say that a process $Y = \{y_t : t \in T, y_t \in \Omega\}$ is called a matrix process if there exist matrices $\mathbf{a}$, $\mathbf{R}(t)$, $\mathbf{B}_\alpha$, $\mathbf{b}$ with complex elements such that

$$Pr(y_{t_1} = \alpha_1, y_{t_1+t_2} = \alpha_2, \ldots, y_{t_1+t_2+\ldots+t_m} = \alpha_m) = \mathbf{a} \prod_{i=1}^{m} \mathbf{R}(t_i)\mathbf{B}_{\alpha_i}\mathbf{b} \tag{1.3.1}$$

for every $\alpha \in \Omega$. Here $\mathbf{a}$ is a matrix row and $\mathbf{b}$ is a matrix column. In the following discussion, we consider only matrix processes of a particular form where $y_j = e_j$ is binary, and the square matrix $\mathbf{R}(t) = \mathbf{R}$ does not depend on t and has a finite order n. Obviously, the HMM is a special case of the matrix process model with $\mathbf{a} = \boldsymbol{\pi}, \mathbf{R} = \mathbf{P}$, $\mathbf{B}_0 = \mathbf{I} - \mathbf{E}$, $\mathbf{B}_1 = \mathbf{E}$, $\mathbf{b} = \mathbf{1}$. However, not every matrix process can be represented as a function of a Markov chain with a finite number of states. [10-12,15]

Matrix processes have many properties that make them similar to Markov functions. Indeed, denoting $\mathbf{M}(e) = \mathbf{R}\mathbf{B}_e$, we obtain by (1.3.1)

$$Pr(e_1, e_2, \ldots, e_m) = \mathbf{a} \prod_{i=1}^{m} \mathbf{M}(e_i)\mathbf{b} \tag{1.3.2}$$

which coincides formally with Eq. (1.2.1). Although $\mathbf{M} = \mathbf{M}(0) + \mathbf{M}(1)$ may not be a stochastic matrix, it possesses certain properties of stochastic matrices. Like them, it has an eigenvalue $\lambda = 1$. Indeed, it follows from the relation

$$\sum_{e_1,\ldots,e_m} Pr(e_1,\ldots,e_m) = \mathbf{aM}^m\mathbf{b} = 1$$

that

$$\sum_{m=0}^{\infty} \mathbf{aM}^m\mathbf{b}z^m = \mathbf{a}(\mathbf{I} - \mathbf{M}z)^{-1}\mathbf{b} = 1/(1-z)$$

The inverse matrix $(\mathbf{I} - \mathbf{M}z)^{-1}$ can be expressed using formula (A.5.2) as

$$(\mathbf{I} - \mathbf{M}z)^{-1} = \mathbf{B}(z) / \Delta(z)$$

where $\Delta(z) = \det(\mathbf{I} - \mathbf{M}z)$ is the *characteristic polynomial* and $\mathbf{B}(z)$ is the matrix polynomial called the *adjoint matrix*. From the previous equation we obtain

$$\mathbf{aB}(z)\mathbf{b} / \Delta(z) = 1 / (1-z)$$

or $\Delta(z) = \mathbf{aB}(z)\mathbf{b}(1-z)$. But then $\Delta(1) = 0$, and therefore the matrix $\mathbf{M}$ has an eigenvalue $\lambda = 1$.

The collection $\{ \mathbf{a}, \mathbf{M}(0), \mathbf{M}(1), \mathbf{b} \}$ is called the *matrix process representation*. In the regular case, when the matrix $\mathbf{M}$ has a simple eigenvalue $\lambda = 1$ and the absolute values of the rest of the eigenvalues are smaller than 1, we can prove (similarly to Theorem A.6.1) that

$$\lim_{m\to\infty} \mathbf{M}^m = \mathbf{zy}$$

exists. Vectors $\mathbf{z}$ and $\mathbf{y}$ are the right and left eigenvectors of the matrix $\mathbf{M}$, corresponding to the eigenvalue $\lambda = 1$

$$\mathbf{Mz} = \mathbf{z} \qquad \mathbf{yM} = \mathbf{y} \tag{1.3.3}$$

and satisfying the condition

$$\mathbf{yz} = 1 \tag{1.3.4}$$

Passing to the limit when $s\to\infty$, $j\to\infty$ in the equation

$$Pr(\Omega^s, e_1, e_2,\ldots,e_m, \Omega^j) = \mathbf{aM}^s \prod_{i=1}^{m} \mathbf{M}(e_i)\mathbf{M}^j\mathbf{b}$$

we obtain

$$Pr(e_1, e_2,\ldots,e_m) = \mathbf{y} \prod_{i=1}^{m} \mathbf{M}(e_i)\mathbf{z} \tag{1.3.5}$$

so that the initial and final matrices of a stationary matrix process satisfy conditions (1.3.3) and (1.3.4).

Under our assumptions, the vector-matrix product in equation (1.3.5) does not depend on the choice of $\mathbf{y}$ and $\mathbf{z}$ satisfying conditions (1.3.3) and (1.3.4), since the matrices $\mathbf{y}$ and $\mathbf{z}$ are determined uniquely up to numerical factors by the conditions (1.3.3): $\mathbf{y} = \mu\mathbf{y}_0$, $\mathbf{z} = \nu\mathbf{z}_0$. The condition (1.3.4) implies that $\mathbf{yz} = \mu\nu\mathbf{y}_0\mathbf{z}_0 = 1$ and $\mu\nu = 1$ if $\mathbf{y}_0\mathbf{z}_0 = 1$. Substituting $\mathbf{y} = \mu\mathbf{y}_0$, $\mathbf{z} = \nu\mathbf{z}_0$ in Eq. (1.3.5), we observe that the value of the right-hand side of Eq. (1.3.5) does not change when we replace $\mathbf{y}$ by $\mathbf{y}_0$ and $\mathbf{z}$ by $\mathbf{z}_0$.

Obviously, the distribution of the lengths of intervals between errors and between correct symbols in the matrix process model can be found by formulas (1.2.2) and (1.2.3), in which we replace the matrices $\boldsymbol{\pi}$, $\mathbf{P}(e)$, and $\mathbf{1}$ by $\mathbf{y}$, $\mathbf{M}(e)$, and $\mathbf{z}$, respectively.

Thus, a matrix process is a natural generalization of a Markov function. By recognizing that the matrices $\mathbf{M}(e)$ that define a matrix process may have complex elements, we find more ways to transform these matrices while replacing a matrix process by its equivalent than we do in the case of a Markov function. Repeating the transformations of Sec. 1.1.6, we can prove that if there exists a matrix $\mathbf{T}$ such that

$$\mathbf{TM}_1(e) = \mathbf{M}(e)\mathbf{T} \qquad e=0,1 \tag{1.3.6}$$

then the matrix processes $\{\mathbf{yT},\mathbf{M}_1(0),\mathbf{M}_1(1),\mathbf{z}_1\}$ and $\{\mathbf{y},\mathbf{M}(0),\mathbf{M}(1),\mathbf{Tz}_1\}$ are equivalent.

Notice that if $\mathbf{y}$ is a left invariant vector of the matrix $\mathbf{M} = \mathbf{M}(0) + \mathbf{M}(1)$, then $\mathbf{y}_1 = \mathbf{yT}$ is a left invariant vector of the matrix $\mathbf{M}_1 = \mathbf{M}_1(0) + \mathbf{M}_1(1)$, since $\mathbf{y}_1\mathbf{M}_1 = \mathbf{yTM}_1 = \mathbf{yMT} = \mathbf{yT} = \mathbf{y}_1$. Similarly, if $\mathbf{z}_1$ is a right invariant vector of the matrix $\mathbf{M}_1$, then $\mathbf{z} = \mathbf{Tz}_1$, is a right invariant vector of the matrix $\mathbf{M}$ and $\mathbf{yz} = \mathbf{yTz}_1 = \mathbf{y}_1\mathbf{z}_1$.

Obviously, if a matrix $\mathbf{T}$ is square and nonsingular, then Eq. (1.3.6) is a transformation of the similarity. Therefore, it is possible to reduce any matrix $\mathbf{M}(e)$ to its Jordan normal form (A.5.22). In particular, if the matrix $\mathbf{M}(e)$ has a simple structure, then it can be reduced to a diagonal form (A.5.21). It is conceivable that in certain transformations like in Eq. (1.3.6), we obtain nonnegative matrices $\mathbf{M}_1(0) \geq 0$ and $\mathbf{M}_1(1) \geq 0$, such that the matrix $\mathbf{M}_1 = \mathbf{M}_1(0) + \mathbf{M}_1(1)$ is stochastic and $\mathbf{z}_1 = \mathbf{1}$. In this case, the matrix process is equivalent to a Markov function.

Thus, the matrices representing a matrix process can be reduced to a simpler form than the matrices representing an HMM, since there are no restrictions of Eq. (1.2.12). Therefore, it is more convenient to use them in processing experimental data. However, matrix processes are not convenient for probabilistic interpretations of the process and for error source modeling. We use the matrix processes to represent the HMM in analytic computations transforming the matrices $\mathbf{P}(e)$ by equations (1.2.11) to their simplest forms. In such transformations we use, of course, Eq. (1.3.5) and not Eq. (1.2.1).

1.3.2 Block Matrix Representation

Let us consider a matrix process with matrices

$$\mathbf{M}(0) = \begin{bmatrix} \mathbf{M}_{00} & 0 \\ \mathbf{M}_{10} & 0 \end{bmatrix} \qquad \mathbf{M}(1) = \begin{bmatrix} 0 & \mathbf{M}_{01} \\ 0 & \mathbf{M}_{11} \end{bmatrix} \tag{1.3.7}$$

$$\mathbf{y} = [\ \mathbf{y}_0 \ \ \mathbf{y}_1\ ] \qquad \mathbf{z} = \begin{bmatrix} \mathbf{z}_0 \\ \mathbf{z}_1 \end{bmatrix}$$

We can reduce the orders of the matrices involved in calculations by expressing the distributions with the help of the matrix blocks. The distribution (1.3.5) takes the form

$$Pr(e_1,e_2,\ldots,e_m) = \mathbf{y}_{e_1} \prod_{i=1}^{m-1} \mathbf{M}_{e_i e_{i+1}} \mathbf{z}_{e_m} \tag{1.3.8}$$

and the matrices $\mathbf{y}_e$ and $\mathbf{z}_e$ in the stationary case satisfy the equations

$$\sum_{i=0}^{1} \mathbf{y}_i \mathbf{M}_{ij} = \mathbf{y}_j \quad \sum_{j=0}^{1} \mathbf{M}_{ij} \mathbf{z}_j = \mathbf{z}_i \quad \sum_{i=1}^{1} \mathbf{y}_i \mathbf{z}_i = 1 \quad i,j=0,1 \tag{1.3.9}$$

Notice that because of equation (1.3.7), the probabilities of error and lack of error are given by

$$p_{\text{err}} = Pr(1) = \mathbf{y}_1 \mathbf{z}_1 \quad p_{\text{cor}} = Pr(0) = \mathbf{y}_0 \mathbf{z}_0 \tag{1.3.10}$$

The multidimensional distributions of the intervals between the errors and correct symbols have the form

$$p_0(\lambda_1, \lambda_2, \ldots, \lambda_m) = \mathbf{y}_1 \prod_{i=1}^{m} \{\mathbf{M}_{10} \mathbf{M}_{00}^{\lambda_i - 1} \mathbf{M}_{01}\} \mathbf{z}_1 / \mathbf{y}_1 \mathbf{M}_{10} \mathbf{z}_0 \tag{1.3.11}$$

$$p_1(l_1, l_2, \ldots, l_m) = \mathbf{y}_0 \prod_{i=1}^{m} \{\mathbf{M}_{01} \mathbf{M}_{11}^{l_i - 1} \mathbf{M}_{10}\} \mathbf{z}_0 / \mathbf{y}_0 \mathbf{M}_{01} \mathbf{z}_1 \tag{1.3.12}$$

We can easily show that an arbitrary matrix process is equivalent to some matrix process, the matrices of which have the form (1.2.21). Indeed, it is obvious that the process with the matrices

$$\mathbf{M}_1(0) = \begin{bmatrix} \mathbf{M}(0) & 0 \\ \mathbf{M}(0) & 0 \end{bmatrix} \qquad \mathbf{M}_1(1) = \begin{bmatrix} 0 & \mathbf{M}(1) \\ 0 & \mathbf{M}(1) \end{bmatrix}$$

is equivalent to the process with the matrices $\mathbf{M}(e)$. The formal proof follows immediately from equation (1.3.6) because of the relation $\mathbf{T}\mathbf{M}_1(e) = \mathbf{M}(e)\mathbf{T}$, where $\mathbf{T} = [\ \mathbf{I} \ \ \mathbf{0}\ ]$.

1.3.3 Matrix Processes and Difference Equations

The probabilities of various combinations of symbols governed by a matrix process can be expressed by difference equations. For the sake of simplifying the discussion, we consider the binary matrix process $\{e_t\}$, $(e_t = 0,1)$. For a matrix process, the probability of a series $e,e,\ldots,e = e^k$ can be found from Eq. (1.3.8):

$$p_e(k) = Pr(e^k) = \mathbf{y}_e \mathbf{M}_{ee}^{k-1} \mathbf{z}_e \tag{1.3.13}$$

Let

$$\psi_e(\lambda) = \lambda^{\nu_e} - \sum_{i=1}^{\nu_e} a_{ei} \lambda^{\nu_e - i} \tag{1.3.14}$$

be any annulling polynomial of the matrix $\mathbf{M}_{ee}$, i.e.,

$$\psi_e(\mathbf{M}_{ee}) = \mathbf{M}_{ee}^{\nu_e} - \sum_{i=1}^{\nu_e} a_{ei} \mathbf{M}_{ee}^{\nu_e - i} = 0 \tag{1.3.15}$$

One such polynomial, according to the Cayley-Hamilton theorem (A.5.29), is the characteristic polynomial $\Delta(\lambda) = \det(\lambda \mathbf{I} - \mathbf{M}_{ee})$ of the matrix $\mathbf{M}_{ee}$.

After multiplying Eq. (1.3.15) by $\mathbf{M}_{ee}^{k-1-\nu_e}$ we obtain

$$\mathbf{M}_{ee}^{k-1} = \sum_{i=1}^{\nu_e} a_{ei}\mathbf{M}_{ee}^{k-i-1}$$

Multiplying this equation from the left by $\mathbf{y}_e$ and on the right by $\mathbf{z}_e$, we obtain, because of Eq. (1.3.13),

$$p_e(k) = \sum_{i=1}^{\nu_e} a_{ei} p_e(k-i) \tag{1.3.16}$$

Thus, the distribution $p_e(k)$ satisfies a linear homogeneous difference equation with constant coefficients. The coefficients of this equation are not uniquely defined. However, the coefficients of the equation with the minimal ν_e are uniquely determined. If we assume the opposite that there is an equation

$$p_e(k) = \sum_{i=1}^{\nu_e} b_{ei} p_e(k-i)$$

with at least one coefficient b_i different from a_i, then, subtracting this equation from Eq. (1.3.16), we obtain the equation whose order is smaller than ν_e. This contradicts the ν_e minimality and, therefore, our assumption was incorrect. Thus, the coefficients of the equation with the minimal ν_e are uniquely defined. Obviously, the minimal ν_e does not exceed the order of the matrix $\mathbf{M}_{ee}$.

If the order of the matrix $\mathbf{M}_{ee}$ is greater than the minimal ν_e, it is possible to find an equation of the form of equation (1.3.13) with matrices of the minimal order ν_e. Indeed, let us consider a vector $\mathbf{p}_e(k) = [\, p_e(k) \;\; p_e(k+1) \, \ldots \, p_e(k+\nu_e-1) \,]$ Its coordinates satisfy Eq. (1.3.16) which can be expressed in the following matrix form $\mathbf{p}_e(k) = \mathbf{p}_e(k-1)\mathbf{A}_{ee}$ where

$$\mathbf{A}_{ee} = \begin{bmatrix} 0 & 0 & \ldots & 0 & a_{e\nu_e} \\ 1 & 0 & \ldots & 0 & a_{e\nu_e-1} \\ \ldots & \ldots & \ldots & \ldots & \ldots \\ 0 & 0 & \ldots & 1 & a_{e1} \end{bmatrix}$$

This special structure of a matrix is called the matrix *echelon form*. Repeating the preceding recursion, we obtain

$$\mathbf{p}_e(k) = \mathbf{p}_e(k-2)\mathbf{A}_{ee}^2 = \cdots = \mathbf{p}_e(1)\mathbf{A}_{ee}^{k-1} = \mathbf{y}_e^{(1)}\mathbf{A}_{ee}^{k-1}$$

where $\mathbf{y}_e^{(1)} = [\, p_e(1) \;\; p_e(2) \, \ldots \, p_e(\nu_e) \,]$ Multiplying both sides of this equation from the right by the matrix column $\mathbf{z}_e^{(1)} = [\, 1 \;\; 0 \, \ldots \, 0 \,]'$, we obtain $p_e(k) = \mathbf{y}_e^{(1)}\mathbf{A}_{ee}^{k-1}\mathbf{z}_e^{(1)}$ This equation is analogous to Eq. (1.3.13) with the matrix $\mathbf{M}_{ee}$ of the minimal order.

Example 1.3.1. Let us find the difference equation for the error-free run distribution according to Gilbert's model.

As we saw in Example 1.2.1, the error-free run distribution is presented by Eq. (1.2.8), which has the form of Eq. (1.3.13) with $\mathbf{M}_{00} = \mathbf{P}(0)$. The characteristic polynomial of $\mathbf{P}(0)$ equals

$$\Delta(g) = g^2 - (Q+qh)g + h(Q-p)$$

Thus the difference equation is given by

$$f_\lambda(k) = (Q+qh)\, f_\lambda(k-1) + h(p-Q)\, f_\lambda(k-2)$$

and its matrix has the following form

$$\mathbf{A}_{00} = \begin{bmatrix} 0 & h(p-Q) \\ 1 & Q+qh \end{bmatrix}$$

Similarly, we can show that the two-dimensional distributions $p(0^k 1^m)$, $p(1^k 0^m)$ satisfy two-dimensional difference equations. Indeed, it follows from Eq. (1.3.15) that

$$p(0^k 1^m) = \sum_i a_{0i} p(0^{k-i} 1^m)$$
$$p(0^k 1^m) = \sum_i a_{1i} p(0^k 1^{m-i}) \tag{1.3.17}$$

To represent these equations in the matrix form we introduce the matrix

$$\mathbf{P}_0(k,m) = \left[p(0^{k+i-1} 1^{m+j-1})\right]_{v_0 v_1}$$

which, according to Eq. (1.3.17), can be expressed as

$$\mathbf{P}_0(k,m) = \mathbf{A}'_{00}\mathbf{P}_0(k-1,m) \qquad \mathbf{P}_0(k,m) = \mathbf{P}_0(k,m-1)\mathbf{A}_{11}$$

and therefore $\mathbf{P}_0(k,m) = (\mathbf{A}'_{00})^{m-1}\mathbf{P}_0(1,1)\mathbf{A}_{11}^{k-1}$, where $\mathbf{A}'$ is the transposed matrix $\mathbf{A}$.

Multiplying the matrix $\mathbf{P}_0(k,m)$ from the left by $\mathbf{y}_0^{(2)} = [\ 1\ 0\ 0 \ldots 0\ ]$ and from the right by $\mathbf{z}_1^{(1)}$, we obtain $p(0^k 1^m) = \mathbf{y}_0^{(2)}(\mathbf{A}'_{00})^{k-1}\mathbf{P}_0(1,1)\mathbf{A}_{11}^{m-1}\mathbf{z}_1^{(1)}$. The probability $p(1^k 0^m)$ can be expressed similarly:

$$p(1^k 0^m) = \mathbf{y}_1^{(2)}(\mathbf{A}'_{11})^{k-1}\mathbf{P}_1(1,1)\mathbf{A}_{00}^{m-1}\mathbf{z}_0^{(1)}$$

with

$$\mathbf{P}_1(k,m) = \left[p(1^{k+i-1} 0^{m+j-1})\right]_{v_1 v_0}$$

Since matrices $\mathbf{A}'_{ee}$ and $\mathbf{A}_{ee}$ are similar (that is, $\mathbf{A}'_{ee} = \mathbf{T}_e^{-1}\mathbf{A}_{ee}\mathbf{T}_e$), the previous relations can be rewritten in the form

$$p(0^k 1^m) = \mathbf{y}_0^{(3)}\mathbf{A}_{00}^{k-1}\mathbf{A}_{01}\mathbf{A}_{11}^{m-1}\mathbf{z}_1^{(1)}$$
$$p(1^k 0^m) = \mathbf{y}_1^{(3)}\mathbf{A}_{11}^{k-1}\mathbf{A}_{10}\mathbf{A}_{00}^{m-1}\mathbf{z}_0^{(1)}$$

where $\mathbf{y}_e^{(3)} = \mathbf{y}_e^{(2)}\mathbf{T}_e^{-1}$ and $\mathbf{A}_{e(1-e)} = \mathbf{T}_e\mathbf{P}(0,0)\mathbf{A}_{(1-e)(1-e)}$.

The equations developed in this section establish connections between parameters of a matrix process and the coefficients of the corresponding difference equations. It is convenient to use these representations when estimating parameters of matrix processes.

1.3.4 Matrix Processes and Markov Functions

As mentioned previously, the matrix processes are a direct generalization of Markov functions. The restrictions imposed on the matrices describing a Markov function complicate the solution of the problem of the equivalence of two Markov functions and the problem of simplifying the model description. However, since Markov functions are matrix processes, the conditions of their equivalence can be formulated in terms of matrix processes. For two binary HMM with matrices $\mathbf{P}(e)$ and $\mathbf{P}_1(e)$ to be equivalent, it is necessary and sufficient that they be equivalent as matrix processes. In other words, it is necessary and sufficient that the models' ranks, canonic basis sequences, and their probabilities $p(s_i e t_j)$ coincide (see Appendix 1).

Any matrix process can be replaced by an equivalent process with the matrices $\mathbf{A}(e)$ of the minimal order, equal to the rank of the process (see Appendix 1). The similar replacement is impossible in the case of Markov functions. [11] Not every matrix process is equivalent to a function of a Markov chain with a finite number of states, [12,15] but every matrix process is equivalent to a function of a Markov chain with a countable number of states. [15]

The constraints that separate the class of Markov functions from the class of the matrix processes complicate minimizing the sizes of the matrices that define the Markov functions. The question of minimization of the number of states is closely related to Markov chain state lumpability (see Appendix 1). However, if an HMM is used for analytical computations, then minimization in the class of the matrix processes is possible, since the computation formulas have the same form for matrix processes as for Markov functions. If one uses the HMM for the error source simulation, then it is convenient to use a Markov function description, thus achieving a simple probabilistic interpretation.

Occasionally, especially in the processing of experimental data, it is important to know whether there is a possibility of reducing the number of states in the description of the HMM. A simple sufficient condition for the minimality of the number of states of the Markov chain can be obtained by analyzing the distributions of the lengths of the intervals between errors and the lengths of the intervals between correct symbols.

If the distributions of the length λ of the intervals between the errors and the lengths l of the intervals between correct symbols of a matrix process are polygeometric

$$p_0(\lambda) = \sum_{i=1}^{m_0} a_{0i} q_{0i}^{\lambda-1} \tag{1.3.18}$$

$$p_1(l) = \sum_{i=1}^{m_1} a_{1i} q_{1i}^{l-1} \tag{1.3.19}$$

($|q_{e1}| < |q_{e2}| < \cdots < |q_{em_e}|$), and the matrices $\mathbf{P}_{ee}$ have the orders m_e, then the number of the model states $m_0 + m_1$ is minimal (see Appendix 1). We will use this simple argument often.

1.4 ERROR SOURCE DESCRIPTION BY SEMI-MARKOV PROCESSES

1.4.1 Semi-Markov Processes

Consider now a generalization of the error source model based on semi-Markov processes: The source of errors has a finite number of states; transitions of the process from state to state are governed by a *semi-Markov* [21] or *Markov renewal* [8] process, which may be described as follows.

Suppose that the process was initially in the state c_0. It then enters the state c_1 with the probability

$f_{c_0c_1}^{(0)}(l_0)$ after l_0 steps. From the state c_1 it passes into the state c_2 with the probability $f_{c_1c_2}(l_1)$ after l_1 steps, and so on.

Consider now a set of h binary channels and assume that errors in these channels are generated by the common source, which may be in one of k states forming a semi-Markov process. If the error source transfers from state i to state j, an error $\mathbf{e}=(e_1,e_2,\ldots,e_h)$ appears with probability $\varepsilon_{ij}(\mathbf{e})$, where $e_k=1$ if k-th channel bit is in error and $e_k=0$ otherwise.

Since the model states are not observable, it is called a *hidden semi-Markov model* (HSMM). To describe the model we need a matrix of initial state probabilities $\mathbf{p}_0 = [\ p_i\ ]_{1,k}$, matrix of initial transition probabilities $\mathbf{F}_0(l) = [\ f_{ij}^{(0)}(l)\ ]_{k,k}$, matrix of transition probabilities $\mathbf{F}(l) = [\ f_{ij}(l)\ ]_{k,k}$, and matrices of the state-error probabilities $\mathbf{E}(\mathbf{e}) = [\ \varepsilon_{ij}(\mathbf{e})\ ]_{k,k}$. We showed previously that the Markov model of the error source is general enough to describe a wide variety of error statistics. However, the size of matrices used for defining the model may be quite large. The semi-Markov approach often leads to a significant reduction of the model parameter set.

An alternative description of a semi-Markov process is based on imbedded Markov chains. Let us consider the semi-Markov process only at the moments when it changes its states. The new process is the Markov chain with the transition probabilities $p_{ij}^{(0)} = \sum_{l=1}^{\infty} f_{ij}^{(0)}(l)$ for the first transition from the initial state and $p_{ij} = \sum_{l=1}^{\infty} f_{ij}(l)$ for subsequent transitions. The intervals between the two consecutive transitions have the distributions

$$w_{ij}^{(0)}(l) = f_{ij}^{(0)}(l)\ /\ p_{ij}^{(0)} \qquad w_{ij}(l) = f_{ij}(l)/p_{ij} \tag{1.4.1}$$

which are called the *state-holding time* probabilities. [21] The successive state transitions are governed by the Markov chain with transition matrices $\mathbf{P}^{(0)} = [\ p_{ij}^{(0)}\ ]_{k,k}$, $\mathbf{P} = [\ p_{ij}\ ]_{k,k}$. After a transition $i \to j$ has been selected, the process still holds the previous state i for l steps, according to transition distributions (1.4.1).

Thus, the model can be described by the matrix of the state initial probabilities $\mathbf{p}_0 = [\ p_i\ ]_{1,k}$, matrices of transition probabilities $\mathbf{P}^{(0)} = [\ p_{ij}^{(0)}\ ]_{k,k}$, $\mathbf{P} = [\ p_{ij}\ ]_{k,k}$, matrices of the interval distributions $\mathbf{W}^{(0)}(l) = [\ w_{ij}^{(0)}(l)\ ]_{k,k}$ and $\mathbf{W}(l) = [\ w_{ij}(l)\ ]_{k,k}$, and matrices of the probabilities of errors $\mathbf{E}(\mathbf{e}) = [\ \varepsilon_{ij}(\mathbf{e})\ ]_{k,k}$. We call the described error source model the *semi-Markov model*, and the corresponding channel is called a *symmetric semi-Markov channel.*

If the interval distributions depend only on the current state of the process $w_{ij}(l) = w_i(l)$, $i,j=1,2,\ldots,k$, the process is called *autonomous.* In this case, the matrix $\mathbf{W}(l)$ has identical columns. Any channel model based on a semi-Markov process is equivalent to some model based on an autonomous semi-Markov process. Indeed, consider a process whose states are defined as pairs (i,j) of the previously defined Markov process. The new process is a Markov process with the transition probabilities

$$p_{(ij),(m,n)} = \begin{cases} p_{jn} & \text{for} \quad m=j \\ 0 & \text{for} \quad m\neq j \end{cases}$$

For a process thus defined, the transition distribution $w_{ij}(l)$ depends only on the first state (i,j) if we assume that impossible transitions from the state (i,j) also have the same interval distribution.

Clearly, a Markov chain is a particular case of the semi-Markov process (with the same transition probability matrix and $\mathbf{W}(1)=\mathbf{I}$). Since triplets (i,j,l) constitute a Markov chain (with an infinite number of states), a semi-Markov process may be considered as a function of some Markov chain. We use this result to apply the statistical methods developed for Markov chains to semi-Markov models.

In general, it is possible for a semi-Markov process to transfer from some state i into the same state (with the probability $f_{ii}(l)$) so that no *real* transition of the process occurs. These types of transitions are called *virtual* transitions. [21] If we assume that the conditional probabilities $\varepsilon_{ij} = \varepsilon_i$ depend only on the current state of the process, then we can replace the semi-Markov process with an equivalent process whose virtual transitions are real. In order to do so, we have to replace the interval transition distribution $f_{ij}(l)$ with the real transition distribution

$$f_{ij}^{(1)}(l) = \sum_{m=1}^{\infty} f_{ij}^{*m}(l) \qquad \text{for } i \neq j$$

$$f_{ii}^{(1)}(l) = 0 \tag{1.4.2}$$

where $f_{ij}^{*m}(l)$ denotes the convolution

$$f_{ij}^{*m}(l) = \sum_{l_1+\ldots+l_m=l} f_{ii}(l_1)\cdots f_{ii}(l_{m-1}) f_{ij}(l_m)$$

Since the z-transform of a convolution is equal to the product of the z-transforms of its components, [16] the z-transform of the distribution (1.4.2)

$$\rho_{ij}^{(1)}(z) = \sum_{m,l=1}^{\infty} f_{ij}^{*m}(l) z^{-l} = \sum_{m=1}^{\infty} \rho_{ii}^{m-1}(z)\rho_{ij}(z)$$

has the form

$$\rho_{ij}^{(1)}(z) = \rho_{ij}(z)/[1-\rho_{ii}(z)] \tag{1.4.3}$$

where $\rho_{ij}(z) = \sum_l f_{ij}(l) z^{-l}$ is the z-transform of the original distribution. Distribution (1.4.2) can be found by decomposing its z-transform (1.4.3) into the Laurent series or by using the inverse Fourier transform.

Example 1.4.1. A Markov chain may be treated as a special case of the semi-Markov process whose interval distributions have the form $f_{ij}(1) = p_{ij}$. The z-transform $\rho_{ij}(z) = p_{ij}/z$ and Eq. (1.4.3) then takes the form

$$\rho_{ij}^{(1)}(z) = p_{ij}/(z-p_{ii}) = \sum_{l=1}^{\infty} p_{ii}^{l-1} p_{ij} z^{-l}$$

Therefore the real transition interval distribution $f_{ij}^{(1)}(l) = (1-p_{ii})p_{ii}^{l-1}$ is geometric, and the Markov chain can also be described as a semi-Markov process with geometric interval distributions.

The semi-Markov process with the interval distribution in the form of Eq. (1.4.2) is equivalent to the original process. The transitional probabilities of the modified process are equal to

$$p_{ij}^{(1)} = \sum_l f_{ij}^{(1)}(l) = \rho_{ij}^{(1)}(1) = p_{ij}/(1-p_{ii})$$

for $i \neq j$. Therefore, the i-th row of the modified matrix of transition probabilities has the form

$$(1-p_{ii})^{-1}[\, p_{i1} \;\; p_{i2} \cdots p_{i,i-1} \;\; 0 \;\; p_{i,i+1} \cdots p_{ik} \,] \tag{1.4.4}$$

In the preceding derivations we assume that the state i is not an initial state of the process. For the initial state the transformations can be performed quite analogously.

Thus, if $p_{ij} \neq 0$, the error source model description is not unique. Therefore, we assume that the transformations (1.4.2) and (1.4.4) are performed for every $i=1,2,...,k$ and the diagonal elements of the matrix $\mathbf{P}$ are equal to zero. We call such a representation of a semi-Markov process *canonical.*

In a particular case, when the source has two states ($k=2$), the canonical transition probability matrix has the form

$$\mathbf{P} = \begin{bmatrix} 0 & 1 \\ 1 & 0 \end{bmatrix}$$

we have an *alternating renewal process* with different probabilities of errors in different states. In this case, one of the lengths is often regarded as the length of a cluster of errors and the other as the time between clusters.

In practice, models based on semi-Markov processes or alternating processes are used relatively rarely because of the difficult calculations they entail. However, it is often convenient to represent the HMM in the form of the semi-Markov process for processing experimental data and especially for simulating an error flow. Let us study, then, under what conditions it is possible to describe the HMM by the semi-Markov process.

1.4.2 Semi-Markov Lumping of Markov Chains

Suppose we are given a homogeneous Markov chain with matrix $\mathbf{P} = [\, p_{ij} \,]_{u,u}$ and a matrix of initial probabilities $\mathbf{p}_0 = row\{p_i\} = [\, p_i \,]_{1,u}$. Let $\{A_1, A_2, ..., A_s\}$ be a partition of the set of all states into nonintersecting subsets:

$$\bigcup_{i=1}^{s} A_i = A \qquad A_i \cap A_j = \varnothing \qquad i \neq j$$

Let us consider a process y_t whose states are defined by the condition $y_t = i$ if at time t the Markov chain is in some state from the set A_i. The process y_t is usually called a *Markov function* (function of a Markov chain) or, in the terminology of Ref. 24, a *lumped process.*

To simplify notations, let us assume that the Markov chain states are numbered in such a way that

$$A_1 = \{1,2,...,m_1\},\; A_2 = \{m_1+1,...,m_2\}, ...,\; A_s = \{m_{s-1}+1,...,m_s\}$$

If the assumption is not valid, we need to renumber the states (see Appendix 6). In correspondence with this partition we can write the matrices $\mathbf{P}$ and $\mathbf{p}_0$ in the block form

$$\mathbf{P} = block[\mathbf{P}_{ij}]_{s,s} = \begin{bmatrix} \mathbf{P}_{11} & \mathbf{P}_{12} & \cdots & \mathbf{P}_{1s} \\ \mathbf{P}_{21} & \mathbf{P}_{22} & \cdots & \mathbf{P}_{2s} \\ \cdots & \cdots & \cdots & \cdots \\ \mathbf{P}_{s1} & \mathbf{P}_{s2} & \cdots & \mathbf{P}_{ss} \end{bmatrix}$$

$$\mathbf{p}_0 = block\ row\{\mathbf{p}_i\} = [\, \mathbf{p}_1 \;\; \mathbf{p}_2 \;\; \cdots \;\; \mathbf{p}_s \,]$$

where matrix block $\mathbf{P}_{ij}$ consists of all transition probabilities $p_{\alpha\beta}$, $\alpha \in A_i$, $\beta \in A_j$.

The lumped process y_t is not usually a Markov or semi-Markov process. Several necessary and sufficient conditions for the lumped process to be a Markov chain may be found in Ref. 24 (see also Appendix 1). We find conditions for the lumped process to be a semi-Markov process.

DEFINITION 1.4.1: We say that a Markov chain with the initial distribution $\mathbf{p}_0 = [\ \mathbf{p}_1\ \ \mathbf{p}_2\ \ \cdots\ \ \mathbf{p}_s\ ]$ is $\mathbf{p}_0$-semi-Markov *lumpable* by a partition $\{A_1, A_2, \ldots, A_s\}$, if the lumped process y_t is a homogeneous semi-Markov process.

Theorem 1.4.1: A necessary and sufficient condition for a Markov chain to be semi-Markov lumpable by a partition $\{A_1, A_2, \ldots, A_s\}$ is that there exist a matrix row of the initial probabilities $\hat{\mathbf{p}}_0 = row\{\hat{p}_i\}$ and transition matrix distributions $\mathbf{F}^{(0)}(l) = [\ f_{ij}^{(0)}(l)\ ]_{s,s}$ $\mathbf{F}(l) = [\ f_{ij}(l)\ ]_{s,s}$, such that

$$\mathbf{p}_{y_0} \prod_{i=0}^{m} \mathbf{P}_{y_i y_i}^{l_i - 1} \mathbf{P}_{y_i y_{i+1}} \mathbf{1} = \hat{p}_{y_0} f_{y_0 y_1}^{(0)}(l_0) \prod_{i=1}^{m} f_{y_i y_{i+1}}(l_i) \qquad (1.4.5)$$

for all $y_i = 1, 2, \ldots, s$, $l_i \in \mathbf{N}$, $i = 0, 1, \ldots, m$, $m \in \mathbf{N}$.

The proof of the theorem is obvious, since the previous relation is actually a paraphrase of the semi-Markov process definition. The left-hand side of the equation is the probability that the Markov function starts at y_0, stays in this state for l_0 steps, then enters y_1, stays in this state for l_1 steps, enters y_2, and so on and finally enters y_{m+1}. The right-hand side of the equation is the probability of the same trajectory if one assumes that y_t is a semi-Markov process.

Let us consider the Markov process in the moments of its transitions from a state of one subset A_i into a state of another subset A_j, assuming that the sets A_i are not absorbing. We call a process built in this way a *transition process*, which is obviously a Markov chain. Denote its matrix $\mathbf{Q} = [\ \mathbf{Q}_{ij}\ ]$. Let α and β be two states of the initial Markov chain and $\alpha \in A_i$, $\beta \in A_j$. The probability $q_{\alpha\beta}$ of transition from the state α into state β is a sum of probability $p_{\alpha\beta}$ of transition of the initial process from the state α into the state β in one step, in two steps:

$$\sum_{\gamma \in A_i} p_{\alpha\gamma} p_{\gamma\beta}$$

and so on. Noting that these probabilities are elements of block matrices, $\mathbf{P}_{ij}$, $\mathbf{P}_{ii}\mathbf{P}_{ij}$, and so on; we obtain

$$\mathbf{Q}_{ij} = \sum_{m=0}^{\infty} \mathbf{P}_{ii}^{m} \mathbf{P}_{ij}$$

Since the set A_i is not absorbing, this series converges. Its sum is

$$\mathbf{Q}_{ij} = (\mathbf{I} - \mathbf{P}_{ii})^{-1} \mathbf{P}_{ij} \qquad (1.4.6)$$

It is also obvious that $\mathbf{Q}_{ii} = 0$. Thus, if the initial process has the matrix $\mathbf{P} = [\ \mathbf{P}_{ij}\ ]$, then the transition process has the matrix

$$\mathbf{Q} = \begin{bmatrix} 0 & (\mathbf{I}-\mathbf{P}_{11})^{-1}\mathbf{P}_{12} & \cdots & (\mathbf{I}-\mathbf{P}_{11})^{-1}\mathbf{P}_{1s} \\ (\mathbf{I}-\mathbf{P}_{22})^{-1}\mathbf{P}_{21} & 0 & \cdots & (\mathbf{I}-\mathbf{P}_{22})^{-1}\mathbf{P}_{2s} \\ \cdots & \cdots & \cdots & \cdots \\ (\mathbf{I}-\mathbf{P}_{ss})^{-1}\mathbf{P}_{s1} & (\mathbf{I}-\mathbf{P}_{ss})^{-1}\mathbf{P}_{s2} & \cdots & 0 \end{bmatrix} \tag{1.4.7}$$

If we denote as $p_{\alpha\beta}(l)$ the probabilities that the chain remains in the set A_i during l steps, under the condition that the initial state belongs to this set ($\alpha \in A_i$) and the final state does not belong to the set ($\beta \notin A_i$), then we can verify that the probabilities $p_{\alpha\beta}(l)$ are the elements of the matrix

$$\mathbf{Q}_i(l) = \mathbf{P}_{ii}^{l-1}(\mathbf{I} - \mathbf{P}_{ii}) \tag{1.4.8}$$

Using the transition process notations, we rewrite the multidimensional distributions in the left-hand side of Eq. (1.4.5) as

$$\mathbf{p}_{y_0} \prod_{i=0}^{m} \mathbf{P}_{y_i y_i}^{l_i - 1} \mathbf{P}_{y_i y_{i+1}} \mathbf{1} = \mathbf{p}_{y_0} \prod_{i=0}^{m} \mathbf{Q}_{y_i}(l_i) \mathbf{Q}_{y_i y_{i+1}} \mathbf{1} \tag{1.4.9}$$

and Eq. (1.4.5) becomes

$$\mathbf{p}_{y_0} \prod_{i=0}^{m} \mathbf{Q}_{y_i}(l_i) \mathbf{Q}_{y_i y_{i+1}} \mathbf{1} = \hat{p}_{y_0} f_{y_0 y_1}^{(0)}(l_0) \prod_{i=1}^{m} f_{y_i y_{i+1}}(l_i) \tag{1.4.10}$$

If the Markov chain with the matrix $\mathbf{P}$ is regular, the transition process is not necessarily regular as well. This fact becomes obvious if we consider a Markov chain with two states and matrix

$$\mathbf{P} = \begin{bmatrix} p_{11} & p_{12} \\ p_{21} & p_{22} \end{bmatrix}$$

The matrix of the transition process, according to Eq. (1.4.7), is equal to

$$\mathbf{Q} = \begin{bmatrix} 0 & 1 \\ 1 & 0 \end{bmatrix}$$

and is not regular. However, if the matrix $\mathbf{P}$ is regular and its invariant vector is $\boldsymbol{\pi} = block\ row\{\boldsymbol{\pi}_i\}$, then it is easy to verify that the matrix $\mathbf{Q}$ has the invariant vector $\mathbf{q} = block\ row\{\mathbf{q}_i\}$ of the form

$$\mathbf{q}_i = \boldsymbol{\pi}_i(\mathbf{I} - \mathbf{P}_{ii}) / \sum_{i=1}^{s} \boldsymbol{\pi}_i(\mathbf{I} - \mathbf{P}_{ii})\mathbf{1}$$

which is the unique solution of the system

$$\mathbf{qQ} = \mathbf{q} \quad \mathbf{q1} = 1 \tag{1.4.11}$$

(see Appendix 1.3).

If the process starts at a random moment, it is necessary to use the initial transition distribution $f_{ij}^{(0)}(l)$. However, if the process starts at the moment of the set change (or, equivalently, the Markov function state change), then $f_{ij}^{(0)}(l) = f_{ij}(l)$ [21] and Eq. (1.4.10) takes on the form

$$\mathbf{q}_{y_1} \prod_{i=1}^{m} \mathbf{Q}_{y_i}(l_i)\mathbf{Q}_{y_i y_{i+1}}\mathbf{1} = \hat{q}_{y_1} \prod_{i=1}^{m} f_{y_i y_{i+1}}(l_i) \qquad (1.4.12)$$

where $\mathbf{q} = [\mathbf{q}_1 \ \ \mathbf{q}_2 \ \ldots \ \mathbf{q}_s \ ]$ is the initial distribution of the chain states at the moment of a transition from one of the sets A_i. Equation (1.4.12) can also be obtained from (1.4.10) by performing summation with respect to y_0 and l_0. In this case

$$\mathbf{q}_j = \sum_{i \neq j} \mathbf{p}_i \mathbf{Q}_{ij}$$

or $\mathbf{q} = \mathbf{pQ}$.

Thus, if a Markov chain is semi-Markov **p**-lumpable with respect to $\{A_1, A_2, \ldots, A_s\}$ and sets A_i are not absorbing, then it is **pQ**-lumpable with the initial transition distribution $f_{ij}^{(0)}(l)$ coinciding with the intermediate transition distribution $f_{ij}(l)$. We assume in the sequel that all the necessary conditions are met and use Eq. (1.4.12) as a criterion of a semi-Markov lumpability.

Consider now some properties of lumpable processes. For $m = 1$, Eq. (1.4.12) takes on the form

$$\mathbf{q}_i \mathbf{Q}_i(l) \mathbf{Q}_{ij} \mathbf{1} = \hat{q}_i f_{ij}(l) \qquad (1.4.13)$$

If we perform summation with respect to l, we obtain

$$\mathbf{q}_i \mathbf{Q}_{ij} \mathbf{1} = \hat{q}_i \hat{p}_{ij} \qquad (1.4.14)$$

The summation with respect to j gives

$$\mathbf{q}_i \mathbf{1} = \hat{q}_i$$

since the sum of the elements of each row of a stochastic matrix is equal to one:

$$\sum_{j=1}^{s} \mathbf{Q}_{ij} \mathbf{1} = \mathbf{1}$$

The z-transform of the matrix-geometric distribution

$$f_{ij}(l) = \mathbf{q}_i \mathbf{P}_{ii}^{l-1} \mathbf{P}_{ij} \mathbf{1} / \mathbf{q}_i \mathbf{1} \qquad (1.4.15)$$

is a rational function

$$\rho_{ij}(z) = \sum_{l=1}^{\infty} f_{ij}(l) z^{-l} = \mathbf{q}_i (z\mathbf{I} - \mathbf{P}_{ii})^{-1} \mathbf{P}_{ij} \mathbf{1} / \mathbf{q}_i \mathbf{1} \qquad (1.4.16)$$

so the interval distribution $f_{ij}(l)$ can be found by carrying out a partial-fraction expansion and identifying the inverse z-transform of the partial fractions (see Appendices 2.1 and 5.1.3):

$$f_{ij}(l) = \sum_k [A_{ijk}(l) \cos(a_{ik} l) + B_{ijk}(l) \sin(a_{ik} l)] \, q_{ik}^{l-1} \qquad (1.4.17)$$

where $A_{ijk}(l)$ and $B_{ijk}(l)$ are polynomials in l and $q_{ik}[\cos(a_{ik}) + \mathrm{j}\ \sin(a_{ik})]$ are the roots of the characteristic polynomial $\Delta(z) = \det\,(z\mathbf{I} - \mathbf{P}_{ii})$ or, in other words, eigenvalues of the matrix $\mathbf{P}_{ii}$.

The previous equations are the *necessary* conditions of semi-Markov lumpability.

Consequently, if a semi-Markov process has an interval transition distribution that does not have the form of Eq. (1.4.17), then one cannot represent this process as a function of a Markov chain with a finite number of states.

Equations (1.4.14) and (1.4.15) lead us to the following conclusion. If a Markov chain is semi-Markov lumpable by the partition $\{A_1, A_2, ..., A_s\}$, then the initial distribution and the interval transition distributions are uniquely defined. Thus, to check whether the Markov chain is lumpable, one may first determine the semi-Markov process parameters from equations (1.4.14) and (1.4.15) and then check if they satisfy Eq. (1.4.10). However, the system (1.4.10) contains an infinite number of equations, hence it is necessary to develop some methods to verify the process lumpability in a finite number of steps. These methods are presented in Appendix 1.2. We discuss here the simplest and, from the practical point of view, the most important sufficient conditions of lumpability.

Theorem 1.4.2: If the sets of states $\{A_1, A_2, ..., A_s\}$ of a Markov chain are not absorbing, and the transition matrices have the form

$$\mathbf{P}_{ij} = \mu_{ij}\mathbf{M}_i\mathbf{N}_j \qquad i \neq j \quad (i,j = 1,2,...,s) \tag{1.4.18}$$

where $\mathbf{N}_j$ is a matrix row, $\mathbf{M}_i$ is a matrix column, and μ_{ij} is a number, then the Markov chain is semi-Markov lumpable.

The proof of the theorem is given in Appendix 1. Without loss of generality, we can assume that $\mathbf{N}_j\mathbf{1} = 1$. Then $\mathbf{M}_i = (\mathbf{I} - \mathbf{P}_{ii})\mathbf{1}/\mu_i$, where $\mu_i = \sum_j \mu_{ij}$ [see Eq. (A.1.36) of Appendix 1]. The lumped semi-Markov process transition probability matrix can be written as $\hat{\mathbf{P}} = [\mu_{ij}/\mu_i]_{s,s}$ and the interval distributions are matrix-geometric: $f_{ij}(l) = \mathbf{N}_i\mathbf{P}_{ii}^{l-1}\mathbf{M}_j$. One can easily verify that rows of the product $\mathbf{M}_i\mathbf{N}_j$ are proportional so that condition (1.4.18) is equivalent to proportionality in the rows of the matrices

$$[\mathbf{P}_{i1} \;\; \mathbf{P}_{i2} \; ... \; \mathbf{P}_{i-1,i} \;\; \mathbf{P}_{i+1,i} \;\; \mathbf{P}_{is}]$$

Several examples of this theorem application are presented in the next section.

In the special case, the set of all states is partitioned into two subsets $\{A_1, A_2\}$ and the Markov chain matrix has the form

$$\mathbf{P} = \begin{bmatrix} \mathbf{P}_{11} & \mu_{12}\mathbf{M}_1\mathbf{N}_2 \\ \mu_{21}\mathbf{M}_2\mathbf{N}_1 & \mathbf{P}_{22} \end{bmatrix}$$

We obtain an equivalent semi-Markov process with the matrix

$$\hat{\mathbf{P}} = \begin{bmatrix} 0 & 1 \\ 1 & 0 \end{bmatrix}$$

and the interval transition distributions $f_{12}(l) = \mathbf{N}_1\mathbf{P}_{11}^{l-1}(\mathbf{I} - \mathbf{P}_{11})\mathbf{1}$ and $f_{21}(l) = \mathbf{N}_2\mathbf{P}_{22}^{l-1}(\mathbf{I} - \mathbf{P}_{22})\mathbf{1}$.

Thus, in this case, the Markov process is equivalent to an alternating renewal process. This result coincides with the result presented in Ref. 4. Other conditions for equivalency to an alternating renewal process presented in Ref. 4 are, in effect, conditions of strong and weak lumpability [24] of a Markov chain states by the partition $\{A_1, A_2\}$.

In practice, the opposite problem often arises: Determine whether a model described by a semi-Markov process (and an alternating renewal process in particular) can be described as an HMM. The solution of this problem is especially important in experimental data processing, since it is usually simpler to estimate the semi-Markov model parameters from experimental data, but analytic calculations are simpler for the HMM. As shown here, it is necessary that the interval distributions $f_{ij}(l)$ be matrix-geometric for the equivalence of a semi-Markov process to a Markov chain. The sufficient conditions are given in the following theorem.

Theorem 1.4.3: If the interval distributions of the semi-Markov process with matrices

$$\hat{\mathbf{P}} = [\ \hat{p}_{ij}\ ]_{s,s} \quad \hat{\mathbf{P}}_0 = [\ \hat{p}_i\ ]_{1,s} \tag{1.4.19}$$

have the form (1.4.15), where the matrices $\mathbf{p}_i$, $\mathbf{P}_{ii}$ satisfy the conditions

$$\mathbf{P}_{ii} > 0 \quad \mathbf{p}_i \geq 0 \quad \mathbf{p}_i \neq 0 \quad \mathbf{P}_{ii}\mathbf{1} < \mathbf{1} \quad \mathbf{p}_i\mathbf{P}_{ii} \leq \mathbf{p}_i \tag{1.4.20}$$

then this semi-Markov process is equivalent to the Markov function with the matrix

$$\mathbf{P} = \begin{bmatrix} \mathbf{P}_{11} & \hat{p}_{12}\mathbf{M}_1\mathbf{N}_2 & \cdots & \hat{p}_{1s}\mathbf{M}_1\mathbf{N}_s \\ \hat{p}_{21}\mathbf{M}_2\mathbf{N}_1 & \mathbf{P}_{22} & \cdots & \hat{p}_{2s}\mathbf{M}_2\mathbf{N}_s \\ \cdots & \cdots & \cdots & \cdots \\ \hat{p}_{s1}\mathbf{M}_s\mathbf{N}_1 & \hat{p}_{s2}\mathbf{M}_s\mathbf{N}_2 & \cdots & \mathbf{P}_{ss} \end{bmatrix} \tag{1.4.21}$$

where $\mathbf{M}_i = (\mathbf{I} - \mathbf{P}_{ii})\mathbf{1}$, $\mathbf{N}_j = \mathbf{p}_j(\mathbf{I} - \mathbf{P}_{jj})/\mathbf{p}_j(\mathbf{I} - \mathbf{P}_{jj})\mathbf{1}$, and the vectors of the initial probabilities are bound by the conditions

$$\mathbf{p}_0 = \mathit{block\ row}\{\mu_i\mathbf{p}_i\} \tag{1.4.22}$$

where

$$\mu_i = \frac{\hat{p}_i[\mathbf{p}_i(\mathbf{I}-\mathbf{P}_{ii})\mathbf{1}]^{-1}}{\sum_j \hat{p}_j[\mathbf{p}_j(\mathbf{I}-\mathbf{P}_{jj})\mathbf{1}]^{-1}\mathbf{p}_j\mathbf{1}}$$

Proof: It is sufficient to show that the matrix $\mathbf{P}$ is a stochastic matrix, since the equivalence of a Markov function to a semi-Markov process follows from Theorem 1.4.2. It follows from Eq. (1.4.20) that $\mathbf{P}_{ii} > 0$, but then $\mathbf{P}_{ij} = \hat{p}_{ij}\mathbf{M}_i\mathbf{N}_j \geq 0$. Since

$$\sum_{j=1}^{s}\mathbf{P}_{ij}\mathbf{1} = \mathbf{P}_{ii}\mathbf{1} + \sum_{j\neq i}\hat{p}_{ij}(\mathbf{I} - \mathbf{P}_{ii})\mathbf{1} = \mathbf{1}$$

the matrix $\mathbf{P}$ is stochastic.

If the matrix $\mathbf{P}$ is regular and $\hat{\mathbf{p}}_0$ is an invariant vector of the matrix $\hat{\mathbf{P}}$, then the vector $\mathbf{p}_0$ defined by formula (1.4.22) is an invariant vector of $\mathbf{P}$. Indeed,

$$\sum_{i=1}^{s} \mu_i \mathbf{p}_i \mathbf{P}_{ij} = \mu_j \mathbf{p}_j \mathbf{P}_{jj} + \sum_{i \neq j} \hat{p}_{ij} \mu_i \mathbf{p}_i \mathbf{M}_i \mathbf{N}_j$$

$$= \mu_j \mathbf{p}_j \mathbf{P}_{jj} + \sum_{i \neq j} \hat{p}_{ij} \hat{p}_i \mu_j \mathbf{p}_j (\mathbf{I} - \mathbf{P}_{jj}) / \hat{p}_j = \mu_j \mathbf{p}_j$$

This theorem is very important for fitting an HMM to any process. First we perform the process quantization in such a way that the quantization level transitions are approximated by a Markov chain whose states correspond to the quantization levels. Then we approximate the state durations with the matrix-geometric distributions that satisfy the theorem conditions. The HMM transition probability matrix is constructed using Eq. (1.4.21). Examples of the theorem application are given in Chapter 3.

1.4.2.1 Polygeometric Distribution

A simplest case of the matrix-geometric distribution satisfying the previous theorem conditions is a polygeometric distribution

$$w_i(l) = \sum_{j=1}^{J_i} a_{ij} (1 - q_{ij}) q_{ij}^{l-1} \tag{1.4.23}$$

$a_{ij} \geq 0$, $0 \leq q_{ij} < q_{i,j+1} < 1$, $(i = 1,2,\ldots,s, j = 1,2,\ldots,J_i)$. Indeed, it is easy to verify that

$$w_i(l) = \mathbf{N}_i \mathbf{Q}_i^{l-1} (\mathbf{I} - \mathbf{Q}_i) \mathbf{1}$$

where

$$\mathbf{p}_i = row\{a_{i1}/(1 - q_{i1})\} = \mathbf{N}_i (\mathbf{I} - \mathbf{Q}_i)^{-1} \tag{1.4.24}$$

$$\mathbf{N}_i = row\{a_{ij}\} \qquad \mathbf{Q}_i = \mathbf{P}_{ii} = diag\{q_{ij}\} \tag{1.4.25}$$

It is easy to check that under the imposed constraints these matrices satisfy the conditions of Theorem 1.4.3, and therefore the semi-Markov process with the polygeometric interval distributions can be modeled by an HMM.

This case has a simple probabilistic interpretation. Equation (1.4.23) represents a mixture of geometric distributions. The coefficient a_{ij} can be interpreted as a probability of selecting a particular mixture which can be treated as transitional probabilities of a Markov chain: if the semi-Markov process is in state i, then the j-th component is selected with probability a_{ij}. Thus, we have a composite semi-Markov process with states i,j. Since the composite process state durations are geometric $w_{ij}(l) = (1 - q_{ij}) q_{ij}^{l-1}$, it is a Markov chain. The transition probabilities of this chain have the form

$$Pr((k,m) \mid (i,j)) = \begin{cases} q_{ij} & \text{for } (k,m) = (i,j) \\ (1 - q_{ij}) \hat{p}_{ik} a_{km} & \text{for } (k,m) \neq (i,j) \end{cases}$$

Note that the transition probability matrix of this chain is a special case of Eq. (1.4.21).

If a semi-Markov process has two states (an alternating renewal process) with the polygeometric state holding time distributions, we have

$$\hat{\mathbf{P}} = \begin{bmatrix} 0 & 1 \\ 1 & 0 \end{bmatrix}$$

and therefore the equivalent binary HMM, according to Eq. (1.4.21), has the matrix

$$\mathbf{P} = \begin{bmatrix} \mathbf{Q}_1 & (\mathbf{I} - \mathbf{Q}_1)\mathbf{1N}_2 \\ (\mathbf{I} - \mathbf{Q}_2)\mathbf{1N}_1 & \mathbf{Q}_2 \end{bmatrix} \tag{1.4.26}$$

1.5 SOME PARTICULAR ERROR SOURCE MODELS

In this section we consider several error source models based on experimental data for a large variety of real channels. All of them can be described as HMMs. For the majority of real channels, the probability of staying in an error-free state is high. The better the channel is, the closer these probabilities are to one. Therefore, it is convenient to present these models by the matrix $\mathbf{R} = \mathbf{P} - \mathbf{I}$. Note that the stationary distribution satisfies the equation $\boldsymbol{\pi}\mathbf{R} = 0$ and matrix $\mathbf{R}$ has the same eigenvectors as $\mathbf{P}$ with eigenvalues $\lambda_R = \lambda_P - 1$. We discuss these properties in more detail when we consider continuous-time HMMs.

1.5.1 Single-Error-State Models

Let us consider one of the popular HMMs in which errors occur only in one state.[17,42,43] If the conditional probability of error in this state differs from 1, the model can be replaced with the equivalent model in which this probability is equal to 1 (see Example 1.2.3).

In this case the Markov chain matrix has the form

$$\mathbf{P} = \begin{bmatrix} \mathbf{P}_{11} & \mathbf{P}_{12} \\ \mathbf{P}_{21} & p_{nn} \end{bmatrix} \tag{1.5.1}$$

where $\mathbf{P}_{12}$ is a matrix column and $\mathbf{P}_{21}$ is a matrix row. The conditions (1.4.18) are satisfied with

$$\mathbf{M}_1 = \mathbf{P}_{12} \quad \mathbf{N}_1 = \mathbf{P}_{21} \quad \mu_{12} = \mu_{21} = 1 \quad \mathbf{N}_2 = 1 \quad \mathbf{M}_2 = 1$$

The interval transition distributions are given by

$$f_{12}(l) = \mathbf{P}_{21}\mathbf{P}_{11}^{l-1}\mathbf{P}_{12} \qquad f_{21}(l) = p_{nn}^{l-1}(1-p_{nn})$$

This Markov chain is equivalent to a renewal process with the matrix-geometric interval distribution

$$f(0) = p_{nn} \qquad f(l) = (1-p_{nn})f_{12}(l) \qquad \text{for } l > 0 \tag{1.5.2}$$

If the distribution $f_{12}(l)$ is polygeometric,

$$f_{12}(l) = \sum_{i=1}^{n-1} a_i(1-g_i)g_i^{l-1} \quad a_i > 0 \quad g_i > 0 \quad i = 1,2,\ldots,n-1$$

the matrix $\mathbf{P}_{11}$ can be diagonalized. Then the Markov chain matrix (1.5.1) takes the form

$$\mathbf{P} = \begin{bmatrix} g_1 & 0 & \dots & 0 & 1-g_1 \\ 0 & g_2 & \dots & 0 & 1-g_2 \\ \dots & \dots & \dots & \dots & \dots \\ 0 & 0 & \dots & g_{n-1} & 1-g_{n-1} \\ b_1 & b_2 & \dots & b_{n-1} & p_{nn} \end{bmatrix} \qquad (1.5.3)$$

where $b_j = (1-p_{nn})a_j$. The model whose matrix has this form is called Fritchman's partitioned model and is often used in practice.[17,23,42,43] It is easy to verify that the model's stationary distribution can be expressed as

$$\pi = [\, p_e b_1 (1-g_1)^{-1},\ p_e b_2 (1-g_2)^{-1}, \dots,\ p_e b_{n-1}(1-g_{n-1})^{-1},\ p_e \,]$$

where

$$p_e = \left(1 + \sum_{i=1}^{n-1} b_i/(1-g_i) \right)^{-1}$$

is the bit error probability.

Several models, not equivalent to Fritchman's partitioned model, are based on the random walk between the states.[28] In the case of pure random walk, the transition probability matrix (1.5.1) is given by[16,28]

$$\mathbf{P} = \begin{bmatrix} 0.5 & 0.5 & 0 & 0 & \cdots & 0 & 0 & 0 \\ 0.5 & 0 & 0.5 & 0 & \cdots & 0 & 0 & 0 \\ 0 & 0.5 & 0 & 0.5 & \cdots & 0 & 0 & 0 \\ 0 & 0 & 0.5 & 0 & \cdots & 0 & 0 & 0 \\ \cdots & \cdots & \cdots & \cdots & \cdots & \cdots & \cdots & \cdots \\ 0 & 0 & 0 & 0 & \cdots & 0.5 & 0 & 0.5 \\ 0 & 0 & 0 & 0 & \cdots & 0 & 0.5 & 0.5 \end{bmatrix}$$

The eigenvalues of this matrix are equal to[16]

$$g_i = \cos\left(\frac{2i-1}{2n-1}\pi\right) \qquad i = 1,2,\dots,n-1$$

Some of these eigenvalues are negative, and therefore the model's matrix is not diagonalizable in the class of Markov functions. It may be transferred to its diagonal form in the class of matrix processes (see Sec. 1.3).

It follows from Theorem 1.4.3 that a renewal process with matrix-geometric interval distribution is equivalent to a Markov function. Thus, if intervals between consecutive errors are independent, the model can be constructed by approximating the interval distribution with the matrix-geometric distribution. In particular, an alternating process model[14] with polygeometric interval distribution can be expressed in Fritchman's partitioned form (1.5.3). This result is often used in practice for estimating model parameters.

Since Gilbert's model is a single-error-state model, the previous results are applicable to it. Therefore, the model matrix may be presented in the form of Eq. (1.5.3):

$$\mathbf{P} = \begin{bmatrix} g_1 & 0 & 1-g_1 \\ 0 & g_2 & 1-g_2 \\ b_1 & b_2 & q(1-h) \end{bmatrix} \quad \text{or} \quad \mathbf{R} = \begin{bmatrix} g_1-1 & 0 & 1-g_1 \\ 0 & g_2-1 & 1-g_2 \\ b_1 & b_2 & -p-qh \end{bmatrix} \tag{1.5.4}$$

(We obtained this result independently in Example 1.2.4.)

Because the model has only three independent parameters, it is sufficient to approximate the error-free run distribution with the bigeometric function

$$f_\lambda(\lambda) = a_1(1-g_1)g_1^\lambda + a_2(1-g_2)g_2^\lambda$$

and then obtain the model parameters by solving some algebraic Eq. (see Example 1.2.4 and Ref. 19). For statistical measurements of one of the real telephone channels, [1] the following estimates have been found [19] $a_1 = 0.184$, $a_2 = 0.816$, $g_1 = 0.99743$, and $g_2 = 0.81$. These values give the model parameter estimates $h = 0.84$, $P = 0.003$, and $p = 0.034$. Matrix (1.5.4) is estimated by

$$\mathbf{R} = \begin{bmatrix} -0.00257 & 0 & 0.00257 \\ 0 & -0.19 & 0.19 \\ 0.15556 & 0.68988 & -0.84544 \end{bmatrix} \tag{1.5.5}$$

Using similar reasoning, the following models have been constructed for a high-frequency radio link [17]

$$\mathbf{R} = \begin{bmatrix} -0.34 & 0 & 0.34 \\ 0 & -9\cdot 10^{-4} & 9\cdot 10^{-4} \\ 0.44 & 0.34 & -0.78 \end{bmatrix} \tag{1.5.6}$$

and for a troposcatter channel: [43]

$$\mathbf{R} = \begin{bmatrix} -4.8\cdot 10^{-4} & 0 & 0 & 4.8\cdot 10^{-4} \\ 0 & -4.66\cdot 10^{-3} & 0 & 4.66\cdot 10^{-3} \\ 0 & 0 & 5.804\cdot 10^{-2} & 5.804\cdot 10^{-2} \\ 0.06902 & 0.25289 & 0.63308 & -0.95499 \end{bmatrix} \tag{1.5.7}$$

A typical model for the T1 digital repeatered line [5] has the matrix

$$\mathbf{R} = \begin{bmatrix} -5.072\cdot 10^{-2} & 0 & 5.072\cdot 10^{-2} \\ 0 & -2.211\cdot 10^{-9} & 2.211\cdot 10^{-9} \\ 0.1446 & 0.7688 & -0.9134 \end{bmatrix} \tag{1.5.8}$$

1.5.2 Alternating Renewal Process Models

An alternating renewal process model is a particular case of semi-Markov model with two states. These states alternate and have different interval transition distributions (see Sec. 1.4.1.). Several models satisfy this description.

One of the models of this type was proposed by Bennet and Froelich. [2] According to this model, bursts of errors occur independently with the fixed probability p_b; length l burst appears with the probability $P_b(l)$. Inside bursts, errors occur with some probability ε; there are no errors outside bursts.

Strictly speaking, this model permits the overlapping of several bursts, but, in practice, it is difficult to identify such a case, so overlapping bursts are treated as a single burst. If we assume that bursts do not overlap, the model can be described as a semi-Markov process with the transition probability matrix

$$\mathbf{P} = \begin{bmatrix} q_b & p_b \\ q_b & p_b \end{bmatrix}$$

and the state-holding time distributions

$$w_{11}(1) = 1 \qquad w_{22}(l) = P_b(l)$$

If we transform this description to its canonical form (see Sec. 1.4.1), we obtain an alternating renewal process with the matrix

$$\mathbf{P} = \begin{bmatrix} 0 & 1 \\ 1 & 0 \end{bmatrix}$$

and transition distributions

$$f_{12}(l) = p_b q_b^{l-1} \qquad f_{21}(l) = \sum_{m=1}^{\infty} q_b p_b^{m-1} P_b^{*m}(l)$$

The z-transforms of these transition distributions are defined by Eq. (1.4.3), which in our case take the form

$$\rho_{12}(z) = \frac{p_b}{z-q_b} \qquad \rho_{21}(z) = \frac{q_b \rho_b(z)}{z-p_b\rho_b(z)}$$

If the burst-length distribution is matrix geometric, this model, according to Theorem 1.4.3, can be described as an HMM. If the burst-length distribution is polygeometric,

$$P_b(l) = \sum_{i=1}^{n-1} a_i(1-g_i)g_i^{l-1} \qquad a_i > 0 \quad g_i > 0 \qquad i=1,2,\ldots,n-1$$

then the model matrix can be written in the form of Eq. (1.5.3). However, in contrast to the single-error-state models considered in the previous section, this model has only one state in which errors do not occur. In all the remaining states they occur with the probability ε.

If the burst-length distribution is geometric

$$P_b(l) = (1-g)g^{l-1}$$

then

$$\rho_{21}(z) = \frac{(1-g)p_b}{z-gq_b-p_b}$$

and therefore the transition distributions are also geometric:

$$f_{12}(l) = p_b q_b^{l-1} \qquad f_{21}(l) = (1-q)q^{l-1}$$

where $q = gq_b + p_b$. In this case, this model is Gilbert's model with the matrix

$$\mathbf{P} = \begin{bmatrix} q_b & p_b \\ 1-q & q \end{bmatrix}$$

The error-free runs in this case have the bigeometric distribution

$$f_\lambda(\lambda) = a_1(1-g_1)g_1^\lambda + a_2(1-g_2)g_2^\lambda$$

in which g_1 and g_2 are different positive roots of the quadratic Eq. (see Example 1.2.1)

$$g^2 - (q_b + q(1-\varepsilon))g + (1-\varepsilon)(q_b + q - 1) = 0$$

Analysis of experimental data showed that the assumption that error bursts are independent is not valid in some cases. In many cases the assumption that error bursts occur independently inside larger error clusters agreed with the experimental data. [30] It has been discovered that, for a large variety of telephone channels, error clusters occurred independently with the fixed probability p_c. The error-cluster length distribution was geometric

$$P_c(l) = (1-g_c)g_c^{l-1}$$

and the burst-length distribution was tri-geometric

$$P_b(l) = a_1(1-g_1)g_1^{l-1} + a_2(1-g_2)g_2^{l-1} + a_3(1-g_3)g_3^{l-1}$$

Ignoring the possibility of error-cluster overlapping and error bursts overlapping inside the clusters, we can easily conclude that this model can be described by an alternating renewal process with the transition distributions

$$f_{12}(l) = a_{11}(1-g_{11})g_{11}^{l-1} + a_{12}(1-g_{12})g_{12}^{l-1} \qquad f_{21}(l) = P_b(l)$$

where g_{11} and g_{12} are the roots of the quadratic equation

$$g^2 - [q_c + q(1-p_{bc})]g + (1-p_{bc})(q_c + q - 1) = 0$$

In this equation, $q_c = 1 - p_c$, p_{bc} is the conditional probability that an error burst starts at any bit inside the cluster, and $q = g_c q_c + p_c$.

This model is a particular case of the HMM considered in Sec. 1.4.2.1 and has the matrix (1.4.26), in which

$$\mathbf{Q}_1 = \begin{bmatrix} g_{11} & 0 \\ 0 & g_{11} \end{bmatrix} \qquad \mathbf{Q}_2 = \begin{bmatrix} g_1 & 0 & 0 \\ 0 & g_2 & 0 \\ 0 & 0 & g_3 \end{bmatrix}$$

$$\mathbf{N}_1 = [\, a_1 \;\; a_2 \;\; a_3 \,] \qquad \mathbf{N}_2 = [\, a_{11} \;\; a_{12} \,]$$

A typical model, described in Ref. 4, has the parameter estimates

$$\mathbf{Q}_1 = \begin{bmatrix} 0.99789 & 0 \\ 0 & 0.9999951 \end{bmatrix} \qquad \mathbf{Q}_2 = \begin{bmatrix} 0.26 & 0 & 0 \\ 0 & 0.93 & 0 \\ 0 & 0 & 0.992 \end{bmatrix}$$

$$\mathbf{N}_1 = [\, 0.416 \;\; 0.267 \;\; 0.317 \,] \qquad \mathbf{N}_2 = [\, 0.86 \;\; 0.14 \,]$$

and the probability of errors inside bursts $\varepsilon = 0.67$.

1.5.3 Alternating state HMM

The renewal assumptions made in the previous sections simplify the model fitting to experimental data. However, these assumptions do not agree with experimental data on errors in many real communication channels: long intervals between errors (or error bursts) are more likely to follow long intervals. A more general model with polygeometric interval distributions which does not assume the interval independence is an HMM whose transition probability matrix has the form (1.2.29)

$$\mathbf{P} = \begin{bmatrix} \mathbf{Q}_1 & \mathbf{P}_{12} \\ \mathbf{P}_{21} & \mathbf{Q}_2 \end{bmatrix} \tag{1.5.9}$$

where $\mathbf{Q}_1 = diag\{\gamma_i\}$ and $\mathbf{Q}_2 = diag\{g_i\}$ are diagonal matrices. This model does not allow transfers between states of the same type. Thus, good and bad states alternate. A state transition diagram of the model of this type with two good states (G_1 and G_2) and two bad states (B_1 and B_2) is depicted in Fig. 1.5.

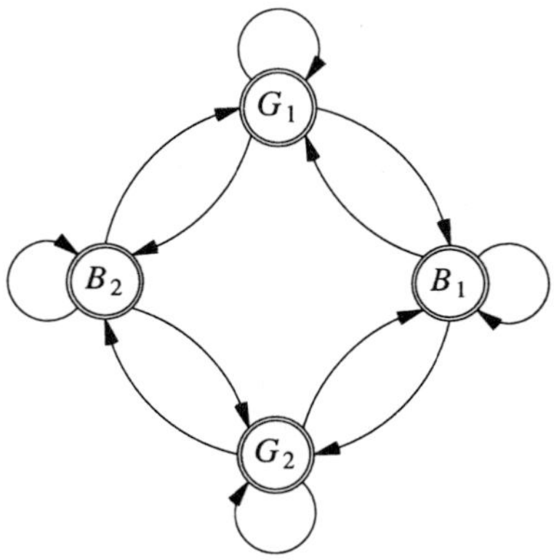

Figure 1.5. Alternating State Model.

This model is more general than the single-error-state model considered above. Indeed, if $\mathbf{Q}_2 = p_{nn}$ is a number(a 1×1 matrix), it is a single-state-error model. If rows of the matrices $\mathbf{P}_{12}$ and $\mathbf{P}_{21}$ are proportional, the model represents an alternating renewal process with the polygeometric interval distributions.

The model is general enough to approximate large varieties of experimental data. On the other hand, the model's special structure simplifies its applications and fitting to experimental data [37,44] (see Chapter 3).

The following canonic model was obtained for the indoor wireless channel.[37]

$$\mathbf{R} = \begin{bmatrix} -10^{-3} & 0 & 0 & 0 & 10^{-3} & 0 \\ 0 & -4.9\cdot 10^{-2} & 0 & 0 & 0.047 & 0.003 \\ 0 & 0 & -0.305 & 0.1 & 0.023 & 0.182 \\ 0 & 0.001 & 0.514 & -0.515 & 0 & 0 \\ 0.077 & 0.76 & 0.112 & 0 & -0.948 & 0 \\ 0 & 0.045 & 0.717 & 0 & 0 & -0.762 \end{bmatrix} \tag{1.5.10}$$

The first three states of this model are good (no errors) and the last three states are bad (errors occur with probability one).

1.5.4 Fading Channel Errors

We described the fading channel model in Sec.1.1.9 and showed that it can be described as a symbol error HMM with a symbol error given by Eq. (1.1.30) or (1.1.31). The description depends on the fading quantized envelope HMM. In practice, the quantization levels are selected in such a way that the sequence of quantized fade amplitudes can be approximated by a semi-Markov model whose states are the quantization levels. Since we assumed that fading is slow, the fade levels of the two consecutive received symbols belong to the same quantization interval or to the adjacent intervals. Therefore, the semi-Markov process transition probability matrix is tridiagonal with zeroes on the main diagonal:

$$\hat{\mathbf{P}} = \begin{bmatrix} 0 & \hat{p}_{12} & 0 & 0 & \cdots & 0 & 0 \\ \hat{p}_{21} & 0 & \hat{p}_{23} & 0 & \cdots & 0 & 0 \\ 0 & \hat{p}_{32} & 0 & \hat{p}_{34} & \cdots & 0 & 0 \\ \cdots & \cdots & \cdots & \cdots & \cdots & \cdots & \cdots \\ 0 & 0 & 0 & 0 & \cdots & 0 & \hat{p}_{m-1,m} \\ 0 & 0 & 0 & 0 & \cdots & \hat{p}_{m,m-1} & 0 \end{bmatrix} \tag{1.5.11}$$

We can use multiple Markov chains described in Sec. 1.1.7 to incorporate the fading trend into our model.

In mobile communications, the following model is very popular. It is assumed that $a(t)$ is a zero-mean complex Gaussian process with independent and identically distributed real and imaginary parts. The PDF of a sequence $\boldsymbol{a} = (a(t_1), a(t_2), \ldots, a(t_k))$ has the form

$$f(\boldsymbol{a}) = (2\pi)^{-k} |\mathbf{D}| \exp(-0.5\mathrm{Re}\{\boldsymbol{a}\mathbf{D}\boldsymbol{a}^H\}) \tag{1.5.12}$$

where $|\mathbf{D}|$ denotes determinant of $\mathbf{D}$, $\boldsymbol{a}^H$ denotes the conjugate transpose of $\boldsymbol{a}$, and $\mathbf{D}^{-1}$ is the process covariance matrix:

$$\mathbf{D}^{-1} = [R(t_j - t_i)]_{k,k} \tag{1.5.13}$$

$R(\tau)$ is the process autocorrelation function. Since the fading envelope $\alpha = |a(t)|$ has a Rayleigh distribution

$$Pr(\alpha < a) = 1 - \exp(-0.5a^2/R(0))$$

it is called a Rayleigh fading. [22] Different models of fading channels are based on different assumptions about the fading power spectral density $S(f)$ (or autocorrelation function $R(\tau)$). The most popular model assumes that

$$S(f) = \mu / \pi\sqrt{f_D^2 - f^2} \qquad R(\tau) = \mu J_0(2\pi f_D|\tau|) \tag{1.5.14}$$

where $J_0(\cdot)$ is the Bessel function of the first kind and f_D is the maximal Doppler frequency. The other models [25,33] include rational functions, simple irrational functions, and Gaussian PDF. Note that an HMM autocorrelation function has a form given by Eq. (2.1.18) which can be written as

$$R(\tau) = \sum_j [P_j(\tau)\cos\omega_j\tau + Q_j(\tau)\sin\omega_j\tau)q_j^\tau$$

where $P_j(\tau)$ and $Q_j(\tau)$ are polynomials.

In the majority of papers [35,41,43,46] the quantized Rayleigh fading is modeled as a Markov chain. As we know, this means that the fade level durations have geometrical distributions. However, our study showed that this assumption is not valid, [45] but it is possible to fit matrix-geometric distributions to the fade level duration histograms and obtain an HMM whose state transition probability matrix has the form of Eq. (1.4.21) with $\hat{\mathbf{P}}$ of equation (1.5.11).

1.6 CONCLUSION

We have considered various methods of modeling digital stochastic processes and proved that the HMM is general enough to approximate a wide variety of experimental data. The actual model building on the basis of the experimental data is presented in Chapter 3.

We discussed advantages and disadvantages of different descriptions of the model and presented several methods for simplifying the model. It turned out that to minimize the number of the model parameters it is necessary to generalize the model allowing its parameters to be complex numbers. This generalization preserves the matrix-train form of multidimensional distributions. The set of probabilities of all possible error sequences is a finite-dimensional linear space, therefore, it is possible to express the probability of any sequence as a linear combination of probabilities of some basis sequences. We showed in Appendix 1 that the basis selection can simplify the model description. The major part of this material is an adaptation of results published in mathematical papers and monographs.

In Appendix 1, we prove basic statements made in this chapter and show the relationships between the basis sequence selection and the problems of lumping the states of a Markov chain.

Problems

1. Gilbert-Elliott model [13,19] is defined as a two-state HMM with the transition probability matrix

$$\mathbf{P} = \begin{bmatrix} p_{00} & p_{01} \\ p_{10} & p_{11} \end{bmatrix}$$

and state bit error probabilities ε_0 and ε_1.

— Find matrix probabilities of bit errors $\mathbf{P}(e)$

— Find conditions for the matrix $\mathbf{P}$ regularity and the stationary state probability vector $\boldsymbol{\pi}$.

— Find conditions for the matrices $\mathbf{P}(0)$ and $\mathbf{P}(1)$ commutativity.

— Find the canonical representation for the model.

— Find the conditions for the model to be equivalent to the channel without memory.

— For the stationary model, find bit-error probability, probability distribution of intervals between consecutive errors, and probability that a block of n bits does not have errors.

2. For a binary stationary HMM

— Show that $Pr(0^m 1) = Pr(10^m) = Pr(0^m) - Pr(0^{m+1})$.

— Show that probabilities $f_\lambda(0,0,...,0) = Pr(1^m \mid 1)$ and $f_l(m) = Pr(1^m 0 \mid 0)$ are dependent.

3. For the Gilbert-Elliott model (see Problem 1) in its canonical form, find the transition process matrix (1.4.7) and the set-holding distributions (1.4.8).

4. Let y_t be a quantized autoregressive process

$$y_k = Q(y_{k-1} - y_{k-2} + n_k)$$

where n_k is the white Gaussian noise with the power spectral density $N_0 = 3$ and equation

$$Q(y) = \begin{cases} 0 & \text{if } y \le 0 \\ 1 & \text{if } 0 < y \le 2 \\ 3 & \text{if } y > 2 \end{cases}$$

defines the quantizer. Find MPs of y_t.

5. Consider the HMM whose MPDF is given by

$$\mathbf{P} = \begin{bmatrix} p_{11} & p_{12} \\ p_{21} & p_{22} \end{bmatrix}, \quad \mathbf{F}(x) = \begin{bmatrix} \lambda_1 e^{-\lambda_1 x} & 0 \\ 0 & \lambda_2 e^{-\lambda_2 x} \end{bmatrix}$$

Present this model as an DTS. [Write equations similar to Eq. (1.1.7).]

6. We say that a fading channel is in a good state if the fading amplitude is above a certain threshold h and in bad state otherwise. Suppose that the good state duration and bad state duration are independent and have the following polygeometric probability distributions

$$p_g(x) = \mu(1-q_{11})q_{11}^{x-1} + (1-\mu)(1-q_{12})q_{12}^{x-1}$$
$$p_b(x) = \nu(1-q_{21})q_{21}^{x-1} + (1-\nu)(1-q_{22})q_{22}^{x-1}$$

Explain why this model is equivalent to an HMM and find the canonical HMM which is equivalent to this model.

7. Describe the system shown in Fig. 1.3 (including the differential encoder and decoder) as an FSC.

References

1 A. A. Alexander, R. M. Gryb, and D. W. Nast, "Capabilities of the telephone network for data transmission," *Bell Syst. Tech. J.*, **39**(3), 471-476, May (1963).

2 W. R. Bennett and F. E. Froelich, "Some results on the effectiveness of error control procedures in digital data transmission," *IRE Trans. on Comm. Syst.*, **CS-9**(1), 58-65, (1961).

3 D. Blackwell and L. Koopmans, "On the identifiability problem for functions of finite Markov chains," *Ann. Math. Stat.*, **28**(4), 1011-1015, (1957).

4 E. L. Bloch, O. V. Popov, W. Ya. Turin, *Models of Error Sources in Channels for Digital Information Transmission* (in Russian), (Sviaz Publishers, Moscow, 1971).

5 M. B. Brilliant, "Observations of errors and error rates on T1 digital repeatered lines," *Bell Syst. Tech. J.*, **57**(3), 711-746, March (1978).

6 R. G. Bucharaev, *Probabilistic Automata* (in Russian), (Kazan University Press, Kazan, 1970).

7 J. W. Carlyle, "Reduced forms for stochastic sequential machines," *J. Math. Anal. and Appl.*, **7**(2), 167-175, (1963).

8 E. Cinlar, *Introduction to Stochastic Processes,* (Prentice Hall, Englewood Cliffs, New Jersey, 1975).

9 E. L. Cohen and S. Berkovits, "Exponential distributions in Markov chain models for communication channels," *Inform. Control,* **13**, 134 -139, (1968).

10 S. W. Dharmadhikary, "Functions of finite Markov chains," *Ann. Math. Stat.*, **34**(3), 1022-1032, (1963).

11 S. W. Dharmadhikary, "Sufficient conditions for a stationary process to be a function of a finite Markov chain," *Ann. Math. Stat.*, **34**(3), 1033-1041, (1963).

12 S. W. Dharmadhikary, "Characterization of class of functions of finite Markov chains," *Ann. Math. Stat.*, **35**(2), 524-528, (1965).

13 E. O. Elliott, "Estimates of error rates for codes on burst-noise channels," *Bell Syst. Tech. J.*, **42**, 1977-1997, Sept. (1963).

14 E. O. Elliott, "A model for the switched telephone network for data communications," *Bell Syst. Tech. J.*, **44**, 89-119, Jan. (1965).

15 R. V. Ericson, "Functions of Markov chains," *Ann. Math. Stat.*, **41**(3), 843-850, (1970).

16 W. Feller, *An Introduction to Probability Theory and Its Applications*, **1**, (John Wiley & Sons, New York, 1962).

17 B. D. Fritchman, "A binary channel characterization using partitioned Markov chains," *IEEE Trans. Inform. Theory,* **IT-13**, 221-227, Apr. (1967).

18 I. I. Gichman and A. V. Scorochod, *The Theory of Stochastic Processes* (in Russian), **1**, (Science Publishers, Moscow, 1971).

19 E. N. Gilbert, "Capacity of a burst-noise channel," *Bell Syst. Tech. J.*, **39**, 1253-1266, Sept. (1960).

20 A. Graham, *Kronecker Products and Matrix Calculus with Applications*, (Ellis Horwood Ltd., 1981).

21 R. A. Howard, *Dynamic Probabilistic Systems,* **II:** *Semi-Markov and Decision Processes.* (John Wiley & Sons, New York, 1971).

22 W.C. Jakes, *Microwave Mobile Communications*, (John Wiley & Sons, New York, 1974).

24 J. G. Kemeny and J. L. Snell, *Finite Markov Chains,* (Van Nostrand, Princeton, New Jersey, 1960).

23 L. N. Kanal and A. R. K. Sastry, "Models for channels with memory and their applications to error control," *Proc. of the IEEE,* **66**(7), 724-744, July (1978).

25 W. C. Y. Lee, *Mobile Communications Engineering*, 2nd ed., (McGraw-Hill, New York, 1997).

26 M. F. Neuts, *Matrix-Geometric Solutions in Stochastic Models*, (Johns Hopkins, Baltimore, 1981).

27 R. H. McCullough, "The binary regenerative channel," *Bell Syst. Tech. J.*, **47**, 1713-1735, Oct. (1968).

28 K. Müller, "Simulation Buschelatiger Storimpulse," *Nachrichtechn. Z.*, **21**(11), 688-692, (1968).

29 A. Paz, *Introduction to Probabilistic Automata*, (Academic Press, New York, 1971).

30 O. V. Popov and W. Ya. Turin, "On the nature of errors in binary communication over standard telephone channels" (in Russian), *Second All-Union Conf. on Coding Theory and Its Applications*, sec. 3, part II, (1965).

31 J.G. Proakis, *Digital Communications,* (McGraw-Hill, New York, 1989).

32 L. Rabiner and B.-H. Juang, *Fundamentals of Speech Recognition*, (Prentice Hall, Englewood Cliffs, New Jersey, 1993).

33 S.O. Rice, "Distribution of the duration of fades in radio transmission: Gaussian noise model," *Bell Syst. Tech Journ.*, **37**, 581-635, May (1958).

34 R. Y. Rubinstein, *Simulation and the Monte Carlo Method,* (Wiley, 1981, New York).

35 M. Sajadieh, F.R. Kschischang, and A. Leon-Garcia, "A block memory model for correlated Rayleigh fading channels," *roc. IEEE Int. Conf. Commun.*, 282-286, June (1996).

36 M. Schwartz, W. R. Bennett and S. Stein, *Communication Systems and Techniques,* (McGraw-Hill, New York, 1966).

37 S. Sivaprakasam and K.S. Shanmugan, "An equivalent Markov model for burst errors in digital channels," *IEEE Trans. Commun.*, **43**, 1347-1355, April (1995).

38 C. E. Shannon, "A Mathematical Theory of Communication," *Bell Syst. Tech. J.*, **27**, 379-423, 623-656, July, October, (1948). (Also in *Claude Elwood Shannon Collected Papers*, N. J. A. Sloan and A. D. Wyner Eds, IEEE Press, Piscataway, New Jersey, 1993.)

39 J. J. Spilker, Jr., *Digital Communications by Satellite,* (Prentice-Hall, Englewood Cliffs, New Jersey, 1977).

40 M. L. Steinberger, P. Balaban, K. S. Shanmugan, "On the effect of uplink noise on a nonlinear digital satellite channel," in *Conf. Rec., 1981 Int. Conf. Commun.*, paper 20.2, Denver, Colorado, June (1981).

41 F. Swarts and H.C. Ferreira, "Markov characterization of channels with soft decision outputs," *IEEE Trans. Commun.*, **41**, 678-682, May (1993).

42 J. Swoboda, "Ein Statistischen Modell für die Fehler bei Binarer Datenübertragung auf Fernsprechknälen," *Arch. Elektr. Ubertag.*, no. 6, (1969).

43 S. Tsai, "Markov characterization of the HF channel," *IEEE Trans. Commun. Technol.*, **COM-17**, 24-32, Feb. (1969).

44 W. Turin, and M. M. Sondhi, "Modeling error sources in digital channels," *IEEE Journ. Sel. Areas in Commun.*, **11**(3), 340-347, Apr. (1993).

45 W. Turin and R. VanNobelen, "Hidden Markov modeling of fading channels," *48th Vech,. Tech. Conf.*, 1234-1238, May, (1998).

46 H.S. Wang and P.-C. Chang, "On verifying the first-order Markovian assumption for a Rayleigh fading channel model," *IEEE Trans. Veh. Technol.*, **45**, 353-357, May (1996).

47 H.S. Wang and N. Moayeri, "Finite-state Markov channel - a useful model for radio communication channels," *IEEE Trans. Veh. Technol.*, **44**, 163-171, Feb. (1995).

48 M. Zorzi and R.R. Rao, "On the statistics of block errors in bursty channels," *IEEE Trans. Commun.*, **45**, 660-667, Jun. (1997).

2

MATRIX PROBABILITIES

2.1 MATRIX PROBABILITIES AND THEIR PROPERTIES

A communication system performance analysis includes the calculation of performance characteristics such as received symbol-error probability, the average interval between errors, the average length of an error burst, error number distribution in a message, etc. The convenience of calculating these and similar characteristics often determines the choice of an error source model. In this chapter, we develop methods of calculation based on the HMM.

With this model, the probabilities of various events depend on the HMM states. The conditional probabilities of an event for different HMM states form a matrix that we call the matrix probability (MP) of the event. Usage of matrix probabilities simplifies calculations and allows us to find solutions in the matrix form that are applicable to all HMMs. Elementary examples of using MP were considered in Chapter 1. In this chapter, we present some theoretical aspects of MP application which are necessary to understand the rest of the book.

2.1.1 Basic Definitions

The MP of a sequence of errors $\mathbf{e}_1^m$ is defined by the following equation

$$\mathbf{P}(\mathbf{e}_1^m) = \prod_{i=1}^{m} \mathbf{P}(\mathbf{e}_i) \tag{2.1.1}$$

Using this notation we can rewrite Eq. (1.1.9) as

$$Pr(\mathbf{e}_1^m) = \mathbf{p}_0 \mathbf{P}(\mathbf{e}_1^m)\mathbf{1} \tag{2.1.2}$$

It is convenient to use MPs, because they do not depend on the HMM initial state distribution. The transitive property of the multiplication of matrix probabilities permits the construction of a "matrix probability theory" whose methods, as we will show later, are useful in calculating HMM characteristics. The corresponding scalar probabilities are found from formula (2.1.2).

Denote $\Omega_{t_0}^m$ a set of all possible ordered sequences of the type $(\mathbf{e}_{t_0+1}, \mathbf{e}_{t_0+2}, \ldots, \mathbf{e}_{t_0+m})$. An event A is defined as a subset of the set $\Omega_{t_0}^m$.

Consider now a system $N_{t_0}^m$ of all possible subsets of the set $\Omega_{t_0}^m$, including also the set $\Omega_{t_0}^m$ and the impossible event $\varnothing$. This system is called a σ-algebra if it is closed under the complementation (if $A \in N_{t_0}^m$, then its complement $\bar{A} = \Omega_{t_0}^m \setminus A$ is also $\in N_{t_0}^m$) and countable unions:

$$\bigcup_{i=1}^{\infty} A_i \in N_{t_0}^m \quad \text{if} \quad A_i \in N_{t_0}^m, \quad i=1,2,...$$

Suppose that events A and B belong to $N_{t_0}^m$, the union of these sets $A \cup B$ is called the events sum, the intersection of the sets $A \cap B$ is called the events product, and the complement of A is called its opposite event $\bar{A}$.

Let A be an event that depends only on values $\mathbf{e}_t$ in the interval $(t_{0A}, t_{1A}]$ and B be an event that depends only on the values $\mathbf{e}_t$ in the interval $(t_{0B}, t_{1B}]$. If these intervals do not coincide, then the events A and B belong to different σ-algebras. Denote $\tau_0 = \min(t_{0A}, t_{0B})$, $\tau_1 = \max(t_{1A}, t_{1B})$. Then the interval $(\tau_0, \tau_1]$ contains both intervals $(t_{0A}, t_{1A}]$ and $(t_{0B}, t_{1B}]$. The events A and B belong to $N_{t_0}^m$ and, therefore, the operations with these events are defined.

Example 2.1.1. Let $A = \{1\}$ be an error on the first position and $B = \{1\}$ be an error on the second position of a two-bit block. The event A belongs to the σ-algebra $N_{t_0}^1$ and B belongs to the different σ-algebra $N_{t_0+1}^1$, but both of them belong to the same σ-algebra $N_{t_0}^2$. In this σ-algebra, $A = \{(1,0), (1,1)\}$ and $B = \{(0,1), (1,1)\}$. The event $A \cap B$ is the occurrence of two errors in the block which is defined as $A \cap B = (1,1)$.

Suppose that an event $A \in N_{t_0}^m$, then the matrix

$$\mathbf{P}(A;t_0,t_1) = \sum_{\mathbf{e}_1^m \in A} \prod_{i=1}^{m} \mathbf{P}(\mathbf{e}_i) = \sum_{A} \prod_{i=1}^{m} \mathbf{P}(\mathbf{e}_i) \tag{2.1.3}$$

is called an MP of the event A. In this equation, the summation is performed over all sequences $\mathbf{e}_1^m$ that constitute the set A, and $\mathbf{P}(\mathbf{e}_i)$ are the parameters of the HMM (or matrix process). The ij-th element of this matrix is the conditional probability

$$p_{ij}(A;t_0,t_1) = Pr(A, s_{t_1} = j \mid s_{t_0} = i)$$

of the event A and transferring from the state $s_{t_0} = i$ (at the moment t_0) into the state $s_{t_1} = j$ (at the moment t_1):

$$\mathbf{P}(A;t_0,t_1) = [\, p_{ij}(A;t_0,t_1) \,]_{u,u}$$

Indeed, the matrix product in Eq. (2.1.3) can be written as (see Appendix 5)

$$p_{ij}(A;t_0,t_1) = \sum_{A} \sum_{s_1^{m-1}} Pr(\mathbf{e}_1, s_1 \mid i) Pr(\mathbf{e}_2, s_2 \mid s_1) \cdots Pr(\mathbf{e}_m, j \mid s_{m-1})$$

which is equal to $Pr(A, s_{t_1} = j \mid s_{t_0} = i)$ due to the theorem of total probability.

According to Eq. (2.1.2), the usual (scalar) probability of A is defined as

$$p(A;t_0,t_1) = \mathbf{p}_0 \mathbf{P}(A;t_0,t_1)\mathbf{1} \tag{2.1.4}$$

For the sake of simplicity, we will often omit the symbols t_0 and t_1 from the MP notation. If the Markov chain is stationary, then this probability does not depend on t_0:

$$p(A;t_0,t_1) = p(A) = \boldsymbol{\pi}\mathbf{P}(A)\mathbf{1} \tag{2.1.5}$$

where $\boldsymbol{\pi}$ is the chain stationary distribution.

Example 2.1.2. Consider Gilbert's model (see Sec. 1.1.5) with the matrices

$$\mathbf{P}(0) = \begin{bmatrix} Q & Ph \\ p & qh \end{bmatrix} \qquad \mathbf{P}(1) = \begin{bmatrix} 0 & P(1-h) \\ 0 & q(1-h) \end{bmatrix}$$

Let us find the probability of two errors in a block of three symbols.

The event A is composed of the union of sequences (110), (101), (011). Therefore, according to formula (2.1.3),

$$\mathbf{P}(A) = \mathbf{P}(0)\mathbf{P}^2(1) + \mathbf{P}(1)\mathbf{P}(0)\mathbf{P}(1) + \mathbf{P}^2(1)\mathbf{P}(0)$$

After multiplication and addition of the matrices in the right-hand side of this equation, we obtain

$$\mathbf{P}(A) = (1-h)^2 \begin{bmatrix} Ppq & (Pp+Qq)P+3Pq^2h \\ pq^2 & 2Ppq+3q^3h \end{bmatrix}$$

The corresponding scalar probability is defined by Eq. (2.1.5). Multiplying the matrices

$$p(A) = \boldsymbol{\pi}\mathbf{P}(A)\mathbf{1} = \frac{1}{P+p}[\, p,\ P\,]\ \mathbf{P}(A)\begin{bmatrix} 1 \\ 1 \end{bmatrix}$$

we find that

$$p(A) = (1-h)^2 P(Pp+2pq+3q^2h)/(P+p)$$

2.1.2 Properties of Matrix Probabilities

Consider now the basic properties of MP that play the main role in applications.

1. *The Unity Theorem.* The following formula holds

$$\mathbf{P}(\Omega^m) = \mathbf{P}^m \tag{2.1.6}$$

where $\mathbf{P} = \sum_{\Omega} \mathbf{P}(\mathbf{e})$. The proof of this formula follows immediately from Eq. (2.1.3):

$$\mathbf{P}(\Omega^m) = \sum_{\Omega^m} \prod_{i=1}^{m} \mathbf{P}(\mathbf{e}_i) = \prod_{i=1}^{m} \sum_{\mathbf{e}_i \in \Omega} \mathbf{P}(\mathbf{e}_i) = \mathbf{P}^m$$

The matrix $\mathbf{P}^m$ plays the role of unity in the matrix theory of probabilities; no MP for the events in the σ-algebra N^m can exceed this probability:

$$0 \le \mathbf{P}(A) \le \mathbf{P}^m \qquad A \in N^m$$

2. *Interval Expansion.* If $A \in N_{t_0}^m$ depends only on values $\mathbf{e}_t$ in the interval $(t_0, t_1]$, then it can also be considered as an event from an σ-algebra containing $N_{t_0}^m$. In this case we can write

$$\mathbf{P}(A;t_0-\tau_0,t_1+\tau_1) = \mathbf{P}^{\tau_0}\mathbf{P}(A;t_0,t_1)\mathbf{P}^{\tau_1} \tag{2.1.7}$$

The proof of this property is similar to the previous proof. It is important to note that $\boldsymbol{\pi}\mathbf{P}^{\tau} = \boldsymbol{\pi}$ and $\mathbf{P}^{\tau}\mathbf{1} = \mathbf{1}$, in many cases there is no need to use this equation when calculating scalar probabilities.

If the Markov chain is regular (see Appendix 6), then, letting $\tau_0 \to \infty$ and $\tau_1 \to \infty$, we obtain

$$\mathbf{P}(A;-\infty,\infty) = \mathbf{1}\boldsymbol{\pi}\mathbf{P}(A;t_0,t_1)\mathbf{1}\boldsymbol{\pi}$$

Since

$$\boldsymbol{\pi}\mathbf{P}(A;t_0,t_1)\mathbf{1} = p(A;t_0,t_1)$$

is a number, we can rewrite the previous equation as

$$\mathbf{P}(A;-\infty,\infty) = p(A;t_0,t_1)\mathbf{1}\boldsymbol{\pi}$$

3. *The Theorem of Addition of Probabilities.* If $A \in N^m$ and $B \in N^m$, then

$$\mathbf{P}(A \cup B) = \mathbf{P}(A) + \mathbf{P}(B) - \mathbf{P}(A \cap B) \tag{2.1.8}$$

The proof of this theorem immediately follows from the theorem of addition of scalar probabilities and the rule of addition of matrices.

4. *The Theorem of Multiplication of Probabilities.* Let $A \in N^m$ and $B \in N^m$. Then, according to the theorem of multiplication of scalar probabilities, we have

$$Pr(A \cap B, s_{t_1}=j \mid s_{t_0}=i) = Pr(A, s_{t_1}=j \mid s_{t_0}=i)Pr(B \mid A, s_{t_1}=j, s_{t_0}=i) \tag{2.1.9}$$

This formula permits us to determine all the elements of the MP $\mathbf{P}(A \cap B;t_0,t_1)$. Let us consider several particular cases of this equation.

If an event A depends only on $\mathbf{e}_t$ in the interval $(t_0,\tau]$ for $t_0<\tau<t_1$, then

$$\begin{aligned} p_{ij}(A \cap B;t_0,t_1) &= \sum_{k=1}^{u} Pr(A \cap B, s_\tau=k, s_{t_1}=j \mid s_{t_0}=i) \\ &= \sum_{i=1}^{u} Pr(A, s_\tau \mid s_{t_0}=i)Pr(B, s_{t_1}=j \mid A, s_\tau=k, s_{t_0}=i) \end{aligned} \tag{2.1.10}$$

If for any $k=1,2,\ldots,u$ the condition

$$Pr(B, s_{t_1}=j \mid A, s_\tau=i, s_{t_0}=k) = Pr(B, s_{t_1}=j \mid A, s_\tau=i)$$

holds, then we call the matrix

$$\mathbf{P}(B \mid A;\tau,t_1) = [\ Pr(B, s_{t_1}=j \mid A, s_\tau=i)\ ]_{u,u}$$

the conditional MP of event B given event A. In this case formula (2.1.10) can be rewritten in the matrix form

$$\mathbf{P}(A \cap B;t_0,t_1) = \mathbf{P}(A;t_0,\tau)\mathbf{P}(B \mid A;\tau,t_1) \tag{2.1.11}$$

If we assume that $(Pr(B, s_{t_1}=j \mid A, s_\tau=i) = Pr(B, s_{t_1}=j \mid s_\tau=i)$ or, in other words, that A and B are conditionally independent, then

$$\mathbf{P}(A \cap B;t_0,t_1) = \mathbf{P}(A;t_0,\tau)\mathbf{P}(B;\tau,t_1) \tag{2.1.12}$$

This formula is similar to the usual probability of a product of two independent events.

However, in this case the events A and B are not independent (they depend on the states of the Markov chain).

5. *The Formula of Total Probability.* The formula of total probability is an obvious corollary of the theorems of addition and multiplication of probabilities. If $A \in N^m$ and the events H_k are pairwise disjoint and form a complete group of events:

$$\bigcup_k H_k = \Omega^m \qquad H_i \cap H_j = \varnothing \quad \text{for } i \neq j$$

then, obviously,

$$\mathbf{P}(A;\, t_0, t_1) = \sum_k \mathbf{P}(A \cap H_k; t_0, t_1)$$

The MP $\mathbf{P}(A \cap H_k; t_0, t_1)$ in the last formula can be obtained from the theorem of multiplication of probabilities. In particular, if formula (2.1.11) is valid, then

$$\mathbf{P}(A;\, t_0, t_1) = \sum_k \mathbf{P}(H_k; t_0, \tau_k)\mathbf{P}(A \mid H_k; \tau_k, t_1) \tag{2.1.13}$$

Example 2.1.3. Let $\mathbf{P}_n(t)$ be the matrix probability that t errors occur in a block of length n. Then according to the formula of total probability we have

$$\mathbf{P}_n(t) = \mathbf{P}(0)\mathbf{P}_{n-1}(t) + \mathbf{P}(1)\mathbf{P}_{n-1}(t-1)$$

2.1.3 Random Variables

Consider an HMM whose observations $\xi \in R$ are real numbers. If ξ is discrete, we define its MP as a matrix $\mathbf{P}_\xi(x;t)$ whose ij-th element is the $Pr(\xi = x, s_t = j \mid s_{t-1} = i)$. If ξ is continuous, we define similarly its matrix probability density function (MPDF) $\mathbf{p}_\xi(x;t)$. MP and MPDF associated with time intervals are defined similarly to the MP defined by Eq. (2.1.3) and are denoted $\mathbf{P}_\xi(x; t_0, t_1)$ and $\mathbf{p}_x(x; t_0, t_1)$, respectively. Most of the time we will be dealing with the nonnegative discrete random variables. We use the term MPDF infrequently to emphasize that we are considering a continuous random variable.

Matrix expected value of a function $\phi(\xi)$ are defined in a usual way:

$$\mathbf{E}\{\xi; t_0, t_1\} = \sum_x \phi(x)\mathbf{P}_\xi(x; t_0, t_1) \tag{2.1.14}$$

for a discrete variable or

$$\mathbf{E}\{\xi; t_0, t_1\} = \int_x \phi(x)\mathbf{p}_\xi(x; t_0, t_1)\,dx \tag{2.1.15}$$

for a continuous variable. The corresponding scalar expected value $E\{\xi\} = \mathbf{p}_0\mathbf{E}\{\xi\}\mathbf{1}$.

It follows from this definition that the elementary properties of the expectations of scalar random variables can be extended to the matrix expectations:

1. If c is a constant, then $\mathbf{E}\{c\xi\} = c\mathbf{E}\{\xi\}$ and $\mathbf{E}\{c\} = c\mathbf{P}$.
2. $\mathbf{E}\{\sum_{i=1}^k \xi_i\} = \sum_{i=1}^k \mathbf{E}\{\xi_i\}$.

It follows from the second property that the mean number of errors in a block of n symbols is equal to np_e which is the same as for the channels with independent errors.

Let us now calculate an HMM autocorrelation function $\mathbf{R}(\tau) = \mathbf{E}\{\xi_t \xi'_{t+\tau}\}$. We have

$$\mathbf{R}(\tau) = \sum_{x_t, x_{t+\tau}} x_t x'_{t+\tau} \mathbf{P}(x_t) \mathbf{P}^{\tau-1} \mathbf{P}(x_{t+\tau}) = \mathbf{E}\{\xi\} \mathbf{P}^{\tau-1} \mathbf{E}\{\xi'\} \tag{2.1.16}$$

Thus, the autocorrelation function of a stationary HMM has the form

$$R(\tau) = \mathbf{a} \mathbf{P}^{\tau-1} \mathbf{b} \tag{2.1.17}$$

where $\mathbf{a} = \boldsymbol{\pi} \mathbf{E}\{\xi\}$ and $\mathbf{b} = \mathbf{E}\{\xi'\}\mathbf{1}$. Using matrix $\mathbf{P}^{\tau-1}$ spectral decomposition [Eq. (A.6.7) of Appendix 6], we obtain the autocorrelation function general form

$$R(\tau) = m_\xi^2 + \sum_{j=2}^{r} \sum_{i=1}^{k_j} d_{ij} \binom{\tau-1}{i-1} \lambda_j^{\tau-i} \tag{2.1.18}$$

where $m_\xi = E\{\xi\} = \mathbf{a1} = \boldsymbol{\pi}\mathbf{b}$ is the distribution mean and λ_j are the matrix $\mathbf{P}$ eigenvalues. It follows from this equation that if $\mathbf{P}$ is regular, the process autocovariance function $C(\tau) = R(\tau) - m_\xi^2$ tends to 0 with the exponential speed as $\tau \to \infty$.

The previous results can be generalized for the n-dimensional HMM observations. In particular, we can prove that the process autocovariance matrix decays exponentially as $\tau \to \infty$.

2.2 MATRIX TRANSFORMS

2.2.1 Matrix Generating Functions

The generating function of a matrix distribution $\mathbf{P}_\xi(x)$ of discrete nonnegative random variable ξ is defined as the power series

$$\boldsymbol{\Phi}_\xi(z) = \sum_x \mathbf{P}_\xi(x) z^x = E\{z^\xi\} \tag{2.2.1}$$

which converges inside the unit disk ($|z| \le 1$) of the complex z-plane. Its elements are defined by

$$\phi_{ij}(z) = \sum_x p_{ij}(x) z^x \tag{2.2.2}$$

If the distribution-generating function is known, then the matrix probability $\mathbf{P}_\xi(x)$, being a coefficient of z^x of the Taylor series (2.2.1), is given by the inversion formulae [7]

$$\mathbf{P}_\xi(x) = \frac{1}{x!} \frac{d^x}{dz^x} \boldsymbol{\Phi}_\xi(z)\big|_{z=0} \tag{2.2.3}$$

or

$$\mathbf{P}_\xi(x) = \frac{1}{2\pi j} \oint_\gamma \boldsymbol{\Phi}_\xi(z) z^{-x-1} dz \tag{2.2.4}$$

where γ is a counterclockwise contour that encircles the origin and lies entirely in the region of convergence of $\boldsymbol{\Phi}_\xi(z)$, which contains $|z| \le 1$.

It is convenient to use generating functions of matrix distributions when dealing with HMM, because, as we will see, they have a comparatively simple form (in contrast to matrix distributions). Besides, it is easier to find distribution moments and asymptotics using generating functions.

Example 2.2.1. Let us calculate the matrix distribution of the number of errors $\mathbf{P}_3(t)$ in a block of length 3 for a binary HMM error source.

By the formula of total probability, we have

$$\mathbf{P}_3(0) = \mathbf{P}^3(0), \quad \mathbf{P}_3(1) = \mathbf{P}^2(0)\mathbf{P}(1)+\mathbf{P}(0)\mathbf{P}(1)\mathbf{P}(0)+\mathbf{P}(1)\mathbf{P}^2(0)$$

$$\mathbf{P}_3(2) = \mathbf{P}^2(1)\mathbf{P}(0)+\mathbf{P}(1)\mathbf{P}(0)\mathbf{P}(1)+\mathbf{P}(0)\mathbf{P}^2(1), \quad \mathbf{P}_3(3) = \mathbf{P}^3(1)$$

The generating function of this distribution is

$$\mathbf{\Phi}_t(z) = \mathbf{P}_3(0) + \mathbf{P}_3(1)z + \mathbf{P}_3(2)z^2 + \mathbf{P}_3(3)z^3$$

Substituting the values of matrix probabilities in this formula and simplifying the result, we obtain

$$\mathbf{\Phi}_\xi(z) = [\mathbf{P}(0) + \mathbf{P}(1)z]^3$$

This generating function has the same form as for independent events [4] occurring with the probability $p(1)$ [binomial distribution $\binom{3}{t}p^t(1)p^{3-t}(0)$]. We encounter this fact very often. However, the matrix distribution $\mathbf{P}_3(t)$ cannot generally be written in the form $\binom{3}{t}\mathbf{P}^t(1)\mathbf{P}^{3-t}(0)$ because the product of matrices is noncommutative.

The matrix distribution $\mathbf{P}_3(t)$ can be found from the generating function by using the inversion formula (2.2.3). For example, let us find $\mathbf{P}_3(1)$. We have (see Sec. 5.1.7 of Appendix 5)

$$\mathbf{\Phi}_t'(z) = \frac{d}{dz}[\mathbf{P}(0)+z\mathbf{P}(1)]^3 = \mathbf{P}(1)[\mathbf{P}(0)+z\mathbf{P}(1)]^2$$

$$+ [\mathbf{P}(0)+z\mathbf{P}(1)]\mathbf{P}(1)[\mathbf{P}(0)+z\mathbf{P}(1)] + [\mathbf{P}(0)+z\mathbf{P}(1)]^2\mathbf{P}(1)$$

Therefore,

$$\mathbf{P}_3(1) = \mathbf{\Phi}_t'(0) = \mathbf{P}(1)\mathbf{P}(0)^2 + \mathbf{P}(0)\mathbf{P}(1)\mathbf{P}(0) + \mathbf{P}^2(0)\mathbf{P}(1)$$

The expansion of the function $\mathbf{\Phi}_t(z)$ in the vicinity of the origin can be obtained by the matrix polynomial expansion. The simplest way of doing it is by performing matrix multiplication and gathering the same power terms:

$$\mathbf{\Phi}_t(z) = (\mathbf{P}(0) + \mathbf{P}(1)z)(\mathbf{P}(0) + \mathbf{P}(1)z)(\mathbf{P}(0) + \mathbf{P}(1)z) = \mathbf{P}^3(0)$$

$$+ [\mathbf{P}^2(0)\mathbf{P}(1)+\mathbf{P}(0)\mathbf{P}(1)\mathbf{P}(0)+\mathbf{P}(1)\mathbf{P}^2(0)]z$$

$$+ [\mathbf{P}^2(1)\mathbf{P}(0)+\mathbf{P}(1)\mathbf{P}(0)\mathbf{P}(1)+\mathbf{P}(0)\mathbf{P}^2(1)]z^2 + \mathbf{P}^3(1)z^3$$

The generating function of an ordinary distribution has the form

$$\phi_\xi(z) = \mathbf{p}_0\mathbf{\Phi}_\xi(z)\mathbf{1} \tag{2.2.5}$$

Several obvious properties of the generating functions are listed here:

1. If $\mathbf{p}_0$ is an arbitrary probabilistic vector, then we obtain from Eq. (2.2.5)

$$\phi_\xi(1) = \mathbf{p}_0\mathbf{\Phi}_\xi(1)\mathbf{1} = 1 \tag{2.2.6}$$

2. If a random variable ξ depends on $\mathbf{e}_t$ in the interval $(t_0,t_1]$, then

$$\mathbf{\Phi}_\xi(1) = \sum_x \mathbf{P}(x;t_0,t_1) = \mathbf{P}^{t_1-t_0} \tag{2.2.7}$$

3. It is convenient to use generating functions to find matrix binomial moments

$$\mathbf{B}_m = \sum_n \binom{n}{m} \mathbf{P}_\xi(n)$$

Indeed, it is easy to check that if we differentiate Eq. (2.2.1) m times and then let $z=1$, we obtain the binomial moment

$$\mathbf{B}_m = \frac{1}{m!} \mathbf{\Phi}_\xi^{(m)}(1)$$

In particular, for $m=1$ this formula gives the matrix mean $\mathbf{B}_1$, and consequently the scalar mean is given by

$$\bar{\xi} = \phi_\xi'(1) = \mathbf{p}_0 \mathbf{B}_1 \mathbf{1} = \mathbf{p}_0 \mathbf{\Phi}_\xi'(1)\mathbf{1} \tag{2.2.8}$$

4. We say that the random variables ξ and η are conditionally independent if, for any x and y from the region of their definition, events $\xi=x$ and $\eta=y$ are conditionally independent. If the random variables ξ and η are conditionally independent, then according to the formula of total probability (2.1.13), the matrix distribution of their sum is the convolution

$$\mathbf{P}_{\xi+\eta}(y) = \sum_{x=0}^{y} \mathbf{P}_\xi(x)\mathbf{P}_\eta(y-x) \tag{2.2.9}$$

If we denote the generating functions of the ξ and η distributions as $\mathbf{\Phi}_\xi(z)$ and $\mathbf{\Phi}_\eta(z)$, then we can easily verify that the generating function of their sum is

$$\mathbf{\Phi}_{\xi+\eta}(z) = \mathbf{\Phi}_\xi(z)\mathbf{\Phi}_\eta(z) \tag{2.2.10}$$

In the general case of n conditionally independent variables, the generating function of the probability distribution of their sum has the form

$$\mathbf{\Phi}_{\xi_1+\xi_2+\ldots+\xi_n}(z) = \prod_{i=1}^{n} \mathbf{\Phi}_{\xi_i}(z) \tag{2.2.11}$$

Since the distribution of the sum is the convolution of distributions of the addends, the previous equation is often called the *convolution theorem:* the generating function of a convolution is equal to the product of its terms' generating functions. The converse is also true, because of the uniqueness of the Taylor series. [7]

Example 2.2.2. Let us find the matrix generating function $\mathbf{\Phi}_t(z;n)$ of the distribution of the number of errors in a block of length n for a binary HMM error source.

For a block of length $n=1$ we have

$$\mathbf{\Phi}_e(z) = \mathbf{P}(0) + \mathbf{P}(1)z$$

The number of errors in the block is equal to the sum $t = e_1+e_2+\cdots+e_n$ so that, according to formula (2.2.11),

$$\mathbf{\Phi}_t(z;n) = \mathbf{\Phi}_e^n(z) = [\mathbf{P}(0) + \mathbf{P}(1)z]^n$$

The average number of errors in the block can be found using equation (2.2.8). The generating function derivative with respect to z is

$$\phi_t'(z) = \sum_{i=0}^{n-1} \mathbf{p}_0[\mathbf{P}(0)+z\mathbf{P}(1)]^i\mathbf{P}(1)[\mathbf{P}(0)+z\mathbf{P}(1)]^{n-i-1}\mathbf{1}$$

With $z=1$, we have

$$\bar{t} = \sum_{i=0}^{n-1} \mathbf{p}_0 \mathbf{P}^i \mathbf{P}(1) \mathbf{P}^{n-i-1} \mathbf{1}$$

Since $\mathbf{P} = \mathbf{P}(0)+\mathbf{P}(1)$ is a stochastic matrix, sums of the elements in each of its rows are equal to 1 (that is, $\mathbf{P1} = 1$), and therefore

$$\bar{t} = \sum_{i=0}^{n-1} \mathbf{p}_0 \mathbf{P}^i \mathbf{P}(1) \mathbf{1}$$

If we also assume that the chain is stationary (that is, $\mathbf{p}_0 = \boldsymbol{\pi} = \boldsymbol{\pi}\mathbf{P}$), then this formula can be further simplified:

$$\bar{t} = \sum_{i=0}^{n} \boldsymbol{\pi}\mathbf{P}(1)\mathbf{1} = np_e$$

where $p_e = \boldsymbol{\pi}\mathbf{P}(1)\mathbf{1}$ is the bit-error probability. We see that the mean number of errors in the block for an HMM error sources is the same as for independent errors with the bit error probability p_e.

The previous results can be generalized for multidimensional distributions. Let $\{\mathbf{A}_{m_1,m_2,\ldots,m_k}\}$ be a sequence of matrices. Then the matrix

$$\boldsymbol{\Phi}(z_1,z_2,\ldots,z_k) = \sum_{m_1,m_2,\ldots,m_k} \mathbf{A}_{m_1,m_2,\ldots,m_k} z_1^{m_1} z_2^{m_2} \cdots z_k^{m_k} \tag{2.2.12}$$

is called the sequence multidimensional generating function.

The elements of the sequence $\{\mathbf{A}_{m_1,m_2,\ldots,m_k}\}$ are found from the generating function by inversion formulas similar to Eq. (2.2.3) and (2.2.4):

$$\mathbf{A}_{m_1,m_2,\ldots,m_k} = \frac{1}{m_1! \cdots m_k!} \frac{\partial^N \boldsymbol{\Phi}(z_1,\ldots,z_k)}{\partial z_1^{m_1} \cdots \partial z_k^{m_k}} \Big|_{\mathbf{0}} \tag{2.2.13}$$

$$\mathbf{A}_{m_1,m_2,\ldots,m_k} = \frac{1}{(2\pi \mathrm{j})^k} \int_\gamma \cdots \int \frac{\boldsymbol{\Phi}(z_1,z_2,\ldots,z_k)}{z_1^{m_1+1} \cdots z_k^{m_k+1}} dz_1 \cdots dz_k \tag{2.2.14}$$

where $\mathbf{0}=(0,\ldots,0)$ is the k-dimensional space origin. The integral representation (2.2.14) is convenient in many calculations related to matrix probabilities. It is particularly true for summing matrix series and for obtaining asymptotic formulas (see Appendices 2 and 5).

2.2.2 Matrix z-Transforms

In some applications it is convenient to use the so-called z-transform of a matrix distribution:

$$\boldsymbol{\Psi}_\xi(z) = \boldsymbol{\Phi}_\xi(1/z) = \sum_x \mathbf{P}_\xi(x) z^{-x} \tag{2.2.15}$$

Since the series on the right-hand side of the $\boldsymbol{\Phi}_\xi(z)$ definition (2.2.1) converges inside the unit disk of the complex z-plane, the series on the right-hand side of Eq. (2.2.15) converges outside the unit disk ($|z| \geq 1$).

As we see, the z-transform of a distribution is only notationally different from the generating function. To find the z-transform inversion we may use Eq. (2.2.4) with the substitution $z = 1/\zeta$:

$$\mathbf{P}_\xi(x) = \frac{1}{2\pi \mathrm{j}} \oint_{\gamma_1} \boldsymbol{\Psi}_\xi(\zeta) \zeta^{x-1} d\zeta \tag{2.2.16}$$

where γ_1 is a counterclockwise contour that encircles the origin and lies entirely in the region of convergence of $\boldsymbol{\Psi}_\xi(z)$.

The basic reason for using $\boldsymbol{\Psi}_\xi(z)$ instead of $\boldsymbol{\Phi}_\xi(z)$ is that often it simplifies the residue theorem application. In the case of Eq. (2.2.4), we must find all the residues of the function $\boldsymbol{\Phi}_\xi(z)$ at the singularities located outside the unit circle, including the point of infinity $z=\infty$. The residue at the infinity is calculated differently than all the other residues. To simplify the residue theorem application, it is convenient to transform ∞ into 0 by the inversion $1/z$ of the complex plane. Thus, the integral (2.2.16) may be computed using all the residues at the singularities of $\boldsymbol{\Psi}_\xi(z)$ located inside the unit circle.

It is also convenient to use z-transforms to find the random variable ξ matrix negative-binomial moments

$$\mathbf{N}_m = \sum_n \binom{m+n-1}{m} \mathbf{P}_\xi(n)$$

Indeed, if we differentiate Eq. (2.2.15) m times and then let $z=1$, we obtain

$$\mathbf{N}_m = \frac{(-1)^m}{m!} \boldsymbol{\Psi}_\xi^{(m)}(1) \tag{2.2.17}$$

In the particular case of $m=1$, this equation gives the matrix mean $\mathbf{N}_1$. The scalar mean is given by

$$\bar{\xi} = -\psi'_\xi(1) = \mathbf{p}_0 \mathbf{N}_1 \mathbf{1} = -\mathbf{p}_0 \boldsymbol{\Psi}'_\xi(1)\mathbf{1}$$

2.2.3 Matrix Fourier Transform

The matrix Fourier transform (or characteristic function) of a matrix distribution $\mathbf{P}_\xi(x)$ is defined as

$$\mathbf{F}(\omega) = \boldsymbol{\Phi}(\mathrm{e}^{-\mathrm{j}\omega}) = \sum_x \mathbf{P}_\xi(x)\mathrm{e}^{-\mathrm{j}\omega x} \tag{2.2.18}$$

Since $|\mathrm{e}^{-\mathrm{j}\omega}|=1$, the matrix Fourier transform of the distribution coincides with its generating function or z-transform on the unit circle. Hence, all basic properties of generating functions are applicable to Fourier transforms.

The Fourier transform inversion formula may be obtained from Eq. (2.2.4) by replacing z with $\mathrm{e}^{-\mathrm{j}\omega}$:

$$\mathbf{P}_\xi(x) = \frac{1}{2\pi}\int_0^{2\pi} \mathbf{F}_\xi(\omega)\mathrm{e}^{\mathrm{j}\omega x} d\omega \tag{2.2.19}$$

Although it is convenient to find binomial moments using the generating functions, the central moments

$$\mathbf{M}_k = \sum_x \mathbf{P}_\xi(x)x^k$$

are usually computed using the Fourier transform derivatives:

$$\mathbf{M}_k = \mathrm{j}^k \mathbf{F}_\xi^{(k)}(0) \tag{2.2.20}$$

The corresponding scalar moments are given by

$$m_k = \mathbf{p}_0 \mathbf{M}_k \mathbf{1}$$

Example 2.2.3. Let us compute the standard deviation of the number of errors in a block for a binary HMM error source.

The error number distribution

$$p_n(m) = \boldsymbol{\pi} \mathbf{P}_n(m) \mathbf{1}$$

has the Fourier transform

$$f(\omega) = \boldsymbol{\pi}(\mathbf{P}(0) + \mathbf{P}(1)e^{-j\omega})^n \mathbf{1}$$

which can be found from the generating function of Example 2.2.2. Using Eq. (2.2.20), we obtain

$$m_1 = j f'(0) = n\boldsymbol{\pi}\mathbf{P}(1)\mathbf{1} = np_e$$

$$m_2 = -f''(0) = (n-1)(n-2)p_e^2 + 2\boldsymbol{\pi}\mathbf{P}(1)\mathbf{U}_n\mathbf{P}(1)\mathbf{1} + np_e$$

where

$$\mathbf{U}_n = [(n-1)\mathbf{I} - n\mathbf{Q} + \mathbf{Q}^n](\mathbf{I}-\mathbf{Q})^{-2} \qquad \mathbf{Q} = \mathbf{P} - \mathbf{1}\boldsymbol{\pi}$$

The distribution variance is expressed as [4]

$$\sigma_n^2 = m_2 - m_1^2$$

Since the matrix $\mathbf{P}$ is regular, $\mathbf{Q}^n \to 0$ when $n \to \infty$, and we obtain

$$\sigma_n^2 \sim n[p_e - 3p_e^2 + 2\boldsymbol{\pi}\mathbf{P}(1)(\mathbf{I}-\mathbf{Q})^{-1}\mathbf{P}(1)\mathbf{1}]$$

2.2.4 Matrix Discrete Fourier Transform

Consider now the distribution Fourier transform in the discrete points uniformly distributed on the unit circle of the complex z-plane $z_k = e^{-2\pi jk/p}$:

$$\mathbf{F}_k = \mathbf{\Phi}(z_k) = \mathbf{F}(2\pi k/p) \tag{2.2.21}$$

This transform is called the *distribution discrete Fourier transform*. In general, this transform is not invertible. That is, we cannot uniquely determine the matrix distribution $\mathbf{P}_\xi(x)$ from its discrete Fourier transform.

Indeed, because of the exponential function periodicity ($z_k = z_{k+p}$), the number of different Eq. (2.2.21) cannot exceed p. So we have a system of p linear equations with an infinite number of unknowns that in general does not have a unique solution. However, if the generating function is a polynomial whose degree is less than p, this system always has a unique solution because its determinant (being Vandermonde's determinant) is not equal to 0.

If we denote $\alpha = z_1 = e^{-2\pi j/p}$, the system (2.2.21) takes the form

$$\mathbf{F}_k = \sum_{i=0}^{p-1} \alpha^{ik} \mathbf{P}_\xi(i) \qquad k = 0, 1, \ldots, p-1 \tag{2.2.22}$$

In order to find the unique solution of this system, we multiply its k-th equation by α^{-mk} and then sum all the equations:

$$\sum_{k=0}^{p-1} \alpha^{-mk} \mathbf{F}_k = \sum_{i=0}^{p-1} \mathbf{P}_\xi(i) \sum_{k=0}^{p-1} \alpha^{k(i-m)} \tag{2.2.23}$$

If we make use of the easily verifiable identity

$$\sum_{k=0}^{p-1} \alpha^{k(i-m)} = \begin{cases} 0 & \text{if } i \neq m+lp \\ p & \text{if } i = m+lp \end{cases} \tag{2.2.24}$$

where l is an arbitrary integer, then Eq. (2.2.23) becomes

$$\sum_{k=0}^{p-1} \alpha^{-mk} \mathbf{F}_k = p\mathbf{P}_\xi(m)$$

So the unique solution of system (2.2.22) can be written as

$$\mathbf{P}_\xi(m) = p^{-1} \sum_{k=0}^{p-1} \alpha^{-mk} \mathbf{F}_k \tag{2.2.25}$$

This formula is extremely useful for the matrix distribution calculations because of the utility of the fast Fourier transform (FFT).

Consider now the general case when the generating function degree may be greater than p. In this case, replacing the upper limit of summation in Eq. (2.2.22) by ∞ and repeating the previous transformations, we obtain

$$\sum_{l=0}^{\infty} \mathbf{P}_\xi(m+lp) = p^{-1} \sum_{k=0}^{p-1} \alpha^{-mk} \mathbf{F}_k \tag{2.2.26}$$

This equation is obviously a generalization of Eq. (2.2.25), which is obtained when $\mathbf{P}_\xi(m+lp)=0$ for $m=0,1,\ldots,p-1$, and $l>0$. Equation (2.2.26) may be used to calculate sums of the elements whose indices have the form $m+lp$.

Example 2.2.4. Let us find the probability of an odd number of errors in a block of length n for a binary HMM error source.

The matrix probability of an odd number of errors is the probability that the number of errors is $2l+1$:

$$\mathbf{P}_{odd} = \sum_l \mathbf{P}_n(2l+1)$$

This sum can be found from equation (2.2.26) when $p=2$ and $m=1$:

$$\mathbf{P}_{odd} = 2^{-1}(\mathbf{F}_0 - \mathbf{F}_1)$$

Using the results of Example 2.2.2, we obtain

$$\mathbf{F}_0 = \mathbf{\Phi}(1) = (\mathbf{P}(0) + \mathbf{P}(1))^n = \mathbf{P}^n$$

$$\mathbf{F}_1 = \mathbf{\Phi}(-1) = (\mathbf{P}(0) - \mathbf{P}(1))^n$$

Thus

$$\mathbf{P}_{odd} = [\mathbf{P}^n + (\mathbf{P}(0) - \mathbf{P}(1))^n]/2$$

In the future, we will need the multidimensional generalization of Eq. (2.2.25). Let us consider an n-dimensional discrete Fourier transform

$$\mathbf{F_k} = \sum_{\mathbf{i}=0}^{\mathbf{q}} \boldsymbol{\alpha}^{\mathbf{ik}} \mathbf{P}_\xi(\mathbf{i}) \tag{2.2.27}$$

where

$$\mathbf{k}=(k_1,k_2,\ldots,k_n) \quad \mathbf{i}=(i_1,i_2,\ldots,i_n) \quad \mathbf{q}=(p_1-1,p_2-1,\ldots,p_n-1)$$
$$\boldsymbol{\xi}=(\xi_1,\xi_2,\ldots,\xi_n) \quad \boldsymbol{\alpha}^{\mathbf{ik}}=\alpha_1^{i_1k_1}\alpha_2^{i_2k_2}\cdots\alpha_n^{i_nk_n} \quad \text{and} \quad \alpha_i=e^{-2\pi j/p_i}$$

As with the one-dimensional Fourier transform, we derive the transform inversion formula

$$\mathbf{P}_{\xi}(\mathbf{m}) = V^{-1}\sum_{\mathbf{k}=0}^{\mathbf{q}}\boldsymbol{\alpha}^{-\mathbf{mk}}\mathbf{F}_{\mathbf{k}} \tag{2.2.28}$$

where $V=p_1p_2\cdots p_n$. The generalized Eq. (2.2.26) is given by

$$\sum_{\mathbf{l}=0}^{\infty}\mathbf{P}_{\xi}(\boldsymbol{\mu}(\mathbf{l})) = V^{-1}\sum_{\mathbf{k}=0}^{\mathbf{q}}\boldsymbol{\alpha}^{-\mathbf{km}}\mathbf{F}_{\mathbf{k}} \tag{2.2.29}$$

where $\boldsymbol{\mu}(\mathbf{l})=(m_1+l_1p_1,m_2+l_2p_2,\ldots,m_n+l_np_n)$. In the particular case $p_1=p_2=\cdots=p_n=p$, this formula can be simplified:

$$\sum_{\mathbf{l}=0}^{\infty}\mathbf{P}_{\xi}(\boldsymbol{\mu}(\mathbf{l})) = p^{-n}\sum_{\mathbf{k}=0}^{\mathbf{q}}\alpha^{-\mathbf{k}\cdot\mathbf{m}}\mathbf{F}_{\mathbf{k}} \tag{2.2.30}$$

where $\mathbf{k}\cdot\mathbf{m}=k_1m_1+k_2m_2+\cdots+k_nm_n$ is a vector dot product.

2.2.5 Matrix Transforms and Difference Equations

In many cases it is convenient to obtain generating functions with the help of recursive equations. For example, formula (2.2.10) was obtained using Eq. (2.2.9). Consider a difference equation

$$\mathbf{W}_n = \sum_{m=1}^{n}\mathbf{A}_m\mathbf{W}_{n-m} + \mathbf{B}_n \tag{2.2.31}$$

When the matrices $\mathbf{W}_0$, $\mathbf{A}_1,\mathbf{A}_2,\ldots$, and $\mathbf{B}_1,\mathbf{B}_2,\ldots$, are given, the matrices

$$\mathbf{W}_1 = \mathbf{A}_1\mathbf{W}_0 + \mathbf{B}_1 \quad \mathbf{W}_2 = \mathbf{A}_1\mathbf{W}_1 + \mathbf{A}_2\mathbf{W}_0 + \mathbf{B}_2 \cdots$$

can be found by sequential substitution, so that the equation has a unique solution.

Define formally $\mathbf{A}_0=0$. Then Eq. (2.2.31) can be rewritten in the form

$$\mathbf{W}_n = \sum_{m=0}^{n}\mathbf{A}_m\mathbf{W}_{n-m} + \mathbf{B}_n \tag{2.2.32}$$

The latter equation has the following form in terms of generating functions:

$$\mathbf{W}(z) = \mathbf{A}(z)\mathbf{W}(z) + \mathbf{B}(z)$$

If $\mathbf{A}(z)$ is a square matrix, then we obtain the solution-generating function

$$\mathbf{W}(z) = [\,\mathbf{I} - \mathbf{A}(z)\,]^{-1}\mathbf{B}(z) \tag{2.2.33}$$

The converse problem often interests us in calculations of matrix distributions. Given generating functions of unknown matrix probabilities, find the recursive equations for their computation. Usually, these equations are obtained by factoring the generating function and using the convolution theorem.

Example 2.2.5. Let us find the recursive equations for the matrix distribution $\mathbf{P}_n(t)$ of t errors in a block of n bits for a binary HMM error source.

As we saw in example 2.2.2, the generating function of this distribution has the form

$$\Phi_t^n(z) = [\,\mathbf{P}(0) + z\mathbf{P}(1)\,]^n$$

Since $[\,\mathbf{P}(0) + z\mathbf{P}(1)\,]^n = [\,\mathbf{P}(0) + z\mathbf{P}(1)\,]^{n-m}[\,\mathbf{P}(0) + z\mathbf{P}(1)\,]^m$, we obtain the equation

$$\mathbf{P}_n(t) = \sum_{\tau} \mathbf{P}_{n-m}(\tau)\mathbf{P}_m(t-\tau) \tag{2.2.34}$$

This equation is useful for fast computation of the distribution.

Let us consider closely the case when the elements of the generating function are rational fractions. As is shown in Appendix 2.1, the generating function can be written in the form of (A.2.2):

$$\mathbf{W}(z) = \mathbf{V}(z)/q(z) \tag{2.2.35}$$

where

$$q(z) = \sum_{i=0}^{k} q_i z^i \qquad \mathbf{V}(z) = \sum_{i=0}^{r} \mathbf{V}_i z^i$$

Bringing the previous relation to the common denominator, we obtain $q(z)\mathbf{W}(z) = \mathbf{V}(z)$. The comparison of the coefficients at the same powers of z leads us to the desired recursive equations

$$\mathbf{W}_m = (1/q_0)\mathbf{V}_m - \sum_{i=1}^{k} (q_i/q_0)\mathbf{W}_{m-i} \tag{2.2.36}$$

where $\mathbf{W}_i = 0$ when $i < 0$. The corresponding scalar distribution $w_m = \mathbf{p}_0\mathbf{W}_m\mathbf{1}$ obviously satisfies the equation

$$w_m = (1/q_0)v_m - \sum_{i=1}^{k} (q_i/q_0)w_{m-i} \tag{2.2.37}$$

where $v_m = \mathbf{p}_0\mathbf{V}_m\mathbf{1}$. The results of this section can be trivially generalized to recursive equations and generating functions of several variables (see Sec. 2.6.2.2 for an example).

Consider now some matrix distributions that generalize the most often used scalar distributions.

2.3 MATRIX DISTRIBUTIONS

2.3.1 Matrix Multinomial Distribution

Suppose that at every moment t, one and only one event from a set $A_1, A_2, \ldots, A_r$ can occur. Each event depends only on $\mathbf{e}_t$ and has constant matrix probability:

$$\mathbf{P}_i = \mathbf{P}(A_i; t-1, t)$$

Let us find the matrix probability $\mathbf{P}_n(m_1, m_2, \ldots, m_r)$ that, in the time interval $(t, t+n]$, the event A_1 occurs m_1 times, A_2 occurs m_2, ..., and A_r occurs m_r times $(m_1 + m_2 + \ldots + m_r = n)$.

In general, the direct calculation of the distribution $\mathbf{P}_n(m_1, m_2, \ldots, m_r)$ is complicated by the fact that the matrices $\mathbf{P}_i$ and $\mathbf{P}_j$ do not commute. For example, the direct formula becomes complex even for $n=5$, $r=2$, $m_1=3$, $m_2=2$:

$$\mathbf{P}_5(3,2) = \mathbf{P}_1^3\mathbf{P}_2^2+\mathbf{P}_1^2\mathbf{P}_2^2\mathbf{P}_1+\mathbf{P}_1\mathbf{P}_2^2\mathbf{P}_1^2+\mathbf{P}_2^2\mathbf{P}_1^3+\mathbf{P}_2\mathbf{P}_1^3\mathbf{P}_2$$
$$+\mathbf{P}_2\mathbf{P}_1^2\mathbf{P}_2\mathbf{P}_1+\mathbf{P}_2\mathbf{P}_1\mathbf{P}_2\mathbf{P}_1^2+\mathbf{P}_1\mathbf{P}_2\mathbf{P}_1\mathbf{P}_2\mathbf{P}_1$$
$$+\mathbf{P}_1\mathbf{P}_2\mathbf{P}_1^2\mathbf{P}_2+\mathbf{P}_1^2\mathbf{P}_2\mathbf{P}_1\mathbf{P}_2$$

One can propose methods to calculate matrix probabilities based on recursive equations which are obtained using the formula of total probability: [2,12,15]

$$\mathbf{P}_n(m_1,m_2,\ldots,m_r) = \sum_{\mu_1,\ldots,\mu_r} \mathbf{P}_{n-k}(m_1-\mu_1,m_2-\mu_2,\ldots,m_r-\mu_r)\mathbf{P}_k(\mu_1,\mu_2,\ldots,\mu_r) \tag{2.3.1}$$

For $k=1$, we obtain the important particular case of this formula

$$\mathbf{P}_n(m_1,m_2,\ldots,m_r) = \sum_{j=1}^{r}\mathbf{P}_{n-1}(m_1,\ldots,m_j-1,\ldots,m_r)\mathbf{P}_j \tag{2.3.2}$$

All the recursive equations need initial conditions to start the recursion. In the previous equation, we assume that $n>0$, $\mathbf{P}_n(m_1,m_2,\ldots,m_r) = 0$ if m_i is negative or $m_1+m_2+\ldots+m_r>n$, $\mathbf{P}_1(\mathbf{e}_j) = \mathbf{P}_j$ where $\mathbf{e}_j = (0,\ldots,0,1,0,\ldots,0)$ is a unit vector whose j-th coordinate is one. If the initial conditions are obvious, we do not mention them to avoid writing something like the previous sentence after each recursive equation. The recursive Eq. (2.3.1) and (2.3.2) permit us to calculate the distribution $\mathbf{P}_n(m_1,m_2,\ldots,m_r)$ for a not very large number of HMM states and n. Therefore, it is important to obtain approximate formulas and asymptotic estimations of this distribution. To solve this problem we use several generating functions of the matrix multinomial distribution.

The simplest generating function is given by the convolution theorem, or by using the recursive Eq. (2.3.1):

$$\mathbf{\Phi}_n(z_1,z_2,\ldots,z_r) = \Big(\sum_{i=1}^{r}\mathbf{P}_i z_i\Big)^n \tag{2.3.3}$$

where

$$\mathbf{\Phi}_n(z_1,z_2,\ldots,z_r) = \sum_{m_1,\ldots,m_r}\mathbf{P}_n(m_1,m_2,\ldots,m_r)z_1^{m_1}z_2^{m_2}\cdots z_r^{m_r}$$

The elements of the matrix $\mathbf{\Phi}_n(z_1,z_2,\ldots,z_r)$ are polynomials in $z_1,z_2,\ldots,z_r$. The matrix $\mathbf{P}_n(m_1,m_2,\ldots,m_r)$ can be found as a coefficient of $z_1^{m_1}z_2^{m_2}\cdots z_r^{m_r}$ in the expansion of this polynomial. In order to find the asymptotic formula when $n\to\infty$, it is convenient to use the generating function

$$\mathbf{\Psi}(w;z_1,z_2,\ldots,z_r) = \sum_{n=0}^{\infty}\mathbf{\Phi}_n(z_1,z_2,\ldots,z_r)w^n$$

where we assume $\mathbf{\Phi}_0=\mathbf{I}$ for the convenience of summation. Obviously, the series on the right-hand side of the previous equation converges if $|z_i|<1$, $(i=1,2,\ldots,r)$, $|w|<1$, and by Eq. (2.3.3), its sum is

$$\mathbf{\Psi}(w;z_1,z_2,\dots,z_r) = (\mathbf{I}-w\sum_{i=1}^{r}\mathbf{P}_i z_i)^{-1} \tag{2.3.4}$$

2.3.2 Matrix Binomial Distribution

2.3.2.1 Generating Functions

Let us study more closely a particular case, $r=2$, of the matrix multinomial distribution, the matrix binomial distribution, that plays the most important role in applications.

Denote $\mathbf{P}_1(0)=\mathbf{P}(0)$, $\mathbf{P}_1(1)=\mathbf{P}(1)$, $\mathbf{P}_n(m)=\mathbf{P}_n(n-m,m)$, $\mathbf{\Phi}_n(1,z)=\mathbf{\Phi}_n(z)$, and $\mathbf{\Psi}(w;1,z)=\mathbf{\Psi}(w;z)$. Using these simplified notations, we can rewrite the basic equations from the previous section. Formula (2.3.1) now has the form

$$\mathbf{P}_n(m) = \sum_{\mu}\mathbf{P}_{n-\nu}(m-\mu)\mathbf{P}_{\nu}(\mu) \tag{2.3.5}$$

and formula (2.3.2) can be rewritten as [2]

$$\mathbf{P}_n(m) = \mathbf{P}_{n-1}(m)\mathbf{P}(0) + \mathbf{P}_{n-1}(m-1)\mathbf{P}(1) \tag{2.3.6}$$

Generating functions (2.3.3) and (2.3.4) become

$$\mathbf{\Phi}_n(z) = [\mathbf{P}(0) + \mathbf{P}(1)z]^n \tag{2.3.7}$$

and

$$\mathbf{\Psi}(w;z) = [\mathbf{I} - \mathbf{P}(0)w - \mathbf{P}(1)wz]^{-1} \tag{2.3.8}$$

These equations have exactly the same form as in the case of the simple binomial distribution

$$P_n(m) = \binom{n}{m}p^m q^{n-m}$$

This is obviously a particular case of the matrix distribution where all the matrices are scalars (their sizes are equal to one). However, in the case of matrix probabilities, this simple binomial formula is not generally valid.

One can take advantage of the relative simplicity of the generating functions and calculate the matrix distribution using the inverse transforms described in Sec. 2.2. In particular, the inversion formulas (2.2.4) and (2.2.14) yield

$$\mathbf{P}_n(m) = \frac{1}{2\pi \mathrm{j}}\oint_{\gamma}[\mathbf{P}(0)+\mathbf{P}(1)z]^n z^{-m-1}dz \tag{2.3.9}$$

$$\mathbf{P}_n(m) = \frac{1}{(2\pi \mathrm{j})^2}\int\int_{\gamma_1}[\,\mathbf{I}-\mathbf{P}(0)w-\mathbf{P}(1)wz]^{-1}z^{-m-1}w^{-n-1}dwdz \tag{2.3.10}$$

One more generating function may be obtained if we expand the generating function $\mathbf{\Psi}(w;z)$ into a power series:

$$\mathbf{\Psi}(w;z) = \sum_{m=0}^{\infty}[(\mathbf{I} - \mathbf{P}(0)w)^{-1}\mathbf{P}(1)w]^m(\mathbf{I}-\mathbf{P}(0)w)^{-1}z^m$$

The coefficient at z^m in this expansion is a generating function

$$\boldsymbol{\Psi}_m(w) = \sum_{n=m}^{\infty} \mathbf{P}_n(m)w^n = [(\mathbf{I} - \mathbf{P}(0)w)^{-1}\mathbf{P}(1)w]^m(\mathbf{I} - \mathbf{P}(0)w)^{-1} \quad (2.3.11)$$

which is convenient to use for asymptotic expansions when $n \to \infty$.

One can obtain recursive equations that differ from the previous ones by using this generating function. Indeed, rewriting the function $\boldsymbol{\Psi}_m(w)$ in the form of the product $\boldsymbol{\Psi}_m(w) = \boldsymbol{\Psi}_k(w)\mathbf{P}(1)w\boldsymbol{\Psi}_{m-k-1}(w)$, we obtain from the convolution theorem

$$\mathbf{P}_n(m) = \sum_{\nu} \mathbf{P}_\nu(k)\mathbf{P}(1)\mathbf{P}_{n-\nu-1}(m-k-1) \quad (2.3.12)$$

and, in particular,

$$\mathbf{P}_n(m) = \sum_{\nu=0}^{n-m} \mathbf{P}^\nu(0)\mathbf{P}(1)\mathbf{P}_{n-\nu-1}(m-1) \quad (2.3.13)$$

This equation may be used to calculate $\mathbf{P}_n(m)$ for small m. For example,

$$\mathbf{P}_n(0) = \mathbf{P}^n(0) \qquad \mathbf{P}_n(1) = \sum_{\nu=0}^{n-1} \mathbf{P}^\nu(0)\mathbf{P}(1)\mathbf{P}^{n-\nu-1}(0)$$

2.3.2.2 Fourier Transform

The relative simplicity of the generating functions of the matrix binomial distribution makes the transform method an attractive tool for calculating the distribution. Since generating function (2.3.7) of the distribution is a polynomial, we may use inversion formula (2.2.25) of the discrete Fourier transform:

$$\mathbf{P}_n(m) = p^{-1} \sum_{k=0}^{p-1} \alpha^{-mk}[\mathbf{P}(0) + \mathbf{P}(1)\alpha^k]^n \quad (2.3.14)$$

where $\alpha = e^{-2\pi j/p}$ and $p > n$. Taking into account the symmetry of the function α^k, one may decrease the upper limit of summation in Eq. (2.3.14):

$$\mathbf{P}_n(m) = p^{-1}\mathbf{P}^n + \eta_{mp}[\mathbf{P}(0)-\mathbf{P}(1)]^n + 2p^{-1}\operatorname{Re}\sum_{k=1}^{\nu} [\mathbf{P}(0)+\mathbf{P}(1)\alpha^k]^n\alpha^{-mk} \quad (2.3.15)$$

where $\eta_{mp} = 0.5(-1)^m[1+(-1)^p]p^{-1}$ and $\nu = \lfloor 0.5(p-1) \rfloor$ is the integer part of $0.5(p-1)$. A dramatic increase in the speed of computation can be achieved by using the FFT. [14] If we select $p = 2^r$, then Eq. (2.3.14) becomes

$$\mathbf{P}_n(m) = 2^{-r} \sum_{k=0}^{p/2-1} [\mathbf{P}(0)+\mathbf{P}(1)\alpha^k]^n\alpha^{-mk} + 2^{-r} \sum_{k=p/2-1}^{p-1} [\mathbf{P}(0)+\mathbf{P}(1)\alpha^k]^n\alpha^{-mk}$$

which, after obvious simplifications, takes the form

$$\mathbf{P}_n(m) = 2^{-r} \sum_{k=0}^{p/2-1} [(\mathbf{P}(0)+\mathbf{P}(1)\alpha^k)^n + (-1)^m(\mathbf{P}(0)-\mathbf{P}(1)\alpha^k)^n]\alpha^{-mk}$$

If $m = 2r$ is even, the previous equation becomes

$$\mathbf{P}_n(2l) = 2^{-r}\sum_{k=0}^{p/2-1}[(\mathbf{P}(0)+\mathbf{P}(1)\alpha^k)^n + (\mathbf{P}(0)-\mathbf{P}(1)\alpha^k)^n]\beta^{-lk} \quad (2.3.16)$$

where $\beta=\alpha^2$. The right-hand side of this equation is a $p/2$ point Fourier transform, and therefore we may use the $p/2$ point transform (instead of the original p point transform) to calculate the probability $\mathbf{P}_n(m)$ for even m. A similar result is valid for an odd m:

$$\mathbf{P}_n(2l+1) = 2^{-r}\sum_{k=0}^{p/2-1}[(\mathbf{P}(0)+\mathbf{P}(1)\alpha^k)^n - (\mathbf{P}(0)-\mathbf{P}(1)\alpha^k)^n]\alpha^k\beta^{-lk} \quad (2.3.17)$$

Thus, instead of having to compute the p point DFT for the function $2^{-r}(\mathbf{P}(0)+\mathbf{P}(1)\alpha^k)^n$, we may use the $p/2$ point DFTs for the function $2^{-r}[(\mathbf{P}(0)+\mathbf{P}(1)\alpha^k)^n + (\mathbf{P}(0)-\mathbf{P}(1)\alpha^k)^n]$ to calculate $\mathbf{P}_n(m)$ for an even m and $p/2$ point DFT for the function $2^{-r}\alpha^k[(\mathbf{P}(0)+\mathbf{P}(1)\alpha^k)^n - (\mathbf{P}(0)-\mathbf{P}(1)\alpha^k)^n]$ to calculate $\mathbf{P}_n(m)$ for an odd m. Using similar reasoning, we may break each of the two $p/2$ point DFTs into two $p/4$ point DFTs, and so on.

This method of calculating the DFT is called the fast Fourier transform (FFT). Generally, when p is large the FFT substantially increases the efficiency of computation. However, it requires a significant amount of computer memory to store p matrices $(\mathbf{P}(0)+\mathbf{P}(1)\alpha^k)^n$ and their transformations.

In many applications, we need to compute $\mathbf{P}_n(m)$ for only several values of m. In this case, we may prefer to use the original formula (2.3.14), which allows us to calculate $(\mathbf{P}(0)+\mathbf{P}(1)\alpha^k)^n$ on the fly rather than storing it for further use as required by the FFT. This method is especially useful for obtaining the matrix probabilities $\mathbf{P}_n(m_1<m\le m_2) = \sum_{m=m_1+1}^{m_2}\mathbf{P}_n(m)$ that the number of occurrences of the event A_2 is greater than m_1 but not greater than m_2.

If we replace $\mathbf{P}_n(m)$ with the right-hand side of Eq. (2.3.15) and sum the geometric progressions, then

$$\mathbf{P}_n(m_1<m\le m_2) = \xi_{m_1,m_2}\mathbf{P}^n + \eta_{m_1m_2}(\mathbf{P}(0)-\mathbf{P}(1))^n + \mathrm{Re}\sum_{k=1}^{v}(\mathbf{P}(0)+\mathbf{P}(1)\alpha^k)^n\zeta_{m_1m_2k} \quad (2.3.18)$$

where

$$\xi_{m_1m_2} = (m_2 - m_1)/p$$

$$\eta_{m_1m_2} = [(-1)^{m_2}-(-1)^{m_1}][1+(-1)^p]/4p$$

$$\zeta_{m_1m_2k} = 2(\alpha^{-(m_1+1)k}-\alpha^{-(m_2+1)k})/p(1-\alpha^{-k})$$

This formula not only allows one to calculate $\mathbf{P}_n(m_1<m\le m_2)$ for large m_2-m_1 faster than with the FFT, but also delivers higher accuracy than does the FFT. It is implemented in the M-file `fpnm18`.

Alternatively, the probability in equation (2.3.18) can also be computed using the formula $\mathbf{P}_n(m_1<m\le m_2)=\mathbf{P}_n(\le m_2)-\mathbf{P}_n(\le m_1)$, where the cumulative matrix probability distribution $\mathbf{P}_n(\le t) = \mathbf{P}_n(m\le t)$ satisfies the following equation

$$\mathbf{P}_n(\le t) = \mathbf{P}_{n-1}(\le t)\mathbf{P}(0) + \mathbf{P}_{n-1}(\le t-1)\mathbf{P}(1)$$

with the initial conditions $\mathbf{P}_1(\leq 0)=\mathbf{P}(0)$ and $\mathbf{P}_1(\leq 1)=\mathbf{P}$. This equation is obtained by summing Eq. (2.3.6), with respect to m, from $m=0$ to $m=t$.

2.3.2.3 Asymptotic Expansion

As we mentioned previously, it is convenient to use z−transforms when dealing with asymptotic expansions. If we replace $w=1/z$ in equation (2.3.11), we obtain the z-transform

$$\mathbf{\Gamma}_m(z) = z[(\mathbf{I}z - \mathbf{P}(0))^{-1}\mathbf{P}(1)]^m[\mathbf{I}z - \mathbf{P}(0)]^{-1} \tag{2.3.19}$$

Since this is a rational function, we may use the results of Appendix 2 to find the asymptotic relations. To transform this function to the form of Eq. (A.2.2), we express the inverse characteristic matrix (see Appendix 5) as

$$[\,\mathbf{I}z - \mathbf{P}(0)\,]^{-1} = \mathbf{B}(z)/\Delta(z) \tag{2.3.20}$$

where $\Delta(z) = \det(\mathbf{I}z - \mathbf{P}(0))$ is the characteristic polynomial of the matrix $\mathbf{P}(0)$ and $\mathbf{B}(z)$ is the characteristic adjoint matrix; that is, its elements are adjuncts of the corresponding elements of the matrix $\mathbf{I}z - \mathbf{P}(0)$. Equation (2.3.19) becomes

$$\mathbf{\Gamma}_m(z) = z[\mathbf{B}(z)\mathbf{P}(1)]^m\mathbf{B}(z)/\Delta^{m+1}(z) \tag{2.3.21}$$

The right-hand side of this equation has the form of Eq. (A.2.2), and therefore the distribution $\mathbf{P}_n(m)$ is given by formula (A.2.6).

Consider the most important particular case when the eigenvalues of the matrix $\mathbf{P}(0)$ are different or, in other words, the roots $z_1, z_2, \ldots, z_k$ of the polynomial $\Delta(z)$ are all different. In this case

$$q(z) = \Delta^{m+1}(z) = (z-z_1)^{m+1}(z-z_2)^{m+1} \cdots (z-z_k)^{m+1} \tag{2.3.22}$$

By formula (A.2.6), we obtain

$$\mathbf{P}_n(m) = \sum_{j=1}^{k}\sum_{\nu=0}^{m} \mathbf{A}_{j\nu}\binom{n}{\nu} z_j^{n-\nu} \tag{2.3.23}$$

where

$$\mathbf{A}_{j\nu} = \mathbf{\Theta}_m^{(m-\nu)}(z_j)/(m-\nu)! \qquad \mathbf{\Theta}_m(z) = (z-z_j)^{m+1}\mathbf{\Gamma}_m(z)z^{-1}$$

It is convenient to use these equations if the dimensions of the matrices and m are not large.

Since the matrix z-transform $\mathbf{\Gamma}_m(z)$ has the form of Eq. (A.2.2), we can use formula (A.2.10) for the $\mathbf{P}_n(m)$ asymptotic expansion. Suppose that the matrix $\mathbf{P}(0)$ has a simple positive eigenvalue z_1 with the largest absolute value. In this case, the asymptotic formula (A.2.10) becomes

$$\mathbf{P}_n(m) \sim \mathbf{A}_{1m}\binom{n}{m} z_1^{n-m} \tag{2.3.24}$$

where

$$\mathbf{A}_{1m} = [\mathbf{B}(z_1)\mathbf{P}(1)]^m\mathbf{B}(z_1)/[\Delta'(z_1)]^{m+1}$$

This formula is exact if the matrices $\mathbf{P}(0)=p$ and $\mathbf{P}(1)=q$ are scalars (first-order matrices). Indeed, in this case the characteristic equation $\Delta(z) = z-q$, has only one root,

$z_1 = q$. Since $\mathbf{B}(z) = 1$ and $\mathbf{A}_{1m} = p^m$, we obtain, as expected, the usual binomial distribution

$$\mathbf{P}_n(m) = \binom{n}{m} p^m q^{n-m}$$

Using Theorem A.2.1, we obtain the asymptotic formula

$$\mathbf{P}_n(m) \sim [\mathbf{B}(z_1)/\Delta'(z_1)][n\mathbf{P}(1)\mathbf{B}(z_1)/\Delta'(z_1)]^m z_1^{n-m}/m! \qquad (2.3.25)$$

which is analogous to the Poisson formula for approximating the binomial distribution. [4] Switching the matrices $\mathbf{P}(1)$ and $\mathbf{P}(0)$, as is usually done in the study of binomial distribution, we can obtain a formula analogous to Eq. (2.3.20) if m is close to n.

2.3.2.4 The Central Limit Theorem

As in the case of the ordinary binomial distribution,

$$p_n(m) = \mathbf{p}_0 \mathbf{P}_n(m)\mathbf{1}$$

tends to have a normal distribution when $n \to \infty$ under some mild regularity conditions. More precisely, if a Markov chain with the matrix $\mathbf{P} = \mathbf{P}(0) + \mathbf{P}(1)$ is regular and $\mathbf{P}(0) \neq 0$ and $\mathbf{P}(1) \neq 0$, then $(m - a_n)/\sigma_n$ is asymptotically (0,1) normal.

Indeed, according to Eq. (2.3.7), the distribution characteristic function (see also Example 2.2.2) is equal to

$$\phi_n(t) = \mathbf{p}_0 \mathbf{Q}^n(t)\mathbf{1} \qquad (2.3.26)$$

where

$$\mathbf{Q}(t) = \mathbf{P}(0) + \mathbf{P}(1)e^{-jt}$$

Since $\mathbf{Q}(0) = \mathbf{P}$ is a regular matrix, $\mathbf{Q}(t)$ has a simple eigenvalue $\lambda(t)$ for small t such that $\lambda(0) = 1$. Then, asymptotically, when $n \to \infty$, we obtain from the matrix $\mathbf{Q}(t)$ spectral representation (see Appendix 5.1.5)

$$\phi_n(t) \sim q(t)\lambda_1^n(t)$$

This means (see Ref. 1, p. 37) that, asymptotically, $\phi_n(t)$ may be regarded as a characteristic function of a sum of independent equally distributed variables, which, as is well known, [4] is asymptotically normal.

2.3.3 Matrix Pascal Distribution

Along with matrix binomial distribution, we often use matrix Pascal distribution and matrix-geometric distribution in particular.

Assume that two events A_1 and A_2 are mutually exclusive, depend only on the value $\mathbf{e}_t$, and have matrix probabilities $\mathbf{P}(A_1) = \mathbf{P}(0)$ and $\mathbf{P}(A_2) = \mathbf{P}(1)$. Let us find the probability distribution $\mathbf{P}(l,r)$ that event A_2 occurs for the r-th time on the Markov chain's l-th step.

If $r = 1$, then, clearly, the matrix probability that event A_2 occurs for the first time on the l-th step is equal to

$$\mathbf{P}(l,1) = \mathbf{P}^{l-1}(A_1)\mathbf{P}(A_2) = \mathbf{P}^{l-1}(0)\mathbf{P}(1) \tag{2.3.27}$$

This distribution, called the *matrix-geometric distribution,* has the generating function

$$\mathbf{\Theta}_1(z) = \sum_{l=1}^{\infty} \mathbf{P}^{l-1}(0)\mathbf{P}(1)z^l = (\mathbf{I} - \mathbf{P}(0)z)^{-1}\mathbf{P}(1)z \tag{2.3.28}$$

In general, the variable l can be represented as a sum of r conditionally independent variables $l = l_1 + l_2 + \cdots + l_r$, each having geometric distribution (2.3.27). By the convolution theorem, the generating function $\mathbf{\Theta}_r(z)$ of the matrix distribution $\mathbf{P}(l,r)$ is equal to the product of the generating functions of the addends:

$$\mathbf{\Theta}_r(z) = \mathbf{\Theta}_1^r(z) = [(\mathbf{I} - \mathbf{P}(0)z)^{-1}\mathbf{P}(1)z]^r \tag{2.3.29}$$

The distribution $\mathbf{P}(l,r)$ is called the *matrix Pascal distribution*. Since

$$\mathbf{\Theta}_r(z) = \mathbf{\Theta}_{r_1}(z)\mathbf{\Theta}_{r-r_1}(z)$$

this distribution satisfies the recursive equations

$$\mathbf{P}(l,r) = \sum_x \mathbf{P}(x,r_1)\mathbf{P}(l-x,r-r_1) \tag{2.3.30}$$

and, in particular,

$$\mathbf{P}(l,r) = \sum_x \mathbf{P}^{x-1}(0)\mathbf{P}(1)\mathbf{P}(l-x,r-1) \tag{2.3.31}$$

The generating function $\mathbf{\Theta}_r(z)$ differs from $\mathbf{\Psi}_r(z)$, defined by formula (2.3.11), by only a factor:

$$\mathbf{\Theta}_r(z) = \mathbf{\Psi}_r(z)[\,\mathbf{I} - \mathbf{P}(0)z\,] \tag{2.3.32}$$

and, therefore, satisfies the similar asymptotic relations

$$\mathbf{P}(l,r) \sim [\mathbf{B}(z_1)\mathbf{P}(1)/\Delta'(z_1)]^r l^{r-1} z_1^{l-r}/(r-1)! \tag{2.3.33}$$

where z_1 is the greatest eigenvalue of the matrix $\mathbf{P}(0)$.

2.4 MARKOV FUNCTIONS

2.4.1 Block Matrix Probabilities

As was shown in the previous chapter, it is sometimes convenient to describe the HMM as a Markov function. In this case, the set of the Markov chain states can be partitioned into subsets corresponding to the events, and the transition probability matrix can be presented in the block form

$$\mathbf{P} = block[\mathbf{P}_{ij}]_{s,s} = \begin{bmatrix} \mathbf{P}_{11} & \mathbf{P}_{12} & \cdots & \mathbf{P}_{1s} \\ \mathbf{P}_{21} & \mathbf{P}_{22} & \cdots & \mathbf{P}_{2s} \\ \cdots & \cdots & \cdots & \cdots \\ \mathbf{P}_{s1} & \mathbf{P}_{s2} & \cdots & \mathbf{P}_{ss} \end{bmatrix} \tag{2.4.1}$$

The case when events A_i do not depend on the states deterministically can be transformed into a case of the deterministic dependency by increasing the number of the Markov chain states. Indeed, consider a new Markov chain whose states are defined as vectors (i,k), where i is the state of the original chain and k is the event A_k index. If the so-defined chain is in the state (i,k), then the event A_k occurs. The new chain matrix has the form

$$\mathbf{P}_1 = \begin{bmatrix} \mathbf{P}(A_1) & \mathbf{P}(A_2) & \dots & \mathbf{P}(A_s) \\ \mathbf{P}(A_1) & \mathbf{P}(A_2) & \dots & \mathbf{P}(A_s) \\ \cdots & \cdots & \cdots & \cdots \\ \mathbf{P}(A_1) & \mathbf{P}(A_2) & \dots & \mathbf{P}(A_s) \end{bmatrix}$$

The price that we pay for such a description is an increase in the number of states by a factor of s. Sometimes, however, by using specifics of the process it is possible to achieve the equivalent description with fewer states. We illustrate this statement in the following example.

Example 2.4.1. As we saw in Example 1.2.3, Gilbert's model with two states can be described as a function of the Markov chain with three states and the matrix

$$\mathbf{P}_1 = \begin{bmatrix} Q & Ph & P(1-h) \\ p & qh & q(1-h) \\ p & qh & q(1-h) \end{bmatrix}$$

when errors occur only in the third state. If A_2 denotes an error event and A_1 its complement $(e_t = 0)$, then this matrix $\mathbf{P}_1$ can be written in form (2.4.1)

$$\mathbf{P}_1 = \begin{bmatrix} \mathbf{P}_{11} & \mathbf{P}_{12} \\ \mathbf{P}_{21} & \mathbf{P}_{22} \end{bmatrix}$$

where

$$\mathbf{P}_{11} = \begin{bmatrix} Q & Ph \\ p & qh \end{bmatrix} \quad \mathbf{P}_{12} = \begin{bmatrix} P(1-h) \\ q(1-h) \end{bmatrix} \quad \mathbf{P}_{21} = \begin{bmatrix} p & qh \end{bmatrix} \quad \mathbf{P}_{22} = q(1-h)$$

Let us express the principal distributions using the transition matrix blocks. Since the Markov function is a particular case of the HMM (see Sec. 1.2.4), the corresponding formulas are corollaries of the formulas of the previous sections. The matrices $\mathbf{P}(A_i)$ have the form

$$\mathbf{P}(A_i) = \begin{bmatrix} 0 & \dots & 0 & \mathbf{P}_{1i} & 0 & \dots & 0 \\ 0 & \dots & 0 & \mathbf{P}_{2i} & 0 & \dots & 0 \\ \cdots & & & & & & \cdots \\ 0 & \dots & 0 & \mathbf{P}_{si} & 0 & \dots & 0 \end{bmatrix} \tag{2.4.2}$$

2.4.2 Matrix Multinomial Distribution

All the formulas from section 2.3 are preserved if the matrices $\mathbf{P}(A_i)$ have the form of Eq. (2.4.2). Noting that the matrices have many zero elements, we can rewrite the formulas using block matrices whose size is considerably smaller than the size of the matrix $\mathbf{P}(A_i)$. Let $\mathbf{P}_{ij}(m_1,m_2,...,m_r;n)$ be the blocks of the matrix $\mathbf{P}_n(m_1,m_2,...,m_r)$. Then it follows from Eq. (2.4.1) that

$$\mathbf{P}_{ij}(m_1,m_2,...,m_r;n) = \sum_{k,\mu_1,...,\mu_r} \mathbf{P}_{ik}(m_1-\mu_1,m_2-\mu_2,...,m_r-\mu_r;n-m)\mathbf{P}_{kj}(\mu_1,\mu_2,...,\mu_r;m) \tag{2.4.3}$$

and in particular

$$\mathbf{P}_{ij}(m_1,m_2,...,m_r;n) = \sum_{k=1}^{s} \mathbf{P}_{ik}(m_1,...,m_j-1,...,m_r;n-1)\mathbf{P}_{kj} \tag{2.4.4}$$

The corresponding generating functions can also be rewritten in the block form

$$\mathbf{\Phi}_{ij}(z_1,z_2,...,z_r;n) = \sum_{k=1}^{s} \mathbf{\Phi}_{ik}(z_1,z_2,...,z_r;n-m)\mathbf{\Phi}_{kj}(z_1,z_2,...,z_r;m) \tag{2.4.5}$$

where $\mathbf{\Phi}_{ij}(z_1,z_2,...,z_r;n)$ are the blocks of the generating function (2.3.3), which has the form

$$\mathbf{\Phi}_n(z_1,z_2,...,z_r) = [\ \mathbf{PT}(z_1,z_2,...,z_r)\ ]^n \tag{2.4.6}$$

where

$$\mathbf{T} = block\ diag\{\mathbf{I}_i z_i\} = \sum_{k=1}^{s}{}^{(\cdot)} \mathbf{I}_k z_k \tag{2.4.7}$$

The symbol $\sum^{(\cdot)}$ denotes a direct sum of matrices which is defined by Eq. (2.4.7).

In the particular case of matrix binomial distribution, we have

$$\mathbf{P}(A_1) = \mathbf{P}(0) = \begin{bmatrix} \mathbf{P}_{11} & 0 \\ \mathbf{P}_{21} & 0 \end{bmatrix} \qquad \mathbf{P}(A_2) = \mathbf{P}(1) = \begin{bmatrix} 0 & \mathbf{P}_{12} \\ 0 & \mathbf{P}_{22} \end{bmatrix} \tag{2.4.8}$$

and the recursive Eq. (2.4.4) become

$$\begin{aligned} \mathbf{P}_{i1}(m;n) &= \mathbf{P}_{i1}(m;n-1)\mathbf{P}_{11} + \mathbf{P}_{i2}(m;n-1)\mathbf{P}_{21} \\ \mathbf{P}_{i2}(m;n) &= \mathbf{P}_{i1}(m-1;n-1)\mathbf{P}_{12} + \mathbf{P}_{i2}(m-1;n-1)\mathbf{P}_{22} \end{aligned} \tag{2.4.9}$$

Generating function (2.3.7) is given by

$$\mathbf{\Phi}_n(z) = \begin{bmatrix} \mathbf{P}_{11} & \mathbf{P}_{12}z \\ \mathbf{P}_{21} & \mathbf{P}_{22}z \end{bmatrix}^n \tag{2.4.10}$$

Finally, using Eq. (2.3.19) and (2.4.8), we have

$$\mathbf{\Gamma}_m(z) = z[(\mathbf{I}z - \mathbf{P}(0))^{-1}\mathbf{P}(1)]^m(\mathbf{I}z - \mathbf{P}(0))^{-1}$$

where

$$(\mathbf{I}z - \mathbf{P}(0))^{-1} = \begin{bmatrix} (\mathbf{I}z-\mathbf{P}_{11})^{-1} & 0 \\ z^{-1}\mathbf{P}_{21}(\mathbf{I}z-\mathbf{P}_{11})^{-1} & z^{-1}\mathbf{I} \end{bmatrix}$$

After some straightforward transformations, we obtain

$$\boldsymbol{\Gamma}_m(z) = z^{-m}\Delta_1^{-m-1}(z)\mathbf{A}(z)\mathbf{C}^{m-1}(z)\mathbf{D}(z)$$

where

$$\mathbf{A}(z) = \begin{bmatrix} (\mathbf{I}z - \mathbf{P}_{11})^{-1}\mathbf{P}_{12}z \\ \mathbf{C}(z) \end{bmatrix}$$

$$\mathbf{C}(z) = \mathbf{P}_{21}\mathbf{B}_1(z)\mathbf{P}_{12} + \Delta_1(z)\mathbf{P}_{22} \qquad \mathbf{D}(z) = \begin{bmatrix} \mathbf{P}_{21}\mathbf{B}_1(z) & \mathbf{I} \end{bmatrix}$$

$\Delta_1(z) = \det(\mathbf{I}z - \mathbf{P}_{11})$ is the characteristic polynomial of the matrix $\mathbf{P}_{11}$, and $\mathbf{B}_1(z) = \Delta_1(z)(\mathbf{I}z - \mathbf{P}_{11})^{-1}$ is its adjoint matrix. Since $\boldsymbol{\Gamma}_m(z)$ is a rational function, the matrices $\mathbf{P}_{ij}(m;n)$ can be expressed in a form similar to that of Eq. (2.3.23). However, the sizes of the matrices in these equations are normally significantly smaller than the sizes of matrices in Eq. (2.3.23).

Suppose that the characteristic polynomial $\Delta_1(z)$ has a simple root z_1 whose absolute value is greater than the absolute values of its remaining roots. For this to be true, by the theorem of Frobenius, it is sufficient that the matrix $\mathbf{P}_{11}$ is irreducible and acyclic (see Appendix 6). (If, in addition, $\mathbf{P}_{12}\mathbf{1} > 0$, then $z_1 < 1$.) Then the asymptotic representations of the matrices $\mathbf{P}_{ij}(m;n)$ have the form of Eq. (2.3.24).

Analyzing the results of this section, we conclude that block matrix distributions have a more complex form than in terms of matrices $\mathbf{P}(A_i)$. However, since the orders of the blocks are smaller than the order of the whole matrix, it is worthwhile to use the formulas in the block form in practical calculations.

2.4.3 Interval Distributions of Markov Functions

Let $\mathbf{P}_{ij}(l)$ be the matrix probability that a Markov chain leaving the states of the set A_i for the first time enters the states of the set A_j on the l-th step. Denote $B_j = \bar{A}_j$ as the complement of the subset A_j with respect to the whole set of states. Then, the distribution $\mathbf{P}_{ij}(l)$ is given by

$$\begin{aligned} \mathbf{P}_{ij}(1) &= \mathbf{P}_{ij} \\ \mathbf{P}_{ij}(l) &= \mathbf{P}(B_j \mid A_i)\mathbf{P}^{l-2}(B_j \mid B_j)\mathbf{P}(A_j \mid B_j) \quad \text{for } l > 1 \end{aligned} \tag{2.4.11}$$

where the matrix $\mathbf{P}(B_j \mid A_i)$ is obtained from matrix $\mathbf{P}$ by eliminating all the columns corresponding to the states of A_j and all the rows corresponding to the states that do not belong to the set A_i, matrix $\mathbf{P}(B_j \mid B_j)$ is the matrix obtained from matrix $\mathbf{P}$ by eliminating all the rows and columns corresponding to the states of A_j, and matrix $\mathbf{P}(A_j \mid B_j)$ is the matrix obtained from matrix $\mathbf{P}$ by eliminating all the columns corresponding to the states that do not belong to A_j and all the rows corresponding to the states of the set A_j.

For example, if $i = 1, j = 2$, $s = 3$, and

$$\mathbf{P} = \begin{bmatrix} \mathbf{P}_{11} & \mathbf{P}_{12} & \mathbf{P}_{13} \\ \mathbf{P}_{21} & \mathbf{P}_{22} & \mathbf{P}_{23} \\ \mathbf{P}_{31} & \mathbf{P}_{32} & \mathbf{P}_{33} \end{bmatrix}$$

then

$$\mathbf{P}(B_2 \mid A_1) = \begin{bmatrix}\mathbf{P}_{11} & \mathbf{P}_{13}\end{bmatrix} \quad \mathbf{P}(B_2 \mid B_2) = \begin{bmatrix}\mathbf{P}_{11} & \mathbf{P}_{13} \\ \mathbf{P}_{31} & \mathbf{P}_{33}\end{bmatrix} \quad \mathbf{P}(A_2 \mid B_2) = \begin{bmatrix}\mathbf{P}_{12} \\ \mathbf{P}_{32}\end{bmatrix}$$

The generating function of the matrices $\mathbf{P}_{ij}(l)$ is given by

$$\boldsymbol{\theta}_{ij}(z) = \sum_{l=1}^{\infty} \mathbf{P}_{ij}(l) z^l = z\mathbf{P}_{ij} + \sum_{l=2}^{\infty} z^2 \mathbf{P}(B_j \mid A_i)[z\mathbf{P}(B_j \mid B_j)]^{l-2} \mathbf{P}(A_j \mid B_j)$$

After summing the matrix-geometric progression (see Appendix 5.1.3), we obtain

$$\boldsymbol{\theta}_{ij}(z) = z\mathbf{P}_{ij} + z^2 \mathbf{P}(B_j \mid A_i)[\mathbf{I} - z\mathbf{P}(B_j \mid B_j)]^{-1} \mathbf{P}(A_j \mid B_j) \qquad (2.4.12)$$

In the particular case when $i = j$, the previous expression gives the generating function of the distribution of the time of the first return to the subset A_i:

$$\boldsymbol{\theta}_{ii}(z) = z\mathbf{P}_{ii} + z^2 \mathbf{P}(B_i \mid A_i)[\mathbf{I} - z\mathbf{P}(B_j \mid B_j)]^{-1} \mathbf{P}(A_j \mid B_j) \qquad (2.4.13)$$

Quite analogously, one can find the generating functions of the so-called tabu distributions. [3] Let ${}_h\mathbf{P}_{ij}(l)$ be the matrix probability that the Markov chain leaving the states of the set A_i arrives at a state of the set A_j for the first time in l steps without visiting the states of the set A_h or returning to A_i. If we denote $B_g = \bar{A}_i \cap \bar{A}_j \cap \bar{A}_h$, then this distribution can be written in the form

$$\begin{aligned} {}_h\mathbf{P}_{ij}(1) &= \mathbf{P}_{ij} \\ {}_h\mathbf{P}_{ij}(l) &= \mathbf{P}(B_g \mid A_i)\mathbf{P}^{l-2}(B_g \mid B_g)\mathbf{P}(A_j \mid B_g) \quad \text{for } l > 1 \end{aligned} \qquad (2.4.14)$$

But then the generating function is given by

$${}_h\boldsymbol{\theta}_{ij}(z) = z\mathbf{P}_{ij} + z\mathbf{P}(B_g \mid A_i)[\mathbf{I} - z\mathbf{P}(B_g \mid B_g)]^{-1} z\mathbf{P}(A_j \mid B_g) \qquad (2.4.15)$$

Using these generating functions and applying the convolution theorem, one can find the generating function of the interval sum distributions. For example, the generating function of the matrix probabilities that the Markov chain leaving the states of the set A_i enters a state of the set A_j for the r-th time on the l-th step is equal to the product $\boldsymbol{\theta}_{ij}(z)\boldsymbol{\theta}_{jj}^{r-1}(z)$ of the generating functions defined previously.

It is convenient to use the generating functions of this section to find mean values for the intervals. For example, the mean interval of entering the states of A_j can be found using the generating function (2.4.12):

$$\mathbf{m}_{ij} = \boldsymbol{\theta}'_{ij}(1)\mathbf{1}$$

After taking the derivative, we obtain

$$\begin{aligned} \boldsymbol{\theta}'_{ij}(1) &= \mathbf{P}_{ij}\mathbf{1} + 2\mathbf{P}(B_j \mid A_i)[\mathbf{I} - \mathbf{P}(B_j \mid B_j)]^{-1}\mathbf{P}(A_j \mid B_j)\mathbf{1} \\ &+ \mathbf{P}(B_j \mid A_i)[\mathbf{I} - \mathbf{P}(B_j \mid B_j)]^{-2}\mathbf{P}(A_j \mid B_j)\mathbf{1} \end{aligned}$$

Since $\mathbf{P}$ is a stochastic matrix, $\mathbf{P1} = \mathbf{1}$ and, therefore,

$$\mathbf{P}(B_j \mid B_j)\mathbf{1} + \mathbf{P}(A_j \mid B_j)\mathbf{1} = \mathbf{1}$$

This can be rewritten as

$$[\mathbf{I} - \mathbf{P}(B_j \mid B_j)]^{-1}\mathbf{P}(A_j \mid B_j)\mathbf{1} = \mathbf{1}$$

Using this formula we obtain

$$\mathbf{m}_{ij} = \mathbf{1} + \mathbf{P}(B_j \mid A_i)[\,\mathbf{I} - \mathbf{P}(B_j \mid B_j)\,]^{-1}\mathbf{1}$$

2.4.4 Signal-Flow Graph Applications

When dealing with Markov functions, we find it convenient to compute matrix probabilities and generating functions with the help of signal-flow graphs.[6,10] We assume that the events depend deterministically on Markov chain states and that the transition matrix has the form of Eq. (2.4.1). Accordingly, we consider a block stochastic graph (see Appendix 2.3) whose matrix branch transmissions are $\mathbf{T}_{ij} = \mathbf{P}_{ij}$. In this section, we consider problems that can be solved with the help of signal-flow graphs.

2.4.4.1 Exclusion of Unobserved States

Suppose that we observe a Markov chain only when it is in a subset A_i. It is obvious that the new process is the Markov chain. If we denote, as in the previous section, $B_i = \overline{A}_i$ as the subset A_i complement, then the transition matrix of the new chain is given by[6]

$$\overline{\mathbf{P}}_i = \mathbf{P}_{ii} + \sum_{m=2}^{\infty} \mathbf{P}(B_i \mid A_i)\mathbf{P}(B_i \mid B_i)^{m-2}\mathbf{P}(A_i \mid B_i)$$

or

$$\overline{\mathbf{P}}_i = \mathbf{P}_{ii} + \mathbf{P}(B_i \mid A_i)[\,\mathbf{I} - \mathbf{P}(B_i \mid B_i)\,]^{-1}\mathbf{P}(A_i \mid B_i) \qquad (2.4.16)$$

(provided, of course, that the set B_i is not absorbing).

This expression has exactly the same form as formula (A.2.23). Therefore, according to the results of Appendix 2, we conclude that a stochastic signal-flow graph of the observed process is the *residue*[10] of the original graph after absorption of the nodes that correspond to the unobserved states. The transfer matrix of the residue graph is the transfer matrix of the observed process.

2.4.4.2 The Generating Function of the Distribution of Transition Time

If we compare Eq. (2.4.15) with expression (A.2.23), we conclude that the matrix generating function ${}_h\boldsymbol{\theta}_{ij}(z)$ can be computed as the transfer matrix from the subgraph corresponding to the states of the set A_i to the subgraph corresponding to the states of the set A_j in the residue graph obtained from the original stochastic graph with the matrix $z\mathbf{P}$, after absorption of the subgraph corresponding to the states of the set B_g.[8]

2.4.4.3 Multinomial Distribution Graph

We have shown in Appendix 2 that the elementary operations of matrix algebra can be performed using transformation of some signal-flow graphs. Since the generating function of the matrix multinomial distribution may be expressed as a matrix product by Eq. (2.4.6), it can be presented as a transfer matrix from subgraph G_1 to subgraph G_n of the graph shown in Fig. 2.1.

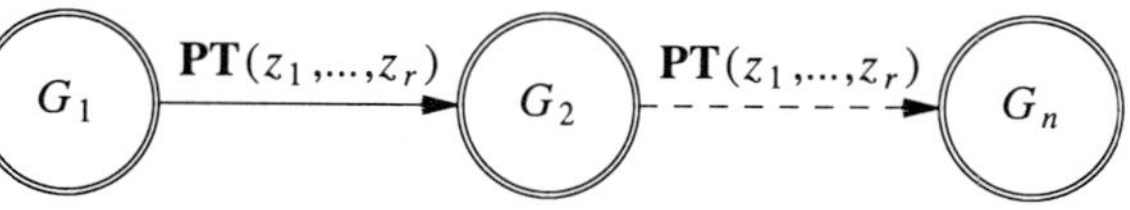

Figure 2.1. Matrix multinomial distribution transition graph.

2.4.4.4 Computing Scalar Probabilities and Generating Functions

As shown previously, scalar probabilities and generating functions can be found from their matrix counterparts as matrix products

$$Pr(A) = \mathbf{p}_0\mathbf{P}(A)\mathbf{1} \qquad \phi(z) = \mathbf{p}_0\mathbf{\Phi}(z)\mathbf{1}$$

Since matrix products may be interpreted using signal-flow graphs, the preceding equations can be presented as transfers from node S to node F of the graph depicted in Fig. 2.2.

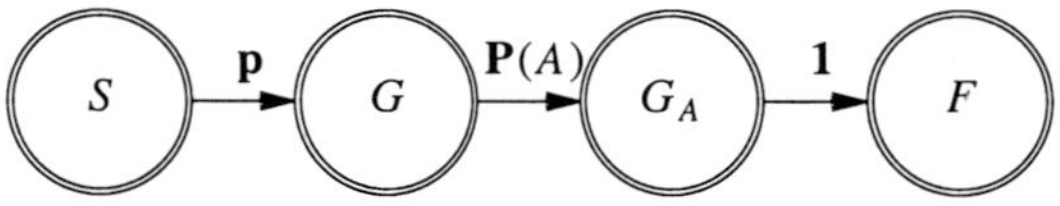

Figure 2.2. Scalar probability calculation.

2.4.4.5 Stationary Probabilities

We can find stationary probabilities of a Markov chain using signal-flow graphs, [11] since the stationary distribution satisfies the equation

$$\boldsymbol{\pi}\mathbf{P} = \boldsymbol{\pi} \qquad \boldsymbol{\pi}\mathbf{1} = 1 \tag{2.4.17}$$

However, Mason's formula [10] cannot be applied directly to the equation $\boldsymbol{\pi}\mathbf{P}=\boldsymbol{\pi}$ because of the matrix $\mathbf{I}-\mathbf{P}$ singularity [det($\mathbf{I} - \mathbf{P}$) = 0]. To solve the system, the graph with the matrix $\mathbf{P}$ should be replaced by another graph, or subgraphs of the graph are considered.

We can propose other methods for computing the stationary probabilities. One of them consists of computing the matrix $(\mathbf{I}z - \mathbf{P})^{-1}$ when $|z|>1$. This transformation is performed by absorbing the loop in the graph of Fig. 2.3.

A stationary distribution can be found as (see Appendix 6)

$$\lim_{k\to\infty} \mathbf{P}^k = \mathbf{1}\boldsymbol{\pi}$$

On the other hand, the same limit may be obtained from the sequence-generating function [4]

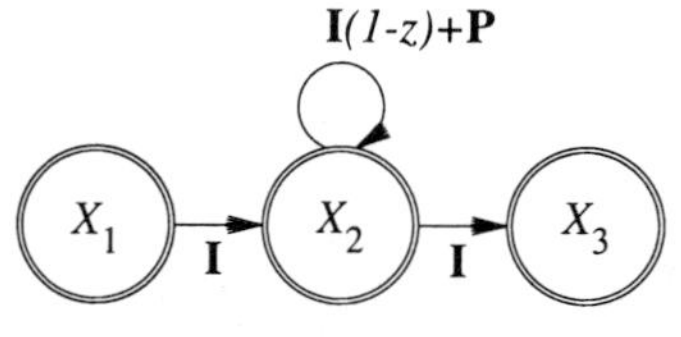

Figure 2.3. Characteristic matrix graph.

$$\mathbf{1P} = \lim_{z \downarrow 1}(z-1)(\mathbf{I}z - \mathbf{P})^{-1} = \mathbf{B}(1)/\Delta'(1)$$

where $\Delta(z)$ is the matrix characteristic polynomial and $\mathbf{B}(z)$ is its adjoint matrix (see Appendix 5).

Since all the rows of the matrix $\mathbf{1\pi}$ are identical, it is sufficient to find only one row of the matrix $(\mathbf{I}z - \mathbf{P})^{-1}$, as is illustrated in the following example.

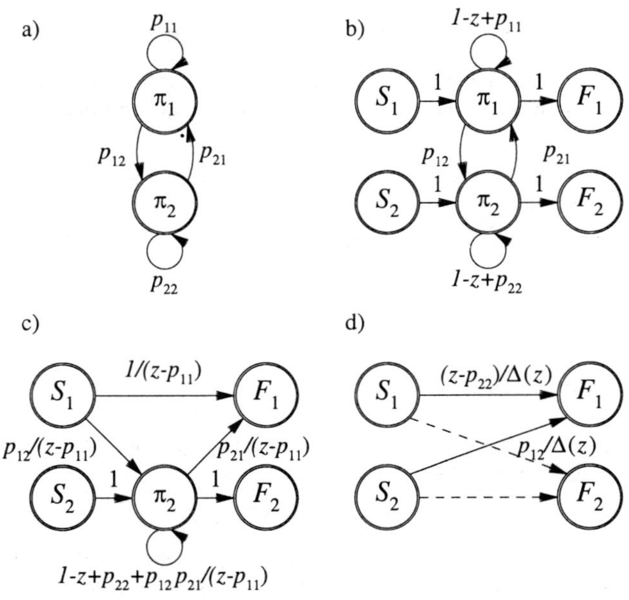

Figure 2.4. Reduction of a characteristic matrix graph.

Example 2.4.2. Consider a Markov chain with the matrix

$$\mathbf{P} = \begin{bmatrix} p_{11} & p_{12} \\ p_{21} & p_{22} \end{bmatrix}$$

whose graph is shown in Fig. 2.4a.

In order to invert the matrix $\mathbf{I}z - \mathbf{P}$ we create the block graph shown in Fig. 2.3. This graph is shown in Fig. 2.4b. After the absorption of the nodes π_1 and π_2 using methods described in Ref. 10, we subsequently obtain the graphs shown in Figs. 2.4c and 2.4d.

The first line of the matrix $(\mathbf{I}z - \mathbf{P})^{-1}$ is presented by the transitions from vertices S_1, S_2 to the node F_1 which is

$$\frac{1}{\Delta(z)}\begin{bmatrix} z-p_{22} & 1 \end{bmatrix}$$

where $\Delta(z)=(z-p_{22})(z-p_{11})-p_{12}p_{21}$. After multiplying this row matrix by $z-1$, we obtain, when $z\to\infty$,

$$\boldsymbol{\pi} = \frac{1}{\Delta'(1)}\begin{bmatrix} 1-p_{22} & 1 \end{bmatrix} = \begin{bmatrix} \dfrac{p_{21}}{p_{12}+p_{21}} & \dfrac{p_{12}}{p_{12}+p_{21}} \end{bmatrix}$$

It is also possible to find the stationary probabilities by representing system (2.4.17) in the form of a graph. All the vertices of the stochastic graph with the matrix $\mathbf{P}$ have to be connected with an additional node that is ascribed the value of 1, and all the transfers to this node should be equal to 1. One has to take into account that, in this case, Mason's formula is inapplicable and the graph should be solved by absorption of the nodes. [10]

Example 2.4.3. Using the second method, let us find the stationary distribution for the Markov chain considered in Example 2.4.2.

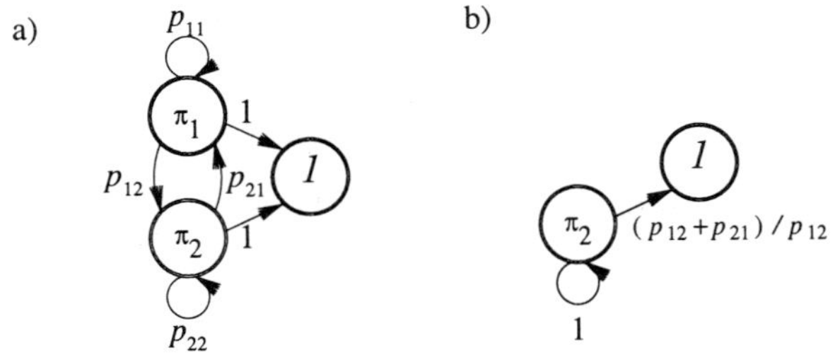

Figure 2.5. Stationary Distribution Calculation.

The graph for the system (2.4.17) is depicted in Fig. 2.5a. After absorbing the node π_1, we obtain the graph shown in Fig. 2.5b. From this graph it follows that $\pi_2(p_{21}+p_{12})/p_{12} = 1$. Thus $\pi_2 = p_{12}/(p_{21}+p_{12})$.

2.5 MONTE CARLO METHOD

The analytical methods for calculating basic HMM characteristics often lead to complex expressions. Sometimes it is convenient to use computer simulation (the Monte Carlo method) to calculate these characteristics. If the sample size is large, the difference between a characteristic and its estimated value obtained by simulation will be small.

The advantage of the Monte Carlo method is its relative simplicity: in many cases it is easier to program the system operation than to evaluate its performance analytically. We can also use raw data obtained from channel-error statistics measurements to evaluate system performance characteristics.

The main disadvantage of the Monte Carlo method is its low degree of accuracy when it is used to estimate small values and the amount of statistical data is limited. To achieve adequate accuracy, the simulation time may have to be very long.

In this section, we compare the Monte Carlo method of error number distribution calculation with the analytical methods described in the previous sections. But first we consider methods of HMM simulation using a random or pseudorandom number generator.

2.5.1 Error Source Simulation

Let us assume that an HMM error source model is presented by its canonical form: its states are governed by the Markov chain with the transition matrix **P**, and conditional probabilities of errors are either 0 or 1 (see Sec. 1.2.3). The error source simulation may be performed using the Markov chain definition: simulate its initial state s_0, according to the initial distribution $\mathbf{p}_0$, then simulate the next state s_1 with the transition probability $p_{s_0 s_1}$, and so on. This method uses one random number to generate the next state and would be very inefficient for real channel models, since errors appear infrequently and, therefore, the Markov chain does not change its states frequently. One may take advantage of these error distribution specifics by simulating intervals between errors.

According to the results of Sec. 1.4.2, a Markov chain with the matrix P is equivalent to the semi-Markov process with the matrix

$$\mathbf{Q} = [q_{ij}]_{m,m} \qquad q_{ii} = 0 \qquad q_{ij} = p_{ij}/(1-p_{ii}) \quad \text{for } i \neq j \tag{2.5.1}$$

and geometric interval distributions

$$w_i(l) = (1-p_{ii})p_{ii}^{\,l-1} \qquad i = 1,2,\ldots,m \tag{2.5.2}$$

Using this description, one may perform the simulation in the following way: According to the initial distribution $\boldsymbol{\pi} = (\pi_1,\pi_2,\ldots,\pi_n)$, select an initial state s_0, then, according to distributions (2.5.1) and (2.5.2), select the next state and the transition interval, and so on.

To be more specific, let us consider a sequence $\xi_1,\xi_2,\ldots,\xi_i,\ldots$ of samples of the [0,1] uniformly distributed random numbers. Then the initial state s_0 is selected if

$$\pi_0 + \pi_1 + \cdots + \pi_{s_0 - 1} \le \xi_1 < \pi_1 + \cdots + \pi_{s_0}$$

where $\pi_0 = 0$. The solution of this inequality can be also written as

$$s_0 = \min\, \{s: \sum_{i=1}^{s} \pi_i > \xi_1\} \tag{2.5.3}$$

Next we simulate the state-holding time, which is a series of $l_1 = \lceil \log\,(1-\xi_2)/\log\, p_{s_0 s_0} \rceil$ states s_0. If the probability of the error **e** is equal to 1 in this state, a series of l_1 errors (including the initial error) is generated.

We select s_1 as the next state of the semi-Markov process as

$$s_1 = \min\, \{s: \sum_{i=1}^{s} q_{s_0 i} > \xi_3\} \quad \text{where} \quad q_{ij} = p_{ij}/(1-p_{ii}) \tag{2.5.4}$$

the state s_1 holding time is simulated as $l_2 = \lceil \log\,(1-\xi_4)/\log\, p_{s_1,s_1} \rceil$ and so on.

If a Markov chain is semi-Markov lumpable by a partition $\{A_1,A_2,\ldots,A_s\}$ and we have a generator that produces random numbers whose distribution has the form

$$w_{ij}(l) = \mathbf{q}_i \mathbf{P}_{ii}^{l-1} \mathbf{P}_{ij} \mathbf{1} / \mathbf{q}_i \mathbf{Q}_{ij} \mathbf{1} \tag{2.5.5}$$

then it is possible to speed up the simulation process by considering the lumped process, which has fewer states than the original chain. However, often we need to solve the opposite problem: creating a random-number generator for a distribution of the form

$$p(l) = \mathbf{a}\mathbf{B}^{l-1}\mathbf{c} \tag{2.5.6}$$

The standard method of solving this problem requires the inversion of the cumulative distribution function

$$F(x) = \sum_{l=1}^{x} p(l) = \mathbf{a}(\mathbf{I} - \mathbf{B}^x)(\mathbf{I}-\mathbf{B})^{-1}\mathbf{c} \tag{2.5.7}$$

since $\zeta = \lceil F^{-1}(\xi) \rceil$ has a cumulative distribution $F(x)$ if ξ is [0,1] uniformly distributed. In general, inverting the function (2.5.7) is a complex problem and cannot be solved analytically. However, it is possible to solve the problem indirectly. We know that the similarity transformation

$$\mathbf{a}_1 = \mathbf{a}\mathbf{T} \qquad \mathbf{T}\mathbf{B}_1 = \mathbf{B}\mathbf{T} \qquad \mathbf{T}\mathbf{c}_1 = \mathbf{c} \tag{2.5.8}$$

does not change the distribution (see Theorem 1.4.1). If we can find nonnegative matrices $\mathbf{a}_1$, $\mathbf{B}_1$, and $\mathbf{c}_1$ that satisfy the conditions

$$\mathbf{c}_1 = (\mathbf{I} - \mathbf{B}_1)\mathbf{1} \quad \text{and} \quad \mathbf{a}_1\mathbf{1} = 1 \tag{2.5.9}$$

then the random variable l can be modeled as an interval between two consecutive returns into the last state of the Markov chain with the matrix

$$\begin{bmatrix} \mathbf{B}_1 & \mathbf{c}_1 \\ \mathbf{a}_1 & 0 \end{bmatrix} \tag{2.5.10}$$

As an example, consider a variable l with the polygeometric probability distribution

$$p(l) = \sum_{i=1}^{m} a_i(1-q_i)q_i^{l-1} \tag{2.5.11}$$

with $a_i > 0$ and $q_i > 0$. In this case we satisfy the preceding conditions by selecting $\mathbf{a}_1 = (a_1, a_2, \ldots, a_m)$ and

$$\mathbf{B}_1 = diag\{q_i\} = \begin{bmatrix} q_1 & 0 & \ldots & 0 \\ 0 & q_2 & \ldots & 0 \\ \cdots & \cdots & \cdots & \cdots \\ 0 & 0 & \ldots & q_m \end{bmatrix} \tag{2.5.12}$$

Therefore, the polygeometrically distributed variable can be simulated as an interval between two consecutive returns into the last state of the chain with the matrix

$$\mathbf{P} = \begin{bmatrix} q_1 & 0 & \cdots & 0 & 1-q_1 \\ 0 & q_2 & \cdots & 0 & 1-q_2 \\ \cdots & \cdots & \cdots & \cdots & \cdots \\ 0 & 0 & \cdots & q_m & 1-q_m \\ a_1 & a_2 & \cdots & a_m & 0 \end{bmatrix} \tag{2.5.13}$$

Using the previously described method of simulating the Markov chain, we select a transition state i with the probability a_i and then select the state-holding time according to geometric distribution

$$w_i(l) = (1-q_i)q_i^{l-1} \tag{2.5.14}$$

2.5.2 Performance Characteristic Calculation

As we mentioned before, sometimes it is simpler to find system performance characteristics using the Monte Carlo method. For example, the error number distribution

$$p_n(m) = \mathbf{p}_0 \mathbf{P}_n(m) \mathbf{1} \tag{2.5.15}$$

where $\mathbf{P}_n(m)$ is the matrix binomial distribution defined in Sec. 2.3.2 may be obtained by direct simulation of the Markov chain with the matrix

$$\begin{bmatrix} \mathbf{P}(0) & \mathbf{P}(1) \\ \mathbf{P}(0) & \mathbf{P}(1) \end{bmatrix} \tag{2.5.16}$$

The probability that the number of errors in a block of length n is equal to m is estimated by

$$p_n(m) \approx \frac{\nu_m}{N} \tag{2.5.17}$$

where ν_m is the number of length n blocks that contain exactly m errors and Nn is the total number of simulated symbols. We give more examples of applying this method in the following chapters.

Since any model is an approximation of the real error statistics and random generator quality may be poor, it is important to verify the agreement between experimental and simulated data. This problem is a particular case of statistical hypothesis testing, which we consider in the next chapter.

It can be shown that the number of times a regular Markov chain enters its state is asymptotically normal. Since an error number is a sum of the numbers of times the chain enters "bad" states, it is also asymptotically normal. Therefore, assuming that the regularity conditions are satisfied, we can use the standard χ^2 or information criteria of closeness between the experimental data and the result of simulation.

Let $p_i = p_n(m_{i-1} \le m < m_i)$ be the probability that the number of errors in a block satisfies the condition $m_{i-1} \le m < m_i$:

$$p_i = \sum_{m=m_{i-1}}^{m_i-1} p_n(m)$$

Denote

$$\hat{p}_{i,exp} = \nu_{i,exp}/N_{exp}$$

as the corresponding experimental frequencies and

$$\hat{p}_{i,sim} = \nu_{i,sim}/N_{sim}$$

as the estimates obtained by simulation.

If the estimates $\hat{p}_{i,exp}$ and $\hat{p}_{i,sim}$ are asymptotically normal, then [9]

$$\chi^2 = N_{sim}N_{exp}\sum_{i=1}^{r}(\hat{p}_{i,exp} - \hat{p}_{i,sim})^2/(\nu_{i,exp}+\nu_{i,sim})$$

and

$$2I = 2N_{exp}N_{sim}\sum_{i=1}^{r}(\hat{p}_{i,exp} - \hat{p}_{i,sim})\ln(\hat{p}_{i,exp}/\hat{p}_{i,sim})/(N_{exp}+N_{sim})$$

have asymptotically the χ^2 distribution with $r-1$ degrees of freedom.

These criteria values allow us to decide whether the experimental data agree with the simulation results. In the next chapter we discuss how the values should be selected.

2.6 COMPUTING SCALAR PROBABILITIES

In the previous sections we considered methods for calculating MPs. However, evaluating MPs is an intermediate step; the main objective is to obtain scalar probabilities. Equations (2.1.2), (2.1.4), and (2.1.5) connect matrix probabilities with their scalar counterparts. In this section we consider computational aspects of this relationship.

2.6.1 The Forward and Backward Algorithms

If we calculate the $Pr(\mathbf{e}_1^m)$ using Eq. (2.1.1) and (2.1.2), it would require $(m-1)u^2$ multiplications to evaluate the MP $\mathbf{P}(\mathbf{e}_1^m)$ plus u multiplications by $\mathbf{p}_0$. Thus, in general we need about $(m-1)u^2+u$ multiplications (multiplications by $\mathbf{1}$ are in effect additions).

Alternatively, we can evaluate Eq. (2.1.2) recursively starting from the left:

$$\boldsymbol{\alpha}(\mathbf{e}_1^0) = \mathbf{p}_0,\quad \boldsymbol{\alpha}(\mathbf{e}_1) = \boldsymbol{\alpha}(\mathbf{e}_1^0)\mathbf{P}(\mathbf{e}_1),\ \dots\ ,\boldsymbol{\alpha}(\mathbf{e}_1^m) = \boldsymbol{\alpha}(\mathbf{e}_1^{m-1})\mathbf{P}(\mathbf{e}_m)$$

where

$$\boldsymbol{\alpha}(\mathbf{e}_1^k) = \mathbf{p}_0\mathbf{P}(\mathbf{e}_1^k)$$

This algorithm is called a *forward algorithm.* [13] It requires $m \cdot u$ multiplications which is about a factor of u less than the previous algorithm. Finally, we obtain $Pr(\mathbf{e}_1^m) = \boldsymbol{\alpha}(\mathbf{e}_1^m)\mathbf{1}$ without multiplications.

The same complexity has the following *backward algorithm*

$$\boldsymbol{\beta}(\mathbf{e}_{m+1}^m) = \mathbf{1} \qquad \boldsymbol{\beta}(\mathbf{e}_k^m) = \mathbf{P}(\mathbf{e}_k)\boldsymbol{\beta}(\mathbf{e}_{k+1}^m), \qquad k=m,m-1,...,1$$

and $Pr(\mathbf{e}_1^m) = \mathbf{p}_0\boldsymbol{\beta}(\mathbf{e}_1^m)$.

We see that, in general, the forward or backward algorithms are more efficient than calculation of MPs. If we have two parallel processors, we can apply the forward algorithm on one processor to obtain $\boldsymbol{\alpha}(\mathbf{e}_1^k)$ and the backward algorithm on the other processor to obtain $\boldsymbol{\beta}(\mathbf{e}_{k+1}^m)$ and then obtain the probability $Pr(\mathbf{e}_1^m) = \boldsymbol{\alpha}(\mathbf{e}_1^k)\boldsymbol{\beta}(\mathbf{e}_{k+1}^m)$. If we have many parallel processors and m is large, we can break the interval $(1,m)$ into subintervals $1<m_1<m_2<...<m_k<m$ and compute MPs of the form $\mathbf{P}(\mathbf{e}_{m_i}^{m_{i+1}-1})$ on separate processors as well as $\boldsymbol{\alpha}(\mathbf{e}_1^{m_1-1})$ and $\boldsymbol{\beta}(\mathbf{e}_{m_k}^m)$ to obtain

$$Pr(\mathbf{e}_1^m) = \boldsymbol{\alpha}(\mathbf{e}_1^{m_1-1})\mathbf{P}(\mathbf{e}_{m_1}^{m_2-1})\mathbf{P}(\mathbf{e}_{m_2}^{m_3-1}) \cdots \mathbf{P}(\mathbf{e}_{m_{k-1}}^{m_k-1})\boldsymbol{\beta}(\mathbf{e}_{m_k}^m)$$

The same idea of replacing square matrices with vectors can be applied to compute other probabilities. For example, multiplying both sides of equation (2.3.6) by $\boldsymbol{\pi}$, we obtain

$$\boldsymbol{p}(m,n) = \boldsymbol{p}(m,n-1)\mathbf{P}(0) + \boldsymbol{p}(m-1,n-1)\mathbf{P}(1)$$

where $\boldsymbol{p}(m,n) = \boldsymbol{\pi}\mathbf{P}_{(m)}$ is a row vector. This forward recursive equation requires u times less computations and memory than Eq. (2.3.6). We can obtain the backward equations similarly:

$$\boldsymbol{q}(m,n) = \mathbf{P}(0)\boldsymbol{q}(m,n-1) + \mathbf{P}(1)\boldsymbol{q}(m-1,n-1)$$

where $q(m,n) = \mathbf{P}_n(m)\mathbf{1}$ is a column vector. We can use the vectors $\boldsymbol{p}(m-\mu,n-\nu)$ and $\boldsymbol{q}(\mu,\nu)$ to compute $p_n(m) = \boldsymbol{\pi}\mathbf{P}_n(m)\mathbf{1}$ according to Eq. (2.3.5):

$$p_n(m) = \sum_{\mu}\boldsymbol{p}(m-\mu,n-\nu)\boldsymbol{q}(\mu,\nu)$$

2.6.2 Scalar Generating Functions

Scalar generating functions are obtained from the corresponding matrix generating functions by premultiplying them by $\mathbf{p}_0 = \boldsymbol{\pi}$ and postmultiplying by $\mathbf{1}$. These multiplications can significantly simplify the application (see Example 2.2.2). To illustrate this point, consider the applications of the scalar generating function

$$\psi(w;z) = \boldsymbol{\pi}\boldsymbol{\Psi}(w;z)\mathbf{1} = \boldsymbol{\pi}[\mathbf{I} - \mathbf{P}(0)w - \mathbf{P}(1)wz]^{-1}\mathbf{1}$$

of the matrix binomial distribution.

The generating function

$$\phi_n(z) = \boldsymbol{\pi}\boldsymbol{\Phi}_n(z)\mathbf{1} = \boldsymbol{\pi}[\mathbf{P}(0) + \mathbf{P}(1)z]^n\mathbf{1}$$

is the coefficient of w^n in the $\psi(w;z)$ expansion:

$$\psi(w;z) = \sum_{n=0}^{\infty} w^n\phi_n(z)$$

2.6.2.1 Moment-Generating Functions

The function $\phi_n(z)$ derivatives at $z=1$ give us the number of error binomial moments (see Sec. 2.2.1). Therefore, the function $\psi(w;z)$ derivatives, with respect to z, give us the binomial moment-generating functions:

$$\zeta_k(w) = \frac{1}{k!}\frac{\partial^k \psi(w;z)}{\partial z^k}\Big|_{z=1} = \sum_{n=0}^{\infty} w^n b_{n,k}$$

where $b_{n,k} = \mathbf{E}\{\binom{m}{k}\} = \sum_{m=0}^{n}\binom{m}{k}p_n(m)$ It is easy to see that (see Problem 6).

$$\frac{\partial^k \psi(w;z)}{\partial z^k} = k!\boldsymbol{\pi}[\boldsymbol{\Psi}(w;z)\mathbf{P}(1)w]^k\boldsymbol{\Psi}(w;z)\mathbf{1}$$

Thus, we have

$$\zeta_k(w) = [\boldsymbol{\Psi}(w;1)\mathbf{P}(1)w]^k\boldsymbol{\Psi}(w;1)\mathbf{1}$$

where, according to Eq. (2.3.8), $\boldsymbol{\Psi}(w;1) = (\mathbf{I} - \mathbf{P}w)^{-1}$. Since $\boldsymbol{\pi}\boldsymbol{\Psi}(w;1) = \boldsymbol{\pi}(1-w)^{-1}$ and $\boldsymbol{\Psi}(w;1)\mathbf{1} = = \mathbf{1}(1-w)^{-1}$ (see Problem 6), the moment-generating function can be written as

$$\zeta_k(w) = \boldsymbol{\pi}\mathbf{P}(1)[\boldsymbol{\Psi}(w;1)\mathbf{P}(1)]^{k-1}w^k(1-w)^{-2}$$

In particular, if $k=1$, we have the generating function of means

$$\zeta_1(w) = \boldsymbol{\pi}\mathbf{P}(1)\mathbf{1}w/(1-w)^2 = p_e w/(1-w)^2$$

The mean number of errors in the block of n bits is a coefficient of w^n in this function expansion. Since

$$\frac{w}{(1-w)^2} = \sum_{n=0}^{\infty} nw^n$$

the mean number of errors is np_e (we obtained the same result in Example 2.2.2).

If $k=2$, we have the generating function of the moments $b_{n,2} = \mathbf{E}\{m(m-1)/2\}$

$$\zeta_2(w) = \boldsymbol{\pi}\mathbf{P}(1)(\mathbf{I} - \mathbf{P}w)^{-1}\mathbf{P}(1)\mathbf{1}w^2/(1-w)^2$$

$b_{n,2}$ is the coefficient of w^n in this function expansion. To find the coefficient we can use decomposition of this generating function into partial fractions as we did before. Alternatively, one can use the matrix $\mathbf{P}$ spectral decomposition $\mathbf{P} = \mathbf{1}\boldsymbol{\pi} + \mathbf{Q}$, which gives us

$$(\mathbf{I} - \mathbf{P}w)^{-1} = (\mathbf{I} - \mathbf{Q}w)^{-1} + w(1-w)^{-1}\mathbf{1}\boldsymbol{\pi}$$

(see Problem 6). Using this formula, we obtain

$$\zeta_2(w) = p_e^2 w^3(1-w)^{-3} + \boldsymbol{\pi}\mathbf{P}(1)(\mathbf{I} - \mathbf{Q}w)^{-1}\mathbf{P}(1)\mathbf{1}w^2/(1-w)^2$$

Next, we use the following matrix partial fraction decomposition (see Appendix 5)

$$(\mathbf{I}-\mathbf{Q}w)^{-1}w^2/(1-w)^2 = (\mathbf{I}-\mathbf{Q})^{-1}w/(1-w)^2 - (\mathbf{I}-\mathbf{Q})^{-2}w/(1-w) + (\mathbf{I}-\mathbf{Q})^{-2}\mathbf{Q}(\mathbf{I}-\mathbf{Q}w)^{-1}w$$

to obtain

$$b_{n,2} = \frac{(n-1)(n-2)}{2} p_e^2 + \boldsymbol{\pi}\mathbf{P}(1)[n(\mathbf{I}-\mathbf{Q})^{-1} - (\mathbf{I}-\mathbf{Q})^{-2} + (\mathbf{I}-\mathbf{Q})^{-2}\mathbf{Q}^n]\mathbf{P}(1)\mathbf{1}$$

Since $m_2 = \mathbf{E}\{m^2\} = 2b_{n,2} + \mathbf{E}\{m\}$, we have

$$m_2 = (n-1)(n-2)p_e^2 + 2\boldsymbol{\pi}\mathbf{P}(1)[n(\mathbf{I}-\mathbf{Q})^{-1} - (\mathbf{I}-\mathbf{Q})^{-2} + (\mathbf{I}-\mathbf{Q})^{-2}\mathbf{Q}^n]\mathbf{P}(1)\mathbf{1} + np_e$$

which agrees with the result of Example 2.2.3.

2.6.2.2 Finite Difference Equations

As we pointed out in Sec. 2.2.5, if a probability generating function is a rational fraction, we can write a finite difference equation for the probabilities. Obviously, we can use the same method if the generating function has several variables. The generating function $\psi(w;z)$ is a rational fraction and can be written as [Eq. (A.5.10) of Appendix 5]

$$\psi(w;z) = \boldsymbol{\pi}[\mathbf{I} - \mathbf{P}(0)w - \mathbf{P}(1)wz]^{-1}\mathbf{1} = \boldsymbol{\pi}\mathbf{B}(w;z)\mathbf{1} / \Delta(w;z)$$

On the other hand, the probability $p_n(m)$ of m errors in the block of n bits is the coefficient of $w^n z^m$ of this function expansion into a power series:

$$\sum_{n}^{\infty} \sum_{m=0}^{\infty} p_n(m) w^n z^m = b(w;z) / \Delta(w;z) \tag{2.6.1}$$

where $b(w;z) = \boldsymbol{\pi}\mathbf{B}(w;z)\mathbf{1}$ is a polynomial. Multiplying both sides of this equation by $\Delta(w;z)$ and comparing coefficients of the same powers on both sides of this equation, we obtain the desired recursive equations. Let us illustrate this technique in the following example.

Example 2.6.1. For the Gilbert's model we have

$$\psi(w;z) = \boldsymbol{\pi} \begin{bmatrix} 1-Qw & -Pwh(z) \\ -pw & 1-qwh(z) \end{bmatrix}^{-1} \mathbf{1}$$

where $h(z) = h + z(1-h)$. We have

$$\begin{bmatrix} 1-Qw & -Pwh(z) \\ -pw & 1-qwh(z) \end{bmatrix}^{-1} = \mathbf{B}(w;z) / \Delta(w;z) = \begin{bmatrix} 1-qwh(z) & Pwh(z) \\ pw & 1-Qw \end{bmatrix} \frac{1}{\Delta(w;z)}$$

where $\Delta(w;z) = 1-Qw-qwh(z)+(Q-p)w^2h(z) = 1-\delta_1 w-\delta_2 zw-\delta_3 w^2-\delta_4 zw^2$ and $b(w;z) = \boldsymbol{\pi}\mathbf{B}(w;z)\mathbf{1} = 1-(Q-p)w(\pi_1 h(z)+\pi_2) = 1-b_1 w-b_2 zw$. Multiplying both sides of Eq. (2.6.1) by $\Delta(w;z)$, we obtain

$$(1-\delta_1 w-\delta_2 zw-\delta_3 w^2-\delta_4 zw^2)\sum_{n=0}^{\infty}\sum_{m=0}^{\infty} p_n(m) w^n z^m = 1-b_1 w-b_2 zw$$

Comparing coefficients of the same powers on both sides of this equation, we obtain the following difference equations:

$$p_n(m) = \delta_1 p_{n-1}(m)+\delta_2 p_{n-1}(m-1)+\delta_3 p_{n-2}(m)+\delta_4 p_{n-2}(m-1), \quad n=2,3,\ldots$$

with the boundary conditions

$$p_n(m) = 0 \quad \text{for} \quad m<0 \quad \text{or} \quad m>n$$

$$p_0(0) = 1, \; p_1(0) = \delta_1 - b_1 = 1-p_e, \; p_1(1) = \delta_2 - b_2 = p_e$$

2.7 CONCLUSION

The matrix-train form of error sequence probabilities enables us to introduce matrix probabilities by stripping off the initial state probability matrix and the final matrix **1**. The matrix probability of a sequence of conditionally independent events can be expressed similarly to the probability of independent events. However, since matrix products are noncommutative, matrix probabilities of combinations of events have a much more complex form than their scalar counterparts.

In this chapter, we have generalized methods of classical probability theory to matrix probabilities. We analyzed the most commonly used matrix distributions and developed numerous methods for their calculation. In particular, the transform methods proved to be the most efficient in many cases. We showed that signal-flow graphs with block transitions can be used to solve various problems related to matrix probabilities. In particular, by using matrix probabilities as transitions on a graph describing a finite state machine, we were able to calculate probabilities and distributions that characterize the machine's performance. These methods are applied to performance analysis of communication protocols in Chapter 5. Because of the generality of the finite state machine model, these methods can be successfully used in applications different from those considered in this book (such as queueing theory and speech and image recognition).

We illustrated the methods by calculating the probability distribution of the numbers of errors in a block. The results of the calculations agree with experimental data and computer simulation.

Some additional material, like the derivation of asymptotic and approximate formulae for matrix probabilities, is given in Appendix 2. We have also analyzed basic operations with block graphs in this appendix to make the book self-contained.

Problems

1. Let m be the number of errors in a block of n bits. For the stationary Gilbert's model (Sec. 1.1.5)

   — Find the characteristic function of m (see Example 2.2.2).

   — Find the mean and standard deviation of m (see Examples 2.2.2 and 2.2.3). (Hint: Use the matrix **P** spectral decomposition

   $$\mathbf{P} = \mathbf{1}\boldsymbol{\pi} + \lambda\mathbf{Q} = \begin{bmatrix} \pi_1 & \pi_1 \\ \pi_2 & \pi_2 \end{bmatrix} + (Q-p)\begin{bmatrix} \pi_2 & -\pi_1 \\ -\pi_1 & \pi_2 \end{bmatrix}$$

   to find the sum

   $$\sum_{i=0}^{k} \mathbf{P}^i = (k+1)\mathbf{1}\boldsymbol{\pi} + \frac{1-\lambda^{k+1}}{P+p}\mathbf{Q}$$

   where $\lambda = Q-p = 1-P-p$.)

   — Find the probability $p_n(m=1)$ of one error in the block.

   — Find the conditional probability of one error in the block given that the previous block does not have errors.

   — Find the probability that the number of errors is divisible by 3.

— Find the probability that the number of errors is even, but not divisible by 4.

— Find the conditional mean of m given that the previous block does not have errors.

— Find the closed-form expression for the MP distribution of the number of errors if $h=0$ (see Sec. 1.1.5). Find the the distribution asymptotic formula and the normal approximation. Compare results for the model parameters numerical values given in Sec. 1.1.5.

2. For the stationary Gilbert's model, find the mean number of steps of the r-th return into the bad state.

3. Show that the matrix $\bar{\mathbf{P}}_i$ in Eq. (2.4.16) is a stochastic matrix ($\bar{\mathbf{P}}_i \geq 0$ and $\bar{\mathbf{P}}_i \mathbf{1} = \mathbf{1}$). Find the relation between the stationary distribution $\boldsymbol{\pi}_i$ of $\bar{\mathbf{P}}_i$ and the stationary distribution $\boldsymbol{\pi}$ of the original matrix $\mathbf{P}$.

4. For a regular binary HMM show that the matrix $(\mathbf{I} - \mathbf{P}(0))^{-1}\mathbf{P}(1)$ and its stationary vector is $\boldsymbol{\pi}(\mathbf{I} - \mathbf{P}(0))$, where $\boldsymbol{\pi}$ is a stationary vector of $\mathbf{P} = \mathbf{P(0)} + \mathbf{P}(1)$.

5. Show that

$$\frac{d}{dz}(\mathbf{I} - \mathbf{A}z)^{-1} = \mathbf{A}(\mathbf{I} - \mathbf{A}z)^{-2} = (\mathbf{I} - \mathbf{A}z)^{-2}\mathbf{A}$$

[Hint: Differentiate the identity $(\mathbf{I} - \mathbf{A}z)^{-1}(\mathbf{I} - \mathbf{A}z) = \mathbf{I}$.]

6. Use the equation of the previous problem to prove that

$$\frac{d^k}{dz^k}(\mathbf{I} - \mathbf{A}z)^{-1} = k!(\mathbf{I} - \mathbf{A}z)^{-k-1}\mathbf{A}^k$$

7. Show that for $\boldsymbol{\Psi}(w;z)$ defined by Eq. (2.3.8)

— $\frac{\partial}{\partial z}\boldsymbol{\Psi}(w;z) = \boldsymbol{\Psi}(w;z)\mathbf{P}(1)w\boldsymbol{\Psi}(w;z)$
[Hint: Differentiate the identity $\Psi(w;z)(\mathbf{I} - \mathbf{P}(0)w - \mathbf{P}(1)zw) = \mathbf{I}$.]

— Using the previous equation, prove that

$$\frac{\partial^k \boldsymbol{\Psi}(w;z)}{\partial z^k} = k![\boldsymbol{\Psi}(w;z)\mathbf{P}(1)w]^k\boldsymbol{\Psi}(w;z)$$

— Show that $\boldsymbol{\pi}(\mathbf{I} - \mathbf{P}w)^{-1} = \boldsymbol{\pi}(1-w)^{-1}$ and $(\mathbf{I} - \mathbf{P}w)^{-1} = \mathbf{1}(1-w)^{-1}$. [Hint: Use the identities $\boldsymbol{\pi}(\mathbf{I} - \mathbf{P}w) = \boldsymbol{\pi} - \boldsymbol{\pi}w$ and $(\mathbf{I} - \mathbf{P}w)\mathbf{1} = \mathbf{1} - \mathbf{1}w$.]

— For the matrix binomial distribution, find recursive equations for calculating the mean and the variance.

References

1 M. S. Bartlett, *Introduction to Stochastic Processes with Special Reference to Methods and Applications,* (Cambridge Univ. Press, Cambridge, 1978).

2 E. L. Bloch, O. V. Popov, and W. Ya. Turin, "Error number distribution for the stationary symmetric channel with memory," *Second International Symposium on Information Theory,* Academiai Kiado, Budapest, 1972.

3 K. L. Chung, *Markov Chains with Stationary Transition Probabilities,* (Springer-Verlag, Berlin, 1960).

4 W. Feller, *An Introduction to Probability Theory and Its Applications*, **1**, (John Wiley & Sons, New York, 1962).

5 I. I. Gichman and A. V. Scorochod, *The Theory of Stochastic Processes* (in Russian), **1**, (Science Publishers, Moscow, 1971).

6 J. G. Kemeny and J. L. Snell, *Finite Markov Chains,* Van Nostrand, (Princeton, New Jersey, 1960).

7 B. A. Fuchs and B. V. Shabat, *Functions of a Complex Variable*, **1**, (Addison-Wesley Publishing Co., Reading, Massachusetts, 1964).

8 W. Huggins, "Signal flow graphs and random signals," *Proc. IRE,* **45**, 74-86, 1957.

9 S. Kullback, *Information Theory and Statistics*, (John Wiley & Sons, New York, 1959).

10 S. J. Mason and H. J. Zimmermann, *Electronic Circuits, Signals, and Systems,* (John Wiley & Sons, New York, 1965).

11 G. A. Medvedev and V. P. Tarasenko, *Probabilistic Methods in Extremal Systems Investigation* (in Russian), (Science Publishers, Moscow, 1967).

12 O. V. Popov and W. Ya. Turin, "The probability distribution law of different numbers of errors in a combination," *Telecommunications and Radioengineering,* (5), 1967.

13 L. Rabiner and B.-H. Juang, *Fundamentals of Speech Recognition*, (Prentice Hall, Englewood Cliffs, New Jersey, 1993).

14 L. R. Rabiner and C. M. Rader, *Digital Signal Processing*, (IEEE Press, New York, 1972).

15 W. Ya. Turin, "Probability distribution laws for the number of errors in several combinations,' *Telecommunications and Radioengineering*, **27**(12), 1973.

3

MODEL PARAMETER ESTIMATION

In this chapter, we develop methods for approximating a stochastic process with an HMM and, in particular, for fitting HMMs to experimental data. We present the iterative *expectation maximization* (EM) algorithm for the process approximation which generalizes the statistical EM algorithm and, in particular, the *Baum-Welch algorithm* (BWA) for approximating the process with an HMM. This generalized algorithm can be also applied to curve fitting and finding a maximum of a nonnegative function of several variables. The EM algorithm is iterative and converges slowly. Therefore, it is important to select a good initial model. We describe several methods for choosing the initial HMM. The question of speeding up the algorithm convergence is also addressed.

Since an HMM can be described in many different ways, we consider several approaches to model building. We consider the model approximation based on multiple Markov chains and semi-Markov processes. Then we examine the question of the model building on the basis of matrix processes. And, finally, we indicate the modifications in experimental data processing methodology for the development of a model of the source of errors in several channels.

A brief statistical introduction, which is necessary to understand the chapter material, is given in Appendix 3.

3.1 THE EM ALGORITHM

3.1.1 Kullback-Leibler Divergence

In order to approximate a stochastic process with some other process, we need to select the approximation quality measure $L(\tau) = \Delta(F(x), G(x,\tau))$ where $F(x)$ is a cumulative probability distribution of some variable x for the original process, $G(x,\tau)$ is the corresponding probability distribution for its model, and τ is the model parameter vector. The optimal fit is achieved by minimizing $\Delta(F(x), G(x,\tau))$ with respect to τ, which we denote as

$$\hat{\tau} = \arg\min_{\tau} \Delta(F(x), G(x,\tau)) = \arg\min_{\tau} L(\tau) \qquad \tau \in \Omega \tag{3.1.1}$$

To simplify notation, we assume that the random variable x is discrete and has probabilities $f(x)$ for the original process and $g(x,\tau)$ for its model. There are many different approximation quality measures. In statistical literature, two measures are most frequently used:

$$\chi^2(\tau) = \sum_x [f(x) - g(x,\tau)]^2 / g(x,\tau) \tag{3.1.2}$$

and the Kullback-Leibler divergence (KLD) [30]

$$D(F \,\|\, G_\tau) = \sum_x f(x)\log(f(x) / g(x,\tau)) = E\{\log(f(x) / g(x,\tau) \mid f(x)\} \tag{3.1.3}$$

If x is a continuous random variable, the sums are replaced with integrals and probabilities are replaced with the PDFs. These two measures are closely related (see Appendix 3). The measures are equal to zero if and only if the distributions are identical (in the continuous case almost everywhere).

3.1.2 Minimization Algorithms

In order to find the optimal fit, we need to solve the minimization problem described by Eq. (3.1.1). In the majority of cases, it is impossible to solve this problem analytically and the minimum is found iteratively: we find a sequence of parameters $\tau_0, \tau_1, \ldots$ that decrease the function $L(\tau)$. The majority of methods are based on the function Taylor expansion. The first-order approximation is

$$L(\tau) \approx L(\tau_p) + \frac{\partial L(\tau_p)}{\partial \tau_p}(\tau - \tau_p)$$

(We assume that the gradient is a row vector and τ is a column vector.) It follows from this equation that the direction of the function's fastest decrease at the point τ_p is opposite to its gradient $\partial L(\tau_p)/\partial \tau$. Thus we have the gradient algorithm

$$\tau'_{p+1} = \tau'_p - \lambda_p \frac{\partial L(\tau_p)}{\partial \tau_p} \qquad p = 0,1,2,\ldots \tag{3.1.4}$$

$L(\tau_p) \geq L(\tau_{p+1})$ if λ_p is sufficiently small. A difficult problem is to find a good λ_p at each step.

We can achieve better results by using the second-order approximation:

$$L(\tau) \approx L(\tau_p) + \frac{\partial L(\tau_p)}{\partial \tau_p}(\tau - \tau_p) + (\tau - \tau_p)' \frac{\partial^2 L(\tau_p)}{\partial \tau_p^2}(\tau - \tau_p)$$

where $\partial^2 L(\tau)/\partial \tau^2 = [\partial^2 L(\tau)/\partial \tau_i \partial \tau_j]$ is the Hessian (the matrix of the function second derivatives). [43] It is not difficult to see that the extremum of the second-order approximation has the form

$$\tau'_{p+1} = \tau'_p - \left[\frac{\partial^2 L(\tau_p)}{\partial \tau_p^2}\right]^{-1} \frac{\partial L(\tau_p)}{\partial \tau_p} \qquad p = 0,1,2,\ldots \tag{3.1.5}$$

It is a minimum if the Hessian is positive definite (see Appendix 5). This equation defines the Newton-Raphson algorithm.

In general, we develop an iterative algorithm by approximating $L(\tau)$ with some function $Q(\tau, \tau_p)$ in the vicinity of τ_p and minimizing $Q(\tau, \tau_p)$ with respect to τ:

$$\tau_{p+1} = \arg\min_{\tau} Q(\tau,\tau_p) \tag{3.1.6}$$

with the hope that $L(\tau_p) \geq L(\tau_{p+1})$. To decide if it is true, we need to analyze the residual $H(\tau,\tau_p) = L(\tau) - Q(\tau,\tau_p)$.

We have

$$L(\tau) = Q(\tau,\tau_p) + H(\tau,\tau_p) \tag{3.1.7}$$

If the approximation is good, the residual $H(\tau,\tau_p)$ is much smaller than $Q(\tau,\tau_p)$. However, it does not mean that the same is true for their increments and we cannot guarantee that $L(\tau)$ decreases whenever $Q(\tau,\tau_p)$ decreases. On the other hand, if both $Q(\tau,\tau_p)$ and $H(\tau,\tau_p)$ decrease, then $L(\tau)$ also decreases. Moreover, if $H(\tau,\tau_p)$ always decreases [in other words, $H(\tau_p,\tau_p)$ is its global maximum]:

$$H(\tau,\tau_p) \leq H(\tau_p,\tau_p) \qquad \forall \tau \in \Omega \tag{3.1.8}$$

then a decrease of $Q(\tau,\tau_p)$ causes a decrease of $L(\tau)$. Thus, in this case, the sequence given by Eq. (3.1.6) decreases $L(\tau)$. the EM algorithm [15] is a special case of this algorithm. Obviously, a similar algorithm can be applied for a function maximization: If both $Q(\tau,\tau_p)$ and $H(\tau,\tau_p)$ increase, then $L(\tau)$ also increases. In particular, if $H(\tau_p,\tau_p)$ is a global minimum of $H(\tau,\tau_p)$, we have the following generalized EM algorithm

$$\tau_{p+1} = \arg\max_{\tau} Q(\tau,\tau_p) \tag{3.1.9}$$

The important question that we need to address when dealing with any iterative algorithm is its convergence. There are many papers and books addressing this problem. The conditions of convergence are discussed in Ref. 61 and 62. If a minimization algorithm converges ($\lim_{k\to\infty} \tau_p = \hat{\tau}$ exists in Ω), in the limit the function may attain its local minimum or it can be its saddle point. In many cases, we do not worry about the algorithm convergence because if the initial approximation obtained by other methods is good, we perform only several iterations to improve the initial approximation.

The other important parameter of the algorithm is the speed of its convergence. The EM algorithm is robust, but slow. We can use it in combination with faster, but unstable algorithms (such as Newton-Raphson algorithm) to increase the convergence rate. We will discuss the algorithm acceleration in Section 3.1.5.

Example 3.1.1. For example, let $L(\tau) = 4 - \tau^2$. This function has a global maximum at $\tau = 0$. To illustrate the convergence of the algorithm, we choose $H(\tau,\tau_p) = -a(\tau - \tau_p)^2$ where $a > 0$. Then $Q(\tau,\tau_p) = L(\tau) + H(\tau,\tau_p)$ has a maximum $\tau = (a/(a+1))\,\tau_p$ and we have

$$\tau_{p+1} = (a/(a+1))\tau_p$$

Obviously, this sequence converges to 0 for any τ_0 but the convergence is slow for large a. The rate of convergence $a/(a+1)$ is equal to the "missing information" ratio $(\partial^2 H/\partial\tau^2)/(\partial^2 Q/\partial\tau^2)$. [34]

3.1.3 The EM algorithm

Let us now address the question of fitting a probability distribution by minimizing the KLD given by Eq. (3.1.3). Obviously, the minimization of $D(F \parallel G_\tau)$ is equivalent to maximization of the likelihood function

$$L(\tau) = \sum_x f(x) \log g(x,\tau) = E\{\log g(x,\tau) \mid f(x)\} \tag{3.1.10}$$

which has a simpler form. If x is a continuous variable, the sum is replaced with the corresponding integral. If some of the variable components are discrete and the others are continuous, we have both sums and integrals. In all the cases we can write $L(\tau) = E\{\log g(x,\tau) \mid f(x)\}$. Because of the universality of this notation, we will use most often. We derive the EM algorithm for the discrete case, however, the derivation is similar for all the cases (see Ref. 57).

Suppose that we have a function $\kappa(z;x,\tau)>0$ whose sum (or an integral) over z is a constant. We can always assume that the constant is equal to one by normalizing this function. In this case, $\kappa(z;x,\tau)$ can be interpreted as a conditional probability (or a PDF in the continuous case) of z. Denote $\psi(z;x,\tau) = g(x,\tau)\kappa(z;x,\tau)$. Taking the logarithm on both sides of this equation, we obtain

$$\log g(x,\tau) = \log\psi(z;x,\tau) - \log\kappa(z;x,\tau)$$

Since the left hand side of this equation does not depend on z, taking the expected value with respect to $\kappa(z;x,\tau_p)$ on both sides of the equation yields

$$\log g(x,\tau) = E\{\log\psi(z;x,\tau) \mid \kappa(z;x,\tau_p)\} - E\{\log\kappa(z;x,\tau) \mid \kappa(z;x,\tau_p)\}$$

The expected value of the left hand side of this equation with respect to $f(x)$ gives us the log-likelihood function $L(\tau)$ according to Eq. (3.1.1)

$$L(\tau) = Q(\tau,\tau_p) + H(\tau,\tau_p) \tag{3.1.11}$$

where

$$Q(\tau,\tau_p) = E\{E\{\log\psi(z;x,\tau) | \kappa(z;x,\tau_p)\} \mid f(x)\} \tag{3.1.12}$$

$$H(\tau,\tau_p) = -E\{E\{\log\kappa(z;x,\tau) | \kappa(z;x,\tau_p)\} \mid f(x)\}$$

To apply the EM algorithm we need to prove that $H(\tau_p,\tau_p)$ is a global minimum of $H(\tau,\tau_p)$. Indeed, according to Jensen's inequality for any convex function $u(x)$ [30]

$$E\{u(x)\} \geq u(E\{x\})$$

We leave the proof of this inequality to the reader (see Problem 1). Since $-\log(x)$ is a convex function, we have

$$-E\{\log \frac{\kappa(z;x,\tau)}{\kappa(z;x,\tau_p)} | \kappa(z;x,\tau_p)\} \geq -\log(E\{\frac{\kappa(z;x,\tau_p)}{\kappa(z;x,\tau_p)} | \kappa(z;x,\tau_p)\}) = 0$$

assuming that $\kappa(z;x,\tau)$ and $\kappa(z;x,\tau_p)$ are equal to zero for the same values of their arguments (are absolutely continuous). It follows from the previous inequality that $-E\{\log(\kappa(z;x,\tau) | \kappa(z;x,\tau_p)\} \geq -E\{\log\kappa(z;x,\tau_p) | \kappa(z;x,\tau_p)\}$. Taking expectations on both sides of this inequality with respect to $f(x)$ we obtain $H(\tau,\tau_p) \geq H(\tau_p,\tau_p)$, which proves that $H(\tau_p,\tau_p)$ is a global minimum. Therefore, we can use the EM algorithm (3.1.9) to find the likelihood maximum.

In summary, we can find

$$\hat{\tau} = \arg\max_{\tau} E\{\log g(x,\tau) \mid f(x)\} \tag{3.1.13}$$

iteratively by the EM algorithm

$$\tau_{p+1} = \arg\max_{\tau} E\{q(x,\tau,\tau_p) \mid f(x)\} \quad p=0,1,2,...$$

where

$$q(x,\tau,\tau_p) = E\{\log\psi(z;x,\tau)|\kappa(z;x,\tau_p)\} \tag{3.1.14}$$

Combining the PDFs, we can also rewrite this equation as

$$\tau_{p+1} = \arg\max_{\tau} E\{\log\psi(z;x,\tau) \mid r(z;x,\tau_p)\} \quad p=0,1,2,...$$

where

$$r(z;x,\tau) = \kappa(z;x,\tau)f(x) = \psi(z;x,\tau)f(x)/g(x,\tau)$$

Each iteration increases the likelihood $E\{\log g(x,\tau) \mid f(x)\}$ and consists of two steps:

1. *Expectation*: Calculating $E\{E\{\log\psi(z;x,\tau)|\kappa(z;x,\tau_p)\} \mid f(x)\}$
2. *Maximization*: Finding $\tau_{p+1} = \arg\max_{\tau} E\{E\{\log\psi(z;x,\tau)|\kappa(z;x,\tau_p)\} \mid f(x)\}$

Note that we can choose different auxiliary probability distributions $\kappa(z;x,\tau)$ to obtain different EM algorithms. Usually, $\kappa(z;x,\tau)$ is selected in such a way so that $\psi(z;x,\tau)$ is not too complex allowing us to find analytically a maximum of the auxiliary function $Q(\tau,\tau_p) = E\{E\{\log\psi(z;x,\tau)|\kappa(z;x,\tau_p)\} \mid f(x)\}$.

Let us discuss the algorithm convergence and the number of necessary iterations. Usually, the first step of the algorithm significantly increases the likelihood $L(\tau)$. This property of the algorithm is illustrated in the next section. Subsequent steps deliver smaller increments. Since $L(\tau_p)$ is monotonically increasing and, according to Jensen's inequality, is bounded by $E\{\log f(x) \mid f(x)\}$, [30] it converges to some value L^*. The conditions of the algorithm's convergence can be found in Ref. 61 and 62. The KLD [which is the difference between $L(\tau_p)$ and the upper bound $E\{\log f(x) \mid f(x)\}$] gives us an idea of the approximation quality. Note that this bound can be achieved if $f(x)$ belongs to the family $\{g(x,\tau)\}$. It follows from Eq. (3.1.26) that the limiting point could be a local minimum or a saddle point of $L(\tau)$. However, the saddle point is not stable and small rounding errors usually throw us on the right track. This phenomenon is sometimes observed in practice: it seems that we are close to the solution because $L(\tau_p)$ practically does not change but then it starts increasing and we find a better solution. Thus, the number of iterations should not be decided on the basis of the function change but rather on the value of the KLD.

In conclusion, we would like to note that the requirement that $f(x)$ and $g(x,\tau)$ are probability distributions is not necessary. Indeed, analyzing the algorithm development, we observe that it is sufficient to require that $f(x)>0$, $g(x,\tau)>0$, and $\kappa(z;x,\tau)>0$ only within the region of summation (or integration) and that all the sums (or integrals) used in the derivation exist. In this case we can always normalize $f(x)$ and $\kappa(z;x,\tau)$ so that they can be treated as probability distributions. In the future, we will perform these normalizations as necessary. In particular, if $f(x)$ is a continuous function and the region of integration is X, we replace it with its normalized version

$$\bar{f}(x) = f(x)/\lambda, \quad \lambda = \int_X f(x)\,dx$$

so that $\int_X \bar{f}(x)\,dx = 1$. This normalization condition can be written for both discrete and continuous variables as

$$E\{1 \mid \bar{f}(x)\} = 1 \tag{3.1.15}$$

We will often use this equation to avoid consideration of discrete and continuous variables separately.

We assume that $f(x)$ and $\kappa(z;x,\tau)$ are normalized, but $g(x,\tau)$ does not have to be normalized. Then the maximization problem

$$\hat{\tau} = \underset{\tau}{argmax}\; E\{\log g(x,\tau) \mid f(x)\} \tag{3.1.16}$$

can be solved iteratively by the EM algorithm

$$\tau_{p+1} = \underset{\tau}{argmax}\; E\{q(x,\tau,\tau_p) \mid f(x)\} \tag{3.1.17}$$

where $q(x,\tau,\tau_p)$ is given by Eq. (3.1.14).

3.1.3.1 Fitting Distribution Mixtures

Suppose that we want to approximate a PDF $f(x)$, $x \in R^n$ with a finite mixture of densities

$$f(x) \approx g(x,\tau) = \sum_{z=1}^{m} a_z\, p_z(x,\phi_z) \tag{3.1.18}$$

To apply the EM algorithm, we define

$$\kappa(z;x,\tau) = a_z\, p_z(x,\phi_z) \,/\, g(x,\tau) \tag{3.1.19}$$

so that $\sum_{z=1}^{m} \kappa(z;x,\tau) = 1$ and, therefore,

$$\psi(z;x,\tau) = a_z\, p_z(x,\phi_z) \tag{3.1.20}$$

The auxiliary function $q(x,\tau,\tau_p) = E\{\log\psi(z;x,\tau) | \kappa(z;x,\tau_p)\}$ has the form

$$q(x,\tau,\tau_p) = \sum_{z=1}^{m} \kappa(z;x,\tau_p)\log a_z + \sum_{z=1}^{m} \kappa(z;x,\tau_p)\log p_z(z;x,\tau)$$

Taking the expectation with respect to $f(x)$ on both sides of the previous equation yields

$$Q(\tau,\tau_p) = \sum_{z=1}^{m} \log a_z E\{\kappa(z;x,\tau_p) \mid f(x)\} + \sum_{z=1}^{m} E\{\log[p_z(x,\phi_z)]\,\kappa(z;x,\tau_p) \mid f(x)\}$$

Now we need to take the maximization step. According to Jensen's inequality (see Problem 1), the function

$$\sum_{1}^{m} a_i \log x_i \quad \text{with } a_i > 0 \text{ and } \sum_{1}^{m} x_i = 1 \tag{3.1.21}$$

attains its global maximum at

$$x_i = a_i / \sum_{k=1}^{m} a_k \tag{3.1.22}$$

Therefore, the maximum of $Q(\tau,\tau_p)$ is attained at

$$a_{z,p+1} = E\{\kappa(z;x,\tau_p) \mid f(x)\} = a_{z,p} E\{p_z(x,\phi_{z,p})/g(x,\tau_p) \mid f(x)\} \tag{3.1.23}$$

$$\phi_{z,p+1} = \underset{\phi_z}{argmax} E\{\log[p_z(x,\phi_z)] \mid C_z(x,\tau_p)\} \tag{3.1.24}$$

where we denoted as

$$C_z(x,\tau_p) = \lambda\kappa(z;x,\tau_p)f(x) = \lambda r(z;x,\tau_p)$$

The coefficient of proportionality λ is obtained from the normalization condition (3.1.15) which in our case takes the form $E\{1 \mid C_z(x,\tau_p)\} = 1$ or

$$E\{1 \mid C_z(x,\tau_p)\} = \lambda E\{\kappa(z;x,\tau_p) \mid f(x)\} = \lambda a_{z,p+1} = 1$$

Thus,

$$C_z(x,\tau_p) = a_{z,p+1}^{-1} a_{z,p} p_z(x,\phi_{z,p}) f(x)/g(x,\tau_p) = a_{z,p+1}^{-1} r(z;x,\tau_p)$$

In summary, the EM algorithm allows us to separate the component densities and fit them independently. Note that Eq. (3.1.24) has the form of Eq. (3.1.13) and, therefore, it is possible to develop the EM algorithm for solving it. In many applications of interest, Eq. (3.1.24) can be solved uniquely and explicitly.

Example 3.1.2. As an illustration, consider mixtures of multidimensional Gaussian distributions:

$$p_i(x,\phi_i) = \mathrm{N}(x,\mu_i,\Sigma_i) = (2\pi)^{-m/2}|\Sigma_i|^{-1/2}\exp[-0.5(x-\mu_i)\Sigma_i^{-1}(x-\mu_i)']$$

where the row vectors μ_i are the mixture mean vectors and $\mathbf{\Sigma}_i$ are the variance matrices, $|\mathbf{\Sigma}_i|$ is the determinant of $\mathbf{\Sigma}_i$. In this case, Eq. (3.1.23) takes the form

$$a_{i,p+1} = \int_X r(i;x,\tau_p)\,dx$$

where $r(i;x,\tau_p) = a_{z,p} p_i(x,\phi_i) f(x)/g(x,\tau_p)$. Equation (3.1.24) takes the form

$$R(\phi_i,\phi_{i,p}) = -0.5\int_x [(x-\mu_i)\Sigma_i^{-1}(x-\mu_i)' + \log|\Sigma_i|]\,C_i(x,\tau_p)\,dx + c$$

where $\phi_i = (\mu_i,\Sigma_i)$ and $c = -0.5m\log(2\pi)$ is the constant which can be neglected in the maximization step. We can also remove the coefficient 0.5 in the previous equation and denote $\mathbf{D}_i = \Sigma_i^{-1}$ to simplify notation. With some abuse of notation, we can write

$$R(\phi_i,\phi_{i,p}) = -\int_x tr[(x-\mu_i)'(x-\mu_i)\mathbf{D}_i]\,C_i(x,\tau_p)\,dx + \log|\mathbf{D}_i|$$

where we used the fact that $C_i(x,\tau_p)$ is normalized and $tr(a) = a$ for any scalar. Using the properties of the trace operator (see Appendix 5), we can write

$$R(\phi_i,\phi_{i,p}) = -tr[\int_x (x-\mu_i)'(x-\mu_i)\,C_i(x,\tau_p)\,dx\mathbf{D}_i] + \log|\mathbf{D}_i|$$

To find a maximum of this function we compute the gradient with respect to μ_i using Eq. (A.5.51) of Appendix 5:

$$\frac{\partial R(\phi_i,\phi_{i,p})}{\partial \mu_i} = -\int_x 2(x-\mu_i)\,C_i(x,\tau_p)\,dx\mathbf{D}_i = 0$$

Solving this equation with respect to μ_i we obtain

$$\mu_{i,p+1} = \int_X x\, C_i(x,\tau_p)\,dx = a_{i,p+1}^{-1} \int_X x\, r(i;x,\tau_p)\,dx$$

To find the matrix $\mathbf{D}_i$, we need to find the derivatives with respect to its elements. However, the derivative with respect to $\mathbf{D}_i$ is not defined since this matrix is symmetric and, therefore its elements are dependent (see Sec. 5.1.9 of Appendix 5). To circumvent this problem we use the theory of the conditional extrema.

Denote as $d_{km}^{(i)}$ the k,m-th element of $\mathbf{D}_i$. Since this matrix is symmetric, $d_{km}^{(i)} - d_{mk}^{(i)} = 0$. It follows from the theory of conditional extrema that the conditional maximum of $R(\phi_i, \phi_{i,p})$ can be found as unconditional maximum of the auxiliary function

$$R(\phi_i,\phi_{i,p}) + \sum_{k=1}^{n} \sum_{m=k+1}^{n} \lambda_{km}(d_{km}^{(i)} - d_{mk}^{(i)})$$

where n is the dimension of the vector x and, therefore, the size of the matrix $\mathbf{D}_i$. This means that for this auxiliary function we can apply the derivatives with respect to the matrix $\mathbf{D}_i$ with independent elements. Using Eq. (A.5.49) and (A.5.50) of Appendix 5, we can express the derivative as

$$\mathbf{G} - \mathbf{D}_i^{-1} - \mathbf{\Lambda} = 0$$

where

$$\mathbf{G} = \int_x (x-\mu_i)'(x-\mu_i)\,C_i(x,\tau_p)\,dx \tag{3.1.25}$$

and

$$\mathbf{\Lambda} = \begin{bmatrix} 0 & \lambda_{12} & \cdots & \lambda_{1n} \\ -\lambda_{12} & 0 & \cdots & \lambda_{2n} \\ \cdots & \cdots & \cdots & \cdots \\ -\lambda_{1n} & -\lambda_{2n} & \cdots & 0 \end{bmatrix}$$

The derivatives with respect to λ_{km} give us $d_{km}^{(i)} - d_{km}^{(i)} = 0$ which means that we are looking for a symmetric matrix $\mathbf{D}_i$.

It is clear from Eq. (3.1.25) that $\mathbf{G}$ is symmetric. Therefore, $\mathbf{D}_i$ can be symmetric only if $\mathbf{\Lambda} = 0$. Thus, $\mathbf{D}_i^{-1} = \mathbf{G}$ or $\mathbf{\Sigma}_i = \mathbf{G}$ which after substitutions can be written as

$$\mathbf{\Sigma}_{i,p+1} = a_{i,p+1}^{-1} \int_X (x-\mu_{i,p+1})'(x-\mu_{i,p+1})\,r(i;x,\tau_p)\,dx$$

In the one-dimensional case, the previous equations take the form

$$p_i(x,\phi_i) = \mathrm{N}(x,\mu_i,\sigma_i) = (2\pi)^{-0.5}\sigma_i^{-1} \exp[-0.5(x-\mu_i)^2\sigma_i^{-2}]$$

$$a_{i,p+1} = \int_{-\infty}^{\infty} r(i;x,\tau_p)\,dx \qquad \mu_{i,p+1} = a_{i,p+1}^{-1} \int_{-\infty}^{\infty} x\, r(i;x,\tau_p)\,dx$$

$$\sigma_{i,p+1}^2 = a_{i,p+1}^{-1} \int_{-\infty}^{\infty} (x-\mu_{i,p+1})^2\, r(i;x,\tau_p)\,dx$$

The MATLAB M-file *wgauss* implements this algorithm. To test the algorithm, we selected a true mixture

$$f(x) = 0.75\,\mathrm{N}(x, -1, 2) + 0.25\,\mathrm{N}(x, 1, 0.5)$$

and then applied the fitting algorithm with the initial function

$$g(x,\tau_0) = 0.5\,\mathrm{N}(x, -6, 4) + 0.5\,\mathrm{N}(x, 6, 4)$$

After 40 iterations, we obtained

$$g(x,\tau_{40}) = 0.743\,\mathrm{N}(x, -1.009, 1.987) + 0.257\,\mathrm{N}(x, 0.997, 0.515)$$

This test was implemented i the MATLAB M-file `tstgauss` The successive approximations $g(x,\tau_p)$ are depicted in Figure 3.1. We see that the original approximation is far from $f(x)$, but the first iteration gives us $g(x,\tau_1)$, which is reasonably close to the target. The last approximation $g(x,\tau_{40})$ is marked by 0s. We would like to note that if a problem is ill conditioned, a good curve fit does not necessarily lead to good model parameter estimates. In order to obtain good estimates, the problem should be regularized by changing the fit criterion and limiting the solution set. An example of such an approach can be found in Ref. 58.

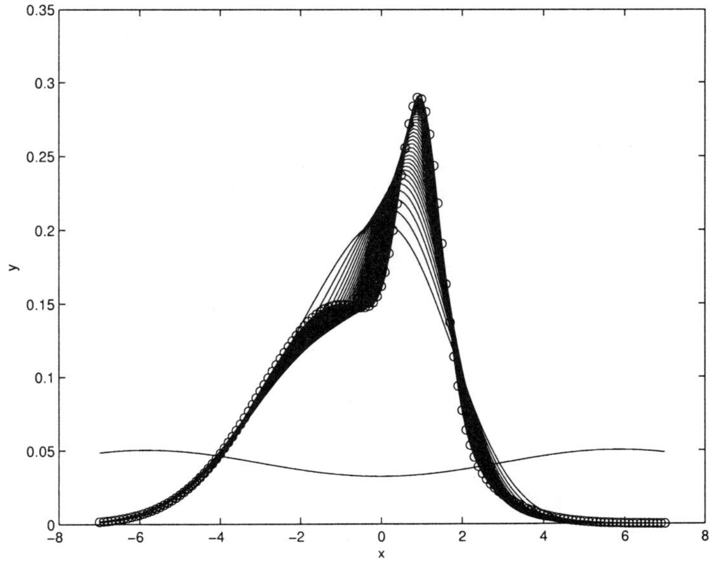

Figure 3.1. Successive approximations.

3.1.4 Maximization of a Function

The EM algorithm can be used to find a maximum of a positive function $g(x)$. Indeed, since $\log(x)$ is a monotonically increasing function

$$\arg\max_x g(x) = \arg\max_x \log g(x) \tag{3.1.26}$$

which is a special case of Eq. (3.1.16) if $f(x)$ is concentrated at x ($Pr(X=x) = 1$) so that $E\{\log g(X) \mid f(X)\} = \log g(x)$. In this case the expectation with respect to $f(x)$ disappears and the EM algorithm takes the form

$$x_{p+1} = \arg\max_x q(x, x_p), \quad p = 0, 1, 2, \ldots \tag{3.1.27}$$

where

$$q(x, x_p) = E\{\log \psi(z; x) \mid \kappa(z; x_p)\} \tag{3.1.28}$$

Example 3.1.3. To illustrate the algorithm, consider the problem of finding a maximum of a mixture of Gaussian PDFs:

$$g(x) = \sum_{z=1}^{m} a_z \mathrm{N}(x, \mu_z, \Sigma_z)$$

As we can see, finding a maximum of this highly nonlinear function is a difficult task. To apply the EM algorithm we define the auxiliary function $\kappa(z;x) = a_z \mathrm{N}(x, \mu_z, \Sigma_z)/g(x)$ as in Eq. (3.1.19). Then, according to (3.1.20), we have

$$\log\psi(z;x) = \log[a_z(2\pi)^{-m/2}|\Sigma_z|^{-1/2}] - 0.5(x-\mu_z)\Sigma_z^{-1}(x-\mu_z)'$$

Therefore, the expectation step given by Eq. (3.1.28) can be expressed as

$$q(x,x_p) = \sum_{z=1}^{m} d(z,x_p)(x-\mu_z)\Sigma_z^{-1}(x-\mu_z)' + const$$

where $d(z,x_p) = -0.5\,\kappa(z,x_p)$ and the constant does not depend on x and, therefore, can be neglected. As we can see, the auxiliary function represents a quadratic form whose global maximum can be found analytically. Since the auxiliary function is a scalar, we can write $q(x,x_p) = tr[q(x,x_p)]$. Using the properties of the trace operator (see Appendix 5), we can write

$$q(x,x_p) = \sum_{z=1}^{m} d(z,x_p)\,tr[\Sigma_z^{-1}(x-\mu_z)'(x-\mu_z)]$$

The derivative (gradient) of this function can be found using Eq. (A.5.51) of Appendix 5:

$$\frac{\partial q(x,x_p)}{\partial x} = 2\sum_{z=1}^{m} d(z,x_p)(x-\mu_z)\Sigma_z^{-1} = 0$$

The solution of this *linear* equation gives us the the result of the maximization step

$$x_{p+1} = \sum_{z=1}^{m} d(z,x_p)\mu_z\Sigma_z^{-1}[\sum_{z=1}^{m} d(z,x_p)\Sigma_z^{-1}]^{-1}$$

In the one-dimensional case, this formula takes the form

$$x_{p+1} = \sum_{z=1}^{m} d(z,x_p)\mu_z\sigma_z^{-2} / \sum_{z=1}^{m} d(z,x_p)\sigma_z^{-2}$$

The MATLAB M-file `emmax` implements this algorithm. To test the algorithm, we selected the same mixture as in example 3.1.2. The results of successive iterations are shown in Figure 3.2

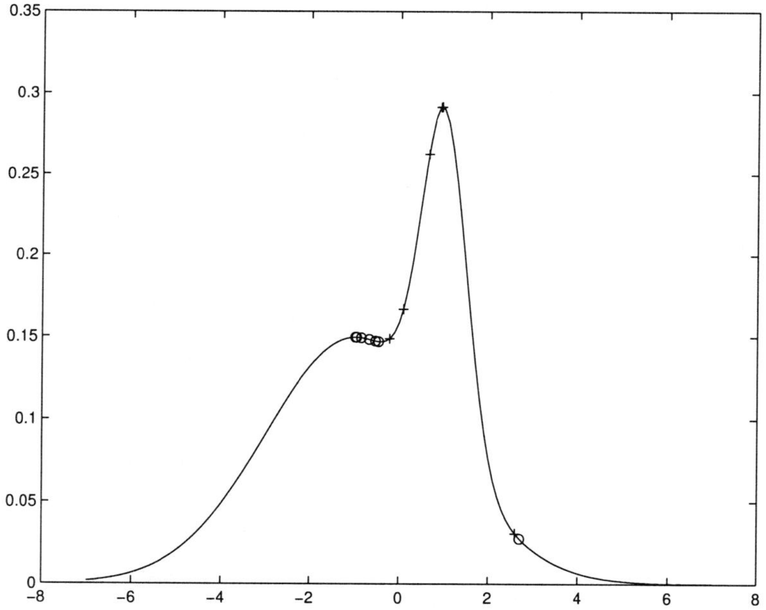

Figure 3.2. Successive Iterations.

For the sequence marked with "+" with the initial value $x_0 = 2.6$ the algorithm converges to the global maximum while for the sequence marked with "o" with the initial value $x_0 = 2.7$ it converges to the local maximum. It is interesting to note that both sequences start from the right side of the global maximum but converge to different maxima. Thus, the regions of attraction for the maxima might be disconnected.

3.1.5 EM Algorithm Acceleration

The algorithm speed of convergence depends on the selection of the auxiliary functions $\kappa(z;x,\tau)$ and $\psi(z;x,\tau)$. If $\psi(z;x,\tau) = g(x;\tau)$, we have the fastest algorithm because, in this case, $H(\tau,\tau_p) = 0$ and the problem is solved in one iteration. However, $\psi(z;x,\tau)$ is usually selected simple enough so that we can perform the maximization step analytically. To satisfy this requirement, we might obtain a very slow algorithm. In order to accelerate the convergence, we can combine the algorithm with faster, but less robust, algorithms in the following way: After each iteration for all the algorithms, select the best result (or a weighted sum of several good results).

For example, we can combine the EM algorithm with the Newton-Raphson algorithm. As we have seen, initially the EM algorithm does a good job and, after several iterations when it slows down, the Newton-Raphson algorithm will quickly finish the job. The price of this approach is the algorithm complexity. However, it is possible to calculate all the derivatives associated with the Newton-Raphson algorithm as a by-product of the EM algorithm. Indeed, since $H(\tau_p,\tau_p)$ is a global minimum of $H(\tau,\tau_p)$, we have $\partial H/\partial\tau = 0$ at $\tau=\tau_p$. Therefore, differentiating Eq. (3.1.11), we obtain

$$\frac{\partial L(\tau_p)}{\partial \tau_p} = \frac{\partial Q(\tau,\tau_p)}{\partial \tau}\Big|_{\tau=\tau_p} = \mathbf{E}\{\frac{\partial \zeta(z;x,\tau_p)}{\partial \tau_p} \mid r(z;x,\tau_p)\} \qquad (3.1.29)$$

where $\zeta(z;x,\tau) = \log\psi(z;x,\tau)$ and $r(z;x,\tau) = f(x)\psi(z;x,\tau)/g(x,\tau)$. This equation allows us to express the gradient $\partial L(\tau_p)/\partial\tau_p$ as an expectation of a simpler gradient $\partial\zeta(z;x,\tau_p)/\partial\tau_p$.

Higher derivatives are obtained by differentiating equation (3.1.29). For example, it is not difficult to see that the Hessian $\partial^2 L(\tau_p)/\partial\tau_p^2$ can be expressed as

$$\frac{\partial^2 L(\tau_p)}{\partial \tau_p^2} = \frac{\partial^2 Q(\tau,\tau_p)}{\partial \tau^2}\Big|_{\tau=\tau_p} + \mathbf{V}\{\zeta(z;x,\tau_p) \mid r(z;x,\tau_p)\}$$

where

$$\mathbf{V}\{\zeta(z;x,\tau) \mid r(z;x,\tau)\} = \mathbf{E}\{\mu(z;\tau) \mid r(z;x,\tau_p)\} - \mathbf{E}\{\nu(x;\tau)\nu(x;\tau)' \mid F(x)\}$$

and

$$\mu(z;\tau) = \frac{\partial\zeta(z;x,\tau)}{\partial\tau}\left(\frac{\partial\zeta(z;x,\tau)}{\partial\tau}\right) \qquad \nu(x;\tau) = \mathbf{E}\{\frac{\partial\zeta(z;x,\tau)}{\partial\tau} \mid \mathrm{K}(z;x,\tau)\}$$

The problem of inverting the Hessian in Eq. (3.1.5) can be solved iteratively using the matrix inversion lemma (see Appendix 5).

3.1.6 Statistical EM Algorithm

Let $\boldsymbol{x}_1^T = x_1, x_2, \ldots, x_T$ be a sequence of independent samples of a random variable x with the probability distribution $g(x,\tau)$. The observation log likelihood is given by (see

Appendix 3)

$$L(\tau) = \sum_{i=1}^{T} \log g(x_i,\tau) \tag{3.1.30}$$

The maximum likelihood (ML) estimate of the model parameter is obtained by maximizing this function. Since Eq. (3.1.30) has the form of Eq. (3.1.16) with $f(x)=1$, the statistical EM algorithm [15] is a special case of the deterministic algorithm given by Eq. (3.1.17) in which we need to replace $f(x) = 1$ with its normalized version $\overline{f}(x)=1/T$ given by Eq. (3.1.15). Thus, the statistical EM algorithm takes the form

$$\tau_{p+1} = \underset{\tau}{argmax} \sum_i q(x_i,\tau,\tau_p)$$

where $q(x_i,\tau,\tau_p)$ is given by Eq. (3.1.14). Therefore, we can apply the previously developed methods to fitting probability distributions to experimental data. For example, using the notation of Sec. 3.1.3.1, we obtain the following EM algorithm for fitting the distribution mixtures to experimental data

$$a_{z,p+1} = (1/T)\sum_{i=1}^{T} \kappa(z;x_i,\tau_p)$$

$$\phi_{z,p+1} = \underset{\phi_z}{argmax} \sum_{i=1}^{T} \log[p_z(x_i;\phi_z)]\, \kappa(z;x_i,\tau_p) \tag{3.1.31}$$

If we denote as n_i the number of the same observation values X_i where all X_i are different, we can rewrite Eq. (3.1.30) as

$$L(\tau) \propto \sum_{i=1}^{N} \nu_i \log g(X_i,\tau)$$

where $\nu_i = n_i/T$ is the observation frequency. This equation has the form of Eq. (3.1.10) with $f(X_i)=\nu_i$. Therefore, we can say that the statistical EM algorithm fits the probability distribution $g(x,\tau)$ to the observation frequencies.

Example 3.1.4. As an illustration, consider fitting a mixture of one-dimensional Gaussian PDFs of Example 3.1.2 to experimental data. As in Example 3.1.2, we can solve analytically Eq. (3.1.31) and obtain [15]

$$a_{z,p+1} = (1/T)\sum_{i=1}^{T} \kappa(z;x_i,\tau_p) \qquad m_{z,p+1} = a_{z,p+1}^{-1} \sum_{i=1}^{T} x_i\, \kappa(z;x_i,\tau_p)$$

$$\sigma_{z,p+1}^2 = a_{z,p+1}^{-1} \sum_{i=1}^{T} (x_i - m_{z,p+1})^2\, \kappa(z;x_i,\tau_p)$$

where

$$\kappa(z;x,\tau_p) = a_{z,p} \mathrm{N}(x,m_{z,p},\sigma_{z,p}) \Big/ \sum_i a_{i,p} \mathrm{N}(x,m_{i,p},\sigma_{i,p})$$

3.2 BAUM-WELCH ALGORITHM

The Baum-Welch algorithm (BWA) [4] is a special case of the EM algorithm for fitting an HMM to a stochastic process. Before we consider a general case, let us develop an EM algorithm for approximating a probability distribution $f(x)$ with a *phase-type* (PH) distribution because this algorithm is the simplest case of the BWA. The other reason for considering this case separately from the general BWA is because this distribution plays an important role in modeling intervals of the same observation values such as error-free intervals. The detailed derivation of the BWA for this simplest case helps to understand the derivation of a more general BWA.

3.2.1 Phase-Type Distribution Approximation

PH distribution is a special case of a matrix-geometric distribution of the form

$$g(x,\tau) = \boldsymbol{\pi}\mathbf{A}^{x-1}\mathbf{b} \quad x=1,2,...$$

where the parameter τ consists of a row vector $\boldsymbol{\pi}$, a column vector $\mathbf{b}$, and an $u\times u$ square matrix $\mathbf{A}$ such that

$$\mathbf{P} = \begin{bmatrix} \mathbf{A} & \mathbf{b} \\ \boldsymbol{\pi} & 0 \end{bmatrix}$$

is a stochastic matrix (its elements are nonnegative and $\mathbf{P1} = \mathbf{1}$). It follows from this definition that $\boldsymbol{\pi}\mathbf{1} = 1$ and $\mathbf{b} = (\mathbf{I} - \mathbf{A})\mathbf{1}$.

Consider a binary HMM whose state transition probability matrix is $\mathbf{P}$. The HMM produces "1" in the $u+1$-th state and "0" in the rest of the states $(1,2,...,u)$. The probability $g(x,\tau)$ can be interpreted as a probability distribution $p(0^x 1 \mid 1)$ of intervals between consecutive returns into the $u+1$-th state.

Using this interpretation, we introduce the auxiliary variable as the sequence of hidden states $z = (i_0, i_1, \ldots, i_{x-1}) = \boldsymbol{i}_0^{x-1}$ and define as $\kappa(\boldsymbol{i}_0^{x-1};x,\tau)=p(\boldsymbol{i}_0^{x-1} \mid x,\tau)$ the conditional probability of the sequence of states given the state duration x. This conditional probability can be written as

$$p(\boldsymbol{i}_0^{x-1} \mid x,\tau) = p(\boldsymbol{i}_0^{x-1},x,\tau) \,/\, p(x,\tau) = \psi(\boldsymbol{i}_0^{x-1};x,\tau) \,/\, g(x,\tau)$$

where we denoted as

$$\psi(\boldsymbol{i}_0^{x-1};x,\tau)=p(\boldsymbol{i}_0^{x-1},x,\tau)=\pi_{i_0}\, a_{i_0 i_1} \cdots a_{i_{x-2} i_{x-1}} a_{i_{x-1},u+1}$$

$a_{i,u+1} = b_i$ and $i_x = u+1$.

To obtain the auxiliary function $Q(\tau,\tau_p)$ of the EM algorithm according to equation (3.1.12), we need to find $q(\tau,\tau_p) = E\{\log\psi(\boldsymbol{i}_0^{x-1};x,\tau) \mid \kappa(z;x,\tau)\}$. Replacing the $\log\psi(\boldsymbol{i}_0^{x-1};x,\tau)$ as a sum of logarithms:

$$\log\psi(\boldsymbol{i}_0^{x-1};x,\tau) = \log\pi_{i_0} + \sum_{t=1}^{x} \log a_{i_{t-1},i_t}$$

we can write

$$q(\tau,\tau_p) = \sum_{i=1}^{u} \gamma_0(i,x,\tau_p)\log\pi_i + \sum_{t=1}^{x} \sum_{i=1}^{u} \sum_{j=1}^{u+1} \gamma_{t-1}(i,j,x,\tau_p)\log a_{ij}$$

where

$$\gamma_{t-1}(i,j,x,\tau) = p(i_{t-1}=i,i_t=j \mid x,\tau) = p(i_{t-1}=i,i_t=j,x,\tau) \,/\, g(x,\tau) \qquad (3.2.1)$$

is the conditional probability of states $i_{t-1}=i, i_t=j$ given the state duration x. If this probability is known, the probability $\gamma_t(i,x,\tau) = p(i_t=i \mid x,\tau)$ can be obtained as

$$\gamma_t(i,x,\tau) = \sum_{i=1}^{u} \gamma_{t-1}(i,j,x,\tau) = \sum_{j=1}^{u+1} \gamma_t(i,j,x,\tau)$$

The probability $p(i_{t-1}=i,i_t=j,x,\tau)$ needed to evaluate $\gamma_{t-1}(i,j,x,\tau)$ in Eq. (3.2.1) can be obtained as

$$p(i_{t-1}=i,i_t=j,x,\tau) = p(i_{t-1}=i,t-1,\tau)p(i_t=j \mid i_{t-1}=i,\tau)p(x-t \mid i_t=j,\tau)$$

To simplify notation, define as $\alpha_i(t,\tau) = p(i_t=i,t,\tau)$ the probability of the HMM observation sequence 0^t and transferring into state $i_t=i$. It is easy to show (see Appendix 6) that this probability is the i-th element of the vector

$$\boldsymbol{\alpha}(t,\tau) = \boldsymbol{\pi}\mathbf{A}^t \tag{3.2.2}$$

The probability $\beta_j(x-t,\tau) = p(x-t \mid i_t=j,\tau) = p(0^{x-t-1} \mid i_t=j,\tau)$ for $j \le u$ is the j-th element of the vector

$$\boldsymbol{\beta}(x-t,\tau) = \mathbf{A}^{x-t-1}\mathbf{b}$$

Using these definitions, we can rewrite Eq. (3.2.1) as

$$\gamma_{t-1}(i,j,x,\tau) = \frac{\alpha_i(t-1,\tau_p)a_{ij}\beta_j(x-t,\tau_p)}{g(x,\tau)} \tag{3.2.3}$$

and, therefore,

$$\gamma_{t-1}(i,x,\tau) = \sum_{j=1}^{u+1} \gamma_{t-1}(i,j,x,\tau) = \frac{\alpha_i(t-1,\tau_p)\beta_i(x-t+1,\tau_p)}{g(x,\tau)}$$

In particular,

$$\gamma_0(i,x,\tau) = \pi_i\beta_i(x,\tau_p) \,/\, g(x,\tau)$$

$$\gamma_{x-1}(i,x,\tau) = \alpha_i(x-1,\tau_p)b_i \,/\, g(x,\tau)$$

The auxiliary function $Q(\tau,\tau_p) = E\{q(\tau,\tau_p) \mid f(x)\}$ of the EM algorithm takes the form

$$Q(\tau,\tau_p) = \sum_{i}^{u} \Pi_i(\tau_p)\log\pi_i + \sum_{i=1}^{u}\sum_{j=1}^{u+1} V_{ij}(\tau_p)\log a_{ij}$$

where

$$\Pi_i(\tau_p) = \sum_x f(x)\gamma_0(i,x,\tau_p) = \sum_x \pi_{i,p}\beta_i(x,\tau_p)\rho(x,\tau_p) \tag{3.2.4a}$$

$$V_{ij}(\tau_p) = \sum_x f(x)\sum_{t=1}^{x}\gamma_{t-1}(i,j,x,\tau_p) = \sum_x A_{ij}(x,\tau_p)\rho(x,\tau_p) \tag{3.2.4b}$$

Here we denoted as $\rho(x,\tau_p) = f(x) \,/\, g(x,\tau_p)$ and

$$A_{ij}(x,\tau) = \sum_{k=1}^{x-1} \alpha_i(k-1,\tau)a_{ij}\beta_j(x-k,\tau) \tag{3.2.5}$$

As we can see the function $Q(\tau,\tau_p)$ has the form of Eq. (3.1.21), therefore, its maximum is found using Eq. (3.1.22) which in our case leads to the following EM algorithm

$$\pi_{i,p+1} = \Pi_i(\tau_p) \Big/ \sum_{i=1}^{u} \Pi_i(\tau_p) = \Pi_i(\tau_p) \tag{3.2.6a}$$

$$a_{ij,p+1} = V_{ij}(\tau_p) \Big/ \sum_{j=1}^{u} V_{ij}(\tau_p) \tag{3.2.6b}$$

Substituting Eq. (3.2.4) into Eq. (3.2.6) we obtain the BWA for fitting the PH distribution to $f(x)$:

$$\pi_{i,p+1} = \sum_x f(x)\gamma_0(i,x,\tau_p) = \pi_{i,p} \sum_x \rho(x,\tau_p)\beta_i(x,\tau_p) \tag{3.2.7a}$$

$$a_{ij,p+1} = a_{ij,p} \frac{\sum_x \rho(x,\tau_p) \sum_{k=1}^{x-1} \alpha_i(k-1,\tau_p)\beta_j(x-k,\tau_p)}{\sum_x \rho(x,\tau_p) \sum_{k=1}^{x} \alpha_i(k-1,\tau_p)\beta_i(x-k+1,\tau_p)} \tag{3.2.7b}$$

$$b_{i,p+1} = b_{i,p} \frac{\sum_x \rho(x,\tau_p)\alpha_i(x-1,\tau_p)}{\sum_x \rho(x,\tau_p) \sum_{k=1}^{x} \alpha_i(k-1,\tau_p)\beta_i(x-k+1,\tau_p)} \tag{3.2.7c}$$

The last equation is not necessary since we can find $\mathbf{b}$ from the normalization conditions: $\mathbf{b} = \mathbf{1} - \mathbf{A1}$. However, as a practical matter, it is convenient to compute recursively just the numerators of the reestimation equations and, in the end, to normalize the results.

It follows from Eq. (3.2.2) that $\boldsymbol{\alpha}(x,\tau)$ can be evaluated recursively by the forward algorithm:

$$\boldsymbol{\alpha}(0,\tau) = \boldsymbol{\pi} \quad \boldsymbol{\alpha}(x,\tau) = \boldsymbol{\alpha}(x-1,\tau)\mathbf{A} \quad x=1,2,... \tag{3.2.8}$$

while $\boldsymbol{\beta}(x,\tau)$ can be evaluated recursively by the backward algorithm:

$$\boldsymbol{\beta}(1,\tau) = \mathbf{b} \quad \boldsymbol{\beta}(x+1,\tau) = \mathbf{A}\boldsymbol{\beta}(x,\tau) \quad x=1,2,...$$

The sum in Eq. (3.2.5) can be evaluated by the *forward-backward* algorithm which is a combination of the forward and backward algorithms: in the forward part, we calculate and save $\boldsymbol{\alpha}(k,\tau)$ for $k=1,2,...,x-1$; in the backward part, we use the saved values together with the calculated $\boldsymbol{\beta}(x-k,\tau)$ to obtain $A_{ij}(x,\tau)$. The forward-backward algorithm is not the optimal algorithm in the general case. The sum in Eq. (3.25) can be evaluated analytically using its z-transform. For this purpose, we rewrite Eq. (3.2.5) in the following matrix form:

$$A_{ij}(x,\tau) = a_{ij} \sum_{k=1}^{x-1} \boldsymbol{\alpha}(k-1,\tau)\mathbf{U}_{ij}\boldsymbol{\beta}(x-k,\tau) = a_{ij}\boldsymbol{\pi} \sum_{k=1}^{x-1} \mathbf{A}^{k-1}\mathbf{U}_{ij}\mathbf{A}^{x-k-1}\mathbf{b} \tag{3.2.9}$$

where $\mathbf{U}_{ij} = \boldsymbol{e}_i' \, \boldsymbol{e}_j$ is the matrix whose ij-th element is one and the rest of its elements are zeroes. Multiplying both sides of this equation by z^{-x} and summing with respect to x, we obtain

$$S_{ij}(z) = \sum_{x=2}^{\infty} A_{ij}(x,\tau)z^{-x} = a_{ij}\boldsymbol{\pi}(\mathbf{I}z - \mathbf{A})^{-1}\mathbf{U}_{ij}(\mathbf{I}z - \mathbf{A})^{-1}\mathbf{b} \tag{3.2.10}$$

By inverting this transform, we find $A_{ij}(x,\tau)$ (as a coefficient of z^{-x}). The inverse matrix $(\mathbf{I}z-\mathbf{A})^{-1}$ can be evaluated as $\mathbf{B}(z)/\Delta(z)$, where $B(z)$ is a polynomial matrix and $\Delta(z) = \det(\mathbf{I}z-\mathbf{A})$ is the characteristic polynomial (see Appendix 5). Hence, $S_{ij}(z)$ is a rational function and its expansion into a power series can be found using its partial fraction decomposition (see Appendix 5).

Example 3.2.1. As an illustration, let us construct the EM algorithm for approximating some probability distribution $f(x)$ with the polygeometric distribution

$$g(x,\tau) = \sum_{i=1}^{u} \pi_i(1-q_i)q_i^{x-1} \quad \pi_i>0 \quad 0<q_i<1 \quad i=1,2,\ldots,u$$

This distribution is a special case of PH with the matrices

$$\boldsymbol{\pi} = row\{\pi_i\} \quad \mathbf{A} = diag\{q_i\} \quad \mathbf{b} = col\{1-q_i\}$$

We have

$$\boldsymbol{\alpha}(x,\tau) = row\{\pi_i q_i^x\} \quad \boldsymbol{\beta}(x,\tau) = col\{q_i^{x-1}(1-q_i)\}$$

Since $\mathbf{I}z - \mathbf{A} = diag\{(z-q_i)\}$ is a diagonal matrix, its inverse is $(\mathbf{I}z - \mathbf{A})^{-1} = diag\{1/(z-q_i)\}$. Therefore, according to Eq. (3.2.10), $S_{ij}(z) = 0$ for $i \neq j$ and

$$S_{ii}(z) = \pi_i q_i(1-q_i)(z-q_i)^{-2}$$

Using the well-known formula (see Appendix 5)

$$(z-q_i)^{-2} = \sum_{x=2}^{\infty} (x-1)q_i^{x-1}z^{-x}$$

we obtain

$$A_{ii}(x,\tau) = \pi_i(x-1)q_i^{x-1}(1-q_i) \quad \text{and} \quad A_{i,u+1} = \pi_i q_i^{x-1}(1-q_i)$$

Therefore,

$$V_{ii}(x,\tau) = \pi_i(1-q_i)\sum_x (x-1)q_i^{x-1}\rho(x,\tau) \quad \text{and} \quad V_{i,u+1} = \pi_i(1-q_i)\sum_x q_i^{x-1}\rho(x,\tau)$$

and the BWA (3.2.7) can be written as

$$\pi_{i,p+1} = \pi_{i,p}(1-q_{i,p})\sum_x \rho(x,\tau_p)q_{i,p}^{x-1}$$

$$1 - q_{i,p+1} = \sum_x \rho(x,\tau_p)q_{i,p}^{x-1} \Big/ \sum_x x\rho(x,\tau_p)q_{i,p}^{x-1}$$

3.2.1.1 *Fitting PH Distribution to Experimental Data*

If we have a sequence $\boldsymbol{x}_1^T$ of independent samples of the PH distributed variables, we can apply the statistical EM algorithm of Sec. 3.1.6 for estimating the distribution parameters. We can obtain the statistical EM algorithm from the algorithm developed in the previous section by replacing $f(x)$ with $1/T$:

$$\pi_{i,p+1} = \frac{\pi_{i,p}}{T}\sum_{m=1}^{T} \beta_i(x_m,\tau_p) \,/\, g(x_m,\tau_p) \tag{3.2.11a}$$

$$a_{ij,p+1} = a_{ij,p} \frac{\sum_{m=1}^{T} \sum_{k=1}^{x_m - 1} \alpha_i(k-1,\tau_p)\beta_j(x_m-k,\tau_p) / g(x_m,\tau_p)}{\sum_{m=1}^{T} \sum_{k=1}^{x_m} \alpha_i(k-1,\tau_p)\beta_i(x_m-k+1,\tau_p) / g(x_m,\tau_p)} \tag{3.2.11b}$$

$$b_{i,p+1} = b_{i,p} \frac{\sum_{m=1}^{T} \alpha_i(x_m-1,\tau_p) / g(x_m,\tau_p)}{\sum_{m=1}^{T} \sum_{k=1}^{x_m} \alpha_i(k-1,\tau_p)\beta_i(x_m-k+1,\tau_p) / g(x_m,\tau_p)} \tag{3.2.11c}$$

Example 3.2.2. In a single-error-state model (see Sec. 1.5.1), error-free intervals have a polygeometric distribution and the error series duration has a geometric distribution. Thus, in order to estimate parameters of this model, we need to fit a polygeometric distribution to the observed error-free intervals. This simplicity of fitting the model to experimental data explains its popularity in applications. Let us write the BWA for estimating parameters of the polygeometric distribution.

The BWA for fitting a polygeometric distribution to experimental data can be obtained from its deterministic counterpart considered in Example 3.2.1 by replacing $f(x)$ with $1/T$:

$$\pi_{i,p+1} = \pi_{i,p}(1-q_{i,p}) \sum_{m=1}^{T} q_{i,p}^{x_m-1} / [T/g(x_m,\tau_p)]$$

$$q_{i,p+1} = 1 - \frac{\sum_{m=1}^{T} q_{i,p}^{x_m-1} / g(x_m,\tau_p)}{\sum_{m=1}^{T} x_m q_{i,p}^{x_m-1} / g(x_m,\tau_p)}$$

In many papers, the fitting is performed by some heuristic methods. The BWA can significantly improve the accuracy of the fit after several iterations.

When applying the BWA, it is important to perform scaling because $\lim_{x\to\infty} \boldsymbol{\alpha}(x,\tau) = 0$ so that for large x the dynamic range of $\boldsymbol{\alpha}(x,\tau)$ is small which leads to the loss of accuracy. To reduce the accuracy loss it is necessary to scale $\boldsymbol{\alpha}(x,\tau)$ by some coefficient μ_x. It is possible to use the scaled variables because the reestimation equations represent fractions, so that multiplying their numerators and denominators by the same numbers does not change the fractions.

It is recommended to normalize [46] $\boldsymbol{\alpha}(x,\tau)$ and compute

$$\bar{\boldsymbol{\alpha}}(x,\tau) = \boldsymbol{\alpha}(x,\tau) / \mu_x, \quad \text{where } \mu_x = \boldsymbol{\alpha}(x,\tau)\mathbf{1} \tag{3.2.12}$$

so that the normalized $\bar{\boldsymbol{\alpha}}(x,\tau)\mathbf{1} = 1$ and, therefore does not tend to zero. Substituting $\boldsymbol{\alpha}(x,\tau) = \mu_x \bar{\boldsymbol{\alpha}}(x,\tau)$ into Eq. (3.2.8) we obtain the normalized forward algorithm

$$\bar{\boldsymbol{\alpha}}(0,\tau) = \boldsymbol{\pi} \quad \bar{\boldsymbol{\alpha}}(x,\tau) = \bar{\boldsymbol{\alpha}}(x-1,\tau)\mathbf{A} / c_{x-1} \quad x=0,1,... \tag{3.2.13}$$

where the normalization coefficients $c_k = \bar{\boldsymbol{\alpha}}(k,\tau)\mathbf{A}\mathbf{1} = \mu_k / \mu_{k-1}$ must be remembered in the forward path and used in the backward path to ensure that the numerators and denominators of the reestimation equations are multiplied by the same numbers.

To find the appropriate scaling for the backward algorithm, consider a typical element of the sum in Eq. (3.2.5) which is multiplied by some common factor C

$$C\alpha_i(k-1,\tau)a_{ij}\beta_j(x-k,\tau) = \bar{\alpha}_i(k-1,\tau)a_{ij}\bar{\beta}_j(x-k,\tau) \quad (3.2.14)$$

where we defined as $\bar{\beta}_j(x-k,\tau) = C\mu_{k-1}\beta_j(x-k,\tau)$. It is easy to see that for any selection of C, the scaled backwards algorithm takes the form

$$\bar{\boldsymbol{\beta}}(1,\tau) = C\mu_{x-2}\mathbf{b}, \quad \bar{\boldsymbol{\beta}}(x-k,\tau) = \mathbf{A}\bar{\boldsymbol{\beta}}(x-k-1,\tau)\,/\,c_k$$

Comparing Eq. (3.2.14) and (3.2.3), we conclude that it is convenient to choose $C = 1/g(x,\tau)$ so that Eq. (3.2.3) is simplified:

$$\gamma_{t-1}(i,j,x,\tau) = \bar{\alpha}_i(t-1,\tau_p)a_{ij}\bar{\beta}_j(x-t,\tau_p)$$

3.2.1.2 *Unidirectional BWA*

The forward-backward algorithm requires a lot of memory if the probabilities of large x are not small (the distribution has "fat tails"), because for large x we need to save in memory all the forward vectors $\boldsymbol{\alpha}(1), \boldsymbol{\alpha}(2), \ldots, \boldsymbol{\alpha}(x-1)$. Let us show that it is possible to use instead a forward-only or backward-only algorithms.

The forward-backward algorithm is used to calculate $A_{ij}(x,\tau)$ according to Eq. (3.2.9). To develop a unidirectional algorithm, we introduce the following matrix

$$\mathbf{W}(x,\tau) = \sum_{k=1}^{x-1}\mathbf{A}^{x-k-1}\mathbf{b}\boldsymbol{\pi}\mathbf{A}^{k-1} = \sum_{k=1}^{x}\boldsymbol{\beta}(x-k,\tau)\boldsymbol{\alpha}(k-1,\tau)$$

Since $\boldsymbol{\beta}(x-k,\tau) = \mathbf{A}^{x-k-1}\mathbf{b}$ is a column vector and $\boldsymbol{\alpha}(k-1,\tau) = \boldsymbol{\pi}\mathbf{A}^{k-1}$ is a row vector, the product $\boldsymbol{\beta}(x-k,\tau)\boldsymbol{\alpha}(k-1,\tau)$ is a square matrix whose ij-th element is

$$\boldsymbol{e}_i\mathbf{A}^{x-k-1}\mathbf{b}\boldsymbol{\pi}\mathbf{A}^{k-1}\boldsymbol{e}_j' = \alpha_j(k-1,\tau)\beta_i(x-k,\tau)$$

Therefore, $A_{ij}(x,\tau) = a_{ij}w_{ji}(x,\tau)$, where $w_{ji}(x,\tau)$ is the ji-th element of $\mathbf{W}(x,\tau)$.

Matrix $\mathbf{W}(x,\tau)$ can be evaluated recursively. Indeed, we can write

$$\mathbf{W}(x+1,\tau) = \sum_{k=1}^{x}\mathbf{A}^{x-k}\mathbf{b}\boldsymbol{\pi}\mathbf{A}^{k-1} = \mathbf{A}\mathbf{W}(x,\tau) + \mathbf{b}\boldsymbol{\alpha}(x-1,\tau)$$

which gives us the following forward-only algorithm

$$\boldsymbol{\alpha}(0,\tau) = \boldsymbol{\pi} \quad \mathbf{W}(1,\tau) = 0$$

$$\mathbf{W}(k+1,\tau) = \mathbf{A}\mathbf{W}(k,\tau) + \mathbf{b}\boldsymbol{\alpha}(k-1,\tau) \quad \boldsymbol{\alpha}(k,\tau) = \boldsymbol{\alpha}(k-1,\tau)\mathbf{A} \quad k=1,2,\ldots,x-1$$

In the end, we obtain $\mathbf{W}(x,\tau_p)$ and $\boldsymbol{\alpha}(x-1,\tau_p)$. Then, we compute $A_{ij}(x,\tau_p) = a_{ij,p}w_{ji}(x,\tau_p)$. We use these values to reestimate the distribution parameters. For example, Eq. (3.2.11b) and (3.2.11c) can be written as

$$a_{ij,p+1} = C_i\sum_{m=1}^{T}A_{ij}(x_m,\tau_p)\,/\,g(x_m,\tau_p)$$

$$b_{i,p+1} = C_i\sum_{m=1}^{T}\alpha_i(x_m-1,\tau_p)\,/\,g(x_m,\tau_p)$$

where C_i is the normalization factor:

$$C_i^{-1} = \sum_{m=1}^{T} [A_{ij}(x_m,\tau_p)\,/\,g(x_m,\tau_p) + \alpha_i(x_m-1,\tau_p)\,/\,g(x_m,\tau_p)]$$

Similarly, we obtain the backward-only algorithm

$$\boldsymbol{\beta}(1,\tau) = \mathbf{b} \quad \mathbf{V}(1,\tau) = 0$$

$$\mathbf{V}(k+1,\tau) = \mathbf{V}(k,\tau)\mathbf{A} + \boldsymbol{\beta}(k,\tau)\boldsymbol{\pi} \quad \boldsymbol{\beta}(k+1,\tau) = \mathbf{A}\boldsymbol{\beta}(k,\tau) \quad k=1,2,\ldots,x-1$$

In contrast with the forward-backward algorithms, these algorithms' memory requirements do not depend on the sample size. The unidirectional algorithms require a fixed amount of memory to store matrices $\mathbf{W}(k,\tau)$ and $\boldsymbol{\alpha}(k,\tau)$ [or $\boldsymbol{\beta}(k,\tau)$]. The other advantage of using the unidirectional algorithms is that if we need to compute several sums $A_{ij}(x_m,\tau)$, which is the case in Eq. (3.2.11), we can reuse the previous results: Sort $\{x_m\}$ in the increasing order, compute $\mathbf{W}(x,\tau)$ for the smallest x_m, and use it as the initial value for the next larger x and so on.

If x is large, it is possible to speed up the unidirectional algorithms. Indeed, it is easy to verify that

$$\mathbf{W}(x+y,\tau) = \mathbf{A}^x\mathbf{W}(y,\tau) + \mathbf{W}(x,\tau)\mathbf{A}^y$$

This equation allows us to find $\mathbf{W}(x+y,\tau)$ if we know $\mathbf{W}(x,\tau)$ and $\mathbf{W}(y,\tau)$. It can be used similarly to the matrix fast exponentiation (see Algorithm A5.2 of Appendix 5). Denote $REM(m,2)$ as a remainder after dividing m by 2. Then, the algorithm can be written as

Algorithm 3.1.

`Initialize:`

$$b_0 = REM(x,2) \quad x_0 = x \quad \mathbf{V}_0 = b_0\mathbf{b}\boldsymbol{\pi} \quad \mathbf{S}_0 = \mathbf{b}\boldsymbol{\pi} \quad \mathbf{Q}_0 = \mathbf{A}^{b_0} \quad \mathbf{R}_0 = \mathbf{A}$$

`While` $x_i > 0$

`Begin`

$$i+1 \rightarrow i$$

$$\mathbf{S}_i = \mathbf{R}_{i-1}\mathbf{S}_{i-1} + \mathbf{S}_{i-1}\mathbf{R}_{i-1} \quad \mathbf{R}_i = \mathbf{R}_{i-1}^2$$

$$x_i = (x_{i-1} - b_{i-1})\,/\,2 \quad b_i = REM(x_i,2)$$

$$\mathbf{V}_i = \mathbf{R}_i^{b_i}\mathbf{V}_{i-1} + b_i\mathbf{S}_i\mathbf{Q}_{i-1} \quad \mathbf{Q}_i = \mathbf{Q}_{i-1}\mathbf{R}_i^{b_i}$$

`End`

At the end, we obtain $\mathbf{W}(x) = \mathbf{V}_k$ and $\mathbf{A}^x = \mathbf{Q}_k$, where k is the number of iterations.

3.2.1.3 Modeling Fade Durations

The following approach is often used for modeling a stochastic process with an HMM. The process values are quantized to several levels whose selection is application dependent. It is usually possible to select the quantization levels in such a way that the level change process is a Markov chain. In this case, we have a semi-Markov process. To model this process with an HMM, we need to approximate the state durations with the PH distributions and then construct the HMM according to Eq. (1.4.21).

We applied this method to modeling fading envelope of the Rayleigh fading described in Sec. 1.5.4. We performed simulation of the fading with the maximal Doppler frequency $f_D = 100Hz$. A typical fade duration PDF is depicted in Fig. 3.3. It resembles a Gamma distribution. [48]

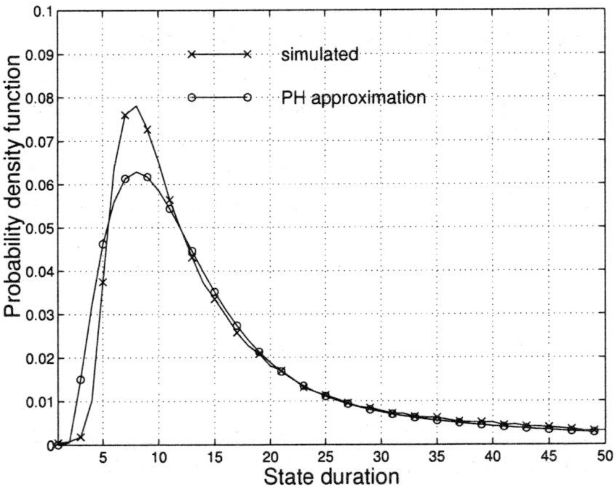

Figure 3.3. A typical state duration distribution and its approximation.

As we can see, this distribution is not geometrical. Since this distribution is not monotonic, it cannot be approximated by the polygeometric distribution. However, since the family of PH distributions is dense, it can be approximated by a PH distribution. The result of fitting an PH distribution to the simulated data is shown in Fig. 3.3 for comparison.

3.2.2 HMM Approximation

Let $f(\boldsymbol{x}_1^T)$ be a process multidimensional distribution. The corresponding HMM distribution, according to Eq. (1.1.9), has the form

$$g(\boldsymbol{x}_1^T,\tau) = \boldsymbol{\pi}(\tau)\prod_{i=1}^{T}\mathbf{P}(x_i,\tau)\mathbf{1} \tag{3.2.15}$$

where τ is the model parameter vector. (Sometimes, to simplify notation, we omit τ in the equations in which its actual value is not important.) It is clear that maximization of the likelihood function in Eq. (3.1.10) for this distribution is a difficult problem. In the following sections we show that this problem can be solved by the EM algorithm in many important practical cases. The EM algorithm for fitting HMMs is called the Baum-Welch algorithm (BWA) because this version of the algorithm was introduced in Ref. [4] before its generalization appeared in Ref. [15].

3.2.2.1 Fundamental formula

Let us develop a formula for computing the auxiliary function $Q(\tau,\tau_p)$ for the general HMM. As in the previous section, the auxiliary random variable $z = \boldsymbol{i}_0^T$ is the sequence of the HMM hidden states and choose as $\kappa(\boldsymbol{i}_0^T;\boldsymbol{x}_1^T,\tau) = p(\boldsymbol{i}_0^T \mid \boldsymbol{x}_1^T,\tau)$ the conditional probability of the sequence given the observation sequence $\boldsymbol{x}_1^T$. This probability can be written as

$$p(\boldsymbol{i}_0^T \mid \boldsymbol{x}_1^T,\tau) = \psi(\boldsymbol{i}_0^T;\boldsymbol{x}_1^T,\tau)/g(\boldsymbol{x}_1^T,\tau)$$

where we denoted as $\psi(\boldsymbol{i}_0^T;\boldsymbol{x}_1^T,\tau) = p(\boldsymbol{i}_0^T,\boldsymbol{x}_1^T,\tau)$ the probability of the state sequence $z = \boldsymbol{i}_0^T$ and the corresponding observation sequence $\boldsymbol{x}_1^T$:

$$\psi(\boldsymbol{i}_0^T;\boldsymbol{x}_1^T,\tau) = \pi_{i_0}(\tau)\prod_{t=1}^{T} p_{i_{t-1}i_t}(x_t,\tau)$$

To obtain the auxiliary function $Q(\tau,\tau_p)$ according to Eq. (3.1.12), we need to find the expected value of $\psi(\boldsymbol{i}_0^T;\boldsymbol{x}_1^T,\tau)$ with respect to $\kappa(\boldsymbol{i}_0^T;\boldsymbol{x}_1^T,\tau)$. Expanding the logarithm of the product, we can write

$$\log\psi(\boldsymbol{i}_0^T;\boldsymbol{x}_1^T,\tau) = \log\pi_{i_0}(\tau) + \sum_{t=1}^{T} \log p_{i_{t-1}i_t}(x_t,\tau)$$

Therefore, the auxiliary function

$$q(\boldsymbol{x}_1^T,\tau,\tau_p) = E\{\log\psi(\boldsymbol{i}_0^T;\boldsymbol{x}_1^T,\tau) \mid \kappa(\boldsymbol{i}_0^T;\boldsymbol{x}_1^T,\tau_p)\} = q_0(\boldsymbol{x}_1^T,\tau,\tau_p) + q_1(\boldsymbol{x}_1^T,\tau,\tau_p)$$

where

$$q_0(\boldsymbol{x}_1^T,\tau,\tau_p) = E\{\log\pi_{i_0} \mid \kappa(\boldsymbol{i}_0^T;\boldsymbol{x}_1^T,\tau_p)\}$$

and

$$q_1(\boldsymbol{x}_1^T,\tau,\tau_p) = \sum_{t=1}^{T} E\{\log p_{i_{t-1}i_t}(x_t,\tau) \mid \kappa(\boldsymbol{i}_0^T;\boldsymbol{x}_1^T,\tau_p)\}$$

Since $\kappa(\boldsymbol{i}_0^T;\boldsymbol{x}_1^T,\tau) = p(\boldsymbol{i}_0^T \mid \boldsymbol{x}_1^T,\tau)$ we can simplify these expressions and write

$$q(\boldsymbol{x}_1^T,\tau,\tau_p) = q_0(\boldsymbol{x}_1^T,\tau,\tau_p) + q_1(\boldsymbol{x}_1^T,\tau,\tau_p) \tag{3.2.16}$$

where

$$q_0(\boldsymbol{x}_1^T,\tau,\tau_p) = E\{\log\pi_{i_0}(\tau) \mid p(i_0 \mid \boldsymbol{x}_1^T,\tau) = \sum_i \gamma_0(i,\boldsymbol{x}_1^T,\tau_p)\log\pi_i(\tau)$$

and

$$q_1(\boldsymbol{x}_1^T,\tau,\tau_p) = \gamma_{t-1}(i,j,\boldsymbol{x}_1^T,\tau_p)\log p_{ij}(x_t,\tau)$$

In these equations, we used the standard in technical publications notation for the probability $p(i_{t-1}=i,i_t=j \mid \boldsymbol{x}_1^T,\tau)$:

$$\gamma_{t-1}(i,j,\boldsymbol{x}_1^T,\tau) = p(i_{t-1}=i,i_t=j,\boldsymbol{x}_1^t;\tau)/g(\boldsymbol{x}_1^T,\tau) \tag{3.2.17}$$

and probability

$$p(i_t=j \mid \boldsymbol{x}_1^T) = \gamma_t(j,\boldsymbol{x}_1^T,\tau) = \sum_i \gamma_{t-1}(i,j,\boldsymbol{x}_1^T,\tau) = \sum_i \gamma_t(j,i,\boldsymbol{x}_1^T,\tau) \qquad (3.2.18)$$

The auxiliary function $Q(\tau,\tau_p)$ of the EM algorithm is obtained by taking the expectation with respect to $f(\boldsymbol{x}_1^T)$ on both sides of Eq. (3.2.16):

$$Q(\tau,\tau_p) = Q_0(\tau,\tau_p) + Q_1(\tau,\tau_p) \qquad (3.2.19)$$

where

$$Q_0(\tau,\tau_p) = E\{\sum_i \gamma_0(i,\boldsymbol{x}_1^T,\tau_p)\log\pi_i(\tau) \mid f(\boldsymbol{x}_1^T)\}$$

and

$$Q_1(\tau,\tau_p) = E\{\sum_{t=1}^{T}\sum_{ij}\gamma_{t-1}(i,j,\boldsymbol{x}_1^T,\tau_p)\log p_{ij}(x_t,\tau) \mid f(\boldsymbol{x}_1^T)\}$$

We call Eq. (3.2.19) a fundamental formula for computing the auxiliary function because all the special cases of the HMM approximation using the EM algorithm that we will consider are derived from this equation.

3.2.2.2 Forward-Backward Algorithm

Equation (3.2.16) shows that we need to develop an algorithm for computing $\gamma_t(i,j,\boldsymbol{x}_1^T,\tau_p)$ to evaluate the auxiliary function $q(\tau,\tau_p)$. According to Eq. (3.2.17), $\gamma_t(i,j,\boldsymbol{x}_1^T,\tau_p)$ is proportional to $p(i_{t-1},i_t,\boldsymbol{x}_1^T;\tau)$ which can be evaluated, using matrix probabilities, as

$$p(i_{t-1},i_t,\boldsymbol{x}_1^T;\tau) = \boldsymbol{\pi}(\tau)\prod_{k=1}^{t-1}\mathbf{P}(x_k,\tau)\mathbf{P}_t(i_{t-1},i_t,x_t,\tau)\prod_{k=t+1}^{T}\mathbf{P}(x_k,\tau)\mathbf{1}$$

where $\mathbf{P}_t(i,j,x_t,\tau)$ is the matrix probability of the event $(i_{t-1}=i,i_t=j,x_t)$ which is the matrix whose i,j-th element is equal to $p_{ij}(x_t,\tau)$ and the rest of its elements are zeroes. As we can see, this equation can be expressed using the forward and backward probability vectors (see Sec. 2.6.1) as

$$p(i_{t-1}=i,i_t=j,\boldsymbol{x}_1^T;\tau) = \boldsymbol{\alpha}(\boldsymbol{x}_1^{t-1},\tau)\mathbf{P}_t(i_{t-1},i_t,x_t,\tau)\boldsymbol{\beta}(x_{t+1}^T,\tau) \qquad (3.2.20)$$

where

$$\boldsymbol{\alpha}(\boldsymbol{x}_1^t,\tau) = \boldsymbol{\pi}(\tau)\prod_{i=1}^{t}\mathbf{P}(x_i,\tau) \quad \text{and} \quad \boldsymbol{\beta}(x_t^T,\tau) = \prod_{i=t}^{T}\mathbf{P}(x_i,\tau)\mathbf{1} \qquad (3.2.21)$$

Equation (3.2.20) can be also written as

$$p(i_{t-1}=i,i_t=j,\boldsymbol{x}_1^t;\tau) = \alpha_i(\boldsymbol{x}_1^{t-1},\tau)p_{ij}(x_t,\tau)\beta_j(x_{t+1}^T,\tau)$$

where $\alpha_i(\boldsymbol{x}_1^{t-1},\tau)$ is the i-th element of the vector $\boldsymbol{\alpha}(\boldsymbol{x}_1^{t-1},\tau)$ and $\beta_j(x_{t+1}^T,\tau)$ is the j-th element of the vector $\boldsymbol{\beta}(x_{t+1}^T,\tau)$. Thus, Eq. (3.2.17) can be written as

$$\gamma_{t-1}(i,j,\boldsymbol{x}_1^T;\tau) = p(i_{t-1}=i,i_t=j \mid \boldsymbol{x}_1^T,\tau) = \frac{\alpha_i(\boldsymbol{x}_1^{t-1},\tau)p_{ij}(x_t,\tau)\beta_j(x_{t+1}^T,\tau)}{g(\boldsymbol{x}_1^T,\tau)} \qquad (3.2.22)$$

It follows from Eq. (3.2.18) that the conditional state density $\gamma_{t-1}(i,\boldsymbol{x}_1^T,\tau)$

$= p(i_{t-1}=i \mid \boldsymbol{x}_1^T,\tau)$ is equal to

$$\gamma_{t-1}(i,\boldsymbol{x}_1^T,\tau) = \sum_j \gamma_{t-1}(i,j,\boldsymbol{x}_1^T,\tau) = \frac{\alpha_i(\boldsymbol{x}_1^{t-1})\beta_i(\boldsymbol{x}_t^T)}{g(\boldsymbol{x}_1^T,\tau)} \tag{3.2.23}$$

and in particular

$$\gamma_0(i,\boldsymbol{x}_1^T,\tau) = p(i_0=i \mid \boldsymbol{x}_1^T,\tau) = \frac{\pi_i(\tau)\beta_i(\boldsymbol{x}_1^T,\tau)}{g(\boldsymbol{x}_1^T,\tau)} \tag{3.2.24}$$

As we can see, in order to find $\gamma_{t-1}(i,j,\boldsymbol{x}_1^T;\tau)$ we need to know both the forward and backward probability vectors. The forward probability vectors can be evaluated recursively by the forward algorithm

$$\boldsymbol{\alpha}(\boldsymbol{x}_1^0,\tau) = \boldsymbol{\pi}(\tau), \quad \boldsymbol{\alpha}(\boldsymbol{x}_1^t,\tau) = \boldsymbol{\alpha}(\boldsymbol{x}_1^{t-1},\tau)\mathbf{P}(x_t,\tau) \quad t=1,2,...,T \tag{3.2.25}$$

and the backward probability vectors can be evaluated by the backward algorithm

$$\boldsymbol{\beta}(\boldsymbol{x}_{T+1}^T,\tau) = \mathbf{1} \quad \boldsymbol{\beta}(\boldsymbol{x}_t^T,\tau) = \mathbf{P}(x_t,\tau)\boldsymbol{\beta}(\boldsymbol{x}_{t+1}^T,\tau) \quad t=T,T-1,...,1 \tag{3.2.26}$$

To compute $\gamma_{t-1}(i,j,\boldsymbol{x}_1^T,\tau_p)$ we can use the following *forward-backward algorithm* (FBA):

Algorithm 3.2.

`Initialize:` *(Forward part)*

$$\boldsymbol{\alpha}(\boldsymbol{x}_1^0,\tau) = \boldsymbol{\pi}(\tau)$$

`For` $t=1,2,...,T$
`Begin`
`Compute and save`

$$\boldsymbol{\alpha}(\boldsymbol{x}_1^t,\tau) = \boldsymbol{\alpha}(\boldsymbol{x}_1^{t-1},\tau)\mathbf{P}(x_t,\tau)$$

`End`
`Initialize:` *(Backward part)*

$$\boldsymbol{\beta}(\boldsymbol{x}_{T+1}^T,\tau) = \mathbf{1}, \quad g(\boldsymbol{x}_1^T,\tau) = \boldsymbol{\alpha}(\boldsymbol{x}_1^T)\mathbf{1}$$

`For` $t=T,T-1,...,1$
`Begin`
`Compute`

$$\gamma_t(i,\boldsymbol{x}_1^T,\tau) = \alpha_i(\boldsymbol{x}_1^t)\beta_i(\boldsymbol{x}_{t+1}^T)\,/\,g(\boldsymbol{x}_1^T,\tau)$$

$$\gamma_{t-1}(i,j,\boldsymbol{x}_1^T;\tau) = \alpha_i(\boldsymbol{x}_1^{t-1},\tau)p_{ij}(x_t,\tau)\beta_j(x_{t+1}^T,\tau)\,/\,g(\boldsymbol{x}_1^T,\tau)$$

$$\boldsymbol{\beta}(\boldsymbol{x}_t^T,\tau) = \mathbf{P}(x_t,\tau)\boldsymbol{\beta}(\boldsymbol{x}_{t+1}^T,\tau)$$

`End`

The maximization of $Q(\tau,\tau_p)$ can be a difficult problem in the general case. Consider now several cases in which the maximization step can be expressed in a closed form.

3.2.2.3 Independent transitional probabilities

In the simplest case, the observation set is finite and the parameter vector τ consists of the Markov chain initial probability vector $\boldsymbol{\pi}$ and the observation matrix probabilities $\mathbf{P}(x)$. In this case, the fundamental Eq. (3.2.19)can be written as

$$Q(\tau,\tau_p) = \sum_i \Pi_i(\tau_p) \log\pi_i + \sum_x \sum_{ij} V_{ij}(x,\tau_p)\log p_{ij}(x) \tag{3.2.27}$$

where

$$\Pi_i(\tau_p) = E\{\gamma_0(i,\mathbf{x}_1^T,\tau_p) \mid f(\mathbf{x}_1^T)\} \tag{3.2.28a}$$

To find $V_{ij}(x,\tau_p)$, we introduce the conditional PDF $f_t(\mathbf{x}_1^{t-1},\mathbf{x}_{t+1}^T \mid x_t) = f(\mathbf{x}_1^T)\,/\,f_t(x_t)$ where

$$f_t(x_t) = \int_{X^{T-1}} f(\mathbf{x}_1^T)\,d\mathbf{x}_1^{t-1}\,d\mathbf{x}_{t+1}^T$$

which allows us to write

$$V_{ij}(x,\tau_p) = f_t(x) \sum_{t=1,x_t=x}^{T} E\{\gamma_{t-1}(i,j,\mathbf{x}_1^T,\tau_p) \mid f_t(\mathbf{x}_1^{t-1},\mathbf{x}_{t+1}^T \mid x_t)\} \tag{3.2.28b}$$

The summation is performed over the sequences in which $x_t = x$.

As we can see, Eq. (3.2.27) has the form of Eq. (3.1.21), therefore, the maximum of $Q(\tau,\tau_p)$ can be found using Eq. (3.1.22) as

$$\pi_{i,p+1} = \Pi_i(\tau_p)$$

$$p_{ij,p+1}(x) = V_{ij}(x,\tau_p)/\sum_x \sum_j V_{ij}(x,\tau_p)$$

These equations represent the closed form expressions for the maximization step for the considered model. However, this model is rarely used because it has too many parameters.

Let us consider now the model in which the parameter vector τ consists of the state initial probability vector $\boldsymbol{\pi}$, state transition probability matrix $\mathbf{P}$, and the parameters ϕ_j of the state observation probabilities (or PDFs) $b_j(x,\phi_j)$. This model has, in general, less parameters than the previous one and also covers continuous observations.

For this model, the fundamental Eq. (3.2.19) takes the form

$$Q(\tau,\tau_p) = \sum_i \Pi_i(\tau_p) \log\pi_i + \sum_{ij} V_{ij}(\tau_p)\log p_{ij} + \sum_j B_j(\phi_j,\tau_p) \tag{3.2.29}$$

where

$$V_{ij}(\tau_p) = E\{\sum_{t=1}^{T} \gamma_{t-1}(i,j,\mathbf{x}_1^T,\tau_p) \mid f(\mathbf{x}_1^T)\} = \int_X V_{ij}(x,\tau_p)\,dx \tag{3.2.30b}$$

$$B_j(\phi_j,\tau_p) = E\{\sum_{t=1}^{T} \gamma_t(j,\mathbf{x}_1^T,\tau_p)\log b_j(x_t,\phi_j) \mid f(\mathbf{x}_1^T)\} \tag{3.2.30c}$$

and we used Eq. (3.2.18) to derive this expression for $B_j(\phi_j,\tau_p)$.

Because of the assumed parameter independence, the three sums in Eq. (3.2.29) can be maximized independently. Using Eq. (3.1.22), we find

$$\pi_{i,p+1} = \Pi_i(\tau_p) \tag{3.2.31a}$$

$$p_{ij,p+1} = V_{ij}(\tau_p)/\sum_j V_{ij}(\tau_p) = V_{ij}(\tau_p)/V_{i.}(\tau_p) \tag{3.2.31b}$$

$$\phi_{j,p+1} = \arg\max_{\phi_j} B_j(\phi_j,\tau_p) \tag{3.2.31c}$$

Equations (3.2.31a) and (3.2.31b) represent closed form expressions for reestimating $\boldsymbol{\pi}$ and $\mathbf{P}$ while the solutions of Eq. (3.2.31c) depend on the form of the conditional PDFs $b_j(x,\phi_j)$. These solutions may be found in a closed form in some cases or they can be found using iterative algorithms and, in particular, by using an EM algorithm.

In order to apply the EM algorithm we transform Eq. (3.2.31c) to the form of equation (3.1.10). Using the conditional PDF $f_t(\mathbf{x}_1^{t-1},\mathbf{x}_{t+1}^T \mid x_t) = f(\mathbf{x}_1^T)\,/\,f_t(x_t)$, we can rewrite Eq. (3.2.30c) as

$$B_j(\phi_j,\tau_p) = E\{\log b_j(x,\phi_j) D_j(x,\tau_p) \mid f_t(x)\} \tag{3.2.32}$$

where

$$D_j(x,\tau_p) = \sum_{t=1,x_t=x}^{T} E\{\gamma_t(j,\mathbf{x}_1^T,\tau_p) \mid f_t(\mathbf{x}_1^{t-1},\mathbf{x}_{t+1}^T \mid x_t)\}$$

Comparing this expression with (3.2.27) we conclude that

$$f_t(x) D_j(x,\tau_p) = \sum_i V_{ij}(x,\tau_p) = V_{.j}(x,\tau_p)$$

If we normalize this function so that it can be considered as a PDF $C_j(x,\tau_p) = \lambda_j(\tau_p) V_{.j}(x,\tau_p)$, we can present Eq. (3.2.32) as

$$B_j(\phi_j,\tau_p) = E\{\log b_j(x,\phi_j) \mid C_j(x,\tau_p)\}$$

The normalization factor $\lambda_j(\tau_p)$ is selected by the normalization condition $E\{1 \mid C_j(x)\} = 1$ which can be written as

$$\lambda_j(\tau_p) E\{\sum_{t=1}^{T} \gamma_t(j,\mathbf{x}_1^T,\tau_p) \mid f(\mathbf{x}_1^T)\} = \lambda_j(\tau_p) \sum_i V_{ij}(\tau_p) = 1$$

Therefore, $\lambda_j(\tau_p) = 1/V_{.j}(\tau_p)$ where $V_{.j}(\tau_p) = \sum_i V_{ij}(\tau_p)$.

Using thus defined PDF $C_j(x,\tau_p)$ we can rewrite Eq. (3.2.31c) in the standard form of Eq. (3.1.10) as

$$\phi_{i,p+1} = \arg\max_{\phi_j} E\{\log b_j(x,\phi_j) \mid C_j(x,\tau_p)\}$$

which allows us to apply the EM algorithm to find ϕ_j. Let us consider several important cases of this algorithm.

3.2.2.4 Nonparametric distribution

In the simplest case the observations are discrete random variables whose state conditional probabilities $b_j(x)$ do not depend on ϕ_j (nonparametric distribution). In this case, the solution of Eq. (3.2.31c) can be found using formula (3.1.22) as

$$b_{j,p+1}(x) = C_j(x,\tau_p) \tag{3.2.33}$$

Thus, if the state observation probabilities do not depend on ϕ_j, Eq. (3.2.31a), (3.2.31b), and (3.2.33) represent the closed form expressions for the maximization step.

3.2.2.5 Mixture Density HMMs

Since Eq. (3.2.31c) has the form of Eq. (3.1.13), we can apply the EM algorithm to find parameters ϕ_j. For example, if the observation PDF is a mixture of distributions

$$b_j(x,\phi_j) = \sum_{k=1}^{k_j} d_{jk} p_{jk}(x,\theta_{jk}) \qquad \sum_k d_{jk} = 1$$

we can apply the algorithm developed in Sec. 3.1.3.1. According to Eq. (3.1.23),

$$d_{jk,p+1} = d_{jk,p} E\{p_{jk}(x,\theta_{jk,p})/b_j(x,\phi_{j,p}) \mid C_j(x,\tau_p)\}$$

and Eq. (3.1.24) leads to

$$\theta_{jk,p+1} = arg\max_{\theta_{jk}} E\{\log p_{jk}(x,\theta_{jk}) \mid C_{jk}(x,\tau_p)\}$$

where

$$C_{jk}(x,\tau_p) = d_{jk,p} p_{jk}(x_t;\theta_{jk}) C_j(x,\tau_p)/[b_j(x,\phi_{j,p}) d_{jk,p+1}] \tag{3.2.34}$$

Example 3.2.3. Consider an HMM in which the observation conditional PDFs are mixtures of multidimensional Gaussian distributions (see Example 3.1.2)

$$b_j(x,\phi_j) = \sum_{k=1}^{k_j} d_{jk} \mathrm{N}(x,\mu_{jk},\mathbf{D}_{jk})$$

where μ_{jk} and $\mathbf{D}_{jk}$ are the mean vectors and covariance matrices. These models are very popular in modeling speech signals.[46]

In this case, Eq. (3.2.34) can be solved analytically (see Example 3.1.2) and we obtain

$$d_{jk,p+1} = \int_{X^T} r_{jk}(\mathbf{x}_1^T,\tau_p)\, d\mathbf{x}_1^T$$

$$\mu_{jk,p+1} = d_{jk,p+1}^{-1} \int_{X^T} x_t C_{jk}(t,\mathbf{x}_1^T;\tau_p)\, d\mathbf{x}_1^T \Big/ \int_{\mathbf{x}_1^T} \sum_{t=1}^{T} C_{jk}(t,\mathbf{x}_1^T;\tau_p)\, d\mathbf{x}_1^T$$

$$\mathbf{D}_{jk} = \int_{\mathbf{x}_1^T} \sum_{t=1}^{T} (x-\mu_{jk,p+1})'(x-\mu_{jk,p+1}) C_{jk}(t,\mathbf{x}_1^T;\tau_p)\, d\mathbf{x}_1^T \Big/ \int_{\mathbf{x}_1^T} \sum_{t=1}^{T} C_{jk}(t,\mathbf{x}_1^T;\tau_p)\, d\mathbf{x}_1^T$$

We would like to note that by expanding the model state space it is possible to construct an equivalent model without the mixtures. Indeed, we can split j-th state into k_j substates $(j,1),(j,2),\ldots,(j,k_j)$, $j=1,2,\ldots,u$. Define the expanded Markov chain transition probabilities as

$$p_{(i,l),(j,m)} = p_{ij} d_{jm} \qquad i,j = 1,2,\ldots,u, \quad m=1,2,\ldots,k_j$$

It is easy to show, using the methods in Sec. 1.1.6, that the HMM with the state transition probabilities $p_{(i,l),(j,m)}$ and observation PDFs $b_{(j,m)}(x) = p_{jm}(x,\theta_{jm})$ is equivalent to the original HMM whose state transitional probabilities are p_{ij} and observation PDF are the distribution mixtures $b_j(x,\phi_j)$.

3.2.3 Fitting HMM to Experimental Data

Suppose that we have multiple independent observation sequences $\{\boldsymbol{x}_1\}_1^{T_1}=(x_{11},x_{12},\ldots,x_{1,T_1})$, $\{\boldsymbol{x}_2\}_1^{T_2}=(x_{21},x_{22},\ldots,x_{2,T_2})$, $\{\boldsymbol{x}_K\}_1^{T_K}=(x_{K1},x_{K2},\ldots,x_{K,T_K})$ of the same HMM. The maximum likelihood estimate of the model parameters is

$$\hat{\tau} = \underset{\tau}{argmax} \sum_{k=1}^{K} \log g(\{\boldsymbol{x}_k\}_1^{T_k},\tau)$$

where $g(\{\boldsymbol{x}_k\}_1^{T_k},\tau)$ is given by Eq. (3.2.15). As we explained in Sec. 3.1.6, the statistical EM algorithm is a special case of the HMM approximation with $f(\{\boldsymbol{x}_k\}_1^{T_k}) = 1/K$

$$\tau_p = (1/K)\sum_i q(\{\boldsymbol{x}_i\}_1^{T_i},\tau,\tau_p)$$

where $q(\{\boldsymbol{x}_i\}_1^{T_i},\tau,\tau_p)$ is given by Eq. (3.2.16). Therefore, we can apply the EM in which the expectation $E\{(\cdot)\mid f(\boldsymbol{x}_1^T)\}$ with respect to $f(\boldsymbol{x}_1^T)$ is replaced by $(1/K)\sum_1^K(\cdot)$. For example, Eq. (3.2.31) take the form

$$\pi_{i,p+1} = \Pi_i(\tau_p) = \frac{1}{K}\sum_{k=1}^{K}\gamma_0(i,\{\boldsymbol{x}_k\}_1^{T_k},\tau_p) \tag{3.2.35a}$$

$$p_{ij,p+1}(x) = V_{ij}(x,\tau_p)\,/\sum_x\sum_j V_{ij}(x,\tau_p) \tag{3.2.35b}$$

where

$$V_{ij}(x,\tau_p) = (1/K)\sum_{k=1}^{K}\sum_{t=1,x_{k,t}=x}^{T_k}\gamma_{t-1}(i,j,\{\boldsymbol{x}_k\}_1^{T_k},\tau_p) \tag{3.2.35c}$$

If we have only one long observation sequence $\boldsymbol{x}_1^T$, then $K=1$ and the previous equations can be simplified:

$$\pi_{i,p+1} = \gamma_0(i,\boldsymbol{x}_1^T,\tau_p) = \pi_{i,p}\beta_i(\boldsymbol{x}_1^T,\tau_p)\,/\,g(\boldsymbol{x}_1^T,\tau_p)$$

$$p_{ij,p+1}(x) = \sum_{t=1,x_t=x}^{T}\gamma_t(i,j,\boldsymbol{x}_1^T,\tau_p)\Big/\sum_{t=1}^{T}\gamma_{t-1}(i,\boldsymbol{x}_1^T,\tau_p)$$

or, more explicitly,

$$p_{ij,p+1}(x) = \sum_{t=1,x_t=x}^{T}\alpha_i(\boldsymbol{x}_1^{t-1},\tau_p)p_{ij,p}(x_t)\beta_j(\boldsymbol{x}_{t+1}^T,\tau_p)\Big/\sum_{t=1}^{T}\alpha_i(\boldsymbol{x}_1^{t-1},\tau_p)\beta_i(\boldsymbol{x}_t^T,\tau_p)$$

Note that if we know the numerator of this and the following equations, the denominator can be obtained using the normalization conditions. Thus, it is sufficient to write

$$\pi_{i,p+1} \propto \pi_{i,p}\beta_i(\boldsymbol{x}_1^T,\tau_p)$$

$$p_{ij,p+1}(x) \propto \sum_{t=1,x_t=x}^{T} \alpha_i(\boldsymbol{x}_1^{t-1},\tau_p)p_{ij,p}(x_t)\beta_j(\boldsymbol{x}_{t+1}^T,\tau_p)$$

If the HMM is described by $\boldsymbol{\pi}$, $\mathbf{P}$, and conditional probabilities $b_j(x,\phi)$ of observations, then, in addition to (3.2.35a), we use Eq. (3.2.31b) and (3.2.31c) which in our case take the form

$$p_{ij,p+1} = \sum_{k=1}^{K}\sum_{t=1}^{T_k}\gamma_{t-1}(i,j,\{\boldsymbol{x}_k\}_1^{T_k},\tau_p)/\sum_{k=1}^{K}\sum_{t=1}^{T_k}\gamma_{t-1}(i,\{\boldsymbol{x}_k\}_1^{T_k},\tau_p)$$

$$\phi_j = \underset{\phi_j}{arg\max}\sum_{k=1}^{K}\sum_{t=1}^{T_k}\gamma_t(j,\{\boldsymbol{x}_k\}_1^{T_k},\tau_p)\log b_j(x_{k,t},\phi_j)$$

In particular, if the observation probability distributions $b_j(x)$ are nonparametric, the solution of this equation is given by Eq. (3.2.33) which in our case can be written as

$$b_{j,p+1}(x) = \sum_{k=1}^{K}\sum_{t=1,x_{kt}=x}^{T_k}\gamma_t(j,\{\boldsymbol{x}_k\}_1^{T_k},\tau_p)/\sum_{k=1}^{K}\sum_{tp=1}^{T}\gamma_t(j,\{\boldsymbol{x}_k\}_1^{T_k},\tau_p)$$

If we have only one long observation sequence $\boldsymbol{x}_1^T$, the combined reestimation equations take the following form

$$\pi_{i,p+1} = \pi_{i,p}\beta_i(\boldsymbol{x}_1^T,\tau_p)\,/\,g(\boldsymbol{x}_1^T,\tau_p) \tag{3.2.36a}$$

$$p_{ij,p+1} = p_{ij,p}\frac{\sum_{t=1}^{T}\alpha_i(\boldsymbol{x}_1^{t-1},\tau_p)b_{j,p}(x_t)\beta_j(\boldsymbol{x}_{t+1}^T,\tau_p)}{\sum_{t=1}^{T}\alpha_i(\boldsymbol{x}_1^{t-1},\tau_p)\beta_i(\boldsymbol{x}_t^T,\tau_p)} \tag{3.2.36b}$$

$$b_{j,p+1}(x) = \sum_{t=1,x_t=x}^{T}\alpha_j(\boldsymbol{x}_1^t,\tau_p)\beta_j(\boldsymbol{x}_{t+1}^T,\tau_p)/\sum_{t=1}^{T}\alpha_j(\boldsymbol{x}_1^t,\tau_p)\beta_j(\boldsymbol{x}_{t+1}^T,\tau_p) \tag{3.2.36c}$$

These equations represent the original BWA. [4] We have derived it as a special case of the statistical EM algorithm. However, the BWA was invented before the EM algorithm. [15]

Example 3.2.4. Let us consider a simple example of estimating parameters of the HMM when $\boldsymbol{x}_1^4 = 0101$ is the received error sequence. We assume that the model parameters $\boldsymbol{\pi}_0$, $\mathbf{P}_0(0)$, and $\mathbf{P}_0(1)$ have been selected originally and after p iterations we have $\boldsymbol{\pi}_p$, $\mathbf{P}_p(0)$, and $\mathbf{P}_p(1)$ (In the sequel we omit the subscript p to simplify notation.)

In the next $p+1$-th step of the BWA we apply the forward-backward algorithm. In the forward part, we compute and store

$$\boldsymbol{\alpha}(0) = \boldsymbol{\pi}\mathbf{P}(0) \quad \boldsymbol{\alpha}(01) = \boldsymbol{\alpha}(0)\mathbf{P}(1) \quad \boldsymbol{\alpha}(010) = \boldsymbol{\alpha}(01)\mathbf{P}(0) \quad \boldsymbol{\alpha}(0101) = \boldsymbol{\alpha}(010)\mathbf{P}(1)$$

In the backward part, we compute

$$\boldsymbol{\beta}(1) = \mathbf{P}(1)\mathbf{1} \quad \boldsymbol{\beta}(01) = \mathbf{P}(0)\boldsymbol{\beta}(1) \quad \boldsymbol{\beta}(101) = \mathbf{P}(1)\boldsymbol{\beta}(01) \quad \boldsymbol{\beta}(0101) = \mathbf{P}(0)\boldsymbol{\beta}(101)$$

and simultaneously accumulate the following sums

$$W_{ij}(0) = \alpha_i(01)p_{ij}(0)\beta_j(1) + \pi_i p_{ij}(0)\beta_j(101)$$

$$W_{ij}(1) = \alpha_i(010)p_{ij}(1) + \alpha_i(0)p_{ij}(1)\beta_j(01)$$

The reestimation Eq. (3.2.36a) gives us

$$\pi_{i,p+1} = \pi_i\beta_i(0101)/g(\boldsymbol{x}_1^4)$$

where the error sequence probability $g(\boldsymbol{x}_1^4) = \boldsymbol{\pi}\mathbf{P}(0)\mathbf{P}(1)\mathbf{P}(0)\mathbf{P}(1)\mathbf{1}$ can be obtained using the normalization condition $\sum_i \pi_{i,p+1} = 1$ which gives us

$$g(x_1^4) = \sum_i \pi_i \beta_i(0101) = \boldsymbol{\pi}\boldsymbol{\beta}(0101)$$

According to Eq. ()

$$p_{ij,p+1}(0) = W_{ij}(0)/W_i \quad p_{ij,p+1}(1) = W_{ij}(1)/W_i$$

The coefficient of proportionality can be obtained using the normalization condition:

$$\sum_j [p_{ij,p+1}(0) + p_{ij,p+1}(1)] = 1$$

which leads to

$$W_i = \sum_j [W_{ij}(0) + W_{ij}(1)]$$

3.2.3.1 Using Conditional Expectations

The BWA has an interesting interpretation. If the Markov chain states were observable, the state transition probabilities could be estimated by the transition frequencies $\hat{p}_{ij} = n_{ij} / n_{i.}$, where n_{ij} is the number of transitions from state i to state j and $n_{i.} = \sum_j n_{ij}$ is the number of transitions from state i (see Appendix 3). The BWA replaces the number of transitions by their expected values. Indeed, according to Eq. (3.2.17)

$$Pr(s_{t-1} = i, s_t = j \mid \boldsymbol{x}_1^T) = \gamma_{t-1}(i, j, \boldsymbol{x}_1^T, \tau_p)$$

Therefore, the expected number of transitions, given the observation sequence, can be expressed as

$$E\{n_{ij} \mid \boldsymbol{x}_1^T, \tau_p) = \sum_{t=1}^{T} \gamma_{t-1}(i, j, \boldsymbol{x}_1^T, \tau_p)$$

which is proportional to the numerator of Eq. (3.2.36b). Similarly, we obtain

$$E\{n_{i.} \mid \boldsymbol{x}_1^T, \tau_p) = \sum_{t=1}^{T} \gamma_{t-1}(i, \boldsymbol{x}_1^T, \tau_p)$$

which is proportional to the denominator of Eq. (3.2.36b). Thus, Eq. (3.2.36b) takes the form

$$p_{ij,p+1} = E\{n_{ij} \mid \boldsymbol{x}_1^T, \tau_p\} / E\{n_{i.} \mid \boldsymbol{x}_1^T, \tau_p\}$$

Equation (3.2.36c) may be presented as

$$b_{j,p+1}(x) = E\{m_{jx} \mid \boldsymbol{x}_1^T, \tau_p\} / E\{n_{.j} \mid \boldsymbol{x}_1^T, \tau_p\}$$

where m_{jx} is the number of times of observing x in state j.

Using a similar reasoning, we can estimate the Markov chain stationary distribution as a by-product of the EM algorithm. Indeed, if the chain states were not hidden, the stationary probability of being in state i can be estimated as (see Problem 4)

$$\hat{\pi}_i = n_{i.} / T$$

As before, we estimate this probability by the corresponding expectations: [56]

$$\hat{\pi}_i = E\{n_{i.} \mid \boldsymbol{x}_1^T, \tau_p\} / T = \sum_{t=1}^{T} \gamma_{t-1}(i, \boldsymbol{x}_1^T, \tau_p) / T \tag{3.2.37}$$

3.2.4 Unidirectional and Parallel Algorithms

Since all the BWAs are based on the fundamental formula given by Eq. (3.2.19), they require to compute sums of the form

$$S(\boldsymbol{x}_1^T) = \sum_{t=1}^{T} \boldsymbol{\alpha}(\boldsymbol{x}_1^{t-1})\mathbf{U}(t,x_t)\boldsymbol{\beta}(\boldsymbol{x}_{t+1}^T) \tag{3.2.38}$$

where $\mathbf{U}(t,x_t)$ are some matrices. Our goal is to develop efficient numerical methods for evaluating these sums. The standard method for computing the sum is the forward-backward algorithm. However, depending on the application, other algorithms might be more efficient. We analyze these algorithms in Sec. 3.3.

3.2.4.1 Parallel BWA

It follows from the forward and backward probability definition (3.2.21) that Eq. (3.2.38) can be written as

$$S = \boldsymbol{\pi}\mathbf{S}(\boldsymbol{x}_1^T)\mathbf{1} \tag{3.2.39}$$

where

$$\mathbf{S}(\boldsymbol{x}_1^T) = \sum_{t=1}^{T} \mathbf{P}(\boldsymbol{x}_1^{t-1})\mathbf{U}(t,x_t)\mathbf{P}(\boldsymbol{x}_{t+1}^T) \tag{3.2.40}$$

and

$$\mathbf{P}(\boldsymbol{x}_u^v) = \prod_{i=u}^{v} \mathbf{P}(x_i)$$

are the corresponding matrix probabilities.

As we mentioned before, the matrix form of any equation usually requires more computations. However, it allows us to use matrix algebra methods which might lead to more efficient algorithms. Unlike its scalar counterpart, the matrix sum can be evaluated recursively. Indeed, denote

$$\mathbf{S}(\boldsymbol{x}_u^v) = \sum_{t=u}^{v} \mathbf{P}(\boldsymbol{x}_1^{t-1})\mathbf{U}(t,x_t)\mathbf{P}(\boldsymbol{x}_{t+1}^v) \tag{3.2.41}$$

It is easy to verify that for any $u \le v \le w$

$$\mathbf{P}(\boldsymbol{x}_u^w) = \mathbf{P}(\boldsymbol{x}_u^v)\mathbf{P}(\boldsymbol{x}_{v+1}^w) \tag{3.2.42a}$$

$$\mathbf{S}(\boldsymbol{x}_u^w) = \mathbf{S}(\boldsymbol{x}_u^v)\mathbf{P}(\boldsymbol{x}_{v+1}^w) + \mathbf{P}(\boldsymbol{x}_u^v)\mathbf{S}(\boldsymbol{x}_{v+1}^w) \tag{3.2.42b}$$

These equations form the basis of the recursive algorithms.

If we can break the interval $[1,T]$ into several subintervals $1 < t_1 < t_2 < \ldots < t_K < T$ and compute $\mathbf{S}(\boldsymbol{x}_{t_k+1}^{t_{k+1}})$ together with $\mathbf{P}(\boldsymbol{x}_{t_k+1}^{t_{k+1}})$ independently (possibly on parallel processors or using a network computing), then we can combine the results in many different ways using Eq. (3.2.42) recursively to obtain the total sum $\mathbf{S}(\boldsymbol{x}_1^T)$. For example, we can apply Eq. (3.2.42) in the forward direction:

$$\mathbf{P}(\boldsymbol{x}_1^{t_2}) = \mathbf{P}(\boldsymbol{x}_1^{t_1})\mathbf{P}(\boldsymbol{x}_{t_1+1}^{t_2}) \qquad \mathbf{S}(\boldsymbol{x}_1^{t_2}) = \mathbf{S}(\boldsymbol{x}_1^{t_1})\mathbf{P}(\boldsymbol{x}_{t_1+1}^{t_2}) + \mathbf{P}(\boldsymbol{x}_1^{t_1})\mathbf{S}(\boldsymbol{x}_{t_1+1}^{t_2})$$

Applying Eq. (3.2.42) again, we obtain

$$\mathbf{P}(\boldsymbol{x}_1^{t_3}) = \mathbf{P}(\boldsymbol{x}_1^{t_2})\mathbf{P}(\boldsymbol{x}_{t_2+1}^{t_3}) \qquad \mathbf{S}(\boldsymbol{x}_1^{t_3}) = \mathbf{S}(\boldsymbol{x}_1^{t_2})\mathbf{P}(\boldsymbol{x}_{t_2+1}^{t_3}) + \mathbf{P}(\boldsymbol{x}_1^{t_2})\mathbf{S}(\boldsymbol{x}_{t_2+1}^{t_3})$$

and so on. This method is valid for any partition of the interval $[1,T]$. However, if there are repeated sequences, the partition should take advantage of this property by sharing common matrices.

In the special case in which $t_k = t$ and $t_{k+1} = t+1$, Eq. (3.2.42) take the form

$$\mathbf{P}(\boldsymbol{x}_1^{t+1}) = \mathbf{P}(\boldsymbol{x}_1^{t})\mathbf{P}(x_{t+1})$$

$$\mathbf{S}(\boldsymbol{x}_1^{t+1}) = \mathbf{S}(\boldsymbol{x}_1^{t})\mathbf{P}(x_{t+1}) + \mathbf{P}(\boldsymbol{x}_1^{t})\mathbf{U}(t,x_{t+1})$$

Similarly, we can evaluate $\mathbf{S}(\boldsymbol{x}_1^T)$ using the backward algorithm starting with $\mathbf{S}(\boldsymbol{x}_{t_K}^T)$, $\mathbf{P}(\boldsymbol{x}_{t_K}^T)$ and recursively computing

$$\mathbf{P}(\boldsymbol{x}_{t_k}^T) = \mathbf{P}(\boldsymbol{x}_{t_k}^{t_{k+1}-1})\mathbf{P}(\boldsymbol{x}_{t_{k+1}}^T) \tag{3.2.43a}$$

$$\mathbf{S}(\boldsymbol{x}_{t_k}^T) = \mathbf{P}(\boldsymbol{x}_{t_k}^{t_{k+1}-1})\mathbf{S}(\boldsymbol{x}_{t_{k+1}}^T) + \mathbf{S}(\boldsymbol{x}_{t_k}^{t_{k+1}-1})\mathbf{P}(\boldsymbol{x}_{t_{k+1}}^T) \tag{3.2.43b}$$

and, in particular,

$$\mathbf{P}(\boldsymbol{x}_t^T) = \mathbf{P}(x_t)\mathbf{P}(\boldsymbol{x}_{t+1}^T)$$

$$\mathbf{S}(\boldsymbol{x}_t^T) = \mathbf{P}(x_t)\mathbf{S}(\boldsymbol{x}_{t+1}^T) + \mathbf{U}(t,x_t)\mathbf{P}(\boldsymbol{x}_{t+1}^T)$$

Unlike its scalar counterpart, the matrix BWA allows us to evaluate $\mathbf{S}(\boldsymbol{x}_1^T)$ in many different ways: we can compute part of the matrix products by the forward algorithm and part of the products by the backward algorithm. In the parallel implementation, the direction is defined by a trie, in which the value $\mathbf{S}(\boldsymbol{x}_u^w)$ at the parent node is evaluated using the values $\mathbf{S}(\boldsymbol{x}_u^v)$ and $\mathbf{S}(\boldsymbol{x}_{v+1}^w)$ of its children according to Eq. (3.2.42b) and $\mathbf{S}(\boldsymbol{x}_1^T)$ is evaluated at the root. The price that we pay for the gained freedom of computing is the increased number of operations because we need to evaluate the whole matrix instead of just one number. We can reduce the computation load by introducing a vector form of the algorithm.

3.2.4.2 Forward-Only BWA

It follows from Eq. (3.2.39) that we can convert the matrix algorithm into a vector algorithm by multiplying equations (3.2.42) from the left by $\boldsymbol{\pi}$ to obtain

$$\boldsymbol{\alpha}(\boldsymbol{x}_1^{t_{k+1}}) = \boldsymbol{\alpha}(\boldsymbol{x}_1^{t_k})\mathbf{P}(\boldsymbol{x}_{t_k+1}^{t_{k+1}}) \tag{3.2.44a}$$

$$\boldsymbol{s}(\boldsymbol{x}_1^{t_{k+1}}) = \boldsymbol{s}(\boldsymbol{x}_1^{t_k})\mathbf{P}(\boldsymbol{x}_{t_k+1}^{t_{k+1}}) + \boldsymbol{\alpha}(\boldsymbol{x}_1^{t_k})\mathbf{S}(\boldsymbol{x}_{t_k+1}^{t_{k+1}}) \tag{3.2.44b}$$

where $\boldsymbol{s}(\boldsymbol{x}_1^t) = \boldsymbol{\pi}\mathbf{S}(\boldsymbol{x}_1^t)$. We can still evaluate $\mathbf{S}(\boldsymbol{x}_{t_k+1}^{t_{k+1}})$ in parallel but $\boldsymbol{s}(\boldsymbol{x}_1^{t_{k+1}})$ are computed sequentially.

If $t_k = t$ and $t_{k+1} = t+1$, we have the following forward algorithm [50,55]

$$\boldsymbol{\alpha}(\boldsymbol{x}_1^{t+1}) = \boldsymbol{\alpha}(\boldsymbol{x}_1^{t})\mathbf{P}(x_{t+1}) \tag{3.2.45a}$$

$$s(x_1^{t+1}) = s(x_1^t)\mathbf{P}(x_{t+1}) + \boldsymbol{\alpha}(x_1^t)\mathbf{U}(t,x_{t+1}) \tag{3.2.45b}$$

The sum in Eq. (3.2.39) is found as $S = s(x_1^T)\mathbf{1}$ We still need to calculate $\mathbf{P}(x_1^T)$ [using forward Eq. (3.2.43a)] because we need $\boldsymbol{\beta}(x_1^T) = \mathbf{P}(x_1^T)\mathbf{1}$ for reestimating $\boldsymbol{\pi}_{p+1}$.

3.2.4.3 *Backward-Only BWA*

Multiplying both sides of Eq. (3.2.42) from the right by **1** we convert the matrix BWA into a column-vector BWA:

$$\boldsymbol{\beta}(x_{t_k}^T) = \mathbf{P}(x_{t_k}^{t_{k+1}-1})\boldsymbol{\beta}(x_{t_{k+1}}^T)$$

$$\boldsymbol{\sigma}(x_{t_k}^T) = \mathbf{P}(x_{t_k}^{t_{k+1}-1})\boldsymbol{\sigma}(x_{t_{k+1}}^T) + \mathbf{S}(x_{t_k}^{t_{k+1}-1})\boldsymbol{\beta}(x_{t_{k+1}}^T)$$

where $\boldsymbol{\sigma}(x_{t_k}^T) = \mathbf{S}(x_{t_k}^T)\mathbf{1}$ and, for $t_k = t$, $t_{k+1} = t+1$, we have

$$\boldsymbol{\beta}(x_t^T) = \mathbf{P}(x_t)\boldsymbol{\beta}(x_{t+1}^T) \tag{3.2.46a}$$

$$\boldsymbol{\sigma}(x_t^T) = \mathbf{P}(x_t)\boldsymbol{\sigma}(x_{t+1}^T) + \mathbf{U}(t,x_t)\boldsymbol{\beta}(x_{t+1}^T) \tag{3.2.46b}$$

The sum in Eq. (3.2.39) is $S = \boldsymbol{\pi\sigma}(x_1^T)$.

Since Eq. (3.2.46) compute $\boldsymbol{\beta}(x_1^T)$, we do not need to compute $\mathbf{P}(x_1^T)$ as in the forward-only algorithm.

We have outlined several methods for computing $S(x_t^T)$, which are different from the standard forward-backward algorithms. In the standard implementation, the algorithm requires the amount of memory which is proportional to the data size T which can be very large. In contrast, the unidirectional algorithms require the amount of memory which is independent of T. The other advantage is that we can apply the forward-only and parallel algorithms on-line without waiting till the whole sequence of length T has been received because the algorithm does not have a backward part.

Example 3.2.5. To illustrate the forward-only algorithm, consider the BWA presented by Eq. (3.2.36). The numerator of Eq. (3.2.36a) can be written as

$$S_{ij}(x_1^T) = \sum_{t=1}^{T} \boldsymbol{\alpha}(x_1^{t-1})\mathbf{U}_{ij}p_{ij,p}b_{j,p}(x_t)\boldsymbol{\beta}(x_{t+1}^T)$$

where $\mathbf{U}_{ij}$ is the matrix whose ij-th element is one and the rest of its elements are zeroes. Since this equation has the form of Eq. (3.2.38), Eq. (3.2.45b) for computing S_{ij}, has the form

$$s_{ij}(x_1^{t+1}) = s_{ij}(x_1^t)\mathbf{P}(x_{t+1}) + \boldsymbol{\alpha}(x_1^t)\mathbf{U}_{ij}p_{ij,p}b_{j,p}(x_t) \tag{3.2.47}$$

The numerator of Eq. (3.2.36c) can be written as

$$S_j(x) = \sum_{t=1}^{T} \boldsymbol{\alpha}(x_1^t)\mathbf{U}_{jj}\delta(x,x_t)\boldsymbol{\beta}(x_{t+1}^T)$$

where $\delta(x,x_t) = 1$ if $x = x_t$ and $\delta(x,x_t) = 0$ otherwise. Equation (3.2.45b), for computing $S_j(x)$, has the form

$$s_j(x_1^{t+1},x) = s_j(x_1^t,x)\mathbf{P}(x_{t+1}) + \boldsymbol{\alpha}(x_1^t)\delta(x,x_{t+1})\mathbf{U}_{ii} \tag{3.2.48}$$

Equations (3.2.47) and (3.2.48) together with Eq. (3.2.45a) and (3.2.43a) represent the forward-only BWA. The initial conditions for the algorithm are

$$s_{ij}(x_1^0) = s_j(x_1^0,x) = 0, \quad \boldsymbol{\alpha}(x_1^0) = \boldsymbol{\pi}_p$$

The reestimation Eq. (3.2.36) can be written as

$$\pi_{i,p+1} = \pi_{i,p}\beta_i(x_1^T) / \pi_p\beta(x_1^T) \qquad \beta(x_1^T) = \mathbf{P}(x_1^T)\mathbf{1}$$

$$p_{ij,p+1} = s_{ij}(x_1^T)\mathbf{1} / \sum_j s_{ij}(x_1^T)\mathbf{1}$$

$$b_{j,p+1}(x) = s_j(x_1^T,x)\mathbf{1} / \sum_x s_j(x_1^T,x)\mathbf{1}$$

3.2.5 Scaling

We would like to note two practical aspects of reestimation. As can be seen from Eq. (3.2.36b), the reestimation algorithm preserves the matrix structure: if $p_{ij,p} = 0$, then $p_{ij,p+1} = 0$. Thus, if we improperly selected the initial matrix, its power might be a machine zero. To avoid this, a preliminary estimation of some mean values (like error-free mean interval) is necessary.

The other aspect of reestimation is scaling.[46] It follows from Eq. (3.2.5) that the forward and backward probability vectors tend to zero exponentially and for sufficiently large sample size (which is always the case in modeling error sources) the dynamic range of the $\alpha(x_1^t)$ computation exceeds the precision of any computer. To improve the computation accuracy, it is necessary to incorporate a scaling procedure. Since all the reestimation formulae are fractions, we can improve the accuracy by multiplying the numerators and denominators by common factors. This can be achieved by normalizing $\alpha(x_1^t)$:

$$\bar{\alpha}(x_1^t) = \alpha(x_1^t) / \mu_t, \quad \mu_t = \alpha(x_1^t)\mathbf{1}$$

Note that the normalized forward vectors represent the HMM state *a posteriori* probabilities:

$$Pr(s_t \mid x_1^t) = Pr(s_t = i, x_1^t) / Pr(x_1^t) = \alpha_i(x_1^t) / \alpha(x_1^t)\mathbf{1}$$

The forward algorithm for the normalized probability vectors can be obtained by substituting $\alpha(x_1^T) = \mu_t \bar{\alpha}(x_1^T)$ into Eq. (3.2.25) to obtain

$$\bar{\alpha}(x_1^t) = c_t^{-1}\bar{\alpha}(x_1^{t-1})\mathbf{P}(x_t)$$

where

$$c_t = \bar{\alpha}(x_1^{t-1})\mathbf{P}(x_t)\mathbf{1} = \mu_t / \mu_{t-1} \tag{3.2.49}$$

It is not difficult to show (see Problem 5) that $\mu_t = \prod_{i=1}^{t} c_i$ so that

$$\alpha(x_1^t) = \bar{\alpha}(x_1^t)\prod_{i=1}^{t} c_i$$

and, therefore,

$$g(x_1^T) = \alpha(x_1^T)\mathbf{1} = \prod_{i=1}^{T} c_i$$

Thus, if we apply the normalized forward algorithm, we obtain the likelihood function as its by-product.

To use the normalized forward probabilities in the BWA, we need to scale the backward probabilities in such a way so that each element of the sum in Eq. (3.2.38) is multiplied by the common factor C and the sum is also multiplied by this factor. The BWA must be compensated for this factor. Most of the reestimation equations need not to be compensated because they are ratios which do not depend on this common factor. If we replace

$\boldsymbol{\alpha}(\boldsymbol{x}_1^{t-1}) = \mu_{t-1}\bar{\boldsymbol{\alpha}}(\boldsymbol{x}_1^{t-1})$ in a typical element of the sum in Eq. (3.2.38) multiplied by C, we obtain

$$C\boldsymbol{\alpha}(\boldsymbol{x}_1^{t-1})\mathbf{U}(t,x_t)\boldsymbol{\beta}(\boldsymbol{x}_{t+1}^T) = \bar{\boldsymbol{\alpha}}(\boldsymbol{x}_1^{t-1})\mathbf{U}(t,x_t)\bar{\boldsymbol{\beta}}(\boldsymbol{x}_{t+1}^T) \tag{3.2.50}$$

where we defined as $\bar{\boldsymbol{\beta}}(\boldsymbol{x}_{t+1}^T) = C\mu_{t-1}\boldsymbol{\beta}(\boldsymbol{x}_{t+1}^T)$ It is easy to see that for any selection of C, the scaled backwards algorithm takes the form

$$\bar{\boldsymbol{\beta}}(\boldsymbol{x}_{T+1}^T) = C\mu_T\mathbf{1}, \quad \bar{\boldsymbol{\beta}}(\boldsymbol{x}_t^T) = c_{t-1}^{-1}\mathbf{P}(x_t)\bar{\boldsymbol{\beta}}(\boldsymbol{x}_{t+1}^T)$$

Comparing Eq. (3.2.50) and (3.2.22), we find that it is convenient to select $C = 1/\mu_T = 1/g(\boldsymbol{x}_1^T)$ because in this case Eq. (3.2.22) and (3.2.23) needed for computing the fundamental formula have especially simple form:

$$\gamma_{t-1}(i,j,\boldsymbol{x}_1^T;\tau) = p(i_{t-1}=i, i_t=j \mid \boldsymbol{x}_1^T,\tau) = \bar{\alpha}_i(\boldsymbol{x}_1^{t-1},\tau)p_{ij}(x_t,\tau)\bar{\beta}_j(x_{t+1}^T,\tau)$$

$$\gamma_{t-1}(i,\boldsymbol{x}_1^T,\tau) = p(i_{t-1}=i \mid \boldsymbol{x}_1^T,\tau) = \bar{\alpha}_i(\boldsymbol{x}_1^{t-1})\bar{\beta}_i(\boldsymbol{x}_t^T)$$

In the FBA, the scaling coefficients c_t must be remembered in the forward part and then used in the backward part. In the forward-only BWA, there is no need to save them because there is no backward part. For example, the scaled version of the BWA described by Eq. (3.2.45) has the form

$$\bar{\boldsymbol{\alpha}}(\boldsymbol{x}_1^{t+1}) = c_{t+1}^{-1}\bar{\boldsymbol{\alpha}}(\boldsymbol{x}_1^t)\mathbf{P}(x_{t+1})$$

$$\bar{\boldsymbol{s}}(\boldsymbol{x}_1^{t+1}) = c_{t+1}^{-1}\bar{\boldsymbol{s}}(\boldsymbol{x}_1^t)\mathbf{P}(x_{t+1}) + c_{t+1}^{-1}\bar{\boldsymbol{\alpha}}(\boldsymbol{x}_1^t)\mathbf{U}(t,x_{t+1})$$

where c_t is computed according to Eq. (3.2.49). The MATLAB M-file implementing this algorithm is called `forw`.

3.3 MARKOV RENEWAL PROCESSES

The direct application of the BWA developed in the previous section to estimating parameters of error sources in real communication channels requires prohibitive amounts of computations, because the intervals between two consecutive bursts are quite large. In some cases, their mean durations are on the order of 10^9 bits. The observation sequences required to estimate parameters of such a model must of necessity be even much longer. In general, if an observation sequence $\boldsymbol{x}_1^T$ contains long stretches of identical observations (for example, sequences of correct bits), we need to develop an estimation algorithm that takes advantage of this property. The main idea of the modified algorithms is to use fast methods of matrix exponentiation and analytical expressions for the matrix sums corresponding to state durations, and then use these expressions in the BWA. The analytical expressions are simplified for the models that do not allow transitions between states corresponding to the same observation values.

The observation sequence $\boldsymbol{x}_1^T$ can be compactly described by the sequence of different observation values $X_1, X_2, \ldots$ and the numbers of their repetitions $m_1, m_2, \ldots$. For example, a sequence 110000111 can be written as $1^2 0^4 1^3$. This method of compression is called a *run-length* coding. Using these notations, we can rewrite the likelihood function (3.2.18) as

$$g(\mathbf{x}_1^T,\tau) = h(\mathbf{v}_1^N,\tau) = \boldsymbol{\pi}\prod_{i=1}^{N}\mathbf{P}^{m_i}(X_i,\tau)\mathbf{1} \tag{3.3.1}$$

where we denoted $v_i = (X_i, m_i)$ and N is the number of subsegments containing identical observations. This method of describing HMMs is called a *Markov renewal process* (MRP) [13] and is popular in queueing theory in which m_i are interpreted as interarrival or service times.

3.3.1 Fast Forward and Backward Algorithms

If an interval m_i is large, we can use the matrix fast exponentiation algorithm which requires on the order of $\log m_i$ matrix multiplications to calculate $\mathbf{P}^{m_i}(X_i,\tau)$ (see Appendix 5) instead of applying m_i times the forward algorithm of Eq. (3.2.25) and memorizing m_i vectors. To calculate $\mathbf{P}^m(X)$ using the fast exponentiation algorithm (see Algorithm A.5.1 of Appendix 5), we express the power as a binary number

$$m = b_1 + 2b_2 + \cdots + 2^{k-1}b_k \tag{3.3.2a}$$

$\mathbf{P}^m(X)$ is found as a result of the following recursion:

Algorithm 3.3.

`Initialize:`

$$\mathbf{R}_1 = \mathbf{P}(X) \qquad \mathbf{Q}_0 = \mathbf{I} \tag{3.3.2b}$$

`For` $i=1,2,\ldots,k$

`Begin`

$$\mathbf{Q}_i = \mathbf{Q}_{i-1}\mathbf{R}_i^{b_i} \qquad \mathbf{R}_{i+1} = \mathbf{R}_i^2 \tag{3.3.2c}$$

`End`

At the end, we have $\mathbf{P}^m(X) = \mathbf{Q}_k$.

If we multiply these equations by the row vector $\boldsymbol{\alpha}_0$ from the left, we obtain the fast forward algorithm

$$\boldsymbol{\alpha}_i = \boldsymbol{\alpha}_{i-1}\mathbf{R}_i^{b_i}, \quad \mathbf{R}_{i+1} = \mathbf{R}_i^2 \quad i=1,2,\ldots,k \tag{3.3.3}$$

which can be used to calculate the forward probabilities

$$\boldsymbol{\alpha}(\mathbf{v}_1^{n_t},\tau) = \boldsymbol{\pi}\prod_{i=1}^{n_{t-1}}\mathbf{P}^{m_i}(X_i,\tau) \tag{3.3.4}$$

where n_t is the number of different observation series up to time t. Initially, $\boldsymbol{\alpha}_0 = \boldsymbol{\pi}$. Then, we calculate $\boldsymbol{\alpha}(v_1,\tau) = \boldsymbol{\pi}\mathbf{P}^{m_1}(X_1,\tau)$ by the fast forward algorithm and use it as $\boldsymbol{\alpha}_0$ for calculating

$$\boldsymbol{\alpha}(\mathbf{v}_1^2,\tau) = \boldsymbol{\alpha}(v_1,\tau)\mathbf{P}^{m_2}(X_2,\tau)$$

and so on. Similarly, we obtain the fast backward algorithm

$$\boldsymbol{\beta}_i = \mathbf{R}_i^{b_i}\boldsymbol{\beta}_{i-1} \qquad \mathbf{R}_{i+1} = \mathbf{R}_i^2 \quad i=1,2,\ldots,k$$

for calculating the backward probabilities

$$\boldsymbol{\beta}(\boldsymbol{v}_{n_t}^N,\tau) = \prod_{i=n_t}^{N} \mathbf{P}^{m_i}(X_i,\tau)\mathbf{1}$$

3.3.1.1 Markov Functions

As we have seen in Sec. 2.4, any HMM with a finite observation range can be expressed as a deterministic function of the Markov chain whose transition probability matrix is given by Eq. (2.4.1). The corresponding MRP distribution can be written in block-matrix form

$$h(\boldsymbol{v}_1^N,\tau) = \boldsymbol{\pi}_{X_1} \prod_{i=1}^{N-1} [\mathbf{P}_{X_iX_i}^{m_i-1} \mathbf{P}_{X_iX_{i+1}}] \mathbf{P}_{X_NX_N}^{m_N-1}\mathbf{1} \tag{3.3.5}$$

To express the forward and backward algorithms for computing $h(\boldsymbol{v}_1^N,\tau)$ in block-vector form, we define

$$\boldsymbol{\alpha}_0(\boldsymbol{v}_1^t,\tau) = \boldsymbol{\pi}_{X_1} \prod_{i=1}^{t-1} [\mathbf{P}_{X_iX_i}^{m_i-1} \mathbf{P}_{X_iX_{i+1}}] \qquad \boldsymbol{\alpha}(\boldsymbol{v}_1^t,\tau) = \boldsymbol{\pi}_{X_1} \prod_{i=1}^{t-1} [\mathbf{P}_{X_iX_i}^{m_i-1} \mathbf{P}_{X_iX_{i+1}}] \mathbf{P}_{X_tX_t}^{m_t-1}$$

These vectors can be evaluated recursively by the following forward algorithm

$$\begin{aligned} \boldsymbol{\alpha}_0(\boldsymbol{v}_1^0,\tau) = \boldsymbol{\pi}_{X_1} \quad \boldsymbol{\alpha}(\boldsymbol{v}_1^t,\tau) &= \boldsymbol{\alpha}_0(\boldsymbol{v}_1^t,\tau)\mathbf{P}_{X_tX_t}^{m_t-1} \\ \boldsymbol{\alpha}_0(\boldsymbol{v}_1^{t+1},\tau) &= \boldsymbol{\alpha}(\boldsymbol{v}_1^t,\tau)\mathbf{P}_{X_tX_{t+1}} \end{aligned} \tag{3.3.6}$$

Similarly, if we define

$$\boldsymbol{\beta}_0(\boldsymbol{v}_t^N,\tau) = \prod_{i=t}^{T} [\mathbf{P}_{X_{i-1}X_i} \mathbf{P}_{X_iX_i}^{m_i-1}]\mathbf{1} \qquad \boldsymbol{\beta}(\boldsymbol{v}_t^N,\tau) = \mathbf{P}_{X_tX_t}^{m_t-1} \prod_{i=t+1}^{T} [\mathbf{P}_{X_{i-1}X_i} \mathbf{P}_{X_iX_i}^{m_i-1}]\mathbf{1}$$

we obtain the block-form backward algorithm

$$\begin{aligned} \boldsymbol{\beta}_0(\boldsymbol{v}_{N+1}^N,\tau) = \mathbf{1} \quad \boldsymbol{\beta}(\boldsymbol{v}_t^N,\tau) &= \mathbf{P}_{X_tX_t}^{m_t-1}\boldsymbol{\beta}_0(\boldsymbol{v}_{t+1}^N,\tau) \\ \boldsymbol{\beta}_0(\boldsymbol{v}_t^N,\tau) &= \mathbf{P}_{X_tX_{t+1}}\boldsymbol{\beta}(\boldsymbol{v}_{t+1}^N,\tau) \end{aligned} \tag{3.3.7}$$

We can apply the fast matrix exponentiation algorithm to block matrices whose sizes are smaller than the size of the whole matrix. If some of the block matrices $\mathbf{P}_{XX}$ are diagonal ($\mathbf{P}_{XX} = diag\{p_{i_Xi_X}\}$), then the matrix power calculation is especially simple: $\mathbf{P}_{XX}^m = diag\{p_{i_Xi_X}^m\}$. This property is useful in error source modeling.[49,50,56]

3.3.2 Fitting MRP

Suppose that we have a distribution $f(\boldsymbol{v}_1^N)$ and we would like to approximate it with an HMM. For this purpose we need to develop the BWA for the HMM MRP distribution $h(\boldsymbol{v}_1^N,\tau)$ given by Eq. (3.3.1).

Consider first the simplest case in which the observation conditional probabilities $p_{ij}(x)$ are the model independent parameters. In accordance with Eq. (3.3.1), we define

$$\psi(z;\boldsymbol{v}_1^N,\tau) = \pi_{i_0}\prod_{t=1}^{N}\prod_{k=u_{t-1}+1}^{u_t} p_{i_{k-1}i_k}(X_t)$$

where u_t are the boundaries of the series of identical observations: $u_0=0$, $u_k=m_1+m_2+\cdots+m_k$, $k=1,2,\ldots,N$.

Using these definitions, we derive the BWA which is an MRP adaptation of Eq. (3.2.29):

$$\pi_{i,p+1} = \pi_{i,p}\sum_{\boldsymbol{v}_1^N}\beta_i(\boldsymbol{v}_1^N,\tau_p)\rho(\boldsymbol{v}_1^N,\tau_p)$$

where

$$\rho(\boldsymbol{v}_1^N,\tau) = f(\boldsymbol{v}_1^N)\,/\,h(\boldsymbol{v}_1^N,\tau)$$

$$p_{ij,p+1}(X) = V_{ij}(X,\tau_p)/\sum_{j,X}V_{ij}(X,\tau_p)$$

and

$$V_{ij}(X,\tau) = \sum_{\boldsymbol{v}_1^N}\rho(\boldsymbol{v}_1^N,\tau_p)p_{ij,p}(X)\partial h(\boldsymbol{v}_1^N,\tau_p)\,/\,\partial p_{ij,p}(X)$$

or

$$V_{ij}(X,\tau) = \sum_{t=1}^{N}\sum_{\boldsymbol{v}_1^N}\boldsymbol{\alpha}(\boldsymbol{v}_1^{t-1},\tau)\mathbf{W}_{ij}(X,v_t,\tau)\boldsymbol{\beta}(\boldsymbol{v}_{t+1}^N,\tau)\rho(\boldsymbol{v}_1^N,\tau)$$

$\mathbf{W}_{ij}(X,v_t,\tau) = 0$ for $X \neq X_t$ and

$$\mathbf{W}_{ij}(X,v_t,\tau) = \sum_{u=1}^{m_t}\mathbf{P}^{u-1}(X_t,\tau)\mathbf{U}_{ij}p_{ij}(X)\mathbf{P}^{m_t-u}(X_t,\tau) \qquad \text{for } X = X_t$$

In the most popular case, the observations are discrete and their state conditional probabilities depend only on the current state (see Sec. 1.1.4). In this case, $p_{ij}(X) = p_{ij}b_j(X)$ and we have the following BWA:

$$\pi_{i,p+1} = \pi_{i,p}\sum_{\boldsymbol{v}_1^N}\beta_i(\boldsymbol{v}_1^N,\tau_p)\rho(\boldsymbol{v}_1^N,\tau_p) \tag{3.3.8a}$$

$$p_{ij,p+1} = V_{ij}(\tau_p)/\sum_{j}V_{ij}(\tau_p) \tag{3.3.8b}$$

$$b_{j,p+1}(X) = C_j(X,\tau_p)\,/\,\sum_{X}C_j(X,\tau_p) \tag{3.3.8c}$$

where

$$V_{ij}(\tau) = \sum_{X}V_{ij}(X,\tau) \quad \text{and} \quad C_j(X,\tau) = \sum_{i}V_{ij}(X,\tau)$$

which can be written as

$$V_{ij}(\tau) = \sum_{t=1}^{N}\sum_{\boldsymbol{v}_1^N}\boldsymbol{\alpha}(\boldsymbol{v}_1^{t-1},\tau)\mathbf{W}_{ij}(v_t,\tau)\boldsymbol{\beta}(\boldsymbol{v}_{t+1}^N,\tau)\rho(\boldsymbol{v}_1^N,\tau) \tag{3.3.8d}$$

$$\mathbf{W}_{ij}(v_t,\tau) = \sum_{u=1}^{m_t} \mathbf{P}^{u-1}(X_t,\tau)\mathbf{U}_{ij}p_{ij}b_j(X_t,\phi_j)\mathbf{P}^{m_t-u}(X_t,\tau) \tag{3.3.8e}$$

$$C_j(X,\tau) = \sum_{t=1}^{N}\sum_{v_1^N} \boldsymbol{\alpha}(\mathbf{v}_1^t,\tau)\mathbf{W}_j(X,v_t,\tau)\boldsymbol{\beta}(\mathbf{v}_{t+1}^N,\tau)\rho(\mathbf{v}_1^N,\tau) \tag{3.3.8f}$$

$$\mathbf{W}_j(X,v_t,\tau) = \sum_{u=1}^{m_t} \mathbf{P}^{u}(X,\tau)\mathbf{U}_{jj}\mathbf{P}^{m_t-u}(X,\tau) \quad \text{for } X = X_t \tag{3.3.8g}$$

otherwise $\mathbf{W}_j(X,v_t,\tau) = 0$

In Sec. 3.3.1 we presented the fast forward and backward algorithms for computing $\boldsymbol{\alpha}(\mathbf{v}_1^t,\tau)$ and $\boldsymbol{\beta}(\mathbf{v}_{n_t}^N,\tau)$. We need to develop fast algorithms for computing sums (3.3.8e) and (3.3.8g) of the form

$$\mathbf{S}_m(X) = \sum_{u=1}^{m} \mathbf{P}^{u-1}(X)\mathbf{W}(X)\mathbf{P}^{m-u}(X) \tag{3.3.9}$$

to develop a fast BWA for fitting an MRP.

3.3.2.1 Analytical Method

As we can see, Eq. (3.3.9) represents a matrix convolution. According to the convolution theorem [Eq. (2.2.10)] and Eq. (A.5.3), the z-transform of the sequence $\{\mathbf{S}_m(X)\}$ has the form

$$\mathbf{S}(z) = (\mathbf{I}z - \mathbf{P}(X))^{-1}\mathbf{W}(X)(\mathbf{I}z - \mathbf{P}(X))^{-1}$$

Matrix $\mathbf{S}_m(X)$ is a coefficient of z^{-m-1} in the expansion of this z-transform. Using Eq. (A.5.4), we can write

$$\mathbf{S}(z) = \mathbf{B}(z)\mathbf{W}(X)\mathbf{B}(z)\Delta^{-2}(z)$$

where $\mathbf{B}(z)$ is a polynomial matrix and $\Delta(z)=\det(\mathbf{I}z-\mathbf{P}(X))$ is the matrix $\mathbf{P}(X)$ characteristic polynomial, which can be factored as $\Delta(z)=(z-\lambda_1)^{\nu_1}(z-\lambda_2)^{\nu_2}\cdots(z-\lambda_r)^{\nu_r}$. Since $\mathbf{S}(z)$ is a rational function, it can be expanded into partial fractions

$$\mathbf{S}(z) = \sum_{j=1}^{r}\sum_{i=1}^{2\nu_j} \mathbf{E}_{ij}\,(z-\lambda_j)^{-i}$$

which gives us (see Appendix 5)

$$\mathbf{S}_m(X) = \sum_{j=1}^{r}\sum_{i=1}^{2\nu_j} \mathbf{E}_{ij}\binom{m}{i-1}\lambda_j^{m-i+1}$$

This equation allows us to calculate $\mathbf{S}_m(X)$ for large m. However, it is usually difficult to find the eigenvalues $\{\lambda_j\}$ and the partial fraction decomposition. The other problem with this approach is that it is necessary to evaluate some of the matrix $\mathbf{P}(X)$ eigenvalues and corresponding matrices $\mathbf{E}_{ij}$ with very high precision which is often impossible to obtain using commercially available software packages. However, if the HMM has a special structure, the analytical approach can be quite useful.

If an HMM is a deterministic Markov function whose state transition probability matrix has the form of equation (2.4.1), then the likelihood function is given by equation (3.3.5) and can be evaluated by the forward and backward algorithms (3.3.6) and (3.3.7) in the block-matrix form. The BWA is given by Eq. (3.3.8a) and (3.3.8b). To estimate the model state transition probability matrix according to Eq. (3.3.8b), we need to calculate $p_{ij}\partial h(\mathbf{v}_1^N,\tau)/\partial p_{ij}$, which is given by Eq. (3.3.5).

If p_{iXjX} is an element of the matrix $\mathbf{P}_{XX}$, we have

$$p_{iXjX}\partial h(\mathbf{v}_1^N,\tau)/\partial p_{iXjX} = \sum_{X_t=X} \boldsymbol{\alpha}_0(\mathbf{v}_1^t,\tau)\mathbf{B}_{iXjX}(m_t)\boldsymbol{\beta}_0(\mathbf{v}_t^N,\tau) \tag{3.3.10}$$

where

$$\mathbf{B}_{iXjX}(m_t) = \sum_{u=1}^{m_t-1} \mathbf{P}_{XX}^{u-1}\mathbf{U}_{iXjX}p_{iXjX}\mathbf{P}_{XX}^{m_t-u-1} \tag{3.3.11}$$

If p_{iXjY} is an element of the matrix $\mathbf{P}_{XY}$ and $X \neq Y$,

$$p_{iXjY}\partial h(\mathbf{v}_1^N,\tau)/\partial p_{iXjY} = \sum_{X_t=X,X_{t+1}=Y} p_{iXjY}\alpha_{iX}(\mathbf{v}_1^t,\tau)\beta_{jY}(\mathbf{v}_{t+1}^N,\tau)$$

where $\alpha_{iX}(\mathbf{v}_1^t,\tau)$ and $\beta_{jY}(\mathbf{v}_{t+1}^N,\tau)$ are the elements of $\boldsymbol{\alpha}(\mathbf{v}_1^t,\tau)$ and $\boldsymbol{\beta}(\mathbf{v}_{t+1}^N,\tau)$, respectively.

The matrix sum $\mathbf{B}_{iXjX}(m_t)$ of Eq. (3.3.11) can be obtained from its z-transform

$$\boldsymbol{\Phi}_{iXjX}(z) = p_{iXjX}(z\mathbf{I} - \mathbf{P}_{XX})^{-1}\mathbf{U}_{ij}(z\mathbf{I} - \mathbf{P}_{XX})^{-1}$$

If matrices $\mathbf{P}_{XX} = diag\{p_{iXiX}\}$ are diagonal, this z-transform is given by

$$\boldsymbol{\Phi}_{iXiX} = p_{iXiX}\mathbf{U}_{iXiX}(z-p_{iXiX})^{-2}$$

$\mathbf{B}_{iXjX}(m)$ is obtained using the inverse z-transform (see Appendix 5):

$$\mathbf{B}_{iXiX}(m) = (m-1)p_{iXiX}^{m-1}\mathbf{U}_{iXiX} = (m-1)\mathbf{P}_{XX}^{m-1}\mathbf{U}_{iXiX}$$

Substituting this expression into Eq. (3.3.10) yields

$$p_{iXiX}\partial h(\mathbf{v}_1^N,\tau)/\partial p_{iXiX} = \sum_{X_t=X} (m_t-1)\alpha_{iX}(\mathbf{v}_1^t,\tau)\beta_{iX}^{(0)}(\mathbf{v}_{t+1}^T,\tau)$$

where $\beta_{iX}^{(0)}(\mathbf{v}_{t+1}^T,\tau)$ is the element of $\boldsymbol{\beta}_0(\mathbf{v}_{t+1}^T,\tau)$.

Using these results, we derive from Eq. (3.3.8b) the following algorithm for estimating the transition probabilities from experimental data

$$p_{iXiX,p+1} = \sum_{X_t=X} (m_t-1)\alpha_{iX}(\mathbf{v}_1^t,\tau_p)\beta_i^{(0)}(\mathbf{v}_{t+1}^T,\tau_p)/w_{iX} \tag{3.3.12a}$$

$$p_{iXjY,p+1} = p_{iXjY,p} \sum_{X_{t-1}=X,X_t=Y} \alpha_{iX}(\mathbf{v}_1^t,\tau_p)\beta_{jY}^{(0)}(\mathbf{v}_t^T,\tau_p)/w_{iX} \tag{3.3.12b}$$

where

$$w_{iX} = \sum_{X_t=X} m_t\alpha_{iX}(\mathbf{v}_1^t,\tau_p)\beta_i^{(0)}(\mathbf{v}_{t+1}^T,\tau_p) \tag{3.3.12c}$$

The stationary distribution is estimated by Eq. (3.2.38), which takes the form

$$\hat{\pi}_{i_X,p} = w_{i_X} / \sum_{i_X} w_{i_X}$$

This form of the BWA not only speeds up the computation by allowing us to use the analytical expression for the matrix powers but also saves memory. In the forward path, we need to save only $\boldsymbol{\alpha}(\boldsymbol{v}_1^t,\tau)$ whose number is equal to the number of identical observation stretches, while the standard BWA needs to save $\boldsymbol{\alpha}(\boldsymbol{x}_1^T)$ whose number is equal to the total number of observations. However, the requirement that matrices $\mathbf{P}_{XX}$ are diagonal is quite restrictive because, in this case, the identical observation series lengths have polygeometric distributions which are monotonically decreasing functions of the length. In the case of an BSC described in Sec. 1.2.4, it requires that matrices $\mathbf{P}_{00}$ and $\mathbf{P}_{11}$ be diagonal. The vast majority of the binary error source models described in the literature possess this property. For example, the Fritchman's partitioned model described in Sec. 1.5.1 satisfies this requirement.

To test the algorithm, we decided to "prove" by simulation that Gilbert's model described in Sec. 1.1.5 is equivalent to the single-error-state model with the matrix (1.5.4). For that we simulated 1,567,289 bits for the Gilbert's model of Sec. 1.1.5 and then fitted to this data a single-error-state model. We selected

$$\mathbf{P}_0 = \begin{bmatrix} 0.999 & 0 & 0.001 \\ 0 & 0.99 & 0.01 \\ 0.4 & 0.1 & 0.5 \end{bmatrix}$$

as the initial approximation and obtained after 10 iterations

$$\hat{\mathbf{P}} = \begin{bmatrix} 0.99741 & 0 & 0.00258 \\ 0 & 0.81274 & 0.18725 \\ 0.18853 & 0.6577 & 0.15377 \end{bmatrix}$$

which agrees favorably with the theoretical result given by equation (1.5.5):

$$\mathbf{P} = \mathbf{I} + \mathbf{R} = \begin{bmatrix} 0.99743 & 0 & 0.00257 \\ 0 & 0.81 & 0.19 \\ 0.15556 & 0.68988 & 0.15456 \end{bmatrix}$$

Note that we needed to save only 20,283 forward vectors $\boldsymbol{\alpha}(\boldsymbol{v}_1^t,\tau)$, while the standard BWA would save 1,567,288 forward vectors.

The requirement that matrices $\mathbf{P}_{00}$ and $\mathbf{P}_{11}$ are diagonal can be relaxed by artificially segmenting experimental data according to error densities. In this case, we have more than two observation types. A typical example occurs if there are intervals with loss of synchronization during which the bit-error rate is close to 0.5. This channel can be described by three types of observations: correct bit, "normal" error, and synchronization loss. For this description, in order to apply Eq. (3.3.12), we require that the matrices corresponding to transitions between the same observation types be diagonal. However, this approach might lead to a very complex model. The algorithm described in the next section can be applied in the general case.

3.3.2.2 Fast BWA

Fast matrix exponentiation is based on doubling the matrix power according to Eq. (3.3.2c). We would like to develop a similar algorithm for computing $\mathbf{S}_m(X)$. Equation (3.3.9) is a special case of Eq. (3.2.43), hence, we can write according to (3.2.45b)

$$\mathbf{S}_{m+n}(X) = \mathbf{S}_m(X)\mathbf{P}^n(X) + \mathbf{P}^m(X)\mathbf{S}_n(X)$$

Thus, similarly to doubling the power, we have

$$\mathbf{S}_{2m}(X) = \mathbf{S}_m(X)\mathbf{P}^m(X) + \mathbf{P}^m(X)\mathbf{S}_m(X)$$

Similarly to the fast exponentiation, we derive the following forward algorithm for calculating $\mathbf{S}_m$ (see Algorithm A5.2 of Appendix 5):

Algorithm 3.4.

`Initialize:`

$$b_0 = REM(m,2) \quad m_0 = m \quad \mathbf{V}_0 = b_0\mathbf{W}(X)$$
$$\mathbf{R}_0 = \mathbf{P}(X) \quad \mathbf{Q}_0 = \mathbf{P}^{b_0}(X) \quad \mathbf{G}_0 = \mathbf{W}(X) \tag{3.3.13a}$$

`While` $m_i > 0$

`Begin`

$$i+1 \to i$$

$$\mathbf{G}_i = \mathbf{G}_{i-1}\mathbf{R}_{i-1} + \mathbf{R}_{i-1}\mathbf{G}_{i-1} \quad \mathbf{R}_i = \mathbf{R}_{i-1}^2 \tag{3.3.13b}$$

$$m_i = (m_{i-1} - b_{i-1})/2 \quad b_i = REM(m_i,2) \tag{3.3.13c}$$

$$\mathbf{V}_i = \mathbf{V}_{i-1}\mathbf{R}_i^{b_i} + b_i\mathbf{Q}_{i-1}\mathbf{G}_i \quad \mathbf{Q}_i = \mathbf{R}_i^{b_i}\mathbf{Q}_{i-1} \tag{3.3.13d}$$

`End`

At the end, we obtain

$$\mathbf{P}^m(X) = \mathbf{Q}_k \quad \text{and} \quad \mathbf{S}_m(X) = \mathbf{V}_k \tag{3.3.13e}$$

where k is the number of iterations.

The backward algorithm has a similar form (see Problem 9).

Both algorithms can be used by the parallel BWA described in Sec. 3.2.3.1. As in Sec 3.2.3.2, we can convert the parallel BWA into the forward-only algorithm. Indeed, if we segment the observation values as in Eq. (3.3.1), then Eq. (3.2.48) can be written as

$$\boldsymbol{\alpha}(\boldsymbol{x}_1^{t_{k+1}}) = \boldsymbol{\alpha}(\boldsymbol{x}_1^{t_k})\mathbf{P}^{m_{t+1}}(X) \tag{3.3.14a}$$

$$\boldsymbol{s}(\boldsymbol{x}_1^{t_{k+1}}) = \boldsymbol{s}(\boldsymbol{x}_1^{t_k})\mathbf{P}^{m_{t+1}}(X) + \boldsymbol{\alpha}(\boldsymbol{x}_1^{t_k})\mathbf{S}_{m_{t+1}}(X) \tag{3.3.14b}$$

Thus, we need to compute row vectors of the form

$$\boldsymbol{\alpha}(m,X) = \boldsymbol{\alpha}_0\mathbf{P}^m(X) \quad \boldsymbol{\gamma}(m,X) = \boldsymbol{\gamma}_0\mathbf{P}^m(X) \quad \mathbf{v}(m,X) = \boldsymbol{\alpha}_0\mathbf{S}_m(X)$$

where we denoted $\boldsymbol{\alpha}_0 = \boldsymbol{\alpha}(\boldsymbol{x}_1^{t_k})$ and $\boldsymbol{\gamma}_0 = \boldsymbol{s}(\boldsymbol{x}_1^{t_k})$. We obtain the forward-only fast algorithm by multiplying the corresponding equations of the matrix algorithm (3.3.13) from the left by $\boldsymbol{\alpha}_0$ or $\boldsymbol{\gamma}_0$:

Algorithm 3.5.

`Initialize:`

$$b_0 = REM(m_{t+1},2) \quad k_0 = m_{t+1} \quad \mathbf{v}_0 = b_0 \boldsymbol{\alpha}_0$$
$$\mathbf{R}_0 = \mathbf{P}(X) \quad \mathbf{G}_0 = \mathbf{W}(X) \quad \boldsymbol{\gamma}_0 = \boldsymbol{\alpha}_0 \tag{3.3.15a}$$

`While` $k_i > 0$

`Begin`

$$i+1 \rightarrow i$$

$$\mathbf{G}_i = \mathbf{R}_{i-1}\mathbf{G}_{i-1} + \mathbf{G}_{i-1}\mathbf{R}_{i-1} \quad \mathbf{R}_i = \mathbf{R}_{i-1}^2 \tag{3.3.15b}$$

$$k_i = (k_{i-1} - b_{i-1})/2 \quad b_i = REM(k_i,2) \tag{3.3.15c}$$

$$\mathbf{v}_i = \mathbf{v}_{i-1}\mathbf{R}_i^{b_i} + b_i \boldsymbol{\alpha}_{i-1}\mathbf{G}_i \quad \boldsymbol{\alpha}_i = \boldsymbol{\alpha}_{i-1}\mathbf{R}_i^{b_i} \quad \boldsymbol{\gamma}_i = \boldsymbol{\gamma}_{i-1}\mathbf{R}_i^{b_i} \tag{3.3.15d}$$

`End`

We obtain

$$\boldsymbol{\alpha}(\boldsymbol{x}_1^{t_{k+1}}) = \boldsymbol{\alpha}_l \quad \text{and} \quad \boldsymbol{s}(\boldsymbol{x}_1^{t_{k+1}}) = \boldsymbol{\gamma}_l + \mathbf{v}_l \tag{3.3.15e}$$

The backward-only algorithm can be derived similarly.

3.4 MATRIX-GEOMETRIC DISTRIBUTION PARAMETER ESTIMATION

As we have seen in the previous chapters, matrix-geometric distribution plays an important role in the HMM and matrix process modeling. This distribution can be written as

$$p(x) = \mathbf{S}\mathbf{M}^{x-1}\mathbf{F} \tag{3.4.1}$$

where $\mathbf{S}$ is order u matrix row, $\mathbf{F}$ is order u matrix column, and $\mathbf{M}$ is order u square matrix. All the elements of these matrices are assumed to be complex numbers. The absolute values of all the matrix $\mathbf{M}$ eigenvalues are assumed to be less than 1, so that the series

$$\sum_{x=1}^{\infty} p(x) = \mathbf{S}(\mathbf{I} - \mathbf{M})^{-1}\mathbf{F} = 1$$

converges for any $\mathbf{S}$ and $\mathbf{F}$. Matrix-geometric distribution is more general than the PH distribution considered in Sec. 3.2.1 and, therefore, is more flexible in data processing.

It is possible to derive necessary and sufficient conditions for a distribution to be matrix-geometric. These conditions can be obtained as a special case of the conditions for the matrix processes if we note that the process with the matrices

$$\mathbf{M}_0 = \begin{bmatrix} \mathbf{M} & 0 \\ \mathbf{S} & 0 \end{bmatrix} \quad \mathbf{M}_1 = \begin{bmatrix} 0 & \mathbf{F} \\ 0 & 0 \end{bmatrix}$$

is a matrix process.

3.4.1 Distribution Simplification

The distribution parameters are the elements of the matrices $\mathbf{S}$, $\mathbf{M}$, and $\mathbf{F}$. We must estimate u elements of the matrix $\mathbf{S}$, u^2 elements of the matrix $\mathbf{M}$, and u elements of the matrix $\mathbf{F}$. This makes u^2+2u-1 parameters to be estimated, since one parameter can be found from the condition that the total sum of the probabilities is equal to 1. It is possible to decrease significantly the number of parameters by using the similarity transformation

$$\mathbf{M}_1 = \mathbf{T}^{-1}\mathbf{M}\mathbf{T}$$

since the distribution has the same form for all similar matrices:

$$\mathbf{S}\mathbf{M}^{x-1}\mathbf{F} = \mathbf{S}_1\mathbf{M}_1^{x-1}\mathbf{F}_1$$

where $\mathbf{S}_1 = \mathbf{S}\mathbf{T}$ $\mathbf{F}_1 = \mathbf{T}^{-1}\mathbf{F}$ and any nonsingular $\mathbf{T}$. The Jordan normal form of the matrix (see Ref. 24 and Appendix 5.1.5)

$$\mathbf{J} = block\ diag\{\mathbf{J}_i\} \tag{3.4.2}$$

where

$$\mathbf{J}_i = \mu_i\mathbf{I} + \mathbf{B}_i \qquad \mathbf{B}_i = \begin{bmatrix} 0 & 1 & 0 \ldots 0 \\ 0 & 0 & 1 \ldots 0 \\ \ldots & \ldots & \ldots \\ 0 & 0 & 0 \ldots 1 \\ 0 & 0 & 0 \ldots 0 \end{bmatrix}_{s_i,s_i} \qquad i = 1,2,\ldots,m$$

is similar to $\mathbf{M}$ and has the minimal number of parameters. Its parameters μ_i are the eigenvalues of the matrix $\mathbf{M}$.

Let us assume that the matrix $\mathbf{M}$ in representation (3.4.1) already has the Jordan form. Then, Eq. (3.4.1) can be written as

$$p(x) = \sum_{i=1}^{m} \mathbf{S}_i\mathbf{J}_i^{x-1}\mathbf{F}_i$$

where $\mathbf{S}_i$ and $\mathbf{F}_i$ are the corresponding subblocks of the matrices $\mathbf{S}$ and $\mathbf{F}$. Since

$$\mathbf{J}_i^r = (\mu_i\mathbf{I} + \mathbf{B}_i)^r = \sum_{k=0}^{s_i-1} \binom{r}{k}\mu_i^{r-k}\mathbf{S}_i\mathbf{B}_i^k\mathbf{F}_i$$

We assume here that $\binom{0}{0} = 1$ and $\binom{r}{k} = r(r-1)\cdots(r-k+1)/k!$, so that $\binom{r}{k} = 0$ for $0 \le r < k$ (see Appendix 4.1). Distribution (3.4.1) takes the form

$$p(x) = \sum_{i=1}^{m}\sum_{k=0}^{s_i-1} \binom{x-1}{k}\mu_i^{x-k-1}\mathbf{S}_i\mathbf{B}_i^k\mathbf{F}_i \tag{3.4.3}$$

Without loss of generality, we can replace $\mathbf{F}$ with $\mathbf{1}$. Thus, the simplified representation of the distribution is given by the following equation:

$$p(x) = \sum_{i=1}^{m}\sum_{k=0}^{s_i-1} \binom{x-1}{k}\mu_i^{x-k-1}c_{ik} \tag{3.4.4}$$

where

$$c_{ik} = \mathbf{S}_i\mathbf{B}_i^k\mathbf{1} \tag{3.4.5}$$

If some of the eigenvalues that belong to different Jordan blocks are equal ($\mu_i = \mu_j$ for

$i \neq j$), then the corresponding terms in the distribution can be factored out. We assume that all the μ_i in Eq. (3.4.4) are different. It is important to note that, if the coefficients c_{ik} have been found, the elements of the matrix $\mathbf{S}_i$ can be determined from linear Eq. (3.4.5) that always have a unique solution. This solution can be expressed in the matrix form:

$$\mathbf{S}_i = \mathbf{C}_i \mathbf{R}(\mathbf{I} - \mathbf{B}_i) \tag{3.4.6}$$

where $\mathbf{C}_i = row\{c_{ik}\}$

$$\mathbf{R} = \begin{bmatrix} 0 & 0 & \dots & 0 & 1 \\ 0 & 0 & \dots & 1 & 0 \\ \cdot & \cdot & \cdot & \cdot & \cdot \\ 1 & 0 & \dots & 0 & 0 \end{bmatrix}$$

Let $\mu_i \neq 0$ for $i = 1,2,\dots,m-1$ and $\mu_m = 0$; then, using the complex number trigonometric form

$$\mu_i = q_i\,(\cos\omega_i + \mathrm{j}\sin\omega_i) \qquad \text{for } i = 1,2,\dots,m-1$$

we can rewrite the distribution $p(x)$ explicitly in its real form:

$$\begin{aligned} p(x) &= \mathbf{S}\mathbf{J}^{x-1}\mathbf{1} \\ &= \sum_{i=1}^{m-1}\sum_{k=0}^{s_i-1}\binom{x-1}{k}\,[\,a_{ik}\cos\omega_i(x-k-1) - b_{ik}\sin\omega_i(x-k-1)\,]\,q_i^{x-k-1} + a_{m,x-1} \end{aligned} \tag{3.4.7}$$

where $a_{ik} = \mathrm{Re}\,c_{ik}$ and $b_{ik} = \mathrm{Im}\,c_{ik}$. If the matrix $\mathbf{M}$ does not have zero eigenvalues, then $a_{m,x-1} = 0$.

In the most important (for applications) case, all the nonzero eigenvalues μ_i are simple ($s_i = 1$ for $i = 1,2,\dots,m-1$), and the distribution is given by

$$p(x) = \sum_{i=1}^{m-1} a_{i0}\,q_i^{x-1} + a_{m,x-1} \tag{3.4.8}$$

This distribution is polygeometric except for the term $a_{m,x-1}$, which can be used to reach better agreement with the experimental data for small x.

Let us consider some methods for estimating distribution (3.4.7) parameters.

3.4.2 ML Parameter Estimation

The logarithmic likelihood function, according to (3.1.8), has the form

$$L(\mathbf{S},\boldsymbol{\mu},\mathbf{z}) = \sum_{i=1}^{r} \nu_i \ln P_i(\mathbf{S},\boldsymbol{\mu},\mathbf{z})$$

where $\boldsymbol{\mu} = (\mu_1,\mu_2,\dots,\mu_m)$, $\mathbf{z} = (s_1,s_2,\dots,s_m)$, and $P_i(\mathbf{S},\boldsymbol{\mu},\mathbf{z}) = Pr(x_{i-1} \leq x < x_i)$ is the probability that x satisfies the condition $x_{i-1} \leq x < x_i$:

$$P_i(\mathbf{S},\boldsymbol{\mu},\mathbf{z}) = \mathbf{S}\mathbf{b}_i(\boldsymbol{\mu},\mathbf{z}) \qquad \mathbf{b}_i(\boldsymbol{\mu},\mathbf{z}) = (\mathbf{J}^{x_{i-1}-1} - \mathbf{J}^{x_i})(\mathbf{I} - \mathbf{J})^{-1}\mathbf{1}$$

ν_i is the corresponding empirical frequency.

The ML parameter estimates are found by maximization of $L(\mathbf{S},\boldsymbol{\mu},\mathbf{z})$, subject to the restriction that the total sum of the probabilities is equal to 1. The matrix order u and the sizes s_i of its Jordan blocks are discrete parameters. Normally, the maximization with respect to these variables is carried out by an exhaustive search: for a selected set of the discrete variables the function is maximized with respect to the rest of the parameters. Then a new set of the variables is selected and the maximum with respect to the continuous parameters is found, and so on. The discrete variables that deliver the global maximum over all possible sets give the ML estimates. In the future, we will assume that the discrete variables are fixed and will consider the problem of estimating the rest of the parameters.

The ML estimates can be found from the necessary conditions of the function $L(\mathbf{S},\boldsymbol{\mu})$ maximum:

$$\frac{\partial}{\partial S_j}\left[\sum_{i=1}^{r} \nu_i \ln P_i(\mathbf{S},\boldsymbol{\mu}) + \lambda \sum_{i=1}^{r} P_i(\mathbf{S},\boldsymbol{\mu}) - \lambda\right] = 0 \quad j=1,2,\ldots,n$$

$$\frac{\partial}{\partial \mu_k}\left[\sum_{i=1}^{r} \nu_i \ln P_i(\mathbf{S},\boldsymbol{\mu}) + \lambda \sum_{i=1}^{r} P_i(\mathbf{S},\boldsymbol{\mu}) - \lambda\right] = 0 \quad k=1,2,\ldots,m$$

$$\sum_{i=1}^{r} P_i(\mathbf{S},\boldsymbol{\mu}) - 1 = 0$$

where λ is a Lagrange's multiplier. After taking the derivatives, we obtain

$$\sum_{i=1}^{r}\left[\frac{\nu_i}{\mathbf{S}\mathbf{b}_i(\boldsymbol{\mu})} + \lambda\right] b_{ij} = 0 \qquad j=1,2,\ldots,n$$

$$\sum_{i=1}^{r}\left[\frac{\nu_i}{\mathbf{S}\mathbf{b}_i(\boldsymbol{\mu})} + \lambda\right] d_{ik} = 0 \qquad k=1,2,\ldots,m \tag{3.4.9}$$

$$\mathbf{S}(\mathbf{I} - \mathbf{J})^{-1}\mathbf{1} - 1 = 0$$

where b_{ij} is the j-th element of the matrix-column $\mathbf{b}_i(\boldsymbol{\mu})$

$$d_{ik} = \mathbf{S}_k[\mathbf{G}_k(x_{i-1}-1) - \mathbf{G}_k(x_i)](\mathbf{I}-\mathbf{J}_k)^{-2}\mathbf{1}$$

$$\mathbf{G}_k(x) = [x(\mathbf{I}-\mathbf{J}_k) + \mathbf{J}_k]\mathbf{J}_k^{x-1}$$

We see that the ML parameter estimates satisfy a system of nonlinear equations. This system can be solved using one of the standard numerical methods. However, these methods converge very slowly when estimating parameters of the distribution of intervals between errors. Some of the eigenvalues μ_i are very close to 1, and small variations of these parameters significantly change the direction of the next iteration. We consider several alternative techniques that use the specifics of the distribution and converge faster than direct methods of solving system (3.4.9).

3.4.3 Sequential Least-Mean-Square Method

The estimation of parameters can be carried out using the minimum χ^2 method, which is asymptotically equivalent to the ML method considered in the previous section. The χ^2 criterion (3.1.2) takes the form

$$\chi^2 = \sum_i [\nu_i - NP_i(\mathbf{S},\boldsymbol{\mu})]^2 / NP_i(\mathbf{S},\boldsymbol{\mu})$$

Normally, the χ^2 minimization is carried out using the gradient methods. However, the standard gradient methods applied directly to χ^2 are unstable for the majority of practical cases. The optimal solution may be found by sequential application of the least-squares (LS) techniques which can be considered as an alternative to the EM algorithm.

Suppose that $\boldsymbol{\theta}^{(s)} = (\mathbf{S}^{(s)}, \boldsymbol{\mu}^{(s)})$ is the initial value of the parameter vector. Then the next approximation can be found by minimization of

$$\chi_{c+1}^2 = \sum_i [\nu_i - NP_i(\boldsymbol{\theta})]^2 / NP_i(\boldsymbol{\theta}^{(s)})$$

Thus, starting with some initial value $\boldsymbol{\theta}^{(0)}$, we find recursively $\boldsymbol{\theta}^{(1)}$, $\boldsymbol{\theta}^{(2)}$, and so on.

Further simplification may be achieved if we use some specifics of the distribution (3.4.7). If $\boldsymbol{\mu}$ is fixed, then the minimization of χ_{c+1}^2 is a linear LS problem and the elements of the matrix **S** can be found by solving a system of linear equations or a linear matrix equation. [47] Suppose that $\boldsymbol{\theta}^{(c)}$ is the parameter approximation. The next approximation of the parameter $\mathbf{S}^{(c+1)}$ can be found in two steps. First, we find the next approximation of matrix **S** by minimizing

$$\chi_{c+1}^2 = \sum_i [\nu_i - N\mathbf{S}b_i(\boldsymbol{\mu}^{(c)})]^2 / NP_i(\boldsymbol{\theta}^{(c)})$$

The necessary condition of minimum

$$\frac{\partial \chi_{c+1}^2}{\partial S_i} = 0$$

gives the following linear equation

$$\mathbf{SA} = \mathbf{c} \tag{3.4.10}$$

where

$$\mathbf{A} = \left[\sum_k b_{ki}(\boldsymbol{\mu}^{(c)}) b_{kj}(\boldsymbol{\mu}^{(c)}) / NP_k(\boldsymbol{\theta}^{(c)}) \right] \qquad \mathbf{c} = \left[\sum_k \nu_k b_{kj}(\boldsymbol{\mu}^{(c)}) / NP_k(\boldsymbol{\theta}^{(c)}) \right]$$

This system is usually called the *normal equation.* [47] Thus, the optimal **S** satisfies some normal equation.

Second, we find the optimal $\boldsymbol{\mu}$ minimizing

$$\chi_{c+1}^2 = \sum_i [\nu_i - N\mathbf{S}^{(c+1)} b_i(\boldsymbol{\mu})]^2 / NP_i(\boldsymbol{\theta}^{(c)})$$

with respect to $\boldsymbol{\mu}$. Then, the associated value **S** is found from the new normal equation and so on until a minimum for χ^2 is found. This method can be used effectively when the number of parameters is small. However, if the number of parameters is large or unknown, the computations become quite complex. Since direct χ^2 minimization is complex, it is very important to find some quasi-optimal methods of parameter estimation. These methods are based on various linearization techniques of the matrix-geometric function of the type in Eq. (3.4.7).

3.4.4 Piecewise Logarithmic Transformation

If we assume that the distribution $p(x)$ has form (3.4.8)

$$p(x) = \sum_{i=1}^{m-1} a_i(1-q_i)q_i^{x-1} + a_{m,x-1}$$

then $p(x) \approx a_1(1-q_1)q_1^{x-1}$ for $q_1 = \max(q_i)$ and large x. This means that

$$\ln p(x) \approx \ln [a_1(1-q_1)] + (x-1) \ln q_1$$

can be approximated with a straight line. After we approximate $\ln p(x)$ for large x with a linear function $k(x-1)+b$, the distribution parameters are found from obvious equations: $\ln q_1 = k$ and $\ln[a_1(1-q_1)] = b$. Therefore, $a_1 = e^b/(1-e^k)$ and $q_1 = e^k$.

Consider now the function $p^{(1)}(x) = p(x) - a_1(1-q_1)q_1^{x-1}$. Using a similar approach, we can extract the next term, $a_2(1-q_2)q_2^{x-1}$ (if $q_2 = \max(q_i)$, $i \neq 1$), and so on. Finally, we estimate $a_{m,x-1}$ by the differences between the approximation and the empirical frequencies for small x.

This method may be used in conjunction with the method of moments which gives additional equations for the parameters

$$\sum_x p(x) = \sum_{i=1}^{m-1} a_i + \sum_x a_{m,x-1} = 1$$

$$\bar{x} = \sum_x xp(x) = \sum_{i=1}^{m-1} a_i/(1-q_i) + \sum_x xa_{m,x-1}$$

The most basic deficiency of this method is the arbitrariness in selecting the distribution form: we can approximate by polygeometric distributions, by polynomials, by arbitrary sequences $a_{m,x-1}$, or by any combination of these. Naturally, in this situation one should choose the approximation which contains the least number of parameters. We also note an instability of the χ^2 minimization in estimating distribution (3.4.7) parameters. This especially concerns the parameter q_1 (small variation of the parameter significantly changes the magnitude of χ^2). We will show a different way to estimate distribution (3.4.7) parameters which in a certain sense is devoid of these deficiencies.

3.4.5 Prony's Method

The other method is based on the fact that the distribution $p(x)$ satisfies a *linear* difference Eq. (see Sec. 1.3.3). Therefore, if we can estimate the coefficients of this equation, then the distribution $p(x)$ may be found as a solution of the difference equation.

In order to find the correspondence between parameters of the distribution and its linear equation, we use the z-transform

$$\rho(z) = \sum_{x=1}^{\infty} p(x) z^{-x} \tag{3.4.11}$$

which, according to (3.4.1), is equal to

$$\rho(z) = \mathbf{S}(\mathbf{I}z - \mathbf{M})^{-1}\mathbf{F}$$

If we replace the inverse matrix in this equation with $(\mathbf{I}z - \mathbf{M})^{-1} = \mathbf{B}(z) / \Delta(z)$, where $\mathbf{B}(z)$ is the adjoint matrix and $\Delta(z)$ is the characteristic polynomial of the matrix $\mathbf{M}$ (see Appendix 5), then $\rho(z)$ may be represented by the rational function

$$\rho(z) = u(z) / \Delta(z) \tag{3.4.12}$$

where

$$u(z) = \mathbf{SB}(z)\mathbf{F} = \sum_{k=1}^{c} u_k z^k \qquad c < r$$

According to Eq. (3.4.2), the characteristic polynomial $\Delta(z) = \det(\mathbf{I}z - \mathbf{M})$ has the form

$$\Delta(z) = \prod_{i=1}^{m} (z - \mu_i)^{s_i} = z^r + \sum_{i=1}^{r} \alpha_i z^{r-i} \tag{3.4.13}$$

Now the problem of estimating the distribution $p(x)$ parameters can be stated in terms of its z-transform: Given the estimates $\hat{p}(x)$, find an approximation of the distribution z-transform with rational function (3.4.12). The estimates of μ_i are found from characteristic Eq. (3.4.13), and the elements of the matrix $\mathbf{S}$ can be found by decomposition of the rational function into partial fractions. Indeed, let

$$\rho(z) = \sum_{i=1}^{m} \sum_{k=0}^{s_i - 1} c_{ik} (z - \mu_i)^{-k-1}$$

be the decomposition of the rational function $\rho(z)$ into partial fractions. Using an easily verifiable formula (see Appendix 5)

$$(z - \mu)^{-k-1} = \sum_{x=1}^{\infty} \binom{x-1}{k} \mu^{x-k-1} z^{-x} \tag{3.4.14}$$

and Eq. (3.4.11), we obtain the equation

$$p(x) = \sum_{i=1}^{m} \sum_{k=0}^{s_i - 1} \binom{x-1}{k} \mu_i^{x-k-1} c_{ik}$$

which coincides with Eq. (3.4.4). This result proves that Eq. (3.4.4) always has a unique solution which can be found by the distribution z-transform decomposition into partial fractions. (The direct proof can be obtained by calculating the system determinant, which is the generalized Vandermonde's determinant.) The matrix $\mathbf{S}$ is found from Eq. (3.4.6), where $\mathbf{F} = \mathbf{1}$, so that we can construct distribution (3.4.1) from its z-transform, which is a rational function.

The problem of finding rational function approximations for a given analytical function has been intensively studied in the mathematical literature. [23 31 38 42 59]

To obtain the difference equation for $p(x)$, we rewrite Eq. (3.4.12) in the following form:

$$\Delta(z)\rho(z) = U(z) \quad degU(z) < r \tag{3.4.15}$$

Comparing the coefficients of the same powers of z on both sides of this equation, we obtain

$$p(x) + \sum_{i=1}^{r} \alpha_i p(x-i) = u_x \tag{3.4.16}$$

where $u_x = 0$ for $x > r$ and $p(x) = 0$ for $x < 1$.

From the mathematical point of view, Eq. (3.4.4) represents the solution of the linear Eq. (3.4.16), which satisfies the zero initial conditions. We can also treat the parameters of the system (3.4.16) as an alternative set of parameters that describe Eq. (3.4.1). As is well known in the theory of linear systems, we can treat Eq. (3.4.1) as a solution of the homogeneous system ($u_x = 0$), which satisfies nonzero initial conditions. Therefore, the initial conditions $p(1), p(2), ..., p(r)$ can replace the parameters $u_1, u_2, ..., u_r$ in describing distribution (3.4.1). The initial r equations of the system (3.4.16) represent the relationship between these two sets of parameters.

If one can estimate the coefficients of this equation and the initial conditions, then distribution (3.4.4) can be found as the unique solution of this equation. The estimates of μ_i are found as the roots of the equation characteristic polynomial (3.4.13). The coefficients c_{ik} of the polynomials can be found from the first r equations of the system of linear Eq. (3.4.4), in which μ_i and $p(x)$ have been replaced by their estimates. In the particular case where the distribution is polygeometric, this method was introduced by Prony [45] in 1795.

The relationship between the system (3.4.16) and the rational function (3.4.13) is well known in the theory of linear systems. The rational function (3.4.12) is interpreted as a linear filter transfer function, the system (3.4.16) describes the filter operation in the time domain, and the sequence $p(x)$ represents the signal on the filter output. The problem of estimating the distribution parameters is equivalent to the linear system identification.

There are many methods for estimating Eq. (3.4.16) parameters. [28,35,36] The simplest method, suggested by Prony, [45] is to replace $p(x)$ with their estimates $\hat{p}(x) = \nu_x / N$ and then treat Eq. (3.4.16) as a system of linear equations with $r+c$ unknowns $(\alpha_1, \alpha_2, ..., \alpha_r, u_1, ..., u_c)$. The unknowns α_i can be found independently of u_i if we consider the part of the system (3.4.16) for $x > c$:

$$\hat{p}(x) + \sum_{i=1}^{r} \alpha_i \hat{p}(x-i) = 0 \tag{3.4.17}$$

If this system has a solution, then the unknowns u_i can be found directly from the part of the system (3.4.16) for $x \le c$. System (3.4.17) is called a *Toeplitz system*.

There are many efficient methods for solving Toeplitz systems. These methods have been developed independently in the following theories: Padé approximations, [42,31,38] orthogonal polynomials, [26,54] algebraic codes, [5,9,37,52] and linear prediction and spectrum analysis. [9,19,28,32,35,53] Normally, the solution of a set of r equations would require $O(r^3)$ operations. However, the special structure of the system (3.4.17) allows us to reduce the number of operations. Levinson, [32] Durbin, [19] and Berlekamp [5] developed methods that require $O(r^2)$ operations. The method based on Euclid's algorithm [10,52,53] requires only $O(r\log^2 r)$ operations. The main feature of these algorithms is that they are *order recursive*. This means that if we find the best fit for the linear system (3.4.16) of an order r and an approximation criterion is not satisfied, we can increase the system's order and calculate new parameters by a simple modification of the previous parameters without the necessity of solving the system from scratch.

Example 3.4.1. To illustrate Prony's method, we first assume that the equation order is

$r=1$. Equation (3.4.16) parameter estimate is found from the first equation of (3.4.17)

$$\hat{\alpha}_1 = -\hat{p}(2)/\hat{p}(1)$$

The characteristic equation

$$\Delta(z) = z+\hat{\alpha}_1 = 0$$

root is estimated by $\hat{\mu}_1 = -\hat{\alpha}_1 = \hat{p}(2)/\hat{p}(1)$. The matrix $\mathbf{S}=S_1$ is found from Eq. (3.4.4)

$$\hat{p}(1) = S_1\hat{\mu}_1^0$$

so that $\hat{S}_1 = \hat{p}(1)$. If the null hypothesis that the equation has the order $r=1$ is true, then the remaining probabilities are approximated by

$$\hat{q}_i = \hat{S}_1\hat{\mu}_1^{i-1} = \hat{p}(1)(\hat{p}(2)/\hat{p}(1))^{i-1}$$

Note that, in this case, we are estimating parameters of the geometric distribution (3.1.5). The estimate that we found is different from the ML estimate given by Eq. (3.1.6). For the grouped data, the χ^2 criterion is given by

$$\chi^2 = N\sum_{i=1}^{t} [\hat{P}(i)-\hat{Q}(i)]^2/\hat{Q}(i) \tag{3.4.18}$$

where

$$\hat{P}(i) = \sum_{j=x_{i-1}}^{x_i-1} \hat{p}(j) \qquad \hat{Q}(i) = \sum_{j=x_{i-1}}^{x_i-1} \hat{q}(j)$$

If the null hypothesis is rejected by this criterion, we try to test the hypothesis that the difference equation has the order $r=2$.

The equation parameters are estimated from the first two equations of system (3.4.17):

$$\hat{p}(3) = \alpha_1\hat{p}(2) + \alpha_2\hat{p}(1)$$

$$\hat{p}(4) = \alpha_1\hat{p}(3) + \alpha_2\hat{p}(2)$$

Solving this equation, we obtain

$$\hat{\alpha}_1 = \frac{\hat{p}(3)\hat{p}(2) - \hat{p}(4)\hat{p}(1)}{\hat{p}(2)\hat{p}(2) - \hat{p}(3)\hat{p}(1)} \qquad \hat{\alpha}_2 = \frac{\hat{p}(4)\hat{p}(2) - \hat{p}(3)\hat{p}(3)}{\hat{p}(2)\hat{p}(2) - \hat{p}(3)\hat{p}(1)}$$

The characteristic polynomial is estimated by

$$\Delta(z) = z^2 + \hat{\alpha}_1 z + \hat{\alpha}_2$$

Suppose that this equation has different roots $\hat{\mu}_1$ and $\hat{\mu}_2$. Then we can select $\mathbf{F} = \mathbf{1}$, and the elements of the matrix $\mathbf{S} = [S_1 \;\; S_2]$ can be estimated from the first equations of (3.4.4),

$$\hat{p}(1) = S_1 + S_2$$
$$\hat{p}(2) = S_1\hat{\mu}_1 + S_2\hat{\mu}_2$$

which have the following solution:

$$\hat{S}_1 = \frac{\hat{p}(1)\hat{\mu}_2-\hat{p}(2)}{\hat{\mu}_2-\hat{\mu}_1} \qquad \hat{S}_2 = \frac{\hat{p}(2)-\hat{p}(1)\hat{\mu}_1}{\hat{\mu}_2-\hat{\mu}_1}$$

Next, we calculate the approximation of the remaining probabilities using Eq. (3.4.4):

$$q_i = \hat{S}_1\hat{\mu}_1^{i-1} + \hat{S}_2\hat{\mu}_2^{i-1}$$

If criterion (3.4.18) rejects the hypothesis, we try to fit an order $r=3$ equation, and so on. Finally, if an order r equation agrees with the experimental data, we refine the approximation, using the minimum χ^2 method.

Prony's method of estimating is analogous to decoding Reed-Solomon code. [60] The eigenvalues $\boldsymbol{\mu}$ serve as error locators, while $\mathbf{S}$ gives the errors values. This method approximates well the initial part of the distribution (which is good in the case of error correction, since a small number of errors is most probable), but it performs poorly in approximating the distribution tail. To improve the approximation agreement with the experimental data, it is better to estimate the matrix $\mathbf{S}$ using the normal Eq. (3.4.10) rather than Eq. (3.4.4). Initially, we replace $NP_i(\boldsymbol{\theta}^{(0)})$ in Eq. (3.4.10) with ν_i.

3.4.6 Matrix Echelon Parametrization

The parameters μ_i of distribution (3.4.4) are found as the roots of the characteristic polynomial. There is no need to find these roots if we use the alternative matrix-echelon parametrization described in Sec. 1.3.4. In this parametrization, the distribution (3.4.1) can be expressed explicitly through the coefficients $\boldsymbol{\alpha} = (\alpha_1, \alpha_2, ..., \alpha_r)$ of equation (3.4.16) and the initial conditions $\mathbf{S} = [\ \hat{p}(1)\ \ \hat{p}(2)\ \ ...\ \ \hat{p}(r)\ ]$:

$$p(x) = \mathbf{SM}^{x-1}\mathbf{F}$$

where $\mathbf{F} = [\ 1\ \ 0\ \ 0\ \ \cdots\ \ 0\ ]'$ and

$$\mathbf{M} = \begin{bmatrix} 0 & 0 & ... & 0 & -\alpha_r \\ 1 & 0 & ... & 0 & -\alpha_{r-1} \\ . & . & . & . & . \\ 0 & 0 & ... & 1 & -\alpha_1 \end{bmatrix}$$

Indeed, it is obvious that $\mathbf{SF} = p(1)$. Since $\mathbf{MF} = [\ 0\ \ 1\ \ 0\ \ \cdots\ \ 0\ ]'$, we obtain $\mathbf{SMF} = p(2)$. Analogously, we prove that Eq. (3.4.1) is valid for all $x \le r$.

For $x = r+1$, the product in the right-hand side of Eq. (3.4.1) can be expressed as

$$\mathbf{SM}^r\mathbf{F} = \mathbf{SMM}^{r-1}\mathbf{F} = \mathbf{SM}\ [\ 0\ \ 0\ \ 0\ \ \cdots\ \ 0\ \ 1]'$$

Since

$$\mathbf{SM} = [\ p(2)\ \ p(3)\ \ \cdots\ \ p(r)\ \ \ -\sum_{i=1}^{r}\alpha_i p(r+1-i)\ ]$$

we obtain

$$\mathbf{SM}^r\mathbf{F} = \mathbf{SM}\ [\ 0\ \ 0\ \ 0\ \ \cdots\ \ 0\ \ 1]' = -\sum_{i=1}^{r}\alpha_i p(r+1-i) = p(r+1)$$

so that Eq. (3.4.1) is satisfied for $x = r+1$. Repeating these transformations for $x = r+2,\ r+3, ...$, we prove that the previously defined matrices satisfy Eq. (3.4.1).

The ML estimates of the alternative set of parameters can be found from Eq. (3.4.9), where $\boldsymbol{\mu}$ is replaced with $\boldsymbol{\alpha}$. This system structure is more complex than Eq. (3.4.9) because it is difficult to express powers of matrices using the matrix-echelon form. Similar problems arise when we use the minimum χ^2 method. However, if we do not group the experimental data and use the sequential LS method, the nonlinear problem can be reduced to a sequence of linear ones.

Indeed, in this case, $\boldsymbol{\alpha}^{(k+1)}$ and $\mathbf{S}^{(k+1)}$ approximations can be found by minimizing the following function with respect to $\alpha_1, \alpha_2, ..., \alpha_r, u_1, ..., u_c$:

$$\chi^2_{k+1} = \sum_x [\ \hat{p}(x) + \sum_{i=1}^{r}\alpha_i \hat{p}(x-i) - u_x\ ]^2 w_x(k)$$

where $w_x(k) = 1 / Np(x; \mathbf{S}^{(k)}, \boldsymbol{\alpha}^{(k)})$ and $p(x; \mathbf{S}^{(k)}, \boldsymbol{\alpha}^{(k)})$ is the probability approximation found after k-th iteration. Since χ^2_{k+1} is a quadratic function of α_i and u_x, the problem of finding $\boldsymbol{\alpha}^{(k+1)}$ is an LS problem and the parameter value satisfies the following normal equation:

$$\boldsymbol{\alpha}\mathbf{B} = \mathbf{c} \tag{3.4.19}$$

where

$$\mathbf{B} = \left[\sum_x w_x \hat{p}(x-i)\hat{p}(x-j)\right]_{r,r} \qquad \mathbf{c} = -\left[\sum_x w_x \hat{p}(x)\hat{p}(x-j)\right]_{1,r}$$

The parameter $\mathbf{u}^{(k+1)}$ satisfies Eq. (3.4.16), which becomes

$$\hat{u}_x^{(k+1)} = \hat{p}(x) + \sum_{i=1}^{r} \alpha_i^{(k+1)} \hat{p}(x-i) \qquad x = 1,2,\ldots,r$$

It follows from these equations that the initial conditions $\hat{p}(1), \hat{p}(2), \ldots, \hat{p}(r)$ represent the LS estimate of the parameter $\mathbf{S}^{(k+1)}$.

Generally, $\mathbf{B}$ is not a Toeplitz matrix, however, if the weights w_x are constant, this matrix becomes a symmetrical Toeplitz matrix. This symmetry property is important in solving system (3.4.19) because, in this case, we can apply the classical Levinson-Durbin [19,32] algorithm.

When using the linear system (3.4.16) for the distribution estimation, it is necessary to have the experimental data $\hat{p}(x)$ at the adjacent points ($x = 1,2,3,\ldots$). This is not normally available. At the same time, Prony's method is very sensitive to the initial part of the distribution, which leads to overestimating the equation order. We discuss here several methods of data compression to overcome this deficiency.

3.4.7 Utilization of the Cumulative Distribution Function

If $p(x)$ satisfies a difference equation, the gap function $p(mx+1)$ with fixed m also satisfies a difference equation. To find this equation, we first consider the gap function z-transform:

$$r_m(z) = \sum_{x=0}^{\infty} p(mx+1) z^{-x-1}$$

which, according to Eq. (3.4.7), is equal to

$$r_m(z) = \mathbf{S}(\mathbf{I}z - \mathbf{M}^m)^{-1}\mathbf{F}$$

It follows from this equation that the gap sequence $p(mx+1)$ characteristic polynomial is $\det(\mathbf{I}z - \mathbf{M}^m)$ and the difference equation for the gap sequence can be obtained by the characteristic polynomial decomposition similar to Eq. (3.4.13). The roots of this characteristic polynomial are equal to μ_i^m, and, since

$$\lim_{m\to\infty} \mu_i^m = 0$$

for large m, only the terms of the distribution that correspond to the largest $q_i = |\mu_i|$ should be taken into account. Therefore, for large m, the gap sequence can be approximated by a function of form (3.4.7) with a smaller number of parameters than the original distribution

$p(x)$. Then we can extract the component that corresponds to these parameters from the original distribution, thereby, reducing the number of parameters to be estimated.

One of the problems with this approach is that the there may not be enough data for estimating the gap function, especially for large m. More reliable estimates can be found if we make use of the fact that the cumulative interval distribution and $p(x)$ satisfy the same difference equation. Indeed, the cumulative distribution $P(l) = Pr(x \geq l)$, according to Eq. (3.4.7), has the form

$$P(l) = \sum_{i=l}^{\infty} p(x) = \mathbf{S}\mathbf{M}^{l-1}(\mathbf{I}-\mathbf{M})^{-1}\mathbf{F}$$

It has the same characteristic polynomial $\Delta(z) = \det(\mathbf{I}z - \mathbf{M})$ as the sequence $\{p(x)\}$, which means that they satisfy the same difference equation. If we find parameters of the distribution tails $P(l)$, then the parameters of the distribution $p(x)$ are found from the identity $p(x) = P(x) - P(x+1)$.

Now we can outline a method for estimating the distribution parameters, which is similar to the piecewise logarithmic transformation. First, we find the approximation of the gap sequence of the cumulative distribution $P(ml+1)$ for large m, using Prony's method or piecewise logarithmic transformation by a function of type (3.4.7):

$$Q_1(ml+1) = \mathbf{B}_1 \mathbf{C}_1^l \mathbf{D}_1$$

The natural interpolation of this function is given by

$$Q_1(l) = \mathbf{B}_1 \mathbf{A}_1^{l-1} \mathbf{D}_1$$

where $\mathbf{A}_1^m = \mathbf{C}_1$. [24] The eigenvalues of the matrix $\mathbf{A}_1$ are equal to the order m roots of the eigenvalues of the matrix $\mathbf{C}_1$. Next, we subtract this component from the original cumulative distribution to obtain the residual function $P^{(1)}(l) = P(l) - Q_1(l)$. Repeating similar transformations with respect to this residual function using a smaller gap, we extract from the original distribution the components that correspond to intermediate length intervals, and so on. In the end, we obtain the cumulative function approximation in the form

$$P(l) \approx \sum_i \mathbf{B}_i \mathbf{A}_i^{l-1} \mathbf{D}_i$$

This may be rewritten as

$$P(l) \approx \hat{\mathbf{S}}\hat{\mathbf{M}}^{l-1}\hat{\mathbf{F}}_i$$

where

$$\hat{\mathbf{S}} = block\ row\{\mathbf{B}_i\} \quad \hat{\mathbf{M}} = block\ diag\{\mathbf{A}_i\} \quad \hat{\mathbf{F}} = block\ col\ \{(\mathbf{I}-\mathbf{A}_i)^{-1}\mathbf{D}_i\}$$

If the criterion χ^2 for this approximation is not small enough, we may use the estimated parameter vector as an initial vector for a more powerful but complex method of χ^2 minimization.

In practice it is convenient to select a sequence of gaps that grow exponentially. For example, from 1 to 10 use gap $m = 1$ (that is, consider $P(1), P(2), \ldots, P(10)$); from 10 to 100 use gap $m = 10$ (that is, consider $P(10), P(20), \ldots, P(100)$); from 100 to 1000 use gap $m = 100$, and so on. Then start the distribution parameter estimation from the interval with the largest gap. Often the estimates from the adjacent intervals with different gaps do not interfere with each other.

3.4.8 Method of Moments

Let us consider another method for data compression, which also leads to a Toeplitz system for the distribution parameter estimation. Because of the matrix-geometric structure of distribution (3.4.7), its moments also have such a structure and therefore satisfy some Toeplitz difference equations. Since in moment calculation we utilize all the available data, this lets us overcome certain ambiguity of gap selection in the previous method. The other advantage of using moments is the possibility of calculating them sequentially so that we can use the efficient methods of recursive estimation discussed previously.

Consider first the negative binomial moments

$$\nu(m) = \sum_x \binom{m+x-1}{m} p(x)$$

which, according to Eq. (2.2.17), can be determined by differentiation of the distribution z-transform:

$$\nu(m) = \frac{(-1)^m}{m!}\rho^{(m)}(1) = \mathbf{S}(\mathbf{I}-\mathbf{M})^{-m-1}\mathbf{F}$$

It follows from this equation that the negative binomial moments of the interval distribution satisfy a linear difference equation. Indeed, the z-transform of the sequence $\{\nu(m)\}$

$$\zeta(z) = \sum_{m=0}^{\infty} \nu(m) z^{-m} = \mathbf{S}(\mathbf{I}(1-z^{-1}) - \mathbf{M})^{-1}\mathbf{F} \qquad (3.4.20)$$

is a rational function and therefore the sequence $\{\nu(x)\}$ satisfies the difference equation

$$\nu(x) + \sum_{i=1}^{r} \beta_i \nu(x-i) = 0 \quad \text{for } x > r$$

whose coefficients are determined by the z-transform $\zeta(z)$ poles.

If we compare Eq. (3.4.20) with Eq. (3.4.11), we can observe that $\zeta(z) = \rho(1-z^{-1})$, so that the the function $\zeta(z)$ denominator is equal to $z^r\Delta(1-z^{-1})$, where $\Delta(z)$ is defined by Eq. (3.4.13). Because of this relationship, the coefficients of the difference equation for $\{\nu(x)\}$ may be expressed as linear combinations of the coefficients of Eq. (3.4.17).

To find the relations between the coefficients, let us decompose the denominator $z^r\Delta(1-z^{-1})$ of the z-transform $\zeta(z)$ into a power sum. After simple algebraic transformations, we obtain

$$z^r\Delta(1-z^{-1}) = \sum_{m=0}^{r} z^m(-1)^{r-m} \sum_{i=0}^{m} \binom{r-i}{r-m} \alpha_i$$

so that

$$\beta_m = (-1)^m \sum_{i=0}^{r-m} \binom{r-i}{m} \alpha_i$$

where $\alpha_0 = 1$.

The inverse equations

$$\alpha_i = (-1)^{r-i} \sum_{m=r-i}^{r} \binom{m}{r-i} \beta_m \qquad (i=0,1,\ldots,m)$$

can be found quite analogously or by direct inversion of the previous equations.

Thus, if we can estimate coefficients of the difference equation for the sequence of negative binomial moments $\{\nu(x)\}$, then we can find the coefficients of the difference equation of the distribution $\{p(x)\}$ and vice versa.

The roots of the moment sequence characteristic polynomial

$$z^r \Delta(1-z^{-1}) = \det [\, \mathbf{I}(z-1) - z\mathbf{M} \,]$$

have the form $1/(1-\mu_i)$, so that the model parameters μ_i can be calculated directly from the negative binomial moment characteristic equation.

Similar results may be obtained by using binomial moments

$$\mathbf{S}(\mathbf{I}-\mathbf{M})^{-n-1}\mathbf{M}^n\mathbf{F} = \sum_x \binom{x-n}{n} p(x)$$

whose characteristic equation roots have the form $(2-\mu_i)/(1-\mu_i)$.

3.4.9 Transform Domain Approximation

Another method with similar merits for estimating the parameters of distributions (3.4.7) is the method of the distribution z-transform approximation. Since interval distributions take the form of Eq. (3.4.7), if and only if their z-transforms are rational fractions (with real coefficients), to estimate their parameters one can approximate the z-transforms of the empirical distribution $p(x)$ with rational fractions.

An important role is played here by the choice of the approximation error measure. Because the χ^2 criterion is more sensitive to the divergence of the tails of distributions, one should choose a measure of proximity of z-transforms which would have this property. In this sense, the arithmetic mean of deviations of z-transforms on the circle of radius $r \ll 1$ has the necessary property. Indeed, let $\rho(z)$ and $\hat{\rho}(z)$ be the z-transforms of the distribution $p(x)$ and its empirical $\hat{p}(x)$ approximation, respectively. If γ is a circle of radius r such that all the poles of the z-transforms are inside the circle, then Cauchy's formula gives

$$p(x)-\hat{p}(x) = \frac{1}{2\pi j} \oint_\gamma [\rho(z)-\hat{\rho}(z)] z^{x-1} dz$$

This leads to the inequality

$$|p(x)-\hat{p}(x)| \le \varepsilon r^x$$

where

$$\varepsilon = \frac{1}{2\pi r} \oint_\gamma |\rho(z)-\hat{\rho}(z)| |dz|$$

This ensures the necessary property of estimations.

Questions of approximating functions with rational fractions in a complex plane are studied intensively in mathematical literature. [59] However, there are several problems in constructing the empirical z-transform $\hat{\rho}(z)$ because the interval distribution $\hat{p}(x)$ is not available for all possible values of its argument. Therefore, one must produce one or another interpolation and extrapolation, since empirical frequencies usually come in a grouped form. It is possible to use geometric interpolation $\hat{p}(x) = a_i q_i^{x-1}$, where $a_i = a_{i-1} q_{i-1}^{x_{i-1}}$ for $x_{i-1} < x \leq x_i$ and q_i is found from the condition

$$\hat{p}(x_{i-1} < x \leq x_i) = \frac{a_i}{1-q_i}(q_i^{x_{i-1}} - q_i^{x_i})$$

The other alternative is to produce smoothing with the help of geometric progression on longer intervals, taking approximately equal parameters of geometric progression for several adjacent intervals, which is especially important in extrapolation of distribution tails. It is also possible to use a different method of interpolation, specifically the spline method, which in a certain sense generalizes the previous method.

3.5 MATRIX PROCESS PARAMETER ESTIMATION

3.5.1 ML Parameter Estimation

Let $e_1, e_2, \ldots, e_n$ be a sequence of errors in a binary channel. We want to test a hypothesis H_0 that the source of errors in this channel can be described as a matrix process $\{\mathbf{Y}(\boldsymbol{\tau}), \mathbf{M}(0,\boldsymbol{\tau}), \mathbf{M}(1,\boldsymbol{\tau}), \mathbf{Z}(\boldsymbol{\tau})\}$ whose matrices depend on k independent parameters $\boldsymbol{\tau} = (\tau_1, \tau_2, \ldots, \tau_k)$. Within this hypothesis the likelihood function is given by

$$L = \mathbf{Y}(\boldsymbol{\tau}) \prod_{i=1}^{n} \mathbf{M}(e_i, \boldsymbol{\tau}) \mathbf{Z}(\boldsymbol{\tau}) \tag{3.5.1}$$

The ML parameter estimates can be found as a solution of the system

$$\frac{\partial L}{\partial \tau_i} = 0 \qquad i = 1, 2, \ldots, k$$

However, because of the matrix noncommutativity, the derivative in the left-hand side of this equation has a very complex form:

$$\frac{\partial L}{\partial \tau_i} = \frac{\partial \mathbf{Y}(\boldsymbol{\tau})}{\partial \tau_i} \prod_{j=1}^{n} \mathbf{M}(e_j, \boldsymbol{\tau}) \mathbf{Z}(\boldsymbol{\tau})$$

$$+ \sum_{m=1}^{n} \prod_{j=1}^{m-1} \mathbf{M}(e_j, \boldsymbol{\tau}) \frac{\partial \mathbf{M}(e_m, \boldsymbol{\tau})}{\partial \tau_i} \prod_{j=m+1}^{n} \mathbf{M}(e_j, \boldsymbol{\tau}) \mathbf{Z}(\boldsymbol{\tau})$$

$$+ \mathbf{Y}(\boldsymbol{\tau}) \prod_{i=1}^{n} \mathbf{M}(e_i, \boldsymbol{\tau}) \frac{\partial \mathbf{Z}(\boldsymbol{\tau})}{\partial \tau_i}$$

It is difficult to find the ML parameter estimates using these equations when n is large. Similar difficulties arise if we use the information criterion to test the null hypothesis. Therefore, we must find alternative ways of estimating parameters and testing H_0.

One of the methods consists of using the process basis sequences (see Appendix 1). If the process has finite rank, there exist so-called basis sequences of errors. The probability of any sequence of errors can be uniquely expressed as a linear combination of the probabilities of these basis sequences (see Appendix 1). The description of the method of estimating the process parameters is given in Appendix 3.

3.5.2 Interval Distribution Estimation

In practice, it is convenient to use methods of estimation of parameters of the matrix process by interval distributions because it is possible to use known methods of mathematical statistics. We will see that the two-dimensional interval distributions enable us to estimate all the model parameters.

Consider first the distribution $p_0(\lambda)$ of intervals between errors and the distribution $p_1(l)$ of lengths of series of consecutive errors. It follows from Eq. (1.3.11) and (1.3.12) that these distributions have the following form:

$$p_0(\lambda) = \mathbf{Y}_1\mathbf{M}_{10}\mathbf{M}_{00}^{\lambda-1}\mathbf{M}_{01}\mathbf{Z}_1 / \mathbf{Y}_1\mathbf{M}_{10}\mathbf{Z}_0 \tag{3.5.2}$$

$$p_1(l) = \mathbf{Y}_0\mathbf{M}_{01}\mathbf{M}_{11}^{l-1}\mathbf{M}_{10}\mathbf{Z}_0 / \mathbf{Y}_0\mathbf{M}_{01}\mathbf{Z}_1 \tag{3.5.3}$$

It is convenient to use the general form for both previous equations

$$p_e(x) = \mathbf{S}_e\mathbf{M}_{ee}^{x-1}\mathbf{F}_e$$

where

$$\mathbf{S}_e = \mathbf{Y}_{1-e}\mathbf{M}_{1-e,e} / \mathbf{Y}_{1-e}\mathbf{M}_{1-e,e}\mathbf{Z}_e \quad \mathbf{F}_e = \mathbf{M}_{e,1-e}\mathbf{Z}_{1-e}$$

This equation has the matrix-geometric form considered in Sec. 3.5.1. Using the methods described in this section, we can estimate parameters $\mathbf{S}_e$, $\mathbf{M}_{ee}$, and $\mathbf{F}_e$. If these distributions do not contradict experimental data, then the hypothesis H_0 is not rejected, since for two given distributions in form Eq. (3.5.2) and (3.5.3) it is always possible to construct a matrix process which has the same distributions of lengths of intervals between errors and lengths of series of errors. Using interval distributions (3.5.2) and (3.5.3), matrices $\mathbf{M}_{00}$ and $\mathbf{M}_{11}$ can be determined in their canonical Jordan form (3.4.2). The estimated sizes of these matrices $\hat{r}_e$ can serve as estimates of ranks r_e of the matrix process. Thus, estimating parameters of distributions (3.5.2) and (3.5.3) using experimental data, we estimate ranks r_e of the process, matrices $\mathbf{M}_{ee}$, $\mathbf{S}_e$, and $\mathbf{F}_e$.

If intervals between consecutive errors and correct symbols are independent, then the matrix process is equivalent to an alternating renewal process and can be constructed using formula (1.4.26), in which all matrices are replaced by their estimates. If the hypothesis of the independence of intervals is rejected, then to estimate the remaining parameters it is necessary to examine some other distributions related to the matrix process.

3.5.3 Utilization of Two-Dimensional Distributions

The parameters that we need to estimate are elements of the matrices $\mathbf{Y}_e$, $\mathbf{Z}_e$, $\mathbf{M}_{01}$, and $\mathbf{M}_{10}$. We can estimate these parameters by solving a system of the method of moments for the probabilities that are independent of the interval distributions used before. When constructing a system of equations, one must, for the sake of simplicity of calculation, try to choose such basis sequences whose probabilities depend on unknown parameters in the most simple way (see Appendix 1). Specifically, by choosing sequences in the form 000...011...11, we obtain linear equations for the elements of the matrix $\mathbf{M}_{01}$; by choosing sequences in the form 11...100...0, we obtain linear equations for the elements of the matrix $\mathbf{M}_{10}$. The two-dimensional distributions of these sequences are given by

$$p_{01}(\lambda,l) = \mathbf{S}_0 \mathbf{M}_{00}^{\lambda-1} \mathbf{M}_{01} \mathbf{M}_{11}^{l-1} \mathbf{F}_1 \quad (3.5.4)$$

$$p_{10}(l,\lambda) = \mathbf{S}_1 \mathbf{M}_{11}^{l-1} \mathbf{M}_{10} \mathbf{M}_{00}^{\lambda-1} \mathbf{F}_0 \quad (3.5.5)$$

These two equations can be rewritten in a common general form:

$$p_{e,1-e}(x,y) = \mathbf{S}_e \mathbf{M}_{ee}^{x-1} \mathbf{M}_{e,1-e} \mathbf{M}_{1-e,1-e}^{y-1} \mathbf{F}_{1-e} \quad (3.5.6)$$

If we replace in these equations the probabilities $p_{e,1-e}(x,y)$ with their estimates for several different values of x and y, we obtain a system of linear equations with unknown elements of the matrix $\mathbf{M}_{e,1-e}$. By solving this system we obtain the matrix estimate. The matrix $\mathbf{M}_{e,1-e}$ has $r_0 \times r_1$ elements, so that we need at least $r_0 \times r_1$ different estimates $\hat{p}_{e,1-e}(x,y)$ to be able to solve this system. If the matrices $\mathbf{M}_{ee}$ in the interval distributions (3.4.7) have the minimal order (or, in other words, r_e are ranks of the matrix process), then the initial $r_0 \times r_1$ equations let us find the unique solution.

It is easy to verify that these initial equations may be written in the following matrix form:

$$\mathbf{G}_e \mathbf{M}_{e,1-e} \mathbf{H}_{1-e} = \mathbf{W}_{e,1-e} \quad (3.5.7)$$

where

$$\mathbf{W}_{e,1-e} = \left[p_{e,1-e}(i,j) \right]_{r_e, r_{1-e}}$$

$$\mathbf{G}_e = block\ col\ \{ \mathbf{S}_e \mathbf{M}_{ee}^{i-1} \} = \left[\mathbf{S}_e \quad \mathbf{S}_e \mathbf{M}_{ee} \quad \cdots \quad \mathbf{S}_e \mathbf{M}_{ee}^{r_e - 1} \right]'$$

and

$$\mathbf{H}_e = block\ row \{ \mathbf{M}_{ee}^{j-1} \mathbf{F}_e \} = \left[\mathbf{F}_e \quad \mathbf{M}_{ee} \mathbf{F}_e \ \ldots \ \mathbf{M}_{ee}^{r_e - 1} \mathbf{F}_e \right]$$

This system has a unique solution, because matrices $\mathbf{G}_e$ and $\mathbf{H}_e$ are nonsingular. We will prove that these matrices are nonsingular by showing that their rows (or columns) are linearly independent.

Indeed, let us assume that vectors $\mathbf{S}_e$, $\mathbf{S}_e \mathbf{M}_{ee}, \ldots, \mathbf{S}_e \mathbf{M}_{ee}^{r_e - 1}$ are linearly dependent. Then there exist numbers $\sigma_1, \sigma_2, \ldots, \sigma_{r_e}$ such that not all of them are equal to zero and

$$\sum_{i=1}^{r_e} \sigma_i \mathbf{S}_e \mathbf{M}_{ee}^{i-1} = 0$$

so that we can express $\mathbf{S}_e \mathbf{M}_{ee}^k$ with $k < r_e$ as a linear combination of the vectors corresponding to powers less than k:

$$\mathbf{S}_e \mathbf{M}_{ee}^k = \sum_{i=1}^{k} \beta_i \mathbf{S}_e \mathbf{M}_{ee}^{i-1}$$

Multiplying this identity by $\mathbf{M}_{ee}^{x-k-1} \mathbf{F}_e$, we obtain a difference equation for distribution (3.4.7)

$$p_e(x) - \sum_{i=1}^{k} \beta_{k-i+1} p_e(x-i) = 0$$

whose order k is less than r_e. This contradicts the assumption that r_e is the minimal order of this Eq. (see Sec. 1.3.3).

Similarly, we can prove that vectors $\mathbf{Z}_e$, $\mathbf{M}_{ee}\mathbf{Z}$, ..., $\mathbf{M}_{ee}^{r_e-1}$ are also linearly independent. This completes the proof of matrix $\mathbf{G}_e$ and $\mathbf{H}_e$ nonsingularity.

The unique solution of system (3.5.7) can be expressed in the following matrix form:

$$\mathbf{M}_{e,1-e} = \mathbf{G}_e^{-1}\mathbf{W}_{e,1-e}\mathbf{H}_e^{-1}$$

We would like to point out that for estimating the matrix $\mathbf{M}_{e,1-e}$ it is possible to use estimates of two-dimensional cumulative distributions. According to Eq. (3.5.6), the probability

$$p_{e,1-e}(x_{i-1} < x \le x_i, y_{j-1} < y \le y_j) = \sum_{x=x_{i-1}+1}^{x_i} \sum_{y=x_{j-1}+1}^{y_j} p_{e,1-e}(x,y)$$

takes the form

$$p_{e,1-e}(x_{i-1} < x \le x_i, y_{j-1} < y \le y_j) = \mathbf{L}_e(x_{i-1},x_i)\mathbf{M}_{e,1-e}\mathbf{R}_e(y_{j-1},y_j)$$

where

$$\mathbf{L}_e(x_{i-1},x_i) = \mathbf{S}_e(\mathbf{I}-\mathbf{M}_{ee})^{-1}(\mathbf{M}_{ee}^{x_{i-1}} - \mathbf{M}_{ee}^{x_i}) \quad \mathbf{R}_e(y_{j-1},y_j) = (\mathbf{I}-\mathbf{M}_{ee})^{-1}(\mathbf{M}_{ee}^{y_{j-1}} - \mathbf{M}_{ee}^{y_j})\mathbf{F}_e$$

This equation is similar to Eq. (3.5.6) and, therefore, $\mathbf{M}_{e,1-e}$ satisfies a system of linear equations similar to Eq. (3.5.7),

$$\mathbf{G}_e^{(1)}\mathbf{M}_{e,1-e}\mathbf{H}_{1-e}^{(1)} = \mathbf{W}_{e,1-e}^{(1)} \tag{3.5.8}$$

where

$$\mathbf{W}_{e,1-e}^{(1)} = \left[p_{e,1-e}(x_{i-1} < x \le x_i, y_{i-1} < y \le y_i)\right]_{r_e,r_{1-e}}$$

$$\mathbf{G}_e^{(1)} = block\ col\,\{\mathbf{L}_e(x_{i-1},x_i)\} \quad \text{and} \quad \mathbf{H}_e^{(1)} = block\ row\{\mathbf{R}_e(y_{j-1},y_j)\}$$

We can also use the distribution tails by replacing upper bounds of the intervals with ∞.

3.5.4 Polygeometric Distributions

Let us examine the important special case where distributions (3.4.7) are polygeometric:

$$p_e(x) = \sum_{i=1}^{r_e} a_{ei}(1-q_{ei})q_{ei}^{x-1} \tag{3.5.9}$$

In this case, matrices $\mathbf{M}_{ee}$ in formulation (3.4.2) are diagonal:

$$\mathbf{M}_{ee} = diag\{q_{ei}\} \tag{3.5.10}$$

Matrices $\mathbf{S}_e$ and $\mathbf{F}_e$ may be expressed as

$$\mathbf{S}_e = row\{a_{ei}(1-q_{ei})\} \quad \mathbf{F}_e = \mathbf{1}$$

Matrices $\mathbf{G}_e$ and $\mathbf{H}_e$ take the form

$$\mathbf{G}_e = \begin{bmatrix} a_{e1}(1-q_{e1}) & a_{e2}(1-q_{e2}) & \cdots & a_{er_e}(1-q_{er_e}) \\ a_{e1}(1-q_{e1})q_{e1} & a_{e2}(1-q_{e2})q_{e2} & \cdots & a_{er_e}(1-q_{er_e})q_{er_e} \\ \cdots & \cdots & \cdots & \cdots \\ a_{e1}(1-q_{e1})q_{e1}^{r_e-1} & a_{e2}(1-q_{e2})q_{e2}^{r_e-1} & \cdots & a_{er_e}(1-q_{er_e})q_{er_e}^{r_e-1} \end{bmatrix}$$

$$\mathbf{H}_e = \left[q_{ei}^{j-1} \right]_{r_e,r_e} = \begin{bmatrix} 1 & q_{e1} & q_{e1}^2 & \cdots & q_{e1}^{r_e-1} \\ 1 & q_{e2} & q_{e2}^2 & \cdots & q_{e2}^{r_e-1} \\ \cdots & \cdots & \cdots & \cdots & \cdots \\ 1 & q_{er_e} & q_{er_e}^2 & \cdots & q_{er_e}^{r_e-1} \end{bmatrix}$$

Thus, $\mathbf{H}_e$ becomes Vandermonde's matrix and $\mathbf{G}_e = [diag\{a_{ei}(1-q_{ei})\}\mathbf{H}_e]'$.

Matrices $\mathbf{G}_e^{(1)}$ and $\mathbf{H}_e^{(1)}$ can be found quite analogously. In the case of the distributions tails $[p_{e,1-e}(x_{i-1} < x, y_{j-1} < y)]$, they are given by

$$\mathbf{G}_e^{(1)} = \left[a_{ej} q_{ej}^{x_i-1} \right]_{r_e,r_e} \qquad \mathbf{H}_e^{(1)} = \left[q_{ei}^{y_j-1} \right]_{r_e,r_e}$$

3.5.5 Error Bursts

In practice, the following situation is typical: Very long intervals between groups in which error density is relatively high are encountered but series of consecutive errors are rare. Therefore, the procedure described here does not provide a good estimate for the matrix $\mathbf{M}_{11}$. To overcome this deficiency, it is convenient to combine the groups with high error density into a single formation—a burst of errors. Single errors (or errors of small multiplicity) belong to the intervals between bursts. Let us denote the resultant process as $\{g_t\}$ ($g_t = 1$, if the symbol at the moment t belongs to the burst, and $g_t = 0$ if the symbol at the moment t belongs to the interval between bursts). We assume that errors within the intervals and bursts are independent and occur with the constant conditional probabilities ε_0 and ε_1, respectively. In addition, let us assume that the process $\{g_t\}$ is a matrix process $\{\mathbf{Y}^{(1)}, \mathbf{M}_0^{(1)}, \mathbf{M}_1^{(1)}, \mathbf{Z}^{(1)}\}$ where $\mathbf{Y}^{(1)} = [\mathbf{Y}_0^{(1)} \;\; \mathbf{Y}_1^{(1)}]$

$$\mathbf{M}_0^{(1)} = \begin{bmatrix} \mathbf{M}_{00}^{(1)} & 0 \\ \mathbf{M}_{10}^{(1)} & 0 \end{bmatrix} \qquad \mathbf{M}_1^{(1)} = \begin{bmatrix} 0 & \mathbf{M}_{01}^{(1)} \\ 0 & \mathbf{M}_{11}^{(1)} \end{bmatrix}$$

In this case, the original process $\{e_t\}$ is a matrix process and can be presented as $\{\mathbf{Y}, \mathbf{M}_0, \mathbf{M}_1, \mathbf{Z}\}$ with

$$\mathbf{Y} = \left[\mathbf{Y}_0^{(1)}(1-\varepsilon_0) \quad \mathbf{Y}_1^{(1)}(1-\varepsilon_1) \quad \mathbf{Y}_0^{(1)}\varepsilon_0 \quad \mathbf{Y}_1^{(1)}\varepsilon_1 \right]$$

$$\mathbf{Z} = \begin{bmatrix} \mathbf{Z}^{(1)} \\ \mathbf{Z}^{(1)} \end{bmatrix} \qquad \mathbf{M}_0 = \begin{bmatrix} \mathbf{M}_{00} & 0 \\ \mathbf{M}_{11} & 0 \end{bmatrix} \qquad \mathbf{M}_1 = \begin{bmatrix} 0 & \mathbf{M}_{11} \\ 0 & \mathbf{M}_{11} \end{bmatrix}$$

$$\mathbf{M}_{00} = \begin{bmatrix} \mathbf{M}_{00}^{(1)}(1-\varepsilon_0) & \mathbf{M}_{01}^{(1)}(1-\varepsilon_1) \\ \mathbf{M}_{10}^{(1)}(1-\varepsilon_0) & \mathbf{M}_{11}^{(1)}(1-\varepsilon_1) \end{bmatrix} \qquad \mathbf{M}_{11} = \begin{bmatrix} \mathbf{M}_{00}^{(1)}\varepsilon_0 & \mathbf{M}_{01}^{(1)}\varepsilon_1 \\ \mathbf{M}_{10}^{(1)}\varepsilon_0 & \mathbf{M}_{11}^{(1)}\varepsilon_1 \end{bmatrix}$$

Indeed, the probability of the sequence of errors $e_1, e_2, \ldots, e_n$ is given by

$$Pr(e_1, e_2, \ldots, e_n) = \sum_{g_1, \ldots, g_n} Pr(g_1, g_2, \ldots, g_n) \prod_{i=1}^{n} Pr(e_i \mid g_i)$$

or, since $\{g_t\}$ is a matrix process,

$$Pr(e_1, e_2, \ldots, e_n) = \sum_{g_1, \ldots, g_n} \mathbf{Y}_{g_1} Pr(e_1 \mid g_1) \prod_{i=1}^{n-1} \{\mathbf{M}_{g_i, g_{i+1}}^{(1)} Pr(e_{i+1} \mid g_{i+1})\} \mathbf{Z}_{g_n}^{(1)}$$

where $Pr(e \mid g) = \varepsilon_g^e (1-\varepsilon_g)^{1-e}$. The process $\{e_t\}$ is a matrix process because the previous formula can be written in the following matrix form:

$$Pr(e_1, e_2, \ldots, e_n) = \mathbf{Y}_{e_1} \prod_{i=1}^{n-1} \mathbf{M}_{e_i e_{i+1}} \mathbf{Z}_{e_n}$$

if we use the previously defined matrices $\mathbf{Y}_e$, $\mathbf{Z}_e$, and $\mathbf{M}_{e_i e_{i+1}}$.

Thus, having introduced the concept of bursts of errors and the intervals between them, it is possible to produce some smoothing of the experimental data, which simplifies model building.

3.6 HMM NESTED STRUCTURES

3.6.1 Multilayered Error Clusters

As noted in Chapter 1, describing the HMM based on matrix processes is easier from the point of view of minimizing the dimensions of matrices used in this description and minimizing the number of parameters. However, by so doing, we lose the clarity in the model description that is characteristic of the interpretation of Markov functions.

The BWA considered in Sec. 3.2 converges slowly to a local maximum of the likelihood. The rate of convergence and the closeness to the global maximum depend on the initial values of the model parameters. It is also important to select a proper HMM structure. Therefore, we need to develop good heuristical methods that allow us to choose the HMM structure and the initial values of its parameters.

Let us consider the problem of estimating parameters of an HMM, assuming that the states of the underlying chain are semi-Markov lumpable. We partition the experimental sequence into segments of different categories, depending on the intensity of the error flow. The classification can be based on one parameter, for instance, by identifying an error burst as a sequence in which a distance between any errors does not exceed a selected value ρ while the distance between errors outside the burst is greater than ρ. We can also have a classification based on several parameters, for instance, by choosing a sequence of gaps $\rho_1 < \rho_2 < \cdots < \rho_s$ and identifying bursts of type i as a sequence of bits in which the distance λ between adjacent errors lies within the limits $\rho_{i-1} < \lambda \le \rho_i$. So-defined bursts are sometimes called *experimental bursts* in contrast with *theoretical bursts*, which are defined as sequences of bits in which errors occur independently with constant conditional probability ε_c.

If so-defined error bursts can be described by a semi-Markov process whose interval transition probabilities have matrix-geometric form and errors within bursts are i. i. d., then we can create the HMM error source model as shown in Sec. 1.4. To test the validity of these assumptions, we must verify the following hypotheses:

I. The errors within bursts of the same i-th category are independent and occur with constant conditional probability ε_i.

II. Bursts of different categories constitute a semi-Markov process with the matrix-geometric interval transition distributions of form (1.4.15).

If experimental data do not reject any of these hypotheses, then by estimating parameters ε_i and parameters of distribution (1.4.15), we can, according to Theorem 1.4.3, estimate HMM parameters. If, for a selected method of burst error classification, any of the hypotheses discussed here is rejected, then we conduct a more detailed classification based on a larger number of parameters.

Consider now the problem of verification of hypotheses I and II in more detail.

I. *Distribution of errors inside bursts.* We would like to test the hypothesis

H_0: errors within bursts are i. i. d. against H_1: H_0 is not true.

The verification of this hypothesis is performed with the procedures of random number testing.[51]

Preliminarily, during error burst identification, it is better to test the hypothesis using some nonparametric criterion. Let $e_1, e_2, \ldots, e_n$ be a sequence of errors belonging to bursts of the same category. If hypothesis H_0 is true, then $\xi_i = e_{2i-1} - e_{2i}$ is also i. i. d. and $Pr(\xi_i = 1) = Pr(\xi_i = -1) = \varepsilon_c(1-\varepsilon_c)$. By deleting from this sequence all the elements that are equal to zero, we create a sequence $\{\eta_i\}$ of i. i. d. random numbers whose probabilities $Pr(\eta_i = 1)$ and $Pr(\eta_i = -1)$ are equal to 0.5 and do not depend on the error source parameters.

Thus, we transformed the test of the hypothesis with unknown error probability ε_c to the similar test with probability equal to 0.5, which was considered in Example A3.4. The two-sided acceptance region is given by Eq. (A3.1.20) with $p_0 = q_0 = 0.5$. If the sample size n is large, we may use the acceptance region (A3.1.24), which takes the form

$$A(0.5) = (0.5 - 2Z_\alpha\sqrt{2/n},\ 0.5 + 2Z_\alpha\sqrt{2/n})$$

We reject the hypothesis if the frequency of ones in the sequence $\{\eta_i\}$ lies outside this interval. This criterion is not sensitive to data dependency. The so-called series criteria are better suited to this test.[51]

We call a sequence $\eta_{i+1} = 1, \eta_{i+2} = 1, \ldots, \eta_{i+j} = 1$ a length j series of pluses if $\eta_i = -1$ and $\eta_{i+j+1} = -1$. A series of minuses is defined similarly. If the hypothesis H_0 is true, the number of series should be large and there should not be very long series of pluses or minuses. Therefore, one may use a number of series and a maximal series length criteria for the hypothesis testing.

It is easy to show that the number of series r in the sequence of n samples $\{\eta_i\}$ has a binomial distribution $\binom{n-1}{r-1}0.5^{n-1}$ so that the two-sided acceptance region for the number of series may be expressed similarly to Eq. (A3.1.20) and (A3.1.22). For large n, the acceptance region is similar to Eq. (A3.1.24): $(0.5n - 2Z_\alpha 2^{1/2} n^{1/2},\ 0.5n + 2Z_\alpha 2^{1/2} n^{1/2})$.

It may be shown [18] that the probability that there is at least one series whose length is greater than y is given by $Pr(l_{\max} > y) \sim 1 - \exp(-n2^{-y-1})$ so that the significance level α acceptance interval may be expressed as $[0\ ,\ \log(-n/\ \ln(1-\alpha)) - 1]$

The considered nonparametric criteria are simple, but to use them we need to transform the original sequence $\{e_i\}$ into the sequence $\{\eta_i\}$, which leads to a significant loss of information about error dependencies. After we have roughly outlined burst boundaries, using these simple criteria, we need to perform a more detailed analysis of the bursts, using more powerful parametric criteria.

If H_0 is true, then the error number inside bursts of the category c has a binomial distribution $P_{c,n}(t) = \binom{n}{t} \varepsilon_c^t (1-\varepsilon_c)^{n-t}$ and the hypothesis can be tested using the ML estimate of the parameter ε_c and the χ^2 criterion (A3.1.11) for the binomial distribution. The ML parameter estimate can be expressed as the ratio $\hat{\varepsilon}_c = t_c / N_c$ of the total number t_c of errors inside the bursts of c-th categories to the total number of bits in the bursts. The χ^2 criterion is given by

$$\chi^2 = \sum_i (\nu_i - Kp_i)^2 / Kp_i$$

where ν_i is the total number of blocks of length n that contain more than t_i, but less or equal to t_{i+1} errors, $K = \lfloor N_c / n \rfloor$ is the total number of blocks in the sequence of N_c bits, and

$$p_i = \sum_{k=t_i+1}^{t_{i+1}} \binom{n}{t} \hat{\varepsilon}_c^k (1-\hat{\varepsilon}_c)^{n-t}$$

Another distribution that we may use for hypothesis H_0 testing is the distribution of intervals between errors inside bursts. Again, replacing ε_c with its estimate, we can use the χ^2 criterion to verify the agreement of the distribution $p_{0,c}(\lambda) = \hat{\varepsilon}_c(1 - \hat{\varepsilon}_c)^{\lambda-1}$ with the experimental data.

We have examined just the simplest criteria for testing the hypothesis I. If this hypothesis is not true, then we must perform a more thorough sorting of bursts of the category under consideration and partition them into several types of bursts. For this we can use approximation (by polygeometric distributions) of the distribution of lengths of the intervals between adjacent errors within the bursts of the category under consideration.

Sorting can also be achieved using the following method. Let us establish level α acceptance region $(\varepsilon_{cL}, \varepsilon_{cH})$ for the hypothesis that error probability within the bursts of c-th category is equal to ε_c. If the relative frequency of errors in an actual burst falls within that acceptance region, then we assume that the burst belongs to this category. By changing the α, we can decrease or expand the acceptance interval changing the method of sorting of errors by categories.

II. *Semi-Markov process hypothesis testing*. We are ready now to test a hypothesis H_0 that bursts of different categories constitute a semi-Markov process whose interval transition distributions have form (1.4.15). The criteria developed in Sec. A3.3 enable us to perform the testing. In our case, the interval transition distributions have independent parameters. The parameters of each individual distribution can be estimated by one of the methods developed in Sec. A3.1.

However, the fact that the theoretical bursts of errors of a given category are not directly observed in an experiment complicates their identification. We can observe the experimental bursts of errors, that is, sequences beginning and ending with errors, in which the distance between adjacent errors lies within the given limits $\rho_{i-1} < \lambda \le \rho_i$. If the conditional probability of error ε_c is large enough, then by identifying the theoretical burst with the experimental, we make a relatively small error. If, on the other hand, the probability ε_c is small, then the error in determining the boundaries of the theoretical burst with respect to the experimental burst can be large.

The solution lies in approximating distributions of lengths of experimental bursts using matrix-geometric functions and subsequent calculation of the parameters of the distribution of lengths of theoretical bursts. As noted earlier, most difficulty in estimating the parameters of a model is due to the fact that it is impossible to determine exactly a burst's boundaries (especially for small values of conditional probability ε_c of the error within the burst). We can resolve this ambiguity by considering a model in which conditional probabilities of errors ε_i are equal 1 or 0 (see Sec. 1.2.3). Then the model can be described by matrices of type (1.2.21), and the necessity to verify hypothesis I no longer exists.

The essence of the method for estimating the parameters of the model represented by matrices of type (1.2.21) consists of the following. Using certain rules, we relate some nonbinary sequence Y_t with the binary sequence e_t: $e_t = f(Y_t)$. If the constructed sequence Y_t is a Markov chain, then by estimating its parameters we can build an HMM. In practice, it is possible to combine the these methods using Markovian nesting of the sequence of models.

3.6.2 Nested Markov Chains

Let us consider a Markov chain with s states and matrix of transition probabilities $\mathbf{P}$; we will call them macrostates. In addition, let us assume that the i-th macrostate $(i=1,2,\ldots,s)$ consists of s_i microstates n_i $(n_i=1,2,\ldots,s_i)$. If the macrostate does not change, then, within the limits of the time of stay in that macrostate, the microstates form a (conditional) Markov chain with a matrix

$$\boldsymbol{\varepsilon}_{ii} = \left[\varepsilon_{n_i m_i}\right]_{s_i, s_i} \qquad i=1,2,\ldots,s \tag{3.6.1}$$

If, on the other hand, macrostate i changes into macrostate j, then with the probability $\varepsilon_{n_i m_j}$ and independently from past history, the chain passes from microstate n_i to microstate n_j. We denote the matrix of these conditional probabilities as

$$\boldsymbol{\varepsilon}_{ij} = \left[\varepsilon_{n_i m_j}\right]_{s_i, s_j} \qquad i,j=1,2,\ldots,s \tag{3.6.2}$$

It is obvious that the sequence of microstates forms a Markov chain with transition probabilities of the form $q_{n_i m_j} = p_{ij}\varepsilon_{n_i m_j}$ In matrix form, this relationship can be written as follows:

$$\mathbf{Q} = \begin{bmatrix} p_{11}\boldsymbol{\varepsilon}_{11} & p_{12}\boldsymbol{\varepsilon}_{12} & \cdots & p_{1s}\boldsymbol{\varepsilon}_{1s} \\ p_{21}\boldsymbol{\varepsilon}_{21} & p_{22}\boldsymbol{\varepsilon}_{22} & \cdots & p_{2s}\boldsymbol{\varepsilon}_{2s} \\ \cdots & \cdots & \cdots & \cdots \\ p_{s1}\boldsymbol{\varepsilon}_{s1} & p_{s2}\boldsymbol{\varepsilon}_{s2} & \cdots & p_{ss}\boldsymbol{\varepsilon}_{ss} \end{bmatrix} \tag{3.6.3}$$

The matrices $\boldsymbol{\varepsilon}_{ij}$, defined by formulas (3.6.1) and (3.6.2), obviously satisfy the normalization conditions $\boldsymbol{\varepsilon}_{ij}\mathbf{1} = \mathbf{1}$. In a more general case, when the same set of microstates corresponds to different macrostates, the matrix of transition probabilities is given by

$$\mathbf{Q} = \begin{bmatrix} \mathbf{P}_{11}\boldsymbol{\varepsilon}_{11} & \mathbf{P}_{12}\boldsymbol{\varepsilon}_{12} & \cdots & \mathbf{P}_{1s}\boldsymbol{\varepsilon}_{1s} \\ \mathbf{P}_{21}\boldsymbol{\varepsilon}_{21} & \mathbf{P}_{22}\boldsymbol{\varepsilon}_{22} & \cdots & \mathbf{P}_{2s}\boldsymbol{\varepsilon}_{2s} \\ \cdots & \cdots & \cdots & \cdots \\ \mathbf{P}_{s1}\boldsymbol{\varepsilon}_{s1} & \mathbf{P}_{s2}\boldsymbol{\varepsilon}_{s2} & \cdots & \mathbf{P}_{ss}\boldsymbol{\varepsilon}_{ss} \end{bmatrix} \tag{3.6.4}$$

It is obvious that any stochastic matrix can be expressed in the form of Eq. (3.6.3) or (3.6.4) in innumerable ways. If the elements of matrices $\mathbf{P}_{ii}\mathbf{1}$ are large, then transitions from macrostates occur after relatively long time intervals. This construction of a Markov chain using the concept of macrostates is called *Markov chain nesting*. If this operation is repeated m times, we call it m-th order Markov chain nesting.

Let us assume that we have been able to build an HMM for relatively short segments of the experimental data. Then, if we approximate the segment transition process by a Markov chain and the transition probabilities (elements of matrices $\boldsymbol{\varepsilon}_{ij}$ for $i \neq j$) satisfy the conditions of independence from past history, then Eq. (3.6.4) gives the matrix of transition probabilities of the combined process. The elements of matrices $\mathbf{P}_{ij}$ are estimated through the method described here only with application to a conditional Markov function. To estimate the model parameters by multiple Markov nesting, let us partition the sequence of errors into segments of different categories and interpret them as times of stay in respective macrostates. Generally speaking, the chosen segments are different from the bursts we considered earlier (the mean lengths of these segments are usually large, and therefore they are sometimes called clusters of bursts). Next, let us partition the selected segments of each category into (second-order) segments of smaller length. We try to find a partition, such that the distribution of lengths of segments (of second order) of each type can be approximated by a geometric distribution (for polygeometric approximation one must try to extract each geometric component). We estimate matrices $\boldsymbol{\varepsilon}_{ij}$ and $\mathbf{P}_{ij}$ using distributions of lengths of segments of the first and second order and, through the process of their change, we estimate the matrix (3.6.4). If, by so doing, it appears that errors within each second-order segment are independent and occur with constant conditional probability, then we obtain estimates of the parameters of the HMM. If the hypothesis of the independence of errors is rejected, then we must conduct a more detailed sorting, or partition the segments of the second order into segments of the third order, and so on.

If we assume that during the change of the macrostates, the microstates of the category that follows do not depend on the microstates of the previous category, then the matrices $\boldsymbol{\varepsilon}_{ij}$ have the special structure outlined in Sec. 1.4. The sufficient condition of this is the factorization:

$$\boldsymbol{\varepsilon}_{ij} = \mathbf{M}_i\mathbf{N}_j \qquad \mathbf{M}_i = col\ \{m_{ij}\} \qquad \mathbf{N}_j = row\ \{n_{ij}\}$$

Example 3.6.1. Suppose that we have partitioned the error sequence into two categories of error clusters that constitute an alternating renewal process. Let $p_{ch}(\lambda) = p_1 q_1^{\lambda-1}$ be the distribution of intervals between the clusters and $p_{ch}(l) = p_2 q_2^{l-1}$ be the distribution of lengths of clusters. Then, according to Eq. (1.4.26), the macrostate transition probability matrix is equal to

$$\mathbf{P} = \begin{bmatrix} q_1 & p_1 \\ p_2 & q_2 \end{bmatrix}$$

Suppose that inside the clusters error bursts are independent and their lengths have a polygeometric distribution $p_b(l) = a_{21}p_{21}q_{21}^{l-1} + a_{22}p_{22}q_{22}^{l-1}$ and intervals between bursts inside the clusters have geometric distribution $p_b(\lambda) = p_{11}q_{11}^{\lambda-1}$. Then, according to Eq. (1.4.26), we obtain

$$\boldsymbol{\varepsilon}_{11} = 1 \quad \boldsymbol{\varepsilon}_{12} = \begin{bmatrix} a_{21}p_{11} & a_{22}p_{11} & q_{11} \end{bmatrix}$$

$$\boldsymbol{\varepsilon}_{21} = \begin{bmatrix} 1 \\ 1 \\ 1 \end{bmatrix} \quad \boldsymbol{\varepsilon}_{22} = \begin{bmatrix} q_{21} & 0 & p_{21} \\ 0 & q_{22} & p_{22} \\ a_{21}p_{11} & a_{22}p_{11} & q_{11} \end{bmatrix}$$

Finally, by Eq. (3.6.3), the model transition probability matrix is given by

$$\mathbf{P} = \begin{bmatrix} q_1\boldsymbol{\varepsilon}_{11} & p_1\boldsymbol{\varepsilon}_{12} \\ p_2\boldsymbol{\varepsilon}_{21} & q_2\boldsymbol{\varepsilon}_{22} \end{bmatrix}$$

3.6.3 Single-Error-State Models

A single-error-state model is a particular case of a renewal process. Therefore, the intervals between consecutive errors are independent, and all we need is to estimate parameters of the matrix-geometric distributions of the intervals. If this distribution is polygeometric

$$f_\lambda(\lambda) = Pr(0^\lambda 1 \mid 1) = \sum_{i=1}^{n-1} b_i(1-q_i)q_i^\lambda \quad \text{for } \lambda > 0$$

with positive parameters, then the process can be described by Fritchman's partitioned model (see Sec. 1.5.1) with the matrix

$$\mathbf{P} = \begin{bmatrix} q_1 & 0 & \dots & 0 & 1-q_1 \\ 0 & q_2 & \dots & 0 & 1-q_2 \\ \dots & \dots & \dots & \dots & \dots \\ 0 & 0 & \dots & q_{n-1} & 1-q_{n-1} \\ b_1 & b_2 & \dots & b_{n-1} & p_{nn} \end{bmatrix}$$

where $p_{nn} = f_\lambda(0)$ and $b_j = (1-p_{nn})a_j$. This method is broadly used in practice.[5,11,12,19]

However, more stable estimates can be obtained by the experimental data sorting discussed in the previous section. Depending on error density, we define different types of error bursts, thus constructing a nonbinary sequence out of binary error statistics. Then we can apply the methods considered in Sec. 3.2 for estimating model parameters. Suppose that the intervals between consecutive errors are partitioned into $n-1$ categories. Let $\lambda_1^{(i)}$, $\lambda_2^{(i)}, \dots, \lambda_{k(i)}^{(i)}$ be a sequence of intervals of i-th category. If we associate state i with the correct symbols within the intervals of i-th category, then the total number of transitions from the state i to the only error-state n is equal to the number of these intervals $k(i)$. We obtain estimates of the transition probabilities as

$$\hat{p}_{in} = 1 - \hat{q}_i = k(i) \,/ \sum_{m=1}^{k(i)} \lambda_m^{(i)} \qquad i \neq n$$

The number of transitions from the error state n to the state i is equal to $k(i)$. And, finally,

$$\hat{p}_{nn} = 1 - \sum_{m=1}^{n-1} \hat{p}_{ni}$$

$$\hat{p}_{ni} = \hat{b}_i = k(i) \,/\, \sum_{m=1}^{n} k(m)$$

We use these estimates as initial values for the BWA. In the majority of practical cases only three iterations were necessary to obtain good estimates.

To complete the statistical analysis, we must test the hypothesis that the experimental data do not contradict the single-error-state model. One method of testing this hypothesis is outlined in Sec. A3.2.2. The other method consists of testing the hypothesis that the intervals between consecutive errors are statistically independent and their length distribution is polygeometric. Using this simple approach, we were able to build models for the source of errors in the majority of the T1 digital repeatered lines. [11]

3.7 MONTE CARLO METHOD OF MODEL BUILDING

The idea of the Monte Carlo method is simple: find model parameters through computer simulations. If the results of the simulation are close to the experimental data, the computer model parameters can be used as the estimates. To test the hypothesis that the computer model describes the real process, we need some criteria of closeness between the experimental data and the result of simulation.

Normally we can select several distributions of some random variables and compare the difference between their measured and simulated estimates. Let $p(x)$ be a variable x probability distribution and

$$\hat{P}_{i,exp} = m_{i,exp} \,/\, N_{exp} \qquad \hat{P}_{i,sim} = m_{i,sim} \,/\, N_{sim}$$

be the probability that the random variable belongs to $[x_{i-1}, x_i)$ estimated from the experimental data and simulation, respectively.

If the null hypothesis is true and the estimates $\hat{P}_{i,exp}$ and $\hat{P}_{i,sim}$ are asymptotically normal, then [30]

$$\chi^2 = N_{sim} N_{exp} \sum_{i=1}^{r} (\hat{P}_{i,exp} - \hat{P}_{i,sim})^2 \,/\, (\hat{m}_{i,exp} + \hat{m}_{i,sim})$$

and

$$2\,I = 2N_{exp} N_{sim} \sum_{i=1}^{r} (\hat{P}_{i,exp} - \hat{P}_{i,sim}) \ln (\hat{P}_{i,exp} \,/\, \hat{P}_{i,sim}) \,/\, (N_{exp} + N_{sim})$$

are asymptotically χ^2 distributed with $r-1$ degrees of freedom. The null hypothesis is rejected if the calculated values χ^2 or $2\,I$ exceed the selected critical value. We can use similar criteria for the two-dimensional distributions described in Sec 3.5.3.

We would like to point out that the Monte Carlo method is less efficient than the majority of the previously considered methods. However, this method is comparatively easy to implement. The method should also be used as an additional test of a model agreement with the experimental data, if the model parameters have been estimated using some other method. It is especially important to do this test if we intend to use the model for computer simulation.

3.8 ERROR SOURCE MODEL IN SEVERAL CHANNELS

Let us now address the question of error source model building in several communication channels. As we mentioned before, the methodology of model parameter estimation in this case does not differ significantly from the case of a single channel. We just outline the steps in model building and emphasize only the features that are specific to multiple channels.

Consider a matrix process model of error source in h binary symmetrical channels. There are $\sigma = 2^h$ possible values of the vector error $\mathbf{e} = (e_1, e_2, ..., e_h)$ in the set of h channels. The matrix process can be described by the following block matrices:

$$\mathbf{M}_{e_j} = \begin{bmatrix} 0 & \cdots & 0 & \mathbf{M}_{\mathbf{e}_1\mathbf{e}_j} & 0 & \cdots & 0 \\ 0 & \cdots & 0 & \mathbf{M}_{\mathbf{e}_2\mathbf{e}_j} & 0 & \cdots & 0 \\ \cdots & \cdots & \cdots & \cdots & \cdots & \cdots & \cdots \\ 0 & \cdots & 0 & \mathbf{M}_{\mathbf{e}_\sigma\mathbf{e}_j} & 0 & \cdots & 0 \end{bmatrix}$$

We need to estimate the elements of these matrices. As in the case of a single channel, the diagonal blocks $\mathbf{M}_{\mathbf{e}_i\mathbf{e}_i}$ of the matrix

$$\mathbf{M} = \sum_{i=1}^{\sigma} \mathbf{M}_{\mathbf{e}_i}$$

can be estimated using the estimated parameters of the error series distributions which have the form of Eq. (3.4.7):

$$p_{\mathbf{e}_i\mathbf{e}_j}(l) = \mathbf{S}_{\mathbf{e}_i}\mathbf{M}_{\mathbf{e}_i\mathbf{e}_i}^{l-1}\mathbf{F}_{\mathbf{e}_j}$$

As in the case of a single channel, the parameters of these distributions can be estimated by the difference Eq. (3.4.16). The rest of the matrices can be estimated using the error series two-dimensional distributions, as described in Sec. 3.5.3.

The methodology of model parameter estimation that has been proposed in Sec. 3.5 can be modified for the case of multiple channels. The model based on the Markov chain with the matrix $\mathbf{P}$ and matrix conditional probabilities of errors $\mathbf{E}_i$, described in Sec. 1.1.2, has a special interest for applications. For this interpretation it is necessary to test the hypothesis that errors inside a burst are independent and occur with constant conditional probabilities so that the conditional probability $P_{c,n}(t_1, t_2, ..., t_h)$ that in a block of n symbols inside the burst of the c-th category there are t_1 errors in the first channel, t_2 errors in the second channel, and so on, is given by

$$P_{c,n}(t_1, t_2, ..., t_h) = \prod_{i=1}^{h} \binom{h}{t_i} \varepsilon_{ci}^{t_i}(1-\varepsilon_{ci})^{n-t_i}$$

where ε_{ci} is the probability of an error in the i-th channel inside the burst of the c-th category. If this hypothesis does not contradict the experimental data, then we test the hypothesis that bursts of different categories constitute a semi-Markov process with the state-holding distributions of form (3.4.7).

In general, the complexity of the model of error source in multiple channels grows exponentially with the number of channels. However, model parameter estimation can be simplified if we make some assumptions about the model's structure. In many practical cases we can neglect the adjacent channel interference and therefore assume that error sources in these channels are independent. If this hypothesis is true, it is possible to build the error source models independently for each channel.

To test the hypothesis of independence, consider two communication channels. Let $\mathbf{P}_1(e)$ and $\mathbf{P}_2(e)$ be the matrix probabilities of errors in the first and the second channels, respectively. If the hypothesis is true, the error source in the two channels can be described by the model with the matrix probabilities

$$\mathbf{P}(e_1,e_2) = \mathbf{P}_1(e_1)\otimes\mathbf{P}_2(e_2)$$

This equation may be treated as a matrix generalization of the condition of independence (A3.1.25) of random variables. In the particular case where the model is described by multiple Markov chains, this condition was used [1] to test the hypothesis that two subsets of states of the chain are independent. However, since the matrix probabilities are not defined uniquely, the preceding condition is not necessarily true when the hypothesis of independence is true. So this condition cannot be used for hypothesis testing in the general case.

In order to test the hypothesis in the general case, we use distributions of different sequences of errors. If the hypothesis is true, then the intervals between errors $\mathbf{e}_i$ and $\mathbf{e}_j$ have the following distribution:

$$p_{\mathbf{e}_i\mathbf{e}_j}(l) = \mathbf{F}_{\mathbf{e}_i}\mathbf{P}^{l-1}(0,0)\mathbf{S}_{\mathbf{e}_j}$$

If the chain is stationary, then, using the properties of the Kronecker product, we can rewrite the previous equation as

$$p_{\mathbf{e}_i\mathbf{e}_j}(l) = p_{e_{1i}e_{1j}}(l)p_{e_{2i}e_{2j}}(l)$$

where

$$p_{e_{ki}e_{kj}}(l) = \mathbf{F}_{\mathbf{e}_{ki}}\mathbf{P}_k^{l-1}(0)\mathbf{S}_{\mathbf{e}_{kj}}$$

Therefore, if the null hypothesis is true, then intervals between errors in different channels are independent. We can prove similarly that for the stationary models all sequences of intervals in different channels are independent.

3.9 CONCLUSION

We have considered the questions of statistical inference concerning the HMM. Since the model has numerous representations, we proposed a variety of methods of model parameter estimation. Major part of this chapter is devoted to the EM algorithm and its special case, the BWA for fitting HMMs. This algorithm can be used for approximating a general random process with HMM. We illustrate this by considering the HMM approximation of the envelope of the Rayleigh fading. This method can be also used for reducing the number of an HMM parameters by fitting a smaller HMM. We consider many forms of the BWA for the most popular model structures. Simple models usually have an enormous number of parameters. The limited amount of experimental data does not allow us to estimate reliably the large number of parameters. To solve this problem, more complex structures with smaller number of parameters are used. The algorithm structural complexity reflects the complexity of the model.

Fitting a matrix-geometric distribution and PH distribution plays a central role in HMM modeling. It allows us to estimate the model parameters and select a proper model structure which impacts the BWA convergence rate and the model usage.

We consider fitting matrix processes to experimental data and show that the two-dimensional distributions of the series of errors and intervals between errors can be used to identify uniquely the canonical matrix process that is equivalent to a model. This canonical matrix process can be viewed as the kernel of any HMM. If two models are equivalent, their kernels must coincide. Therefore, we can translate statistical problems related to the HMM into the corresponding problems of two-dimensional distributions.

The problem of estimating interval distribution parameters is a problem of the nonlinear function minimization. It can be reduced to a problem of a linear equation parameter estimation, which has been studied by many authors. However, this approach is very formal and destroys the clear relationship between the model parameters and statistical data. The alternative approach, which uses the notion of error bursts and clusters of bursts, allows us not only to have a "physical" model but also is more efficient. In many cases, a model based on bursts of errors and clusters of bursts better suits applications. For example, if we calculate the error number distribution in a short block, we may make an assumption that this block either lies inside a cluster of bursts or outside the clusters, which simplifies the distribution calculation.

An important part of modeling is testing the model. Usually, some parameters are evaluated using the model and compared with their experimental values. The other method of testing is based on the computer simulation. If simulated data differ significantly from the corresponding experimental data, the model can be adjusted using the Monte Carlo method.

Problems

1. A function $u(x)$ defined on some interval I is called convex if for any $x_1, x_2 \in I$ and any $p_1 \geq 0$ and $p_2 = 1 - p_1 \geq 0$

$$p_1 u(x_1) + p_2 u(x_2) \geq u(p_1 x_2 + p_2 x_2)$$

If p_1 and p_2 are interpreted as the probability distribution $f(x)$, then this inequality can be written as

$$E\{u(x) \mid f(x)\} \geq u(E\{x \mid f(x)\})$$

This inequality is called a *Jensen's inequality*. Prove it for any discrete distribution $p_1, p_2, \ldots, p_n$ and then generalize it for the continuous random variables.

2. Let $a_i > 0$, $b_i > 0$ for $i = 1, 2, \ldots, n$ and $\sum_{i=1}^{n} b_i = B$

— Show that

$$\sum_{i=1}^{n} a_i \log b_i \leq \log(A/B) \sum_{i=1}^{n} a_i \log a_i$$

where $A = \sum_{i=1}^{n} a_i$. (Hint: use the convexity of -log (x) and Jensen's inequality for $E\{(f_2(x)/f_1(x) \mid f_1(x).)$

— Show that the function

$$L(b_1^n) = \sum_{i=1}^{n} a_i \log b_i$$

attains its global maximum at $b_i = (B/A)a_i$. (Hint: use the previous inequality.)

3. Let

$$\mathrm{N}(x,\mu,\mathbf{D}) = (2\pi)^{-m/2}|\mathbf{D}|^{-1/2} \exp[-0.5(x-\mu)\mathbf{D}^{-1}(x-\mu)']$$

be a multidimensional Gaussian PDF and denote $\tau = (\mu,\mathbf{D})$ the distribution parameter vector. Show that the solution to the following equation

$$\hat{\tau} = \arg\max_{\tau} \int_X \log \mathrm{N}(x,\mu,\mathbf{D})\, r(x)\, dx$$

has the form

$$\hat{\mu} = \int_X x r(x)\, dx \Big/ \int_X r(x)\, dx$$

$$\hat{\mathbf{D}} = \int_X (x-\hat{\mu})(x-\hat{\mu})' r(x)\, dx \Big/ \int_X r(x)\, dx$$

[Hint: denote $\mathbf{A} = \mathbf{D}$ and use the formula $\partial|\mathbf{A}|/\partial a_{ij} = A_{ij}$, which follows from the determinant expansion $|\mathbf{A}| = \sum a_{ij}A_{ij}$ (see Appendix 5)].

4. Suppose that $g(\boldsymbol{x}_1^T,\tau)$ is defined by Eq. (3.2.18)

— Show that the reestimation Eq. (3.2.32a) can be written as

$$\pi_{i,p+1} = \pi_{i,p} E\{\frac{\partial \log g(\boldsymbol{x}_1^T,\tau)}{\partial \pi_i} \mid f(\boldsymbol{x}_1^T)\}$$

(Hint: $\partial g(\boldsymbol{x}_1^T,\tau)\ /\ \partial \boldsymbol{\pi} = \boldsymbol{\beta}(\boldsymbol{x}_1^T,\tau)$.)

— Using the formula of Appendix 5.1.9 for differentiating matrix products, show that Eq. (3.2.30b) can be written as

$$V_{ij}(\tau_p) = E\{\frac{p_{ij,p}}{g(\boldsymbol{x}_1^T,\tau)} \frac{\partial g(\boldsymbol{x}_1^T,\tau)}{\partial p_{ij,p}} \mid f(\boldsymbol{x}_1^T)\} = p_{ij,p} E\{\frac{\partial \log g(\boldsymbol{x}_1^T,\tau)}{\partial p_{ij,p}} \mid f(\boldsymbol{x}_1^T)\}$$

5. Show that if ν_{ij} is the number of transitions from state i to state j, the Markov chain stationary distribution can be estimated as

$$\hat{\pi}_i = \nu_i \Big/ \sum_i \nu_i$$

where $\nu_i = \sum_j \nu_{ij}$. (Hint: the chain transition probabilities are estimated as $\hat{p}_{ij} = \nu_{ij}/\nu_i$.)

6. Prove Eq. (3.2.44) by induction:

$$\bar{\boldsymbol{\alpha}}(\boldsymbol{x}_1^t,\tau) = \bar{\boldsymbol{\alpha}}(\boldsymbol{x}_1^{t-1},\tau)\mathbf{P}(x_t,\tau)/c_t = \boldsymbol{\alpha}(\boldsymbol{x}_1^{t-2},\tau)\mathbf{P}(x_{t-1},\tau)\mathbf{P}(x_t,\tau)/c_{t-1}c_t = \cdots$$

7. Derive a backward-only BWA for the case described in example 3.3.1 and test the algorithm on simulated data.

8. Derive a forward-only BWA for estimating parameters of an HMM whose observations have state conditional Gaussian distributions and test the algorithm on simulated data.

9. Derive the following backward algorithm for computing $\mathbf{S}_m(X)$ given by Eq. (3.3.9):

Algorithm 3.6.

Initialize:

$$b_0 = REM(m,2) \quad m_0 = m \quad \mathbf{V}_0 = b_0\mathbf{W}(X) \quad \mathbf{R}_0 = \mathbf{P}(X) \quad \mathbf{Q}_0 = \mathbf{P}^{b_0}(X),\ \mathbf{G}_0 = \mathbf{W}(X)$$

While $m_i > 0$

```
Begin
```

$$i+1 \to i$$

$$\mathbf{G}_i = \mathbf{R}_{i-1}\mathbf{G}_{i-1} + \mathbf{G}_{i-1}\mathbf{R}_{i-1} \qquad \mathbf{R}_i = \mathbf{R}_{i-1}^2$$

$$m_i = (m_{i-1} - b_{i-1})/2 \qquad b_i = REM(m_i, 2)$$

$$\mathbf{V}_i = \mathbf{R}_i^{b_i}\mathbf{V}_{i-1} + b_i\mathbf{G}_i\mathbf{Q}_{i-1} \qquad \mathbf{Q}_i = \mathbf{Q}_{i-1}\mathbf{R}_i^{b_i}$$

```
End
```

$$\mathbf{P}^m(X) = \mathbf{Q}_k \quad \text{and} \quad \mathbf{S}_m(X) = \mathbf{V}_k$$

References

1 T. W Anderson and L. A. Goodman, "Statistical inference about Markov chains," *Ann. Math. Statist.*, **28**, 89-110, (1957).

2 M. S. Bartlett, *Introduction to Stochastic Processes with Special Reference to Methods and Applications*, (Cambridge Univ. Press, Cambridge, 1978).

3 M. S. Bartlett, "The frequency goodness of fit test for probability chains," *Proc. Cambridge Philos. Soc.*, **4a**7, 86-95, (1951).

4 L. E. Baum, T. Petrie, G. Soules, and N. Weiss, "A maximization technique occurring in the statistical analysis of probabilistic functions of Markov chains," *Ann. Math. Statist.*, **41**, 164-171, (1970).

5 E. R. Berlekamp, *Algebraic Coding Theory*, (McGraw-Hill, New York, 1968).

6 P. Billingsley, "Statistical methods in Markov chains," *Ann. Math. Statist.*, **32**, 12-40, (1961).

7 P. Billingsley, *Statistical Inference for Markov Processes*, (University of Chicago Press, Chicago, 1961).

8 R. E. Blahut, *Fast Algorithms for Digital Signal Processing*, (Addison-Wesley, Reading, Massachusetts, 1984).

9 R. E. Blahut, *Theory and Practice of Error Control Codes*, (Addison-Wesley, Reading, Massachusetts, 1983).

10 R. P. Brent, F. G. Gustavson, and D. Y. Y. Yun, "Fast solution of Toeplitz systems of equations and computation of Pade approximants," *J. Algorithms*, **1**, 259-295, (1980).

11 M. B. Brilliant, "Observations of errors and error rates on T1 digital repeatered lines," *Bell Syst. Tech. J.*, **57**(3), 711-746, March (1978).

12 C. Chatfield, "Statistical inference regarding Markov chain models," *Appl. Statist.*, **22**, 7-20, (1973).

13 E. Cinlar, *Introduction to Stochastic Processes*, Prentice Hall, (Englewood Cliffs, New Jersey, 1975).

14 H. Cramer, *Mathematical Methods of Statistics*, (Princeton Univ. Press, Princeton, New Jersey, 1946).

15 A. P. Dempster, N. M. Laird, and D. B. Rubin, "Maximum likelihood from incomplete data via the EM algorithm," *J. R. Statist. Soc.*, **76**, 341-353, (1977).

16 S. W. Dharmadhikary, "Functions of finite Markov chains," *Ann. Math. Stat.*, **34**(3), 1022-1032, (1963).

17 S. W. Dharmadhikary, "Sufficient conditions for a stationary process to be a function of a finite Markov chain," *Ann. Math. Stat.*, **34**(3), 1033-1041, (1963).

18 I. V. Dunin-Barkovsky and K. V. Smirnov, *Probability Theory and Mathematical Statistics* (in Russian), (Gostechizdat, Moscow, 1955).

19 J. Durbin, "The fitting of time-series models," *Rev. Inst. Int. Stat.*, **28**(3), 233-243, (1960).

20 R. V. Ericson, "Functions of Markov chains," *Ann. Math. Stat.*, **41**(3), 843-850, (1970).

21 W. Feller, *An Introduction to Probability Theory and Its Applications*, **1**, (John Wiley & Sons, 1962).

22 R. Fortret, "Random determinants," *J. Res. Nat. Bureau Standards*, **47**, 465-470, (1951).

23 E. Frank, "Corresponding type continued fractions," *Amer. Jour. of Math.*, **68**, 89-108, (1946).

24 F. R. Gantmacher, *The Theory of Matrices*, (Chelsea Publishing Co., New York, 1959).

25 V. L. Girco, *Random Matrices* (in Russian), (Vischa Shcola Publishers, Kiev, 1975).

26 U. Grenander and G. Szego, *Toeplitz Forms and Their Applications,"* (University of California Press, Berkeley, 1958).

27 P. G. Hoel, "A test for Markoff chains," *Biometrica*, **41**, 430-433, (1954).

28 T. Kailath, "A view of three decades of linear filtering theory," *IEEE Trans. Inform. Theory*, **IT-20**(2), 146-181, (1974).

29 J. Komlós, "On the Determinant of Random Matrices," *Studia Sci. Math. Hungary*, **3**(4), (1968).

30 S. Kullback, *Information Theory and Statistics*, (John Wiley & Sons, New York, 1959).

31 W. Leighton and W. T. Scott, "A general continued fraction expansion," *Bull. Am. Math. Soc.*, **45**, 596-605, (1935).

32 N. Levinson, "The Wiener RMS (Root Mean Square) error criterion in filter design and prediction," *J. Math. Phys.*, **25**(4), 261-278, 1947). (Also Appendix B in N. Wiener, *Extrapolation, Interpolation and Smoothing of Stationary Time Series*, MIT Press, Cambridge, MA, 1964).

33 F. W. Leysieffer, "Functions of finite Markov chains," *Ann. Math. Stat.*, vol. 38(1), 206-212, (1967).

34 T. A. Louis, "Finding the observed information matrix when using the EM algorithm," *J. R. Statist. Soc.*, **44**, 226-233, (1982).

35 J. Makhoul, "Linear prediction, a tutorial review," *Proc. IEEE*, **63**(4), 561-580, April (1975).

36 H. B. Mann and A. Wald, "On the statistical treatment of linear stochastic difference equations," *Econometrica*, **11**(3), 173-220, (1943).

37 J. L. Massey, "Shift register synthesis and BCH decoding," *IEEE Trans. Inform. Theory*, **IT-15**, 122-127, (1969).

38 W. H. Mills, "Continued fractions and linear recurrences," *Math. of Comp.*, **29**(129), 173-180, (1975).

39 C. Moller and C. Van Loan, "Nineteen dubious ways to compute the exponential of a matrix," *SIAM Review* **20**, 801-836, (1978).

40 M. F. Neuts, *Matrix-Geometric Solutions in Stochastic Models*, (Johns Hopkins, Baltimore, 1981).

41 H. Nyquist, S. Rice, and J. Riordan, "The distribution of random determinants," *Quart. Appl. Math.*, **12**(2), (1954).

42 H. Padé, "Sur la Representation Aprochee d'une Fonction par des Fractions Rationelles," *Ann. Ecole Norm.*, supp., **3**(9), 1-93 1892 (supplement).

43 W. H. Press, S. A. Teukolsky, W. T. Vetterling, and B. P. Flannery, *Numerical Recipes in C*, (Cambridge University Press, Cambridge, 1992).

45 R. de Prony, "Essai experimentale et analytique," *J. Ecole Polytechnique* 24-76, (1795).

46 L. Rabiner and B.-H. Juang, *Fundamentals of Speech Recognition*, (Prentice Hall, Englewood Cliffs, New Jersey, 1993).

47 C. R. Rao, *Linear Statistical Inference and Its Applications*, (John Wiley & Sons, New York, 1965).

48 M. Sajadieh, F.R. Kschischang, and A. Leon-Garcia, "A block memory model for correlated Rayleigh fading channels," *roc. IEEE Int. Conf. Commun.*, 282-286, June (1996).

49 S. Sivaprakasam and K. S. Shanmugan, "A forward-only recursion based HMM for modeling burst errors in digital channels," GLOBECOM, **2**, 1054-1058, (1995).

50 S. Sivaprakasam and K. S. Shanmugan, "An equivalent Markov model for burst errors in digital channels," *IEEE Trans. Commun.*, **43**, 1347-1355, (1995).

51 Yu. A. Shreider, *Method of Statistical Testing*, (Elsevier Publishing Co., Amsterdam, 1964).

52 Y. Sugiyama, M. Kasahara, S. Hirsawa, and T. Namekawa, "A method for solving key equations for decoding Goppa codes," *Inform. Control*, **27**, 87-99, (1975).

53 Y. Sugiyama, "An algorithm for solving discrete-time Wiener-Hopf equations based upon Euclid's algorithm," *IEEE Trans. Inform. Theory,* **IT-32**, 394-409, (1986).

54 G. Szegö, "Orthogonal Polynomials," Colloquium Publications(23), *Amer. Math. Society,* (1939).

55 N. Tan, *Adaptive Channel/Code Matching*, Ph. D. Dissertation, (University of Southern California, November, 1993).

56 W. Turin and M. M. Sondhi, "Modeling error sources in digital channels," *IEEE Journ. Sel. Areas in Commun.*, **11**(3), 340-347, (1993).

57 W. Turin, "Fitting probabilistic automata via the EM algorithm," *Stoch. Models,* **12**(3), 405-424, (1996).

58 W. Turin and R. A. Boie, "Bar code recovery via the EM algorithm," *IEEE Trans. Sig. Proc.,* **46.**(2), Feb. (1998).

59 J. L. Walsh, *Interpolation and Approximation by Rational Functions in the Complex Domain*, Amer. Math. Soc., **20**, 354-363 (1969).

60 J. K. Wolf, "Decoding of Bose-Chaudhuri-Hocquenghem codes and Prony's method of curve fitting," *IEEE Trans. Inform. Theory,* **IT-13**, 608, Oct. (1967).

61 C. F. J. Wu, "On the convergence properties of the EM algorithm," *Ann. of Statistics,* **11**(1), 95-103 (1983).

62 W. I. Zangwill, *Nonlinear Programming: A Unified Approach*, (Prentice Hall, Englewood Cliffs, New Jersey (1969).

4

PERFORMANCE OF FORWARD ERROR-CORRECTION SYSTEMS

4.1 BASIC CHARACTERISTICS OF ONE-WAY SYSTEMS

In one-way systems the information flow is strictly unidirectional: from transmitter to receiver. Such systems can be described in terms of their input and output process characterization. However, in practical applications, only certain characteristics of this process are usually considered. Some basic performance characteristics that are used for comparing these systems are:

p_s	the symbol-error probability on a decoder output
p_*	the symbol-erasure probability (the probability of receiving a symbol with detected errors)
p_c	the probability of receiving a message without errors
p_u	the probability of receiving a message with undetected errors
p_d	the probability of receiving a message with detected errors (obviously $p_c+p_u+p_d = 1$)
t_d	the average path delay
R	the average information rate (the average ratio of the number of information symbols to the total number of transmitted symbols)
$P(EFS)$	the average percent of error-free seconds (the average percentage of the one-second intervals that do not have errors)

These parameters, along with the system's complexity and implementation cost, are taken into consideration when the system is designed. The method of comparison depends upon the system's designation.

Selecting a method of forward error correction (FEC) or error detection is the important part of system design. The optimal FEC scheme takes into account real-channel error statistics and therefore corrects the most probable sequences of errors. Normally, it is assumed that a channel is the random-error channel or memoryless channel when errors occur independently with constant bit-error probability. This assumption often leads to selecting an inappropriate code for the channels with memory. For example, a very popular extended Hamming cyclic code with the generator polynomial [27] $x^{16} + x^{15} + x^2 + 1$, broadly known as sixteen bit Cyclical Redundancy Check (CRC-16), is able to detect up to three errors or correct a single error in a block. [31] If channel errors are bursty, then the probability of having more than one error in a block might be higher than the probability of a single error. In this case, the redundancy of 16 bits is not used for the blocks without errors, but this redundancy is not sufficient to correct the most probable multiple errors.

To be able to select an optimal code, one must have an adequate error source description. As we saw in Chapter 1, the HMM is general enough to describe error statistics in real channels. In this chapter, we show that the model is also convenient to use for calculating system performance parameters. We consider, first, a general approach to computing FEC efficiency, then analyze particular coding algorithms and derive the numerical results.

4.2 ELEMENTS OF ERROR-CORRECTING CODING

Our goal is to demonstrate the application of methods developed in the previous chapters to evaluating error correction and error-detection performance. We assume that the reader is familiar with the basics of coding theory; however, we present here the background necessary for understanding performance evaluation of communication systems. We present the basic results without proofs, since excellent treatments of the subject can be found in texts by E. R. Berlekamp, [2] R. G. Gallager, [13] F. J. MacWilliams and N. J. A. Sloane, [19] and W. W. Peterson and E. J. Weldon. [27]

4.2.1 Field Galois Arithmetic

Digital communication devices perform operations on bits using finite length registers. The logical operations between these registers may be interpreted as some mathematical operations between the integers that those registers represent. If we impose certain rules on these operations, we may call them addition, subtraction, multiplication, and division. The "usual" mathematical operations performed by digital computers suffer from a deficiency that is not tolerable in data communications: The result of an operation may not fit into the finite length register, and thus register overflow and loss of accuracy occur. We want the operations to be defined in such a way that for any two numbers of the set, the result of any operation (except for division by zero) is a number of the same set that fits into a finite length register. Such number sets are called *finite fields* if the operations satisfy some standard conditions (axioms) that are familiar from elementary mathematics. We present here a method of finite field construction.

Let p be a prime number. A *ground field*, or *Galois, field* of p elements GF(p) may be defined as a set of nonnegative integers that are less than p with the modulo p operations between them. The modulo p operation on an integer gives the remainder of division of the number by p. The modulo p addition of two numbers $(a+b) \bmod p$ is equal to the remainder of division of $(a+b)$ by p. The product $ab \bmod p$ is also the remainder of division of ab by p. The subtraction and division modulo p are defined as the inverse operations to the addition and multiplication, respectively.

Example 4.2.1. Let us illustrate the operations in GF(5), which is the set of numbers 0, 1, 2, 3, 4 with the modulo 5 operations:

addition	(2 + 3) mod 5 = 5 mod 5 = 0, (4 + 3) mod 5 = 7 mod 5 = 2, (1 + 2) mod 5 = 3.
subtraction	is inverse to addition. To find $(2-4)\bmod 5 = x$, we must solve the equation $(x+4)\bmod 5 = 2$. By testing all five elements of the field, we see that $x = 3$ is the only solution to this equation. Therefore, (2 – 4) mod 5 = 3. This result may be obtained faster if we make use of the identity 2 = 7 mod 5, so that (2 – 4) mod 5 = (7 – 4) mod 5 = 3.
multiplication	$2 \cdot 3$ mod 5 = 6 mod 5 = 1 , 4^2 mod 5 = 16 mod 5 = 1.
division	is inverse to multiplication. To find (2/3) mod 5 = x, we must solve the Eq. $3x$ mod 5 = 2. By testing all five elements of the field we determine that $x = 4$ is the only solution of this equation. Therefore, (2/3) mod 5 = 4. It is convenient to find the inverse of each field element (except for 0): 1^{-1} = 1/1 = 1, 2^{-1} = 1/2 mod 5 = 3, 3^{-1} = 1/3 mod 5 = 2, 4^{-1} = 1/4 mod 5 = 4. Using these results as a table and the identity a/b mod 5 = $a \cdot b^{-1}$ we find that 2/3 mod 5 = $2 \cdot 3^{-1} = 2 \cdot 2 = 4$, etc.

Using the ground field we can define a more general GF(p^n). Let $q(x)$ be n-th degree polynomial with the coefficient from the ground field GF(p). Then, the field GF(p^n) can be constructed as the set of all polynomials whose degree is less than n

$$a(x) = \sum_{i=0}^{n-1} a_i x^i$$

with the coefficients a_i from the ground field GF(p), and modulo *field generator polynomial* $q(x)$ operations between the polynomials. The field generator polynomial must be *primitive* [that is, not factorable and not divide any polynomial $x^k - 1$ for $k < 2^n - 1$ over GF(p)]. According to this definition, the sum (or difference) of two polynomials is the polynomial whose coefficients are modulo p sums (or differences) of their corresponding coefficients.

The definition of the product of two polynomials is more complex. The formal product of two polynomials can be presented as

$$\sum_{i=0}^{n-1} a_i x^i \cdot \sum_{i=0}^{n-1} b_i x^i = \sum_{i=0}^{2n-2} c_i x^i$$

where

$$c_i = \sum_{k=0}^{i} a_k b_{i-k} \bmod p$$

The degree of the product $c(x)$ may be greater than $n-1$, so that it may not belong to the set of polynomials with degrees less than n. However, the remainder after division of $a(x)b(x)$ by $q(x)$ has a degree less than n, which we define as the product $a(x)b(x) \bmod q(x)$ in GF(p^n).

Since a polynomial

$$a(x) = \sum_{i=0}^{n-1} a_i x^i$$

is uniquely represented by its coefficients, we sometimes use vector notation to represent field elements: $\mathbf{a} = (a_0, a_1, ..., a_{n-1}) = row\{a_i\} \in \text{GF}(p^n)$. The correspondence between these two equivalent representations is denoted as $a(x) \sim \mathbf{a}$.

The most important fields used in practice are GF(2^k) because binary operations are the basic operations of the vast majority of digital devices. In this case, the ground field FG(2) operations have special notations. The modulo 2 addition or subtraction is called *exclusive-*

OR and is denoted as $\oplus$: $0 \oplus 0 = 0$, $1 \oplus 0 = 1$, $0 \oplus 1 = 1$, $1 \oplus 1 = 0$. The modulo 2 multiplication is normally called the *AND* operation and is denoted as & : $0\&0 = 0\&1 = 1\&0 = 0$, $1\&1 = 1$. The field GF(2^k) addition or subtraction is usually called *bitwise exclusive-OR*. However, we will not use these special notations in the future if it is clear that the operations are carried out in GF(2^k).

Example 4.2.2. Consider the field GF(2^3) with the generator polynomial $q(x) = 1 + x + x^3$. This polynomial is primitive. Indeed, it is not factorable (which can be proved by trying to divide it by all possible linear and quadratic polynomials having binary coefficients) and it does not divide x^3+1, x^4+1, x^5+1, and x^6+1. The field consists of all polynomials whose degree is less than 3: $a(x) = a_0 + a_1 x + a_2 x^2$ or, equivalently, all possible three-bit-long sequences $\mathbf{a} = (a_0, a_1, a_2)$.

Let us illustrate the field GF(2^3) operations between two numbers $\mathbf{a} = (0,1,0)$ and $\mathbf{b} = (0,1,1)$. Employing the bitwise exclusive-OR, we obtain $(0,1,0) + (0,1,1) = (0,0,1)$. To find the product of these numbers, we represent them as polynomials: $a(x) = x$, $b(x) = x + x^2$. Then

$$a(x)b(x) = x \cdot (x + x^2) \bmod q(x) = (x^2 + x^3) \bmod q(x)$$

The polynomial division

$$\begin{array}{r|l} & \overline{1 \quad \text{(quotient)}} \\ x^3 + x + 1 & \; x^3 + x^2 \\ & \underline{x^3 \qquad\quad + x + 1} \\ & \qquad x^2 + x + 1 \quad \text{(remainder)} \end{array}$$

gives the remainder $1 + x + x^2$, and we can write $(0,1,0) \cdot (0,1,1) = (1,1,1)$.

In every GF(p^n) there is a so-called *primitive element* α such that any nonzero element of the field can be expressed as a power of α.

Let us illustrate this using the Galois field of Example 4.2.2. Indeed, let $\boldsymbol{\alpha} = (0,1,0)$ which corresponds to the polynomial $\alpha = x$.

$$\begin{array}{ll}
\alpha^1 = x \sim & (0,1,0) \\
\alpha^2 = x^2 \sim & (0,0,1) \\
\alpha^3 = x^3 \bmod (1 + x + x^3) = 1 + x \sim & (1,1,0) \\
\alpha^4 = \alpha^3 x = x + x^2 \sim & (0,1,1) \\
\alpha^5 = \alpha^4 x \bmod (1 + x + x^3) = 1 + x + x^2 \sim & (1,1,1) \\
\alpha^6 = 1 + x^2 \sim & (1,0,1) \\
\alpha^7 = 1 \sim & (1,0,0)
\end{array}$$

Galois fields are important in practice because they enable us to apply mathematical methods to optimize the performance of digital communication systems. Many familiar mathematical methods are directly applicable to these fields. For example, we can solve a system of linear equations by using the standard Gaussian algorithm; a linear system has a unique solution if its determinant is not equal to zero. In the next section, we illustrate applications of matrix theory and linear algebra to channel-coding algorithms based on the Galois field notion.

4.2.2 Linear Codes

Suppose that elements of a finite field represent symbols $s_1, s_2, \ldots, s_k$ that are transmitted over a noisy channel. Let us also assume that any element of the field can be transmitted so that there is no redundancy in the alphabet. Then it is impossible to detect an error, since a corrupted symbol is still a legitimate element of the alphabet. In order to be able to detect and/or correct errors we need to transmit some redundant symbols $s_{k+1}, s_{k+2}, \ldots, s_n$ that depend on the previously transmitted symbols. The initial k symbols in this case are called the *information* symbols and the added $r = n - k$ symbols are called the *parity check* symbols. The set of all possible combinations $s_1, s_2, \ldots, s_n$ of information and parity symbols is called a *systematic* (n,k) *code.*

We consider only the case where information and parity symbols satisfy a system of *linear* equations

$$\sum_{i=1}^{n} s_i h_{ji} = 0 \pmod{p} \qquad j = 1,2,..,r$$

This system can be rewritten in matrix form:

$$\mathbf{sH}' = 0 \qquad \mathbf{s} = row\{s_i\} \qquad \mathbf{H} = \left[h_{ij}\right]_{r,n} \tag{4.2.1}$$

The matrix **H** is called a *parity-check matrix.* The set of all possible solutions of this system is called the *linear* (n,k) *code* that corresponds to the parity-check matrix **H**; any element **s** of the code is called a *code word.*

A well-known result from linear algebra says that all possible solutions of a linear homogeneous system constitute a linear vector space. [13] Thus, this code represents a linear vector subspace of the n-dimensional vector space. If we denote $\mathbf{g}_1, \mathbf{g}_2, \ldots, \mathbf{g}_k$ as the code basis, then any code word can be represented as a linear combination of these vectors:

$$\mathbf{s} = \sum_{i=1}^{k} c_i \mathbf{g}_i = \mathbf{cG} \quad \text{where} \quad \mathbf{G}' = \begin{bmatrix} \mathbf{g}_1 & \mathbf{g}_2 & \cdots & \mathbf{g}_k \end{bmatrix}$$

is the matrix whose rows are the k basis vectors of the code, which is called the *code generator matrix.* From the definitions of **G** and **H** it follows that

$$\mathbf{GH}' = 0 \tag{4.2.2}$$

Suppose that, because of the channel errors, the transmitted code word **s** is received as $\tilde{\mathbf{s}}$. If $\tilde{\mathbf{s}}$ is not a code word, that is

$$\tilde{\mathbf{s}}\mathbf{H}' = \mathbf{S} \neq 0 \tag{4.2.3}$$

then the received information contains errors. We may try to correct the errors by selecting a code word that is closest to the received word $\tilde{\mathbf{s}}$, or we may just decide not to try to recover the information but instead ask to retransmit the whole message.

The vector **S** in the right-hand side of the previous equation is called the received message $\tilde{\mathbf{s}}$ *syndrome* because if $\mathbf{S} \neq 0$, it indicates the presence of errors. However, when $\mathbf{S} = 0$ the received information is not necessarily correct: the error pattern might be such that $\tilde{\mathbf{s}}$ is also a valid code word that differs from the transmitted **s**. In this case, the code is not able to detect the error pattern.

The difference between the received and transmitted words is called an *error word* $\mathbf{e} = \tilde{\mathbf{s}} - \mathbf{s}$. Since $\mathbf{sH} = 0$, the syndrome can be expressed through the error word

$$\mathbf{eH}' = \mathbf{S} \tag{4.2.4}$$

If errors are additive, the syndrome probability distribution depends only on the error word distribution. In this case, for each syndrome we can define the most probable error pattern $\mathbf{e}(\mathbf{S})$. Then the most probable transmitted word can be obtained by subtracting the most probable error pattern from the received word $\hat{\mathbf{s}} = \tilde{\mathbf{s}} - \mathbf{e}(\mathbf{S})$. This technique is called *maximum likelihood syndrome decoding*.

Example 4.2.3. Consider the binary (7,4) Hamming code whose parity-check matrix is given by

$$\mathbf{H} = \begin{bmatrix} 1 & 1 & 1 & 0 & 1 & 0 & 0 \\ 0 & 1 & 1 & 1 & 0 & 1 & 0 \\ 0 & 0 & 1 & 1 & 1 & 0 & 1 \end{bmatrix}$$

This code consists of all possible solutions of Eq. (4.2.1) and is the four-dimensional subspace of the seven-dimensional vector space. The basis of the code consists of four linearly independent vectors. The vectors

$$\mathbf{s}_1 = [1\ 0\ 1\ 1\ 0\ 0\ 0]$$
$$\mathbf{s}_2 = [0\ 1\ 0\ 1\ 1\ 0\ 0]$$
$$\mathbf{s}_3 = [0\ 0\ 1\ 0\ 1\ 1\ 0]$$
$$\mathbf{s}_4 = [0\ 0\ 0\ 1\ 0\ 1\ 1]$$

constitute the code basis, since the matrix has full rank (see Appendix 5):

$$rank \begin{bmatrix} 1 & 0 & 1 & 1 & 0 & 0 & 0 \\ 0 & 1 & 0 & 1 & 1 & 0 & 0 \\ 0 & 0 & 1 & 0 & 1 & 1 & 0 \\ 0 & 0 & 0 & 1 & 0 & 1 & 1 \end{bmatrix} = 4$$

This is true because the determinant of the leftmost square subblock equals 1:

$$\det \begin{bmatrix} 1 & 0 & 1 & 1 \\ 0 & 1 & 0 & 1 \\ 0 & 0 & 1 & 0 \\ 0 & 0 & 0 & 1 \end{bmatrix} = 1$$

Thus, the matrix whose rows are $\mathbf{s}_1$, $\mathbf{s}_2$, $\mathbf{s}_3$, and $\mathbf{s}_4$ can be selected as the code-generator matrix

$$\mathbf{G} = \begin{bmatrix} 1 & 0 & 1 & 1 & 0 & 0 & 0 \\ 0 & 1 & 0 & 1 & 1 & 0 & 0 \\ 0 & 0 & 1 & 0 & 1 & 1 & 0 \\ 0 & 0 & 0 & 1 & 0 & 1 & 1 \end{bmatrix}$$

Every code word can be represented as a linear combination (with binary coefficients) of the basis vectors

$$\mathbf{s} = c_1\mathbf{s}_1 + c_2\mathbf{s}_2 + c_3\mathbf{s}_3 + c_4\mathbf{s}_4 = \mathbf{cG}$$

Let us consider now the problem of error correction. Suppose that single errors in the block of seven bits have the highest probabilities. The syndrome of the first bit error $\mathbf{e}_1 = (1\ 0\ 0\ 0\ 0\ 0\ 0)$ is equal to $\mathbf{e}_1\mathbf{H}' = (1\ 0\ 0)$, which is the transposed first column of $\mathbf{H}$. Similarly, one can easily see that the i-th bit-error syndrome is the transposed i-th column of the matrix $\mathbf{H}$. Therefore, each syndrome uniquely identifies the position of a single error. If we know the position of the error, we can correct it by inverting the corresponding received bit. More specifically, if the syndrome of the received word is equal to the i-th column of the matrix **H**, then the i-th bit of the received word should be inverted.

By permuting columns of **H**, it is possible to obtain the parity-check matrix, whose i-th column is the binary representation of i:

$$\mathbf{H}_p = \begin{bmatrix} 0 & 0 & 0 & 1 & 1 & 1 & 1 \\ 0 & 1 & 1 & 0 & 0 & 1 & 1 \\ 1 & 0 & 1 & 0 & 1 & 0 & 1 \end{bmatrix}$$

then the decoding rule can be simplified: Invert i-th bit of the received word if the syndrome's binary representation is i. For instance, if $\mathbf{S} = (1 \ 0 \ 1)$, then $i = 5$ and $\mathbf{e}(\mathbf{S}) = (0 \ 0 \ 0 \ 0 \ 1 \ 0 \ 0)$. We can use a so-called 3-to-8 decoder to implement this algorithm in hardware.

An important requirement of code design consists of finding a parity-check matrix **H** such that the most probable error patterns have different syndromes. Another important property of a code is the complexity of its implementation. From this point of view, the class of cyclic codes is the most attractive class of codes.

4.2.3 Cyclic Codes

A linear code is called *cyclic* if a cyclic shift of any code word results in another code word. It is convenient to use polynomial notation when dealing with cyclic codes. The cyclic shift, for example, can be represented analytically as modulo $x^n - 1$ multiplication of a polynomial by x: $xa(x) \bmod (x^n - 1) = a_0 x + a_1 x^2 + \cdots + a_{n-2} x^{n-1} + a_{n-1}$, since $x^n = 1 \bmod (x^n - 1)$. Thus, if $a(x)$ is a code word, then modulo $x^n - 1$ products $xa(x)$, $x^2 a(x), \ldots, x^{n-1} a(x)$ are code words. Since any code is a linear vector space, the product of a code word by a number from the ground field is a code word, and the sum of any two code words is a code word. In terms of polynomials, this means that, if $a(x)$ is a code word and $b(x)$ is an arbitrary polynomial, then $b(x)a(x) \bmod (x^n - 1)$ is also a code word.

Every cyclic code has a code word $g(x) = g_0 + g_1 x + \ldots + g_{r-1} x^{r-1} + x^r$ of the minimal degree, which is called the *code-generator polynomial*, such that the whole code can be generated by multiplying this polynomial by arbitrary polynomials modulo $x^n - 1$:

$$c(x) = b(x) g(x) \bmod (x^n - 1) \tag{4.2.5}$$

Since any code word can be expressed as a linear combination of $g(x)$, $xg(x), \ldots,$ $x^{n-r-1} g(x)$, these polynomials represent a basis of the code, and the code-generator matrix can be written as

$$\mathbf{G} = \begin{bmatrix} x^{n-r-1} g(x) \\ x^{n-r-2} g(x) \\ \cdots \\ x g(x) \\ g(x) \end{bmatrix} = \begin{bmatrix} 1 & g_{r-1} & g_{r-2} & \cdots & g_0 & 0 & \cdots & 0 \\ 0 & 1 & g_{r-1} & \cdots & g_1 & g_0 & \cdots & 0 \\ \cdots & \cdots & \cdots & \cdots & \cdots & \cdots & \cdots & \cdots \\ 0 & 0 & \cdots & 1 & g_{r-1} & \cdots & g_0 & 0 \\ 0 & 0 & \cdots & 0 & 1 & \cdots & g_1 & g_0 \end{bmatrix} \tag{4.2.6}$$

The rank of this matrix is equal to $n - r$ because the determinant of its leftmost square subblock is equal to 1.

The corresponding parity-check matrix can be found by solving Eq. (4.2.2). One of the solutions may be expressed in the polynomial form

$$\mathbf{H}' = \left[x^{r-1}h(x) \;\; x^{r-2}h(x) \;\; \cdots \;\; xh(x) \;\; h(x) \right]$$

where

$$h(x) = (x^n - 1)/g(x) \tag{4.2.7}$$

is called the *parity-check polynomial*. Indeed, the product of the i-th row of the matrix $\mathbf{G}$ by the j-th column of the matrix $\mathbf{H}'$ is equal to zero:

$$x^{n-r-i}g(x)x^{r-j}h(x) = x^{n-i-j}(x^n - 1) = 0 \mod (x^n - 1)$$

Cyclic codes are popular because the previously described polynomial operations can easily be implemented in hardware by using shift registers.

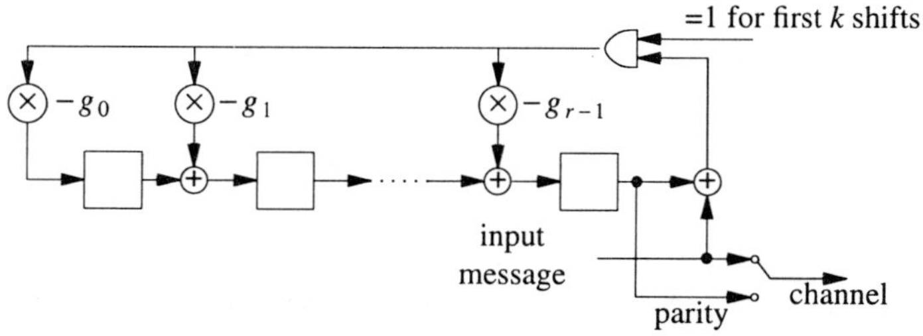

Figure 4.1. Encoder for cyclic (n,k) code.

For example, we can use the shift register shown in Fig. 4.1 to encode the information with the cyclic code whose generator polynomial is $g(x) = g_0 + g_1x + \cdots + g_{k-1}x^{k-1} + x^k$ This encoder works as follows: Initially the shift registers are empty. The AND gate is open, and k information symbols are shifted simultaneously into the circuit and the communication channel. After that, the AND gate is closed. The shift register contents are the r parity symbols, which are shifted into the channel.

It is not difficult to prove [27] that the shift register contents after k shifts can be expressed as $-r(x)$, where $r(x) = b_0+b_1x+...+b_{r-1}x_{r-1}$ is the remainder of division of the information polynomial $i(x) = s_1x^{n-1} + s_2x^{n-2} + \cdots + s_kx^{n-k}$ by the generator polynomial $g(x)$. If we denote $q(x)$ the quotient of the division, then the remainder is given by $r(x) = i(x) - q(x)g(x)$. The sequence transmitted over the channel consists of the information symbols that were shifted when the AND gate was open and the remainder multiplied by (-1) when the gate was closed. Using polynomial notation, the transmitted sequence can be expressed as $i(x) - r(x) = q(x)g(x)$ which, according to Eq. (4.2.5), is a code word.

In the case of the binary cyclic code, the encoder implementation is a fairly simple task because the ground field elements are either 0 or 1. The circuit shown in Fig. 4.1 consists only of binary adders and binary shift registers.

Example 4.2.4. Consider the (7,4) Hamming code with the generator polynomial $g(x)=1+x+x^3$. According to Eq. (4.2.6), its generator matrix is given by

$$\mathbf{G} = \begin{bmatrix} 1&0&1&1&0&0&0 \\ 0&1&0&1&1&0&0 \\ 0&0&1&0&1&1&0 \\ 0&0&0&1&0&1&1 \end{bmatrix}$$

The parity-check polynomial equals $h(x) = (x^7 - 1)/(x^3 + x + 1) = 1 + x + x^2 + x^4$. Parity-check matrix (4.2.7) has the following form

$$\mathbf{H} = \begin{bmatrix} 1&1&1&0&1&0&0 \\ 0&1&1&1&0&1&0 \\ 0&0&1&1&1&0&1 \end{bmatrix}$$

The encoder can be implemented using the circuit shown in Fig. 4.2.

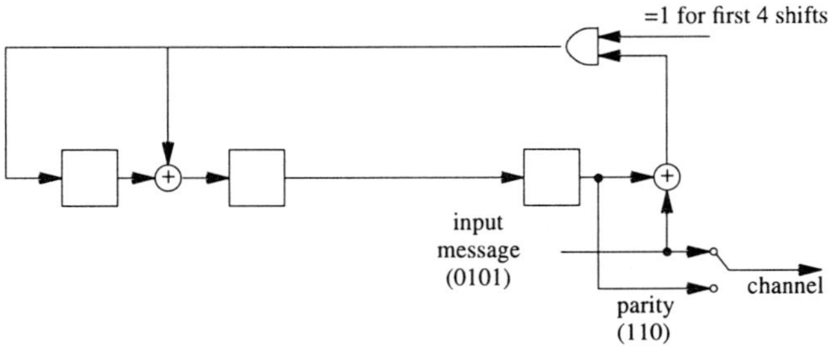

Figure 4.2. Encoder for the (7,4) Hamming code.

Suppose that the information sequence is 1010. After this sequence is shifted simultaneously into the channel and the shift register, the AND gate is closed and the shift register's contents, 110, are shifted into the channel. Thus, the transmitted code word is 1010011. The first four bits of this word represent the information, and the remaining three bits represent the parity.

Let us now address the problem of detecting and correcting errors by cyclic codes. If the transmitted code word is $s(x) = s_1 x^{n-1} + s_2 x^{n-2} + \cdots + s_n$ then errors are detected if the received word $\tilde{s}(x)$ does not belong to the code, or, in other words, it is not divisible by the code-generator $g(x)$. The same circuit that we used for encoding can be used for detecting errors, since it produces the division remainder. The information portion of the received word is shifted into the shift register, where its contents are compared with the parity-check portion of the received word. If they are different, the received word contains errors.

We can also shift the whole received word $\tilde{s}(x)$ into the shift register. Since the register's contents represent the remainder, the received polynomial is not divisible by the code-generator polynomial if the register's contents are not equal to zero. The polynomial that corresponds to the shift register's contents, after the whole received block is shifted into it is called the *syndrome polynomial.* The syndrome polynomial represents the syndrome vector defined by Eq. (4.2.3).

If channel errors $e(x)$ are additive $\tilde{s}(x) = s(x) + e(x)$ then the syndrome polynomial $S(x) = \text{remainder}\{\tilde{s}(x) / g(x)\} = \text{remainder}\{e(x) / g(x)\}$ depends only on errors, since the code word $s(x)$ is divisible by $g(x)$.

Thus, cyclic codes that are used only for detecting errors can be easily implemented. The same shift registers can perform encoding and decoding. The decoder implementation is usually much more complex when the codes are used for correcting errors. One method of error correction consists of selecting (decoding) a code word whose conditional probability, given the received word $\tilde{s}(x)$, is highest: $Pr\ [\ \hat{s}(x)\ |\ \tilde{s}(x)\ ] = \max$. In general, this method can be implemented only for short codes, because the number of code words grows exponentially with the code length. Let us now consider a very attractive class of codes that are both powerful and implementable in a simple manner.

4.2.4 Bose-Chaudhuri-Hocquenghem Codes

Let channel symbols be elements of GF(q) and α be an element of GF(q^m). Then the Bose-Chaudhuri-Hocquenghem (BCH) codes are defined by the lowest degree generator polynomial $g(x)$ over GF(q) whose roots are $\alpha^{m_0}, \alpha^{m_0+1}, ..., \alpha^{m_0+d-2}$, where m_0 and d are fixed numbers.

Any code word also has these roots, since it is divisible by $g(x)$:

$$s(\alpha^j) = \sum_{i=1}^{n} s_i \alpha^{(n-i)j} = 0 \qquad j = m_0, m_0+1, ..., m_0+d-2 \tag{4.2.8}$$

These equations suggest that the code parity-check matrix is

$$\mathbf{H} = \begin{bmatrix} \alpha^{(n-1)m_0} & \alpha^{(n-2)m_0} & \cdots & \alpha^{m_0} & 1 \\ \alpha^{(n-1)(m_0+1)} & \alpha^{(n-2)(m_0+1)} & \cdots & \alpha^{m_0+1} & 1 \\ \cdots & \cdots & \cdots\ \cdots & & \cdots \\ \alpha^{(n-1)(m_0+d-2)} & \alpha^{(n-2)(m_0+d-2)} & \cdots & \alpha^{m_0+d-2} & 1 \end{bmatrix} \tag{4.2.9}$$

Suppose that channel errors $e(x)$ are additive: $\tilde{s}(x) = s(x) + e(x)$. Then the received message syndrome may be expressed in the polynomial form:

$$\tilde{s}(\alpha^j) = e(\alpha^j) = S_j \qquad j = m_0, m_0+1, ..., m_0+d-2$$

These relationships can be treated as equations for the unknown errors that are the coefficients of the polynomial $e(x)$:

$$e(\alpha^j) = \sum_{i=1}^{n} e_i \alpha^{(n-i)j} = S_j \qquad j = m_0, m_0+1, ..., m_0+d-2 \tag{4.2.10}$$

If the number of errors t is not greater than $(d-1)/2$, Eq. (4.2.10) has a unique solution that determines the error values. In this case only t coefficients of the polynomial $e(x)$ differ from zero. If we denote them $Y_k = e_{i_k}$, $k = 1, 2, ..., t$, the previous system can be rewritten as

$$\sum_{k=1}^{t} Y_k X_k^j = S_j \qquad j = m_0, m_0+1, ..., m_0+d-2 \tag{4.2.11}$$

where $X_k = \alpha^{(n-i_k)}$. This system contains $2t$ unknowns $X_1, X_2, ..., X_t$, $Y_1, Y_2, ..., Y_t$ and is equivalent to the previous system. The values X_k uniquely identify error locations i_k and are called the *error locators*, while Y_k give the corresponding *error values*.

The problem of solving system (4.2.11) is similar to the problem of approximating a probability distribution by the polygeometric distribution (3.4.8). Applying Prony's method (described in Sec 3.4.5) to the system solution, [2,37] we obtain a linear difference equation

$$S_l + \sum_{j=1}^{t} \sigma_j S_{l-j} = 0 \qquad (4.2.12)$$

whose characteristic equation

$$x^t + \sum_{j=1}^{t} \sigma_j x^{t-j} = 0 \qquad (4.2.13)$$

roots are the error locators X_k. The corresponding error values are found by solving system (4.2.11).

These results can be summarized as the following algorithm [2] of BCH code decoding:

- Find the coefficients σ_i of difference Eq. (4.2.12) that generates the syndrome sequence.
- Find the error locators X_k as the roots of characteristic Eq. (4.2.13).
- Find the error values Y_k by solving Eq. (4.2.11). This step is not needed in decoding the binary BCH codes, since then $Y_k = 1$.
- Correct errors by subtracting the error values Y_k from the received symbols $\tilde{s}_{n-i_k}$.

System (4.2.12) is a Toeplitz system, therefore, the fast recursive algorithms discussed in the previous chapter can be used to solve it. However, not all the recursive methods are applicable in this case because the majority of them assume that the Toeplitz matrix principal determinants are not equal to zero. This is not a significant restriction with complex numbers, but it is intolerable with finite fields. Berlekamp's algorithm [2] or Euclid's algorithm [30] are free of these restrictions.

We can also improve the algorithm's performance if we notice that Eq. (4.2.10) can be treated as a part of the discrete Fourier transform of the received sequence. For the complete Fourier transform we need n equations rather than the $d-1$ equations that we have. If the number of errors is not greater than $(d-1)/2$, we can obtain as many syndromes as we need by making use of difference Eq. (4.2.12). Therefore, instead of solving the characteristic equation to find the error locators and subsequently solving system (4.2.11) to find the error values, we may proceed as follows:

- Find the linear difference Eq. (4.2.12) coefficients σ_i by using Berlekamp's or Euclid's algorithms.
- Compute the error sequence Fourier transform by using recurrence Eq. (4.2.12) for $l = m_0 + d - 1, \ldots, m_0 + n - 1$.
- Find the error polynomial estimate $\hat{e}(x)$ by applying the inverse Fourier transform.
- Subtract the estimated error sequence from the received sequence to correct errors.

The complexity of this method is proportional to $n \log^2 n$, whereas the complexity of the previous method is proportional to n^2. [17,26]

Example 4.2.5. The Hamming code considered in Example 4.2.4 can be treated as a BCH code. Its generator polynomial $g(x) = 1 + x + x^3$ is the lowest degree polynomial whose root α is the primitive element of GF(2^3) of Example 4.2.2. Indeed,

$$g(\alpha) = 1 + \alpha + \alpha^3 = (1,0,0) + (0,1,0) + (1,1,0) = 0$$

and no linear or quadratic polynomial with binary coefficients has a root α. The code is capable of correcting a single bit error because the condition $\alpha = \alpha^{m_0 + d - 2}$ is satisfied by

$m_0 = 1$ and $d = 2$. Suppose that a single bit error occurred in the first information bit. Equation (4.2.11) then becomes $\alpha^6 = S_1$, so that the syndrome uniquely identifies the error location: bit $i = 1$ must be complemented, since $\alpha^{7-i} = S_1$.

If we decide to use the Fourier transform method, we first find difference Eq. (4.2.12), which in our case takes the form $S_l + \sigma_1 S_{l-1} = 0$ with $\sigma_1 = \alpha^6$. Using this equation, we calculate $S_2 = \alpha^{12} = \alpha^5$, $S_3 = \alpha^{18} = \alpha^4$, $S_4 = \alpha^{24} = \alpha^3$, $S_5 = \alpha^{30} = \alpha^2$, $S_6 = \alpha^{36} = \alpha^1$, and $S_7 = \alpha^{42} = 1$. We find the error bit pattern by using the inverse Fourier transform $e_k = \sum_{j=1}^{7} S_j \alpha^{-(7-k)j}$ which is equal to 1 when $k = 1$ and to 0 otherwise. By adding (1 0 0 0 0 0 0) to the received word, we correct the error.

4.2.5 Reed-Solomon Codes

The Reed-Solomon (RS) code is the particular case of the BCH codes when $\alpha \in \mathrm{GF}(q)$. These codes are becoming increasingly popular because of their efficiency in correcting bursts of errors.

It is obvious that the RS code generator polynomial is given by

$$g(x) = (x - \alpha^{m_0})(x - \alpha^{m_0+1}) \cdots (x - \alpha^{m_0+d-2})$$

since this polynomial is the minimum degree polynomial that has the roots $\alpha^{m_0}, \alpha^{m_0+1}, \ldots, \alpha^{m_0+d-2}$. Being a particular case of the BCH codes, this code is able to correct up to $\lfloor (d-1)/2 \rfloor$. Since the generator polynomial degree $r = d - 1$ is equal to the number of the parity symbols, this code requires only $d - 1$ parity check symbols to be able to correct up to $\lfloor (d-1)/2 \rfloor$ symbol errors, whereas a general BCH code requires more than $d - 1$ parity-check symbols to be able to correct up to $\lfloor (d-1)/2 \rfloor$ symbol errors.

4.2.6 Convolutional Codes

In contrast with the block (n,k) codes, convolutional codes continuously encode information so that parity symbols of any block depend not only on the information symbols of this block but also on some symbols of previous blocks.

A *convolutional* (n,k,m) *code* can be defined by a difference equation

$$\mathbf{y}_j = \sum_{i=0}^{m} \mathbf{x}_{j-i} \mathbf{G}_i \tag{4.2.14}$$

where $\mathbf{G}_i$ are the $k \times n$ matrices, vectors $\mathbf{x}_i = [x_{i1} \; x_{i2} \; \ldots \; x_{ik}]$ represent the information symbols, and $\mathbf{y}_j = [y_{j1} \; y_{j2} \; \ldots \; y_{jn}]$ represent the encoded symbols. The matrices $\mathbf{G}_i$ used to perform the encoding called the *code-generator matrices*. We assume also that the elements of the matrices belong to a Galois field GF(q).

It is convenient to describe convolutions using polynomials. Denote the encoder input sequence

$$\mathbf{x}(D) = \sum_{i=0}^{\infty} \mathbf{x}_i D^i \tag{4.2.15}$$

the encoder output sequence

$$\mathbf{y}(D) = \sum_{i=0}^{\infty} \mathbf{y}_i D^i \tag{4.2.16}$$

and the code-generator polynomial

$$\mathbf{G}(D) = \sum_{i=0}^{m} \mathbf{G}_i D^i \tag{4.2.17}$$

Then, the convolution in Eq. (4.2.14) can be expressed as

$$\mathbf{y}(D) = \mathbf{x}(D)\mathbf{G}(D) \tag{4.2.18}$$

This equation is merely a compressed version of equation (4.2.14). If we *formally* multiply the polynomials and equate the coefficients of the same powers of D, we obtain Eq. (4.2.14).

Equation (4.2.14) defines nonsystematic codes in the general case, since the encoded information $\mathbf{y}_j$ might not contain the original information $\mathbf{x}_j$. In the particular case, when the code-generator polynomial has the special form

$$\mathbf{G}(D) = \begin{bmatrix} \mathbf{I} & \mathbf{G}_P(D) \end{bmatrix} \tag{4.2.19}$$

the code is systematic because the encoder output vector can be decomposed into the information part

$$\mathbf{y}_I(D) = \mathbf{x}(D) \tag{4.2.20}$$

and the parity part

$$\mathbf{y}_P(D) = \mathbf{x}(D)\mathbf{G}_P(D) \tag{4.2.21}$$

According to Eq. (4.2.14), convolutional encoders can be implemented using Galois field multipliers and adders. In the case of binary convolutional codes we need only binary adders.

Example 4.2.6. Let $\mathbf{G}(D) = [1+D+D^2 \quad 1+D^2]$ be a convolutional code-generator polynomial. Then the encoder input–output relationship can be expressed by the convolution in Eq. (4.2.14), which takes the form

$$\mathbf{y}_j = x_j[1 \quad 1] + x_{j-1}[1 \quad 0] + x_{j-2}[1 \quad 1] \tag{4.2.22}$$

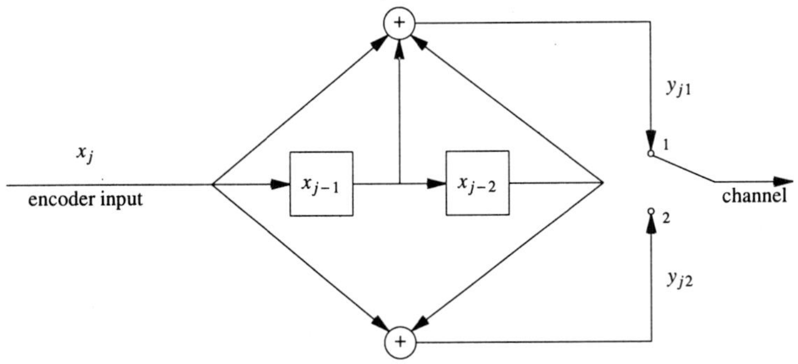

Figure 4.3. Encoder for the (2,1,2) convolutional code

and can be rewritten as

$$y_{j1} = x_j + x_{j-1} + x_{j-2}$$
$$y_{j2} = x_j + x_{j-2} \tag{4.2.23}$$

This encoder is shown in Fig. 4.3. One bit at a time is shifted into the shift register. For each input bit x_j, the encoder outputs two bits: y_{j1} and y_{j2} when the switch is in positions 1 and 2, respectively.

A convolutional encoder can be described as a linear sequential circuit. [14] Indeed, it is easy to verify that the convolution (4.2.14) can be expressed as

$$\mathbf{s}_{j+1} = \mathbf{s}_j\mathbf{A} + \mathbf{x}_j\mathbf{B}$$
$$\mathbf{y}_{j+1} = \mathbf{s}_j\mathbf{C} + \mathbf{x}_j\mathbf{G}_0 \tag{4.2.24}$$

where $\mathbf{s}_j = [\mathbf{x}_j \;\; \mathbf{x}_{j-1} \;\; \ldots \;\; \mathbf{x}_{j-m+1}]$ is the *encoder state* vector. [Sometimes the encoder state is defined as the number of shift registers used in its implementation: only the components of $\mathbf{s}_j$ that are actually used in the system (4.2.24) constitute the state.]

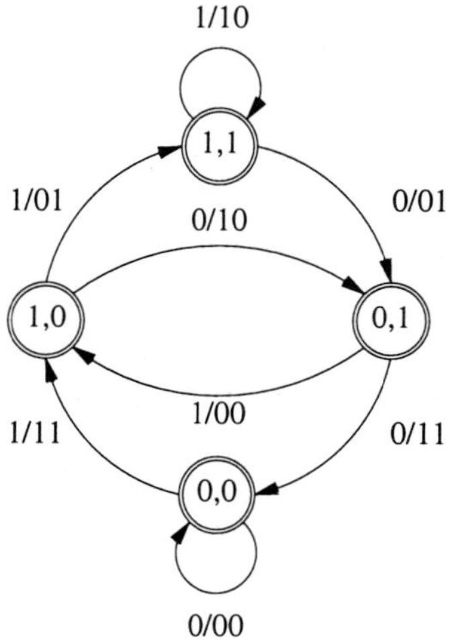

Figure 4.4. Encoder state diagram.

Matrices **A**, **B**, and **C** are given by

$$\mathbf{A} = \begin{bmatrix} 0 & \mathbf{I} & 0 \ldots & 0 \\ 0 & 0 & \mathbf{I} \ldots & 0 \\ \ldots & \ldots & \ldots & \ldots \\ 0 & 0 & 0 \ldots & \mathbf{I} \\ 0 & 0 & 0 \ldots & 0 \end{bmatrix} \quad \mathbf{B} = \begin{bmatrix} \mathbf{I} & 0 \ldots & 0 \end{bmatrix} \quad \mathbf{C} = \begin{bmatrix} \mathbf{G}_1 \\ \mathbf{G}_2 \\ \ldots \\ \mathbf{G}_m \end{bmatrix} \tag{4.2.25}$$

In this interpretation the encoder output $\mathbf{y}_j$ and the next state $\mathbf{s}_j$ depend on its input $\mathbf{x}_j$ and the previous state $\mathbf{s}_{j-1}$. The encoder state diagram is a flow graph whose nodes represent the encoder states, and the symbols I/O along the edges represent the encoder input and output symbols. The state diagram of the encoder of Fig. 4.3 is shown in Fig. 4.4. Sometimes we want to show a convolutional code state transition in time. In this case we treat states $\mathbf{s}_j$ and $\mathbf{s}_{j-1}$ as elements of different sets of states. The state diagram of this process is called a *trellis diagram*. The trellis diagram that corresponds to the state diagram of Fig. 4.4 is shown in Fig. 4.5.

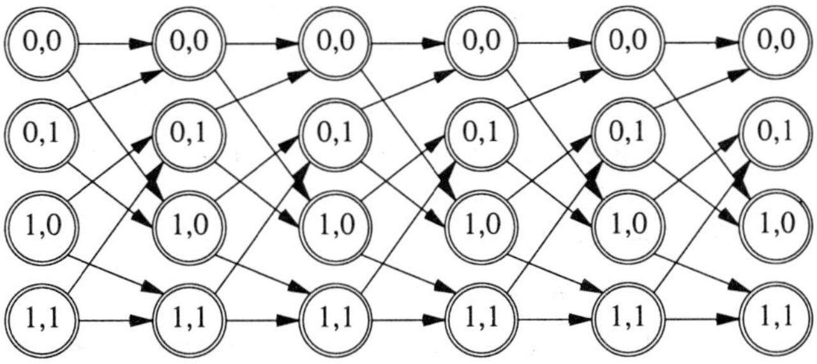

Figure 4.5. Encoder trellis diagram.

State and trellis diagrams are useful in describing convolutional code operation and evaluating its performance.

Let us now address the problem of decoding cyclic codes. We consider here only the algorithms whose performance will be evaluated later in this chapter.

4.2.6.1 Syndrome Decoding

Let us consider syndrome decoding of convolutional codes. Suppose that we have a systematic convolutional code and that the channel errors are additive. Then, the transmitted information is expressed by Eq. (4.2.20) and (4.2.21) and the received information is given by $\mathbf{r}(D) = \mathbf{x}(D)\mathbf{G}(D) + \mathbf{e}(D)$. where $\mathbf{e}(D)$ is the channel error polynomial. Since the code is systematic, the information part and the parity-check information of the received data are given by $\mathbf{r}_I(D) = \mathbf{x}_I(D) + \mathbf{e}_I(D)$ and $\mathbf{r}_P(D) = \mathbf{x}_P(D)\mathbf{G}_P(D) + \mathbf{e}_P(D)$, respectively. The difference

$$\mathbf{s}(D) = \mathbf{r}_I(D)\mathbf{G}_P(D) - \mathbf{r}_P(D) \tag{4.2.26}$$

between the parity check of the received information and the received parity check indicates the presence of errors and is called the *error syndrome polynomial.* This syndrome does not depend on the transmitted information but only on errors:

$$\mathbf{s}(D) = \mathbf{e}_I(D)\mathbf{G}_P(D) - \mathbf{e}_P(D) \tag{4.2.27}$$

since $\mathbf{y}_I(D)\mathbf{G}_P(D) - \mathbf{y}_P(D) = 0$.

If $\hat{\mathbf{e}}(D)$ is a solution to Eq. (4.2.27), then the received information is corrected by subtracting the estimated information error sequence from the received information: $\hat{\mathbf{x}}(D) = \mathbf{y}_I(D) - \hat{\mathbf{e}}_I(D)$. Denote $\mathbf{s}_j$, $\mathbf{e}_{Ij}$, and $\mathbf{e}_{Pj}$ as the coefficients of $\mathbf{s}(D)$, $\mathbf{e}_I(D)$, and $\mathbf{e}_p(D)$, respectively. Then, Eq. (4.2.27) can be rewritten as a system of linear equations for the unknown channel errors,

$$\mathbf{s}_j = \sum_{i=0}^{m} \mathbf{e}_{I(j-i)} \mathbf{G}_{Pi} - \mathbf{e}_{Pj} \tag{4.2.28}$$

This system may be rewritten as $\mathbf{s}_j = \sum_{i=0}^{m} \mathbf{e}_{j-i} \mathbf{H}_i$ where $\mathbf{e}_i = (\mathbf{e}_{Ii}, \mathbf{e}_{Pi})$ and $\mathbf{H}_i$ are the rows of the matrix $\mathbf{H}' = [\mathbf{G}_P \quad -\mathbf{I}]$ which is called a parity-check matrix. Any method of solving Eq. (4.2.28) represents an algorithm of error correction.

Usually, this system is broken down into subsystems. The subsystem that checks $\mathbf{e}_{Ij}$ has the form

$$\mathbf{s}_{j+k} = \sum_{i=0}^{m} \mathbf{e}_{I(j+k-i)} \mathbf{G}_{Pi} - \mathbf{e}_{P(j+k)} \qquad k = 1,2,\ldots,m \tag{4.2.29}$$

The decoding algorithm that estimates

$$\hat{\mathbf{e}}_{Ij} = \mathbf{f}(\mathbf{s}_j, \mathbf{s}_{j+1}, \ldots, \mathbf{s}_{j+m}) \tag{4.2.30}$$

from this subsystem is called *definite decoding*. It can be implemented using shift registers and a circuit that computes the preceding function (see Fig. 4.6).

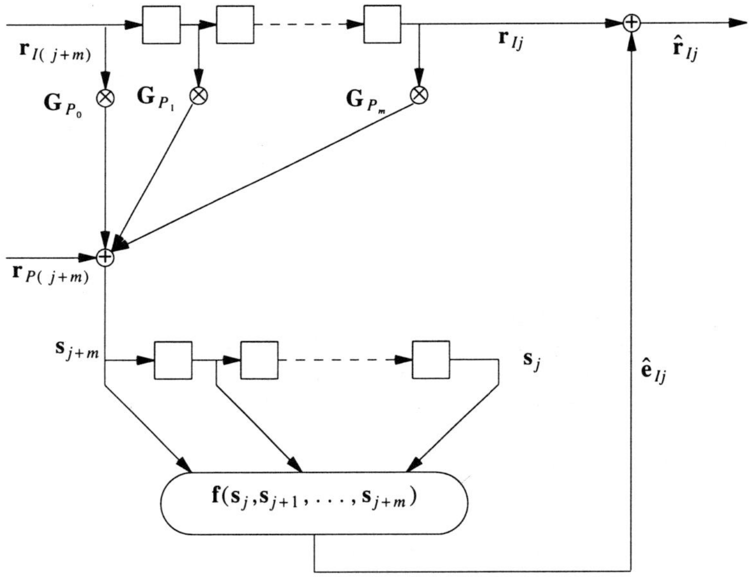

Figure 4.6. Decoder for the convolutional (n,k,m) code.

The received information sequence $\mathbf{r}_I(D)$ is re-encoded by using the encoder replica shown in upper part of Fig. 4.6 to produce $\mathbf{r}_I(D)\mathbf{G}_P(D)$. Then the received parity sequence $\mathbf{r}_P(D)$ is subtracted from $\mathbf{r}_I(D)\mathbf{G}_P(D)$, which, according to Eq. (4.2.26), yields the syndrome sequence $\mathbf{s}(D)$. The syndrome registers remember m most recent syndromes, and the decision circuit estimates error sequence according to Eq. (4.2.30).

The other technique, which is called *feedback decoding*, uses the estimates from the previous subsystems. If error estimates from the previous subsystems are subtracted from the actual errors, the system (4.2.29) becomes

$$\boldsymbol{\sigma}_j^{(k)} = \sum_{i=0}^{m} \boldsymbol{\varepsilon}_{I(j-i)}^{(k)} \mathbf{G}_{Pi} - \mathbf{e}_{P(j+k)} \qquad k=1,2,\ldots,m$$

where

$$\boldsymbol{\varepsilon}_{Ij}^{(k)} = \begin{cases} \mathbf{e}_{Ij} - \hat{\mathbf{e}}_{Ij} & \text{for } j<k, \\ \mathbf{e}_{Ij} & \text{for } j \ge k \end{cases}$$

If the estimated error values from the previous subsystems were correct, then $\boldsymbol{\varepsilon}_{Ij}^{(k)} = 0$ for $j<k$ and the system becomes significantly simpler.

The feedback decoding rule $\hat{\mathbf{e}}_{Ij} = \mathbf{f}(\boldsymbol{\sigma}_j^{(j)}, \boldsymbol{\sigma}_{j+1}^{(j)}, \ldots, \boldsymbol{\sigma}_{j+m}^{(j)})$ is based on the values $\boldsymbol{\sigma}_j^{(k)}$ that are called the *modified syndromes*. Note that $\boldsymbol{\sigma}_j^{(k)} = \mathbf{s}_j$ for $j \ge k+m$.

This technique is called feedback decoding because it can be implemented by feeding the estimated errors back into the syndrome registers. Indeed, one can easily verify by direct substitutions that the modified syndromes used for the $\hat{\mathbf{e}}_{I(j+1)}$ estimation satisfy the following recurrent equations:

$$\begin{aligned} \boldsymbol{\sigma}_{j+1}^{(j+1)} &= \boldsymbol{\sigma}_{j+1}^{(j)} - \hat{\mathbf{e}}_{Ij} \mathbf{G}_{P_1}, \\ \boldsymbol{\sigma}_{j+2}^{(j+1)} &= \boldsymbol{\sigma}_{j+2}^{(j)} - \hat{\mathbf{e}}_{Ij} \mathbf{G}_{P_2}, \\ &\cdots\cdots\cdots\cdots \\ \boldsymbol{\sigma}_{j+m}^{(j+1)} &= \mathbf{s}_{j+m} - \hat{\mathbf{e}}_{Ij} \mathbf{G}_{P_m}. \end{aligned} \tag{4.2.31}$$

These equations can be implemented by adding a feedback to the syndrome registers of the definite decoder, as shown in Fig. 4.7.

Like Eq. (4.2.24), Eq. (4.2.31) may be rewritten in the more compact form

$$\mathbf{u}_{j+1} = \mathbf{u}_j \mathbf{A} + \mathbf{s}_{j+m} \mathbf{B} - \hat{\mathbf{e}}_j \mathbf{G}_P \tag{4.2.32}$$

where $\mathbf{u}_j = (\boldsymbol{\sigma}_{j+m-1}^{(j)}, \boldsymbol{\sigma}_{j+m-2}^{(j)}, \ldots, \boldsymbol{\sigma}_j^{(j)})$ is the syndrome register state, $\mathbf{G}_P = (\mathbf{G}_{P_m}, \mathbf{G}_{P_{m-1}}, \ldots, \mathbf{G}_{P_1})$, and matrices **A** and **B** are defined by Eq. (4.2.25).

If the code has a special structure, the decoder decision element can be implemented by using simple majority-logic circuits. The most popular codes of this type are the *convolutional self-orthogonal codes* (CSOCs). [21,39,40,41]

In the case of a binary convolutional code, elements of system (4.2.28) are zeroes and ones. We say that e_{Ij} is checked by J equations of system (4.2.28) if this system has j equations of the form

$$s_{j_k} = e_{Ij} + \delta_{j_k} \qquad k=1,2,\ldots,J \tag{4.2.33}$$

where δ_{j_k} denotes the sum of other than e_{Ij} terms. Let us assume now that δ_{j_k} and δ_{j_l} for $k \ne l$ do not have any common terms e_{Ik}. The convolutional codes that possess this property are called *self-orthogonal*. They can be decoded by a simple majority-logic rule.

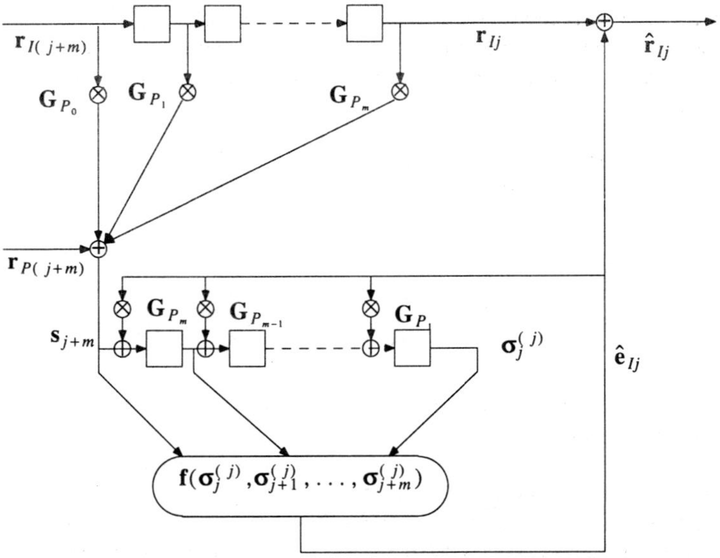

Figure 4.7. Feedback decoder for the (n,k,m) code.

If the number of errors that are checked by Eq. (4.2.33) is not greater than $\lfloor J/2 \rfloor$ and $e_{Ij}=1$, then $\delta_{j_k} = 0$ in more than $\lfloor J/2 \rfloor$ of the equations so that the majority of them have the form $s_{j_k} = 1$. If the number of errors that are checked by Eq. (4.2.33) is not greater than $\lfloor J/2 \rfloor$ and $e_{Ij}=0$, then $\delta_{j_k} = 1$ in no more than $\lfloor J/2 \rfloor$, so that the majority of Eq. (4.2.33) cannot have the form $s_{j_k} = 1$. Therefore, if the number of errors that are checked by the equations (4.2.33) is not greater than $\lfloor J/2 \rfloor$, the following algorithm corrects the error: Decode $e_{Ij} = 1$ when the majority of the syndromes in Eq. (4.2.33) are equal to 1, and decode 0 otherwise.

Example 4.2.7. Consider decoding of the systematic code with $\mathbf{G} = \begin{bmatrix} 1 & 1+D+D^4+D^6 \end{bmatrix}$. The consecutive equations of system (4.2.28) containing e_{Ij} are

$$
\begin{aligned}
s_j &= e_{I(j-6)} + e_{I(j-4)} + e_{I(j-1)} + e_{Ij} + e_{Pj} \\
s_{j+1} &= e_{I(j-5)} + e_{I(j-3)} + e_{Ij} + e_{I(j+1)} + e_{P(j+1)} \\
s_{j+2} &= e_{I(j-4)} + e_{I(j-2)} + e_{I(j+1)} + e_{I(j+2)} + e_{P(j+2)} \\
s_{j+3} &= e_{I(j-3)} + e_{I(j-1)} + e_{I(j+2)} + e_{I(j+3)} + e_{P(j+3)} \\
s_{j+4} &= e_{I(j-2)} + e_{Ij} + e_{I(j+3)} + e_{I(j+4)} + e_{P(j+4)} \\
s_{j+5} &= e_{I(j-1)} + e_{I(j+1)} + e_{I(j+4)} + e_{I(j+5)} + e_{P(j+5)}
\end{aligned}
$$

$$s_{j+6} = e_{Ij} + e_{I(j+2)} + e_{I(j+5)} + e_{I(j+6)} + e_{P(j+6)}$$

The part of this system that checks e_{Ij} is given by

$$\begin{aligned} s_j &= e_{I(j-6)} + e_{I(j-4)} + e_{I(j-1)} + e_{Ij} + e_{Pj} \\ s_{j+1} &= e_{I(j-5)} + e_{I(j-3)} + e_{Ij} + e_{I(j+1)} + e_{P(j+1)} \\ s_{j+4} &= e_{I(j-2)} + e_{Ij} + e_{I(j+3)} + e_{I(j+4)} + e_{P(j+4)} \\ s_{j+6} &= e_{Ij} + e_{I(j+2)} + e_{I(j+5)} + e_{I(j+6)} + e_{P(j+6)} \end{aligned}$$

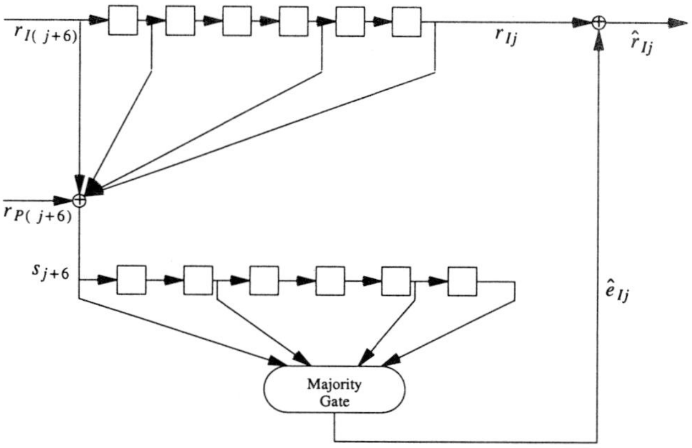

Figure 4.8. Decoder for the convolutional (2,1,6) code.

Each of these equations contains e_{Ij}, but they don't have any other common elements. Therefore, the code is self-orthogonal. If the number of errors in the total of seventeen symbols that are checked by these equations is not greater than two, then $e_{Ij} = 1$ only when the majority of the syndromes s_j, s_{j+1}, s_{j+4} and s_{j+6} are equal to 1. This method does not take into account the solution of the rest of the system and therefore represents definite decoding.

The decoder diagram is shown in Fig. 4.8. The received information symbols r_{Ik} are re-encoded, and the received parity symbols r_{Pk} are subtracted from the calculated parity of the received information symbols, thus producing syndromes s_k according to Eq. (4.2.28). The syndromes are shifted into the syndrome registers. Then the majority-logic element selects $\hat{e}_{Ij} = 1$ if at least three out of four syndromes that enter it are equal to 1, otherwise it selects $\hat{e}_{Ij} = 0$. The estimated error is added to the received information bit, thus correcting the error. To illustrate feedback decoding, let us consider the first four equations of the modified system:

$$\begin{aligned} s_0 &= e_{I0} + e_{P0} \\ s_1 &= e_{I0} + e_{I1} + e_{P1} \\ s_4 &= e_{I0} + e_{I3} + e_{I4} + e_{P4} \\ s_6 &= e_{I0} + e_{I2} + e_{I5} + e_{I6} + e_{P6} \end{aligned}$$

If the number of errors in the total of eleven symbols that are checked by these equations is not greater than two, the error e_{I0} is corrected. If the estimated error $\hat{e}_{I0}$ is equal to the actual error ($e_{I0} = \hat{e}_{I0}$), subtracting of the error from the previous equations eliminates it from the rest of the system.

The subtraction can be performed by feeding the estimated error $\hat{e}_{I0}$ back into syndrome registers (see Fig. 4.9). The contents of the syndrome registers after subtraction are the modified syndromes. The system that checks e_{I1} then becomes

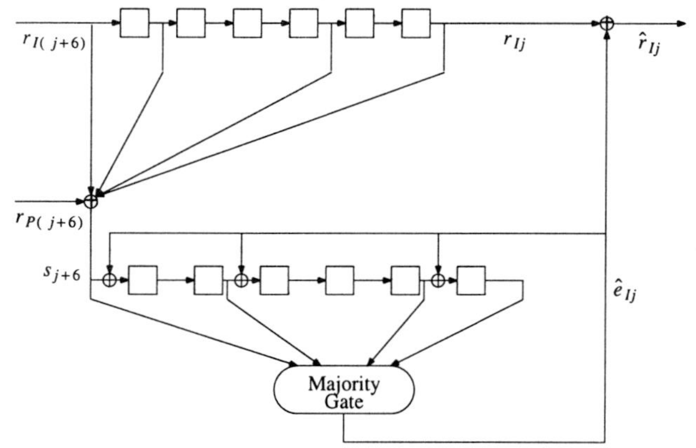

Figure 4.9. Feedback decoder for the (2,1,6) code.

$$\sigma_1 = e_{I1} + e_{P1}$$
$$s_2 = e_{I1} + e_{I2} + e_{P2}$$
$$s_5 = e_{I1} + e_{I4} + e_{I5} + e_{P5}$$
$$s_7 = e_{I1} + e_{I3} + e_{I6} + e_{I7} + e_{P7}$$

where $\sigma_1 = s_1 + e_{I0} + \hat{e}_{I0}$ is the modified syndrome. Again, if no more than two out of eleven symbols have errors and our previous estimation $\hat{e}_{I0}$ was correct, then e_{I1} is corrected. It is obvious that the process can be repeated. If the probability of two errors in eleven is smaller than the probability of having two errors in seventeen symbols, feedback decoding might be better than definite decoding.

The problem arises when an error is estimated incorrectly. Then the incorrect estimate is shifted into the syndrome registers, which could lead to additional decoding errors (error propagation of feedback decoding), even in the absence of channel errors. We investigate the performance merits of feedback decoding later in this chapter.

4.2.7 Trellis-Code Modulation

Trellis codes can be defined as the output X_t of the finite state machine

$$\begin{aligned} \mathbf{s}_{t+1} &= f_t(\mathbf{s}_t, \mathbf{x}_t) \\ \mathbf{y}_{t+1} &= g_t(\mathbf{s}_t, \mathbf{x}_t). \end{aligned} \tag{4.2.34}$$

whose input $\mathbf{x}_t$ is an information symbol and output $\mathbf{y}_t$ is the encoded symbol. The codes are called trellis codes because the encoder state transition can be presented by a trellis whose nodes represent the states $\mathbf{s}_t$ and edges represent the state transitions similar to the one shown in Fig. 4.5. Convolutional codes, according to Eq. (4.2.24), are trellis codes. The trellis codes are more general than the convolutional since Eq. (4.2.34) are nonlinear and time varying.

Symbols transmitted over a communication channel are usually different from the symbols produced by an encoder. A modulator performs the mapping of the encoder output to the channel symbols. Depending on the modulator, a part of the symbol or several encoded symbols are mapped onto a single modulated symbol. Since the mapping is deterministic, we can present the combined encoder-modulator as finite-state machine:

$$\begin{aligned} \boldsymbol{S}_{t+1} &= F_t(\boldsymbol{S}_t, \mathbf{X}_t) \\ \mathbf{Z}_{t+1} &= H_t(\boldsymbol{S}_t, \mathbf{X}_t). \end{aligned} \tag{4.2.35}$$

where $\mathbf{X}_t$ is a part or several input symbols that produce one modulated symbol $\mathbf{Z}_t$ and $\boldsymbol{S}_t$ is the corresponding state. The process described by these equations is called a *trellis code modulation*.

It follows from Eq. (4.2.35) that if the source symbols can be modeled by an HMM (and, in particular, are independent), the modulator output process is an HMM. In particular, if the source symbols are i.i.d. with the probability distribution $p(x_t)$, the HMM state transition probability has the form

$$p_{S_t S_{t+1}}(X_t) = \begin{cases} p(X_t), & \text{if } S_{t+1} = F_t(S_t, X_t) \\ 0 & \text{otherwise} \end{cases}$$

On this state transition the modulator outputs $Z_{t+1} = H(S_t, X_t)$. Thus,

$$Pr(S_{t+1}, Z_{t+1} \mid S_t) = \begin{cases} p(X_t), & \text{if } S_{t+1} = F_t(S_t, X_t) \text{ and } Z_{t+1} = H(S_t, X_t) \\ 0 & \text{otherwise} \end{cases} \tag{4.2.36}$$

Example 4.2.8. Consider the convolutional code of example 4.2.6. The state of the encoder in Fig. 4.3 can be described by the contents of its shift registers $\boldsymbol{s}_t = (x_t, x_{t-1})$. The convolutional code can be described by the finite-state machine given by Eq. (4.2.24)

$$\begin{aligned} \boldsymbol{s}_t &= \boldsymbol{s}_{t-1}\mathbf{A} + x_t\mathbf{B} \\ \mathbf{y}_t &= \boldsymbol{s}_{t-1}\mathbf{C} + x_t\mathbf{G}_0 \end{aligned}$$

where, according to (4.2.25),

$$\mathbf{A} = \begin{bmatrix} 0 & 1 \\ 0 & 0 \end{bmatrix}, \quad \mathbf{B} = [1 \ \ 0], \quad \mathbf{C} = \begin{bmatrix} 1 & 0 \\ 1 & 1 \end{bmatrix}, \quad \mathbf{G}_0 = [1 \ \ 1]$$

Let us assume now that the encoded symbols $\mathbf{y}_t = (y_{t,1}, y_{t,2})$ are modulated by the Quadrature-Phase-Shift-Keying (QPSK) modulator producing

$$O_t = W\,[(2y_{t,1} - 1) + j\,(2y_{t,2} - 1)] \tag{4.2.37}$$

The plot of all possible modulator output symbols is usually called the modulator *constellation*. The QPSK modulator constellation consists of four points $\pm W \pm W j$ on the complex plane.

Let us assume now that the binary source is memoryless and symmetric $[Pr(0) = Pr(1) = 0.5]$. Since the source is symmetric, the TCM system is modeled by a Markov chain whose transition probability matrix is computed according to its state transition diagram in Fig. 4.4:

$$\mathbf{P}_s = \begin{bmatrix} 0.5 & 0 & 0.5 & 0 \\ 0.5 & 0 & 0.5 & 0 \\ 0 & 0.5 & 0 & 0.5 \\ 0 & 0.5 & 0 & 0.5 \end{bmatrix} \tag{4.2.38}$$

whose states are ennumerated by the binary values of the encoder output $2y_{t,1} + y_{t,2}$

From the statistical point of view, the TCM can be described as a method of creating highly correlated sequences that are resistant to channel errors allowing us to recover the source sequences with high reliability. For a given encoder, the modulator can be selected to optimize the the system performance.

4.3 MAXIMUM A POSTERIORI DECODING

In the previous sections, we considered the so-called algebraic decoding algorithms which do not use explicitly channel and source statistics. However, the problem of decoding can be formulated in the general framework of pattern recognition where we are trying to recognize an object **S** given its (distorted) image **R**. In telecommunications, **S** is the source message and **R** is the received message. In speech recognition, **R** is what we hear, while **S** is the corresponding meaning.

To recognize an object, the image space is divided into acceptance regions $O_{\mathbf{S}}$, and **S** is recognized if $\mathbf{R} \in O_{\mathbf{S}}$. The regions are selected by optimizing a certain discrimination criterion which is application dependent.

In the model-based (probabilistic) pattern recognition, the two most popular discrimination criteria are: the likelihood and a posteriori probability (APP) criteria. For a given observation **R**, the *maximum likelihood* (ML) recognizer selects $\tilde{\mathbf{S}}$ which maximizes the conditional probability $Pr(\mathbf{R} \mid \mathbf{S})$:

$$\tilde{\mathbf{S}} = arg\max_{\mathbf{S}} Pr(\mathbf{R} \mid \mathbf{S})$$

while the *maximum a posteriori* (MAP) recognizer maximizes the APP

$$\hat{\mathbf{S}} = arg\max_{\mathbf{S}} Pr(\mathbf{S} \mid \mathbf{R})$$

These probabilities are closely related according to the Bayes theorem

$$Pr(\mathbf{S} \mid \mathbf{R}) = Pr(\mathbf{S},\mathbf{R}) / Pr(\mathbf{R}) = Pr(\mathbf{R} \mid \mathbf{S}) Pr(\mathbf{S}) / Pr(\mathbf{R})$$

Since the observation **R** is fixed, the MAP recognition rule can be written as

$$\hat{\mathbf{S}} = arg\max_{\mathbf{S}} Pr(\mathbf{S},\mathbf{R}) = arg\max_{\mathbf{S}} Pr(\mathbf{R} \mid \mathbf{S}) Pr(\mathbf{S}) \tag{4.3.1}$$

In applications to telecommunications, $Pr(\mathbf{S})$ is obtained on the basis of the source model, while $Pr(\mathbf{R} \mid \mathbf{S})$ is defined by the channel model. In speech recognition, $Pr(\mathbf{S})$ and $Pr(\mathbf{R} \mid \mathbf{S})$ represent the language and acoustic models, respectively. The probability $Pr(\mathbf{S},\mathbf{R})$ gives us a composite description of both the source and channel. We assume that they are modeled by an IOHMM described in Sec. 1.1.2. The model parameters are estimated on the basis of experimental data which is usually called the model *training*. It follows from Eq. (4.3.1) that ML and MAP estimates coincide if $Pr(\mathbf{S}) = const$, that is, if the source **S** is uniformly distributed (and, in particular, is not random). In this sense we can think of the ML recognition as a special case of the MAP recognition. Thus, all the methods for the MAP recognition can be applied to the ML recognition.

In the sequel, we consider MAP decoding of discrete information. However, it follows from the previous discussion that these methods are applicable to other recognition tasks (such as speech and handwriting recognition, information decryption, signal quantization, etc.).

If $\mathbf{S} = \mathbf{x}_1^T$ is the transmitted sequence and $\mathbf{R} = \mathbf{y}_1^T$ is the received sequence, then MAP *sequence decoding* is defined by maximizing the sequence APP (or PDF)

$$\hat{\mathbf{x}}_1^T = arg\max_{\mathbf{x}_1^T} Pr(\mathbf{x}_1^T \mid \mathbf{y}_1^T) \tag{4.3.2}$$

while MAP *symbol decoding* is represented by

$$\hat{x}_t = \underset{x_t}{argmax}\ Pr(x_t \mid \mathbf{y}_1^T) \tag{4.3.3}$$

Note that Eq. (4.3.2) and (4.3.3) have, in general, different solutions. The MAP decoded symbols even might represent a sequence that can not be produced by the source.

4.3.1 Map Symbol Decoding

Since the received sequence $\mathbf{y}_1^T$ is fixed, it is sufficient to maximize the unnormalized APP $Pr(x_t,\mathbf{y}_1^T)=Pr(x_t \mid \mathbf{y}_1^T)Pr(\mathbf{y}_1^T)$:

$$\hat{x}_t = \underset{x_t}{argmax}\ Pr(x_t,\mathbf{y}_1^T) \tag{4.3.4}$$

For the IOHMM, we have

$$Pr(x_t,\mathbf{y}_1^T) = \boldsymbol{\pi}\prod_{i=1}^{t-1}\mathbf{P}(y_i)\mathbf{P}(x_t,y_t)\prod_{i=t+1}^{T}\mathbf{P}(y_i)\mathbf{1} \tag{4.3.5}$$

$$\mathbf{P}(y) = \sum_x \mathbf{P}(x,y)$$

is the observation marginal matrix probability. Equation (4.3.5) can be written as

$$Pr(x_t,\mathbf{y}_1^T) = \boldsymbol{\alpha}(\mathbf{y}_1^{t-1})\mathbf{P}(x_t,y_t)\prod_{i=t+1}^{T}\boldsymbol{\beta}(\mathbf{y}_{t+1}^T)$$

and, therefore, can be evaluated by the forward-backward algorithm (FBA). Since scaling does not change the solution of Eq. (4.3.4), we can use the normalized version of the FBA:

Algorithm 4.1.

`Initialize:` *(Forward part).*

$$\boldsymbol{\alpha}(\mathbf{y}_1^0) = \boldsymbol{\pi}$$

`For` $t=1,2,\ldots,T-1$

`Begin`

`compute and save`

$$\boldsymbol{\alpha}(\mathbf{y}_1^t) = \boldsymbol{\alpha}(\mathbf{y}_1^{t-1})\mathbf{P}(y_t)/c_t \qquad c_t = \boldsymbol{\alpha}(\mathbf{y}_1^{t-1})\mathbf{P}(y_t)\mathbf{1} \tag{4.3.6}$$

`End`

`Initialize:` *(Backward part).*

$$\boldsymbol{\beta}(\mathbf{y}_{T+1}^T) = \mathbf{1}$$

`For` $t=T,T-1,\ldots,2$

`Begin`

$$p(x_t,\mathbf{y}_1^T) = \boldsymbol{\alpha}(\mathbf{y}_1^{t-1})\mathbf{P}(x_t,y_t)\boldsymbol{\beta}(\mathbf{y}_{t+1}^T) \tag{4.3.7}$$

$$\boldsymbol{\beta}(\mathbf{y}_t^T) = \mathbf{P}(y_t)\boldsymbol{\beta}(\mathbf{y}_{t+1}^T)/c_{t-1} \tag{4.3.8}$$

`End`

If the input set has small cardinality, the maximum can be found by direct comparison. For example, if symbols x_t are binary, we need to compare only two numbers $p(x_t=0,\mathbf{y}_1^T)$ and $p(x_t=1,\mathbf{y}_1^T)$. Thus, in the binary case we have the following algorithm: Decode $x_t = 0$

if $p(x_t=0,\mathbf{y}_1^T) \geq p(x_t=1,\mathbf{y}_1^T)$, otherwise decode $x_t = 1$. This is the so-called BCJR algorithm. [1]

We see that if the input alphabet is small, the main difficulty in decoding represents the FBA. As we pointed out before, the algorithm requires an enormous amount of memory for saving all the forward vectors. This requirement can be reduced if we know the inverse matrices $\mathbf{P}^{-1}(y_t)$ because, in this case, we can recover $\boldsymbol{\alpha}(\mathbf{y}_1^{t-1})$ from $\boldsymbol{\alpha}(\mathbf{y}_1^t)$ by inverting equation (4.3.6):

$$\boldsymbol{\alpha}(\mathbf{y}_1^{t-1}) = c_t \boldsymbol{\alpha}(\mathbf{y}_1^t)\mathbf{P}^{-1}(y_t)$$

The problem with this approach is that if one of the matrices $\mathbf{P}^{-1}(y_t)$ does not exist, the process stops. The obvious way around this problem is to save only those $\boldsymbol{\alpha}(\mathbf{y}_1^{t-1})$ which are not recoverable from $\boldsymbol{\alpha}(\mathbf{y}_1^t)$.

4.3.1.1 The Forward-only Algorithm

To develop the forward-only algorithm, consider the following probability vectors

$$\mathbf{s}(x_t,\mathbf{y}_1^{t+\tau}) = \boldsymbol{\pi}\prod_{i=1}^{t-1}\mathbf{P}(y_i)\mathbf{P}(x_t,y_t)\prod_{i=t+1}^{\tau}\mathbf{P}(y_i) \tag{4.3.9}$$

It follows from this definition that $Pr(x_t,\mathbf{y}_1^T) = \mathbf{s}(x_t,\mathbf{y}_1^T)\mathbf{1}$ and $\mathbf{s}(x_t,\mathbf{y}_1^{t+\tau})$ can be evaluated using the forward-only algorithm. Indeed, Eq. (4.3.9) can be written as

$$\mathbf{s}(x_t,\mathbf{y}_1^{t+\tau})=\boldsymbol{\alpha}(\mathbf{y}_1^t)\mathbf{P}(x_t,y_t)\mathbf{P}(y_{t+1}^{t+\tau})$$

Consequently, we can write

$$\mathbf{s}(x_t,\mathbf{y}_1^t)=\boldsymbol{\alpha}_{t-1}\mathbf{P}(x_t,y_t), \quad \mathbf{s}(x_t,\mathbf{y}_1^{t+\tau+1})=\mathbf{s}(x_t,\mathbf{y}_1^{t+\tau})\mathbf{P}(y_{t+\tau})$$

This is the forward-only algorithm. It requires to save in memory and update all the vectors $\mathbf{s}(x_t,\mathbf{y}_1^{t+\tau})$.

4.3.1.2 The Fixed-Lag Algorithm

If the received sequence is too long we can use the approximation $Pr(x_t \mid y_1^T) \sim Pr(x_t \mid y_1^{t+\tau})$ where τ is a fixed number. In this case, we can use the forward-only algorithm requiring to save $\tau+1$ vectors. In particular, if $\tau=0$ (this case is called filtering), we need to save only $\boldsymbol{\alpha}(\mathbf{y}_1^t)$. The block-diagram of this algorithm is shown in Fig. 4.10

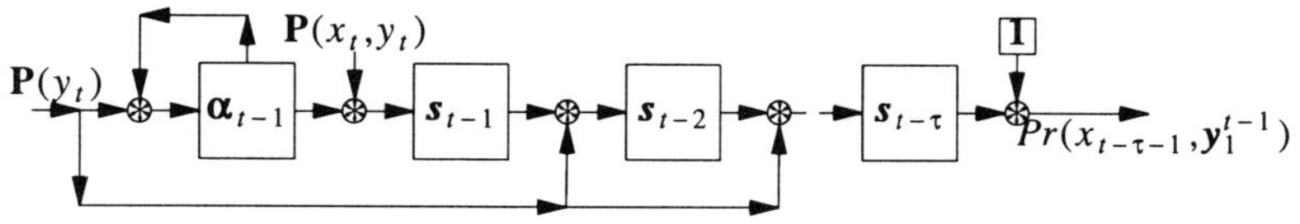

Figure 4.10. Block-diagram of the fixed-lag algorithm.

In this figure, we denoted $\boldsymbol{\alpha}_{t-1} = \boldsymbol{\alpha}(\mathbf{y}_1^{t-1})$ and $\mathbf{s}_{t-k} = \mathbf{s}(x_{t-k}, \mathbf{y}_1^{t-1})$, $*$ denotes the vector-matrix multiplication.

4.3.1.3 TCM Systems

Suppose that a source sequence $\mathbf{X}_1^T$ of i.i.d symbols enters a TCM system described by Eq. (4.2.35) and its output (trellis code modulated) sequence $\mathbf{O}_1^T$ is transmitted over an HMM channel described by the conditional PDF matrices $\mathbf{P}_c(Y_t \mid O_t)$ of receiving Y_t when O_t was transmitted. If we assume that the encoder-modulator and channel are statistically independent, the combined system can be described by Eq. (4.3.36) and the channel conditional PDF matrices as:

$$\mathbf{P}(X_t, Y_t) = [p_{S_{t-1}S_t}(X_t)\mathbf{P}_c(Y_t \mid O_t)]_{uv,uv} \tag{4.3.10}$$

where u is the size of the matrix (4.3.36), v is the size of the matrix $\mathbf{P}_c(Y_t \mid O_t)$, $O_t = H(S_{t-1}, X_t)$ and the only nonzero subblocks of this matrix correspond to $S_t = F(S_{t-1}, X_t)$. The matrix $\mathbf{P}(Y_t)$ has the form

$$\mathbf{P}(Y_t) = \sum_{X_t} \mathbf{P}(X_t, Y_t) = [p_{S_{t-1}S_t}\mathbf{P}_c(Y_t \mid O_t^{(S_{t-1}S_t)})]_{uv,uv}$$

This matrix is sparse which allows us to express the FBA in terms of the matrix subblocks. Direct multiplication of $\boldsymbol{\alpha}(\mathbf{Y}_1^{t-1}) = [\boldsymbol{\alpha}_1(\mathbf{Y}_1^{t-1}), \boldsymbol{\alpha}_2(\mathbf{Y}_1^{t-1}), \ldots, \boldsymbol{\alpha}_u(\mathbf{Y}_1^{t-1})]$ by the previous matrix gives us

$$\boldsymbol{\alpha}_j(\mathbf{Y}_1^t) = \sum_{i \in U(t,j)} \boldsymbol{\alpha}_i(\mathbf{Y}_1^{t-1}) p_{ij} \mathbf{P}_c(Y_t \mid O_t^{(ij)}) \tag{4.3.11}$$

where $U(t,j)$ is the set of solutions for i of the Eq. $j = F_t(i, X_t)$ for all possible X_t (defining all possible states $S_{t-1} = i$ from which transitions into the encoder state $S_t = j$ are possible) and $O_t^{(ij)} = H_t(i, X_t)$ is the corresponding modulator output symbol. These equations can be written using the code trellis diagram on which nodes $i = S_{t-1}$ and $j = S_t$ are connected iff $S_t = F(S_{t-1}, X_t)$. In this case, $U(t,j)$ corresponds to all the nodes S_{t-1} connected to the node $j = S_t$. We convert the trellis into the signal-flow graph by interpreting $p_{ij}\mathbf{P}_c(Y_t \mid O_t^{(ij)})$ as gain on the transition between the nodes. According to Eq. (4.3.11), $\boldsymbol{\alpha}_j(\mathbf{Y}_1^t)$ equals to the sum of products of $\boldsymbol{\alpha}_i(\mathbf{Y}_1^{t-1})$ times the corresponding gains for all the nodes $S_{t-1} = i$ connected to $S_t = j$.

The backward PDFs can be evaluated similarly:

$$\beta_i(Y_t^T) = \sum_{j\in V_{t,i}} p_{ij}\mathbf{P}_c(Y_t \mid O_t^{(ij)})\beta_j(Y_{t+1}^T)$$

Using these block vectors and Eq. (4.3.10), we can express $Pr(X_t, Y_1^T)$ in terms of these block vectors.

Example 4.3.1. Suppose that the binary source is memoryless and symmetric (Pr(0)=Pr(1)=0.5). The information bits are encoded by the rate 1/2 convolutional code of Example 4.2.6, QPSK modulated and transmitted over the flat fading channel with AWGN. Let us describe the MAP symbol decoding for this system.

Denote the source sequence as x_1^T. The encoded symbols $\mathbf{y}_1^T$ are described by the equations

$$s_t = s_{t-1}\begin{bmatrix}0 & 1\\ 0 & 0\end{bmatrix} + \mathbf{x}_t[1 \ \ 0]$$

$$\mathbf{y}_t = s_{t-1}\begin{bmatrix}1 & 0\\ 1 & 1\end{bmatrix} + \mathbf{x}_t[1 \ \ 1]$$

As we pointed out in Example 4. the QPSK modulator output, according to Eq. (4.3.37), is $O_t = A[(2y_{t1}-1) + j(2y_{t2}-1)]$. The encoder output is mapped to the QPSK constellation consisting of four symbols: $O^{(1)} = -A-Aj$, $O^{(2)} = -A+Aj$, $O^{(3)} = A-Aj$, and $O^{(4)} = A+Aj$. The modulator output is a Markov chain with the state transition probability matrix given by Eq. (4.2.38) which we repeat here for the convenience of the reader

$$\mathbf{P}_s = \begin{bmatrix}0.5 & 0 & 0.5 & 0\\ 0.5 & 0 & 0.5 & 0\\ 0 & 0.5 & 0 & 0.5\\ 0 & 0.5 & 0 & 0.5\end{bmatrix}$$

The flat fading channel is described by the following equation

$$r_t = h_t O_t + n_t \tag{4.3.12}$$

where h_t is a complex random process whose real and imaginary parts are independent and n_t is the AWGN. In many papers [25,33,35,36], it is assumed that the quantized fading process can be described by the HMM. Denote the quantized fading levels as $a_1, a_2, \ldots, a_n$. Then the fading HMM is described by the matrix probabilities $\mathbf{P}_f(a_i)$, $i=1,2,\ldots,n$. According to Eq. (4.3.12), the received process is also an HMM with the matrix probabilities

$$\mathbf{P}_c(r_t \mid O_t) = \frac{1}{2\pi\sigma^2}\sum_{k=1}^{n}\exp(-\| r_t - a_k O_t \|^2/2\sigma^2)\mathbf{P}_f(a_k)$$

The composite IOHMM is described by Eq. (4.3.10) which in our case has the form

$$\mathbf{P}(x=0,r) = 0.5\begin{bmatrix}\mathbf{P}_c(r \mid O^{(1)}) & 0 & 0 & 0\\ \mathbf{P}_c(r \mid O^{(4)}) & 0 & 0 & 0\\ 0 & \mathbf{P}_c(r \mid O^{(3)}) & 0 & 0\\ 0 & \mathbf{P}_c(r \mid O^{(2)}) & 0 & 0\end{bmatrix} \tag{4.3.13}$$

$$\mathbf{P}(x=1,r) = 0.5\begin{bmatrix}0 & 0 & \mathbf{P}_c(r \mid O^{(4)}) & 0\\ 0 & 0 & \mathbf{P}_c(r \mid O^{(1)}) & 0\\ 0 & 0 & 0 & \mathbf{P}_c(r \mid O^{(2)})\\ 0 & 0 & 0 & \mathbf{P}_c(r \mid O^{(3)})\end{bmatrix} \tag{4.3.14}$$

The matrix $\mathbf{P}(r) = \mathbf{P}(x=0,r) + \mathbf{P}(x=1,r)$ used in the FBA is given by

$$\mathbf{P}(r) = 0.5\begin{bmatrix}\mathbf{P}_c(r \mid O^{(1)}) & 0 & \mathbf{P}_c(r \mid O^{(4)}) & 0\\ \mathbf{P}_c(r \mid O^{(4)}) & 0 & \mathbf{P}_c(r \mid O^{(1)}) & 0\\ 0 & \mathbf{P}_c(r \mid O^{(3)}) & 0 & \mathbf{P}_c(r \mid O^{(2)})\\ 0 & \mathbf{P}_c(r \mid O^{(2)}) & 0 & \mathbf{P}_c(r \mid O^{(3)})\end{bmatrix}$$

As we can see, these matrices are sparse. Therefore, it is convenient to present the forward and backward algorithms in a block form which corresponds to the block form of the previous equation. We can also disregard the common coefficient 0.5, because it does not change the a posteriori PDF maximization. Equation (4.3.6) takes the form

$$\boldsymbol{\alpha}_1(\boldsymbol{r}_1^t) = \boldsymbol{\alpha}_1(\boldsymbol{r}_1^{t-1})\mathbf{P}_c(r_t \mid O^{(1)}) + \boldsymbol{\alpha}_2(\boldsymbol{r}_1^{t-1})\mathbf{P}_c(r_t \mid O^{(4)}) \tag{4.3.15a}$$

$$\boldsymbol{\alpha}_2(\boldsymbol{r}_1^t) = \boldsymbol{\alpha}_3(\boldsymbol{r}_1^{t-1})\mathbf{P}_c(r_t \mid O^{(3)}) + \boldsymbol{\alpha}_4(\boldsymbol{r}_1^{t-1})\mathbf{P}_c(r_t \mid O^{(2)}) \tag{4.3.15b}$$

$$\boldsymbol{\alpha}_3(\boldsymbol{r}_1^t) = \boldsymbol{\alpha}_1(\boldsymbol{r}_1^{t-1})\mathbf{P}_c(r_t \mid O^{(4)}) + \boldsymbol{\alpha}_2(\boldsymbol{r}_1^{t-1})\mathbf{P}_c(r_t \mid O^{(1)}) \tag{4.3.15c}$$

$$\boldsymbol{\alpha}_4(\boldsymbol{r}_1^t) = \boldsymbol{\alpha}_3(\boldsymbol{r}_1^{t-1})\mathbf{P}_c(r_t \mid O^{(2)}) + \boldsymbol{\alpha}_4(\boldsymbol{r}_1^{t-1})\mathbf{P}_c(r_t \mid O^{(3)}) \tag{4.3.15d}$$

where $\boldsymbol{\alpha}_i(\boldsymbol{r}_1^t)$ is the i-th component vector of $\boldsymbol{\alpha}(\boldsymbol{r}_1^t)$:

$$\boldsymbol{\alpha}(\boldsymbol{r}_1^t) = \begin{bmatrix} \boldsymbol{\alpha}_1(\boldsymbol{r}_1^t) & \boldsymbol{\alpha}_2(\boldsymbol{r}_1^t) & \boldsymbol{\alpha}_3(\boldsymbol{r}_1^t) & \boldsymbol{\alpha}_4(\boldsymbol{r}_1^t) \end{bmatrix}$$

We assume that, initially, the encoder was in the state zero, therefore

$$\boldsymbol{\alpha}(\boldsymbol{r}_1^0) = \begin{bmatrix} \boldsymbol{\pi}_c & 0 & 0 & 0 \end{bmatrix}$$

where $\boldsymbol{\pi}_c$ is the channel HMM initial state probability vector. Equations for $\boldsymbol{\beta}(\boldsymbol{r}_t^T)$ can be written similarly.

$$\boldsymbol{\beta}_1(\boldsymbol{r}_t^T) = \mathbf{P}_c(r_t \mid O^{(1)})\boldsymbol{\beta}_1(\boldsymbol{r}_{t+1}^T) + \mathbf{P}_c(r_t \mid O^{(4)})\boldsymbol{\beta}_3(\boldsymbol{r}_{t+1}^T)$$

$$\boldsymbol{\beta}_2(\boldsymbol{r}_t^T) = \mathbf{P}_c(r_t \mid O^{(4)})\boldsymbol{\beta}_1(\boldsymbol{r}_{t+1}^T) + \mathbf{P}_c(r_t \mid O^{(1)})\boldsymbol{\beta}_3(\boldsymbol{r}_{t+1}^T)$$

$$\boldsymbol{\beta}_3(\boldsymbol{r}_t^T) = \mathbf{P}_c(r_t \mid O^{(3)})\boldsymbol{\beta}_2(\boldsymbol{r}_{t+1}^T) + \mathbf{P}_c(r_t \mid O^{(2)})\boldsymbol{\beta}_4(\boldsymbol{r}_{t+1}^T)$$

$$\boldsymbol{\beta}_4(\boldsymbol{r}_t^T) = \mathbf{P}_c(r_t \mid O^{(2)})\boldsymbol{\beta}_2(\boldsymbol{r}_{t+1}^T) + \mathbf{P}_c(r_t \mid O^{(3)})\boldsymbol{\beta}_4(\boldsymbol{r}_{t+1}^T)$$

Usually the convolutional code is forced into the zero state by the encoder. In this case, we initialize $\boldsymbol{\beta}_1(\boldsymbol{r}_{T+1}^T) = \mathbf{1}$ and the rest of $\boldsymbol{\beta}_i(\boldsymbol{r}_{T+1}^T) = 0$.

As we pointed out before, instead of saving $\boldsymbol{\alpha}_i(\boldsymbol{r}_1^t)$ in memory, we can recover it in the backward part by reversing Eq. (4.3.15).

The symbol x_t unnormalized APP is given by Eq. (4.3.5) which in our case takes the form

$$p(x_t=0,\boldsymbol{r}_1^T) = C[\boldsymbol{\alpha}_1(\boldsymbol{r}_1^t)\boldsymbol{\beta}_1(\boldsymbol{r}_{t+1}^T)+\boldsymbol{\alpha}_2(\boldsymbol{r}_1^t)\boldsymbol{\beta}_2(\boldsymbol{r}_{t+1}^T)]$$

$$p(x_t=1,\boldsymbol{Y}_1^T) = C[\boldsymbol{\alpha}_3(\boldsymbol{r}_1^t)\boldsymbol{\beta}_3(\boldsymbol{r}_{t+1}^T)+\boldsymbol{\alpha}_4(\boldsymbol{r}_1^t)\boldsymbol{\beta}_4(\boldsymbol{r}_{t+1}^T)]$$

where the constant C compensates for the neglected common factor 0.5 at each step of the algorithm. If we apply scaling to improve the numerical stability of the algorithm (see Chapter 3), it will compensate for the neglected common factor automatically.

The symbol APPs are called *soft decisions.* Hard decisions are made by comparing the probabilities: We decode 0 if $p(x_t=0,\boldsymbol{Y}_1^T)>p(x_t=1,\boldsymbol{Y}_1^T)$, otherwise we decode 1.

4.3.1.4 Block Codes

Block codes can be also presented using trellises.[1,38] Indeed, let

$$\mathbf{H} = \begin{bmatrix} h_1 & h_2 & ,..., & h_N \end{bmatrix}$$

be a parity check matrix of a linear (N,K) code. Define the encoder states recursively:

$$S_0 = 0, \qquad S_t = S_{t-1} + x_t h_t$$

Since a codeword $\boldsymbol{x}_1^N$ is in the null space of $\mathbf{H}$:

$$\boldsymbol{x}_1^N\mathbf{H}' = \sum_{t=0}^{N} x_t h_t = 0$$

the trellis encoder needs to keep only the trajectories leading to state $S_N=0$. If the code is systematic, first K symbols are the information symbols x_1^K given by the source, and the parity check symbols x_{K+1}^N are uniquely defined by the path leading from state S_K to state

$S_N = 0$. Since we need to decode only the information symbols, the forward algorithm should be used only within the information sequence [$t = 1, 2, \ldots, K$ in Eq. (4.3.6)].

If the source is binary, there are only two possible transitions from the states corresponding to information bits:

$$S_t = \begin{cases} S_{t-1} & \text{if} \quad x_t = 0 \\ S_{t-1} + h_t & \text{if} \quad x_t = 1 \end{cases}$$

Consider a symmetric i.i.d. binary source which is encoded by an (N,K) systematic block code The encoded bits are transmitted over a binary symmetrical channel whose error source is modeled by symmetric binary HMM with the error matrix probabilities $\mathbf{P}_c(0)$ and $\mathbf{P}_c(1)$. In this case, the system is modeled by the IOHMM with the I/O transition matrix probability whose S_{t-1}, S_t-th block has the form $\mathbf{P}(x_t, y_t, S_t \mid S_{t-1}) = 0.5\mathbf{P}_c(x_t + y_t)$ for the information bits and $\mathbf{P}(x_t, y_t, S_t \mid S_{t-1}) = \mathbf{P}_c(x_t + y_t)$ for the parity bits. Since the coefficient 0.5 does not change the MAP estimate, we ignore it in the sequel. Thus, the I/O transition matrix probability $\mathbf{P}_t(x_t, y_t)$ can be written in a block form whose S_{t-1}, S_t-th block is $\mathbf{P}(x_t, y_t, S_t \mid S_{t-1}) = \mathbf{P}_c(x_t + y_t)$ if $S_t = S_{t-1} + x_t h_t$; otherwise it is a zero matrix. Thus,

$$\mathbf{P}_t(x_t = 0, y_t) = blockdiag\{\mathbf{P}_c(y_t)\} = \mathbf{I} \otimes \mathbf{P}_c(y_t)$$

and

$$\mathbf{P}_t(x_t = 1, y_t) = \mathbf{T}(h_t) \otimes \mathbf{P}_c(y_t + 1)$$

where $\mathbf{T}(h_t)$ is a matrix whose $(S_t, S_t + h_t)$-th and $(S_t + h_t, S_t)$-th elements are equal to 1 for all S_t, and the rest of the elements are equal to 0. Note that $\mathbf{I} + \mathbf{T}(h_t)$ is the trellis transition adjacency matrix: the identity matrix $\mathbf{I}$ corresponds to transitions between the same states $(S_{t-1} = S_t)$, while $\mathbf{T}(h_t)$ corresponds to transitions between different states. The encoder outputs 0 on transitions between the same states and it outputs 1 on transitions between different states (see Fig. 4.11). The matrix $\mathbf{P}_t(y_t) = \mathbf{P}_t(x_t = 0, y_t) + \mathbf{P}_t(x_t = 1, y_t)$ takes the form $\mathbf{P}_t(y_t) = \mathbf{I} \otimes \mathbf{P}_c(y_t) + \mathbf{T}(h_t) \otimes \mathbf{P}_c(y_t + 1)$.

Since the matrix $\mathbf{P}_t(y_t)$ is sparse, it is convenient to represent the forward an backward algorithms in the block form:

$$\boldsymbol{\alpha}(\mathbf{y}_1^t, S_t) = \boldsymbol{\alpha}(\mathbf{y}_1^{t-1}, S_t)\mathbf{P}_c(y_t) + \boldsymbol{\alpha}(\mathbf{y}_1^{t-1}, S_t + h_t)\mathbf{P}_c(y_t + 1) \tag{4.3.16a}$$

where $\boldsymbol{\alpha}(\mathbf{y}_1^t, S_t)$ denotes the S_t-th subblock of the forward probability vector $\boldsymbol{\alpha}(\mathbf{y}_1^t)$. These equations can be written using the code state trellis diagram. We put $\mathbf{P}_c(y_t)$ on the parallel transitions (corresponding to $x_t = 0$) and $\mathbf{P}(_t + 1)$ on the nonparallel transitions and interpret the trellis as the signal flow-graph. The transmission between two nodes is determined as a product of the $\boldsymbol{\alpha}(\mathbf{y}_1^{t-1}, S_{t-1})$ and the gain on the transition between the nodes.

Note, that if we replace S_t with $S_t + h_t$ in this equation, we obtain

$$\boldsymbol{\alpha}(\mathbf{y}_1^t, S_t + h_t) = \boldsymbol{\alpha}(\mathbf{y}_1^{t-1}, S_t + h_t)\mathbf{P}_c(y_t) + \boldsymbol{\alpha}(\mathbf{y}_1^{t-1}, S_t)\mathbf{P}_c(y_t + 1) \tag{4.3.16b}$$

The initial value of $\boldsymbol{\alpha}(\mathbf{y}_1^t)$ has the form

$$\boldsymbol{\alpha}(\mathbf{y}_1^0, S_0 = 0) = \boldsymbol{\pi}_c \qquad \boldsymbol{\alpha}(\mathbf{y}_1^0, S_0 \neq 0) = 0$$

For the subsequent blocks $\boldsymbol{\pi}_c$ should be replaced by $\boldsymbol{\pi}_c \mathbf{P}_c^m$, where m is the number of bits received from the beginning of the first block reception. If m is large, the stationary probability vector ($\boldsymbol{\pi}_c = \boldsymbol{\pi}_c \mathbf{P}_c$) should be used.

The backward algorithm has a similar form:

$$\boldsymbol{\beta}(\mathbf{y}_t^N, S_t) = \mathbf{P}_c(y_t)\boldsymbol{\beta}(\mathbf{y}_{t+1}^N, S_t) + \mathbf{P}_c(y_t+1)\boldsymbol{\beta}(\mathbf{y}_{t+1}^N, S_t+h_t)$$

Instead of saving $\boldsymbol{\alpha}(\mathbf{y}_1^{t-1}, S_t)$ in the forward part, we can reverse Eq. (4.3.16) by solving them with respect to $\boldsymbol{\alpha}(\mathbf{y}_1^{t-1}, S_t)$ and $\boldsymbol{\alpha}(\mathbf{y}_1^{t-1}, S_t+h_t)$ which can be done by inverting its block-form matrix (see Appendix 5)

$$\begin{bmatrix} \mathbf{P}(y_t) & \mathbf{P}(y_t+1) \\ \mathbf{P}(y_t+1) & \mathbf{P}(y_t) \end{bmatrix}^{-1} = \begin{bmatrix} \mathbf{M}(y_t) & \mathbf{M}(y_t+1) \\ \mathbf{M}(y_t+1) & \mathbf{M}(y_t) \end{bmatrix}$$

where

$$\mathbf{M}(y_t) = [\mathbf{P}(y_t) - \mathbf{P}(y_t+1)\mathbf{P}^{-1}(y_t)\mathbf{P}(y_t+1)]^{-1}$$

and

$$\mathbf{M}(y_t+1) = -\mathbf{M}(y_t)\mathbf{P}(y_t+1)\mathbf{P}^{-1}(y_t)$$

The inverted system has the form

$$\boldsymbol{\alpha}(\mathbf{y}_1^{t-1}, S_t) = \boldsymbol{\alpha}(\mathbf{y}_1^t, S_t)\mathbf{M}(y_t) + \boldsymbol{\alpha}(\mathbf{y}_1^t, S_t+h_t)\mathbf{M}(y_t+1)$$

$$\boldsymbol{\alpha}(\mathbf{y}_1^{t-1}, S_t+h_t) = \boldsymbol{\alpha}(\mathbf{y}_1^t, S_t)\mathbf{M}(y_t+1) + \boldsymbol{\alpha}(\mathbf{y}_1^t, S_t+h_t)\mathbf{M}(y_t)$$

If the inverse matrix does not exists, the corresponding $\boldsymbol{\alpha}(\mathbf{y}_1^{t-1}, S_t)$ must be saved in the forward part of the algorithm.

Example 4.3.2. As an example, consider a (5,3) block code with the parity check matrix

$$\mathbf{H} = \begin{bmatrix} 1 & 1 & 0 & 1 & 0 \\ 0 & 1 & 1 & 0 & 1 \end{bmatrix}$$

The code trellis diagram is depicted in Fig. 4.11.

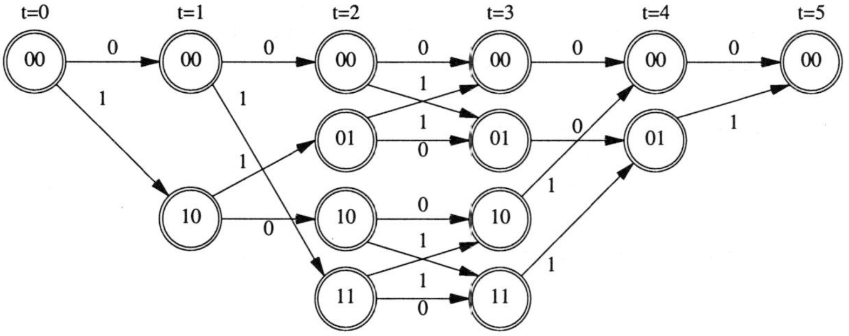

Figure 4.11. The encoder trellis diagram.

This code has four states: 00, 01 , 10, and 11. The encoder output bits mark the corresponding state transition edges: the (horizontal) transitions between the same states correspond to 0 and transitions between different states correspond to 1.

We convert the trellis in Fig. 4.11 into a signal-flow graph by $\mathbf{P}_c(y_t)$ as gain on the parallel transitions between nodes S_{t-1} and S_t (which are marked as 0) and $\mathbf{P}_c(y_t+1)$ on the nonparallel nodes. Using this flow-graph we can write

$$\boldsymbol{\alpha}(y_1, 00) = \boldsymbol{\pi}_c \mathbf{P}_c(y_1), \quad \boldsymbol{\alpha}(y_1, 10) = \boldsymbol{\pi}_c \mathbf{P}_c(y_1+1)$$

$$\boldsymbol{\alpha}(\mathbf{y}_1^2,00) = \boldsymbol{\alpha}(y_1,00)\mathbf{P}_c(y_2)$$
$$\boldsymbol{\alpha}(\mathbf{y}_1^2,01) = \boldsymbol{\alpha}(y_1,10)\mathbf{P}_c(y_2+1)$$
$$\boldsymbol{\alpha}(\mathbf{y}_1^2,10) = \boldsymbol{\alpha}(y_1,10)\mathbf{P}_c(y_2)$$
$$\boldsymbol{\alpha}(\mathbf{y}_1^2,11) = \boldsymbol{\alpha}(y_1,00)\mathbf{P}_c(y_2+1)$$

This ends the forward part of the algorithm because we need to decode only information bits x_1,x_2,x_3. We now apply the backward algorithm using the same signal-flow graph in reverse order:

$$\boldsymbol{\beta}(y_5,00) = \mathbf{P}_c(y_5)\mathbf{1}, \quad \boldsymbol{\beta}(y_5,01) = \mathbf{P}_c(y_5+1)\mathbf{1}$$
$$\boldsymbol{\beta}(\mathbf{y}_4^5,00) = \mathbf{P}_c(y_4)\boldsymbol{\beta}(y_5,00)$$
$$\boldsymbol{\beta}(\mathbf{y}_4^5,10) = \mathbf{P}_c(y_4+1)\boldsymbol{\beta}(y_5,00)$$
$$\boldsymbol{\beta}(\mathbf{y}_4^5,01) = \mathbf{P}_c(y_4)\boldsymbol{\beta}(y_5,01)$$
$$\boldsymbol{\beta}(\mathbf{y}_4^5,11) = \mathbf{P}_c(y_4+1)\boldsymbol{\beta}(y_5,01)$$

We are ready now to calculate the unnormalized APP for the bit x_3 using Eq. (4.3.7) corresponding to parallel transitions between the nodes at $t=2$ and $t=3$ in Fig. 4.11.

$$P(x_3=0,\mathbf{y}_1^5) = \boldsymbol{\alpha}(\mathbf{y}_1^2,00)\mathbf{P}_c(y_3)\boldsymbol{\beta}(\mathbf{y}_4^5,00) + \boldsymbol{\alpha}(\mathbf{y}_1^2,01)\mathbf{P}_c(y_3)\boldsymbol{\beta}(\mathbf{y}_4^5,01)$$
$$+ \boldsymbol{\alpha}(\mathbf{y}_1^2,10)\mathbf{P}_c(y_3)\boldsymbol{\beta}(\mathbf{y}_4^5,10) + \boldsymbol{\alpha}(\mathbf{y}_1^2,11)\mathbf{P}_c(y_3)\boldsymbol{\beta}(\mathbf{y}_4^5,11)$$

$P(x_3=1,\mathbf{y}_1^5)$ can be found similarly using the nonparallel transitions. We decode $x_3=0$ if $P(x_3=0,\mathbf{y}_1^5)>P(x_3=1,\mathbf{y}_1^5)$ or $x_3 = 1$, otherwise. Next, we make another step in the backward algorithm to calculate $P(x_2,\mathbf{y}_1^5)$ and so on. For a memoryless channel, it is possible to perform decoding without using the backward algorithm. [16]

4.3.1.5 Iterative Decoding

The MAP symbol estimation has become very popular because of its applications to iterative decoding of product and concatenated codes with simple component codes and, in particular, the so-called turbo codes. [4] The idea is to perform decoding iteratively by passing the intermediate results of decoding between the component codes. A component decoder estimates the symbol APP with the aid of the additional information obtained from the other components and passes back to them the estimated results. Thus, the component code operates in the "soft-in/soft-out" fashion: it receives soft decisions from the previous component, updates them and passes the updated soft decisions to the next component. [15] The process continues until convergence.

It has been shown by computer simulation that these codes perform remarkably well which stimulated interest in analyzing them. The FBA plays a central role in decoding these codes because the received sequence is modeled by an HMM and the symbol APP is evaluated by Eq. (4.3.3)

4.3.2 MAP Sequence Decoding

Suppose that the source and channel are modeled by an IOHMM according to which

$$Pr(\mathbf{x}_1^T,\mathbf{y}_1^T) = \boldsymbol{\pi}\prod_{i=1}^{T}\mathbf{P}(x_t,y_t)\mathbf{1} \tag{4.3.17}$$

where $\mathbf{x}_1^T$ is the source sequence and $\mathbf{y}_1^T$ is the received sequence.

The MAP sequence decoder maximizes the information sequence APP

$$\hat{\boldsymbol{x}}_1^T = \arg\max_{\boldsymbol{x}_1^T} Pr(\boldsymbol{x}_1^T, \boldsymbol{y}_1^T)$$

A sequence decoding algorithm is a method of solving this equation. Consider first a special IOHMM whose source sequence $\boldsymbol{x}_1^T$ is uniquely defined by its state sequence $\boldsymbol{s}_1^T$.

4.3.2.1 The Viterbi Algorithm

Given a sequence $\boldsymbol{y}_1^T$ of an HMM observations, the Viterbi algorithm performs recursively the MAP estimation of the sequence of its states. Therefore, if the transmitted sequence can be obtained from the sequence of states deterministically, the Viterbi algorithm can be used for the MAP sequence decoding.

To describe the Viterbi algorithm, we assume that the initial state s_0 is fixed. The joined probability distribution (or PDF) of the sequences $\boldsymbol{s}_0^T$ and $\boldsymbol{y}_1^T$ has the form

$$Pr(\boldsymbol{s}_1^T, \boldsymbol{y}_1^T) = \prod_{i=1}^{T} p_{s_{i-1}s_i}(y_i)$$

Our goal is to find a maximum of this function with respect to $\boldsymbol{s}_1^T$ or, equivalently, to minimize

$$L(\boldsymbol{s}_1^T) = -\log Pr(\boldsymbol{s}_1^T, \boldsymbol{y}_1^T) = -\sum_{i=1}^{T} \log p_{s_{i-1}s_i}(y_i)$$

This minimization problem can be solved by a version of dynamic programming which is called the *Viterbi algorithm.* [34]

It is convenient to explain the algorithm using the state trellis diagram (see Fig. 4.5 for an example). A sequence of states $\boldsymbol{s}_0^T$ represents a path on the trellis diagram. If we interpret $m_{s_{i-1}s_i} = -\log p_{s_{i-1},s_i}(y_i)$ as a distance between the nodes s_{i-1} and s_i, then $L(\boldsymbol{s}_1^T)$ is the total length of the path $\boldsymbol{s}_0^T$ in this metric. Thus, the problem of minimizing $L(\boldsymbol{s}_1^T)$ is equivalent to finding a shortest path from the original node s_0 to s_T.

The problem can be solved by comparing the lengths of all possible paths on the trellis diagram which requires an enormous amount of computations because the number of paths grows exponentially with T. The Viterbi algorithm allows us to solve this problem recursively. It is based on the following observation: Any part of the shortest path is also a shortest path. Thus, the shortest path $s_0, s_1^*, \ldots, s_t^*, s_{t+1}$ at s_{t+1} can be found from all the shortest paths $s_0, s_1^*, \ldots, s_{t-1}^*, s_t$ using the following Viterbi algorithm:

$$s_t^* = \arg\min_{s_t} \{L(s_0, s_1^*, \ldots, s_{t-1}^*, s_t) + m_{s_t s_{t+1}}\}$$

$$L(s_0, s_1^*, \ldots, s_t^*, s_{t+1}) = L(s_0, s_1^*, \ldots, s_{t-1}^*, s_t^*) + m_{s_t^* s_{t+1}}$$

If this solution is not unique, one of the optimal solutions is selected at random or using some other fair rule. [5] The shortest path ending at the node s_t is called a *survivor*. The Viterbi algorithm allows us to find the survivors at the moment $t+1$ from the survivors at the moment t recursively. At the end, we will have all the survivors ending at the states s_T. The last state is selected by finding the shortest survivor $s_T^* = \arg\min_{s_T} L(s_0, s_1^*, \ldots, s_{T-1}^*, s_T)$.

The Viterbi algorithm can be applied to the MAP decoding if the HMM states uniquely define the information sequence. If the information symbols on the input to a convolutional encoder are independent and identically distributed (i. i. d.) and channel errors are independent, then, according to Eq. (4.2.24), the encoder output can be described by an HMM whose state vector $\mathbf{s}_j = [\mathbf{x}_{j-1}\ \mathbf{x}_{j-2} \ldots \mathbf{x}_{j-m}]$ is uniquely defined by the source sequence. The encoded symbols $\boldsymbol{y}_1^T$ are transmitted over a channel with independent errors with the symbol error probability $p_c(\mathbf{e})$. Denote the received sequence $\boldsymbol{r}_1^T$. The received sequence $\boldsymbol{r}_1^T$ is modeled by an HMM with the state observation transition probabilities

$$p_{\mathbf{s}_{i-1}\mathbf{s}_i}(\mathbf{r}_i) = p_c(\mathbf{y}_i - \mathbf{r}_i)$$

(For the soft decision decoding, this probability is replaced by the corresponding PDF.) After obtaining the state sequence MAP estimate $\hat{\mathbf{s}}_j$, we find the information symbols $\hat{\mathbf{x}}_{j-1}$ by solving the equation $\hat{\mathbf{s}}_j = [\hat{\mathbf{x}}_{j-1}\ \hat{\mathbf{x}}_{j-2} \ldots \hat{\mathbf{x}}_{j-m}]$:

$$\hat{\mathbf{x}}_j = \hat{\mathbf{s}}_{j+1}[\,\mathbf{I}\ \ 0\ \ 0 \ldots 0\,]'$$

Example 4.3.3. Let us illustrate the Viterbi algorithm for the code of Example 4.2.6, assuming that encoder input is memoryless and symmetric $[p(0)=p(1)=0.5]$, channel errors are independent, and the bit-error rate is p. This communication system can be described as an IOHMM whose states are the encoder states and state-transition observation probabilities are given by the following equation

$$p_{s_{i-1}s_i}(y_i, r_i) = p_c(r_i - y_i) = p^e q^{2-e}$$

$e = HD(y_i, r_i)$ is the Hamming distance (the number of bit-errors) between the transmitted symbol y_i (each consisting of two bits) and received symbol r_i. The branch metric is

$$m_{s_{i-1}s_i} = -\log p_{s_{i-1}s_i}(y_i, r_i) = -2 \log q - e \log (p/q)$$

Since $p < q$, the metric $m_{s_{i-1}s_i}$ is a monotonically increasing function of e. Therefore, we can use the Hamming distance e as a metric which is not only simpler to compute, but also does not depend on the channel parameters. This means that we do not need to know the bit error probability to apply the algorithm.

Suppose that the sequence 0 0 0 0 0 0 entered the encoder when it was in the state (0,0). The encoder sends 00 00 00 00 00 00 (see Fig. 4.4), and the decoder receives the sequence 00 01 00 00 00 00, which contains a single error.

We assume that, initially, the decoder was in the state (0,0). According to the trellis diagram in Fig. 4.5, we have two survivors at time $t=1$: $(0,0)\to(0,0)$ and $(0,0)\to(1,0)$, with the lengths $L((0,0)\to(0,0)) = 0$ and $L((0,0)\to(1,0)) = HD[(0,0),(1,1)] = 2$. The branch $(0,0)\to(1,0)$ has length 2, since the received symbol $r_1 = 00$ is different from the encoder output symbol $y_1 = 11$ in both positions (see Fig. 4.4).

At time $t=2$ we have four survivors with lengths $L((0,0)\to(0,0)\to(0,0)) = 1$, $L((0,0)\to(0,0)\to(1,0)) = 1$, $L((0,0)\to(1,0)\to(0,1)) = 4$, and $L((0,0)\to(1,0)\to(1,1)) = 2$.

At time $t=3$ there are two paths leading to state (0,0). Repeating the calculations we see that the path (0,0)→(0,0)→(0,0)→(0,0) has length 1, whereas the path (0,0)→(1,0)→(0,1)→(0,0) that terminates at the same state has length 6. Therefore, the path (0,0)→(0,0)→(0,0)→(0,0) is a survivor. Similarly, we select survivors for each state in the trellis diagram (see Fig. 4.12). Finally, we discover that the path (0,0)→(0,0)→(0,0)→(0,0)→(0,0)→(0,0)→(0,0) is the shortest (its length is equal to 1), and the corresponding all-zero sequence is decoded, thus correcting the single error.

Ideally, at the end of communication, the survivor $\hat{\mathbf{s}}_0, \hat{\mathbf{s}}_1, \ldots, \hat{\mathbf{s}}_T$ that has the shortest length is selected, and the decoder outputs the corresponding information symbols $\hat{\mathbf{x}}_0, \hat{\mathbf{x}}_1, \ldots, \hat{\mathbf{x}}_T$. This method would require a very large memory to store all the survivors. In practice, it is necessary to perform information decoding based on truncated survivors. If the truncation depth Δ is large enough, then it is highly probable that the initial part (outside the truncation interval) of all the survivors is the same so that we can decode $\hat{\mathbf{x}}_{t-\Delta}$ at the moment

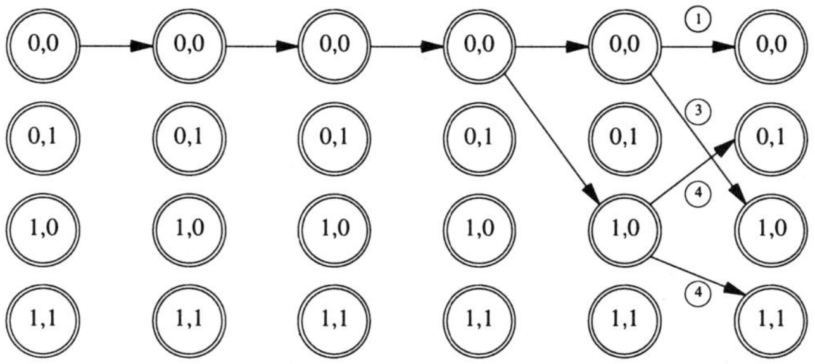

Figure 4.12. The survivors.

t that corresponds to the merged part of all the survivors (the common part of all the survivors in Fig. 4.12). If the survivors disagree outside the truncated interval at the moment $t-\Delta$, then we select $\hat{\mathbf{x}}_{t-\Delta}$, which corresponds to the shortest path.

If the source sequence is not an i.i.d. and/or channel errors are not independent, then the communication system cannot be modeled by an IOHMM whose states are the encoder states. The standard way to work around this problem is by *interleaving*, that is, scrambling the encoded sequence before the transmission and unscrambling it at the receiver to make errors independent. To achieve the desired error decorrelation, the interleaving depth can be very large leading to large delays which cannot be tolerated by some applications. We discuss the interleaver depth selection in Sec. 4.7.1.

4.3.2.2 The EM Algorithm

As we have seen in Sec. 3.2, the HMM structure lends itself to solving the maximization problems of the form (4.3.1) with the $Pr(\boldsymbol{x}_1^T, \boldsymbol{y}_1^T)$ given by equation (4.3.17). The only difference is that now we assume that the model parameters are known and we are maximizing the probability (or a PDF) with respect to $\boldsymbol{x}_1^T$. As in Sec. 3.2, define the auxiliary function as

$$\psi(\boldsymbol{s}_0^T; \boldsymbol{x}_1^T, \boldsymbol{y}_1^T) = \pi_i \prod_{t=1}^{T} p_{s_{t-1}s_t}(x_t, y_t)$$

Then, similarly to the derivations of Sec. 3.2, we can show that the solution of Eq. (4.3.1) can be found iteratively by the following EM algorithm

$$x_{t,p+1} = \underset{x_t}{argmax} Q(\boldsymbol{x}_1^T(p), \boldsymbol{y}_1^T) \tag{4.3.18a}$$

where $\boldsymbol{x}_1^T(p) = x_{1,p}, x_{2,p}, \ldots, x_{T,p}$ is the result of the p-th iteration and

$$Q(\boldsymbol{x}_1^T(p), \boldsymbol{y}_1^T) = \sum_{t=1}^{T} \sum_{i,j} \gamma_{t,ij}(\boldsymbol{x}_1^T(p)) \log p_{ij}(x_t, y_t) \tag{4.3.18b}$$

where

$$\gamma_{t,ij}(\boldsymbol{x}_1^T(p)) = \alpha_i(\boldsymbol{x}_1^{t-1}(p),\boldsymbol{y}_1^{t-1})p_{ij}(x_{t,p},y_t)\beta_j(\boldsymbol{x}_{t+1}^T(p),\boldsymbol{y}_{t+1}^T)$$

$\alpha_i(\boldsymbol{x}_1^{t-1},\boldsymbol{y}_1^{t-1})$ and $\beta_j(\boldsymbol{x}_{t+1}^T,\boldsymbol{y}_{t+1}^T)$ are the i-th and j-th elements of the forward and backward probability vectors, respectively:

$$\boldsymbol{\alpha}(\boldsymbol{x}_1^{t-1},\boldsymbol{y}_1^{t-1}) = \boldsymbol{\pi}\prod_{i=1}^{t-1}\mathbf{P}(x_i,y_i) \qquad \boldsymbol{\beta}(\boldsymbol{x}_{t+1}^T,\boldsymbol{y}_{t+1}^T) = \prod_{i=t+1}^{T}\mathbf{P}(x_i,y_i)\mathbf{1}$$

which can be evaluated by the forward and backward algorithms (see Sec. 3.2).

Generally, it is simpler to find the maximum of the auxiliary function in Eq. (4.3.18b) than to find the maximum of $\mathbf{P}(\boldsymbol{x}_1^T,\boldsymbol{y}_1^T)$ in equation (4.3.17) because in the majority of applications the observation PDFs belong to the exponential family. An example of using this algorithm when the maximum can be found in the closed form is given in Ref. 18.

4.3.2.3 TCM Systems

If the i.i.d. source sequence $\boldsymbol{X}_1^T$ (see Sec. 4.2.7) is trellis code modulated and transmitted over an HMM channel, then the received sequence $\boldsymbol{Y}_1^T$ represents the IOHMM which is described by the matrices $\mathbf{P}(X_t,Y_t)$ given by Eq. (4.3.10) which we rewrite here for the convenience of the reader

$$\mathbf{P}(X_t,Y_t) = [p_{S_{t-1}S_t}(X_t)\mathbf{P}_c(Y_t \mid O_t)]_{uv,uv}$$

where u is the size of the matrix (4.3.36), v is the size of the matrix $\mathbf{P}_c(Y_t \mid O_t)$, $O_t = H_t(S_{t-1},X_t)$ and the only nonzero subblocks of this matrix correspond to $S_t = F_t(S_{t-1},X_t)$.

As in Sec. 4.3.1.3, we can express the APP in terms of the subblocks of these matrices:

$$Pr(\boldsymbol{X}_1^T,\boldsymbol{Y}_1^T) = \boldsymbol{\pi}_c\prod_{t=1}^{T}p_{S_{ss(t-1)}S_{ss(t)}}(X_t)\mathbf{P}_c(Y_t \mid O_t)$$

where

$$p_{S_{t-1}S_t}(X_t) = \begin{cases} Pr(X_t) & \text{if } S_t=F_t(S_{t-1},X_t) \\ 0 & \text{otherwise} \end{cases}$$

In other words, the product of the matrices is taken along the path on the code trellis. To simplify notation, we assume that the source symbols are equiprobable. In this case, the MAP decoding is equivalent to the ML decoding consisting in finding the optimal path on the code trellis that maximizes the likelihood function

$$Pr(\boldsymbol{Y}_1^T \mid \boldsymbol{X}_1^T) = \boldsymbol{\pi}_c\prod_{t=1}^{T}\mathbf{P}_c(Y_t \mid O_t)\mathbf{b} \tag{4.3.19}$$

where $\mathbf{b} = [\mathbf{1}\ \ 0\ \ \ldots\ \ 0]'$ if the code is terminated at the first state.

The auxiliary function of the EM algorithm for maximizing this probability can be written as

$$Q(\boldsymbol{X}_1^T,\boldsymbol{X}_{1,p}^T) = \sum_{t=1}^{T}\sum_{i=1}^{v}\sum_{j=1}^{v}\gamma_{t,ij}(\boldsymbol{X}_{1,p}^T)\log p_{c,ij}(Y_t|O_t) + C \tag{4.3.20}$$

where

$$\gamma_{t,ij}(\boldsymbol{X}_{1,p}^T) = \alpha_i(\boldsymbol{Y}_1^{t-1} \mid \boldsymbol{X}_{1,p}^{t-1}) p_{c,ij}(Y_t | O_t) \; \beta_j(\boldsymbol{Y}_{t+1}^T \mid \boldsymbol{X}_{t+1,p}^T)$$

the constant C does not depend on $\boldsymbol{X}_1^T$ and, therefore, will be neglected, the sub script p indicates that the corresponding sequence is obtained after the p-th iteration, $\alpha_i(\boldsymbol{Y}_1^{t-1} \mid \boldsymbol{X}_{1,p}^{t-1})$ and $\beta_j(\boldsymbol{Y}_{t+1}^T \mid \boldsymbol{X}_{t+1,p}^T)$ are the elements of the forward and backward probability vectors $\boldsymbol{\alpha}(\boldsymbol{Y}_1^{t-1} \mid \boldsymbol{X}_{1,p}^{t-1})$ and $\boldsymbol{\beta}(\boldsymbol{Y}_{t+1}^T \mid \boldsymbol{X}_{t+1,p}^T)$ that can be found using the block form of forward and backward algorithms that in our case are given by

$$\boldsymbol{\alpha}(\boldsymbol{Y}_1^t \mid \boldsymbol{X}_{1,p}^t) = \boldsymbol{\alpha}(\boldsymbol{Y}_1^{t-1} \mid \boldsymbol{X}_{1,p}^{t-1}) = \mathbf{P}_c(Y_t \mid O_{t,p})$$

$$\boldsymbol{\beta}(\boldsymbol{Y}_t^T \mid \boldsymbol{X}_{t,p}^T) = \mathbf{P}_c(Y_t \mid O_{t,p})\boldsymbol{\beta}(\boldsymbol{Y}_{t+1}^T \mid \boldsymbol{X}_{t+1,p}^T)$$

where the products are taken along the code trellis path corresponding to the source sequence $\boldsymbol{X}_{1,p}^T$

The likelihood function in Eq. (4.3.19) is expressed in terms of these subblocks as

$$Pr(Y_1^T \mid \boldsymbol{X}_1^t = \boldsymbol{\alpha}(\boldsymbol{Y}_1^t \mid \boldsymbol{X}_{1,p}^t)\boldsymbol{\beta}(\boldsymbol{Y}_{t+1}^T \mid \boldsymbol{X}_{t+1,p}^T)$$

where $0 \le t \le T$.

It follows from Eq. (4.3.20) that we can apply the Viterbi algorithm with the branch metric

$$m(X_t) = \sum_{i=1}^{v}\sum_{j=1}^{v} \gamma_{t,ij}(\boldsymbol{X}_{1,p}^T)\log p_{c,ij}(Y_t | O_t), \quad t = 1,2,\ldots,T \tag{4.3.21}$$

to find a maximum of $Q(\boldsymbol{X}_1^T, \boldsymbol{X}_{1,p}^T)$.

Thus, the E-step of the algorithm consisting in evaluating the auxiliary function is performed by the FBA while the M-step is realized using the Viterbi algorithm. The forward algorithm computes and saves $\boldsymbol{\alpha}(\boldsymbol{Y}_1^t \mid \boldsymbol{X}_{1,p}^t)$, therefore, it is convenient to perform the Viterbi algorithm together with the backward part of the algorithm which eliminates the need of saving $\gamma_{t,ij}(\boldsymbol{X}_{1,p}^T)$. Or, alternatively, we can perform the backward algorithm by computing and saving $\boldsymbol{\beta}(\boldsymbol{Y}_t^T \mid \boldsymbol{X}_{t,p}^T)$ and combine the Viterbi algorithm with the forward part of the algorithm.

In practice, the received sequence is usually long. Therefore, we need to use the fixed-lag FBA together with the (delayed) Viterbi algorithm.

Example 4.3.4. Let us illustrate the MAP sequence decoding for the system in example 4.3.2 (in which we illustrated the MAP symbol decoding). The matrices $\mathbf{P}(x,r)$ describing the IOHMM are given by Eq. (4.3.13) and (4.3.14). and the auxiliary function $Q(\boldsymbol{x}_1^T(p), \boldsymbol{y}_1^T)$ of the EM algorithm

Suppose that the sequence our initial guess about the transmitted information sequence is $\boldsymbol{x}_{1,0}^3 = 0,1,0$. Assuming that the code is terminated at the 00-state we find that bits 0,0 force the trellis in Fig. 4.5 into the 00-state. Therefore, the encoder input sequence is $\boldsymbol{x}_{1,0}^5 = 0,1,0,0,0$. Therefore, the QPSK modulator output is (see Fig 4.4 and Example 4.3.2) $\boldsymbol{O}_{1,0}^5 = O^{(1)}, O^{(4)}, O^{(3)}, O^{(4)}, O^{(1)}$. Suppose that $r_1^5 = r_1, r_2, r_3, r_4, r_5$ was received then, in the forward part, we compute

$$\boldsymbol{\alpha}(r_1 | 0) = \boldsymbol{\pi}_c \mathbf{P}_c(r_1 \mid O^{(1)}),\ \boldsymbol{\alpha}(r_1^2 | 01) = \boldsymbol{\alpha}(r_1 | 0)\mathbf{P}_c(r_2 \mid O^{(4)}),\ \boldsymbol{\alpha}(r_1^3 | 010)$$
$$= \boldsymbol{\alpha}(r_1^2 | 01)\mathbf{P}_c(r_3 \mid O^{(3)}),\ \boldsymbol{\alpha}(r_1^4 | 0100) = \boldsymbol{\alpha}(r_1^3 | 010)\mathbf{P}_c(r_4 \mid O^{(4)})$$

This completes the forward part. In the backward part, we compute, according to Eq. (4.3.21),

$$L_4(00) = m(X_5=0) = \sum_{i=1}^{v}\sum_{j=1}^{v} \gamma_{5,ij}(01000)\log p_{c,ij}(r_5 \mid O^{(1)})$$

$$L_4(01) = m(X_5=1) = \sum_{i=1}^{v}\sum_{j=1}^{v} \gamma_{5,ij}(01000)\log p_{c,ij}(r_5 \mid O^{(4)})$$

where $\gamma_{5,ij}(01000) = \alpha_i(r_1^4|0100)p_{c,ij}(r_5 \mid O^{(1)}$. Then we compute $\beta(r_5|0) = \mathbf{P}_c(r_5 \mid O^{(1)})\mathbf{1}$ and

$$L_3(IJ) = m(X_4=I)+m(X_5=J),$$

In the previous steps there were no competing transitions, therefore, all possible path were survivors. In the next step, each node has two transition leading to it, therefore, we need to apply the Viterbi algorithm to find the surviovor: Choose the transition 00←00 if

$$L_2(00) = L_3(00)+m(X_3=0) > L_3(10)+m(X_3=1)$$

or choose 00←10 and $L(S_2=00) = L(S_3=01)+m(X_3=1)$, otherwise. We continue this process untill we find the survivor leading to the original 00-state. Then we decode the sequence $\boldsymbol{X}_{1,1}^5$ that corresponds the survivor. The process continues until two consecutive iterations deliver the same result.

4.4 BLOCK CODE PERFORMANCE CHARACTERIZATION

In any one-way communication system, three mutually exclusive events can happen after a message $\mathbf{s} = [s_1, s_2, \ldots, s_n]$ decoding:

$C_\mathbf{s}$ The message has been decoded without errors (that is, the message either did not have errors or the errors have been corrected).

$U_\mathbf{s}$ The message has been decoded with errors (that is, it contains undetected errors or an attempt to correct errors was unsuccessful).

$D_\mathbf{s}$ The message contains errors that the decoder can detect but is not able to correct.

The probabilities of these events characterize the message transmission correctness in one-way systems.

Suppose that messages are encoded by a linear (n,k) block code and that channel errors are additive. Then, the previously described events do not depend on the transmitted message, and their probabilities can be expressed as

$$p_u = \sum_{\mathbf{e}\in U} Pr(\mathbf{e}) \tag{4.4.1}$$

$$p_d = \sum_{\mathbf{e}\in D} Pr(\mathbf{e}) \tag{4.4.2}$$

$$p_c = \sum_{\mathbf{e}\in C} Pr(\mathbf{e}) \tag{4.4.3}$$

where $\mathbf{e}$ is the difference between received and transmitted messages; sets C, U, and D correspond to previously defined events. Obviously, we need to compute only two of these probabilities, since their sum is equal to 1. The use of formulas (4.4.1), (4.4.2), and (4.4.3) is often accompanied by many difficulties due to the necessity of having to perform summation over all the elements of sets C, D, and U. This process can be systematized by using matrix generating functions.

4.4.1 The Probability of Undetected Error

Suppose that a linear (n,k) code with the code-generator matrix $\mathbf{G}$ is used for error detection only. An error $\mathbf{e} = (e_1, e_2, ..., e_n) \neq 0$, where $e_i \in GF(p^m)$ is not detected if it is a code word:

$$\mathbf{e} = \mathbf{cG}$$

Thus, according to Eq. (4.4.1),

$$p_u = \sum_{\mathbf{c}\neq 0} Pr(\mathbf{cG}) = \sum_{\mathbf{c}} Pr(\mathbf{cG}) - p_n(0) \tag{4.4.4}$$

where $p_n(0)$ is the probability that the block does not contain errors.

The probability of undetected errors can also be found by Eq. (4.4.1) if we describe the set U of all undetected errors as the set of all nonzero vectors whose syndromes (4.2.4) are equal to zero: $U = \{\mathbf{e}: \mathbf{eH}' = 0 \quad \mathbf{e}\neq 0\}$.

It is convenient to use generating functions to perform summation in the right-hand side of Eq. (4.4.1). As we have seen in Sec 2.2.1, elements of $GF(p^m)$ may be interpreted as m-tupples with components from the ground field $GF(p)$. Let $(a)_i$ denote an i-th component of an element. Then, error syndrome components can be expressed in the form

$$(S_j)_i = \sum_{k=1}^{n} (e_k h_{jk})_i \pmod p \quad j=1,2,...,r;\ i=1,2,...,m \tag{4.4.5}$$

This equation is the expanded version at the level of the ground field $GF(p)$ of Eq. (4.2.4), which uses matrices whose elements belong to $GF(q)$ with $q = p^m$.

The probability of undetected errors is equal to the sum of probabilities of nonzero errors, for which the right-hand side of Eq. (4.4.5) is equal to zero. To find this sum, we introduce the generating function

$$f(\mathbf{z}) = \sum_{\mathbf{e}} Pr(\mathbf{e})\mathbf{z}^{\mathbf{eH}'} \tag{4.4.6}$$

where $\mathbf{z}^{\mathbf{eH}'} = \prod_{i,j=1}^{m,r} z_{ij}^{w_{ij}}$ and $w_{ij} = \sum_{k=1}^{n} (e_k h_{jk})_i$ for $j=1,2,...,r;\ i=1,2,...,m$.

The probability of undetected errors p_u is equal to the sum of probabilities $Pr(\mathbf{e})$, for which $w_{ij} = 0 \pmod p$ minus the probability $p_n(0)$ of the absence of errors in the block. This probability can be determined from generating function (4.4.6) by using formula (2.2.29):

$$p_u = q^{-r} \sum_{\mathbf{v}} f(\boldsymbol{\zeta}^{\mathbf{v}}) - p_n(0) \tag{4.4.7}$$

where $\mathbf{v}$ is an rm-dimensional vector whose coordinates v_{ij} vary from 0 to $p-1$, $\boldsymbol{\zeta}^{\mathbf{v}}$ denotes a vector with the coordinates $\zeta^{v_{ij}}$, and $\zeta = e^{-2\pi j/p}$. This equation expresses the probability of undetected errors for any additive error source model. It can be simplified if the error source is described by an HMM model, because we can make use of the properties of matrix probabilities. The matrix probability of absence of errors in the block is equal to $\mathbf{P}^n(0)$. The generating function $\mathbf{F}(\mathbf{z})$ that corresponds to (4.4.6) is given by

$$\mathbf{F}(\mathbf{z}) = \sum_{\mathbf{e}} \prod_{k=1}^{n} \mathbf{P}(\mathbf{e})\mathbf{z}^{\mathbf{eH}'} \tag{4.4.8}$$

Since

$$\mathbf{z}^{\mathbf{eH}'} = \prod_{k=1}^{n} \prod_{ij} z_{ij}^{(e_k h_{jk})_i}$$

and the matrix probability

$$\mathbf{P}(\mathbf{e}) = \prod_{k=1}^{n} \mathbf{P}(e_k)$$

generating function (4.4.8) can be expressed as

$$\mathbf{F}(\mathbf{z}) = \prod_{k=1}^{n} \sum_{e_k} \mathbf{P}(e_k)\mathbf{z}^{e_k \mathbf{h}_k} \tag{4.4.9}$$

where

$$\mathbf{z}^{e_k \mathbf{h}_k} = \prod_{ij} z_{ij}^{(e_k h_{jk})_i}$$

Therefore, as in Eq. (4.4.7), we can write the matrix probability of undetected errors

$$\mathbf{P}_u = q^{-r} \sum_{\mathbf{v}} \mathbf{F}(\zeta^{\mathbf{v}}) - \mathbf{P}^n(0) \tag{4.4.10}$$

In the most important case of binary codes ($q = 2$, $\zeta = -1$), generating function (4.4.9) becomes

$$\mathbf{F}(\mathbf{z}) = \prod_{k=1}^{n} [\mathbf{P}(0) + \mathbf{P}(1)\mathbf{z}^{\mathbf{h}_k}] \tag{4.4.11}$$

where $\mathbf{z}^{\mathbf{h}_k} = z_1^{h_{1k}} z_2^{h_{2k}} \cdots z_r^{h_{rk}}$. Thus, the matrix probability of undetected errors (4.4.10) is given by

$$\mathbf{P}_u = 2^{-r} \sum_{\mathbf{v}} \prod_{k=1}^{n} [\mathbf{P}(0) + \mathbf{P}(1)(-1)^{\mathbf{v}\mathbf{h}_k}] - \mathbf{P}^n(0) \tag{4.4.12}$$

where $\mathbf{v}\mathbf{h}_k = v_1 h_{1k} + v_2 h_{2k} + \ldots + v_r h_{rk}$ is a dot product of binary vectors. The scalar probability of undetected errors can be found in the usual way:

$$p_u = \mathbf{p}\mathbf{P}_u\mathbf{1} \tag{4.4.13}$$

Since the code is used only for detecting errors, the matrix probability $\mathbf{P}_c$ of correct reception of a combination and the matrix probability $\mathbf{P}_d$ of error detection are equal to

$$\mathbf{P}_c = \mathbf{P}^n(0) \tag{4.4.14}$$

and

$$\mathbf{P}_d = \mathbf{P}^n - \mathbf{P}^n(0) - \mathbf{P}_u \tag{4.4.15}$$

respectively.

Example 4.4.1. Let us consider the simple $(n, n-1)$ parity-check code. The parity-check matrix of this code has the form $\mathbf{H} = [1\ 1\ 1 \ldots 1]$. Equation (4.4.11) now takes the form $\mathbf{F}(z) = [\mathbf{P}(0) + \mathbf{P}(1)z]^n$ and, according to Eq. (4.4.12), the matrix probability of undetected errors is given by

$$\mathbf{P}_u = 2^{-1}\mathbf{P}^n + 2^{-1}[\mathbf{P}(0) - \mathbf{P}(1)]^n - \mathbf{P}^n(0)$$

The matrix probability of detected errors, according to (4.4.15), is equal to

$$\mathbf{P}_d = 2^{-1}\mathbf{P}^n - 2^{-1}[\mathbf{P}(0) - \mathbf{P}(1)]^n$$

Consider now a more complex case.

Example 4.4.2. Find the probability of undetected errors for the binary (7,4) Hamming code with the following parity-check matrix:

$$\mathbf{H} = \begin{bmatrix} 1\ 1\ 1\ 0\ 1\ 0\ 0 \\ 0\ 1\ 1\ 1\ 0\ 1\ 0 \\ 0\ 0\ 1\ 1\ 1\ 0\ 1 \end{bmatrix}$$

In this case, Eq. (4.4.11) becomes

$$\mathbf{F}(z_1,z_2,z_3) = [\mathbf{P}(0)+\mathbf{P}(1)z_1][\mathbf{P}(0)+\mathbf{P}(1)z_1z_2][\mathbf{P}(0)+\mathbf{P}(1)z_1z_2z_3]$$
$$\times[\mathbf{P}(0)+\mathbf{P}(1)z_2z_3][\mathbf{P}(0)+\mathbf{P}(1)z_1z_3][\mathbf{P}(0)+\mathbf{P}(1)z_2][\mathbf{P}(0)+\mathbf{P}(1)z_3]$$

Then by formula (4.4.12) we can find

$$\mathbf{P}_u = 2^{-3} \sum_{i,j,k=0}^{1} \mathbf{F}((-1)^i,(-1)^j,(-1)^k) - \mathbf{P}^7(0)$$

In the case of the BSC with independent errors and $\mathbf{P}(0) = p_0$ and $\mathbf{P}(1) = p_1$, these equations are simplified: $p_u = 2^{-3}[1+7(p_0-p_1)^4] - p_0^7$.

Note that the MacWilliams formula [20] follows from the results of this section. Indeed, if we substitute $f(\zeta^{\mathbf{v}})$ from equation (4.4.6) into Eq. (4.4.7), we obtain

$$p_u = q^{-r}\sum_{\mathbf{v}} \sum_{\mathbf{e}} Pr(\mathbf{e})\zeta^{(\mathbf{eH}')\mathbf{v}} - p_n(0) \tag{4.4.16}$$

This equation expresses the probability of undetected errors directly through the code parity-check matrix.

Comparing Eq. (4.4.16) and (4.4.4), we obtain the identity

$$\sum_{\mathbf{c}} Pr(\mathbf{cG}) = q^{-r}\sum_{\mathbf{v}} \sum_{\mathbf{e}} Pr(\mathbf{e})\zeta^{(\mathbf{eH}')\mathbf{v}} \tag{4.4.17}$$

which is the generalization of the MacWilliams identity. The latter is obtained if we assume that errors are independent

$$Pr(\mathbf{e}) = \prod_{k=1}^{n} Pr(e_k)$$

and have the same probability $Pr(e) = P/(q-1)$ for $e \neq 0$. [8]

4.4.2 Performance of Forward Error-Correcting Codes

When a block code is used for correcting and detecting errors, the basic probabilities that characterize the system's performance depend not only on the code-generator (or parity-check) matrix but also on the method of correcting errors.

Suppose that maximum likelihood syndrome decoding is being used. Then for each syndrome $\mathbf{S} = \mathbf{eH}'$, we select the most probable error word $\mathbf{e}(\mathbf{S})$ and subtract it from the received word. In this case, the full power of the code is used to correct errors so that the probability of detecting errors is equal to zero. Sometimes not the full code power is used to correct errors. If a syndrome belongs to a subset C, errors are corrected by the maximum likelihood rule; otherwise errors are just detected. The detected error patterns are either suppressed or flagged for subsequent application of some other error-correction scheme.

The probability of correct decoding is the probability that channel error words are equal to $\mathbf{e}(\mathbf{S})$ when syndromes belong to C:

$$p_c = \sum_{\mathbf{S}\in C} Pr(\mathbf{e}(\mathbf{S})) \tag{4.4.18}$$

The probability of detected errors is equal to

$$p_d = \sum_{\mathbf{S}\in \bar{C}} Pr(\mathbf{S}) \tag{4.4.19}$$

where $\bar{C}$ is the complement of C.

The probability of detecting errors can be found using generating function (4.4.6). Indeed, using Eq. (2.2.29) we obtain the probability that a syndrome is equal to $\mathbf{S}$:

$$Pr(\mathbf{S}) = q^{-r} \sum_{\mathbf{v}} \zeta^{-\mathbf{S}\mathbf{v}} f(\zeta^{\mathbf{v}}) \tag{4.4.20}$$

The probability of detected errors is found by substituting these values into Eq. (4.4.19):

$$p_d = q^{-r} \sum_{\mathbf{S}\in \bar{C}} \sum_{\mathbf{v}} \zeta^{-\mathbf{S}\mathbf{v}} f(\zeta^{\mathbf{v}})$$

In calculating characteristics of FEC performance, we can always make use of the identity $p_c + p_u + p_d = 1$ to find one of the probabilities when the other two are known. In many practical cases it is simpler to find p_c and p_u than p_d because it is difficult to describe the set $\bar{C}$. The probability p_u is equal to the probability that an error syndrome $\mathbf{S} = \mathbf{e}\mathbf{H}'$ belongs to C, but the error is not equal to $\mathbf{e}(\mathbf{S})$:

$$p_u = Pr(\mathbf{S}\in C) - p_c \tag{4.4.21}$$

Using Eq. (4.4.20), the previous formula can be rewritten as

$$p_u = q^{-r} \sum_{\mathbf{S}\in C} \sum_{\mathbf{v}} \zeta^{-\mathbf{S}\mathbf{v}} f(\zeta^{\mathbf{v}}) - p_c \tag{4.4.22}$$

Note that this equation is more general than equation (4.4.7), which is obtained when C consists of only one element $\mathbf{S} = 0$.

In the majority of cases, the error patterns that correspond to syndromes are selected according to the *minimum Hamming weight rule*. To describe it we define a block $\mathbf{a}$ of n symbols Hamming weight $|\mathbf{e}|$ as a total number of its nonzero symbols and define the Hamming distance between two vectors $\mathbf{a}$ and $\mathbf{b}$ as the Hamming weight of their difference: $d(\mathbf{a},\mathbf{b}) = |\mathbf{a}-\mathbf{b}|$. Now the minimum Hamming weight rule of decoding defines $\mathbf{e}(\mathbf{S})$ as the word with the smallest number of nonzero components whose syndrome is $\mathbf{S}$:

$$|\mathbf{e}(\mathbf{S})| = \min_{\mathbf{e}\mathbf{H}' = \mathbf{S}} |\mathbf{e}|$$

If a code is able to correct up to t errors and minimum Hamming weight decoding is used, the set C consists of syndromes of all error sequences that contain up to t nonzero components.

The alternative decoding procedure consists of selecting the code word $\mathbf{c}$ whose distance from the received word $\mathbf{e}$ does not exceed t. This procedure is equivalent to the previously described syndrome decoding. Indeed, suppose that $\mathbf{c}$ is the code word whose distance from $\mathbf{e}$ does not exceed t: $|\mathbf{e} - \mathbf{c}| \le t$. Denote $\mathbf{e}_1 = \mathbf{e} - \mathbf{c}$. Since $|\mathbf{e}_1| \le t$ and $\mathbf{e}_1\mathbf{H}' \in C$, the syndrome decoder outputs the same vector $\mathbf{e} + \mathbf{e}_1 = \mathbf{c}$.

The converse statement is also true. If $\mathbf{eH}' \in C$, the syndrome decoder finds a vector $\mathbf{e}_1$ that has the same syndrome as $\mathbf{e}$ (that is, $\mathbf{e}_1\mathbf{H}' = \mathbf{eH}'$) and weight $|\mathbf{e}_1| \le t$. Then the decoder outputs a code word $\mathbf{c} = \mathbf{e} - \mathbf{e}_1$ whose distance from $\mathbf{e}$ does not exceed t: $|\mathbf{e} - \mathbf{c}| = |\mathbf{e}_1| \le t$.

The alternative decoding rule can be interpreted geometrically. A set of points $\mathbf{x}$ that satisfy the equation $|\mathbf{x} - \mathbf{c}| \le r$ is called a sphere of radius r with the center in $\mathbf{c}$. Thus, we may say that a received message $\mathbf{e}$ is decoded as $\mathbf{c}$ if it falls inside the sphere whose center is defined by the code word and has radius t. The code is able to correct up to t errors if and only if these spheres do not have common points, meaning that the Hamming distance between their centers is at least $2t+1$. The minimum distance between code words of a code is called the *code minimum distance*. A code is able to correct up to t errors if and only if its minimum distance $d_{\min} \ge 2t+1$.

If a code corrects up to t errors and does not correct errors of higher multiplicity, the probability p_c of correct decoding is equal to the probability that the number of errors in a word is not greater than t. For the HMM error source model, we have

$$p_c = \mathbf{pP}_c\mathbf{1} \qquad \mathbf{P}_c = \mathbf{P}_n(\le t) \tag{4.4.23}$$

where $\mathbf{P}_n(\le t)$ is the matrix probability that the number of errors does not exceed t and is defined in Sec 2.3.2.2.

The probability of incorrect decoding may be found using Eq. (4.4.22) and generating function (4.4.9):

$$p_u = \mathbf{pP}_u\mathbf{1} \tag{4.4.24}$$

where

$$\mathbf{P}_u = q^{-r} \sum_{\mathbf{S} \in C} \sum_{\mathbf{v}} \zeta^{-\mathbf{Sv}} \mathbf{F}(\zeta^{\mathbf{v}}) - \mathbf{P}_n(\le t) \tag{4.4.25}$$

Since syndromes may be expressed as $\mathbf{S} = \mathbf{eH}'$, we can replace the index of summation $\mathbf{S}$ by $\mathbf{e}$:

$$\mathbf{P}_u = q^{-r} \sum_{|\mathbf{e}| \le t} \sum_{\mathbf{v}} \zeta^{-(\mathbf{eH}')\mathbf{v}} \mathbf{F}(\zeta^{\mathbf{v}}) - \mathbf{P}_n(\le t) \tag{4.4.26}$$

In the case of binary codes these equations become

$$\mathbf{P}_u = 2^{-r} \sum_{\mathbf{S} \in C} \sum_{\mathbf{v}} (-1)^{-\mathbf{Sv}} \prod_{k=1}^{n} [\mathbf{P}(0) + \mathbf{P}(1)(-1)^{\mathbf{vh}_k}] - \mathbf{P}_n(\le t) \tag{4.4.27}$$

and

$$\mathbf{P}_u = 2^{-r} \sum_{|\mathbf{e}| \le t} \sum_{\mathbf{v}} (-1)^{-(\mathbf{eH}')\mathbf{v}} \prod_{k=1}^{n} [\mathbf{P}(0) + \mathbf{P}(1)(-1)^{\mathbf{vh}_k}] - \mathbf{P}_n(\le t) \tag{4.4.28}$$

Example 4.4.3. Suppose that the binary (7,4) Hamming code with the parity-check matrix

$$\mathbf{H} = \begin{bmatrix} 1\,1\,1\,0\,1\,0\,0 \\ 0\,1\,1\,1\,0\,1\,0 \\ 0\,0\,1\,1\,1\,0\,1 \end{bmatrix}$$

is used for correcting errors. For minimum Hamming weight decoding, let us find the probability of undetected errors.

First, we define the decoding rule $\mathbf{e}(\mathbf{S})$ for error words with the smallest Hamming weight. The syndrome for the zero-weight error $\mathbf{e}_0 = (0\ 0\ 0\ 0\ 0\ 0\ 0\ 0\ 0)$ is the zero syndrome $\mathbf{e}_0\mathbf{H}' = \mathbf{S}_0 = (0\ 0\ 0)$. Thus, we decide that no errors occurred, if the received message syndrome equals $\mathbf{S}_0$. Let us now compute syndromes for the single bit errors. The syndrome for $\mathbf{e}_1 = (1\ 0\ 0\ 0\ 0\ 0\ 0\ 0\ 0)$ is $\mathbf{e}_1\mathbf{H}' = \mathbf{S}_1 = (1\ 0\ 0)$, which coincides with the transposed first column of $\mathbf{H}$. Quite analogously we find that the syndrome for $\mathbf{e}_2 = (0\ 1\ 0\ 0\ 0\ 0\ 0\ 0\ 0)$ is $\mathbf{S}_2 = (1\ 1\ 0)$, which is the transposed second column of $\mathbf{H}$, and so on. Since all the columns of matrix $\mathbf{H}$ are different, we obtain all possible syndromes $\mathbf{S}_0, \mathbf{S}_1, ..., \mathbf{S}_7$ because there are only eight different combinations of three bits. This means that the set C consists of all possible sequences of three bits. Thus, the minimum Hamming weight decoding rule can be expressed as $\mathbf{e}(\mathbf{S}_i) = \mathbf{e}_i$. In other words, if a received message syndrome is equal to the i-th column of the matrix $\mathbf{H}$, then the i-th bit of the received block must be corrected (see also Example 4.2.3).

The probability of correct decoding, according to Eq. (4.4.18), is the probability that the number of errors in the block is not greater than 1. For the HMM model, this probability is equal to $p_c = \mathbf{P}_7(\le 1) = \mathbf{p}\mathbf{P}^7(0)\mathbf{1} + \mathbf{p}\mathbf{P}_7(1)\mathbf{1}$ where matrix probabilities $\mathbf{P}_n(t)$ of t errors in a block of n symbols are defined in Sec 2.3. The probability of incorrect decoding is equal to $p_u = 1 - p_c$, since $p_d = 0$ in this case.

4.4.3 Symmetrically Dependent Errors

The exact computations by formulas developed in the previous section are, in general, quite complex, so that we need to find their approximations. For short codes a good approximation can be found by assuming the symmetric dependence of errors, [9] meaning that the probabilities of different permutations of the same combination of errors are equal:

$$Pr(e_1, e_2, ..., e_n) = Pr(e_1^{(1)} e_2^{(1)}, ..., e_n^{(1)}) \tag{4.4.29}$$

where $e_i, e_i^{(1)} \in \mathrm{GF}(q)$ and $(e_1, e_2, ..., e_n)$ is a permutation of $(e_1^{(1)}, e_2^{(1)}, ..., e_n^{(1)})$. This assumption agrees reasonably well with experimental data when n and t are small and, in effect, is very close to the assumption that error source is memoryless. [9] According to this assumption we need not be concerned with the order of matrices in matrix products, and a sum of products of probabilities Eq. (4.4.29) can be replaced by the number of addends times the probability. The matrix probability $\mathbf{P}_c$ in Eq. (4.4.23) is equivalent to

$$\mathbf{P}_c = \sum_{i=0}^{t} \binom{n}{i} \mathbf{P}_0^{n-i} \mathbf{Q}_0^i \tag{4.4.30}$$

where $\mathbf{Q}_0 = \mathbf{P} - \mathbf{P}_0$.

The matrix probability of incorrect decoding is given by the Eq.

$$\mathbf{P}_u = \mathbf{P}(\mathbf{S} \in C) - \mathbf{P}_c \tag{4.4.31}$$

where $\mathbf{P}(\mathbf{S} \in C)$ is the matrix probability that error syndrome belongs to C. This probability is equal to the sum of probabilities that an error vector $\mathbf{e}$ falls into a radius-t sphere with the center at a code vector $\mathbf{c}$:

$$\mathbf{P}(\mathbf{S} \in C) = \sum_{\mathbf{c}} \sum_{|\mathbf{e}-\mathbf{c}| \le t} \mathbf{P}(\mathbf{e}) \tag{4.4.32}$$

This sum can be rearranged into sums according to Hamming weights:

$$\mathbf{P}(\mathbf{S}\in C) = \sum_{w=0}^{n}\sum_{\mathbf{c}_w}\sum_{i=0}^{t}\sum_{j=0}^{n}\sum_{|\mathbf{e}_j-\mathbf{c}_w| = i}\mathbf{P}(\mathbf{e}_j) \tag{4.4.33}$$

where $\mathbf{c}_w$ denotes a weight w code word and $\mathbf{e}_j$ denotes a weight j error vector. After changing the order of summation, this equation can be rewritten in the following form:

$$\mathbf{P}(\mathbf{S}\in C) = \sum_{i=0}^{t}\sum_{j=0}^{n} A_{ij}\mathbf{P}_1^j\mathbf{P}_0^{n-j} \tag{4.4.34}$$

where $\mathbf{P}_1 = (q-1)^{-1}\mathbf{Q}_0$ is the average symbol-error probability, and A_{ij} is the number of vectors of Hamming weight j whose distance from some code word is equal to i. This number is a coefficient of x^iy^j in the generating function expansion

$$\sum_{ij} A_{ij}x^iy^j = A(x+y+(1-\gamma)xy,\ 1+\gamma xy) \tag{4.4.35}$$

where $\gamma = q-1$

$$A(x,y) = \sum_{i=0}^{n} A_i x^i y^{n-i} \tag{4.4.36}$$

and A_i is the number of code words whose Hamming weight is equal to i; $A_i=0$ for $i=1,2,\ldots,2t$.[2,20] The function

$$A(z,1) = A(z) = \sum_{i=0}^{n} A_i z^i \tag{4.4.37}$$

is called the *code-weight generator* polynomial. Obviously, $A(x,y) = y^nA(x/y)$.

It is easy to express probability (4.4.34) in terms of these generating functions. Indeed, it follows from Eq. (4.4.35) that

$$\sum_{ij} A_{ij}\mathbf{P}_1^j\mathbf{P}_0^{n-j}x^iy^j = A(x\mathbf{P}_0+y\mathbf{P}_1+(1-\gamma)xy\mathbf{P}_1,\ \mathbf{P}_0+\gamma xy\mathbf{P}_1) \tag{4.4.38}$$

Comparing this equation with equation (4.4.34), we see that $\mathbf{P}(\mathbf{S}\in C)$ is equal to the sum of the coefficients of x^i for $i=0,1,\ldots,t$ in the expansion of function (4.4.38) when $y=1$:

$$A(x\mathbf{P}_0+\mathbf{P}_1+(1-\gamma)x\mathbf{P}_1,\ \mathbf{P}_0+\gamma x\mathbf{P}_1) \tag{4.4.39}$$

But the probability in Eq. (4.4.30) can also be found by summing the coefficients of x^i for $i=0,1,\ldots,t$ in the expansion of the binomial

$$(\mathbf{P}_0 + \gamma x\mathbf{P}_1)^n$$

Thus, the probability of incorrect decoding equals the sum of the coefficients of x^i for $i=0,1,\ldots,t$ in the expansion of the function

$$\mathbf{V}(x) = A(x\mathbf{P}_0+\mathbf{P}_1+(1-\gamma)x\mathbf{P}_1,\ \mathbf{P}_0+\gamma x\mathbf{P}_1) - (\mathbf{P}_0 + \gamma x\mathbf{P}_1)^n \tag{4.4.40}$$

This equation can be simplified if we use the generator polynomial

$$W(x,y) = y^n[A(x/y)-1] \tag{4.4.41}$$

of all nonzero code weights. It is easy to verify that

$$\mathbf{V}(x) = W(x\mathbf{P}_0+\mathbf{P}_1+(1-\gamma)x\mathbf{P}_1,\ \mathbf{P}_0+\gamma x\mathbf{P}_1) \tag{4.4.42}$$

Using Taylor's formula for the coefficients of a power series, we can express the probability of incorrect decoding as

$$\mathbf{P}_u = \sum_{i=0}^{t} \frac{1}{i!} \frac{d^i}{dx^i} \mathbf{V}(x)|_{x=0} \tag{4.4.43}$$

This probability can also be found as a coefficient of x^t in the expansion of the generating functions of sums:

$$\mathbf{P}_u = \frac{1}{t!} \frac{d^t}{dx^t} \mathbf{R}(x)|_{x=0} \tag{4.4.44}$$

$$\mathbf{R}(x) = W(x\mathbf{P}_0 + \mathbf{P}_1 + (1-\gamma)x\mathbf{P}_1,\ \mathbf{P}_0 + \gamma x \mathbf{P}_1)(1-x)^{-1} \tag{4.4.45}$$

After performing the expansion we obtain

$$\mathbf{P}_u = \sum_{i=1}^{n} A_i \sum_{k=0}^{t} \sum_{h=h_1}^{h_2} \binom{i}{k-h}\binom{n-i}{h} \gamma^h \mathbf{P}_0^{n-i-h} \mathbf{P}_1^{2h+i-k} (\mathbf{P}_0 + (\gamma-1)\mathbf{P}_1)^{k-h} \tag{4.4.46}$$

where $h_1 = \max\{0, k-i\}$ and $h_2 = \min\{n-i, k\}$. In the case of binary codes $\gamma = 1$, the previous expression becomes

$$\mathbf{P}_u = \sum_{i=1}^{n} A_i \sum_{k=0}^{t} \sum_{h=h_1}^{h_2} \binom{i}{k-h}\binom{n-i}{h} \mathbf{P}_0^{n-i+k-2h} \mathbf{P}_1^{2h+i-k} \tag{4.4.47}$$

We would like to emphasize again that the preceding approximation is based on the assumption that errors are symmetrically dependent, which allowed us to change the order in matrix products. In the particular case of the channel with independent errors, the matrices are replaced with the probabilities, and the preceding results coincide with well-known equations. [2,20]

Example 4.4.4. Let us illustrate the calculations for the CRC-16 code with the generator polynomial $x^{16} + x^{15} + x^2 + 1$ assuming that channel errors are independent and occur with the probability p_1.

Since $x^{16} + x^{15} + x^2 + 1 = (x+1)(x^{15}+x+1)$ and the polynomial $x^{15}+x+1$ is irreducible, this code is an extended $(n,n-16)$ Hamming code (see Ref. 27, p. 220) with the block length $n = 2^{15}-1$. Its code-weight generator polynomial (see problem 5 of Ref. 27, p. 142) is equal to $A(x) = 2^{-1}(n+1)^{-1}[(1+x)^n + (1-x)^n + 2n(1-x^2)^{(n-1)/2}]$. Therefore, $W(x,y) = 2^{-1}(n+1)^{-1}[(y+x)^n + (y-x)^n + 2n(y^2-x^2)^{(n-1)/2}] - y^n$. Generating function (4.4.45) then takes the form

$$R(x) = W(xp_0 + p_1, p_0 + xp_1)(1-x)^{-1}$$

where $p_0 = 1 - p_1$.

If the code is used only for detecting errors, then the probability of correct decoding is equal to p_0^n. The probability of undetected errors is equal to the coefficient of x^0 in the generating function expansion $p_u = R(0) = 2^{-1}(n+1)^{-1}[1 + g^n + 2ng^{(n-1)/2}] - p_0^n$, where $g = 1 - 2p_1$.

Suppose now that the code is being used to correct single errors. In this case, the probability of correct decoding is the probability of receiving a block with no more than one error: $p_c = p_0^n + np_1 p_0^{n-1}$. The probability of incorrect decoding is equal to the coefficient of x in the expansion of the generating function $R(x)$:

$$p_u = \frac{d}{dx} R(x)|_{x=0} = R(0) + (1-g^n)n/2(n+1) - np_1 p_0^{n-1}$$

Example 4.4.5. Consider now an (n,k) Reed-Solomon code whose symbols belong to GF(q) where $q = n+1$. Suppose that symbol-errors are independent and the probability of correct symbol reception is equal to P_0.

Let us evaluate the code's performance characteristics when it is used for error detection only. In this case, the probability of correct decoding is equal to $p_c = P_0^n$. The code Hamming weight generator polynomial (see Appendix 4.2) is equal to

$$A(x) = q^{-r}(1+nx)^n + \sum_{i=0}^{r-1} \binom{n}{i} x^i (1-x)^{n-i}(1-q^{i-r})$$

so that

$$W(x,y) = q^{-r}(y+nx)^n + \sum_{i=0}^{r-1} \binom{n}{i} x^i (y-x)^{n-i}(1-q^{i-r}) - y^n$$

where $r=n-k$. Generating function (4.4.42) is given by

$$W(P_1+xQ_1, P_0+nxP_1) = q^{-r}(1+nx)^n - (P_0 + nxP_1)^n$$
$$+ \sum_{i=0}^{r-1} \binom{n}{i}(P_1+xQ_1)^i(1-x)^{n-i}(P_0-P_1)^{n-i}(1-q^{i-r})$$

where $Q_1 = 1-P_1$ and generating function (4.4.45) takes the form

$$R(x) = q^{-r}(1+\gamma x)^n(1-x)^{-1} - (P_0+nxP_1)^n(1-x)^{-1}$$
$$+ \sum_{i=0}^{r-1} \binom{n}{i}(P_1+xQ_1)^i(1-x)^{n-i-1}(P_0-P_1)^{n-i}(1-q^{i-r})$$

The probability of undetected errors is equal to

$$p_u = R(0) = q^{-r} - P_0^n + \sum_{i=0}^{r-1} \binom{n}{i} P_1^i (P_0-P_1)^{n-i}(1-q^{i-r})$$

If the code redundancy r is an even number, then this code is able to correct up to $r/2$ symbol errors (see Sec 4.2.5). Suppose that the code is used to correct up to $t \le r/2$ errors. Then, the probability of correct decoding is given by Eq. (4.4.30):

$$p_c = \sum_{i=0}^{t} \binom{n}{i} P_0^{n-i} P_1^i n^i$$

The probability of incorrect decoding is the coefficient of x^t in the expansion of $R(x)$:

$$p_u = q^{-r} \sum_{i=0}^{t} \binom{n}{i} \gamma^i - \sum_{i=0}^{t} P_0^{n-i} P_1^i n^i$$
$$+ \sum_{i=0}^{r-1} \binom{n}{i}(P_0-P_1)^{n-i}(1-q^{i-r}) \sum_{j=j_1}^{j_2} \binom{i}{j}\binom{n-i-1}{t-j}(-1)^{t-j} P_1^{i-j} Q_1^j$$

where $j_1 = \max\{0, t-i-n\}$ and $j_2 = \min\{i,t\}$.

4.4.4 Upper and Lower Bounds

In many applications of probability theory, it is often easier to find bounds on some probability than to find the probability itself. This observation is especially true for the probability of incorrect decoding, even in the case of independent errors, since the code-weight generator polynomial is known for only a small number of codes.

Upper bounds are often used improperly to compare the performance of different systems. If a performance parameter upper bound of a system A is less than that of a system B, it is not necessarily true for the real values of the parameters. To prove a system's superiority, one must compare the upper bound of the parameter of one system with the lower bound of the parameter of the other system. However, if the upper bounds are tight, they may be used for comparison, but, in this case, they are used not as upper bounds but rather as good approximations of real parameters (since tightness means that they are close to the bounds).

4.4.4.1 Superset Bound

A simple technique for obtaining the upper bound is based on the inequality $Pr(A) \le Pr(B)$ if $A \subset B$. For example, if a code can correct up to t errors and is used for error correction and detection, then the probability of incorrect decoding is upper bounded by the probability that the number of errors is greater than t.

$$p_u \le \mathbf{pP}_n(>t)\mathbf{1} \tag{4.4.48}$$

This upper bound may be exact in some cases.

4.4.4.2 Union Bound

Another commonly used method for finding bounds is based on the fact that the probability of the union of events does not exceed the sum of the probabilities of the events

$$Pr(\bigcup_i A_i) \le \sum_i Pr(A_i)$$

The right-hand side of this inequality is called the *union bound* of its left-hand side.

By using this method, we can obtain a more accurate upper bound than in Eq. (4.4.48). Let us consider the case when the maximum likelihood decoding is used for correcting errors: for each received word the decoder outputs the closest code word and, in the case of a tie, each code word is selected with the same probability by some random number generator. If we assume that the all-zero code word has been transmitted, then the received block $\mathbf{e}$ is decoded incorrectly if its Hamming distance from the transmitted word is larger than from some other code word $|\mathbf{c}-\mathbf{e}|<|\mathbf{e}|$ or $|\mathbf{c}-\mathbf{e}| = |\mathbf{e}|$ but the random-number generator selected $\mathbf{c}$ rather than $\mathbf{e}$. The sum of the probabilities of these inequalities gives us the (union) upper bound of the probability of incorrect decoding:

$$p_u \le \mathbf{pU1}$$
$$\mathbf{U} = \sum_{\mathbf{c}\neq 0}\sum_{\mathbf{e}} \mathbf{P}(|\mathbf{c}-\mathbf{e}|<|\mathbf{e}|) + 0.5\,\mathbf{P}(|\mathbf{c}-\mathbf{e}| = |\mathbf{e}|)$$

For the HMM model, we can calculate this upper bound by using generating functions. Let $\mathbf{c} = (c_1,c_2,...,c_n)$ be a code word and $\mathbf{e} = (e_1,e_2,...,e_n)$ denote an error sequence. Then, the Hamming distance between the code word and the received word can be expressed as $|\mathbf{c}-\mathbf{e}| = \sum_{i=1}^{n} \Delta(c_i,e_i)$ and the distance between the received word and the transmitted word $|\mathbf{e}| = \sum_{i=1}^{n} \Delta(0,e_i)$ where $\Delta(x,y) = 0$ if $x = y$ and otherwise $\Delta(x,y) = 1$. The generating function of the distance difference distribution is given by

$$\mathbf{\Phi}(z) = \sum_{\mathbf{c}\neq 0}\sum_{\mathbf{e}} \mathbf{P}(\mathbf{e})z^{|\mathbf{e}|-|\mathbf{c}-\mathbf{e}|}$$

If we denote $\delta(c_i,e_i) = \Delta(0,e_i)-\Delta(c_i,e_i)$, then this generating function may be expressed as

$$\mathbf{\Phi}(z) = \sum_{\mathbf{c}} \prod_{i=1}^{n} \mathbf{\Psi}(c_i,z) \tag{4.4.49}$$

where $\mathbf{\Psi}(c_i,z) = \sum_{e_i} \mathbf{P}(e_i) z^{\delta(c_i,e_i)}$ Since $\delta(c_i,e_i) = 1$ when $c_i = e_i \neq 0$; $\delta(c_i,e_i) = -1$ when $c_i \neq 0, e_i = 0$, and $\delta(c_i,e_i) = 0$ in all other cases, we have

$$\mathbf{\Psi}(c_i;z) = \begin{cases} \mathbf{P}(c_i)z + \mathbf{P}(0)z^{-1} + \mathbf{P} - \mathbf{P}(c_i) - \mathbf{P}(0) & \text{when } c_i \neq 0 \\ \mathbf{P} & \text{otherwise} \end{cases} \tag{4.4.50}$$

The matrix probability $\mathbf{U}$ is equal to the sum of the coefficients of the positive power terms and one-half the coefficient of the zero power term in the expansion of the generating function $\mathbf{\Phi}(z)$. This sum can be expressed using contour integral representation of the Laurent series [12] coefficients:

$$\mathbf{P}_u \leq \frac{1}{4\pi j} \oint_{\gamma} \frac{z+1}{z(z-1)} \mathbf{\Phi}(z)\, dz$$

This upper bound does not look simpler than exact formula (4.4.25). However, it becomes much simpler when channel errors are symmetrically dependent and, in particular, for channels without memory. Indeed, generating function (4.4.50) in this case takes the form

$$\mathbf{\Psi}(c_i,z) = \mathbf{P}_1 z + (q-2)\mathbf{P}_1 + \mathbf{P}_0 z^{-1}$$

for $c_i \neq 0$. Generating function (4.4.49) is equal to

$$\mathbf{\Phi}(z) = \sum_{i=1}^{n} A_i [\, \mathbf{P}_1 z + (q-2)\mathbf{P}_1 + \mathbf{P}_0 z^{-1} \,]^i \mathbf{P}^{n-i}$$

where A_i is the number of the code words whose Hamming weight is equal to i. Since $\mathbf{P}^{n-i}\mathbf{1} = \mathbf{1}$, this generating function is equivalent to

$$\mathbf{\Phi}(z) = \sum_{i=1}^{n} A_i [\, \mathbf{P}_1 z + (q-2)\mathbf{P}_1 + \mathbf{P}_0 z^{-1} \,]^i = W(\mathbf{P}_1 z + (q-2)\mathbf{P}_1 + \mathbf{P}_0 z^{-1}, 1)$$

In the case of a BSC without memory it takes the form

$$\Phi(z) = \sum_{i=1}^{n} A_i (P_1 z + P_0 z^{-1})^i = W(P_1 z + P_0 z^{-1}, 1) \tag{4.4.51}$$

Collecting all the positive power terms and one-half of the zero power term we obtain

$$p_u \leq \sum_{i=1}^{n} A_i U_i$$

$$U_i = \begin{cases} \displaystyle\sum_{j=(i+1)/2}^{i} \binom{i}{j} P_1^j P_0^{i-j} & \text{when } i \text{ is odd} \\ 0.5 \binom{i}{i/2} (P_0 P_1)^{i/2} + \displaystyle\sum_{j=i/2+1}^{i} \binom{i}{j} P_1^j P_0^{i-j} & \text{when } i \text{ is even} \end{cases}$$

In conclusion, we would like to note that the zero power term in the generating function expansion is needed only when the random number generator is used to resolve a tie. If the code is being used to correct up to t errors, then we can include into the union bound only the terms whose power is greater than $|\mathbf{c}|-2t$.

4.4.4.3 Lower Bound

If we know a sum $S = X+Y$ of two nonnegative variables and an upper bound of one of the addends ($X \le U$), then we can obtain a lower bound of the other ($Y \ge S-U$). Because of the sum's symmetry, we can also find a upper bound on one of the addends if we know a lower bound on the other.

This primitive technique is often applied to the identity $Pr(A) = Pr(A \cup B) - Pr(B)$ where $A \cap B = \varnothing$. If the probability $Pr(A \cup B)$ is known and the probability $Pr(B)$ is upper bounded by some value U, then the lower bound is given by $Pr(A) \ge Pr(A \cup B) - U$ In particular, if A is a complement of B, $Pr(A \cup B) = 1$, and the previous bound takes the form $Pr(A) \ge 1-U$.

To illustrate this technique, suppose that we know a distribution-generating function

$$f(z) = \sum_{i=0}^{\infty} p_i z^{-i}$$

and we want to find a lower bound on the sum $p_t = \sum_{i=0}^{t} p_i$. This sum can be expressed through the distribution tail: $p_t = 1 - \sum_{i=t+1}^{\infty} p_i$, Using the Chernoff inequality (see Appendix 2.2) for the distribution tail

$$\sum_{i=t+1}^{n} p_i \le z^{t+1} f(z)$$

where $z \ge 1$, we obtain the lower bound:

$$p_t \ge 1 - z^{t+1} f(z)$$

The best bound of this form is obtained if z delivers the global minimum to the function $z^{t+1} f(z)$.

4.4.5 Postdecoding Error Probability

In many cases, it is necessary to evaluate the message-processing algorithm after decoding is done. The processing may include additional decoding, descrambling, decryption, and so on. It is, therefore, important to be able to describe the symbol sequence on the decoder output. Normally, only the information part of a message appears on the decoder output, so that errors in the parity portion of a message can be disregarded. However, in some cases, the whole message is passed over to a processor that follows the decoder. Generalized concatenated codes represent an example of such processing. [7] We will consider only this case in the future.

The simplest characteristic of the sequence on the decoder output is the symbol-error probability p_{se}:

$$p_{se} = P \lim_{T\to\infty} \frac{N_{se}}{T}$$

where N_{se} is the total number of errors and T is the total number of symbols on the decoder output. If we assume that the all-zero code word has been transmitted and the received word syndrome $\mathbf{eH}' \in C$, then the decoder outputs the closest to $\mathbf{e}$ code word $\mathbf{c}$. The symbol-error probability on the output of a decoder of a block code used only for correcting errors is

$$p_{se} = q_c + q_e \tag{4.4.52}$$

where

$$q_c = \sum_{\mathbf{eH}' \in C} \frac{|\mathbf{c}(\mathbf{e})|}{n} Pr(\mathbf{e}) \quad q_e = \sum_{\mathbf{eH}' \notin C} \frac{|\mathbf{e}|}{n} Pr(\mathbf{e})$$

In this equation, $\mathbf{c}(\mathbf{e})$ denotes the code word whose distance from the received word $\mathbf{e}$ does not exceed t.

If errors are symmetrically dependent, the above sums can be rearranged according to vectors weights as in Eq. (4.4.34):

$$q_c = \mathbf{pQ}_c\mathbf{1}$$

$$\mathbf{Q}_c = \sum_{w=0}^{n} \sum_{\mathbf{c}_w} \sum_{i=0}^{t} \sum_{j=0}^{n} \sum_{|\mathbf{e}_j - \mathbf{c}_w| = i} \frac{w}{n} \mathbf{P}(\mathbf{e}_j)$$

After changing the order of summation, we obtain

$$\mathbf{Q}_c = \sum_{i=0}^{t} \sum_{j=0}^{n} B_{ij} \mathbf{P}_1^j \mathbf{P}_0^{n-j}$$

where B_{ij} has the generating function

$$\sum_{ij} B_{ij} x^i y^j = B(x+y+(1-\gamma)xy,\ 1+\gamma xy)$$

and

$$B(x,y) = \sum_{i=0}^{n} \frac{i}{n} A_i x^i y^{n-i} = \frac{x}{n} \frac{\partial}{\partial x} A(x,y) \tag{4.4.53}$$

$\mathbf{Q}_c$ is equal to the sum of coefficients of x^i for $i \le t$ in the power series expansion of the function

$$B(x\mathbf{P}_0 + \mathbf{P}_1 + (1-\gamma)x\mathbf{P}_1,\ \mathbf{P}_0 + \gamma x \mathbf{P}_1) \tag{4.4.54}$$

It is equal to the coefficient of x^t in the expansion of the function

$$B(x\mathbf{P}_0 + \mathbf{P}_1 + (1-\gamma)x\mathbf{P}_1,\ \mathbf{P}_0 + \gamma x \mathbf{P}_1)(1-x)^{-1}$$

After performing the expansion, we obtain

$$\mathbf{Q}_c = \sum_{i=1}^{n} iA_i \sum_{k=0}^{t} \sum_{h=h_1}^{h_2} \binom{i}{k-h} \binom{n-i}{h} \gamma^h \mathbf{P}_0^{n-i-h} \mathbf{P}_1^{2h+i-k} [\,\mathbf{P}_0 + (\gamma-1)\mathbf{P}_1\,]^{k-h}$$

where $h_1 = \max\{0, k-i\}$ and $h_2 = \min\{n-i, k\}$.

The second term of Eq. (4.4.52) can be found similarly if we make use of the identity

$$\sum_{\mathbf{e}\mathbf{H}' \notin C} |\mathbf{e}| Pr(\mathbf{e}) = \sum_{\mathbf{e}} |\mathbf{e}| Pr(\mathbf{e}) - \sum_{\mathbf{e}\mathbf{H}' \in C} |\mathbf{e}| Pr(\mathbf{e})$$

This can be rewritten as

$$q_e = \gamma \mathbf{p}\mathbf{P}_1 \mathbf{1} - \mathbf{p}\mathbf{Q}_e \mathbf{1}$$

$$\mathbf{Q}_e = \sum_{\mathbf{e}\mathbf{H}' \in C} \frac{|\mathbf{e}|}{n} \mathbf{P}(\mathbf{e}) = \sum_{w=0}^{n} \sum_{\mathbf{c}_w} \sum_{i=0}^{t} \sum_{j=0}^{n} \sum_{|\mathbf{e}_j - \mathbf{c}_w| = i} \frac{j}{n} \mathbf{P}(\mathbf{e}_j)$$

The sum on the right-hand side of this equation can be found by using the generating function

$$C[\, x\mathbf{P}_0 + \mathbf{P}_1 + (1-\gamma)x\mathbf{P}_1,\ \mathbf{P}_0 + \gamma x \mathbf{P}_1\,](1-x)^{-1}$$

where

$$C(x,y) = \frac{y}{n} \frac{\partial}{\partial y} A(x,y)$$

However, in the majority of cases, $\mathbf{Q}_e$ is small compared to $\mathbf{Q}_c$, and in the particular case when all the syndromes are used for correcting errors $\mathbf{Q}_e = 0$.

Example 4.4.6. Let us calculate the postdecoding error probability for the (7,4) Hamming code (see Example 4.4.3) in the case of BSC with independent errors. We have $A(x,y) = y^7 + 7x^3y^4 + 7x^4y^3 + x^7$. This generating function can be determined directly by listing all 16 code words, or using the general Eq. (see Ref. 27, p. 142). Equation (4.4.53) gives $B(x,y) = 3x^3y^4 + 4x^4y^3 + x^7$. After substituting $\mathbf{P}_0 = p_0$, $\mathbf{P}_1 = p_1$, and $\gamma = 1$ into Eq. (4.4.54), we obtain

$$B(p_1 + p_0x, p_0 + p_1x) = 3(p_1 + p_0x)^3(p_0 + p_1x)^4 + 4(p_1 + p_0x)^4(p_0 + p_1x)^3 + (p_1 + p_0x)^7$$

Finally, the probability of postdecoding error is found by summing the coefficients of x^0 and x^1:

$$p_{se} = 9p_1^2p_0^5 + 19p_1^3p_0^4 + 16p_1^4p_0^3 + 12p_1^5p_0^2 + 7p_1^6p_0 + p_1^7$$

It is easy to prove that, for the general $(n, n-r)$ Hamming code,

$$p_{se} = \frac{1}{n+1}[1 + \frac{n-1}{2} g^{(n+1)/2} - \frac{n+1}{2} g^{(n-1)/2}] + p_1 \frac{(n-1)}{n+1}[1 - g^{(n+1)/2}]$$

where $g = 1 - 2p_1$.

4.4.6 Soft Decision Decoding

So far we have considered the performance of systems in which the transmitted and received symbols belong to the same alphabet GF(q). This happens if the *hard decision demodulator* described in Sec 1.1.8 is used. In the case of a binary channel, the decision circuit in the demodulator outputs bit 1 if, at the sampling instant, the received signal voltage is positive and outputs 0 otherwise. However, a more sophisticated decision circuit can output more than two alternative values that depend on the received signal voltage. The received analog signal is quantized and a decoder receives nonbinary symbols. This type of demodulation is called a *soft decision demodulation*. Soft decision demodulator provides more information about the received signal than hard decision demodulator. The decoder can use this information to improve system performance.

Let $\mathbf{s} = (s_1, s_2, \ldots, s_n)$ be a transmitted block and $\boldsymbol{\theta}(\mathbf{s}) = (\theta_1, \theta_2, \ldots, \theta_n)$ be the received block of n symbols, assuming that there were no errors. Suppose that the transmitted symbols belong to GF(q) and are encoded by an (n,k) code. The received symbols, in general, do not belong to the field of transmitted symbols. We assume also that all the messages are equally probable and that channel errors $\mathbf{e}$ are additive, so that

$$Pr(\boldsymbol{\theta}(\mathbf{s})+\mathbf{e} \mid \mathbf{s}) = Pr(\mathbf{e})$$

The maximum likelihood decoder outputs the code word $\mathbf{s}^*$ for the received message $\mathbf{r}^*$ if the received message has the largest probability:

$$Pr(\mathbf{r}^* \mid \mathbf{s}^*) = \max_{\mathbf{s}} Pr(\mathbf{r}^* \mid \mathbf{s})$$

If this equation has several solutions, a unique solution is selected from the set of solutions with equal probabilities by using a random number generator. For each code word, we define the decision region as the set $\Theta(\mathbf{s})$ of all $\mathbf{r}^*$ that satisfy the previous equation. The decoder outputs $\mathbf{s}$ when the received block $\boldsymbol{\theta} \in \Theta(\mathbf{s})$.

Let us evaluate the performance of a soft decision decoder. Since errors do not depend on the transmitted symbols, we can assume that the all-zero code word $\mathbf{0}$ has been transmitted. The receiver decodes it correctly if the soft decision demodulator outputs $\mathbf{e} \in \Theta(\mathbf{0})$, so that

$$p_c = Pr(\mathbf{e} \in \Theta(\mathbf{0}))$$

The probability of incorrect decoding is equal to

$$p_u = \sum_{\mathbf{s} \neq \mathbf{0}} Pr(\mathbf{e} \in \Theta(\mathbf{s}))$$

The probability of detected errors is equal to zero in this case, since every received block $\boldsymbol{\theta}$ is decoded into a code word $\mathbf{s}$. If we want to correct only the most probable errors, then the decision region can be narrowed by eliminating points $\mathbf{r}^*$ for which the probability $Pr(\mathbf{r}^* \mid \mathbf{s}^*)$ is small. In this case, the probability of detected errors $p_d = 1 - p_c - p_u$.

In many cases, when it is difficult to calculate $Pr(\mathbf{e} \in \Theta(\mathbf{s}))$, the following union bound may be handy:

$$Pr(\mathbf{e} \in \Theta(\mathbf{s})) \leq \sum_{U(\mathbf{s})} Pr(\mathbf{e}) + 0.5 \sum_{T(\mathbf{s})} Pr(\mathbf{e})$$

where $U(\mathbf{s}) = \{\mathbf{e}\colon Pr(\mathbf{e}-\boldsymbol{\theta}(\mathbf{s})) > Pr(\mathbf{e})\}$ and $T(\mathbf{s}) = \{\mathbf{e}\colon Pr(\mathbf{e}-\boldsymbol{\theta}(\mathbf{s})) = Pr(\mathbf{e})\}$. In order to calculate these probabilities we need to have an error source model.

Example 4.4.7. Consider the satellite channel described in Sec 1.1.8. For the sake of simplicity, we assume that the channel is linear, ideally equalized, and that there is no differential encoder. In this case, the signal on the input to the data detector is given by Eq. (1.1.22). Let us assume now that the data detector makes a "soft" decision about the received signal: it does not change the signal. This assumption is valid when the number of quantization levels is large.

Binary information is encoded by the (n,k) code, demultiplexed into the inphase and quadrature sequences, and transmitted over the channel. The receiver multiplexes the samples Re $u_3(iT)$ and Im $u_3(iT)$, which are given by Eq. (1.1.22) and outputs them to the soft decision decoder. Our goal is to evaluate the decoder's performance.

Suppose that a code word $\mathbf{s} = (s_1, s_2, ..., s_n)$ has been transmitted and the block $\boldsymbol{\theta}(\mathbf{s}) = (\theta_1, \theta_2, ..., \theta_n)$ has been received in the absence of noise. Then, according to the channel description, $\boldsymbol{\theta}(\mathbf{s}) = 2\mathbf{s} - \mathbf{1}$, where $\mathbf{1} = (1, 1, ..., 1)$. In the presence of noise, however, we receive $\mathbf{r} = \boldsymbol{\theta} + \mathbf{e}$, where $\mathbf{e}$ is a zero-mean Gaussian variable whose covariance matrix is equal to $diag\{\sigma^2\}$, where σ^2 is the noise power on the output of the receiver filter.

Let us define now the maximum likelihood decoder decision regions. Since **r** is a Gaussian variable whose mean is equal to **θ**, the probability density of receiving **r** if **s** was transmitted is given by

$$f(\mathbf{r} \mid \mathbf{s}) = A\exp(-\|\mathbf{r}-\boldsymbol{\theta}(\mathbf{s})\|^2/2\sigma^2)$$

where $A = (2\pi)^{-n/2}\sigma^{-n}$ and

$$\|\mathbf{r}-\boldsymbol{\theta}(\mathbf{s})\|^2 = \sum_{i=1}^{n}(\theta_i - r_i)^2$$

is the squared Euclidean distance between vectors **r** and **θ**. From this equation, it follows that the maximum of $f(\mathbf{r} \mid \mathbf{s})$ corresponds to the minimum of the Euclidean distance. Therefore, the maximum likelihood decoding rule can be formulated as follows: Decode the code word **s** if $\boldsymbol{\theta}(\mathbf{s})$ is the closest to the received vector **r**. This rule is similar to the minimum Hamming distance decoding considered in the previous section.

In order to evaluate the decoder's performance, suppose that the all-zero code word $\mathbf{0} = (0,0,\ldots,0)$ has been transmitted. In the absence of noise the receiver outputs the sequence $\boldsymbol{\theta}(\mathbf{0}) = -\mathbf{1}$. A received block **r** is decoded correctly if it is closer to $\boldsymbol{\theta}(\mathbf{0})$ than to some other $\boldsymbol{\theta}(\mathbf{s})$, so that

$$p_c = \int_{\Theta(\mathbf{0})} f(\mathbf{r} \mid \mathbf{0})\,d\mathbf{r}$$

where the integration is performed over all **r** that are closer to $\boldsymbol{\theta}(\mathbf{0})$ than to $\boldsymbol{\theta}(\mathbf{s})$ with $\mathbf{s}\neq\mathbf{0}$.

The probability of incorrect decoding is

$$p_u = \sum_{\mathbf{s}\neq\mathbf{0}} \int_{\Theta(\mathbf{s})} f(\mathbf{r} \mid \mathbf{0})\,d\mathbf{r}$$

It is difficult to evaluate the preceding integrals because the integration regions are complex. The union bound can be expressed using simpler integrals:

$$p_u \le \sum_{\mathbf{s}\neq\mathbf{0}} \int_{U(\mathbf{s})} f(\mathbf{r} \mid \mathbf{0})\,d\mathbf{r} \tag{4.4.55}$$

where $U(\mathbf{s}) = \{\mathbf{r}\colon \|\mathbf{r}-\boldsymbol{\theta}(\mathbf{0})\| > \|\mathbf{r}-\boldsymbol{\theta}(\mathbf{s})\|\}$ is the set of all **r** that are closer to $\boldsymbol{\theta}(\mathbf{s})$ than to $\boldsymbol{\theta}(\mathbf{0})$. The integration region $U(\mathbf{s})$ is the semispace bounded by the hyperplane which is equidistant from the points $\boldsymbol{\theta}(\mathbf{0})$ and $\boldsymbol{\theta}(\mathbf{s})$. The integral over $U(\mathbf{s})$ is equal to the probability that the projection of **r** onto $\boldsymbol{\theta}(\mathbf{s})-\boldsymbol{\theta}(\mathbf{0})$ is greater than $0.5\,\|\boldsymbol{\theta}(\mathbf{s})-\boldsymbol{\theta}(\mathbf{0})\|$, which is equal to half of the distance between the points. Since the projection of **r** is a Gaussian variable, we obtain

$$\int_{U(\mathbf{s})} f(\mathbf{r} \mid \mathbf{0})\,d\mathbf{r} = 0.5\,\mathit{erfc}(0.5\beta\,\|\boldsymbol{\theta}(\mathbf{s})-\boldsymbol{\theta}(\mathbf{0})\|)$$

It is easy to verify that the relation between the Euclidean and Hamming distances can be expressed as $0.5\,\|\boldsymbol{\theta}(\mathbf{s})-\boldsymbol{\theta}(\mathbf{0})\| = \sqrt{|\mathbf{s}|}$. If we rearrange the summation in Eq. (4.4.55) according to Hamming weights, we obtain

$$p_u \le \sum_{i=1} A_i\,\mathit{erfc}(\beta\sqrt{i}) \tag{4.4.56}$$

where A_i is the number of Hamming weight i code words.

This bound can be expressed through the code-weight generating function if we make use of the identity [24]

$$\int_a^\infty \frac{e^{-pt}dt}{t\sqrt{t-a}} = \frac{\pi}{\sqrt{a}}\mathit{erfc}(\sqrt{ap})$$

which, after substitutions $\sqrt{a} = \beta$ and $\sqrt{t-a} = v$, takes the form

$$\mathit{erfc}(\beta\sqrt{p}) = \frac{2\beta}{\pi}\int_0^\infty (\beta^2+v^2)^{-1}e^{-p(\beta^2+v^2)}\,dv$$

Using this identity we can perform the summation in Eq. (4.4.56) and obtain the following representation of the union bound for the probability of incorrect decoding:

$$p_u \le \frac{2\beta}{\pi}\int_0^\infty (\beta^2+v^2)^{-1}A_1(e^{-(\beta^2+v^2)})\,dv$$

where $A_1(x) = W(x,1)$ is defined by Eq. (4.4.41).

In particular, for the (7,4) Hamming code (see Example 4.4.6) $A_1(x) = 7x^3+7x^4+x^7$ and the union bound of Eq. (4.4.56) takes the form

$$p_u \le 3.5erfc(\beta\sqrt{3})+3.5erfc(2\beta)+0.5erfc(\beta\sqrt{7})$$

Let us compare the performance of the soft decision decoder with that of the hard decision decoder. For the hard decision decoder, the probability of incorrect decoding is defined by Eq. (4.4.43), which in the case of the (7,4) Hamming code gives

$$p_u = 7(3p_1^2p_0^5 + 5p_1^3p_0^3 + 3p_1^5p_0^2 + p_1^6p_0) + p_1^7 \tag{4.4.57}$$

where $p_1 = 0.5erfc(\beta)$ and $p_0 = 1-p_1$. The union bound for this probability can be obtained by using generating function (4.4.51), which takes the form

$$\Phi(z) = 7(p_1z+p_0z^{-1})^3+7(p_1z+p_0z^{-1})^4+(p_1z+p_0z^{-1})^7$$

Since the code corrects all single errors and does not correct or detect errors of higher multiplicity, we need to collect only the positive power terms of this function expansion:

$$p_u \le 7(3p_1^2p_0+p_1^3+4p_1^3p_0+p_1^4+5p_1^4p_0^3+3p_1^5p_0^2+p_1^6p_0)+p_1^7$$

This bound is very tight for small values of p. It is approximately equal to $7(3p_1^2+5p_1^3)$, whereas the exact value of the probability of incorrect decoding (4.4.61) is approximated by $7(3p_1^2-10p_1^3)$ if p is small.

4.4.7 Correcting Errors and Erasures

Let us now consider a channel with errors and erasures. Erasures can be generated by a soft decision demodulator when it cannot make a decision about received symbol value. They can also appear on the outer decoder input of a concatenated code when the inner decoder detects symbol-errors and passes them along to the outer decoder as erasures. [11] In this case, in addition to symbols belonging to GF(q), we receive a $q+1$-th symbol called an *erasure*. This symbol contains no information about the value of the transmitted symbol; it only marks the location of an indefinite symbol. It is possible to use this information to improve the code's performance.

Suppose that the code minimum distance is equal to $d_{\min}$. Then the code can correct t errors and s erasures if and only if $2t+s<d_{\min}$. It follows from the fact that for correct decoding of a code word with s erased symbols it is necessary and sufficient that this code word be different from any other code word in the remaining $n-s$ nonerased positions. The minimum Hamming distance decoding consists of finding a code word that is closest to the received word in the nonerased positions.

One could use a decoder that corrects errors to correct errors and erasures by trying all possible combinations of symbols in the places of erasures until the syndrome of the combination of the errors and the selected symbols belongs to the set C of the syndromes of correctable errors. Let $\mathbf{r}_0$ be a word obtained from the received word by replacing the erasures with zeroes, and let $\mathbf{e}_s$ be a word that has zeroes on the positions different from erasures. We vary $\mathbf{e}_s$ until we find the one which satisfies the condition $(\mathbf{r}_0+\mathbf{e}_s)\mathbf{H}' = \mathbf{S}\in C$, and then we decode $\mathbf{c} = \mathbf{r}_0+\mathbf{e}(\mathbf{S})$.

In the case of BCH codes, the decoding procedure can be significantly simplified. Indeed, the erasure locations are the roots of the characteristic polynomial in Eq. (4.2.13). We can factor the polynomial into the product of two polynomials $\Lambda(x)\lambda(x)$, where $\Lambda(x)$ is the known polynomial whose roots are the erasure locators and $\lambda(x)$ is the unknown error locator polynomial. Since the syndromes satisfy difference Eq. (4.2.12), the syndrome-generating function is a rational function:

$$S(x) = \phi(x) / (\Lambda(x)\lambda(x)) \tag{4.4.58}$$

Multiplying both sides of this by $\Lambda(x)$, we obtain $S(x)\Lambda(x) = \phi(x)/\lambda(x)$. The coefficients T_j of the generating function $T(x) = S(x)\Lambda(x)$ are called *modified syndromes.* [11] According to the convolution theorem, they can be expressed as

$$T_j = \sum_{i=0}^{s} \Lambda_i S_{j-i} \qquad j=m_0+s, m_0+s+1, \ldots, m_0+d$$

On the other hand, since the generating function $T(x)$ is the rational function given by the right-hand side of Eq. (4.4.58), the modified syndromes satisfy the finite difference equation whose characteristic polynomial is equal to $\lambda(x)$. Thus, by using the modified syndromes T_j instead of the original syndromes S_j, we can use the decoding algorithm described in Sec 4.2.4: Find a difference equation for the modified syndromes; the characteristic polynomial of this equation is the error locator polynomial $\lambda(x)$. The error locators are the roots of this polynomial. Using the error locators and the known erasure locators, we find the error and erasure values by solving system (4.2.11), in which t is replaced with $t+s$.

Let us calculate the code performance characteristics by assuming that the error and erasure source is described by the HMM model with the matrix probability of error $\mathbf{P}_e$ and the matrix probability of erasure $\mathbf{P}_*$. The probability of correct decoding is equal to the sum of the probabilities of all error and erasure sequences that satisfy the condition $2t+s<d$:

$$p_c = \mathbf{p} \sum_{2t+s<d} \mathbf{P}_n(t,s)\mathbf{1}$$

where $\mathbf{P}_n(t,s)$ is the matrix probability that a block of n symbols contains t errors and s erasures. Since it is a matrix multinomial probability, it has generating function (2.3.3), which can be written as

$$\mathbf{\Phi}_n(1,z,w) = \sum_{t,s} \mathbf{P}_n(t,s) z^t w^s = (\mathbf{P}_0 + z\mathbf{P}_e + w\mathbf{P}_*)^n \tag{4.4.59}$$

where $\mathbf{P}_0 = \mathbf{P} - \mathbf{P}_e - \mathbf{P}_*$. The generating function $\mathbf{\Psi}_n(z)$ of the probabilities $\mathbf{\Psi}_{m,n} = \mathbf{P}_n(2t+s = m)$ that $2t+s=m$ can be obtained from the generating function $\mathbf{\Phi}_n(1,z,w)$ with the simple variable substitution

$$\mathbf{\Psi}_n(z) = \mathbf{\Phi}_n(1,z^2,z) = (\mathbf{P}_0 + z^2\mathbf{P}_e + z\mathbf{P}_*)^n \tag{4.4.60}$$

Since $\mathbf{\Psi}_n(z) = \mathbf{\Psi}_k(z)\,\mathbf{\Psi}_{n-k}(z)$, the probabilities $\mathbf{\Psi}_{m,n}$ can be found from the recursive equation

$$\mathbf{\Psi}_{m,n} = \sum_{i=0}^{m} \mathbf{\Psi}_{i,k}\mathbf{\Psi}_{m-i,n-k}$$

For fast calculations, it is advisable to perform the recursion by using the binary representation of n, as in evaluating the matrix binomial distribution discussed in Sec 2.6.

The matrix probability of correct decoding is equal to the sum of the coefficients of power terms z^m for $m=0,1,\ldots,d-1$ in the function $\mathbf{\Psi}_n(z)$ expansion:

$$\mathbf{P}_c = \sum_{m=0}^{d-1} \mathbf{\Psi}_{m,n} \tag{4.4.61}$$

This probability can also be found as a coefficient of z^{d-1} in the expansion of the generating function of sums:

$$\Psi_n(z)/(1-z) = \sum_{i=0}^{\infty} z^i \sum_{m=0}^{i} \Psi_{m,n}$$

Since the generating function (4.4.59) is a polynomial, we may use the Fourier transform inversion formula in Eq. (2.2.25) to calculate the coefficients $\Psi_{m,n}$ and the probability of correct decoding:

$$\Psi_{m,n} = p^{-1} \sum_{k=0}^{p-1} \alpha^{-mk} (\mathbf{P}_0 + \mathbf{P}_e \alpha^{2k} + \mathbf{P}_* \alpha^k)^n$$

$$\mathbf{P}_c = p^{-1} \sum_{k=0}^{p-1} \eta(\alpha^{-k}) (\mathbf{P}_0 + \mathbf{P}_e \alpha^{2k} + \mathbf{P}_* \alpha^k)^n$$

where $\alpha = e^{-2\pi j/p}$, $p > 2n$, and $\eta(x) = (1-x^d)/(1-x)$.

The lower bound on the probability of correct decoding can be obtained using Chernoff's inequality

$$p_c \geq 1 - z_0^{-d} \mathbf{p} \Psi(z_0) \mathbf{1}$$

where $z_0 \geq 1$ minimizes $z^{-d} \mathbf{p} \Psi(z) \mathbf{1}$.

The matrix probability of incorrect decoding can be found similarly to the case when no erasures were present. If channel errors are symmetrically dependent, the equations can be simplified. The probability of incorrect decoding can be expressed using the total probability theorem

$$p_u = \sum_{s=0}^{d-1} Pr(s) \tilde{p}_u(s) \tag{4.4.62}$$

where $\tilde{p}_u(s)$ is the average conditional probability of incorrectly decoding codes that are obtained from the original code by erasing any s symbols. The conditional probability of incorrect decoding can be determined in a manner similar to that of Eq. (4.4.47).

Example 4.4.8. Suppose that the (7,4) Hamming code is used to correct errors and erasures in a channel with independent errors and erasures. This code minimum Hamming weight is equal to $d=3$ (see Example 4.4.6), so that it is able to correct any combination of errors and erasures that satisfy the inequality $2t+s<3$; that is, it can correct a single bit error with no erasures or up to two erasures with no errors. The decoding procedure is simple: Correct a single error according to its syndrome in the absence of erasures or correct no more than two erasures by finding the code word that matches the received word in unerased positions (or by varying bits in erased positions until the word becomes a code word).

Let us now evaluate the code performance characteristics. Generating function (4.4.61) takes the form $\Psi_7(z) = (p_0 + z^2 p_e + z p_*)^7$ and the probability of correct decoding is equal to the sum of the coefficients of z^0, z^1, and z^2 in this function expansion: $p_c = p_0^7 + 7p_0^6(1-p_0) + 21 p_0^5 p_*^2$.

The probability of incorrect decoding can be found using Eq. (4.4.62). The probability of having $s=0$ erasures is $Pr(s=0) = (1-p_*)^7$. The conditional probability $p_u(0)$ of incorrect decoding in this case is given by the ratio of the right-hand side of Eq. (4.4.61), in which p_1 is replaced with p_e, to $(1-p_*)^7$.

The probability of having one erasure is $Pr(s=1) = 7(1-p_*)^6 p_*$. To calculate the corresponding conditional probability of incorrect decoding, we have to find the average weight distribution for all the codes that can be obtained from the original (7,4) code by deleting one position. If we write down all sixteen code words and delete a bit in all of them, we obtain one all-zero word, three words whose weight is 2, eight words whose weight is 3, three words whose weight is 4, and one word whose weight is 6. Thus, the average weight-generating function is $A^{(1)}(x) = 1+3z^2+8z^3+3z^4+z^6$.

As we see, this code has a minimum distance $d=2$ and is not able to correct errors; it can only detect them. Therefore, the incorrect decoding occurs when the code does not detect errors. Using Eq. (4.4.43), we obtain

$$(1-p_*)^6 p_u(1) = W(p_e, p_0) = 3p_e^2 p_0^4 + 8p_e^3 p_0^3 + 3p_e^4 p_0^2 + p_e^6$$

Quite analogously we find $Pr(s=2)=21p_*^2(1-p_*)^5$ and

$$(1-p_*)^5 p_u(2) = p_e p_0^4 + 6p_e^2 p_0^3 + 6p_e^3 p_0^2 + p_e^4 p_0 + p_e^5$$

Finally, the probability of incorrect decoding is given by $p_u = p_u(0)+7p_* p_u(1)+21p_*^2 p_u(2)$

4.4.8 Product Codes

Product (or iterated) codes are long codes that are formed from two or more shorter codes. Suppose that information has been encoded using an (n_1, k_1) code. A superblock consisting of k_2 code words may be viewed as a $k_2 \times n_1$ matrix, each row of which is a code word

$$\mathbf{X}_1 = \begin{bmatrix} \mathbf{X}_{11} & \mathbf{X}_{12} \end{bmatrix}$$

where matrices

$$\mathbf{X}_{11} = \begin{bmatrix} x_{11} & x_{12} & \dots & x_{1,k_1} \\ x_{21} & x_{22} & \dots & x_{2,k_1} \\ \dots & \dots & \dots & \dots \\ x_{k_2 1} & x_{k_2 2} & \dots & x_{k_2 k_1} \end{bmatrix} \quad \text{and} \quad \mathbf{X}_{12} = \begin{bmatrix} x_{1,k_1+1} & \dots & x_{1,n_1} \\ x_{2,k_1+1} & \dots & x_{2,n_1} \\ \dots & \dots & \dots \\ x_{k_2 k_1+1} & \dots & x_{k_2 n_1} \end{bmatrix}$$

represent the information and parity-check symbols, respectively. We can improve the protection of information by encoding each column of the matrix with some (n_2, k_2) block code so that the encoded information is presented by the matrix

$$\mathbf{X} = \begin{bmatrix} \mathbf{X}_{11} & \mathbf{X}_{12} \\ \mathbf{X}_{21} & \mathbf{X}_{22} \end{bmatrix}$$

where $\mathbf{X}_{21}$ represents the parity-check symbols for the columns of the matrix $\mathbf{X}_{11}$, while $\mathbf{X}_{22}$ represents parity-check symbols for columns of $\mathbf{X}_{12}$ or for rows of $\mathbf{X}_{21}$. [27] Thus, we have constructed a $(n_1 n_2,\ k_1 k_2)$ code called a *product code*. We can construct products of more than two codes using multidimensional arrays.

For several reasons, product codes are widely used in practice. One of the reasons is the relative simplicity of their implementation: Very long codes can be constructed out of several short codes. The other reason is that they are capable of fighting against bursts of errors because of their built-in interleaving. If a code matrix row length n_1 is large compared to the average error burst length, then errors in the same column can be treated as independent. Also, if the (n_1, k_1) code detects noncorrectable errors in a row, it may erase the whole row, and the (n_2, k_2) code, will correct errors and erasures.

Let $\mathbf{H}_1$ and $\mathbf{H}_2$ be the parity-check matrices of the (n_1, k_1) and (n_2, k_2) code, respectively. Then, $\mathbf{X}$ is a code word of the product code if and only if

$$\mathbf{X}\mathbf{H}'_1 = 0 \text{ and } \mathbf{H}_2\mathbf{X} = 0$$

Using rectangular matrices simplifies product code description. To present a code word in the conventional vector form, we may scan the matrix **X** and then string all the rows in one long vector. This operation can be expressed in the matrix form [29]

$$\mathbf{s} = (x_{11}, x_{12}, \ldots, x_{1,n_1}, x_{21}, \ldots, x_{n_2,n_1}) = \sum_{i=1}^{n_2} \mathbf{v}_i \mathbf{X} \mathbf{N}_i$$

where $\mathbf{v}_i$ is the row matrix whose i-th element is equal to 1 and the rest of the elements are zeroes, and $\mathbf{N}_i$ is the block-row matrix whose i-th subblock is equal to **I** and the rest of the subblocks are zeroes:

$$\mathbf{v}_i = [\,0 \ldots 0\ 1\ 0 \ldots 0\,]$$

$$\mathbf{N}_i = [\,0 \ldots 0\ \mathbf{I}\ 0 \ldots 0\,]$$

The inverse relation has the form

$$\mathbf{X} = \sum_{i=1}^{n_2} \mathbf{v}'_i \mathbf{s} \mathbf{N}'_i$$

Using these equations, it is easy to verify that the product code parity-check matrix can be expressed as

$$\mathbf{H} = \begin{bmatrix} \mathbf{H}_1 \otimes \mathbf{I} & \mathbf{I} \otimes \mathbf{H}_2 \end{bmatrix}$$

To evaluate product code performance we can use its matrix **H**, which contains $n_1(n_1-k_1)n_2(n_2-k_2)$ elements in all the previously derived equations. However, if we use the code matrix representation, these equations can be simplified, since the component code parity-check matrices have fewer elements $[n_1(n_1-k_1)+n_2(n_2-k_2)]$. It is convenient to use matrix notations for transmitted, received, and error sequences when dealing with product codes. Let

$$\mathbf{E} = \begin{bmatrix} \mathbf{e}_{ij} \end{bmatrix}_{n_1,n_2}$$

be the error matrix that corresponds to the transmitted code matrix **X**. Matrices

$$\mathbf{S}_1 = \mathbf{E}\mathbf{H}'_1 \quad \text{and} \quad \mathbf{S}_2 = \mathbf{H}_2\mathbf{E}$$

are called *received message error syndromes.*

Using these notations, we can rewrite syndrome distribution generating function (4.4.6) as

$$f(\mathbf{z},\mathbf{w}) = \sum_{\mathbf{E}} Pr(\mathbf{E})\, \mathbf{z}^{\mathbf{E}\mathbf{H}'_1}\, \mathbf{w}^{\mathbf{H}'_2\mathbf{E}}$$

All the equations that use generating function (4.4.6) should be modified accordingly. Instead of doing the straightforward modifications, we illustrate their applications in the following example.

Example 4.4.9. Consider the product of two simple parity-check codes that check the code $m \times n$ matrix rows and columns, respectively. For an HMM model, let us find the probability of correct decoding and the probability of undetected errors if we assume that the code is used for error detection only.

The generating function of the syndrome matrix distribution is similar to Eq. (4.4.11) and can be written in the form

$$\mathbf{F}(\mathbf{z},\mathbf{w}) = \prod_{i=1}^{n}\prod_{j=1}^{m}[\mathbf{P}(0) + \mathbf{P}(1)z_i w_j]$$

The matrix probability of correct decoding is equal to the zero power term of the function $\mathbf{P}_c = \mathbf{P}^{mn}(0)$. The matrix probability of undetected errors is given by equation (4.4.12), which, in our case, has the form

$$\mathbf{P}_u = 2^{-n-m}\sum_{\mathbf{v},\boldsymbol{\mu}}\prod_{i=1}^{n}\prod_{j=1}^{m}[\mathbf{P}(0) + \mathbf{P}(1)(-1)^{\nu_i+\mu_j}] - \mathbf{P}^{mn}(0)$$

where $\nu_i, \mu_j = 0,1$. The calculations are simplified if we replace variables z_i with $(-1)^{\nu_i}$ to obtain the generating function

$$\Psi(\mathbf{w}) = 2^{-m}\sum_{\mathbf{v}}\prod_{i=1}^{n}\prod_{j=1}^{m}[\mathbf{P}(0) + \mathbf{P}(1)(-1)^{\nu_i}w_j]$$

$$= [\ \prod_{j=1}^{m}(\ \mathbf{P}(0)+\mathbf{P}(1)w_j\ ) + \prod_{j=1}^{m}(\ \mathbf{P}(0)-\mathbf{P}(1)w_j\ )\ ]^n$$

and then determine the probability of undetected errors using the generating function

$$\mathbf{P}_u = 2^{-n}\sum_{\boldsymbol{\mu}}\Psi((-1)^{\boldsymbol{\mu}}) - \mathbf{P}^{mn}(0)$$

where $(-1)^{\boldsymbol{\mu}} = [(-1)^{\mu_1},(-1)^{\mu_2},...,(-1)^{\mu_m}]$.

4.4.9 Concatenated Codes

Concatenated codes are similar to product codes. The first code that is used to encode the information matrix $\mathbf{X}_{11}$ rows is called the *inner code* and is defined the same way as in product codes. The second code that is used to encode matrix $\mathbf{X}_{11}$ columns is defined differently: The rows of matrix $\mathbf{X}_{11}$ whose elements belong to GF(q) are treated as elements of GF(q^{k_1}) and encoded by some (n_2,k_2) code over this field. This second code is called the *outer code*. In other words, all columns of matrix $\mathbf{X}_{11}$ are treated as one column of elements from GF(q^{k_1}) and encoded as such.

The decoder of the inner code detects and corrects errors. Symbols with detected errors are passed to the decoder of the outer code as erasures, which corrects the erasures and remaining errors. Usually, the RS code is used as the outer code.

The performance of concatenated codes can be evaluated similarly to that of product codes. However, for the HMM, it is possible to evaluate code performance characteristics in two steps: first for the inner code and then for the outer code. The inner encoder output is fully characterized by these three matrix probabilities: the probability $\mathbf{P}_0$ of correct decoding, the probability $\mathbf{P}_e$ of incorrect decoding, and the probability of an erasure which is equal to the probability of detected errors $\mathbf{P}_* = \mathbf{P}^{n_1} - \mathbf{P}_0 - \mathbf{P}_e$. (The methods for calculating these probabilities are described in previous sections.) Then the concatenated code performance can be evaluated by the methods described in Sec 4.4.7.

Example 4.4.10. Let us calculate the performance characteristics of a concatenated code whose inner code is the simple parity-check $(k+1,\ k)$ code and whose outer code is $(N,\ K)$ RS code where $N = 2^k - 1$.

We determine the characteristics of the inner code first. It is obvious that the probability that a block of $k+1$ bits does not have errors $\mathbf{P}_0 = \mathbf{P}^{k+1}(0)$. The probability of undetected errors has been found in Example 4.3.1: $\mathbf{P}_e = 2^{-1}\mathbf{P}^{k+1} + 2^{-1}[\mathbf{P}(0)-\mathbf{P}(1)]^{k+1} - \mathbf{P}^{k+1}(0)$ and the probability of an erasure is given by $\mathbf{P}_* = \mathbf{P}^{k+1} - \mathbf{P}_e - \mathbf{P}_0$ which, after simplification, becomes

$$\mathbf{P}_* = 2^{-1}\mathbf{P}^{k+1} - 2^{-1}[\mathbf{P}(0)-\mathbf{P}(1)]^{k+1}$$

We are ready now to evaluate outer code performance. Since the RS code minimal distance is equal to $D=N-K+1$, this code corrects all the combinations of t errors and s erasures that satisfy the condition $2t+s \le N-K$ and only them. The probability of correct decoding is equal to the sum of the coefficients of z^m for $m=0,1,...,N-K$ in the expansion of generating function (4.4.60) and is given by equations (4.4.61).

4.5 CONVOLUTIONAL CODE PERFORMANCE

Let us consider now methods for evaluating the performance of convolutional codes. Because of the nature of convolutional coding, its performance is usually characterized by some parameters of error sequences on a convolutional decoder output and, in particular, by decoder symbol-error probability.

4.5.1 Viterbi Algorithm Performance

It is possible to find an exact equation for the probability of incorrect decoding when the Viterbi algorithm is used because the decoding procedure itself can be described as an HMM. [5,6] However, the complexity of this approach grows exponentially with the code constraint length and simpler upper (union) bound on this probability is used instead which can be expressed in the closed form. In Appendix 4.3, we derive the closed-form expression for the union bound on Viterbi algorithm˙ performance for a general discrete-time system. According to Eq. (4.2.24), convolutional codes represent a particular case of discrete-time system (A.4.9). Therefore, we can apply the results of Appendix 4.3 to convolutional codes.

Suppose that we want to find the bit-error probability on the output of a decoder that uses the Hamming distance between the received and transmitted symbols as a branch metric

$$m(\mathbf{z}_k,\mathbf{w}_k) = |\mathbf{z}_k-\mathbf{y}_k| = |\mathbf{e}_k| \tag{4.5.1}$$

where $\mathbf{x}_k$ is the encoder input symbol, $\mathbf{y}_k$ is the encoder output symbol, $\mathbf{s}_k = (\mathbf{x}_{k-1},\mathbf{x}_{k-2},...,\mathbf{x}_{k-m+1})$ is the encoder state, and $\mathbf{w}_k = (\mathbf{x}_k,\mathbf{s}_k)$, and $\mathbf{z}_k$ is the decoder input symbol (see Sec 4.2.6 and Appendix 4.3). In this case, the branch metric generating function (A.4.12) is

$$\mathbf{D}(v;\mathbf{w}_k,\hat{\mathbf{w}}_k) = \sum_{\mathbf{e}_k}\mathbf{P}(\mathbf{e}_k)\,v^{|\mathbf{e}_k|-|\hat{\mathbf{e}}_k|} \tag{4.5.2}$$

where $\hat{\mathbf{e}}_k = \mathbf{z}_k-\hat{\mathbf{y}}_k$, $\hat{\mathbf{y}}_k$ is the encoder output for $\hat{\mathbf{w}}_k$.

The distortion measure is the decoder output bit error

$$d(\mathbf{w}_k,\hat{\mathbf{w}}_k) = \begin{cases} 0 & \text{if } \mathbf{x}_k = \hat{\mathbf{x}}_k \\ 1 & \text{if } \mathbf{x}_k \neq \hat{\mathbf{x}}_k \end{cases}$$

This may be rewritten as

$$d(\mathbf{w}_k,\hat{\mathbf{w}}_k) = \delta(\mathbf{x}_k - \hat{\mathbf{x}}_k) \tag{4.5.3}$$

where $\delta(x) = 0.5\,[1-(-1)^{|x|}]$.

Generating function (A.4.17), in our case, depends only on channel errors. If we assume that the encoder input symbols are independent and have the same probability, then we need not average in Eq. (A.4.16) and can assume that the all-zero sequence $\{\mathbf{y}_k = 0\}$ was transmitted.

The system's symmetry allows us to decrease the number of states as compared to the general case presented in Appendix 4.3: Instead of using the product states and considering all possible correct paths, we can assume that there is only one correct path—the all-zero path $\{\mathbf{w}_k = 0\}$. The symbol-error probability union bound (A.4.15) in this case takes the form

$$p_{sb} \le \frac{1}{2\pi j}\frac{d}{du}\{\oint_C (\frac{1}{v-1} - \frac{1}{2v})\mathbf{pG}(u,v)\mathbf{1}dv\}\,|_{u=1} \tag{4.5.4}$$

where

$$\mathbf{G}(u,v) = \sum_{L=1}^{\infty}\sum_{\hat{\sigma}_L}\prod_{k=0}^{L-1}\mathbf{D}(v;0,\hat{\mathbf{w}}_k)u^{\delta(\hat{\mathbf{x}}_k)} \tag{4.5.5}$$

$C\in \{v{:}R_G \cap |v|>1\}$ and R_G is the region of convergence of the sum in Eq. (4.5.5).

Let us now assume that the symbols $\mathbf{y}_k = (y_{1k},y_{2k},...,y_{bk})$ are transmitted serially bit-by-bit over an HMM channel. In this case, according to the HMM model,

$$\mathbf{P}(\mathbf{e}_k) = \prod_{i=1}^{b}\mathbf{P}(e_{ik}) \tag{4.5.6}$$

where e_{ik} denotes an i-th bit error and $\mathbf{e}_k = (e_{1k},e_{2k},...,e_{bk})$. Also,

$$v^{|\mathbf{e}_k| - |\hat{\mathbf{e}}_k|} = \prod_{i=1}^{b} v^{\mu_{ik}}$$

where $\mu_{ik} = -(-1)^{e_{ik}}\hat{y}_{ik}$, so that Eq. (4.5.2) can be rewritten as

$$\mathbf{D}(v;0,\hat{\mathbf{w}}_k) = \prod_{i=1}^{b}[\mathbf{P}(0)v^{-\hat{y}_{ik}} + \mathbf{P}(1)v^{\hat{y}_{ik}}] \tag{4.5.7}$$

The generating function (4.5.5) can be found from the code state transition graph with branch weights

$$\mathbf{D}(v;0,\hat{\mathbf{w}}_k)u^{\delta(\hat{\mathbf{x}}_k)} = \prod_{i=1}^{b}[\mathbf{P}(0)v^{-\hat{y}_{ik}} + \mathbf{P}(1)v^{\hat{y}_{ik}}]u^{\delta(\hat{\mathbf{x}}_k)} \tag{4.5.8}$$

These weights can be composed from the encoder state diagram by using branch input/output symbols ($\hat{\mathbf{x}}$ / $\hat{\mathbf{y}}$).

Example 4.5.1. Let us illustrate the bit error probability union bound calculation for the code of Example 4.2.6.

The encoder state transition graph was shown in Fig. 4.4. To calculate the generating function (4.5.5), we break the loop at the state (0,0), which corresponds to correct decoding. Next, we replace branch I/O symbols with weights (4.5.8). For instance, the branch symbol 1/01, which corresponds to the transition $(1,0) \to (1,1)$ in Fig. 4.4, is replaced with $\mathbf{D}(v;0,\hat{\mathbf{w}})u^{\delta(\hat{\mathbf{x}})} = \mathbf{PD}u$ where $\mathbf{P} = \mathbf{P}(0)+\mathbf{P}(1)$, $\mathbf{D} = \mathbf{P}(0)v^{-1}+\mathbf{P}(1)v$. The generating function $\mathbf{G}(u,v)$ represents the transition weight from the state $\mathbf{0} = (0,0)$ back to it and can be calculated by absorption of the rest of the states. However, it is more convenient to duplicate the state $\mathbf{0}$ so that the edges, which ended at the state $\mathbf{0}$, go instead to its duplicate $\mathbf{0}'$, as shown in Fig. 4.13. After absorption of all the nodes (see Appendix 2.3) except for $\mathbf{0}$ and

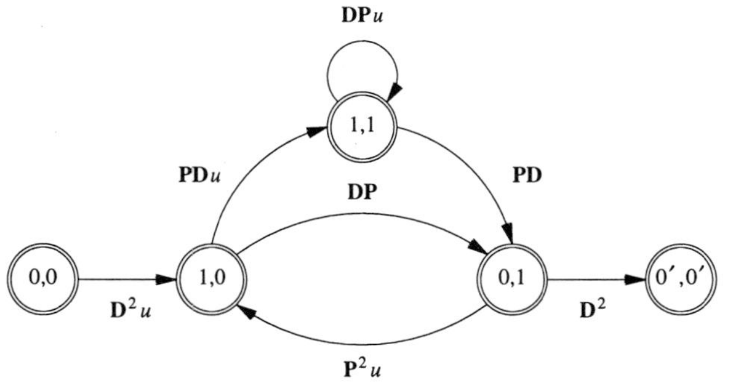

Figure 4.13. Signal flow graph for calculating bit-error probability.

$\mathbf{0}'$, we obtain

$$\mathbf{G}(u,v) = u\mathbf{G}_1(u,v)[\mathbf{I} - u\mathbf{P}^2\mathbf{G}_1(u,v)]^{-1}\mathbf{D}^2$$

where $\mathbf{G}_1(u,v) = \mathbf{DP} + u\mathbf{PD}[\mathbf{I} - u\mathbf{DP}]^{-1}\mathbf{PD}$.

The bit-error probability union bound is obtained from Eq. (4.5.4). The contour integral in Eq. (4.5.4) can be found using the residue theorem. Indeed, $\mathbf{pG}(u,v)\mathbf{1} = a(u,v)/b(u,v)$ is a rational function, and its derivative with respect to u at $u = 1$

$$f(v) = [a'(1,v)b(1,v) - a(1,v)b'(1,v)]/b^2(1,v)$$

is also a rational function. Its poles are the roots of the polynomial $b(1,v)$, which can be written as $b(1,v) = \det[\Delta(v)\mathbf{I} - \mathbf{R}(v)]$ where $\Delta(v) = \det[\mathbf{I} - \mathbf{DP}]$, $\mathbf{R}(v) = \mathbf{DP} + \mathbf{PDB}(v)\mathbf{PD}$, and $\mathbf{B}(v)$ is the adjoint matrix of $\mathbf{I} - \mathbf{DP}$.

In the particular case of BSC with independent errors and the bit-error probability p_1, the calculations are much simpler. In this case, $\mathbf{P} = 1$, $\mathbf{D} = p_1 v + p_0 v^{-1}$, $\mathbf{G}(u,v) = u\mathbf{D}^5/(1 - 2u\mathbf{D})$ and $f(v) = \mathbf{D}^5/(1 - 2\mathbf{D})^2$.

Union bound (4.5.4) can be rewritten as

$$p_{sb} \le \frac{1}{2\pi j}\oint_C \left(\frac{1}{v-1} - \frac{1}{2v}\right)\frac{(p_1 v + p_0 v^{-1})^5}{[1 - 2(p_1 v + p_0 v^{-1})]^2}\,dv$$

where $C \in \{ v : |2(p_1 v + p_0 v^{-1})| < 1 \cap |v| > 1 \}$. Applying the residue theorem, we obtain [28,32]

$$p_b \le 2^{-5}[\,5 - g(5 - 96p_0p_1)(1 - 16p_0p_1)^{-1.5} + 14p_1 + 12p_1^2 - 8p_1^3]$$

where $g = 1 - 2p_1$. As reported in Ref. 28, for $p_1 = 2^{-9}$ the measured error rate equals $\hat{p}_b = 3.3\cdot 10^{-7}$, the preceding upper bound gives $3.9\cdot 10^{-7}$ and the Viterbi upper bound [34] gives $7.9\cdot 10^{-6}$.

4.5.2 Syndrome Decoding Performance

Consider now performance of the syndrome decoding discussed in Sec 4.2.6.3. In the case of definite decoding, error estimate $\hat{\mathbf{e}}_{lj}$ is a function of the syndromes that check this error: $\hat{\mathbf{e}}_{lj} = \mathbf{f}(\mathbf{s}_j, \mathbf{s}_{j+1}, \ldots, \mathbf{s}_{j+m})$. The received symbol $\mathbf{r}_{lj}$ is decoded incorrectly if this estimate is not equal to the actual error $\mathbf{e}_{lj}$. Thus, the probability of incorrect decoding is

$$p_u = Pr(\hat{\mathbf{e}}_{Ij} \neq \mathbf{e}_{Ij}) \tag{4.5.9}$$

Since, according to Eq. (4.2.28), a syndrome is a deterministic function of the finite number of errors, it can be expressed as a Markov function if we assume that the error source is described by an HMM. Therefore, we can calculate the probability of incorrect decoding by using matrix probabilities and generating functions.

Suppose that $j=0$ and consider the generating function

$$\mathbf{F}(\mathbf{t};\mathbf{z}) = \sum_{\mathbf{e}} \prod_{j=-m}^{m} \mathbf{P}(\mathbf{e}_j)\mathbf{z}^{\mathbf{S}}\mathbf{t}^{\mathbf{e}_{I0}}$$

where $\mathbf{e}_j = (\mathbf{e}_{Ij}, \mathbf{e}_{Pj})$, $\mathbf{z} = (\mathbf{z}_1, \mathbf{z}_2, \ldots, \mathbf{z}_m)$, and $\mathbf{z}^{\mathbf{S}} = \prod_{j=1}^{m} \mathbf{z}^{\mathbf{s}_j}$. Since syndromes are linear functions of error values, this generating function can be calculated quite analogously to that of Eq. (4.4.9):

$$\mathbf{F}(\mathbf{t};\mathbf{z}) = \prod_{j=-m}^{m} \sum_{\mathbf{e}_j} \mathbf{P}(\mathbf{e}_j)\mathbf{z}^{\mathbf{e}_j\mathbf{H}_j}\mathbf{t}^{\delta(j)} \tag{4.5.10}$$

where $\boldsymbol{\delta}(0) = \mathbf{e}_{I0}$ and $\boldsymbol{\delta}(j)=0$ for $j\neq 0$. The probability of incorrect decoding is equal to the sum of the coefficients of this function expansion into power series for the powers satisfying the condition $\mathbf{e}_{I0} \neq \mathbf{f}(\mathbf{s}_0, \mathbf{s}_1, \ldots, \mathbf{s}_m)$. The inverse Fourier transform can be used to calculate the probability of the set of errors that satisfy the previous inequality.

Sometimes it is convenient to use conditional generating functions

$$\mathbf{F}_{\mathbf{e}_{I0}}(\mathbf{z}) = \prod_{j=-m}^{m} \sum_{\mathbf{e}_j} \mathbf{P}(\mathbf{e}_j)\mathbf{z}^{\mathbf{e}_j\mathbf{H}_j} \tag{4.5.11}$$

where $\mathbf{e}_{I0}$ is fixed. For each of these generating functions, we calculate the corresponding conditional matrix probabilities of incorrect decoding and then add them to obtain the matrix probability $\mathbf{P}_u$ of incorrect decoding.

Example 4.5.2. To illustrate the probability of incorrect decoding calculation, consider a CSOC with the generator polynomial [23] $\mathbf{G}(D) = [1 \;\; 1+D]$. The Eq. (4.2.28) that check e_{Ij} are

$$\begin{aligned} s_j &= e_{I(j-1)} + e_{Ij} + e_{Pj} \\ s_{j+1} &= e_{Ij} + e_{I(j+1)} + e_{P(j+1)} \end{aligned} \tag{4.5.12}$$

Suppose that we use majority logic for decoding $\hat{e}_{Ij} = s_j s_{j+1}$. Then, the j-th symbol is incorrectly decoded if and only if the estimated error is not equal to the actual error: $e_{Ij} \neq s_j s_{j+1}$. The matrix probability of incorrect decoding is

$$\mathbf{P}_u = \sum_{e_{Ij}\neq s_j s_{j+1}} \mathbf{P}(e_{I(j-1)})\mathbf{P}\mathbf{P}(e_{Ij})\mathbf{P}(e_{Pj})\mathbf{P}(e_{I(j+1)})\mathbf{P}(e_{P(j+1)})$$

This probability can also be found by using generating function (4.5.10), which, in our case, takes the form

$$\mathbf{F}(t;z_1,z_2) = [\mathbf{P}(0)+z_1\mathbf{P}(1)]\mathbf{P}[\mathbf{P}(0)+tz_1z_2\mathbf{P}(1)][\mathbf{P}(0)+z_1\mathbf{P}(1)][\mathbf{P}(0)+z_2\mathbf{P}(1)]^2$$

Conditional generating functions (4.5.11) have the form

$$\mathbf{F}_0(z_1,z_2) = [\mathbf{P}(0)+z_1\mathbf{P}(1)]\mathbf{P}\mathbf{P}(0)[\mathbf{P}(0)+z_1\mathbf{P}(1)][\mathbf{P}(0)+z_2\mathbf{P}(1)]^2$$

$$\mathbf{F}_1(z_1,z_2) = [\mathbf{P}(0)+z_1\mathbf{P}(1)]\mathbf{P}\mathbf{P}(1)z_1z_2[\mathbf{P}(0)+z_1\mathbf{P}(1)][\mathbf{P}(0)+z_2\mathbf{P}(1)]^2$$

An error e_j is estimated incorrectly if $e_{Ij} = 0$ but $s_j s_{j+1} = 1$. In this case, $s_j=1$ and $s_{j+1}=1$ so that the matrix probability of this event can be computed using [as in Eq. (4.4.12)] the inverse Fourier transform

$$\mathbf{P}_u^{(0)} = 2^{-2}[\mathbf{F}_0(1,1)-\mathbf{F}_0(-1,1)-\mathbf{F}_0(1,-1)+\mathbf{F}_0(-1,-1)]$$

The error is also estimated incorrectly if $e_{Ij} = 1$ but $s_j s_{j+1} = 0$. The matrix probability of the complementary event $s_j s_{j+1}$ is given by the previous equation, in which $\mathbf{F}_0(\cdot,\cdot)$ is replaced with $\mathbf{F}_1(\cdot,\cdot)$. Therefore, the matrix probability of $s_j s_{j+1}$ can be expressed as

$$\mathbf{P}_u^{(1)} = \mathbf{P}^2\mathbf{P}(1)\mathbf{P}^2 - 2^{-2}[\mathbf{F}_1(1,1)-\mathbf{F}_1(-1,1)-\mathbf{F}_1(1,-1)+\mathbf{F}_1(-1,-1)]$$

Finally, the matrix probability of incorrect decoding is equal to the sum of the conditional probabilities: $\mathbf{P}_u = \mathbf{P}_u^{(0)} + \mathbf{P}_u^{(1)}$. In particular, for the BSC with independent errors and $\mathbf{P}(0) = p_0$ and $\mathbf{P}(1) = p_1$, this equation takes the form

$$\mathbf{P}_u = p_u = p_1 - 4p_1^3p_0^2 + 4p_1^2p_0^3$$

which coincides with the result of Ref. 23.

Consider now the performance of feedback decoding. The state of syndrome registers of the feedback decoder is given by Eq. (4.2.32)

$$\mathbf{u}_{j+1} = \mathbf{u}_j\mathbf{A} + \mathbf{s}_{j+m}\mathbf{B} - \hat{\mathbf{e}}_j\mathbf{G}_P \tag{4.5.13}$$

and decision function (4.2.52) can be rewritten as

$$\hat{\mathbf{e}}_{Ij} = \mathbf{f}(\mathbf{u}_j,\mathbf{s}_{j+m}) \tag{4.5.14}$$

which depends not only on actual channel errors but also on the previously estimated errors. These equations define an HMM whose output $\hat{\mathbf{e}}_{Ij}$ depends only on its input $\mathbf{s}_{j+m}$ and state $\mathbf{u}_j$. However, this description does not let us calculate the probability of incorrect decoding because the state $\mathbf{u}_j$ and the syndrome $\mathbf{s}_{j+m}$ do not uniquely identify the error $\mathbf{e}_{Ij}$. To find the probability of incorrect decoding according to Eq. (4.5.9), we introduce the decoder extended state $\mathbf{v}_j = (\mathbf{e}_{I(j+m-1)},\mathbf{e}_{I(j+m-2)},\ldots,\mathbf{e}_{I(j-1)},\mathbf{u}_j)$.

The matrix probability of incorrect decoding can be found using several methods. The first method [23] uses the total probability theorem: The probability of incorrect decoding is presented as a sum of the stationary probabilities of the extended states times the conditional probabilities of incorrect decoding at these states.

We can also find the probability of incorrect decoding by introducing states $\mathbf{w}_j = (\mathbf{v}_j,\mathbf{s}_{j+m})$. Decision rule (4.5.14) partitions these states into two subsets: "good" (G) and "bad" (B). A state $\mathbf{w}_j$ belongs to G if and only if $\mathbf{e}_{Ij} = \mathbf{f}(\mathbf{u}_j,\mathbf{s}_{j+m})$; otherwise it belongs to B. The probability of incorrect decoding is equal to the sum of the stationary probabilities of the states that belong to subset B: $p_u = \sum_{\mathbf{w}\in B}\pi_{\mathbf{w}}$

Example 4.5.3. Consider the performance of the code in Example 4.5.2 when feedback decoding is used.

Equations (4.5.13) and (4.5.14) take the form

$$u_{j+1} = s_{j+1} + \hat{e}_j \qquad \hat{e}_j = s_{j+1}u_j \tag{4.5.15}$$

and the syndrome s_{j+1} is defined by Eq. (4.5.12). The decoder extended state $\mathbf{v}_j = (e_{Ij},u_j)$ represents an HMM with transition matrix probabilities

$$\mathbf{P}(\mathbf{v}_{j+1} \mid \mathbf{v}_j) = \mathbf{P}(e_{I(j+1)})\mathbf{P}(u_{j+1} \mid e_{Ij},e_{I(j+1)},u_j)$$

which, after substitutions of Eq. (4.5.12) and (4.5.15), take the following form

$$\mathbf{P}(\mathbf{v}_{j+1} \mid \mathbf{v}_j) = \mathbf{P}(e_{I(j+1)})\mathbf{Pr}\{u_{j+1}=[e_{Ij}+e_{I(j+1)}+e_{P(j+1)}](u_j+1)\}$$

Using this formula, we obtain the HMM transition probability matrix

$$\begin{bmatrix} \mathbf{P}^2(0) & \mathbf{P}(0)\mathbf{P}(1) & \mathbf{P}^2(1) & \mathbf{P}(1)\mathbf{P}(0) \\ \mathbf{P}(0)\mathbf{P} & 0 & \mathbf{P}(1)\mathbf{P} & 0 \\ \mathbf{P}(0)\mathbf{P}(1) & \mathbf{P}^2(0) & \mathbf{P}(1)\mathbf{P}(0) & \mathbf{P}^2(1) \\ \mathbf{P}(0)\mathbf{P} & 0 & \mathbf{P}(1)\mathbf{P} & 0 \end{bmatrix}$$

The HMM state stationary distribution can be found from system (1.1.8), which, in our case, is given by

$$\begin{aligned} \mathbf{v}_{00} &= \mathbf{v}_{00}\mathbf{P}^2(0)+\mathbf{v}_{01}\mathbf{P}(0)\mathbf{P}+\mathbf{v}_{10}\mathbf{P}(0)\mathbf{P}(1)\mathbf{v}_{11}\mathbf{P}(0)\mathbf{P} \\ \mathbf{v}_{01} &= \mathbf{v}_{00}\mathbf{P}(0)\mathbf{P}(1) \quad + \quad \mathbf{v}_{10}\mathbf{P}^2(0) \\ \mathbf{v}_{10} &= \mathbf{v}_{00}\mathbf{P}^2(1)+\mathbf{v}_{01}\mathbf{P}(1)\mathbf{P}+\mathbf{v}_{10}\mathbf{P}(1)\mathbf{P}(0)+\mathbf{v}_{11}\mathbf{P}(1)\mathbf{P} \\ \mathbf{v}_{11} &= \mathbf{v}_{00}\mathbf{P}(1)\mathbf{P}(0) \quad + \quad \mathbf{v}_{10}\mathbf{P}^2(1) \\ &(\mathbf{v}_{00} + \mathbf{v}_{01} + \mathbf{v}_{10} + \mathbf{v}_{11})\mathbf{1} = 1 \end{aligned}$$

The solution of this system can be written as

$$\mathbf{v}_{00} = \boldsymbol{\pi}[\,\mathbf{P}(0)\mathbf{P}-\mathbf{P}(1)\mathbf{P}\mathbf{X}\mathbf{P}^2(0)\,]\mathbf{Y} \quad \mathbf{v}_{01} = \boldsymbol{\pi}\mathbf{P}(0)\mathbf{P}-\mathbf{v}_{00}$$

$$\mathbf{v}_{10} = [\,\boldsymbol{\pi}\mathbf{P}(1)\mathbf{P}-\mathbf{v}_{00}\mathbf{P}(1)\mathbf{P}(0)\,]\mathbf{X} \quad \mathbf{v}_{11} = \boldsymbol{\pi}\mathbf{P}(1)\mathbf{P}-\mathbf{v}_{10}$$

where $\mathbf{X} = [\,\mathbf{I} + \mathbf{P}^2(1)\,]^{-1} \quad \mathbf{Y} = [\,\mathbf{I} + \mathbf{P}(0)\mathbf{P}(1) - \mathbf{P}(1)\mathbf{P}(0)\mathbf{X}\mathbf{P}^2(0)\,]^{-1}$

Here $\boldsymbol{\pi}$ is the stationary distribution of the error source states ($\boldsymbol{\pi}\mathbf{P} = \boldsymbol{\pi}$, $\boldsymbol{\pi}\mathbf{1} = 1$).

The probability of incorrect decoding can be found using the total probability theorem

$$p_u = \sum_{i=0}^{1}\sum_{j=0}^{1} \mathbf{v}_{st}\mathbf{P}_u^{(et)}\mathbf{1}$$

where $\mathbf{P}_u^{(et)}$ is the matrix probability of incorrect decoding, given that the system is in the state (e,t) or, in other words, the probability that the estimated error $\hat{e}_{Ij} = s_{j+1}t$ is not equal to the actual error e:

$$\mathbf{P}_u^{(et)} = \mathbf{Pr}\{t(e_{I(j+1)} + e_{P(j+1)}) = e(t+1)+1\}$$

This equation gives $\mathbf{P}_u^{(00)} = 0$, $\mathbf{P}_u^{(10)} = \mathbf{P}^2$, $\mathbf{P}_u^{(01)} = \mathbf{P}_u^{(11)} = \mathbf{P}(0)\mathbf{P}(1) + \mathbf{P}(1)\mathbf{P}(0)$

In the particular case of the BSC with independent errors and $\mathbf{P}(0) = p_0$ and $\mathbf{P}(1) = p_1$, we find

$$\mathbf{v}_{00} = (1-2p_1+3p_1^2-2p_1^3)R \quad \mathbf{v}_{01} = (p_1-3p_1^3+2p_1^4)R$$

$$\mathbf{v}_{11} = (p_1-3p_1^2+5p_1^3-2p_1^4)R \quad \mathbf{v}_{10} = (3p_1^2-2p_1^3)R$$

where $R = 1/(1+3p_1^2-2p_1^3)$. Thus, $p_u = (7p_1^2-12p_1^3+10p_1^4-4p_1^5)R$ which coincides with the result reported in Ref. 23.

4.6 COMPUTER SIMULATION

Computer simulation of a code operation is a powerful tool for analyzing code performance. As we mentioned in Sec 2.5, a direct bit-by-bit error simulation and decoder operation would be inefficient; the simulation of intervals between errors dramatically increases simulation speed.

Suppose that we want to evaluate a block code performance characteristic. If this code corrects up to t errors and detects up to $d-1$ errors, we do not have to simulate decoding of each block. The blocks that contain less than t errors can be identified immediately as the blocks that decoded correctly. Our HMM error source simulator produces intervals between consecutive errors, as shown in Fig. 4.14. If the code corrects a single error and detects up to two errors, we need to apply the decoding algorithm only to i-th and $i+6$-th blocks. Suppose that the i-th block was incorrectly decoded and that the decoder detected errors in the $i+6$-th block. Then, for the situation depicted in Fig. 4.14, the counter of correct blocks on the decoder output is incremented by 4, the counter of blocks with detected (but not corrected)

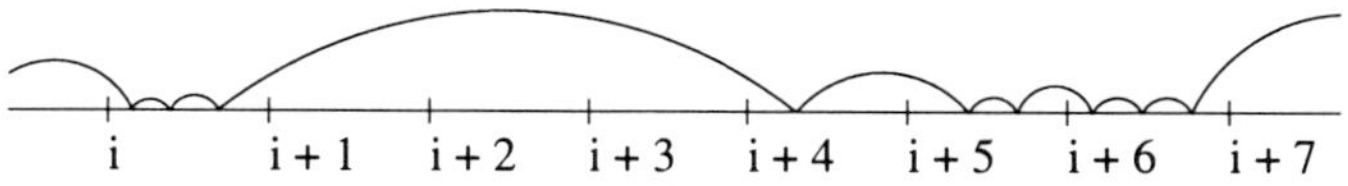

Figure 4.14. Example of an error sequence.

errors is incremented by 2, and the counter of incorrectly decoded blocks is incremented by one.

During the simulation, two possibilities exist: either to memorize the error configuration while errors are occurring in the same block and start calculating error syndromes when the number of errors in the block exceeds $d-1$, or to start calculating syndromes with the first error in the block. It is the usual performance and memory trade-off: In the majority of cases the number of errors in the block is less than $d-1$, so there is no need for calculating syndromes.

Similar methods apply to simulation of a convolutional code with the decoder shift registers playing the role of a block. If the number of errors inside the shift registers does not exceed the number of errors that the code can correct, and if definite decoding is used, we can avoid calculating error syndromes. In the case of feedback decoding our approach is different: As long as the number of errors does not exceed the code error-correcting capability, the simulation proceeds the same way as in definite decoding, but when the number of errors exceeds this limit the syndrome registers' contents must be evaluated and the decoding algorithm applied until all the errors are flushed out.

Consider, for example, simulating (8,7,47) CSOC [41] with the generator polynomial matrix

$$G_P(D) = \begin{bmatrix} 1+D^3+D^{19}+D^{42} \\ 1+D^{21}+D^{34}+D^{43} \\ 1+D^{29}+D^{33}+D^{47} \\ 1+D^{25}+D^{36}+D^{37} \\ 1+D^{15}+D^{20}+D^{46} \\ 1+D^2+D^8+D^{32} \\ 1+D^7+D^{17}+D^{45} \end{bmatrix}$$

This code is popular in satellite communications. [41] The code is majority-logic decodable and is able to correct up to two errors.

To analyze the effect of feedback decoding on error-burst distribution on the decoder output, we simulated the decoder operation in a BSC without memory. The error sequence on the decoder output may be described as an HMM [23] whose states are defined by the contents of the shift registers. Since the number of states is large, direct HMM simulation is not feasible. However, if we simulate the decoder operation only when shift registers contain errors, the simulation may be performed quite successfully. The simulation outcome almost precisely matches (to well within the statistical margin of error) the hardware measurements.

4.7 ZERO-REDUNDANCY CODES

4.7.1 Interleaving

As noted previously, the majority of codes that are actually implemented are not quite effective in an actual channel due to the grouping of errors. Special coding procedures are designed to fight against bursty errors, *interleaving* being one them. This method consists of reordering the symbols before transmitting them over a channel so that the symbols of the same code word are not hit by the same burst. The receiver performs the inverse operation, called *deinterleaving*. If the interleaving depth is large, we may treat errors on the deinterleaver output as independent.

The simplest interleaver separates symbols of the same code word by a fixed number of positions. It can be implemented as a memory array: The encoded symbols are read into the array by columns and transmitted by rows. If the array has j columns, then the adjacent symbols of the same code word (column) are separated by $j-1$ symbols on the interleaver output. When the interleaving depth j grows, the HMM model approaches the memoryless error source model. Indeed, in this case, the probability of having some combination $\mathbf{e}_1, \mathbf{e}_2, \ldots, \mathbf{e}_n$ of errors in the block with interleaved positions is equal to

$$p_j(\mathbf{e}_1, \mathbf{e}_2, \ldots, \mathbf{e}_n) = \boldsymbol{\pi} \prod_{i=1}^{n} \mathbf{P}(\mathbf{e}_i)\mathbf{P}^j \mathbf{1}. \tag{4.7.1}$$

Due to Markov chain regularity (see Appendix 6) $\lim_{j\to\infty} \mathbf{P}^j = \mathbf{1}\boldsymbol{\pi}$.

Therefore, $\mathbf{P}^j$ is approximately equal to $\mathbf{1}\boldsymbol{\pi}$ for large values of j. Asymptotically, relationship (4.7.1) will have the form

$$p_\infty(\mathbf{e}_1, \mathbf{e}_2, \ldots, \mathbf{e}_n) = \prod_{i=1}^{n} \boldsymbol{\pi}\mathbf{P}(\mathbf{e}_i)\mathbf{1} \tag{4.7.2}$$

Since $p(\mathbf{e}_i) = \boldsymbol{\pi}\mathbf{P}(\mathbf{e}_i)\mathbf{1}$ is a number, Eq. (4.7.2) can be rewritten as

$$p_\infty(\mathbf{e}_1, \mathbf{e}_2, \ldots, \mathbf{e}_n) = \prod_{i=1}^{n} p(\mathbf{e}_j)$$

This proves that asymptotically, when the interleaving depth tends to infinity, the interleaved errors are statistically independent. From this immediately follows that for large j the effectiveness of codes with interleaving is close to that of codes in the channel without memory. However, in order to achieve better performance, the interleaving depth might be large, thus requiring it to have a large capacity of transmitter and receiver memory as well as a large delay in delivering information to the customer. Using some sophisticated interleaving techniques, [3] it is possible to improve interleaving efficiency.

It follows from the previously derived equations that we can use all the methods of code performance evaluation by replacing $\mathbf{P}(\mathbf{e})$ with $\mathbf{P}_j(\mathbf{e}) = \mathbf{P}(\mathbf{e})\mathbf{P}^j$ when dealing with interleaving. To evaluate an interleaver performance, one may use the interleaving effectiveness coefficient $\theta_j = p_u^{(j)}/p_u$. This coefficient is equal to the ratio of the probability of incorrect decoding of an interleaved code word to that for the code without interleaving.

The interleaving may be treated as a particular case of coding with zero redundancy. For instance, the product code of the (n_1, n_1) code and (n_2, k_2) code represents the (n_2, k_2) code with depth n_1 canonical interleaving. There are some other zero-redundancy codes that have numerous applications, which are considered in the next section.

4.7.2 Encryptors and Scramblers

Usually, scramblers are implemented as rate 1 convolutional codes. They are used to randomize the transmitted bit sequence, thereby, improving receiver clock recovery. They are also used to eliminate periodic components from a transmitted signal so that it spans more evenly the frequency band allocated to it. Sometimes they are used as inexpensive encryptors to prevent unauthorized reception of the transmitted information.

Encryptors, on the other hand, are viewed as more complex and general transformers than scramblers. They usually represent nonlinear data transformation, which is very difficult to invert unless some additional information (called *key information*) is provided. Block encryption is similar to a block code: The information is transmitted in blocks, which are decrypted independently. Stream encryption is similar to convolutional coding.

Encryptors and scramblers degrade the quality of data communication, and their impact can be evaluated similarly to that of general codes. Usually, they cause error multiplication, and therefore special analysis should be done before they are added to a network. For instance, if a descrambler with three taps precedes a single error correction decoder in BSC with independent errors, then the majority of errors that enter the decoder will not be corrected. For the same reason, an encryptor should be positioned before an encoder, so that the decoder does not have to correct bursts of errors caused by the decryptor.

4.8 CONCLUSION

Using methods developed in Chapter 2, we found basic performance characteristics of error correction and error detection systems. Transform methods proved to be most efficient in calculating matrix probabilities of incorrect decoding, undetected errors, and postdecoding errors. We used characters of GF(q) to calculate these probabilities. In the case of a channel with independent errors, some of the results are well known. The structure of the generating functions developed in this chapter allows us to use fast algorithms for calculating matrix probabilities.

The major part of Sec 4.2 represents an introduction to error control coding and is included to simplify understanding of the coding performance evaluation. In this section, we also propose a modified Viterbi algorithm for channels with memory.

Appendix 4 contains results related to this chapter. In Appendix 4.2, we derive a closed-form expression for the code-weight generating function of the maximum distance separable code. The closed-form expression for the error-event probability of the Viterbi algorithm applied to a general discrete-time communication system with an HMM channel is given in Appendix 4.3.

References

1. L. R. Bahl, J. Cocke, F. Jelinek, and J. Raviv, "Optimal decoding of linear codes for minimizing symbol error rate," *IEEE Trans. Inform. Theory,* **IT-20**, 284-287 (1974).

2. E. R Berlekamp, *Algebraic Coding Theory*, (McGraw-Hill, New York, 1968).

3. E. R. Berlekamp, *Interleaved Coding for Bursty Channels*, Final Project Report, NSF Program SBIR-1982 Phase 1.

4. C. Berrou and A. Glavieux, "Near optimum error correcting coding and decoding: turbo-codes," *IEEE Trans. Commun.*, **COM-44**(10), 1261-1271 (1996).

5. M. R. Best, M. V. Burnashev, Y. Levy, A. Rabinovich, P. C. Fishburn, A. R. Calderbank, and D. J. Costello, Jr., "On a technique to calculate the exact performance of a convolutional code," *IEEE Trans. Inf. Theory*, **41**(2), . 441-447 (1995).

6. A. R. Calderbank and P. C. Fishburn, "The normalized second moment of the binary lattice determined by a convolutional code," *IEEE Trans. Inf. Theory*, **40**(1), 166-174 (1994).

7. E. L. Bloch and V. V. Zyablov, *Generalized Concatenated Codes* (in Russian, Sviaz Publishers, Moscow, 1976).

8. S. C. Chang and J. K. Wolf, "A simple derivation of the MacWilliams' identity for linear codes," *IEEE Trans. Inf. Theory*, **IT-26**(4), 476-477 (1980).

9. W. Feller, *An Introduction to Probability Theory and Its Applications*, **2**, (John Wiley & Sons, New York, 1971).

10. G. D. Forney, Jr.,"The Viterbi algorithm,"*Proc. IEEE*, **61**(3), 268-278 (1973).

11. G. D. Forney, *Concatenated Codes,* Research Monograph 37, (MIT Press, Cambridge, Massachusetts, 1966).

12. B. A. Fuchs and B. V. Shabat, *Functions of a Complex Variable*, **1** (Addison-Wesley Publishing Co., Reading, Massachusetts, 1964).

13. R. G. Gallager, *Information Theory and Reliable Communication,* (John Wiley & Sons, New York, 1968).

14. A. Gill, *Linear Sequential Circuits,* (McGraw-Hill, New York, 1967).

15. J. Hagenauer, E. Offer, and L. Parke, "Iterative decoding of binary block and convolutional codes," *IEEE Trans. Inf. Theory*, **IT-42**(2), pp. (1996).

16. T. Johansson and K. Zigangirov, "A simple one sweep algorithm for optimal APP symbol decoding of linear block codes," In *Proc. IEEE Intern. Symp. on Inf. Theory,* Cambridge, Massachusetts, August 1998, p. 231.

17. J. Justesen, "On the complexity of decoding Reed-Solomon codes," *IEEE Trans. Inform. Theory,* **IT-22**, 237-238 (1976).

18. P. Ligdas, W. Turin, and N. Seshadri, "Statistical methods for speech transmission using hidden Markov models," In *Proc. 31th Conf. on Information Sciences and Systems,* Princeton, New Jersey, 546-551, 1997.

19. F. J. MacWilliams and N. J. A. Sloane, *The Theory of Error-Correcting Codes,* (North-Holland, Amsterdam, 1977).

20. F. J. MacWilliams, "A theorem on the distribution of weights in a systematic code," *Bell System Tech. J.,* **42**, 79-84 (1963).

21. J. L. Massey, *Threshold Decoding,* (MIT Press, Cambridge, Massachusetts, 1963).

22. N. Merhav and Y. Ephraim, "Hidden Markov modeling using a dominant state sequence with applications to speech recognition," *Computer, Speech and Language*, **5**, 327-339 (1991).

23. T. N. Morrissey, Jr., "Analysis of decoders for convolutional codes by stochastic sequential machine methods," *IEEE Trans. Inform. Theory,* **IT-16**, 460-469 (1970).

24. F. Oberhettinger and L. Badii, *Tables of Laplace Transform*, (Springer-Verlag, New York, 1973, p. 16 , eq. 2.34).

25. M. Sajadieh, F.R. Kschischang, and A. Leon-Garcia, "A block memory model for correlated Rayleigh fading channels," In *Proc. IEEE Int. Conf. Commun.*, June 1996, pp. 282-286.

26. D. V. Sarwate, "On the complexity of decoding Goppa codes," *IEEE Trans. Inform. Theory,* **IT-22**, 515-516 (1976).

27. W. W. Peterson and E. J. Weldon, Jr., *Error-Correcting Codes*, (The MIT Press, Cambridge, Massachusetts, 1961).

28. K. A. Post, "Explicit evaluation of Viterbi's union bounds on convolutional code performance for the binary symmetric channel,"*IEEE Trans. Inform. Theory*, **IT-23**, 403-404 (1977).

29. W. K . Pratt, *Digital Image Processing*, (John Wiley & Sons, New York, 1978).

30. Y. Sugiyama, M. Kasahara, S. Hirasawa, and T. Namekawa, "A method of solving key equation for decoding Goppa codes," *Information and Control,* **27**, 173-180 (1975).

31. A. S. Tanenbaum, *Computer Networks*, (Prentice-Hail , Inc., Englewood Cliffs, New Jersey, 1981).

32. W. Turin, "Union bounds on Viterbi algorithm performance,"*AT&T Tech. Journ.*, **64**(10), 2375-2385 (1985).

33. W. Turin and R. van Nobelen, "Hidden Markov modeling of flat fading channels," *IEEE J. Select. Areas Commun.,* **16**(9), 1809-1817 (1998).

34. A. J. Viterbi, "Error bounds for convolutional codes and asymptotically optimum decoding algorithm," *IEEE Trans. Inform. Theory,* **IT-12**, 260-269 (1967).

35. H.S. Wang and P.-C. Chang, "On verifying the first-order Markovian assumption for a Rayleigh fading channel model," *IEEE Trans. Veh. Technol.,* **45**, 353-357 (1996).

36. H.S. Wang and N. Moayeri, "Finite-state Markov channel - a useful model for radio communication channels," *IEEE Trans. Veh. Technol.,* **44**, 163-171 (1995).

37. J. K. Wolf, "Decoding of Bose-Chaudhuri-Hocquenghem codes and Prony's method of curve fitting,"*IEEE Trans. Inform. Theory,* **IT-13**, 608 (1967).

38. J. K. Wolf, "Efficient maximum likelihood decoding od linear block codes using trellis," *IEEE Trans. Inform. Theory,* **IT-24**, 76-80 (1978).

39. W. W. Wu , "New convolutional codes - Part I," *IEEE Trans. Commun.,* **COM-23**, 442-456 (1975).

40. W. W. Wu , "New convolutional codes - Part II," *IEEE Trans. Commun.,* **COM-24**, 19-32 (1976).

41. W. W. Wu , "New convolutional codes - Part III," *IEEE Trans. Commun.,* **COM-24**, 946-955 (1976).

5

PERFORMANCE ANALYSIS OF COMMUNICATION PROTOCOLS

5.1 BASIC CHARACTERISTICS OF TWO-WAY SYSTEMS

Forward error-correction schemes are most effective if the channel error rate and burstiness are low. However, this is not the case in many applications. For example, the wireless channel distortions, such as fading, adjacent channel and co-channel interference, specular reflections, and shadowing, lead to a bursty nature of errors with high error rate. It is difficult to achieve a reasonable reliability of communications using forward error-correction schemes alone because the required code redundancy, decoding delay, and the system complexity are often not acceptable. If a return channel is available, it is possible to achieve high reliability in a simple system by sending the information about the status of the received information back to the transmitter over the return channel.

In this chapter, we give a very brief introduction of the two-way systems (or systems with a feedback) and develop methods of evaluating their performance parameters. There are many excellent textbooks and papers describing these systems (many references to these papers and books can be found in Refs. 10,11,17,18). Most of them analyze the systems performance assuming that channel errors are independent. However, since the two-way systems are specifically designed to combat bursty errors, this assumption is not valid. Our goal is to develop methods for evaluating the two-way systems performance parameters for the channels with bursty errors modeled by HMMs. Therefore, we do not analyze advantages or disadvantages of considered systems; we just show *how* their parameters can be evaluated using methods developed in the previous chapters.

In two-way systems the information between two stations is transmitted in both directions. The set of rules that governs the communication between the stations is usually called the *link control protocol*. These systems can be divided into three categories: systems with repeat-request, systems with comparison, and their combinations.

In systems with automatic repeat-request (ARQ), a receiver instructs the transmitter by sending a negative acknowledgement (NAK) through the return channel to retransmit the same message if it detects noncorrectable errors. Otherwise, the receiver acknowledges receipt of the message by sending a positive acknowledgement (ACK). Thus, in these systems the receiver controls the transmission.

In systems using comparison, a transmitter controls the data flow. In the simplest case, the receiver echoes every message it receives back through the return channel and the transmitter decides whether the message is correct. This method is often used in human-oriented protocols, in which a computer echoes every character it receives and an operator decides whether it is correct. To delete an incorrect character, the operator sends a special erase character.

In combined systems, the decision is made by both receiver and transmitter. For example, if an operator misspelled a command and did not notice the error, the computer might ask the operator to retransmit the command.

We consider here only some aspects of link control protocols, as a detailed description of any of them would require its own book and may be found in the corresponding technical documentation. The protocols can be described using sequential machines. [5] Consider a two-way system whose diagram is depicted in Fig. 5.1. Let $\{\mathbf{u}_k\}$ be the source sequence. The transmitter sends a message $\mathbf{x}_k$, which depends on the transmitter state $\mathbf{s}_k$ and the acknowledgement $\boldsymbol{\rho}_{k-1}$ of the previous message received over the return channel.

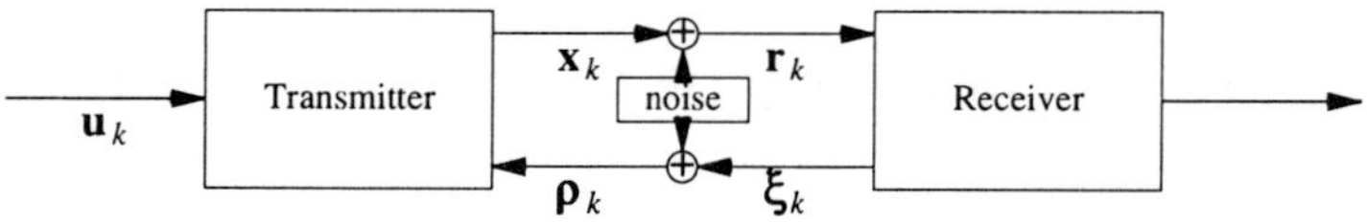

Figure 5.1. Two-way system.

Then the transmitter changes its state to $\mathbf{s}_{k+1}$:

$$\mathbf{x}_k = \mathbf{f}_T(\mathbf{u}_k, \mathbf{s}_k, \boldsymbol{\rho}_{k-1})$$
$$\mathbf{s}_{k+1} = \mathbf{g}_T(\mathbf{u}_k, \mathbf{s}_k, \boldsymbol{\rho}_{k-1})$$

The receiver sends an acknowledgement that depends on the received message $\mathbf{r}_k$ and the receiver state $\boldsymbol{\sigma}_k$:

$$\boldsymbol{\xi}_k = \mathbf{f}_R(\boldsymbol{\sigma}_k, \mathbf{r}_k)$$
$$\boldsymbol{\sigma}_{k+1} = \mathbf{g}_R(\boldsymbol{\sigma}_k, \mathbf{r}_k)$$

In this chapter, we analyze two-way system performance on the basis of an HMM error source model. We assume that the matrix probability of receiving

$$\mathbf{r}_k = \mathbf{x}_k + \mathbf{e}_k^{(d)} \qquad \boldsymbol{\rho}_k = \boldsymbol{\xi}_k + \mathbf{e}_k^{(r)}$$

can be expressed as

$$\mathbf{P}(\mathbf{e}_k^{(d)}, \mathbf{e}_k^{(r)}) = \mathbf{P}(\mathbf{e}_k^{(d)})\mathbf{P}(\mathbf{e}_k^{(r)})$$

This assumption is valid when the same channel is used for both messages and responses (half-duplex systems). It is also valid when the direct and return channels are statistically independent:

$$\mathbf{P}(\mathbf{e}_k^{(d)}, \mathbf{e}_k^{(r)}) = \mathbf{P}_d(\mathbf{e}_k^{(d)}) \otimes \mathbf{P}_r(\mathbf{e}_k^{(r)}) = \mathbf{P}(\mathbf{e}_k^{(d)})\mathbf{P}(\mathbf{e}_k^{(r)})$$

where $\mathbf{P}(\mathbf{e}_k^{(d)}) = \mathbf{P}_d(\mathbf{e}_k^{(d)}) \otimes \mathbf{I}$, $\mathbf{P}(\mathbf{e}_k^{(r)}) = \mathbf{I} \otimes \mathbf{P}_r(\mathbf{e}_k^{(r)})$, and the subscripts d and r

correspond to direct and return-channel models. If errors in the channels are dependent, it is possible to obtain the same result by increasing the number of model states.

We compare different systems and protocols using the following performance characteristics:

- The probability of correct message reception (p_c)
- The probability of incorrect message reception (p_u)
- The probability $p(\gamma)$ that a message was retransmitted γ times
- The probability of receiving a duplicate message (insertion)
- The probability of losing a message
- The maximum possible link utilization (throughput)

We would like to point out that these performance parameters are defined differently in different publications. For example, the probability of correct message reception can be defined among all the transmitted messages or it can be defined among the messages on the receiver output (that is, messages received correctly, with undetected errors, or insertions). The *system throughput* is usually defined as the ratio of the average number of information bits received correctly (per unit time) to the total number of bits that could be transmitted (per unit time). This definition allows us to compare various systems and evaluate the tradeoffs of using some bits for the forward error correction versus using shorter messages with ARQ. There are many other definitions. The most popular definition considers all the received information bits (correctly or incorrectly). Sometimes the throughput is defined as the ratio of messages instead of information bits. Using this definition, it is impossible to compare the effect of using different codes. For example, if no code is used for error detection, the throughput is equal to 100%. However, if we cannot change the message structure and and need only to compare different protocols, the last definition makes sense.

One of the basic goals in designing a communication system is to achieve the information throughput that satisfies customer's needs. Sometimes it is assumed that the information flow rate is the same as that defined by the channel baud rate. However, in noisy channels, the same message might be retransmitted several times, so that the information throughput there would be lower than that in a noiseless channel.

In order to design a system with sufficient throughput, it is important to know the information source statistics. It is often assumed that the source contains an infinite number of messages and that the system is always busy transmitting them. Another popular assumption is that messages or their components arrive randomly with constant probability (binomial or Poisson data flow). Sometimes it is assumed that messages are generated periodically, as in some telemetry systems, in which control messages and status reports are sent regularly.

The other important system parameter that must be taken into account is the maximum tolerable delay in message transmission. This limitation could be "hard" (each message must not stay in the system longer than the specified time period) or "soft" (the limit is imposed on the message delay probability).

The interaction between messages received over the direct channel and the corresponding control messages sent over the return channel is the most important part of a link control protocol. The direct-channel messages are used for information transmission and are similar to those of the one-way systems we discussed before. Let us now consider return-channel messages.

5.2 RETURN-CHANNEL MESSAGES

As noted here, return-channel noise can cause loss of synchronization and may thereby lead to incorrect reception of long sequences of messages. Therefore, return-channel messages should be encoded in such a way that the probability of their mutual transformation is small. Let us consider various methods of coding the acknowledgement.

We consider first the case in which the return channel is used solely for transmitting the acknowledgement messages. Suppose that $\mathbf{f}_0 = (f_{01}, f_{02}, \ldots, f_{0n})$ represents an ACK and $\mathbf{f}_1 = (f_{11}, f_{12}, \ldots, f_{1n})$ represents an NAK. The ACK is decoded if a received message $\mathbf{f} = (f_1, f_2, \ldots, f_n)$ belongs to a vector set V_0, and the NAK is decoded if $\mathbf{f} \in V_1$. If the ACK and NAK are the only control signals that are sent over the return channel, then V_0 is the complement of V_1.

The probability of correctly decoding these messages is equal to

$$r_{uu} = Pr\{\mathbf{f}_u + \mathbf{e} \in V_u\} \qquad u = 0,1 \tag{5.2.1}$$

and the probability of their mutual transformation equals

$$r_{uv} = Pr\{\mathbf{f}_u + \mathbf{e} \in V_v\} \qquad u \neq v \tag{5.2.2}$$

where $\mathbf{e} = (e_1, e_2, \ldots, e_n)$ is the return-channel error.

Equations (5.2.1) and (5.2.2) can be rewritten as

$$r_{uv} = Pr\{\mathbf{e} \in W_{uv}\}$$

where

$$W_{uv} = \{\mathbf{g} : \mathbf{g} = \mathbf{f} - \mathbf{f}_v \qquad \mathbf{f} \in V_u\}$$

is the set of differences between vectors of V_u and f_v. The corresponding matrix probability of decoding $\mathbf{f}_v$ when $\mathbf{f}_u$ was sent is

$$\mathbf{R}_{uv} = \sum_{\mathbf{e} \in W_{uv}} \prod_{i=1}^{n} \mathbf{P}(e_i) \tag{5.2.3}$$

Let us analyze several methods of encoding the acknowledgement. If the minimum Hamming distance decoding is used, then $\mathbf{f}_0$ should be the bitwise complement of $\mathbf{f}_1$. [For instance, $\mathbf{f}_0 = (0,0,\ldots,0)$ and $\mathbf{f}_1 = (1,1,\ldots,1)$.] It is natural to define V_0 as the set of all vectors whose Hamming weight is not greater than h and V_1 as its complement ($V_1 = \overline{V}_0$). According to the results of Sec. 2.3.2,

$$\mathbf{R}_{00} = \mathbf{P}_n(t \leq h) \qquad \mathbf{R}_{01} = \mathbf{P}^n - \mathbf{R}_{00}$$

$$\mathbf{R}_{11} = \mathbf{P}_n(t \leq n-h) \qquad \mathbf{R}_{10} = \mathbf{P}^n - \mathbf{R}_{11}$$

In some full-duplex systems, the acknowledgement messages are encoded together with some other information that is sent over the return channel. The acknowledgement message occupies k_1 symbols out of total k information symbols protected by the (n,k) code. The return-channel decoder corrects up to t_1 and detects up to $d-1$ errors in the block. If the decoder detects noncorrectable errors or the designated k_1 symbols do not represent an ACK, the message is decoded as an NAK; otherwise it is decoded as an ACK.

To calculate the matrix probabilities $\mathbf{R}_{uv}$, let us consider all possible message decoding outcomes:

A The message is decoded correctly

B The decoder detects errors in the message, but does not correct them

C The message is decoded incorrectly

U The acknowledgement portion of the message is correct

V The acknowledgement portion of the message is incorrect

The above-defined error sets W_{uv} may be expressed as

$$W_{00} = A \cup (C \cap U) \qquad W_{01} = B \cup (C \cap V)$$
$$W_{10} = C \cap V \qquad W_{01} = A \cup B \cup (C \cap V)$$

The matrix probabilities $\mathbf{R}_{uv}$ are defined by equation (5.2.3) and can be found using methods developed in Sec. 4.4. Usually, it is possible to neglect the probabilities of events $C \cap U$ and $C \cap V$ as compared to the probabilities of A and B. In this case,

$$\mathbf{R}_{00} \approx \mathbf{P}_n(t \le t_1) \qquad \mathbf{R}_{01} = \mathbf{P}^n - \mathbf{R}_{00}$$
$$\mathbf{R}_{10} = \mathbf{P}_{cv} \qquad \mathbf{R}_{11} = \mathbf{P}^n - \mathbf{R}_{10} \tag{5.2.4}$$

where $\mathbf{P}_{cv}$ is the matrix probability of not detecting errors in the block and of misinterpreting the control subblock for the ACK when the NAK was transmitted. This probability can be calculated using the modified Eq. (4.4.27) in which only the errors that cause the NAK $\to$ ACK transformation are considered. Suppose that the first k_1 bits of a code word are used to encode the acknowledgement information

$$\text{ACK} \sim \mathbf{f}_0 = (0,0,...,0) \quad \text{and} \quad \text{NAK} \sim \mathbf{f}_1 = (1,1,...,1)$$

Then, we obtain from modified Eq. (4.4.27)

$$\mathbf{R}_{00} = 2^{-r} \sum_{|\mathbf{e}| \le t} \sum_{\mathbf{v}} (-1)^{-(\mathbf{eH}')\mathbf{v}} \prod_{j=1}^{n} [\mathbf{P}(0) + \delta_j \mathbf{P}(1)(-1)^{\mathbf{vh}_j}] \tag{5.2.5}$$

$$\mathbf{R}_{10} = 2^{-r} \sum_{|\mathbf{e}| \le t} \sum_{\mathbf{v}} (-1)^{-(\mathbf{eH}')\mathbf{v}} \prod_{j=1}^{n} [\delta_j \mathbf{P}(0) + \mathbf{P}(1)(-1)^{\mathbf{vh}_j}] \tag{5.2.6}$$

where $\delta_j = 0$ for $j \le k_1$ and $\delta_j = 1$ otherwise.

Example 5.2.1. Suppose that the (7,4) Hamming code is used for error detection in the return channel and the first bit of each code word is equal to 0 if it is an ACK; otherwise this bit is set to 1. Let us calculate the return-channel characteristics $\mathbf{R}_{uv}$.

Equation (5.2.5) gives (see Example 4.4.2)

$$\mathbf{R}_{00} = 2^{-3}\mathbf{P}(0) \sum_{i,j,k=0}^{1} \mathbf{F}_1((-1)^i,(-1)^j,(-1)^k)$$

where

$$\mathbf{F}_1(z_1,z_2,z_3) = [\mathbf{P}(0)+\mathbf{P}(1)z_1z_2][\mathbf{P}(0)+\mathbf{P}(1)z_1z_2z_3]$$
$$\times[\mathbf{P}(0)+\mathbf{P}(1)z_2z_3][\mathbf{P}(0)+\mathbf{P}(1)z_1z_3][\mathbf{P}(0)+\mathbf{P}(1)z_2][\mathbf{P}(0)+\mathbf{P}(1)z_3]$$

and Eq. (5.2.6) becomes

$$\mathbf{R}_{10} = 2^{-3}\mathbf{P}(1) \sum_{i,j,k=0}^{1} (-1)^i \mathbf{F}_1((-1)^i,(-1)^j,(-1)^k)$$

From Eq. (5.2.4) we obtain the rest of the matrix probabilities: $\mathbf{R}_{01} = \mathbf{P}^7 - \mathbf{R}_{00}$, $\mathbf{R}_{11} = \mathbf{P}^7 - \mathbf{R}_{10}$.

In the case of a BSC with $\mathbf{P}(0) = q$ and $\mathbf{P}(1) = p$ these equations are simplified:

$$r_{00} = 2^{-3}q[1+4(q-p)^3+3(q-p)^4] \qquad r_{01} = 1 - r_{00}$$
$$r_{10} = 2^{-3}p[1-4(q-p)^3+3(q-p)^4] \qquad r_{11} = 1 - r_{10}$$

Consider now the case when an ACK is encoded by any code vector, while an NAK is represented by a certain noncode vector. The ACK is decoded correctly if the decoder does not detect errors, that is, when a syndrome $\mathbf{S}$ belongs to the set C of correctable errors. As in Eq. (4.4.27), we obtain the matrix probability of decoding the ACK correctly:

$$\mathbf{R}_{00} = 2^{-r} \sum_{|\mathbf{e}|\le t} \sum_{\mathbf{v}} (-1)^{-(\mathbf{eH}')\mathbf{v}} \prod_{k=1}^{n} [\mathbf{P}(0) + \mathbf{P}(1)(-1)^{\mathbf{vh}_k}]$$

The NAK can be encoded either by selecting a predefined noncode vector or by deliberately adding errors into transmitted blocks.

If a fixed noncode vector $\mathbf{f}$ represents the NAK, then the probability of its incorrect decoding is equal to the probability of receiving a word $\mathbf{f}+\mathbf{e}$ whose syndrome belongs to the set of correctable errors

$$\mathbf{R}_{10} = \mathbf{Pr}\{\mathbf{S}(\mathbf{f} + \mathbf{e}) \in C\}$$

Since a syndrome is a linear function, the previous equation may be written as

$$\mathbf{R}_{10} = \mathbf{Pr}\{\mathbf{S}(\mathbf{e}) \in C_\Delta\}$$

where $C_\Delta = \{\mathbf{h} : \mathbf{h} = \mathbf{S}-\mathbf{S}(\mathbf{f}),\ \mathbf{S}\in C\}$. Using Eq. (4.4.25), we can express the probability $\mathbf{R}_{10}$ through generating functions (see Sec. 4.4.2):

$$\mathbf{R}_{10} = 2^{-r} \sum_{\mathbf{S}\in C_\Delta} \sum_{\mathbf{v}} z^{-\mathbf{S}\mathbf{v}} \mathbf{F}(z^{\mathbf{v}})$$

Consider now the case when the NAK is formed by adding to a code vector $\mathbf{c}_i$ of a fixed error configuration $\boldsymbol{\psi}$, that is, the NAK is encoded as $\mathbf{c}_i+\boldsymbol{\psi}$. If all code words are equally probable, we can assume that the all-zero code word ($\mathbf{c}_i = 0$) was transmitted, so the probability $\mathbf{R}_{10}$ can be computed using the previous equation.

5.3 SYNCHRONIZATION

Synchronization is one of the most important tasks in digital communications. The receiver is synchronized with the transmitter if it recognizes the beginning and the end of each block of data. Depending on the size of the blocks, several synchronization levels exist (bit, byte, message, etc). For each level, there are procedures for restoring and maintaining the synchronization that are usually based on block framing: Some redundant symbols are sent before (*preamble*) and after (*postamble*) the block. The framed block of data is called a *frame*.

Synchronization recovery can be done most effectively together with message decoding.

Special codes enable fast synchronization recovery. [16] However, in the majority of systems, only framing symbols are used, and synchronization recovery is performed independently of information decoding. In systems with fixed block size, framing information is transmitted periodically, so that a synchronization algorithm can correlate the framing symbols of several blocks. If messages have variable lengths, then each block is usually synchronized independently of the others, using its own framing symbols.

The simplest technique of synchronization recovery consists of using a predefined sequence of framing symbols. If this sequence does not match the received candidate sequence, the next symbol is tried as a possible framing position, and so on until a satisfactory agreement is reached. In *asynchronous* communications each character (usually eight bits) is framed by so-called start and stop bits, and synchronization is achieved at the beginning of each character. In *synchronous* communications the synchronization is maintained continuously.

In block coding it is important to synchronize at the beginning of each block. Suppose that every code word is framed with a preamble $\mathbf{h} = (h_1, h_2, ..., h_u)$ and a postamble $\mathbf{t} = (t_1, t_2, ..., t_v)$. Except for the first message, we can treat those two sequences as one [$\mathbf{s} = (\mathbf{t}, \mathbf{h})$] and call it a synchronization block (SYNC).

In calculating a one-sided system's performance characteristics, we assumed that code words were perfectly synchronized. Without this assumption, the calculations would have been more complex. We must add the condition of synchronization to every calculation. For instance, if each code word is synchronized independently of the others, then the matrix probability of correct decoding can be expressed as $\mathbf{P}_{\text{syn}}\mathbf{P}_c$, where $\mathbf{P}_{\text{syn}}$ is the matrix probability that the block is synchronized. For example, if the receiver is decoding SYNC that contains less than b errors, then the synchronization matrix probability is given by the matrix binomial probability (defined in Sec 2.3.3): $\mathbf{P}_{\text{syn}} = \mathbf{P}_{u+v}(m<b)$. In the particular case when the receiver is looking for a perfect match $\mathbf{P}_{\text{syn}} = \mathbf{P}^{u+v}(0)$, the matrix probabilities of other events that characterize block code performance can be found in a similar manner.

If the receiver is not able to identify a message start, the whole message and several following messages might be lost or incorrectly decoded. However, the probability of incorrectly decoding a misframed message is usually much smaller than the probability of losing the message, and it therefore can be neglected. The assumption that the synchronization may be lost introduces an additional event—loss of the message—to the set of possible outcomes of the message transmission. The matrix probability of a frame loss is given by $\mathbf{P}_\Lambda = \mathbf{P}^{u+v} - \mathbf{P}_{\text{syn}}$

Example 5.3.1. Binary synchronous communication control (BISYNC) protocol [7] uses two eight-bit characters (SYN SYN) for data-block framing. The frame loss probability is given by

$$p_\Lambda = 1 - \boldsymbol{\pi}\mathbf{P}^{16}(0)\mathbf{1}$$

if we assume that an exact match is required.

In synchronous communications, the receiver usually has two modes of operation: an in-frame mode and a framing search mode. In contrast with asynchronous communications, there is no need to resynchronize each frame independently from the previous frames. If the previous frames were synchronized and the framing pattern of the current frame is slightly different from that expected, the receiver may decide that this difference is caused by channel errors and does not initiate a framing search. The out-of-frame detection algorithm must be robust enough to avoid the interruption of transmission due to unnecessary framing searches. On the other hand, the algorithm must be sensitive enough to avoid the reception of long sequences of misframed information.

Example 5.3.2. Let us calculate the probability distribution of the interval between false out-of-frame detection (misframing) by the algorithm used in the T1 channel format. A T1 frame consists of 192 information bits and one framing bit. The T1 framing algorithm declares an out-of-frame condition when two errors in four consecutive framing bits are detected.

This algorithm can be implemented in a four-bit shift register. Each framing bit is shifted into this register. The receiver compares the register contents with the locally generated framing bit pattern. If they disagree in more than one position, the algorithm declares the out-of-frame condition.

Suppose that $\boldsymbol{\alpha}$ is the probability distribution of HMM model states after synchronization recovery and that the shift register contains no errors; we denote this state of the shift register as $\mathbf{s}_0 = [0\ 0\ 0\ 0]$. If this register contains only one error (that is, $\mathbf{s}_1 = [0\ 0\ 0\ 1]$, or $\mathbf{s}_2 = [0\ 0\ 1\ 0]$, or $\mathbf{s}_3 = [0\ 1\ 0\ 0]$, or $\mathbf{s}_4 = [1\ 0\ 0\ 0]$), there is no misframing. The receiver erroneously declares an out-of-frame condition ($\mathbf{s}_5$) when the shift register contains at least two errors. As we pointed out in the previous chapter, shift register behavior can be described using Markov functions. In our case, the transition probability matrix has the form

$$\begin{bmatrix} \mathbf{Q}(0) & \mathbf{Q}(1) & 0 & 0 & 0 & 0 \\ 0 & 0 & \mathbf{Q}(0) & 0 & 0 & \mathbf{Q}(1) \\ 0 & 0 & 0 & \mathbf{Q}(0) & 0 & \mathbf{Q}(1) \\ 0 & 0 & 0 & 0 & \mathbf{Q}(0) & \mathbf{Q}(1) \\ \mathbf{Q}(0) & 0 & 0 & 0 & 0 & \mathbf{Q}(1) \\ 0 & 0 & 0 & 0 & 0 & \mathbf{I} \end{bmatrix}$$

where $\mathbf{Q}(0) = \mathbf{P}(0)\mathbf{P}^{192}$ and $\mathbf{Q}(1) = \mathbf{P}(1)\mathbf{P}^{192}$. According to Sec. 2.4.3, the interval between misframes has the matrix-geometric distribution

$$Pr(l) = \mathbf{p}\mathbf{P}^{l-1}(A_1 \mid A_1)\mathbf{q}$$

where $\mathbf{p} = [\ \boldsymbol{\alpha}\ \ 0\ \ 0\ \ 0\ \ 0\ ]$, $\mathbf{q}=\mathbf{q}_1\mathbf{1}$, $\mathbf{q}_1 = [\ 0\ \ \mathbf{Q}(1)\ \ \mathbf{Q}(1)\ \ \mathbf{Q}(1)\ \ \mathbf{Q}(1)\ ]'$ is the last column of the previous matrix, and

$$\mathbf{P}(A_1 \mid A_1) = \begin{bmatrix} \mathbf{Q}(0) & \mathbf{Q}(1) & 0 & 0 & 0 \\ 0 & 0 & \mathbf{Q}(0) & 0 & 0 \\ 0 & 0 & 0 & \mathbf{Q}(0) & 0 \\ 0 & 0 & 0 & 0 & \mathbf{Q}(0) \\ \mathbf{Q}(0) & 0 & 0 & 0 & 0 \end{bmatrix}$$

Let us now calculate the mean time between misframes. The generating function of the misframe interval distribution (see Sec. 2.4.3) is equal to

$$\theta_1(z) = z\mathbf{p}[\,\mathbf{I} - z\mathbf{P}(A_1 \mid A_1)\,]^{-1}\mathbf{q}$$

The mean time between misframes can be expressed as a derivative of this generating function:

$$\bar{l} = \theta'_1(1) = \mathbf{p}[\,\mathbf{I} - \mathbf{P}(A_1 \mid A_1)\,]^{-1}\mathbf{q} + \mathbf{p}[\,\mathbf{I} - \mathbf{P}(A_1 \mid A_1)\,]^{-2}\mathbf{P}(A_1 \mid A_1)\mathbf{q}$$

which, after applying the formula

$$[\,\mathbf{I} - \mathbf{P}(A_1 \mid A_1)\,]^{-1}\mathbf{q} = \mathbf{1}$$

becomes

$$\bar{l} = \theta'_1(1) = \mathbf{p}[\,\mathbf{I} - \mathbf{P}(A_1 \mid A_1)\,]^{-1}\mathbf{1}$$

After calculating the inverse matrix, we obtain

$$\bar{l} = \boldsymbol{\alpha}[\,\mathbf{I} + \mathbf{XQ}(1)\,]\,[\,\mathbf{I} - \mathbf{Q}(0)\,]^{-1}\mathbf{1}$$

where

$$\mathbf{X} = [\,\mathbf{I} - \mathbf{Q}(0) - \mathbf{Q}(1)\mathbf{Q}^4(0)\,]^{-1}$$

In particular, for a memoryless channel with bit-error probability p, the previous expression takes the form

$$\bar{l} = \frac{1 + p + pq + pq^2 + pq^3}{p^2(1 + q + q^2 + q^3)}$$

In two-sided systems, the problem of synchronization becomes even more important because we also have to maintain correspondence between the messages that are sent over the channels in both directions. The synchronization between the messages that are sent over the direct channel and the responses received from the return channel we call an MR-sync. Any insertion or loss in the absence of supplementary redundancy may lead to incorrect reception of a group of messages as a result of synchronization loss between the messages and their acknowledgements. Therefore, the ARQ system should have a provision to maintain and recover correspondence between the messages and their acknowledgements.

One method for maintaining synchronization consists of sending each message number as a part of the message and of its acknowledgement. If the same message is received more than once, the receiver recognizes it by its number and the message is discarded. If a message is lost, the receiving station sends a request for its retransmission. In a case when no acknowledgement is received, the station retransmits all the nonacknowledged messages after waiting for an acknowledgement for a certain period of time (timeout). The details of this technique are considered in the following sections.

5.4 ARQ PERFORMANCE CHARACTERISTICS

The matrix probabilities $\mathbf{R}_{uv}$ may be used to determine the probabilities of control message decoding outcomes. However, they alone cannot be used as ARQ system performance characteristics. We must describe the system operation (including the direct channel) and find the probabilities of the system states.

For a frame transmitted over a direct channel, we define the following events, which are related to the results of its decoding:

A The frame is decoded correctly

B The decoder detects but does not correct, errors in the frame

C The frame is decoded incorrectly

We also denote $F = A \cup B$ and $Q = B \cup C$. We defined the events related to decoding of the return-channel control messages as W_{uv} in Sec. 5.2, where

$u = 0$ If an ACK was transmitted

$u = 1$ If an NAK was transmitted

$v = 0$ If the ACK was decoded

$v = 1$ If the NAK was decoded

As mentioned before, we assume that the matrix probabilities of the sequence of events related to both channels can be expressed as products of the matrix probabilities of individual events. For instance, the matrix probability that a frame has been decoded correctly (A) and the corresponding ACK was also decoded correctly (W_{00}) can be expressed as

$$\mathbf{P}_{A0} = \mathbf{Pr}\{A \cap W_{00}\} = \mathbf{P}_A \mathbf{R}_{00}$$

Quite analogously, we define the rest of the matrix probabilities that are needed for describing the system operation:

$$\mathbf{P}_{Av} = \mathbf{Pr}\{A \cap W_{0v}\} = \mathbf{P}_A \mathbf{R}_{0v} \qquad \mathbf{P}_{Cv} = \mathbf{Pr}\{C \cap W_{0v}\} = \mathbf{P}_C \mathbf{R}_{0v}$$

$$\mathbf{P}_{Bv} = \mathbf{Pr}\{B \cap W_{1v}\} = \mathbf{P}_B \mathbf{R}_{1v} \qquad v = 0, 1$$

5.4.1 Stop-and-Wait ARQ

Stop-and-wait is one of the simplest and most broadly used ARQ link controls. [1,2,7,13] With this scheme, after a communication link has been established, the transmitter sends a frame, and then stops and waits for an acknowledgement.

Let us consider first the simplest version of the system, which assumes that frames are perfectly synchronized in each channel and that only an MR-sync loss may happen. There is no frame numbering; every direct-channel frame is acknowledged with an ACK or NAK, depending on the result of decoding the frame. We assume that each message occupies a single frame, so that the frame loss affects only a single message and does not lead to discarding adjacent frames. The system operation is shown in Fig. 5.2.

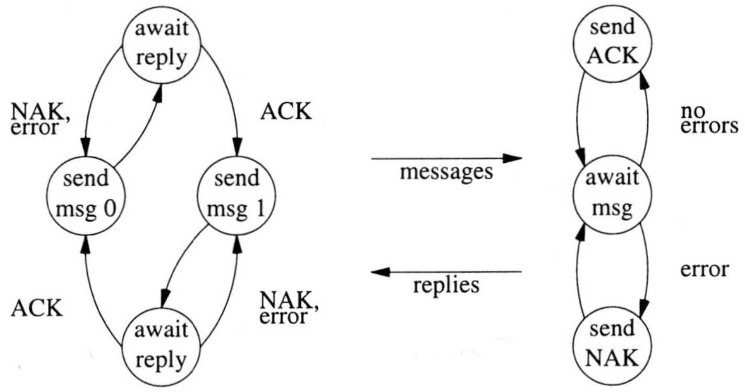

Figure 5.2. Simple ARQ system.

In order to evaluate system performance, let us define all possible outcomes of transmitting a message.

G The message is received correctly the first time. This outcome occurs if the previous message has been positively acknowledged and the current message has been decoded correctly; the matrix probability of the transition $G \to G$ is equal to $\mathbf{P}_{GG} = \mathbf{R}_{00} \mathbf{P}_A$.

U The message is received incorrectly the first time. This outcome occurs if the previous message has been positively acknowledged and the current message has been decoded incorrectly; the matrix probability of the transition $G \rightarrow U$ can be expressed as $\mathbf{P}_{GU} = \mathbf{R}_{00}\mathbf{P}_C$.

I The message is received without detected errors, but not the first time. This outcome occurs if the returning ACK has been improperly interpreted as an NAK and no errors in the direct-channel message have been detected; the matrix probability of the transition $G \rightarrow I$ is given by $\mathbf{P}_{GI} = \mathbf{R}_{01}\mathbf{P}_F$, where $\mathbf{P}_F = \mathbf{P}_A + \mathbf{P}_C$.

S The message is received, but not the first time, and has detectable errors. This outcome occurs if the returning ACK has been improperly interpreted as an NAK and errors in the direct-channel message have been detected; the matrix probability of the transition $G \rightarrow S$ is given by $\mathbf{P}_{GS} = \mathbf{R}_{01}\mathbf{P}_B$.

D The message will be retransmitted next time. This outcome occurs if the previous ACK was decoded correctly, the transmitted message contains detected errors, and its NAK is decoded correctly; the matrix probability of the transition $G \rightarrow D$ is equal to $\mathbf{P}_{GD} = \mathbf{R}_{00}\mathbf{P}_B\mathbf{R}_{11}$.

L The message is lost. This outcome occurs if the previous ACK was decoded correctly, the transmitted message contains detected errors, and its NAK is decoded incorrectly; the matrix probability of the transition $G \rightarrow L$ is equal to $\mathbf{P}_{GL} = \mathbf{R}_{00}\mathbf{P}_B\mathbf{R}_{10}$.

These events make it possible to describe the process on the receiver output as an HMM. For instance, let

$$A_1,\ A_2,\ A_2,\ C_2,\ B_2,\ A_3,\ A_4,\ B_5,\ B_5,\ A_6,$$
$$B_6,\ A_6,\ B_7,\ B_8,\ B_9,\ B_9,\ A_9,\ C_{10},\ A_{10}$$

be a sequence of received messages where capital letters denote the previously defined results of message decoding and indices denote the transmitter output message sequence number. Thus, A_1 means that the first message was decoded correctly, and B_9 means that the decoder detected errors in the ninth source message. Using the previously defined message transmission outcomes, the previous sequence can be transformed into

$$G,\ G,\ I,\ I,\ S,\ G,\ G,\ D,\ L,\ G,\ S,\ I,\ L,\ L,\ D,\ D,\ G,\ U,\ I \qquad (5.4.1)$$

The transformed sequence of events allows us to determine all basic performance characteristics. The probability of receiving a correct message is the probability of G in the sequence of the events, the probability of inserting a message is the probability of I, and so on. The transformed sequence can be described as an HMM because the outcomes of message transition depend deterministically on the sequence of errors in both channels. The transition probability matrix of this HMM has the form

$$\mathbf{T} = [\mathbf{B}_G \quad \mathbf{B}_G \quad \mathbf{B}_G \quad \mathbf{B}_S \quad \mathbf{B}_M \quad \mathbf{B}_M]' \qquad (5.4.2)$$

where

$$\mathbf{B}_G = [\, \mathbf{P}_{GG} \quad \mathbf{P}_{GU} \quad \mathbf{P}_{GI} \quad \mathbf{P}_{GS} \quad \mathbf{P}_{GD} \quad \mathbf{P}_{GL} \,]$$

$$\mathbf{B}_S = [\ \mathbf{R}_{10}\mathbf{P}_A \quad \mathbf{R}_{10}\mathbf{P}_C \quad \mathbf{R}_{11}\mathbf{P}_F \quad \mathbf{R}_{11}\mathbf{P}_B \quad \mathbf{R}_{10}\mathbf{P}_B\mathbf{R}_{11} \quad \mathbf{R}_{10}\mathbf{P}_B\mathbf{R}_{11}\ ]$$

$$\mathbf{B}_M = [\ \mathbf{P}_A \quad \mathbf{P}_C \quad 0 \quad 0 \quad \mathbf{P}_B\mathbf{R}_{11} \quad \mathbf{P}_B\mathbf{R}_{10}\ ]$$

Calculation of the first row $\mathbf{B}_G$ of matrix (5.4.2) was explained during the message outcome definition; the remaining transition probabilities are calculated quite analogously.

The ARQ system performance characteristics are expressed as the stationary probabilities

$$\mathbf{p} = [\ \mathbf{p}(G) \quad \mathbf{p}(U) \quad \mathbf{p}(I) \quad \mathbf{p}(S) \quad \mathbf{p}(D) \quad \mathbf{p}(L)\ ] \tag{5.4.3}$$

of matrix (5.4.2), assuming that this matrix is regular. The stationary probabilities are found from the system $\mathbf{pT} = \mathbf{p}$, $\mathbf{p1} = 1$ which can be simplified because the matrix has several identical rows. If we denote

$$\mathbf{X} = \mathbf{p}(G)+\mathbf{p}(U)+\mathbf{p}(I) \qquad \mathbf{Y} = \mathbf{p}(S) \qquad \mathbf{Z} = \mathbf{p}(D)+\mathbf{p}(L)$$

$$\mathbf{c}_1 = \mathbf{XR}_{00} + \mathbf{YR}_{10} + \mathbf{Z} \qquad \mathbf{c}_2 = \mathbf{XR}_{01} + \mathbf{YR}_{11}$$

then the previous system can be written as

$$\mathbf{p}(G) = \mathbf{c}_1\mathbf{P}_A \quad \mathbf{p}(U) = \mathbf{c}_1\mathbf{P}_C \quad \mathbf{p}(I) = \mathbf{c}_2\mathbf{P}_F$$

$$\mathbf{p}(S) = \mathbf{c}_2\mathbf{P}_B \quad \mathbf{p}(D) = \mathbf{c}_2\mathbf{P}_B\mathbf{R}_{11} \quad \mathbf{p}(L) = \mathbf{c}_2\mathbf{P}_B\mathbf{R}_{10}$$

From this system, it follows that

$$\mathbf{X} = [\ (\mathbf{X} + \mathbf{Y})\mathbf{S} + \mathbf{Z}\ ]\mathbf{P}_F$$

$$\mathbf{Y} = (\mathbf{XR}_{01} + \mathbf{YR}_{11})\mathbf{P}_B$$

$$\mathbf{Z} = (\mathbf{XR}_{00} + \mathbf{YR}_{10} + \mathbf{Z})\mathbf{P}_B\mathbf{S}$$

and $(\mathbf{X} + \mathbf{Y})\mathbf{S} + \mathbf{Z} = \boldsymbol{\pi}$, where $\mathbf{S} = \mathbf{P}_F+\mathbf{P}_B$ and $\boldsymbol{\pi}$ is the stationary distribution of the Markov chain with the matrix $\mathbf{S}$: $\boldsymbol{\pi}\mathbf{S} = \boldsymbol{\pi}$, $\boldsymbol{\pi}\mathbf{1} = 1$. Solving this system, we obtain

$$\mathbf{p}(G) = \mathbf{c}_1\mathbf{P}_A \quad \mathbf{p}(U) = \mathbf{c}_1\mathbf{P}_C \quad \mathbf{p}(I) = \mathbf{c}_2\mathbf{P}_F$$

$$\mathbf{p}(S) = \mathbf{c}_2\mathbf{P}_B \quad \mathbf{p}(D) = \mathbf{c}_2\mathbf{P}_B\mathbf{R}_{11} \quad \mathbf{p}(L) = \mathbf{c}_2\mathbf{P}_B\mathbf{R}_{10}$$

where

$$\mathbf{H}_0 = \mathbf{P}_F\mathbf{R}_{00}+\mathbf{P}_B\mathbf{R}_{10} \tag{5.4.4}$$

$$\mathbf{c}_1 = \boldsymbol{\pi}\mathbf{H}_0(\mathbf{I} - \mathbf{P}_B\mathbf{R}_{11})^{-1}, \quad \mathbf{c}_2 = \boldsymbol{\pi}\mathbf{P}_F\mathbf{R}_{01}(\mathbf{I} - \mathbf{P}_B\mathbf{R}_{11})^{-1}$$

Using stationary distribution (5.4.3), we can find the ARQ system performance characteristics:

$\mathbf{c}_1\mathbf{P}_A\mathbf{1}$ The probability that the message is received correctly (for the first time)

$\mathbf{c}_1\mathbf{P}_C\mathbf{1}$ The probability that the message is received incorrectly (for the first time)

$\mathbf{c}_2\mathbf{P}_F\mathbf{1}$ The probability that the message is an insertion of the previous message

$\mathbf{c}_2\mathbf{P}_B\mathbf{R}_{10}\mathbf{1}$ The probability that the message is lost

To calculate the system throughput, we denote the minimal amount of time between transmissions of two adjacent messages as $n+\Delta$ (measured in the number of symbols that the transmitter could have sent during this period). If there were no retransmissions, insertions, or losses, the throughput would have been $k/(n+\Delta)$. However, only the messages of the type $G \cup U$ should be taken into account, so that the throughput is

$$\tau = [\mathbf{p}(G) + \mathbf{p}(U)]\mathbf{1}k/(n+\Delta)$$

Let us now find interval distributions that characterize sequences of messages on decoder output.

5.4.2 Message Delay Distribution

Let us calculate the probability $Pr(\gamma)$ that the number of message retransmissions until the message is received (correctly or incorrectly) for the first time is equal to γ. This probability can be found as the conditional probability that the previous message has been positively acknowledged and the current message remains in one of the states D for the period γ until it reaches $G \cup U$ for the first time. In other words, $Pr(\gamma)$ is the probability of a series of γ events D in sequence (5.4.1) that ends with G or U.

Using methods described in Sec. 2.4.3, we obtain

$$Pr(\gamma) = \mathbf{c}_1(\mathbf{I} - \mathbf{P}_B\mathbf{R}_{11})(\mathbf{P}_B\mathbf{R}_{11})^{\gamma}\mathbf{P}_F\mathbf{1} / \mathbf{c}_1\mathbf{P}_F\mathbf{1} \tag{5.4.5}$$

The generating function of this distribution has the form

$$\phi(z) = \sum_{\gamma=0}^{\infty} Pr(\gamma)z^{\gamma} = \mathbf{K}(\mathbf{I} - \mathbf{P}_B\mathbf{R}_{11})(\mathbf{I} - \mathbf{P}_B\mathbf{R}_{11}z)^{-1}\mathbf{P}_F\mathbf{1}$$

where $\mathbf{K} = \mathbf{c}_1/\mathbf{c}_1\mathbf{P}_F\mathbf{1}$. This generating function can be used to calculate the distribution moments. The average delay is equal to

$$\bar{\gamma} = \phi'(1) = \mathbf{K}(\mathbf{I} - \mathbf{P}_B\mathbf{R}_{11})^{-1}\mathbf{P}_B\mathbf{R}_{11}\mathbf{P}_F\mathbf{1} \tag{5.4.6}$$

and the variance of the delay is

$$\sigma_{\gamma}^2 = \phi''(1) - \bar{\gamma} - \bar{\gamma}^2$$

where

$$\phi''(1) = 2\mathbf{K}(\mathbf{I} - \mathbf{P}_B\mathbf{R}_{11})^{-2}(\mathbf{P}_B\mathbf{R}_{11})^2\mathbf{P}_F\mathbf{1}$$

The message-delay distribution differs from the period of time that the message occupies the system because it might be received several times (in the case of MR-sync loss). Obviously, the period of time that the system is busy transmitting the message is equal to the length of the series of NAKs received over the return channel. Similarly to Eq. (5.4.5), we obtain

$$Pr(\beta) = \boldsymbol{\pi}\mathbf{H}_0\mathbf{H}_1^{\beta-1}\mathbf{H}_0\mathbf{1} / \boldsymbol{\pi}\mathbf{H}_0\mathbf{1}$$

where

$$\mathbf{H}_1 = \mathbf{P}_F\mathbf{R}_{01} + \mathbf{P}_B\mathbf{R}_{11} \tag{5.4.7}$$

The average busy period is equal to

$$\bar{\beta} = 1/\boldsymbol{\pi}\mathbf{H}_0\mathbf{1}$$

and can be found quite analogously to Eq. (5.4.6).

5.4.3 Insertion and Loss Interval Distribution

It is important to know the distribution of synchronization loss time. According to our assumptions, only MR-sync loss can happen as a result of incorrectly decoding the acknowledgements. Let us first find the distribution of the intervals of the positive synchronization loss (which is a series of events $I \cup S$). The length i of the series is equal to the time that the Markov chain with matrix (5.4.2) stays in the states $I \cup S$. The insertion length distribution can be found using the methods of Sec. 2.4.3. Denote

$$\mathbf{U}_{00} = \begin{bmatrix} \mathbf{R}_{01}\mathbf{P}_F & \mathbf{R}_{01}\mathbf{P}_B \\ \mathbf{R}_{11}\mathbf{P}_F & \mathbf{R}_{11}\mathbf{P}_B \end{bmatrix} = \begin{bmatrix} \mathbf{R}_{01} \\ \mathbf{R}_{11} \end{bmatrix} \begin{bmatrix} \mathbf{P}_F & \mathbf{P}_B \end{bmatrix} \tag{5.4.8}$$

as the subblock of matrix (5.4.2), which corresponds to transitions between states $I \cup S$. Then, the probability distribution of the insertion series length can be expressed as

$$Pr(i) = \mathbf{q}(\mathbf{I}-\mathbf{U}_{00})\,\mathbf{U}_{00}^{i-1}\,(\mathbf{I}-\mathbf{U}_{00})\mathbf{1} \,/\, \mathbf{q}\,(\mathbf{I}-\mathbf{U}_{00})\mathbf{1}$$

where $\mathbf{q} = [\ \mathbf{p}(I) \ \ \mathbf{p}(S)\ ]$. Using factorization (5.4.8), we can simplify this expression:

$$Pr(i) = \mathbf{c}_2\mathbf{H}_1^{i-1}\mathbf{H}_0^2\mathbf{1} \,/\, \mathbf{c}_2\mathbf{H}_0\mathbf{1}$$

The average length of the series is equal to

$$\bar{i} = \mathbf{c}_2\mathbf{1} \,/\, \mathbf{c}_2\mathbf{H}_0\mathbf{1}$$

The probability distribution and the average of the series of consecutive frame losses can be found quite analogously:

$$Pr(l) = \mathbf{c}_3\mathbf{P}_{B0}^{l-1}(\mathbf{I}-\mathbf{P}_{B0})\mathbf{1} \,/\, \mathbf{c}_3\mathbf{1} \qquad \bar{l} = \mathbf{p}(L) \,/\, \mathbf{c}_3\mathbf{1}$$

where $\mathbf{c}_3 = \mathbf{p}(L)(\mathbf{I}-\mathbf{P}_{B0})$ and $\mathbf{P}_{B0} = \mathbf{P}_B\mathbf{R}_{10}$.

5.4.4 Accepted Messages

The HMM with matrix (5.4.2) describes the sequence of messages on decoder output. Some of the messages that contain detected errors have no effect on the messages that are ultimately accepted. The sequence of accepted messages is obtained by the type of D and S messages removed from the decoder output. For example, the sequence (5.4.1), after such removal, becomes

$$G,\ G,\ I,\ I,\ G,\ G,\ L,\ G,\ I,\ L,\ L,\ G,\ U,\ I.$$

Obviously, this sequence represents an HMM. Its matrix is obtained from matrix (5.4.2) by eliminating the unobserved sets of states D and S (see Appendix 6.1.3). After performing the necessary transformations, we obtain

$$\mathbf{T}_1 = \begin{bmatrix} \mathbf{W}_1\mathbf{P}_A & \mathbf{W}_1\mathbf{P}_C & \mathbf{W}_2\mathbf{P}_F & \mathbf{W}_1\mathbf{P}_B\mathbf{R}_{10} \\ \mathbf{W}_1 P_A & \mathbf{W}_1 P_C & \mathbf{W}_2 P_F & \mathbf{W}_1 P_B\mathbf{R}_{10} \\ \mathbf{W}_1 P_A & \mathbf{W}_1 P_C & \mathbf{W}_2 P_F & \mathbf{W}_1 P_B\mathbf{R}_{10} \\ \mathbf{W}_3 P_A & \mathbf{W}_3 P_C & 0 & \mathbf{W}_3 P_B\mathbf{R}_{10} \end{bmatrix} \tag{5.4.9}$$

where $\mathbf{W}_3 = (\mathbf{I}-\mathbf{P}_B\mathbf{R}_{11})^{-1}$, $\mathbf{W}_2 = \mathbf{R}_{01}\mathbf{W}_3$, and $\mathbf{W}_1 = (\mathbf{R}_{00}+\mathbf{R}_{01}\mathbf{W}_3\mathbf{P}_B\mathbf{R}_{01})\mathbf{W}_3$. The corresponding stationary distribution

$$[\ \mathbf{p}_1(G)\ \ \mathbf{p}_1(U)\ \ \mathbf{p}_1(I)\ \ \mathbf{p}_1(L)\ ] \tag{5.4.10}$$

may be obtained from stationary distribution (5.4.3) by the removal of $\mathbf{p}(D)$ and $\mathbf{p}(S)$, with the subsequent normalization:

$$\mathbf{p}_1(G) = \mu\mathbf{p}(G) \qquad \mathbf{p}_1(U) = \mu\mathbf{p}(U)$$
$$\mathbf{p}_1(I) = \mu\mathbf{p}(I) \qquad \mathbf{p}_1(L) = \mu\mathbf{p}(L)$$

where

$$\mu = [\mathbf{c}_1(\mathbf{P}_F+\mathbf{P}_B\mathbf{R}_{10})\mathbf{1}+\mathbf{c}_2\mathbf{P}_F\mathbf{1}]^{-1}$$

Matrices (5.4.9) and (5.4.10) characterize the sequence of accepted messages. Using these matrices, we can obtain the system performance characteristics:

$\mu\mathbf{c}_1\mathbf{P}_A\mathbf{1}$ The probability that the message is received correctly (for the first time)

$\mu\mathbf{c}_1\mathbf{P}_C\mathbf{1}$ The probability that the message is received incorrectly (for the first time)

$\mu\mathbf{c}_2\mathbf{P}_F\mathbf{1}$ The probability that the message is an insertion of the previous message

$\mu\mathbf{c}_1\mathbf{P}_B\mathbf{R}_{10}\mathbf{1}$ The probability that the message is lost

5.4.5 Alternative Assumptions

In some control systems, a message is considered to be accepted correctly if, in a series of its insertions, it appears at least once without errors. To investigate these systems' performance, we introduce the following events (in addition to those previously defined):

G_0 The first correct reception of the message in the series of insertions

U_0 The message in the series of insertions is decoded incorrectly

D_0 The message in the series of insertions contains detected errors

C_0 The message in the series of insertions appears after it has been received correctly at least once

The transition probability matrix for the events G, U, G_0, U_0, D_0, C_0, D, and L has the form

$$[\mathbf{M}_g\ \ \mathbf{M}_g\ \ \mathbf{M}_b\ \ \mathbf{M}_g\ \ \mathbf{M}_g\ \ \mathbf{M}_b\ \ \mathbf{M}_g\ \ \mathbf{M}_g]'$$

where

$$\mathbf{M}_g = [\mathbf{P}_{A0}\ \ \mathbf{P}_{C0}\ \ \mathbf{P}_{A1}\ \ \mathbf{P}_{C1}\ \ 0\ \ 0\ \ \mathbf{P}_{B1}\ \ \mathbf{P}_{B0}]$$
$$\mathbf{M}_b = [\mathbf{H}_0\ \ 0\ \ 0\ \ 0\ \ 0\ \ \mathbf{H}_1\ \ 0\ \ 0]$$

The stationary distribution of the Markov chain with this matrix satisfies the following equations:

$$\mathbf{p}(G) = \mathbf{a}_1\mathbf{P}_{A0} + \mathbf{a}_2\mathbf{H}_0 \quad \mathbf{p}(U) = \mathbf{a}_1\mathbf{P}_{C0} \quad \mathbf{p}(G_0) = \mathbf{a}_1\mathbf{P}_{A1} \quad \mathbf{p}(U_0) = \mathbf{a}_1\mathbf{P}_{C1}$$
$$\mathbf{p}(D_0) = \mathbf{a}_3\mathbf{P}_{B1} \quad \mathbf{p}(C_0) = \mathbf{a}_2\mathbf{H}_1 \quad \mathbf{p}(D) = (\mathbf{a}_1-\mathbf{a}_3)\mathbf{P}_{B1} \quad \mathbf{p}(L) = \mathbf{a}_1\mathbf{P}_{B0}$$

where

$$\mathbf{a}_1 = \mathbf{p}(G) + \mathbf{p}(U) + \mathbf{p}(U_0) + \mathbf{p}(D_0) + \mathbf{p}(D) + \mathbf{p}(L)$$

$$\mathbf{a}_2 = \mathbf{p}(G_0) + \mathbf{p}(C_0) \quad \mathbf{a}_3 = \mathbf{p}(U_0) + \mathbf{p}(D_0)$$

Solving this system, we obtain

$$\mathbf{a}_1 = \pi\mathbf{H}_0(\mathbf{I}-\mathbf{P}\,|\,)^{-1} \quad \mathbf{a}_2 = \pi\mathbf{P}_{A1}(\mathbf{I}-\mathbf{P}\,|\,)^{-1} \quad \mathbf{a}_3 = \mathbf{a}_1\mathbf{P}_{C1}(\mathbf{I}-\mathbf{P}_{B1})^{-1}$$

where $\mathbf{P}\,|\, = \mathbf{P}_{B1} + \mathbf{P}_{C1}$.

For the sequence of accepted messages, the basic probabilities that characterize the system's performance are:

$\lambda\mathbf{p}(G)\mathbf{1}$ The probability that the message is received correctly

$\lambda\mathbf{p}(U)\mathbf{1}$ The probability that the message is received incorrectly

$\lambda\mathbf{p}(I_0)\mathbf{1}$ The probability that the message is a duplicate of the previous message ($I_0 = G_0 \cup U_0 \cup C_0$)

$\lambda\mathbf{p}(L)\mathbf{1}$ The probability that the message is lost

The coefficient λ is obtained by excluding the states corresponding to D_0 and D with the appropriate normalization:

$$\lambda = [1 - \mathbf{p}(D_0)\mathbf{1} - \mathbf{p}(D)\mathbf{1}]^{-1}$$

5.4.6 Modified Stop-and-Wait ARQ

The simple stop-and-wait ARQ considered in the previous section has several deficiencies that may cause a permanent synchronization loss or create a deadlock. If a message or a reply is lost in one of the channels and the loss goes undetected, the system may be deadlocked, with the receiver waiting for the message and the transmitter waiting for an acknowledgement.

The transmission protocol can be modified to avoid this problem. A timeout can be used to break a deadlock: The previous message is retransmitted if no reply has been received during the timeout period. If the direct-channel message has been accepted, but the corresponding ACK has been lost, the retransmitted message duplicates the previous one. To avoid inserting the message, the frames and acknowledgements are modulo 2 numbered (ACK0 and ACK1). The flow control protocol may be described as follows (see Fig. 5.3):

Transmitter:

- If the response was the ACK with the previous message number, the next message is sent.
- If the response was different from the ACK with the previous message number, or no response was received during the timeout period, the previous message is retransmitted.

Receiver:

- If the decoder does not detect errors and the received message number differs from the previous ACK number, the message is accepted and the ACK with this message number is sent to the transmitter.

- If the decoder detects errors in the received message, it sends the NAK.
- If the decoder does not detect errors and the received message number coincides with the previous ACK number, the message is discarded and the ACK with this message number is sent to the transmitter.

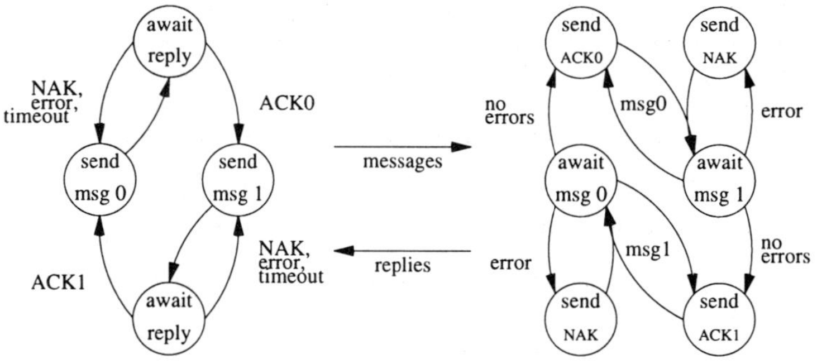

Figure 5.3. Modified ARQ system.

If each channel is perfectly synchronized, the message numbering decreases the probability of the MR-sync loss, because only the ACK with the transmitted message number serves as the positive acknowledgement. Now we have three different acknowledgements that can be sent over the return channel, so that we need to consider the probabilities of their mutual transformations W_{uv} where

$u = a_0$ If an ACK0 was transmitted

$u = a_1$ If an ACK1 was transmitted

$u = 1$ If an NAK was transmitted

$v = a_0$ If the ACK0 was decoded

$v = a_1$ If the ACK1 was decoded

$v = 1$ If the NAK was decoded

The process on the decoder output can be described using an HMM as for the simple stop-and-wait ARQ. However, the message numbering requires consideration of all possible outcomes of the transmission of even-numbered messages G_0, U_0, I_0, S_0, D_0, L_0 and odd-numbered messages G_1, U_1, I_1, S_1, D_1, L_1. The process transition probability matrix can be found analogously to Eq. (5.4.2). Since the matrix contains many zero blocks, it is convenient to describe it in terms of nonzero matrix transitional probabilities:

$$\mathbf{P}_{X_i,G_j} = \mathbf{R}_{a_i a_i}\mathbf{P}_A \quad \mathbf{P}_{X_i,U_j} = \mathbf{R}_{a_i a_i}\mathbf{P}_C \quad \mathbf{P}_{X_i,I_i} = \mathbf{R}_{a_i,1}\mathbf{P}_F$$

$$\mathbf{P}_{X_i,S_i} = \mathbf{R}_{a_i,1}\mathbf{P}_B \quad \mathbf{P}_{X_i,D_i} = \mathbf{R}_{a_i a_i}\mathbf{P}_B\mathbf{R}_{11}^{(i)} \quad \mathbf{P}_{X_i,L_i} = \mathbf{R}_{a_i a_i}\mathbf{P}_B\mathbf{R}_{1,a_i}$$

$$\mathbf{P}_{S_i,G_j} = \mathbf{R}_{1,a_i}\mathbf{P}_A \quad \mathbf{P}_{S_i,U_j} = \mathbf{R}_{1,a_i}\mathbf{P}_C \quad \mathbf{P}_{S_i,I_i} = (\mathbf{R}_{1,a_j}+\mathbf{R}_{11})\mathbf{P}_F$$
$$\mathbf{P}_{S_i,S_i} = (\mathbf{R}_{1,a_j}+\mathbf{R}_{11})\mathbf{P}_B \quad \mathbf{P}_{S_i,D_i} = \mathbf{R}_{1,a_i}\mathbf{P}_B\mathbf{R}_{11}^{(i)} \quad \mathbf{P}_{S_i,L_i} = \mathbf{R}_{1,a_i}\mathbf{P}_B\mathbf{R}_{1,a_i}$$
$$\mathbf{P}_{Y_i,G_i} = \mathbf{P}_A \quad \mathbf{P}_{Y_i,U_i} = \mathbf{P}_C \quad \mathbf{P}_{Y_i,D_i} = \mathbf{P}_B\mathbf{R}_{11}^{(i)} \quad \mathbf{P}_{Y_i,L_i} = \mathbf{P}_B\mathbf{R}_{1,a_i} \tag{5.4.11}$$

In these equations $i,j=0,1$, $i\neq j$, $X_i = G_i, U_i, I_i$, $Y_i = D_i, L_i$, $\mathbf{R}_{a_i,1}$ is the probability of not receiving the ACKi when it was sent, $\mathbf{R}_{11}^{(i)} = \mathbf{R}_{11}+\mathbf{R}_{1,a_j}$ is the probability of not receiving the ACKi when the NAK was sent, $\mathbf{P}_A$ is the probability of correctly decoding the message and its number, and $\mathbf{P}_C$ is the probability of incorrectly decoding the message and correctly decoding its number. We assume that these probabilities do not depend on the message number.

The stationary probabilities of the system states satisfy the following equations:

$$\mathbf{p}(G_i) = \mathbf{c}_j\mathbf{P}_A \quad \mathbf{p}(U_i) = \mathbf{c}_j\mathbf{P}_C \quad \mathbf{p}(I_i) = \mathbf{d}_i\mathbf{P}_F$$
$$\mathbf{p}(S_i) = \mathbf{d}_i\mathbf{P}_B \quad \mathbf{p}(D_i) = \mathbf{c}_i\mathbf{P}_B\mathbf{R}_{11}^{(i)} \quad \mathbf{p}(L_i) = \mathbf{c}_i\mathbf{P}_B\mathbf{R}_{1,a_i}$$

where $i\neq j$ and $i,j=0,1$. One can easily show that the coefficients $\mathbf{c}_i$ and $\mathbf{d}_i$ can be found from the system

$$\mathbf{c}_i = (\mathbf{c}_j+\mathbf{d}_i)\mathbf{P}_F\mathbf{R}_{a_i,a_i} + \mathbf{d}_i\mathbf{P}_B\mathbf{R}_{1,a_i} + \mathbf{c}_i\mathbf{P}_B\mathbf{S}$$
$$\mathbf{d}_i = (\mathbf{c}_j+\mathbf{d}_i)\mathbf{P}_F\mathbf{R}_{a_i,1} + \mathbf{d}_i\mathbf{P}_B\mathbf{R}_{11}^{(i)}$$
$$\mathbf{c}_0 + \mathbf{c}_1 + \mathbf{d}_0 + \mathbf{d}_1 = \boldsymbol{\pi}$$

Solving this system, we obtain

$$\mathbf{c}_0 = \boldsymbol{\pi}(\mathbf{I} + \mathbf{M}_0 + \mathbf{N}_0 + \mathbf{N}_0\mathbf{M}_1)^{-1}$$
$$\mathbf{c}_1 = \mathbf{c}_0\mathbf{N}_0 \quad \mathbf{d}_1 = \mathbf{c}_0\mathbf{M}_0 \quad \mathbf{d}_0 = \mathbf{c}_1\mathbf{M}_1$$
$$\mathbf{M}_j = \mathbf{P}_F\mathbf{R}_{a_i,1}(\mathbf{I} - \mathbf{P}_F\mathbf{R}_{a_i,1} - \mathbf{P}_B\mathbf{R}_{11}^{(i)})^{-1}$$
$$\mathbf{N}_j = [\,\mathbf{I} + \mathbf{M}_j(\mathbf{P}_F\mathbf{R}_{a_i,a_i} + \mathbf{P}_B\mathbf{R}_{1,a_i})\,](\mathbf{I} - \mathbf{P}_B\mathbf{S})^{-1}$$

In the set of accepted messages the ARQ performance characteristics can be expressed using stationary probabilities (5.4.11):

$\nu(\mathbf{c}_0+\mathbf{c}_1)\mathbf{P}_A\mathbf{1}$	The probability that the message is received correctly (the first time)
$\nu(\mathbf{c}_0+\mathbf{c}_1)\mathbf{P}_C\mathbf{1}$	The probability that the message is received incorrectly (the first time)
$\nu(\mathbf{d}_0+\mathbf{d}_1)\mathbf{P}_F\mathbf{1}$	The probability that the message is an insertion of the previous message
$p(L)$	The probability that the message is lost

In these expressions we denoted

$$p(L) = \nu(\mathbf{c}_0\mathbf{P}_B\mathbf{R}_{1,a_0} + \mathbf{c}_1\mathbf{P}_B\mathbf{R}_{1,a_1})\mathbf{1}$$
$$\nu = [1 - (\mathbf{d}_0+\mathbf{d}_1)\mathbf{P}_B\mathbf{1} - (\mathbf{c}_0\mathbf{P}_B\mathbf{R}_{11}^{(0)}+\mathbf{c}_1\mathbf{P}_B\mathbf{R}_{11}^{(1)})\mathbf{1}]^{-1}$$

If the acknowledgements are symmetrical, the probabilities $\mathbf{R}_{a_i,a_i}$, $\mathbf{R}_{a_i,1}$, and $\mathbf{R}_{11}^{(i)}$ do not depend on i and the preceding solutions can be simplified:

$$\mathbf{c}_0 = \mathbf{c}_1 = 0.5\boldsymbol{\pi}(\mathbf{P}_B\mathbf{R}_{1,a_0} + \mathbf{P}_F\mathbf{R}_{a_0,a_0})(\mathbf{I} - \mathbf{P}_B\mathbf{R}_{11}^{(0)})^{-1}$$

$$\mathbf{d}_0 = \mathbf{d}_1 = 0.5\boldsymbol{\pi}\mathbf{P}_F\mathbf{R}_{a_0,1}(\mathbf{I} - \mathbf{P}_B\mathbf{R}_{11}^{(0)})^{-1}$$

Let us consider now the the possibility of losing synchronization in the direct and return channels. To describe the system's operation we introduce more events:

Λ_d The message is lost in the direct channel

Λ_r The message is lost in the return channel

The corresponding matrix probabilities are denoted $\mathbf{P}_{\Lambda_d}$ and $\mathbf{R}_{\Lambda_r}$.

If the message is lost in the direct channel, it will be retransmitted after a timeout so that the system will return to the state in which it was before transmitting the message. To model this situation, we define for each of the previously defined states X_i ($X = G, U, I, S, D, L$) the direct-channel timeout states X_{i+2} with transition probabilities

$$\mathbf{P}_{X_i,X_{i+2}} = \mathbf{P}_{\Lambda_d} \qquad \mathbf{P}_{X_{i+2},X_i} = \mathbf{S}$$

The return-channel message loss can be treated similarly. Suppose that the previous message was in one of the states G_i, U_i, I_i, or S_i. Then, the next message without detected errors represents an insertion of the previous one if the acknowledgement has been lost. The probability of this event is

$$\mathbf{P}_{Y_i,I_{i+4}} = \mathbf{R}_{\Lambda_r}\mathbf{P}_F \qquad (Y = G, U, I, S)$$

If errors are detected, the next message is discarded, with the probability

$$\mathbf{P}_{Y_i,S_{i+4}} = \mathbf{R}_{\Lambda_r}\mathbf{P}_B$$

If the previous message was in one of the states D_i or L_i, then

$$\mathbf{P}_{Z_i,G_i} = \mathbf{R}_{\Lambda_r}\mathbf{P}_A \qquad \mathbf{P}_{Z_i,U_i} = \mathbf{R}_{\Lambda_r}\mathbf{P}_C$$

$$\mathbf{P}_{Z_i,D_i} = \mathbf{R}_{\Lambda_r}\mathbf{P}_B\mathbf{R}_{11}^{(i)} \qquad \mathbf{P}_{Z_i,L_i} = \mathbf{R}_{\Lambda_r}\mathbf{P}_B\mathbf{R}_{1,a_i}$$

The transition probabilities from the states G_{i+4}, U_{i+4}, I_{i+4}, S_{i+4}, D_{i+4}, L_{i+4} are the same as from the states G_i, U_i, I_i, S_i, D_i, L_i. The stationary probabilities of the states of the Markov chain can be found similarly to those in previous cases.

5.4.7 Go-Back-N ARQ

The stop-and-wait method is inefficient because of the idle time that it spends waiting for an acknowledgement. In full-duplex systems, it is possible to send messages and receive acknowledgements continuously. The transmitter sends the next message without waiting for the previous message to be acknowledged. As long as all the messages are positively acknowledged, this method is conceptually the same as the stop-and-wait method. The transmitter continuously transmits messages and simultaneously shifts them into its buffer of length N (whose size is greater or equal to the number of messages transmitted during the round-trip delay) if it receives an ACK. When an NAK is received, the transmitter backs up to the negatively acknowledged message and retransmits it and $N-1$ succeeding messages. The receiver blocks for the period of N messages transmission: It discards the negatively acknowledged message and $N-1$ following messages and waits for their retransmission.

The protocol state transition diagram is shown in Fig. 5.4. In this figure, M_t denotes the transmission of of the message at the moment t and 0_t and 1_t denote the transmission of an ACK and NAK for this message. The edges are marked by the results of the messages decoding which cause the state transfer. The transmitter sends M_{t-N}^{t-1} if it decodes an NAK 1_{t-N}; otherwise it shifts M_t into its buffer and sends it to the receiver. The receiver sends an NAK 1_t if it detects errors in M_t which we denoted $\overline{M}_t$; otherwise it sends an ACK 0_t. $W_1, W_2, \ldots, W_{N-1}$ denote states corresponding to waiting for retransmission. The receiver enters these states after sending an NAK.

Analysis of these systems is complicated by the transmission of the next message before the reception of an acknowledgement for the previous message. Therefore, it is necessary to use multiple Markov chains to describe the system's operation in an HMM channel. The chain order N is defined by the delay between message transmission and decoding of its acknowledgement. Let us consider a piggyback scheme in which messages together with acknowledgements are sent over the channels in both directions. To describe the system functioning, let us introduce the state vector $\mathbf{s} = (m, r, s_1, s_2, \ldots, s_{N-1})$, where

- m is the message sequence number from the beginning of the receiver blocking zone, with $m = 0$ if it is outside the blocking zone
- r is the acknowledgement message sequence number from the beginning of the return-channel blocking zone, with $r = 0$ if it is outside the blocking zone
- s_i is the logical variable that characterizes the message status: $s_1 = 0$ if the message has not been accepted before, $s_1 = 1$ otherwise; $s_2, \ldots, s_{N-1}$ similarly characterize the succeeding messages

These states represent a Markov function whose matrix transitional probabilities $\mathbf{P}(\boldsymbol{\sigma};\ \mathbf{s})$ have the following form:

$$\mathbf{P}(a,b,\mathbf{v};0,0,\boldsymbol{\tau},0) = \mathbf{P}_{F0} \qquad \mathbf{P}(a,b,\mathbf{v};0,1,\boldsymbol{\tau},1) = \mathbf{P}_{F1}$$

$$\mathbf{P}(a,b,\mathbf{v};1,0,\boldsymbol{\tau},0) = \mathbf{P}_{B0} \qquad \mathbf{P}(a,b,\mathbf{v};1,1,\mathbf{w}) = \mathbf{P}_{B1}$$

$$\mathbf{P}(m,b,\mathbf{v};m+1,0,\boldsymbol{\tau},0) = \mathbf{P}_{\Omega 0} \qquad \mathbf{P}(m,b,\mathbf{v};m+1,1,\mathbf{w}) = \mathbf{P}_{\Omega 1}$$

$$\mathbf{P}(a,r,\mathbf{v};0,r+1,\boldsymbol{\tau},1) = \mathbf{P}_{F\Omega} \qquad \mathbf{P}(a,r,\mathbf{v};1,r+1,\mathbf{w}) = \mathbf{P}_{B\Omega}$$

$$\mathbf{P}(m,r,\mathbf{v};m+1,r+1,\mathbf{w}) = \mathbf{P}_{\Omega\Omega} \qquad \mathbf{P}(m,N-1,\mathbf{v};m+1,N,\mathbf{v}) = \mathbf{P}_{\Omega\Omega}$$

$$\mathbf{P}(a,N-1,\mathbf{v};0,N,\mathbf{v}) = \mathbf{P}_{0\Omega} \qquad \mathbf{P}(a,N-1,\mathbf{v};1,N,\mathbf{v}) = \mathbf{P}_{1\Omega}$$

Here $\mathbf{v} = (s_1, s_2, \ldots, s_{N-1})$, $\boldsymbol{\tau} = (s_2, s_3, \ldots, s_{N-1})$, $\mathbf{w} = (s_1, s_2, \ldots, s_{N-1}, s_1)$, $0 < m \le N-1$, $0 < r < N-1$, $a = 0,\ N$, and $b = 0, N$. In these equations, $\mathbf{P}_{Fv}$ is the matrix probability of not detecting errors in a direct-channel message and receiving an ACK ($v = 0$) or NAK ($v = 1$) over the return channel; $\mathbf{P}_{Bv}$ is the probability of detecting errors in a direct-channel message and receiving the reply v; and $\mathbf{P}_{0\Omega}$ and $\mathbf{P}_{1\Omega}$ are the probabilities of decoding an ACK or NAK from the direct-channel message when the NAK was sent. Also we denoted

$$\mathbf{P}_{\Omega v} = \mathbf{P}_{Fv} + \mathbf{P}_{Bv} \quad \mathbf{P}_{F\Omega} = \mathbf{P}_{F0} + \mathbf{P}_{F1} \quad \mathbf{P}_{B\Omega} = \mathbf{P}_{B0} + \mathbf{P}_{B1} \quad v = 0,\ 1,\ \Omega$$

As in the previous sections, we determine first the stationary probabilities of the system states $\mathbf{p}(\mathbf{s}) = \mathbf{p}(m,r,\mathbf{v})$ that satisfy the equations

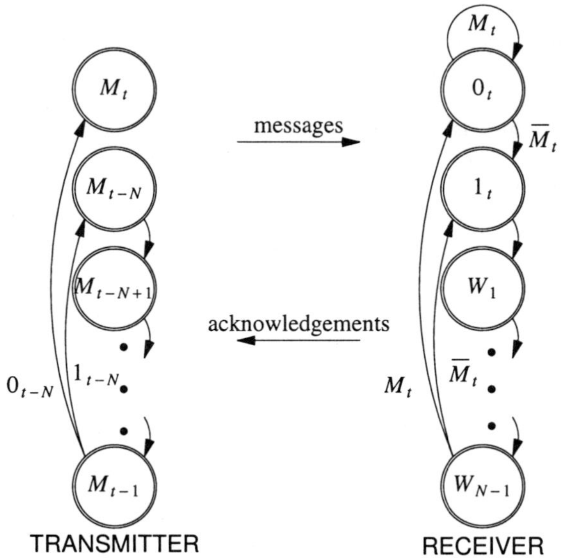

Figure 5.4. Go-Back-N ARQ.

$$\mathbf{p}(\mathbf{s}) = \sum_{\sigma} \mathbf{p}(\sigma)\mathbf{P}(\sigma;\ \mathbf{s}) \qquad \sum_{\mathbf{s}} \mathbf{p}(\mathbf{s}) = \pi \tag{5.4.12}$$

where $\mathbf{P}(\sigma;\ \mathbf{s})$ are the previously defined matrix transition probabilities and π is the stationary distribution of the HMM states. It is also the invariant vector of the matrix $\mathbf{P}_{\Omega\Omega}$. The summation is performed over all the states that are reachable from the all-zero state (0,0, ...,0).

If we know the solution of system (5.4.12), then the system's performance characteristics can be expressed by summing the appropriate stationary probabilities:

$\mathbf{u}_0\mathbf{P}_A\mathbf{1}$ The probability of message correct decoding

$\mathbf{u}_0\mathbf{P}_C\mathbf{1}$ The probability of message incorrect decoding

$\mathbf{u}_1\mathbf{P}_F\mathbf{1}$ The probability of message insertion

$\mathbf{w1}$ The probability of message loss where $\mathbf{w} = \mathbf{u}_0\mathbf{P}_{B0} + \mathbf{v}\mathbf{P}_{\Omega 0}$

In these expressions, we denote

$$\mathbf{u}_s = \sum_{\substack{r \neq N-1 \\ N-1}} \sum_{\tau} \mathbf{p}(0,r,s,\tau) + \mathbf{p}(1,r,s,\tau) \qquad s = 0,1$$

$$\mathbf{v} = \sum_{m=1} \sum_{\tau} \mathbf{p}(m,0,0,\tau) + \mathbf{p}(m,N,0,\tau), \quad \tau = (\mathbf{s}_2,\ldots,\mathbf{s}_{N-1})$$

The system's maximal throughput is greater than that of the stop-and-wait protocol and is equal to

$$\bar{R} = \mathbf{u}_0 \mathbf{P}_F \mathbf{1} k / n$$

To find the probabilities in the set of accepted messages we need to exclude the messages which are retransmitted and perform the appropriate normalization of the previously defined probabilities.

Thus, solving system (5.4.12) is the major difficulty in analyzing the go-back-N scheme. This system has a special structure that simplifies its solution. To simplify demonstration of the method, we assume that the return channel is noiseless. In this case, $\mathbf{P}_{F0} = \mathbf{P}_F$, $\mathbf{P}_{F1} = 0$, $\mathbf{P}_{B1} = \mathbf{P}_B$, $\mathbf{P}_{B0} = 0$, $\mathbf{P}_{\Omega 0} = \mathbf{S}$, $\mathbf{P}_{\Omega 1} = 0$, $r = 0$, $s_i = 0$, and the protocol can be described as an HMM with the state transition probability matrix

$$\mathbf{T} = \begin{bmatrix} \mathbf{P}_F & \mathbf{P}_B & 0 & 0 & \cdots & 0 \\ 0 & 0 & \mathbf{S} & 0 & \cdots & 0 \\ 0 & 0 & 0 & \mathbf{S} & \cdots & 0 \\ \cdots & \cdots & \cdots & \cdots & \cdots & \cdots \\ 0 & 0 & 0 & 0 & \cdots & \mathbf{S} \\ \mathbf{P}_F & \mathbf{P}_B & 0 & 0 & \cdots & 0 \end{bmatrix}$$

System (5.4.12) takes the form

$$\begin{aligned} \mathbf{p}(\mathbf{s}_0) &= [\mathbf{p}(\mathbf{s}_0) + \mathbf{p}(\mathbf{s}_N)]\mathbf{P}_F \\ \mathbf{p}(\mathbf{s}_1) &= [\mathbf{p}(\mathbf{s}_0) + \mathbf{p}(\mathbf{s}_N)]\mathbf{P}_B \\ \mathbf{p}(\mathbf{s}_{i+1}) &= \mathbf{p}(\mathbf{s}_i)\mathbf{S} \quad i = 1,2,..,N \\ \sum_{i=0}^{N} \mathbf{p}(\mathbf{s}_i) &= \boldsymbol{\pi} \end{aligned} \tag{5.4.13}$$

where $\mathbf{s}_i = (i,0,0,...,0)$.

To solve this system, we denote $\mathbf{u}_0 = \mathbf{p}(\mathbf{s}_0) + \mathbf{p}(\mathbf{s}_N)$ and, after the substitution, we obtain the solution in the form

$$\mathbf{p}(\mathbf{s}_0) = \mathbf{u}_0 \mathbf{P}_F \quad \mathbf{p}(\mathbf{s}_i) = \mathbf{u}_0 \mathbf{P}_B \mathbf{S}^{i-1} \quad (i = 1,2,...,N)$$

where the single unknown $\mathbf{u}_0$ is obtained from the last equation of (5.4.13) which becomes

$$\mathbf{u}_0 \mathbf{P}_F + \mathbf{u}_0 \mathbf{P}_B \mathbf{V}_N = \boldsymbol{\pi}$$

where

$$\mathbf{V}_N = \sum_{i=1}^{N} \mathbf{S}^{i-1}$$

Since $\mathbf{P}_F = \mathbf{S} - \mathbf{P}_B$ and $\boldsymbol{\pi}\mathbf{S} = \boldsymbol{\pi}$, this equation follows from

$$\mathbf{u}_0(\mathbf{I} + \mathbf{P}_B \mathbf{V}_{N-1}) = \boldsymbol{\pi}$$

Thus, we have $\mathbf{u}_0 = \boldsymbol{\pi}(\mathbf{I} + \mathbf{P}_B \mathbf{V}_{N-1})^{-1}$. The sum $\mathbf{V}_{N-1}$ can be found using the matrix $\mathbf{S}$ spectral decomposition ($\mathbf{S} = \mathbf{1}\boldsymbol{\pi} + \mathbf{B}$):

$$\mathbf{V}_{N-1} = (N-2)\mathbf{1}\boldsymbol{\pi} + (\mathbf{I} - \mathbf{B}^{N-1})(\mathbf{I} - \mathbf{B})^{-1}$$

Note that there is no need to determine all of the elements of the stationary distribution, since the system performance characteristics depend only on sums of the stationary probabilities. In the previously considered particular case, only the sum $\mathbf{u}_0 = \mathbf{p}(\mathbf{s}_0) + \mathbf{p}(\mathbf{s}_1)$

needs to be found to calculate the probabilities of correct and incorrect decoding and of the system throughput. The probability $p_\beta(x)$ that a frame occupies the system for the time necessary to send x frames without retransmission can also be expressed using this sum:

$$p_\beta(lN+1) = \mathbf{u}_0 \mathbf{P}_F \mathbf{P}_\beta(lN+1)\mathbf{1} / \mathbf{u}_0 \mathbf{P}_F \mathbf{1}$$

where

$$\mathbf{P}_\beta(lN+1) = (\mathbf{P}_B \mathbf{S}^{N-1})^l \mathbf{P}_F$$

is the corresponding matrix distribution whose generating function is

$$\mathbf{\Phi}_\beta(z) = \sum_{l=0}^{\infty} \mathbf{P}_\beta(lN+1) z^{lN} = z(\mathbf{I} - z^N \mathbf{P}_B \mathbf{S}^{N-1})^{-1} \mathbf{P}_F \tag{5.4.14}$$

The go-back-N protocol is more effective than the stop-and-wait protocol, but its implementation is more complex. It requires a buffer to store at least N messages that have not been acknowledged. If N is large and the channel is noisy, the effectiveness of the scheme may suffer significantly because it is necessary to retransmit all N messages. In alternative *selective-repeat systems* only the messages with detected errors are retransmitted. These systems use not only a transmitter buffer but also a receiver buffer, since messages may not be accepted in the same order as they were transmitted and therefore may need to be rearranged on the receiver output.

5.4.7.1 Alternative Assumptions

There are several papers that analyze the throughput performance of the GBN protocol in channels with memory. [3,4,8,9,12,20,22] Most of them assume that the return channel is error-free. This assumption is valid if ACKs and NAKs are well protected. However, it is often not the case (especially when a single bit representing the acknowledgement is a part of the return channel messages). Several papers consider the effects of errors in the return channels [4,8,22] assuming that error sources in the direct and return channel are independent and modeled by a two-state Markov chain. In addition, it is assumed that an ACK cannot be misdecoded as an NAK and vice versa.

If we define the throughput as a ratio of the message length to the mean time that the message occupies the system from the message's first transmission until it leaves the system (received the last time or lost), we can obtain a simple formula for the throughput calculation which generalizes the results of the previous papers.

The message service time is equal to its length L if it has been positively acknowledged after its first reception. We denoted the matrix probability of this event $\mathbf{H}_0 = \mathbf{P}_F \mathbf{R}_{00} + \mathbf{P}_B \mathbf{R}_{10}$. The message service time was $\tau_k = L(kN+1)$ if it has been positively acknowledged after its $k+1$ transmission. The matrix probability of this event is

$$\mathbf{P}(\tau_k) = \mathbf{D}^k \mathbf{H}_0$$

where $\mathbf{D} = \mathbf{H}_1 \mathbf{P}^{N-1}$ is the matrix probability of receiving the same message, $\mathbf{H}_1$ is the matrix probability of decoding an NAK: $\mathbf{H}_1 = \mathbf{P}_F \mathbf{R}_{01} + \mathbf{P}_B \mathbf{R}_{11}$. The sum $\mathbf{S} = \mathbf{H}_0 + \mathbf{H}_1$ of these matrices is the transition probability matrix. The message service time matrix probability generating function is given by

$$\boldsymbol{\Phi}(z) = \sum_{k=0}^{\infty} \mathbf{P}(\tau_k) z^{\tau_k} = \sum_{k=0}^{\infty} \mathbf{D}^k \mathbf{H}_0 z^{L(kN+1)}$$

which, after summing the matrix-geometric series [see Eq. (A.5.9) of Appendix 5], takes the form

$$\boldsymbol{\Phi}(z) = z^L(\mathbf{I} - \mathbf{D}z^{LN})^{-1}\mathbf{H}_0$$

The service time probability distribution has the form

$$p(\tau_k) = \boldsymbol{\pi}_0 \mathbf{P}(\tau_k)\mathbf{1} / \boldsymbol{\pi}_0 \mathbf{1} \qquad \text{for} \quad \tau_k = L(kN+1)$$

where $\boldsymbol{\pi}_0$ is the probability vector of decoding an ACK (for the previous message).

It is convenient to find the distribution mean using its generating function $\phi(z) = \boldsymbol{\pi}_0 \boldsymbol{\Phi}(z)\mathbf{1} / \boldsymbol{\pi}_0 \mathbf{1}$. Thus, the mean service time is

$$\overline{\tau}_{GBN} = \phi'(1) = L\boldsymbol{\pi}_0[N\mathbf{D}(\mathbf{I} - \mathbf{D})^{-2} + (\mathbf{I} - \mathbf{D})^{-1}]\mathbf{H}_0\mathbf{1} / \boldsymbol{\pi}_0 \mathbf{1}$$

and the system throughput is $L / \overline{\tau}_{GBN}$.

If the process is stationary, $\boldsymbol{\pi}_0$ can be found by solving Eq. (5.4.12) which, in our case, can be reduced to

$$\boldsymbol{\pi}_0 = (\boldsymbol{\pi}_0 + \boldsymbol{\pi}_N)\mathbf{H}_0$$
$$\boldsymbol{\pi}_1 = (\boldsymbol{\pi}_0 + \boldsymbol{\pi}_N)\mathbf{H}_1$$
$$\boldsymbol{\pi}_{i+1} = \boldsymbol{\pi}_i \mathbf{S}, \quad i = 1, 2, \ldots, N$$
$$\sum_{i=0}^{N} \boldsymbol{\pi}_i = \boldsymbol{\pi}$$

where the probability vector $\boldsymbol{\pi}_0$ corresponds to decoding an ACK, $\boldsymbol{\pi}_1$ corresponds to decoding an NAK, and $\boldsymbol{\pi}_2, \boldsymbol{\pi}_3, \ldots, \boldsymbol{\pi}_N$ correspond to waiting for an acknowledgment after receiving an NAK. This system has the form of system (5.4.13). Therefore, it solution can be written as

$$\boldsymbol{\pi}_0 = \boldsymbol{\pi}(\mathbf{I} + \mathbf{H}_1\mathbf{V}_{N-1})^{-1}\mathbf{H}_0$$

In this case, after some algebra, we obtain the system throughput

$$\eta = L / \overline{\tau}_{GBN} = \boldsymbol{\pi}(\mathbf{I} + \mathbf{H}_1\mathbf{V}_{N-1})^{-1}\mathbf{H}_0$$

Example 5.3.3. Let us calculate a throughput of a system when the return channel is ideal and the direct channel is modeled by the Markov chain with the matrix

$$\mathbf{S} = \begin{bmatrix} p & q \\ r & s \end{bmatrix}$$

whose first state corresponds to the message reception and second state corresponds to detecting errors in the message. [9]

In this case, we have (see Sec. 1.1.5)

$$\mathbf{H}_0 = \begin{bmatrix} p & 0 \\ r & 0 \end{bmatrix} \quad \mathbf{H}_1 = \begin{bmatrix} 0 & q \\ 0 & s \end{bmatrix}$$

and

$$\boldsymbol{\pi} = [\frac{r}{r+q}, \frac{q}{r+q}]$$

Matrix S spectral decomposition has the form (see Appendix 6, Sec. 6.1.2)

$$\mathbf{S} = \mathbf{1}\boldsymbol{\pi} + \mathbf{D} = \frac{1}{r+q}\begin{bmatrix} r & q \\ r & q \end{bmatrix} + \frac{\lambda}{r+q}\begin{bmatrix} q & -q \\ -r & r \end{bmatrix}$$

$\lambda = 1-r-q$ is the matrix S eigenvalue. Thus,

$$\mathbf{V}_{N-1} = (N-1)\mathbf{1}\boldsymbol{\pi} + \frac{1-\lambda^{N-1}}{(r+q)^2}\begin{bmatrix} q & -q \\ -r & r \end{bmatrix}$$

so that

$$\eta = \frac{r(1-\lambda^N)}{r(1-\lambda^N) + Nq(r+q)}$$

which coincides with Eq. (2.10) of Ref. 13.

Note that if block errors in the direct channel are independent (Markov chain with one state), we have $\mathbf{S} = 1, \mathbf{H}_1 = p, \mathbf{H}_1 = q, \boldsymbol{\pi} = 1, \mathbf{V}_{N-1} = N-1$, and we obtain

$$\eta = (1 + (N-1)q)^{-1}p = \frac{p}{p + Nq}$$

which is the well-known formula. [10,11,17,18]

5.4.8 Multiframe Messages

In the systems considered so far, each frame contained only one message. However, in many cases, a message may occupy several frames. For example, a message may be a file that is transferred from one computer to another. In these cases, long messages are broken up into smaller blocks to fit into transmission frames. The message is transferred correctly if all its blocks are accepted without errors. If at least one of the blocks contains errors, the whole message contains errors.

Suppose that each message is encoded into K blocks which are protected by an (n,k) code. The messages arrive periodically with a transmission period of $\alpha = k+\xi M$ frames. The maximum message transmission delay is equal to the time of transmitting ε frames. Our goal is to calculate the probability of correct and incorrect message reception if the ARQ scheme is used for frame transmission over an HMM channel.

The problem that we are trying to solve is one typical of queueing theory. Messages arrive periodically and the service time is equal to the time that the message occupies the channel, which is equal to the sum of the times taken by the message subblocks. The matrix distribution of the service time can be expressed as a convolution of the addends, and its generating function equals

$$\boldsymbol{\Phi}_\tau(z) = [\boldsymbol{\Phi}_\beta(z)]^K \tag{5.4.15}$$

where $\boldsymbol{\Phi}_\beta(z)$ is the generating function of the time that a single frame occupies the channel. The message service time distribution can be expressed as

$$\mathbf{P}_\tau(m) = \frac{1}{2\pi j}\oint_\gamma \boldsymbol{\Phi}_\tau(z) z^{-m-1}\, dz. \tag{5.4.16}$$

Let us now determine the matrix distribution $\mathbf{Q}(j)$ of the message waiting time, which plays a fundamental role in evaluating the system's performance. According to the HMM, the sequence of waiting times for different messages can be described as a Markov function with the matrix

$$\begin{bmatrix} \mathbf{P}_\tau(\le\alpha) & \mathbf{P}_\tau(\alpha+1) & \cdots & \mathbf{P}_\tau(\varepsilon-1) & \mathbf{P}_\tau(>\varepsilon-1) \\ \mathbf{P}_\tau(\le\alpha-1) & \mathbf{P}_\tau(\alpha) & \cdots & \mathbf{P}_\tau(\varepsilon-2) & \mathbf{P}_\tau(>\varepsilon-2) \\ \cdots & \cdots & \cdots & \cdots & \cdots \\ \mathbf{P}_\tau(1) & \mathbf{P}_\tau(2) & \cdots & \mathbf{P}_\tau(\kappa) & \mathbf{P}_\tau(>\kappa) \\ 0 & \mathbf{P}_\tau(1) & \cdots & \mathbf{P}_\tau(\kappa-1) & \mathbf{P}_\tau(>\kappa-1) \\ \cdots & \cdots & \cdots & \cdots & \cdots \\ 0 & 0 & \cdots & \mathbf{P}_\tau(\alpha-1) & \mathbf{P}_\tau(>\alpha-1) \end{bmatrix} \tag{5.4.17}$$

where $\kappa = \varepsilon-\alpha$,

$$\mathbf{P}_\tau(\le m) = \sum_{j=1}^{m} \mathbf{P}_\tau(j)\mathbf{Q}^{m-j} \tag{5.4.18}$$

$$\mathbf{P}_\tau(>m) = \mathbf{S}^m - \mathbf{P}_\tau(\le m) \tag{5.4.19}$$

The matrix probabilities $\mathbf{Q}(j)$ represent the stationary distribution of the Markov chain with matrix (5.4.17) and can be found from the following system:

$$\begin{aligned} \mathbf{Q}(0) &= \sum_{i=0}^{\alpha-1} \mathbf{Q}(i)\mathbf{P}_\tau(\le\alpha-i) \\ \mathbf{Q}(j) &= \sum_{i=0}^{j+\alpha-1} \mathbf{Q}(i)\mathbf{P}_\tau(\alpha+j-i) \qquad 0<j<\kappa \\ \mathbf{Q}(\kappa) &= \sum_{i=0}^{\kappa} \mathbf{Q}(i)\mathbf{P}_\tau(>\varepsilon-i+1) \\ \sum_{j=0}^{\kappa} \mathbf{Q}(j)\mathbf{1} &= 1 \qquad \mathbf{Q}(j) = 0 \quad \text{for } j>\kappa \end{aligned} \tag{5.4.20}$$

It is easy to verify that

$$\sum_{j=0}^{\kappa} \mathbf{Q}(j)\mathbf{S}^{\kappa-j} = \boldsymbol{\pi} \tag{5.4.21}$$

We find $\mathbf{Q}(\alpha-1)$ from the first equation

$$\mathbf{Q}(\alpha-1) = [\mathbf{Q}(0) - \sum_{i=0}^{\alpha-2} \mathbf{Q}(i)\mathbf{P}_\tau(\le\alpha-i)]\mathbf{P}_\tau^{-1}(1)$$

and substitute it into the rest of the equations. Then, we find $\mathbf{Q}(\alpha)$ from the second equation and replace it in the remaining equations, and so on. Repeating the process, we obtain the system of matrix equations to determine $\mathbf{Q}(0)$, $\mathbf{Q}(1)$,..., $\mathbf{Q}(\alpha-2)$. If α is small, this system can easily be solved. The rest of the matrices $\mathbf{Q}(j)$ are found by directly substituting the solution into the rest of the system.

This system can also be solved using generating functions. We will show that the system can be replaced with an integral equation for the waiting-time distribution generating function

$$\boldsymbol{\rho}(z) = \sum_{j=0}^{\kappa} \mathbf{Q}(j) z^j$$

Indeed, we obtain from system (5.4.20)

$$\boldsymbol{\rho}(z) = \sum_{m=0}^{\infty} \mathbf{Q}(m)[\mathbf{P}_\tau(\le \alpha - m) + \sum_{j=1}^{\kappa-1} \mathbf{P}_\tau(\alpha + j - m) z^j + \mathbf{P}_\tau(> \varepsilon - m + 1) z^\kappa] \quad (5.4.22)$$

Using integral representation (5.4.16) and Eq. (5.4.18) and (5.4.19) we obtain

$$\mathbf{P}_\tau(\le m) = \frac{1}{2\pi j} \oint_\gamma \boldsymbol{\Phi}_\tau(z)(\mathbf{I} - \mathbf{S}z)^{-1} z^{-m-1}\, dz$$

$$\mathbf{P}_\tau(> m) = \frac{1}{2\pi j} \oint_\gamma [\mathbf{I} - \boldsymbol{\Phi}_\tau(z)\mathbf{S}z](\mathbf{I} - \mathbf{S}z)^{-1} z^{-m}\, dz$$

Substituting these expressions into Eq. (5.4.22), we obtain the following integral equation:

$$\boldsymbol{\rho}(z) = z^\kappa \boldsymbol{\pi} + \oint_\gamma \frac{\boldsymbol{\rho}(w)\boldsymbol{\Psi}_\tau(w)(w^\kappa - z^\kappa)\, dw}{2\pi j\, w^\kappa (w - z)} (\mathbf{I} - \mathbf{S}z) \quad (5.4.23)$$

where

$$\boldsymbol{\Psi}_\tau(w) = w^{-\alpha} \boldsymbol{\Phi}_\tau(w)(\mathbf{I} - \mathbf{S}w)^{-1}$$

In deriving this equation we used Eq. (5.4.21), which can be rewritten as

$$\frac{1}{2\pi j} \oint_\gamma \boldsymbol{\rho}(w)(\mathbf{I} - \mathbf{S}w)^{-1} w^{-\varepsilon-1}\, dw = \boldsymbol{\pi} \quad (5.4.24)$$

Integral Eq. (5.4.23) can be solved by the standard methods of the theory of systems of integral equations. For large κ, the solution is close to the case of an infinite delay ($\kappa \to \infty$). Suppose that the message interarrival interval is greater than the average time that the message spends in the system (service time). Then, stationary distribution of waiting time exists. Equation (5.4.23) takes the following form when $\kappa \to \infty$:

$$\boldsymbol{\rho}(z) = \boldsymbol{\rho}(z)\boldsymbol{\Psi}_\tau(z) + \oint_\gamma \frac{\boldsymbol{\rho}(w)\boldsymbol{\Psi}_\tau(w)(\mathbf{I} - \mathbf{S}w)^{-1}\, dw}{2\pi j (w - z)} (\mathbf{I} - \mathbf{S}z)$$

or

$$\boldsymbol{\rho}(z) = \oint_\gamma \frac{\boldsymbol{\rho}(w)\boldsymbol{\Psi}_\tau(w)(\mathbf{I} - \mathbf{S}w)^{-1}\, dw}{2\pi j (w - z)} (\mathbf{I} - \mathbf{S}z)[\mathbf{I} - \boldsymbol{\Psi}_\tau(z)]^{-1} \quad (5.4.25)$$

It is easy to verify that

$$\boldsymbol{\xi}(z) = \oint_\gamma \frac{\boldsymbol{\rho}(w)\boldsymbol{\Psi}_\tau(w)(\mathbf{I} - \mathbf{S}w)^{-1}\, dw}{2\pi j (w - z)}$$

is a degree $\alpha + 1$ polynomial of z^{-1}

$$\boldsymbol{\xi}(z) = \sum_{m=0}^{\alpha+1} \mathbf{B}_m z^{-m}$$

so that the solution of Eq. (5.4.25) has the form

$$\boldsymbol{\rho}(z) = \boldsymbol{\xi}(z)(\mathbf{I} - \mathbf{S}z)[\mathbf{I} - \boldsymbol{\Psi}_\tau(z)]^{-1}$$

The coefficients $\mathbf{B}_m$ of the solution may be obtained by using Eq. (5.4.24) and the conditions of the function $\boldsymbol{\rho}(z)$ regularity inside the unit circle.

According to Eq. (5.4.14) and (5.4.15), $\boldsymbol{\Psi}_\tau(z)$ is a rational function. Therefore, the distribution $\mathbf{Q}(j)$ has the form of Eq. (A.2.6), which in the case of simple roots z_m of the equation

$$\det[\mathbf{I} - \Psi_\tau(z)] = 0$$

that lie outside the unit circle has the form

$$\mathbf{Q}(j) = \sum_m \mathbf{A}_m z_m^{-j}$$

Let us now calculate the system performance characteristics under the assumption that probabilities $\mathbf{Q}(j)$ have been found. The probability of receiving a correct message can be expressed as

$$p_c = \sum_{j=0}^{\kappa} \mathbf{Q}(j)\mathbf{Pr}\{A^K;\ \varepsilon - j - 1\}\mathbf{1} \tag{5.4.26}$$

where $\mathbf{P}\{A^K;\ \varepsilon - j - 1\}$ is the matrix probability that all K subblocks that constitute the message have been received without errors at the time interval of $\varepsilon - j - 1$. This matrix represents the probability that the Markov chain for the K-th time reaches one of the states of the set that corresponds to a frame correct reception for the time $\varepsilon - j$ without entering any state other than a state corresponding to detected errors.

The right-hand side of Eq. (5.4.26) is a convolution, and, therefore, the probability of receiving a correct message can be expressed using transform methods (see Sec. 2.2):

$$p_c = \frac{1}{2\pi j} \oint_\gamma \boldsymbol{\rho}(z)\, \boldsymbol{W}_A(z)\mathbf{1} z^{-\varepsilon+1}\, dz$$

where

$$\boldsymbol{W}_A(z) = \sum_m \mathbf{Pr}\{A^K;\ m\} z^m$$

is the generating function of the matrix probabilities $\mathbf{P}\{A^K;\ m\}$. It is not difficult to show, using the convolution theorem, that this generating function has the form

$$\boldsymbol{W}_A(z) = [\boldsymbol{\Phi}_A(z)]^K (1 - z)^{-1}$$

where $\boldsymbol{\Phi}_A(z)$ is the generating function of the number of frame retransmissions before its first correct reception:

$$\boldsymbol{\Phi}_A(z) = (\mathbf{I} - \mathbf{P}_{B1} z)^{-1} \mathbf{P}_A z$$

The probability of incorrect decoding and message erasure (in the case when not all the message subblocks were accepted correctly) can be found quite analogously.

5.5 DELAY-CONSTRAINED SYSTEMS

In the previous sections, we assumed that the message source has an infinite buffer which always has messages to transmit; the message arrival regime was not included into our analysis. However, in the delay-constrained systems, a message is deleted if it was not received at the destination within the fixed amount of time since its *arrival* at the transmitter. Therefore, we need to include a queueing element into our analysis. If the message generation process can be described by an HMM and the communication channel is modeled by an HMM, then the combined source-channel system is an HMM which we can use for analyzing the system performance.

There are many publications that study the message arrival process. The process depends on many factors and environments. For example, voice communications traffic is completely different from the computer communications sending large files. Thus, we need to consider a mixture of various sources. To describe this situation, we assume that the message arrival and channel are modeled by an IOHMM with the matrix probabilities $\mathbf{P}(X,Y)$ where X is the number of messages arriving at the transmitter during a certain time interval and Y is a parameter characterizing the status of transmitter-receiver communication.

To be more specific and emphasize the queueing element, we consider a system with the stop-and-wait protocol. A time interval between the arrival at the transmitter of two consecutive acknowledgments (ACK or NAK) we call a slot. For this system, we denote as $\mathbf{P}(X,Y)$ the matrix probability of X packets arriving at a slot and receiving an acknowledgement $Y \in (ACK, NAK)$ for the first message in the queue (buffer). This model is general enough to describe large varieties of multimedia traffic: some states may correspond to voice some other states may correspond to data, and so on.

5.5.1 Queue-Length Probability Distribution

Let us obtain the probability distribution of the number m of the packet in the queue waiting for transmission in the stationary state of the system. According to our model, the number of packets in the queue can be described by an HMM whose state transition probability matrix can be computed in the following way.

The matrix probability for the queue to be empty if it was empty in the previous slot is $\mathbf{P}(0|0) = \mathbf{P}(1,ACK)+\mathbf{P}(0,\Omega)$ and $\mathbf{P}(0,\Omega) = \mathbf{P}(0,ACK)+\mathbf{P}(0,NAK)$, where $\Omega = ACK \cup NAK$. We find similarly that $\mathbf{P}(m-1|m) = \mathbf{P}(0,ACK)$ and $\mathbf{P}(m+i|m) = \mathbf{P}(i,NAK)+\mathbf{P}(i+1,ACK)$ for $i = 0,1,\ldots.$ This special form of the state transition probability matrix is called the *structured stochastic matrix of the M/G/1 type* and the stationary distribution of this Markov chain can be obtained by solving the following system of equations [14]

$$\mathbf{x}_m = \sum_{i=0}^{m+1} \mathbf{x}_i \mathbf{P}(m|i),$$

where $\mathbf{x}_i$ is the stationary probability sub-vector of i packets in the queue. We can solve this system using the matrix generating function $\mathbf{\Phi}(z) = \sum_{i=0}^{\infty} \mathbf{x}_i z^i$ which can be easily obtained by multiplying both sides of the previous equation by z^m and summing with respect to m. The solution has the form

$$\mathbf{\Phi}(z) = \mathbf{x}_0 \mathbf{P}(0,A)(z-1)[\mathbf{I}z - \mathbf{\Psi}_{ACK}(z) - z\mathbf{\Psi}_{NAK}(z)]^{-1}$$

where

$$\mathbf{\Psi}_Y(z) = \sum_{X=0}^{\infty} \mathbf{P}(X,Y) z^X$$

$\mathbf{x}_0$ can be obtained from the normalization condition $\lim_{z \to 1} \mathbf{\Phi}(z) = \boldsymbol{\pi}$ where $\boldsymbol{\pi}$ is the model stationary probability vector satisfying $\boldsymbol{\pi}\mathbf{P} = \boldsymbol{\pi}$, $\boldsymbol{\pi}\mathbf{1} = 1$ where

$$\mathbf{P} = \mathbf{\Psi}_{ACK}(1) + \mathbf{\Psi}_{NAK}(1) = \mathbf{H}_0 + \mathbf{H}_1$$

$\mathbf{H}_0$ and $\mathbf{H}_1$ are defined by Eq. (5.4.4) and (5.4.7), respectively.

Alternatively, the solution can be found as [15]

$$\mathbf{x}_0 \mathbf{K} = \mathbf{x}_0$$

where

$$\mathbf{K} = \mathbf{P}(0|0) + \sum_{i=1}^{\infty} \mathbf{P}(0|i)\mathbf{G}, \quad \mathbf{G} = \sum_{i=0}^{\infty} \mathbf{P}(1+i|1)\mathbf{G}^i$$

$$\mathbf{x}_i = [\mathbf{x}_0 \mathbf{B}_i + \sum_{j=0}^{i-1} \mathbf{x}_j \mathbf{C}_{i+1-j}](\mathbf{I} - \mathbf{C}_1)^{-1}$$

$$\mathbf{B}_i = \sum_{j=i}^{\infty} \mathbf{P}(i|0)\mathbf{G}^{j-i}, \quad \mathbf{C}_i = \sum_{j=i}^{\infty} \mathbf{P}(1+i|1)\mathbf{G}^{j-i}$$

If the number of arriving packets is bounded ($\mathbf{P}(X,Y) = 0$ for $X > M$), then $\Phi(z)$ is a rational function and its expansion into a power series can be obtained using its decomposition into partial fractions (see Appendix 5).

If $M = 1$ (only one packet can arrive in a slot) the system has a matrix-geometric solution [14]

$$\mathbf{x}_i = \mathbf{x}_0 \mathbf{R}^i, \quad i = 1,2,...$$

where $\mathbf{x}_0$ can be found from the system

$$\mathbf{x}_0 = \mathbf{x}_0[\mathbf{P}(0|0) + \mathbf{R}\mathbf{P}(0|1)], \quad \mathbf{x}_0(\mathbf{I} - \mathbf{R})^{-1}\mathbf{1} = 1$$

and $\mathbf{R}$ satisfies the following matrix quadratic equation

$$\mathbf{R} = \mathbf{P}(1|0) + \mathbf{R}\mathbf{P}(1|1) + \mathbf{R}^2\mathbf{P}(1|2)$$

This equation can be solved iteratively [15] as

$$\mathbf{R}_{k+1} = \mathbf{P}(1|0) + \mathbf{R}_k\mathbf{P}(1|1) + \mathbf{R}_k^2\mathbf{P}(1|2)$$

with the initial value $\mathbf{R}_0 = \mathbf{I}$. This algorithm is implemented in the M-file `late`.

Since $\sum_{i=0}^{\infty} \mathbf{x}_i = \mathbf{x}_0(\mathbf{I} - \mathbf{R})^{-1} = \boldsymbol{\pi}$ we have $\mathbf{x}_0 = \pi(\mathbf{I} - \mathbf{R})$.

5.5.2 Packet Delay Probability

Let $p_d(D)$ be the probability that a packet is not successfully delivered within D slots from its arrival. This probability can be evaluated as

$$p_d(D) = \sum_{i=0}^{\infty} \mathbf{x}_i \mathbf{P}_D(g \le i)\mathbf{1} \tag{5.5.1}$$

where

$$\mathbf{P}_D(g \le i) = \begin{cases} \sum_{g=0}^{i} \mathbf{P}_D(g) & \text{for } i \le D \\ [\mathbf{H}_0 + \mathbf{H}_1]^D & \text{otherwise} \end{cases}$$

is the matrix probability that there are no more than i successful transmissions in D slots. The sum in Eq. (5.5.1) can be found using the complex convolution theorem as

$$\mathbf{P}_d(D) = \frac{1}{2\pi j} \oint_C \mathbf{\Phi}(1/z)\boldsymbol{\phi}_g(z) z^{-1}\mathbf{1} \tag{5.5.2}$$

where $\boldsymbol{\phi}_g(z)$ is the matrix generating function of $\mathbf{P}_D(g \le i)$.

To find this generating function, we note that the matrix probability $\mathbf{P}_D(g)$ of exactly g successes out of D attempts follows the matrix binomial distribution and its generating function is given by Eq. (2.3.7) which in our case takes the form $(\mathbf{H}_1 + \mathbf{H}_0 z)^D$. The matrix generating function of $\mathbf{P}_D(g \le i)$ can be found using the summation theorem:

$$\boldsymbol{\phi}_g(z) = \sum_{i=0}^{D} \mathbf{P}_D(g \le i) z^i = (1-z)^{-1}[\mathbf{H}_1 + H_0 z]^D \tag{5.5.3}$$

Thus, Eq. (5.5.2) takes the form

$$\mathbf{P}_d(D) = \frac{1}{2\pi j} \oint_C \mathbf{\Phi}(1/z)(\mathbf{H}_1 + \mathbf{H}_0 z)^D (z - z^2)^{-1}\mathbf{1}$$

For the binary arrival process, we can use the matrix-geometric solution to write

$$p_d(D) = \mathbf{x}_0 \sum_{i=0}^{\infty} \mathbf{R}^i \mathbf{P}_D(g \le i)\mathbf{1} \tag{5.5.4}$$

The sum in this equation can be evaluated using the matrix $\mathbf{R}^m$ spectral decomposition (see Appendix 6)

$$\mathbf{R}^m = \sum_{j=1}^{r} \sum_{i=1^{m_j}} \mathbf{B}_{ij} \binom{m}{i-1} \lambda_j^{m-i+1}$$

where λ_j are the matrix $\mathbf{R}$ eigenvalues. Substituting this expression into Eq. (5.5.2) and recalling Eq. (5.5.3), we obtain

$$p_d(D) = \mathbf{x}_0 \sum_{j=1}^{r} \sum_{i=1}^{m_j} \frac{\mathbf{B}_{ij} \boldsymbol{\phi}_g^{(i-1)}(\lambda_j)\mathbf{1}}{(i-1)!},$$

where $\boldsymbol{\phi}_g^{(i)}(z) = \partial^i \boldsymbol{\phi}_g(z) / \partial z^i$. In particular, if the matrix $\mathbf{R}$ has a simple structure (see Appendix 5) we have

$$p_d(D) = \mathbf{x}_0 \sum_{j=1}^{r} \mathbf{B}_{1j} \boldsymbol{\phi}_g(\lambda_{j)}\mathbf{1}$$

which after substituting $\mathbf{x}_0 = \boldsymbol{\pi}(\mathbf{I} - \mathbf{R})$ takes the form

$$p_d(D)=\boldsymbol{\pi}\sum_{j=1}^{r}\mathbf{B}_{1j}[\mathbf{H}_1+\mathbf{H}_0\lambda_j]^D\mathbf{1}$$

The M-file `late` contains the program that computes and plots this probability for various values of D.

5.6 CONCLUSION

In this chapter, we evaluated some communication protocols' performance characteristics. Because these protocols are described by finite state sequential machines, the communication system operation can be described using an HMM. The standard method of analyzing a system is to find its state stationary distribution and use this distribution to express the system's performance characteristics.

When one is calculating these characteristics, it is important to define the set of all possible outcomes of transmitting a message. For instance, when we dealt with message delay distribution and channel throughput, we considered as a possible outcome the retransmission of a message when the decoder detected errors. We excluded this outcome, though, when we computed the probability of correctly accepting the message.

The purpose of this chapter was to demonstrate matrix probability applications for evaluating some protocol performance without considering all the details of the protocols. For example, we did not address call processing (link establishment) and transmission termination. We also omitted the details of the existing protocol frame structure. The inclusion of details into analysis does not change the general approach but only makes the analysis more complicated. This statement was illustrated by considering a simple stop-and-wait scheme and then a more complex, modified stop-and-wait protocol.

The message contents may also change our approach to analyzing a system. For example, in some control systems it is important that a command be received at least once, whereas in others the duplicated command may negate the previously received command. Therefore, correct reception of a message should be defined accordingly. We showed that a change in definition requires some changes in the matrix of the system's state transition probabilities without changing the general approach.

In the delay-constrained environment, it is important to include the message arrival model an consider the corresponding queueing system. The performance of these systems can be analyzed by considering the combined source-channel IOHMM.

References

1 R. J. Benice and A. H. Frey, Jr., "An analysis of retransmission systems," *IEEE Trans. Commun. Technol.*, **COM-12**, 135-145, Dec. (1964).

2 H. O. Burton and D. D. Sullivan, "Errors and error control,"*Proc. IEEE*, **60**, 1293-1301, Nov. (1972).

3 L.F. Chang, "Throughput estimation of ARQ protocols for a Rayleigh fading channels using fade and interfade duration statistics," *IEEE Trans. Veh. Tech.*, **VT-40**, 223-229, Feb. (1991).

4 Y.J Cho and C.K Un, "Performance analysis of ARQ error controls under Markovian block error pattern," *IEEE Trans. Commun.*, **COM-42**, 2051-2061, Feb.-Apr. (1994).

5 J. P. Gray, "Line control procedures," *Proc. IEEE*, **60**, 1301-1312, Nov. (1972).

6 G. J. Holtzman, *Design and Validation of Computer Protocols,* (Prentice-Hall, Englewood Cliffs, New Jersey, 1991).

7 IBM Corp. *General Information - Binary Synchronous Communications*, (File TP-09, Order GA 27-3004-2, 1970).

8 S. R. Kim and C. K. Un, "Throughput analysis for two ARQ schemes using combined transition matrix," *IEEE Trans. Commun.*, **COM-40**, 1679-1683, Nov. (1992).

9 C.H.C Leung, Y. Kikumoto, and S.A. Sorensen, "The throughput efficiency of the Go-Back-N ARQ scheme under Markov and related error structures," *IEEE Trans. Commun.*, **COM-36**, 231-234, Feb. (1988).

10 S. Lin, D. J. Costello, and M. J. Miller, "Automatic-repeat-request error-control scheme," *IIEEE Comm. Mag.*, **72**, 5-17, Dec. (1984).

11 S. Lin and D. J. Costello, *Error Control Coding: Fundamentals and Applications,* (Prentice-Hall, Englewood Cliffs, New Jersey, 1983).

12 D.L. Lu and J.F. Chang, "Performance of ARQ protocols in nonindependent channel errors," *IEEE Trans. Commun.*, **COM-41**, 721-730, May (1993).

13 J. J. Metzner and K. C. Morgan, "Coded binary decision-feedback communication systems," *IRE Trans. Commun. Syst.*, **CS-8**, 101-113, June (1960).

14 M. F. Neuts, *Matrix-Geometric Solutions in Stochastic Models* (Jons Hopkins, Baltimore, 1981).

15 M. F. Neuts, *Structured Stochastic Matrices of M/G/1 Type and Their Applications* (Marcel Dekker, Inc., New York, 1989).

16 W. W. Peterson and E. J. Weldon, Jr., *Error-Correcting Codes*, (The MIT Press, Cambridge, Massachusetts, 1961).

17 M. Schwartz, *Telecommunication Networks Protocols, Modeling and Analysis*, (Addison-Wesley, Reading, Massachusetts, 1987).

18 W. Stallings, *Data and Computer Communications*, (Macmillan, New York, 1988).

19 A. Tanenbaum, *Computer Networks* (Prentice Hall, Englewood Cliffs, New Jersey, 1992).

20 D. Towsley, "A statistical analysis of ARQ protocols operating in nonindependent error environment," *IEEE Trans. Commun.*, **COM-29**, 971-981, Jul. (1981).

21 E. J. Weldon, "The improved selective-repeat ARQ strategy," *IEEE Trans. Commun.*, **COM-29**, 1514-1519, Oct. (1981).

22 M. Zorzi and R.R. Rao, "Throughput analysis of Go-Back-N ARQ in Markov channels with unreliable feedback,"*Proc. IEEE ICC'95,* 1232-1237, Jun (1995).

6

CONTINUOUS TIME HMM

6.1 CONTINUOUS AND DISCRETE TIME HMM

In the previous chapters, we considered HMMs which are probabilistic functions of discrete Markov chains. In the continuous time HMM, observations depend on states of a continuous time Markov process. The continuous time models assume that during a small time interval the probability of a state change is also small. Therefore, if the time interval is small, the state transition probability matrix **P** is close to a unit matrix **I**. When we estimated the model parameters on the basis of experimental data in Sec. 1.5, we discovered that the diagonal elements of the transition probability matrices are close to one. For that reason, it was convenient to represent the models with the matrices $\mathbf{R} = \mathbf{P} - \mathbf{I}$. The other consideration which must be taken into account is the accuracy of all the computations: we would rather express all the equations containing **P** in terms of **R** to avoid computations with numbers having different precision. These and other considerations suggest that we might make another step and consider a limiting case of an HMM when the time unit becomes infinitesimally small. In the limit we obtain a continuous time HMM.

The continuous time HMMs are very popular in queueing theory where they are called *Markov arrival processes.* [12] Of course, the Markov arrival process properties can be studied independently of its discrete counterpart. [2,3,7,8,9] We, however, use the limiting approach so that we can extend all the previous results instead of rederiving them. The other, more important, reason is that we can analyze the conditions of convergence and see if the continuous time model can be used in a real application. From the theoretical point of view, this approach is closely related to the Volterra's multiplicative calculus. [4]

6.1.1 Markov Arrival Processes

To obtain the continuous time HMM as a limiting case of the discrete HMM, we need to impose certain restrictions on the discrete HMM parameters. We select an observation which plays a special role in the model and denote it 0. In modeling error sources it corresponds to the absence of errors or no customer arrival (in the queueing theory). We assume that the probability of this special observation tends to one when the observation interval tends to zero:

$$\lim_{\Delta t \to 0} Pr(X=0 \mid \Delta t) = 1$$

Thus, we assume that discrete time HMM parameters satisfy the following conditions

$$\mathbf{P}_\Delta(0) = \mathbf{I} + \mathbf{Q}_0 \Delta t + o(\Delta t) \tag{6.1.1}$$

and

$$\mathbf{P}_\Delta(x) = \mathbf{Q}_x \Delta t + o(\Delta t) \quad \text{for} \quad x \neq 0 \tag{6.1.2}$$

The HMM interval distribution has the form

$$Pr(\boldsymbol{x}_1^M, \boldsymbol{m}_1^M, \Delta t) = \mathbf{p}_0 \prod_{i=1}^{M} [\mathbf{P}_\Delta^{m_t}(0) \mathbf{P}_\Delta(x_t)] \mathbf{1} \tag{6.1.3}$$

In the error source modeling we call m_t error gaps, while in the queueing theory they are called *interarrival times*.

To obtain the limiting PDF, we denote the interarrival durations as $t_i = m_i \Delta t$ and write the interarrival duration matrix probability as

$$\mathbf{P}_\Delta^{m_i}(0) \mathbf{P}_\Delta(x_i) = [\mathbf{I} + \mathbf{Q}_0 \Delta t + o(\Delta t)]^{t_i/\Delta t} \mathbf{Q}(x_i) \Delta t \tag{6.1.4}$$

Using the well-known formula [4]

$$\lim_{\Delta t \to 0} [\mathbf{I} + \mathbf{Q}_0 \Delta t + o(\Delta t)]^{t/\Delta t} = \exp(\mathbf{Q}_0 t)$$

we obtain the following matrix PDF for the interarrival times:

$$\mathbf{F}(x,t) = \exp(\mathbf{Q}_0 t) \mathbf{Q}_x \tag{6.1.5}$$

The corresponding interval PDF is given by

$$f(\boldsymbol{x}_1^M, \boldsymbol{t}_1^M) = \mathbf{p}_0 \prod_{i=1}^{M} [\exp(\mathbf{Q}_0 t_i) \mathbf{Q}_{x_i}] \mathbf{1} = \prod_{i=1}^{M} \mathbf{F}(x_i, t_i) \mathbf{1} \tag{6.1.6}$$

This limiting HMM was introduced by Neuts as a *Markov arrival process*, [12] and we use Eq. (6.1.6) as its definition. A special case of binary Markov arrival process, in which $\mathbf{Q}_1 = \mathbf{\Lambda} = diag\{\lambda_i\}$, is called a *Markov-modulated Poisson process* (MMPP). These processes can be considered as a natural limiting case of the model with different state error probabilities considered in Sec. 1.1.2. In the MMPP, Poisson arrival rates depend on the states of a continuous time Markov process.

6.1.1.1 Continuous Time Markov Process

The m-step transition probability matrix for a discrete time Markov chain has the form

$$\mathbf{P}_\Delta^m = [\mathbf{I} + \mathbf{Q} \Delta t + o(\Delta t)]^{t/\Delta t} \quad \text{where} \quad \mathbf{Q} = \sum_x \mathbf{Q}_x$$

which in the limit gives us

$$\mathbf{P}(t) = \exp(\mathbf{Q} t) \tag{6.1.7}$$

The matrix $\mathbf{Q}$ is called the Markov chain infinitesimal generator matrix or simply a generator. By differentiating this equation, we conclude that the Markov process transitional probability matrix satisfies the following differential equation

$$\mathbf{P}' = \mathbf{QP} \quad \text{or} \quad \mathbf{P}' = \mathbf{PQ} \quad \text{with} \quad \mathbf{P}|_{t=0} = \mathbf{I} \tag{6.1.8}$$

If the chain is regular, then as for the discrete time Markov chains

$$\lim_{t\to\infty} \mathbf{P}(t) = \mathbf{1}\boldsymbol{\pi} \tag{6.1.9}$$

The stationary distribution $\boldsymbol{\pi}$ can be obtained from the system

$$\boldsymbol{\pi}\mathbf{P}(t) = \boldsymbol{\pi} \qquad \boldsymbol{\pi}\mathbf{1} = 1$$

which can be reduced to

$$\boldsymbol{\pi}\mathbf{Q} = 0 \qquad \boldsymbol{\pi}\mathbf{1} = 1 \tag{6.1.10}$$

by differentiation. The Markov chain is called stationary if its state initial probability distribution is $\boldsymbol{\pi}$.

We see that the vector $\boldsymbol{\pi}$ is a left eigenvector of the chain generator with the eigenvalue $\lambda = 0$. We can prove similarly that $\mathbf{1}$ is a right eigenvector of $\mathbf{Q}$ (see Problem 1).

6.1.1.2 Phase-Type Distribution

It follows from Eq. (6.1.6) that the interarrival time PDF is given by

$$f(t) = \boldsymbol{\pi}\exp(\mathbf{Q}_0 t)\mathbf{b} \tag{6.1.11}$$

where $\mathbf{b} = (\mathbf{Q} - \mathbf{Q}_0)\mathbf{1} = -\mathbf{Q}_0\mathbf{1}$. This PDF is called a *phase-type* (PH) PDF [11] and can be obtained in the limit from its discrete counterpart given by Eq. (3.2.1). The corresponding matrix PDF has the form

$$\mathbf{F}(t) = \exp(\mathbf{Q}_0 t)(\mathbf{Q} - \mathbf{Q}_0) \tag{6.1.12}$$

By differentiating this equation, we obtain the following differential equation for the matrix PH-type PDF.

$$d\mathbf{F}/dt = \mathbf{F}\mathbf{Q}_0 \qquad \mathbf{F}|_{t=0} = \mathbf{Q} - \mathbf{Q}_0 \tag{6.1.13}$$

Example 6.1.1. As an illustration, consider an MMPP whose generator and the observation rate matrices are given by

$$\mathbf{Q} = \begin{bmatrix} -q_1 & q_1 \\ q_2 & -q_2 \end{bmatrix} \qquad \Lambda = \begin{bmatrix} \lambda_1 & 0 \\ 0 & \lambda_2 \end{bmatrix}$$

The chain stationary distribution is found from Eq. (6.1.10), which has the form

$$-q_1\pi_1 + q_2\pi_2 = 0$$
$$\pi_1 + \pi_2 = 1$$

The solution is

$$\pi = [\pi_1 \;\; \pi_2] = \frac{1}{q_1+q_2}[q_2 \;\; q_1]$$

Let us find the Markov chain transition probability matrix $\exp(\mathbf{Q}t)$. To find this matrix, we use Eq. (A.5.37) from Appendix 5 with $f(z) = \exp(zt)$. We need to decompose into partial fractions the following matrix

$$(z\mathbf{I} - \mathbf{Q})^{-1} = \begin{bmatrix} z+q_1 & -q_1 \\ -q_2 & z+q_2 \end{bmatrix}^{-1} = \frac{1}{\Delta(z)}\mathbf{B}(z)$$

where $\Delta(z) = \det(z\mathbf{I} - \mathbf{Q}) = z(z+q_1+q_2)$ and

$$\mathbf{B}(z) = \begin{bmatrix} z+q_2 & q_1 \\ q_2 & z+q_2 \end{bmatrix}$$

is the adjoint matrix.

We have

$$(z\mathbf{I} - \mathbf{Q})^{-1} = \frac{1}{z}\mathbf{1}\boldsymbol{\pi} + \frac{1}{(z+q_1+q_2)}(\mathbf{I} - \mathbf{1}\boldsymbol{\pi})$$

or

$$(z\mathbf{I} - \mathbf{Q})^{-1} = \frac{1}{(q_1+q_2)z}\begin{bmatrix} q_2 & q_1 \\ q_2 & q_1 \end{bmatrix} + \frac{1}{(q_1+q_2)(z+q_1+q_2)}\begin{bmatrix} q_1 & -q_1 \\ -q_2 & q_2 \end{bmatrix}$$

which gives us

$$\exp(\mathbf{Q}t) = \mathbf{1}\boldsymbol{\pi} + (\mathbf{I} - \mathbf{1}\boldsymbol{\pi})\exp[-(q_1+q_2)t]$$

Note that $\lim_{t\to\infty}\exp(\mathbf{Q}t) = \mathbf{1}\boldsymbol{\pi}$ which agrees with Eq. (6.1.9).

The interarrival time PDF

$$\mathbf{F}(t) = \exp[(\mathbf{Q}-\boldsymbol{\Lambda})t]\boldsymbol{\Lambda}$$

is found similarly. It has the following form

$$\mathbf{F}(t) = \mathbf{A}_1\exp(z_1t) + \mathbf{A}_2\exp(z_2t)$$

where

$$z_{1,2} = 0.5(\lambda_1+\lambda_2+q_1+q_2) \pm 0.5\sqrt{(\lambda_1-\lambda_2+q_1-q_2)^2+4q_1q_2}$$

are the roots of

$$\det(z\mathbf{I}-\mathbf{Q}+\boldsymbol{\Lambda}) = \det\begin{bmatrix} z+\lambda_1+q_1 & -q_1 \\ -q_2 & z+\lambda_2+q_2 \end{bmatrix}$$

Both roots are negative so that $\lim_{t\to\infty}\exp[(\mathbf{Q}-\boldsymbol{\Lambda})t] = 0$.

The interarrival time PDF is a PH type distribution and is given by

$$f(t) = \boldsymbol{\pi}\mathbf{F}(t)\mathbf{1} = a_1\exp(z_1t) + a_2\exp(z_2t)$$

where $a_i = \boldsymbol{\pi}\mathbf{A}_i\mathbf{1}$. We leave as an exercise to the reader the derivation of the equation for calculating these coefficients.

6.1.2 Markov Arrival Processes Discrete Skeleton

Most of the results of Chapter 2 can be easily extended to the continuous time HMMs (CTHMMs) by considering them as a limit of the corresponding discrete time HMMs (DTHMMs). To formalize our approach, define a family of DTHMMs which represent uniform samples $x_k = x(k\Delta t)$ of the CTHMM. The DTHMM x_k is called the skeleton of $x(t)$ with the step size Δt. We develop methods of analyzing CTHMMs as a limiting case of the corresponding methods of their skeletons when $\Delta t \to 0$.

The CTHMMs are very popular in queueing theory. In many cases, they are used as an approximation of discrete time processes. For example, in digital communications, packet arrivals are often modeled by the CTHMM. This can be partially explained by the queueing theory traditions: for the Poisson arrivals and exponential distributions of service times, there is a very nice theory for solving queueing problems analytically, while for the discrete time arrivals, the solutions are more complex. CTHMM is a natural generalization of the Poisson processes, which explains their popularity. However, their use as a model cannot always be justified.

For a discrete time HMM with the matrix probabilities $\mathbf{P}(x)$, we define $\mathbf{Q}_0 \Delta t = \ln \mathbf{P}(0)$ so that $\mathbf{P}(0) = \exp(\mathbf{Q}_0 \Delta t)$ and we can write

$$\mathbf{P}^m(0)\mathbf{P}(x) = \exp(\mathbf{Q}_0 t)\mathbf{Q}_x \Delta t \tag{6.1.14}$$

where $t = m\Delta t$ and $\mathbf{Q}_x \Delta t = \mathbf{P}(x)$. This equation is similar to Eq. (6.1.15) and can be used to approximate the DTHMM with the CTHMM. This approximation is effective if the time unit Δt is small enough so that the diagonal elements of $\mathbf{P}(0)$ are close to one (which means that the probability of two adjacent nonzero observations is negligibly small). The other test is the difference between $\mathbf{Q} = \sum_x \mathbf{Q}_x$ and $\ln \mathbf{P} / \Delta t$.

In general, it is difficult to say if the CTHMM approximation is good. This problem is closely related to the problem of identifiability of CTHMM by its discrete skeleton.

In the next section, using the CTHMM skeleton, we derive a limiting case for the matrix binomial distribution considered in Sec. 2.3.2. A similar approach is used in the elementary probability theory to derive the Poisson distribution as a limit of the binomial distribution. By this analogy, we call the limit of the matrix binomial distribution a matrix Poisson distribution.

6.1.3 Matrix Poisson Distribution

For the Markov arrival process, let us find the probability distribution $p(m,t)$ of the number of arrivals during time interval t. We denote the corresponding matrix probability as $\mathbf{P}(m,t)$.

If we divide the interval into $n = t/\Delta t$, small subintervals of length Δt, $\mathbf{P}(m,t)$ can be approximated by the matrix binomial distribution with the parameters

$$\mathbf{P}_\Delta(0) = \exp(\mathbf{Q}_0 \Delta t) = \mathbf{I} + \mathbf{Q}_0 \Delta t + o(\Delta t)$$

and

$$\mathbf{P}_\Delta(1) = \mathbf{Q}_1 \Delta t + o(\Delta t) \qquad \mathbf{Q}_1 = \mathbf{Q} - \mathbf{Q}_0$$

The matrix binomial distribution generating function is given by equation (2.3.7). We obtain the matrix Poisson distribution generating function as a limit

$$\lim_{\Delta t \to 0} [\mathbf{P}_\Delta(0) + z\mathbf{P}_\Delta(1)]^{t/\Delta t} = \exp[(\mathbf{Q}_0 + z\mathbf{Q}_1)t \tag{6.1.15}$$

This generating function [5,6,7,11] has numerous applications in queueing theory.

The probability $\mathbf{P}(m,t)$ is a coefficient of z^m of this function expansion.

Example 6.1.2. As an illustration, consider the one-dimensional case: $\mathbf{Q}_0 = -\lambda$ and $\mathbf{Q}_1 = \lambda$. We have

$$\exp[(\mathbf{Q}_0 + z\mathbf{Q}_1)t = \exp[\lambda t(z-1)] = \exp(-\lambda t) \sum_{m=0}^{\infty} \frac{(\lambda t)^m}{m!} z^m$$

which gives us the Poisson distribution $p(m,t) = \exp(-\lambda t)(\lambda t)^m / m!$.

The matrix Poisson distribution satisfies equations that are similar to those of the matrix binomial distribution. In particular, since

$$\exp[(\mathbf{Q}_0+z\mathbf{Q}_1)t] = \exp[(\mathbf{Q}_0+z\mathbf{Q}_1)(t-\tau)]\exp[(\mathbf{Q}_0+z\mathbf{Q}_1)\tau]$$

according to the convolution theorem, the matrix Poisson distribution $\mathbf{P}(m,t)$ satisfies the equation

$$\mathbf{P}(m,t) = \sum_{\mu=0}^{m} \mathbf{P}(m-\mu,t-\tau)\mathbf{P}(\mu,\tau) \tag{6.1.16}$$

which is similar to Eq. (2.3.5). We cannot use this equation the same way as Eq. (2.3.5) for calculating matrix probabilities recursively because the right-hand side and the left-hand side of it contain matrix probabilities of events at different time intervals. This equation can be used to calculate the probabilities $\mathbf{P}(m,t)$ if the subintervals $t-\tau$ and τ have the same length which is possible if $\tau = t/2$.

It is possible to have the same time intervals on both sides of the equation in the limit when $\tau \to 0$. For small τ, we have

$$\mathbf{P}(0,\tau) = \mathbf{I} + \mathbf{Q}_0\tau + o(\tau) \qquad \mathbf{P}(1,\tau) = \mathbf{Q}_1\tau + o(\tau)$$

and $\mathbf{P}(\mu,\tau) = o(\tau)$ for $\mu > 1$. Hence,

$$\mathbf{P}(m,t) = \mathbf{P}(m,t-\tau)(\mathbf{I} + \mathbf{Q}_0\tau) + \mathbf{P}(m-1,t-\tau)\mathbf{Q}_1\tau + o(\tau)$$

or

$$[\mathbf{P}(m,t) - \mathbf{P}(m,t-\tau)]/\tau = \mathbf{P}(m,t-\tau)\mathbf{Q}_0 + \mathbf{P}(m-1,t-\tau)\mathbf{Q}_1 + o(\tau)/\tau$$

In the limit, we obtain the following system of recursive differential equations: [7]

$$\begin{aligned} \frac{d}{dt}\mathbf{P}(m,t) &= \mathbf{P}(m,t)\mathbf{Q}_0 + \mathbf{P}(m-1,t)\mathbf{Q}_1 \qquad m \ge 1 \quad t \ge 0 \\ \mathbf{P}(0,t) &= \exp(\mathbf{Q}_0 t) \end{aligned} \tag{6.1.17}$$

The analog of generating function (2.3.8) is the Laplace transform of the previously defined matrix Poisson generating function:

$$\Psi(p;z) = \int_0^\infty \exp[(\mathbf{Q}_0+z\mathbf{Q}_1)t]\exp(-pt)\,dt = (\mathbf{I}p - \mathbf{Q}_0 - z\mathbf{Q}_1)^{-1} \tag{6.1.18}$$

The coefficient at z^m in this function's expansion is a Laplace transform

$$\Psi_m(p) = \int_0^\infty \mathbf{P}(m,t)\exp(-pt)\,dt = [(\mathbf{I}p - \mathbf{Q}_0)^{-1}\mathbf{Q}_1]^m(\mathbf{I}p - \mathbf{Q}_0)^{-1} \tag{6.1.19}$$

which, according to the convolution theorem, leads to the recurrence equation

$$\mathbf{P}(m,t) = \int_0^t \mathbf{P}(k,\tau)\mathbf{Q}_1\mathbf{P}(m-k-1,t-\tau)\,d\tau \tag{6.1.20}$$

If $k = m-1$, this equation becomes

$$\mathbf{P}(m,t) = \int_0^t \mathbf{P}(m-1,\tau)\mathbf{Q}_1\exp[\mathbf{Q}_0(t-\tau)]\,d\tau \tag{6.1.21}$$

Several suggestions for solving this system numerically have been given in Ref. 11.

In many applications of matrix Poisson distribution it is convenient to use its integral representations. Using Cauchy's integral, we obtain the equation

$$\mathbf{P}(m,t) = \frac{1}{2\pi j}\oint_{\gamma} \exp[(\mathbf{Q}_0 + z\mathbf{Q}_1)t]\, z^{-m-1}\, dz$$

This equation is similar to Eq. (2.3.9). The matrix probability $\mathbf{P}(m_1 < m \le m_2)$ that the number of occurrences of the event A_2 is greater than m_1 but not greater than m_2 can be expressed as

$$\mathbf{P}(m_1 < m \le m_2, t) = \frac{1}{2\pi j}\oint_{\gamma} \exp[(\mathbf{Q}_0 + z\mathbf{Q}_1)t]\, \frac{z^{-m_1} - z^{-m_2}}{z^2 - z}\, dz$$

By replacing z with $\exp(j\phi)$, we can convert these integrals into Fourier transforms.

The inversion of the Laplace transform yields

$$\mathbf{P}(m,t) = \frac{1}{2\pi j}\int_{a-j\infty}^{a+j\infty} \Psi_m(p)\exp(pt)\, dp$$

Since $\Psi_m(p)$ is a rational function, this integral can be evaluated using the residue theorem.

For a discrete time batch arrival HMM, with the matrix generating function

$$U(z) = \sum_{i=1}^{\infty} \mathbf{P}_i(x) z^x$$

we have, according to the convolution theorem of Eq. (2.2.11), the following generating function of the total number of arrivals in n steps

$$\mathbf{\Phi}_n(z) = U^n(z)$$

In the continuous time case, we obtain similarly to Eq. (6.1.15)

$$\mathbf{\Phi}(z,t) = \exp[\mathbf{Q}(z)t]$$

where $\mathbf{Q}(z)$ is the corresponding generating function

$$\mathbf{Q}(z) = \sum_{i=1}^{\infty} \mathbf{Q}_x z^x$$

6.2 FITTING CONTINUOUS TIME HMM

We develop the Baum-Welch algorithm (BWA) for estimating parameters of a continuous time HMM as a limiting case of its discrete time counterpart.

6.2.1 Fitting Phase-Type Distribution

6.2.1.1 The Baum-Welch Algorithm

The BWA for approximating an PDF $\phi(t)$ with the continuous time PH PDF $f(t,\tau)$ can be obtained as a limit of its discrete counterpart described in Sec. 3.2.1. Let Δt be a small time interval. Denote $x = t/\Delta t$ as the number of such intervals. Then the continuous time PH distribution can be approximated by the discrete time distribution

$$g_\Delta(x,\tau) = \boldsymbol{\pi}\mathbf{A}_\Delta^{x-1}\mathbf{b}_\Delta \tag{6.2.1}$$

where

$$\mathbf{A}_\Delta = \exp(\mathbf{Q}_0\Delta t) = \mathbf{I} + \mathbf{Q}_0\Delta t + o(\Delta t) \qquad \mathbf{b}_\Delta = \mathbf{b}\Delta t \tag{6.2.2}$$

The BWA developed in Sec. 3.2.1 for approximating $\phi(x\Delta t)\Delta t$ with $g_\Delta(x,\tau)$ is given by Eq. (3.2.10), which can be written as

$$\pi_{i,p+1} = \pi_{i,p}\Pi_i(\tau_p,\Delta t)$$

$$a_{ij,p+1} = a_{ij,p}V_{ij}(\tau_p,\Delta t)\,/\,V_i(\tau_p,\Delta t)$$

$$b_{i,p+1} = b_{i,p}B_i(\tau_p,\Delta t)\,/\,V_i(\tau_p,\Delta t)$$

where

$$\Pi_i(\tau,\Delta t) = \sum_x \rho_\Delta(x,\tau)\beta_i(x,\tau) = \sum_x \rho_\Delta(x,\tau)\boldsymbol{e}_i\mathbf{A}_\Delta^{x-1}\mathbf{b}_\Delta$$

$$V_{ij}(\tau,\Delta t) = \sum_x \rho_\Delta(x,\tau)\sum_{k=1}^{x}\boldsymbol{\pi}\mathbf{A}_\Delta^{k-1}\boldsymbol{e}_i'\boldsymbol{e}_j\mathbf{A}_\Delta^{x-k-1}\mathbf{b}_\Delta$$

$$V_i(\tau,\Delta t) = \sum_x \rho_\Delta(x,\tau)\sum_{k=1}^{x}\boldsymbol{\pi}\mathbf{A}_\Delta^{k-1}\boldsymbol{e}_i'\boldsymbol{e}_i\mathbf{A}_\Delta^{x-k}\mathbf{b}_\Delta$$

$$B_i(\tau,\Delta t) = \sum_x \rho_\Delta(x,\tau)\boldsymbol{\pi}\mathbf{A}_\Delta^{x-1}\boldsymbol{e}_i' \qquad \rho_\Delta(x,\tau) = \phi(x\Delta t)\Delta t\,/\,g_\Delta(x,\tau)$$

Let us find the limits of the previous equations when $\Delta t \to 0$. We have

$$\lim_{\Delta t\to 0} g_\Delta(x,\tau)\,/\,\Delta t = \boldsymbol{\pi}\exp(\mathbf{Q}_0 t)\mathbf{b} = f(t,\tau)$$

Thus, $\lim_{\Delta t\to 0}\rho_\Delta(x,\tau) = \rho(x,\tau) = \phi(t)\,/\,f(t,\tau)$. Also $\lim_{\Delta t\to 0} a_{ij}/\Delta t = q_{ij}^0$ for $i\neq j$ (where q_{ij}^0 is the ij-th element of $\mathbf{Q}_0$).

When $\Delta t \to 0$, the sums with respect to x in the reestimation equations converge to integrals and we obtain in the limit the following BWA

$$\pi_{i,p+1} = \pi_{i,p}\Pi_i(\tau_p) \tag{6.2.3a}$$

$$q_{ij,p+1}^0 = q_{ij,p}^0 V_{ij}(\tau_p)\,/\,V_{ii}(\tau_p) \qquad i \neq j \tag{6.2.3b}$$

$$b_{i,p+1} = b_{i,p}B_i(\tau_p)\,/\,V_{ii}(\tau_p) \tag{6.2.3c}$$

where

$$\Pi_i(\tau) = \int_0^\infty \rho(t,\tau)\,\mathbf{e}_i \exp(\mathbf{Q}_0 t)\,\mathbf{b}\,dt$$

$$V_{ij}(\tau) = \int_0^\infty \rho(t,\tau)\,dt \int_0^t \boldsymbol{\pi} \exp(\mathbf{Q}_0 u)\,\mathbf{e}_i'\,\mathbf{e}_j \exp[\mathbf{Q}_0(t-u)\,\mathbf{b}]\,du$$

$$B_i(\tau) = \int_0^\infty \rho(t,\tau)\,\boldsymbol{\pi} \exp(\mathbf{Q}_0 t)\,\mathbf{e}_i'\,dt$$

These equations do not determine $q^0_{ii,p+1}$ which can be obtained from the normalization conditions $(\mathbf{Q}_0 + \mathbf{b})\mathbf{1} = 0$:

$$q^0_{ii,p+1} = -\sum_{j=1,j\neq i} q^0_{ij,p+1} - b_{i,p+1}$$

Example 6.2.1. As an illustration, let us construct the BWA for approximating some PDF $\phi(x)$ with the mixture of exponential PDFs:

$$f(t,\tau) = \sum_{i=1}^{u} \pi_i \lambda_i e^{-\lambda_i t}$$

In this case, $\mathbf{Q}_0 = -diag\{\lambda_i\}$ and $\mathbf{b} = col\{\lambda_i\}$. Since all the off-diagonal elements are zeroes, we do not use Eq. (6.2.3b). We have

$$\Pi_i(\tau) = \int_0^\infty \rho(t,\tau)\lambda_i e^{-\lambda_i t}\,dt$$

$$B_i(\tau) = \int_0^\infty \rho(t,\tau)\pi_i e^{-\lambda_i t}\,dt$$

and

$$V_{ii}(\tau) = \int_0^\infty \rho(t,\tau)\,dt \int_0^t \pi_i e^{-\lambda_i u} e^{-\lambda_i (t-u)} \lambda_i\,du = \lambda_i \int_0^\infty \rho(t,\tau)\,t e^{-\lambda_i t}\,dt$$

Thus, the BWA can be presented by the following equations

$$\pi_{i,p+1} = \pi_{i,p}\lambda_{i,p} \int_0^\infty \rho(t,\tau) e^{-\lambda_{i,p} t}\,dt$$

$$\lambda_{i,p+1} = \int_0^\infty \rho(t,\tau) e^{-\lambda_{i,p} t}\,dt \Big/ \int_0^\infty t\rho(t,\tau) e^{-\lambda_{i,p} t}\,dt$$

The BWA for fitting a PH distribution to experimental data $\mathbf{t}_1^T$ is a special case of the previous algorithm with

$$\phi(t) = \frac{1}{T}\sum_{m=1}^{T} \delta(t-t_m)$$

where $\delta(x)$ is the Dirac δ-function. The BWA is given by Eq. (6.2.3) with

$$\Pi_i(\tau) = \sum_{m=1}^{T} \mathbf{e}_i \exp(\mathbf{Q}_0 t_m)\,\mathbf{b} \,/\, f(t_m,\tau) \tag{6.2.4a}$$

$$V_{ij}(\tau) = \sum_{m=1}^{T} \int_0^{t_m} \boldsymbol{\pi} \exp(\mathbf{Q}_0 u) \boldsymbol{e}_i' \boldsymbol{e}_j \exp[\mathbf{Q}_0(t_m - u)] \mathbf{b} / f(t_m, \tau) \tag{6.2.4b}$$

$$B_i(\tau) = \sum_{m=1}^{T} \boldsymbol{\pi} \exp(\mathbf{Q}_0 t_m) \boldsymbol{e}_i' / f(t_m, \tau) \tag{6.2.4c}$$

Example 6.2.2. The BWA for fitting a mixture of exponential distributions to experimental data is a special case of the algorithm considered in example 6.2.1 and is represented by the following equations

$$\pi_{i,p+1} = \pi_{i,p} \lambda_{i,p} \frac{1}{T} \sum_{m=1}^{T} e^{-\lambda_{i,p} t_m} / f(t_m, \tau)$$

$$\lambda_{i,p+1} = \frac{\sum_{m=1}^{T} e^{-\lambda_{i,p} t_m} / f(t_m, \tau)}{\sum_{m=1}^{T} t_m e^{-\lambda_{i,p} t_m} / f(t_m, \tau)}$$

6.2.1.2 Numerical Aspects of the BWA

Let us consider various methods for evaluating the PH distribution and the integral

$$v_{ij}(t) = \int_0^t \boldsymbol{\pi} \exp(\mathbf{Q}_0 u) \boldsymbol{e}_i' \boldsymbol{e}_j \exp[\mathbf{Q}_0(t-u)] \mathbf{b} \, du$$

There are many methods for computing the matrix exponential. [10] Some of them are listed in Appendix 5. A particular method selection is application dependent. However, if a matrix does not belong to a special class, the following method is recommended: [10] For small Δt compute $\exp(\mathbf{Q}_0 \Delta t)$ with high accuracy and for $t = m\Delta t$ compute $\exp(\mathbf{Q}_0 t) = [\exp(\mathbf{Q}_0 \Delta t)]^m$. This method is preferred basically because of the availability of the fast matrix exponentiation (see Appendix 5).

Most of the methods of finding the matrix exponential can be modified for computing the integrals $v_{ij}(t)$. The transposed matrix of these integrals can be written as

$$\mathbf{W}(t) = \int_0^t \exp[\mathbf{Q}_0(t-u)] \mathbf{U} \exp(\mathbf{Q}_0 u) \, du \tag{6.2.5}$$

where $\mathbf{U} = \mathbf{b}\boldsymbol{\pi}$.

Since this integral represents a convolution, its Laplace transform can be easily found

$$\boldsymbol{\Phi}(p) = \int_0^\infty \mathbf{W}(t) \exp(-pt) \, dt = (\mathbf{I}p - \mathbf{Q}_0)^{-1} \mathbf{U} (\mathbf{I}p - \mathbf{Q}_0)^{-1}$$

To invert this transform, we compute $(\mathbf{I}p - \mathbf{Q}_0)^{-1} = \mathbf{B}(p)/\Delta(p)$, where $\mathbf{B}(p)$ is the adjoint matrix and $\Delta(p) = \det(\mathbf{I}p - \mathbf{Q}_0)$ is the characteristic polynomial (see Appendix 5). As we see, the Laplace transform is a rational function and it can be decomposed into partial fractions:

$$\mathbf{\Phi}(p) = \sum_{j=1}^{r} \sum_{i=1}^{2v_j} \mathbf{E}_{ij} (p-\lambda_j)^{-i}$$

where λ_j are the matrix $\mathbf{Q}_0$ eigenvalues. The Laplace transform inversion gives us

$$\mathbf{W}(t) = \sum_{j=1}^{r} \sum_{i=1}^{2v_j} \mathbf{E}_{ij}\, t^{i-1} \exp(-\lambda_j t)/(i-1)!$$

This formula is useful in theoretical studies. Its practical value is questionable in the general case because finding eigenvalues and partial fraction decomposition is a computationally intensive process. The commercially available software does not provide the required accuracy in the majority of applications. Alternatively, one can use efficient numerical methods for the Laplace transform inversion. [1]

The other method for computing $\mathbf{W}(t)$ is based on its differential equation. Taking the derivative on both sides of Eq. (6.2.5), we obtain

$$\mathbf{W}'(t) = \mathbf{U}\exp(\mathbf{Q}_0 t) + \mathbf{Q}_0 \int_0^t \exp[\mathbf{Q}_0(t-u)]\,\mathbf{U}\exp(\mathbf{Q}_0 u)\,du$$

which, after substitutions, can be written as

$$\mathbf{W}' = \mathbf{b}\boldsymbol{\pi}\exp(\mathbf{Q}_0 t) + \mathbf{Q}_0\mathbf{W}$$

We see that $\mathbf{W}(t)$ can be found as a solution of a linear nonhomogeneous matrix differential equation (a system of linear nonhomogeneous differential equations for the matrix elements) with the initial condition $\mathbf{W}(0) = 0$.

We can obtain a system of homogeneous matrix differential equations if we add to the previous equation the differential equation for the forward probability vector $\boldsymbol{\alpha}(t,\tau) = \boldsymbol{\pi}\exp(\mathbf{Q}_0 t)$ which is derived by the vector differentiation: $\boldsymbol{\alpha}' = \boldsymbol{\alpha}\mathbf{Q}_0$. The initial condition is $\boldsymbol{\alpha}(0,\tau) = \boldsymbol{\pi}$.

Thus, we can simultaneously obtain $\boldsymbol{\alpha}(t,\tau)$ and $\mathbf{W}(t)$ by solving the following system

$$\mathbf{W}' = \mathbf{b}\boldsymbol{\alpha} + \mathbf{Q}_0\mathbf{W}$$

$$\boldsymbol{\alpha}' = \boldsymbol{\alpha}\mathbf{Q}_0$$

with the initial conditions $\mathbf{W}(0) = 0$ and $\boldsymbol{\alpha}(0,\tau) = \boldsymbol{\pi}$.

Solving this system of differential equations by the standard software packages does not allow us to achieve high accuracy if t is large. [10] However, if we can solve this system with high accuracy for small Δt, we can use the fast algorithms described in Sec. 3.2.1.2 for computing $\boldsymbol{\alpha}(t,\tau)$ and $\mathbf{W}(t,\tau)$. Indeed, it is easy to see from Eq. (6.2.5) that

$$\mathbf{W}(t_1+t_2) = \exp(\mathbf{Q}_0 t_1)\mathbf{W}(t_2) + \mathbf{W}(t_1)\exp(\mathbf{Q}_0 t_2)$$

which leads to the following fast algorithm for computing $\mathbf{W}(t,\tau)$ for $t = \sum_{i=0}^{m} b_i 2^i \Delta t$:

Initialize:

$$\mathbf{S}(y_0) = b_0\mathbf{b}\boldsymbol{\pi} \quad \mathbf{G}_0 = \mathbf{b}\boldsymbol{\pi} \quad \mathbf{V}_0 = \exp(\mathbf{Q}_0\Delta t b_0) \quad \mathbf{R}_0 = \exp(\mathbf{Q}_0\Delta t)$$

For i=1,2,...,m

Begin

$$\mathbf{S}(y_i) = \mathbf{R}^{b_i}\mathbf{S}(y_{i-1}) + b_i\mathbf{G}_i\mathbf{V}_{i-1}$$

$$\mathbf{V}_i = \mathbf{V}_{i-1}\mathbf{R}^{b_i}$$

$$\mathbf{R}_{i+1} = \mathbf{R}_i\mathbf{R}_i$$

$$\mathbf{G}_{i+1} = \mathbf{G}_i\mathbf{R}_i + \mathbf{R}_i\mathbf{G}_i$$

End

In the end, we obtain $\mathbf{W}(t) = \mathbf{S}(y_m)$ and $\exp(\mathbf{Q}_0 t) = \mathbf{V}_m$. We use $\mathbf{W}(t)$ to compute $V_{ij}(\tau_p)$ and $\exp(\mathbf{Q}_0 t)$ to compute $\Pi_i(\tau_p)$ and $B_i(\tau_p)$ in reestimation Eq. (6.2.3).

6.2.2 HMM Approximation

The BWA for approximating a point process with the CTHMM can be obtained as a limiting case of the BWA considered in Sec. 3.3.2. As we did before, we replace m_i with $t_i = m_i\Delta t$. In the limit, when $\Delta t \to 0$, the sums of probabilities in the equations of the discrete BWA of Sec. 3.3.2 are replaced with the integrals of the corresponding PDFs.

We would like to approximate a point process with a Markov arrival process by fitting its PDF $f(\boldsymbol{x}_1^M, \boldsymbol{m}_1^M, \tau)$ of Eq. (6.1.6) to the corresponding PDF $\phi(\boldsymbol{x}_1^M, \boldsymbol{m}_1^M)$ of the point process. Consider first the simplest structure in which $\mathbf{Q}_x$ are independent parameters. For the Markov arrival process skeleton, we can write

$$g_\Delta(\boldsymbol{u}_1^M, \tau) = \prod_{i=1}^{M} [\mathbf{P}_\Delta^{m_i}(0)\mathbf{P}_\Delta(x_i)]\mathbf{1} \tag{6.2.6}$$

where we denoted $u_i = (x_i, m_i)$. This equation is a special case of Eq. (3.3.1) and, therefore, for small Δt the skeleton's parameters can be estimated by the BWA given by Eq. (3.3.10) which, in our case, has the form

$$\pi_{i,p+1} = \pi_{i,p} \sum_{\boldsymbol{u}_1^M} \beta_i(\boldsymbol{u}_1^M, \tau_p)\rho_\Delta(\boldsymbol{u}_1^M, \tau_p) \tag{6.2.7a}$$

$$p_{ij,p+1}(x) = V_{ij}(x, \tau_p) / \sum_{x,j} V_{ij}(x, \tau_p) \tag{6.2.7b}$$

where $\rho_\Delta(\boldsymbol{u}_1^M, \tau) = \phi(\boldsymbol{u}_1^M)\Delta t / g_\Delta(\boldsymbol{u}_1^M, \tau)$

$$V_{ij}(x, \tau) = \sum_{l=1}^{M} \sum_{\boldsymbol{u}_1^M} \boldsymbol{\alpha}(\boldsymbol{u}_1^{l-1}, \tau)\mathbf{W}_{ij}(x, u_l, \tau)\boldsymbol{\beta}(\boldsymbol{u}_{l+1}^M, \tau)\rho_\Delta(\boldsymbol{u}_1^M, \tau) \tag{6.2.7c}$$

$$\mathbf{W}_{ij}(0, u_t, \tau) = \sum_{u=1}^{m_t} \mathbf{P}_\Delta^{u-1}(0, \tau)\mathbf{U}_{ij}p_{ij}(0)\mathbf{P}_\Delta^{m_t - u}(0, \tau)\mathbf{P}_\Delta(x_t) \tag{6.2.7d}$$

$$\mathbf{W}_{ij}(x, u_t, \tau) = \mathbf{P}_\Delta^{m_t}(0)\mathbf{U}_{ij}p_{ij}(x) \quad \text{for} \quad x \neq 0 \tag{6.2.7e}$$

$$\boldsymbol{\alpha}(\boldsymbol{u}_1^l, \tau) = \boldsymbol{\pi}\prod_{i=1}^{l} \mathbf{P}_\Delta^{m_i}(0)\mathbf{P}_\Delta(x_i) \tag{6.2.7f}$$

$$\boldsymbol{\beta}(\boldsymbol{u}_l^M, \tau) = \prod_{i=l}^{M} \mathbf{P}_\Delta^{m_i}(0)\mathbf{P}_\Delta(x_i)\mathbf{1} \tag{6.2.7g}$$

To obtain the BWA for the continuous time model, we need to find the limit of Eq.

(6.2.7) when $\Delta t \to 0$. Denote $v_i = (x_i, t_i)$, where t_i is a continuous variable replacing m_i in the limit. The sums with respect to $\boldsymbol{m}_1^l$ tend to corresponding integrals with respect to $\boldsymbol{t}_1^l$ and we obtain in the limit the following BWA

$$\pi_{i,p+1} = \pi_{i,p} \sum_{\boldsymbol{x}_1^M} \int_{\boldsymbol{t}_1^M} \beta_i(\boldsymbol{v}_1^M, \tau_p) \rho(\boldsymbol{v}_1^M, \tau_p) d\boldsymbol{t}_1^M \tag{6.2.8a}$$

$$q_{ijx,p+1} = V_{ij}(x, \tau_p) / \sum_{x,j} V_{ij}(x, \tau_p) \tag{6.2.8b}$$

where $\rho(\boldsymbol{u}_1^M, \tau) = \phi(\boldsymbol{u}_1^M) / f(\boldsymbol{u}_1^M, \tau)$ $i \neq j$ if $x = 0$,

$$V_{ij}(x, \tau) = \sum_{l=1}^{M} \sum_{\boldsymbol{x}_1^M} \int_{\boldsymbol{t}_1^M} \boldsymbol{\alpha}(\boldsymbol{v}_1^{l-1}, \tau) \mathbf{W}_{ij}(x, v_l, \tau) \boldsymbol{\beta}(\boldsymbol{v}_{l+1}^M, \tau) \rho(\boldsymbol{v}_1^M, \tau) d\boldsymbol{t}_1^M \tag{6.2.8c}$$

$$\mathbf{W}_{ij}(0, v_l, \tau) = \int_0^{t_l} \exp(\mathbf{Q}_{0,p} t) \mathbf{U}_{ij} q_{ij0,p} \exp[\mathbf{Q}_{0,p}(t_l - t)] \, dt \mathbf{Q}_{x_l,p} \tag{6.2.8d}$$

$$\mathbf{W}_{ij}(x, v_l, \tau) = \exp(\mathbf{Q}_{0,p} t_l) \mathbf{U}_{ij} q_{ijx,p} \quad \text{for} \quad x \neq 0 \tag{6.2.8e}$$

$$\boldsymbol{\alpha}(\boldsymbol{v}_1^l, \tau) = \boldsymbol{\pi} \prod_{i=1}^{l} [\exp(\mathbf{Q}_{0,p} t_i) \mathbf{Q}_{x_i,p}] \tag{6.2.8f}$$

$$\boldsymbol{\beta}(\boldsymbol{v}_l^M, \tau) = \prod_{i=l}^{M} [\exp(\mathbf{Q}_{0,p} t_i) \mathbf{Q}_{x_i,p}] \mathbf{1} \tag{6.2.8g}$$

These equations do not determine $q_{ii0,p+1}$ which can be found from the normalization conditions

$$\mathbf{Q}_{0,p} = -\sum_{x \neq 0} \mathbf{Q}_{x,p} \tag{6.2.8h}$$

If a Markov arrival process has a special structure, the equations are modified in the same way as for the DTHMMs. For example, the MMPP is a binary process ($x = 0, 1$) with $\mathbf{Q}_0 = \mathbf{Q} - \boldsymbol{\Lambda}$ and $\mathbf{Q}_1 = \boldsymbol{\Lambda}$, where $\boldsymbol{\Lambda} = diag\{\lambda_i\}$. Since $\boldsymbol{\Lambda}$ is a diagonal matrix, the off-diagonal elements of the generator $\mathbf{Q}$ and $\mathbf{Q}_0$ are the same and Eq. (6.2.8) become

$$\pi_{i,p+1} = \sum_{\boldsymbol{x}_1^M} \int_{\boldsymbol{t}_1^M} \beta_i(\boldsymbol{v}_1^M, \tau_p) \rho(\boldsymbol{v}_1^M, \tau_p) d\boldsymbol{t}_1^M \tag{6.2.9a}$$

$$q_{ij,p+1} = V_{ij}(\tau_p) / \sum_j V_{ij}(\tau_p) \quad i \neq j \tag{6.2.9b}$$

$$\lambda_{j,p+1} = V_j(1, \tau_p) / [V_j(0, \tau_p) + V_j(1, \tau_p)] \tag{6.2.9c}$$

where

$$V_{ij}(\tau_p) = V_{ij}(0, \tau) + V_{ij}(1, \tau) \tag{6.2.9d}$$

and

$$V_j(x, \tau_p) = \sum_i V_{ij}(x, \tau) \tag{6.2.9e}$$

The BWA for approximating a PDF $\phi(t)$ with an PH-type PDF considered in the previous section is special case of this algorithm when $M = 1$, $\boldsymbol{\Lambda} = diag\{b_i\}$.

6.3 CONCLUSION

Continuous time HMMs are popular in queueing theory in which they are called the Markov arrival processes. Major equations for the queueing system performance parameters developed for independent arrivals with the exponential interarrival times can be generalized for the PH interarrival times. [8] Hence, it is important to be able to approximate a general interarrival time PDF with a PH distribution. It is also important to model more complex arrival processes. Markov arrival processes represent a rich family of models allowing us to approximate a wide variety of processes.

In this chapter, we have demonstrated that various methods developed for the discrete time HMMs can be extended to the continuous time HMMs. We obtain the continuous time HMM as a limit of its discrete counterpart. This approach allows us to analyze the conditions for the limit existence which are usually ignored in applications. Sometimes a discrete time process approximation by a continuous time process is not accurate. More than that, during the evaluation of the system performance parameters, the continuous time model is approximated with a discrete time model to be able to use a computer.

Problems

1. Prove that $\mathbf{Q1} = 0$. (Hint: take derivative of $\exp(\mathbf{Q}t)\mathbf{1} = \mathbf{1}$.)
2. Prove that if $\mathbf{Q}$ is a generator of a regular Markov chain, then
   — $\int_0^T \exp(\mathbf{Q}t)\,dt = \mathbf{1}\pi T + [\mathbf{I} - \exp(\mathbf{Q}T)](\mathbf{1}\pi - \mathbf{Q})^{-1}$

   [Hint: use Eq. (A.5.28).]

   — For a binary Markov arrival process, prove that $-\mathbf{Q}_0^{-1}\mathbf{Q}_1$ is a stochastic matrix and its stationary vector is $-\boldsymbol{\pi}\mathbf{Q}_0$ where $\boldsymbol{\pi}$ is a stationary vector of $\mathbf{Q} = \mathbf{Q}_0 + \mathbf{Q}_1$.

   — For the PH-type PDF given by Eq. (6.1.11), prove that $\int_0^\infty f(t)\,dt = 1$.

References

1 J. Abate and W. Whitt, "The Fourier-series method for inverting transforms of probability distributions," *Queueing Syst.*, **10**, 5-88, (1992).

2 S. Asmussen and O. Nerman, "Fitting phase-type distributions via the EM algorithm," *Symposium i Anvendt Statistik*, Copenhagen, 335-346, (1991).

3 W. Fisher and K. Meier-Hellstern, "The Markov-modulated Poisson process (MMPP) cookbook," *Performance Evaluation*, **18**, 149-171, (1992).

4 F. R. Gantmacher, *The Theory of Matrices*, (Chelsea Publishing Co., New York, 1959).

5 H. Heffes and D. M. Lucantoni, "A Markov modulated characterization of packetized voice and data traffic and related statistical multiplexer performance," *IEEE J. on Selected Areas in Comm.*, **SAC-4**(6), 856-868, Sept. 1986.

6 D. M. Lucantoni, *An Algorithmic Analysis of a Communication Model with Retransmissions of Flawed Messages*, (Pitman, London, 1983).

7 D. M. Lucantoni, K. S. Meier-Hellstern, and M. F. Neuts, "A single server queue with server vacations and a class of non-renewal arrival processes," *Adv. in Appl. Probability*, **22**(3), 676-705, Sep. (1990).

8 D. M. Lucantoni, "New results on the single server queue with a batch Markovian arrival processes," *Comm. Statist. Stochastic Models,* **7**, 1-46, (1991).

9 K. S. Meier, "A fitting algorithm for Markov-modulated Poisson processes having two arrival rates," *European J. Oper. Res.,* **29**, 370-377, (1987).

10 C. Moler and C. van Loan, "Nineteen dubious ways to compute the exponential of a matrix," *SIAM Rev.,* **20**, 801-836, (1978).

11 M. F. Neuts, *Matrix-Geometric Solutions in Stochastic Models*, (Johns Hopkins University Press, Baltimore, 1981).

12 M. F. Neuts, "A versatile Markovian point process," *J. Appl. Probab.*, **16**, 764-779 (1979).

7

CONTINUOUS STATE HMM

7.1 CONTINUOUS AND DISCRETE STATE HMM

Continuous state HMMs are very popular in many fields that include control theory, signal processing, speech and image recognition, finance and many others. The application of a general continuous time HMM is much more difficult than of the discrete state HMM because it usually requires to compute multidimensional integrals rather than multiply matrices. There is one class of the continuous state HMM — hidden Gauss-Markov processes (HGMM) for which there are closed form expressions for the multidimensional integrals. This class has been studied intensively by many researchers who developed a rich theory related to the so-called state space linear systems. There are many textbooks and monographs devoted to this theory.

In contrast with the traditional approach, we present this theory as a generalization of the theory of the discrete state HMM developed in the previous chapters. We derive the major facts and algorithms of the theory on the basis of computing certain probability density functions similarly to Refs. 1,4 while the traditional approach is more "statistical" and is based on the estimation theory. [2,17] The probabilistic approach allowed us not only to simplify the presentation, but also to explain the relationship between the discrete state and continuous state HMM theories.

We consider first general continuous state HMM and show that the introduction of the operator probability allows us to generalize the results of the matrix probability theory presented in Chapter 2. This general theory is than applied to the HGMM in which case the calculation of the multidimensional integrals is reduced to calculations of the corresponding moments (means and covariance matrices). As an example of this approach, we show that the forward algorithm can be realized using the Kalman filter. The Baum-Welch algorithm considered in Chapter 3 is realized by computing the corresponding moments.

In the case of non-Gaussian continuous state HMMs, the integrals can be computed numerically by one of the numerical methods which can be interpreted as an approximation of the continuous state HMM by the discrete state HMM which allows us to use all the results of the previous chapters directly. We call this discrete state HMM the skeleton of the continuous state HMM. The main reason for using discrete state HMMs is the simplicity of representing various probabilities using matrix algebra. The HGMM play a similar role allowing us to use matrix algebra for computing the corresponding covariance matrices, but at the cost of narrowing its application only to Gauss-Markov processes. It can also be applied to the non-Gaussian HMMs whose state transition PDFs are Gaussian. The theory can be also extended to the combination discrete-continuous state space.

As we mentioned before, an HMM can be defined as a Markov process † whose states are partially observable. Consider a Markov process whose state at the moment t has the form ‡ $\mathbf{x}_t = (\boldsymbol{s}_t, \mathbf{a}_t)$ where $\boldsymbol{s}_t \in S$ and $\mathbf{a}_t \in A$ represent the unobservable and observable portion of the state, respectively; $\boldsymbol{s}_t$ is called the *hidden* state, $\mathbf{a}_t$ is called the *observation*, and $\mathbf{x}_t$ is called the *complete* state. The process is called an HMM if its state transition PDF has the following special form

$$p_{t_1 t_2}(\mathbf{x}_{t_2} \mid \mathbf{x}_{t_1}) = p_{t_1 t_2}(\mathbf{x}_{t_2} \mid \boldsymbol{s}_{t_1}) = p_{t_1 t_2}(\boldsymbol{s}_{t_2}, \mathbf{a}_{t_2} \mid \boldsymbol{s}_{t_1}), \quad t_1 < t_2 \tag{7.1.1}$$

This means that for any $t_1 < t_2 < ... < t_n$ and any sequences of states

$$p(\mathbf{x}_{t_n} \mid \mathbf{x}_{t_1}, \mathbf{x}_{t_2}, \ldots, \mathbf{x}_{t_{n-1}}) = p(\boldsymbol{s}_{t_n}, \mathbf{a}_{t_n} \mid \boldsymbol{s}_{t_{n-1}})$$

and

$$p(\mathbf{x}_{t_1}, \mathbf{x}_{t_2}, \ldots, \mathbf{x}_{t_n} \mid \boldsymbol{s}_{t_0}) = p(\boldsymbol{s}_{t_1}, \mathbf{a}_{t_1} \mid \boldsymbol{s}_{t_0}) p(\boldsymbol{s}_{t_2}, \mathbf{a}_{t_2} \mid \boldsymbol{s}_{t_1}) \cdots p(\boldsymbol{s}_{t_n}, \mathbf{a}_{t_n} \mid \boldsymbol{s}_{t_{n-1}}) \tag{7.1.2}$$

The hidden portion of the state is a Markov process with the transition PDF

$$p_{t_1 t_2}(\boldsymbol{s}_{t_2} \mid \boldsymbol{s}_{t_1}) = \int_A p_{t_1 t_2}(s_{t_2}, \mathbf{a}_{t_2} \mid \boldsymbol{s}_{t_1}) d\mathbf{a}_{t_2}$$

where $\int_A$ is a shorthand for a multiple integral over the region A of the observation space of vectors $\mathbf{a}_t = (a_{1,t}, a_{2,t}, \ldots, a_{k,t})$ and $d\mathbf{a}_t = da_{1,t} da_{2,t} \cdots da_{k,t}$ denotes the element of volume in this space.

The observation sequence $\mathbf{a}_t$ is not a Markov process in the general case. Note, however, that any Markov process whose states are partially observable can be presented as an HMM by increasing the state space. Indeed, suppose that the state transition PDF does not satisfy Eq. (7.1.1). In this case, we can treat the Markov process $\mathbf{x}_t = (\boldsymbol{s}_t, \mathbf{a}_t)$ as nonobservable and the observation $\mathbf{b}_t$ in state $(\boldsymbol{s}_t, \mathbf{a}_t)$ is $\mathbf{a}_t$. In other words, we consider a new process $\mathbf{y}_t = (\mathbf{x}_t, \mathbf{b}_t)$ whose state transition PDF has the form

$$p_{t_1 t_2}(\mathbf{y}_{t_2} \mid \mathbf{y}_{t_1}) = p_{t_1 t_2}(\mathbf{x}_{t_2}, \mathbf{b}_{t_2} \mid \mathbf{x}_{t_1}) = p_{t_1 t_2}(\boldsymbol{s}_{t_2}, \mathbf{a}_{t_2} \mid \mathbf{x}_{t_1}) \delta(\mathbf{b}_2 - \mathbf{a}_2)$$

In the most popular in applications case, $p_{t_1 t_2}(\mathbf{a}_{t_2} \mid \boldsymbol{s}_{t_1}, \boldsymbol{s}_{t_2}) = p_{t_1 t_2}(\mathbf{a}_{t_2} \mid \boldsymbol{s}_{t_2})$ the observation depends only on the current state. In this case, we can write

$$p_{t_1 t_2}(\boldsymbol{s}_{t_2}, \mathbf{a}_{t_2} \mid \boldsymbol{s}_{t_1}) = p_{t_1 t_2}(\boldsymbol{s}_{t_2} \mid \boldsymbol{s}_{t_1}) p_{t_1 t_2}(\mathbf{a}_{t_2} \mid \boldsymbol{s}_{t_2})$$

For a discrete-time HMM, we assume, without loss of generality, that $t_i = i$ and denote the one-step transition PDF as

$$p_{t-1,t}(\boldsymbol{s}_t, \mathbf{a}_t \mid \boldsymbol{s}_{t-1}) = p_t(\boldsymbol{s}_t, \mathbf{a}_t \mid \boldsymbol{s}_{t-1})$$

We consider only the discrete-time HMMs in the sequel. For the discrete-time HMM, the PDF of the observation sequence $\mathbf{a}_1^T$ can be written as

---

† A brief description of the Markov processes properties used in this chapter is given in Appendix 7.

‡ In this chapter, $\mathbf{a} = (a_1, a_2, \ldots, a_k)$ denotes a column vector while $\mathbf{b} = [a_1, a_2, \ldots, a_k] = \mathbf{a}'$ is a row vector.

$$p(\mathbf{a}_1^T) = \int_{S^{T+1}} p_0(\boldsymbol{s}_0) \prod_{i=1}^{T} p_i(\boldsymbol{s}_i, \mathbf{a}_i \mid \boldsymbol{s}_{i-1}) d\boldsymbol{s}_0^T \tag{7.1.3}$$

The process is called *homogeneous* if its state transition PDF depends only on time difference $t_2 - t_1$. A discrete-time continuous-state homogeneous HMM is described by the state initial PDF $p_0(s_0)$ and the one-step transition PDF which does not depend on t: $p_t(\boldsymbol{s}_t, \mathbf{a}_t \mid \boldsymbol{s}_{t-1}) = p(\boldsymbol{s}_t, \mathbf{a}_t \mid \boldsymbol{s}_{t-1})$.

We can define, similarly to Sec. 1.1, the input-output HMM whose observation consists of a pair $(\mathbf{b}_t, \mathbf{a}_t)$ of an input (source) and corresponding output. The output is always observable, while the input is observable during the system identification (training) and unobservable otherwise.

A discrete space HMM is a special case of the continuous space HMM when the transition PDF has the form

$$p(\boldsymbol{s}_t, \mathbf{a}_t \mid \boldsymbol{s}_{t-1} = i) = \sum_j p_{ij} \delta(\boldsymbol{s}_t - j)$$

On the other hand, the continuous state HMM can be viewed as a limiting case of the discrete state HMM. Thus, a problem related to a continuous state HMM can be solved numerically by approximating it with the discrete state HMM. (This approximation is usually called state quantization.) The goal of this chapter is to generalize the methods developed in the previous chapters for the discrete state HMMs to the continuous state HMMs.

7.2 OPERATOR PROBABILITY

In the previous chapters, we demonstrated the usefulness of the matrix probabilities in applications of discrete state HMMs. In this chapter, we generalize the notion of the matrix probability and define the operator probabilities which are applicable to continuous state HMMs.

Let us start by developing a formula for the PDF of the observation sequence $\mathbf{a}_1^T$. Since the sequence of hidden states is a Markov process, the two-step transition PDF can be evaluated using the Chapman-Kolmogorov equation (see Appendix 7).

$$p(\boldsymbol{s}_2, \mathbf{a}_1, \mathbf{a}_2 \mid \boldsymbol{s}_0) = \int_S p_2(\boldsymbol{s}_2, \mathbf{a}_2 \mid \boldsymbol{s}_1) p_1(\boldsymbol{s}_1, \mathbf{a}_1 \mid \boldsymbol{s}_0) d\boldsymbol{s}_1 \tag{7.2.1}$$

The right hand side of this equation is a generalization of the product of matrices (see Appendix 7) and can be formalized using the integral (Fredholm) operators.

Indeed, let us define the operator $\mathbf{P}_t(\mathbf{a})$ as mapping to itself of the set of functions $\{f(x)\}$ for which the following integral exists

$$\boldsymbol{f}\mathbf{P}_t(\mathbf{a}) = \int_S f(\boldsymbol{s}) p_t(\boldsymbol{x}, \mathbf{a} \mid \boldsymbol{s}) d\boldsymbol{s}$$

This definition generalizes the multiplication of a row-vector by the matrix. The mapping

$$\mathbf{P}_t(\mathbf{a})\boldsymbol{f} = \int_S f(\boldsymbol{x}) p_t(\boldsymbol{x}, \mathbf{a} \mid \boldsymbol{s}) d\boldsymbol{x}$$

generalizes the multiplication of a matrix by a column-vector. The function $p_t(\boldsymbol{s}, \mathbf{a} \mid \boldsymbol{x})$ is called a *kernel* of the operator $\mathbf{P}(\mathbf{a})$. Thus, for the same kernel, we defined two different operators depending on the selection of the variable of integration ($\boldsymbol{s}$ or $\boldsymbol{x}$) which is denoted symbolically by the order of the operands.

Equation (7.2.1) defines the kernel of the the two-step operator $\mathbf{P}(\mathbf{a}_1,\mathbf{a}_2) = \mathbf{P}_1(\mathbf{a}_1)\mathbf{P}_2(\mathbf{a}_2)$ which we call the product of operators. The kernel of the operator

$$\mathbf{P}(\mathbf{a}_1^T) = \mathbf{P}_1(\mathbf{a}_1)\mathbf{P}_2(\mathbf{a}_2)\cdots\mathbf{P}_T(\mathbf{a}_T) = \prod_{i=1}^{T}\mathbf{P}_i(\mathbf{a}_i) \qquad (7.2.2)$$

is the transition PDF

$$p(s_T,\mathbf{a}_1^T \mid s_0) = \int_{S^{T-1}} \prod_{i=1}^{T} p_i(s_i,\mathbf{a}_i \mid s_{i-1})\,ds_1^{T-1}$$

This kernel can be evaluated using the following forward algorithm

$$p(s_k,\mathbf{a}_1^k \mid s_0) = \int_S p_k(s_k,\mathbf{a}_k \mid s_{k-1})p(s_{k-1},\mathbf{a}_1^{k-1} \mid s_0)\,ds_{k-1}, \quad \text{for } k=2,3,\ldots,T$$

which can be written in the operator form as

$$\mathbf{P}(\mathbf{a}_1^k) = \mathbf{P}(\mathbf{a}_1^{k-1})\mathbf{P}(\mathbf{a}_k), \quad k=2,3,\ldots,T$$

It can also be evaluated by the backward algorithm

$$\mathbf{P}(\mathbf{a}_k^T) = \mathbf{P}(\mathbf{a}_k)\mathbf{P}(\mathbf{a}_{k+1}^T), \quad k=T-1,T-2,\ldots,1$$

or by the combination of both to obtain $\mathbf{P}(\mathbf{a}_1^T) = \mathbf{P}(\mathbf{a}_1^r)\mathbf{P}(\mathbf{a}_{r+1}^T)$ where $1<r<T$.

The PDF of the observation sequence $\mathbf{a}_1^T$ can be expressed using these operators as

$$p(\mathbf{a}_1^T) = \mathbf{p}_0\mathbf{P}(\mathbf{a}_1)\mathbf{P}(\mathbf{a}_2)\cdots\mathbf{P}(\mathbf{a}_T)\mathbf{1} = \mathbf{p}_0\prod_{i=1}^{T}\mathbf{P}(\mathbf{a}_i)\mathbf{1} \qquad (7.2.3)$$

where $\mathbf{p}_0 = p_0(s_0)$ is the initial state PDF and $\mathbf{1}$ denotes the function $f(x)\equiv 1$. Note that this equation represents the operator form of Eq. (7.1.3). We can rewrite Eq. (7.2.3) as $p(\mathbf{a}_1^T) = \boldsymbol{\alpha}(\mathbf{a}_1^T)\mathbf{1}$ where

$$\boldsymbol{\alpha}(\mathbf{a}_1^T) = \mathbf{p}_0\mathbf{P}(\mathbf{a}_1^T) = p(s_T,\mathbf{a}_1^T) \qquad (7.2.4)$$

is the forward PDF which can be evaluated using the forward algorithm

$$\boldsymbol{\alpha}(\mathbf{a}_1^0) = \mathbf{p}_0, \quad \boldsymbol{\alpha}(\mathbf{a}_1^k) = \boldsymbol{\alpha}(\mathbf{a}_1^{k-1})\mathbf{P}(\mathbf{a}_k) \qquad (7.2.5a)$$

$k=1,2,\ldots,T$. This operator equation is just a shorthand for the following scalar equation

$$\alpha(s_k,\mathbf{a}_1^k) = \int_S \alpha(s_{k-1},\mathbf{a}_1^{k-1})p(s_k,\mathbf{a}_k \mid s_{k-1})\,ds_{k-1} \qquad (7.2.5b)$$

We can also rewrite Eq. (7.2.3) as $p(\mathbf{a}_1^T) = \mathbf{p}_0\boldsymbol{\beta}(\mathbf{a}_1^T)$ where the backward PDF

$$\boldsymbol{\beta}(\mathbf{a}_1^T) = \mathbf{P}(\mathbf{a}_1^T)\mathbf{1} = p(\mathbf{a}_1^T \mid s_0) \qquad (7.2.6)$$

can be evaluated by the following backward algorithm

$$\boldsymbol{\beta}(\mathbf{a}_T^{T+1}) = \mathbf{1}, \quad \boldsymbol{\beta}(\mathbf{a}_k^T) = \mathbf{P}(\mathbf{a}_k)\boldsymbol{\beta}(\mathbf{a}_{k+1}^T) \qquad (7.2.7a)$$

for $k=m,T-1,\ldots,1$. The scalar form of this equation is

$$\beta(s_{k-1},\mathbf{a}_k^T) = \int_S p(s_k,\mathbf{a}_k \mid s_{k-1})\beta(s_k,\mathbf{a}_{k+1}^T)\,ds_k \qquad (7.2.7b)$$

Combining both algorithms we obtain

$$p(\mathbf{a}_1^T) = \boldsymbol{\alpha}(\mathbf{a}_1^k)\cdot\boldsymbol{\beta}(\mathbf{a}_{k+1}^T) = \int_S \alpha(s_k,\mathbf{a}_1^k)\beta(s_k,\mathbf{a}_{k+1}^T)\,ds_k$$

As we can see, Eq. (7.2.3) formally coincides with Eq. (2.1.2) which formed the basis of the matrix probability theory. Thus, we can extend this theory to the *operator probability theory* by defining the operator probability of an event Ξ corresponding to a subset of the observation sequences $\mathbf{a}_1^T$ as

$$\mathbf{P}(\Xi) = \int_\Xi \mathbf{P}(\mathbf{a}_1^T)\,d\mathbf{a}_1^T$$

The (scalar) probability of this event is given by

$$Pr(\Xi) = \mathbf{p}_0\mathbf{P}(\Xi)\mathbf{1} \qquad (7.2.8)$$

Example 7.1. Let the transition PDF have a form

$$p(s_t,a_t \mid s_{t-1}) = \mathrm{N}(s_t - fs_{t-1},\sigma^2)\mathrm{N}(a_t - hs_t,\rho^2)$$

Find the conditional PDF $p(a_t \mid s_{t-1})$.

This conditional PDF is obtained by the marginalization of $p(s_t,a_t \mid s_{t-1})$:

$$p(a_t \mid s_{t-1}) = \int_{-\infty}^{\infty} \mathrm{N}(s_t - fs_{t-1},\sigma^2)\mathrm{N}(a_t - hs_t,\rho^2)\,ds_t$$

The integral in this equation can be evaluated using Eq. (A.7.22) of Appendix 7 which brings us

$$p(a_t \mid s_{t-1}) = \mathrm{N}(a_t - hfs_{t-1},\sigma_1^2) = \frac{\exp[-(a_t - hfs_{t-1})^2/2\sigma_1^2]}{\sqrt{2\pi}\,\sigma_1}$$

where $\sigma_1^2 = \rho^2 + h^2\sigma^2$.

Equations (7.2.2) and (7.2.8) formally coincide with Eq. (2.1.1) and (2.1.5). Therefore, all the results of the matrix probability theory developed in Chapter 2 can be formally extended to the operator probabilities. However, this formalism is not very useful unless we show how to evaluate effectively these operators probabilities.

7.3 FILTERING, PREDICTION, AND SMOOTHING

In many applications (see, for example, Sec. 4.3) it is necessary to compute the conditional PDF $p(s_t \mid \mathbf{a}_1^T)$ of the state s_t given the observation sequence $\mathbf{a}_1^T$. In some applications (such as decoding and channel equalization) it is sufficient to find the most probable state

$$\hat{s}_t = \underset{s_t}{argmax}\,p(s_t \mid \mathbf{a}_1^T) \qquad (7.3.1)$$

If $T=t$, the process of finding this state is called *filtering*, if $T<t$, it is called *prediction*, and, if $T>t$, it is called *smoothing*.

The filtering PDF is called the normalized forward PDF and can be obtained using the forward algorithm since

$$p(\boldsymbol{s}_t \mid \mathbf{a}_1^t) = \bar{\alpha}(\boldsymbol{s}_t,\mathbf{a}_1^t) = \alpha(\boldsymbol{s}_t,\mathbf{a}_1^T)/p(\mathbf{a}_1^t)$$

where

$$p(\mathbf{a}_1^t) = \boldsymbol{\alpha}(\mathbf{a}_1^t)\mathbf{1} = \int_S \alpha(\boldsymbol{s}_t,\mathbf{a}_1^T)\,d\boldsymbol{s}_t$$

The prediction PDF can be also obtained using the modified forward algorithm which consists of the forward algorithm (7.2.5) for the first T steps ($k \le T$) and continues with $p(\boldsymbol{s}_k,\mathbf{a}_k \mid \boldsymbol{s}_{k-1}) = p(\boldsymbol{s}_k \mid \boldsymbol{s}_{k-1})$ for $k > m$. The prediction PDF can be written in the following operator form

$$p(\boldsymbol{s}_t \mid \mathbf{a}_1^T) = \bar{\boldsymbol{\alpha}}(\mathbf{a}_1^T)\mathbf{P}^{t-T}$$

In the fixed-lag prediction $t - T$ is constant and operator $\mathbf{P}^{t-T}$ can be precomputed.

The smoothing PDF of state $\boldsymbol{s}_t$ is usually denoted as $\gamma_t(\boldsymbol{s}_t,\mathbf{a}_1^T) = p(\boldsymbol{s}_t \mid \mathbf{a}_1^T)$ and can be expressed using the Bayes' theorem as

$$\gamma_t(\boldsymbol{s}_t,\mathbf{a}_1^T) = p(\boldsymbol{s}_t,\mathbf{a}_1^t)p(\mathbf{a}_{t+1}^T \mid \boldsymbol{s}_t)/p(\mathbf{a}_1^T)$$

or, using the notations of the previous section, as

$$\gamma_t(\boldsymbol{s}_t,\mathbf{a}_1^T) = \alpha(\boldsymbol{s}_t,\mathbf{a}_1^t)\beta(\boldsymbol{s}_t,\mathbf{a}_{t+1}^T)/p(\mathbf{a}_1^T) \tag{7.3.2}$$

Thus, the smoothing PDF can be obtained using both the forward and backward algorithms.

The *sliding window* smoothing PDF is defined as $\gamma_{t-k}(\boldsymbol{s}_{t-k}^t,\mathbf{a}_1^T) = p(\boldsymbol{s}_{t-k}^t \mid \mathbf{a}_1^T)$ where k is the size of the window. Using the Markovian property, we can write

$$\gamma_{t-k}(\boldsymbol{s}_{t-k}^t,\mathbf{a}_1^T) = p(\boldsymbol{s}_{t-k},\mathbf{a}_1^{t-k})p(\boldsymbol{s}_{t-k+1}^t,\mathbf{a}_{t-k+1}^t \mid \boldsymbol{s}_{t-k})p(\mathbf{a}_{t+1}^T \mid \boldsymbol{s}_t)/p(\mathbf{a}_1^T)$$

This equation can be rewritten using the forward and backward PDFs as

$$\gamma_{t-k}(\boldsymbol{s}_{t-k}^t,\mathbf{a}_1^T) = \alpha(\boldsymbol{s}_{t-k},\mathbf{a}_1^{t-k})\prod_{i=t-k+1}^{t} p(\boldsymbol{s}_i,\mathbf{a}_i \mid \boldsymbol{s}_{i-1})\beta(\boldsymbol{s}_t,\mathbf{a}_{t+1}^T)/p(\mathbf{a}_1^T)$$

If we replace k in this equation by $k-1$, we obtain

$$\gamma_{t-k+1}(\boldsymbol{s}_{t-k+1}^t,\mathbf{a}_1^T)=\alpha(\boldsymbol{s}_{t-k+1},\mathbf{a}_1^{t-k+1})\prod_{i=t-k+2}^{t} p(\boldsymbol{s}_i,\mathbf{a}_i \mid \boldsymbol{s}_{i-1})\beta(\boldsymbol{s}_t,\mathbf{a}_{t+1}^T)/p(\mathbf{a}_1^T)$$

Comparing these equations, we obtain the following relation between $\gamma_{t-k}(\boldsymbol{s}_{t-k}^t,\mathbf{a}_1^T)$ and $\gamma_{t-k+1}(\boldsymbol{s}_{t-k+1}^t,\mathbf{a}_1^T)$

$$\gamma_{t-k}(\boldsymbol{s}_{t-k}^t,\mathbf{a}_1^T) = \gamma_{t-k+1}(\boldsymbol{s}_{t-k+1}^t,\mathbf{a}_1^T)\frac{\alpha(\boldsymbol{s}_{t-k},\mathbf{a}_1^{t-k})p(\boldsymbol{s}_{t-k+1},\mathbf{a}_{t-k+1} \mid \boldsymbol{s}_{t-k})}{\alpha(\boldsymbol{s}_{t-k+1},\mathbf{a}_1^{t-k+1})}$$

This equation allows us to compute the smoothing PDF for the larger window using that of the smaller one and only the filtering PDFs.

$$\gamma_{t-k}(\boldsymbol{s}_{t-k}^{t-1},\mathbf{a}_1^T) = \gamma_{t-k+1}(\boldsymbol{s}_{t-k+1}^t,\mathbf{a}_1^T)\frac{\alpha(\boldsymbol{s}_{t-k},\mathbf{a}_1^{t-k})p(\boldsymbol{s}_{t-k+1},\mathbf{a}_{t-k+1} \mid \boldsymbol{s}_{t-k})}{\alpha(\boldsymbol{s}_{t-k+1},\mathbf{a}_1^{t-k+1})}$$

In particular, for $k = 1$ we have

$$\gamma_{t-1}(\boldsymbol{s}_{t-1},\boldsymbol{s}_t,\mathbf{a}_1^T) = \gamma_t(\boldsymbol{s}_t,\mathbf{a}_1^T)\alpha(\boldsymbol{s}_{t-1},\mathbf{a}_1^{t-1})p(\boldsymbol{s}_t,\mathbf{a}_t \mid \boldsymbol{s}_{t-1})/\alpha(\boldsymbol{s}_t,\mathbf{a}_1^t) \tag{7.3.3}$$

By integrating both sides of the previous equation with respect to $\boldsymbol{s}_t$ (or, in other words, by marginalizing the conditional PDF) we obtain

$$\gamma_{t-1}(\boldsymbol{s}_{t-1},\mathbf{a}_1^T) = \alpha(\boldsymbol{s}_{t-1},\mathbf{a}_1^{t-1})\int_S \gamma_t(\boldsymbol{s}_t,\mathbf{a}_1^T)p(\boldsymbol{s}_t,\mathbf{a}_t \mid \boldsymbol{s}_{t-1})/\alpha(\boldsymbol{s}_t,\mathbf{a}_1^t)\,d\boldsymbol{s}_t$$

This equation allows us to compute the smoothing PDFs using the following forward-backward algorithm:

Algorithm 7.1.

`Initialize` *(Forward part)*

$$\alpha(\boldsymbol{s}_0,\mathbf{a}_0) = p_0(\boldsymbol{s}_0) \tag{7.3.4a}$$

`For` $t=0,1,...,T-1$
`Begin`
`Compute and save`

$$\alpha(\boldsymbol{s}_{t+1},\mathbf{a}_1^{t+1}) = \int_S p(\boldsymbol{s}_{t+1},\mathbf{a}_{t+1} \mid \boldsymbol{s}_t)\alpha(\boldsymbol{s}_t,\mathbf{a}_1^t)\,d\boldsymbol{s}_t \tag{7.3.4b}$$

`End`
`Initialize:` *(Backward Part)*

$$\gamma_T(\boldsymbol{s}_T,\mathbf{a}_1^T) = \bar{\alpha}(\boldsymbol{s}_T,\mathbf{a}_T) = \alpha(\boldsymbol{s}_T,\mathbf{a}_T)/\int_S \alpha(\boldsymbol{s}_T,\mathbf{a}_T)\,d\boldsymbol{s}_T \tag{7.3.4c}$$

`For` $t=T-1,T-2,...,1$
`Begin`
`Compute`

$$\gamma_{t-1}(\boldsymbol{s}_{t-1},\mathbf{a}_1^T) = \alpha(\boldsymbol{s}_{t-1},\mathbf{a}_1^{t-1})\int_S \gamma_t(\boldsymbol{s}_t,\mathbf{a}_1^T)p(\boldsymbol{s}_t,\mathbf{a}_t \mid \boldsymbol{s}_{t-1})/\alpha(\boldsymbol{s}_t,\mathbf{a}_1^t)\,d\boldsymbol{s}_t \tag{7.3.4d}$$

`End`

The *fixed − interval state* smoothing is defined by the solution of Eq. (7.3.1) for $t=1,2,...,T$ and can be achieved by adding the maximization step to the backward part of the previous algorithm. In the case of finite discrete-state HMM, the forward-backward algorithm required to save all the forward probabilities for $t=1,2,...,T$. However, in the case of continuous-state HMM it is, generally, impossible to save the forward PDFs. Nevertheless, if these PDFs depend on the finite number of parameters, it is possible to save the parameters (which is equivalent to saving the PDFs) and, thus, apply the forward-backward algorithm in the parametric form. We will use this approach for the HMM with the Gaussian transition PDF in which case the recursive algorithm for computing the parameters of the forward PDFs is called the Kalman filter while the backward part is the Rauch-Tung-Striebel (RTS) smoother. [14]

Note that thus defined smoothing might deliver a sequence of states $\hat{\boldsymbol{s}}_1^T$ that is not possible for a particular HMM while the *fixed − interval state − sequence* smoothing which is defined as the solution of the following equation

$$\hat{\boldsymbol{s}}_1^T = \underset{\boldsymbol{s}_1^T}{argmax}\, p(\boldsymbol{s}_1^T \mid \mathbf{a}_1^T) = \underset{\boldsymbol{s}_1^T}{argmax}\, p(\mathbf{x}_1^T) \tag{7.3.5}$$

does not have this problem. Because of the Markovian property, presented by Eq. (7.1.2), this equation can be solved using the Viterbi algorithm (see Sec. 4.3.1).

In the forward path, the algorithm is initialized as $\psi_0(\boldsymbol{s}_0) = p_0(\boldsymbol{s}_0)$ and then computes recursively

$$\begin{aligned} \psi_t(\boldsymbol{s}_t) &= \max_{\boldsymbol{s}_{t-1}} p(\boldsymbol{s}_t,\mathbf{a}_t \mid \boldsymbol{s}_{t-1})\psi_{t-1}(\boldsymbol{s}_{t-1}) \\ \eta_{t-1}(\boldsymbol{s}_t) &= \underset{\boldsymbol{s}_{t-1}}{argmax}\, p(\boldsymbol{s}_t,\mathbf{a}_t \mid \boldsymbol{s}_{t-1})\psi_{t-1}(\boldsymbol{s}_{t-1}) \end{aligned}, \quad t=1,2,...,T \tag{7.3.6}$$

Then it finds $\hat{\boldsymbol{s}}_T = \underset{s}{argmax}\,\psi_T(\boldsymbol{s}_T)$ and performs backtracking:

$$\hat{\boldsymbol{s}}_{t-1} = \eta_{t-1}(\hat{\boldsymbol{s}}_t), \quad t=T,T-1,...,2 \tag{7.3.7}$$

As with the forward-backward algorithm, the main problem with this approach is that it is impossible to remember the functions $\psi_t(s)$ and $\eta_t(s)$ unless they depend on the finite number of parameters. Note, however, that instead of evaluating and remembering $\eta_t(s)$ and backtracking, we can can use the following backward algorithm:

$$\hat{\boldsymbol{s}}_{t-1} = \underset{\boldsymbol{s}_{t-1}}{argmax}\, p(\hat{\boldsymbol{s}}_t,\mathbf{a}_t \mid \boldsymbol{s}_{t-1})\psi_{t-1}(\boldsymbol{s}_{t-1}) \tag{7.3.8}$$

In general, Eq. (7.3.1) and (7.3.5) have different solutions. However, if the PDF $p(\boldsymbol{s}_1^T \mid \mathbf{a}_1^T) = \mathrm{N}(\boldsymbol{s}_1^T - \boldsymbol{\mu}_1^T, \boldsymbol{\Sigma}_T)$ is Gaussian, then the solutions of these equations are identical. Indeed, any Gaussian PDF (see Appendix 7) achieves its global maximum at its mean. Therefore, the solution of Eq. (7.3.5) is $E\{\boldsymbol{s}_1^T \mid \mathbf{a}_1^T\} = \boldsymbol{\mu}_1^T$ while the solution of Eq. (7.3.1) is $E\{\boldsymbol{s}_t \mid \mathbf{a}_1^T\} = \boldsymbol{\mu}_t$. In other words, the solutions of (7.3.1) delivered on the basis of the forward-backward algorithm and solutions of (7.3.5) delivered by the Viterbi algorithm are identical for the HMM with the Gaussian PDFs. These HMMs are considered in the next section.

7.4 LINEAR SYSTEMS

Any discrete time HMM can be represented by a nonlinear system of equations. Indeed, let $\mathbf{x}_t = (\boldsymbol{s}_t, \mathbf{a}_t)$ be an HMM with the transition PDF $p(\mathbf{x}_t \mid \boldsymbol{s}_{t-1})$. Denote as $F_t(\mathbf{X}_t \mid \boldsymbol{s}_{t-1}) = Pr(\mathbf{x}_t < \mathbf{X}_t \mid \boldsymbol{s}_{t-1})$ its cumulative probability distribution function and $F_t^{-1}(\mathbf{u}_t \mid \boldsymbol{s}_{t-1}) = inf\{\mathbf{x}_t : F_t(\mathbf{x}_t \mid \boldsymbol{s}_{t-1}) \ge \mathbf{u}_t\}$ its inverse function. Then it is easy to show that the random variable $\mathbf{x}_t$ can be simulated by using the inverse function method: [15]

$$\mathbf{x}_t = F_t^{-1}(\mathbf{u}_t \mid \boldsymbol{s}_{t-1}) \tag{7.4.1}$$

where $\mathbf{u}_t$ is a random vector whose components are independent random variables and are uniformly distributed in the interval [0,1].

The inverse statement is also true: A stochastic process defined by Eq. (7.4.1), in which $\mathbf{u}_t$ are independent vectors, is an HMM. Thus, we can use stochastic recursive equations to describe HMMs. A special class of these HMMs are represented by the linear stochastic equations:

$$\begin{aligned} \boldsymbol{s}_{t+1} &= \mathbf{F}_t \boldsymbol{s}_t + \mathbf{G}_t \boldsymbol{w}_t \\ \mathbf{a}_{t+1} &= \mathbf{H}_{1,t}\boldsymbol{s}_{t+1} + \mathbf{H}_{2,t}\boldsymbol{s}_t + \mathbf{M}_t \boldsymbol{v}_t \end{aligned}, \quad t=0,1,... \tag{7.4.2}$$

where $(\mathbf{w}_t, \mathbf{v}_t)$ are independent random variables. If we substitute $\boldsymbol{s}_{t+1}$ from the first equation of (7.4.2) into its second equation, then (7.4.2) can be rewritten as

$$\begin{aligned} \boldsymbol{s}_{t+1} &= \mathbf{F}_t \boldsymbol{s}_t + \mathbf{u}_{s,t} \\ \mathbf{a}_{t+1} &= \mathbf{H}_t \boldsymbol{s}_t + \mathbf{u}_{a,t} \end{aligned}$$

where $\mathbf{H}_t = \mathbf{H}_{1,t}\mathbf{F}_t + \mathbf{H}_{2,t}$ and

$$\mathbf{u}_{s,t} = \mathbf{G}_t \mathbf{w}_t, \quad \mathbf{u}_{a,t} = \mathbf{M}_t \boldsymbol{v}_t + \mathbf{H}_{1,t}\mathbf{G}_t \boldsymbol{w}_t$$

In other words, a general linear system describing an HMM can be written in terms of its complete states as

$$\mathbf{x}_{t+1} = \mathbf{C}_t \boldsymbol{s}_t + \mathbf{u}_t, \quad t=0,1,... \tag{7.4.3}$$

where $\mathbf{C}_t' = [\mathbf{F}_t', \mathbf{H}_t']$ and $\mathbf{u}_t$ are independent variables.

If we further assume that $\mathbf{u}_t$ are independent Gaussian variables and the initial state $\boldsymbol{s}_0$ is also Gaussian or fixed, then the complete sequence $\mathbf{x}_t$ is a special case of the Gauss-Markov process (see Appendix 7). We call this process a hidden Gauss-Markov model (HGMM). In order to apply the results developed for the Gauss-Markov processes to HGMMs, we rewrite equation (7.4.3) in the standard form presented by Eq. (A.7.38) of Appendix 7 [†]

$$\mathbf{x}_{t+1} = \mathbf{A}_t \mathbf{x}_t + \mathbf{u}_t \tag{7.4.4}$$

where we define $\mathbf{x}_0' = (\boldsymbol{s}_0, 0)$ and

$$\mathbf{A}_t = \begin{bmatrix} \mathbf{F}_t & 0 \\ \mathbf{H}_t & 0 \end{bmatrix} = \begin{bmatrix} \mathbf{C}_t & 0 \end{bmatrix} \tag{7.4.5}$$

We assume that $\mathbf{u}_t \sim N(0, \mathbf{S}_t)$ is zero-mean Gaussian variable whose variance matrix has the following block form corresponding to the hidden states $\boldsymbol{s}_t$ and observations $\mathbf{a}_t$

$$\mathbf{S}_t = \begin{bmatrix} \mathbf{S}_{ss,t} & \mathbf{S}_{sa,t} \\ \mathbf{S}_{as,t} & \mathbf{S}_{aa,t} \end{bmatrix}$$

As we pointed out in Sec. 7.1, any partially observable Markov process can be presented as an HMM by increasing the number of hidden states. In particular, suppose that the process is described by Eq. (7.4.4) where the matrix $\mathbf{A}_t$ has the form

$$\mathbf{A}_t = \begin{bmatrix} \mathbf{A}_{ss,t} & \mathbf{A}_{sa,t} \\ \mathbf{A}_{as,t} & \mathbf{A}_{aa,t} \end{bmatrix}$$

---

† Typically, the linear system is presented in the state space form as

$$\begin{aligned} \boldsymbol{s}_{t+1} &= \mathbf{F}_t \boldsymbol{s}_t + \mathbf{G}_t \boldsymbol{w}_t \\ \mathbf{a}_t &= \mathbf{H}_t \boldsymbol{s}_t + \mathbf{M}_t \boldsymbol{v}_t \end{aligned}, \quad t=0,1,2,...$$

which assumes that there is an observation $\mathbf{a}_0$ in the original state while Eq. (7.4.2) assumes that there is no observations at $t=0$. As we have seen in the previous chapters, the latter assumption is more convenient allowing us to develop the matrix probability theory. Moreover, this concept leads to a single Eq. (7.4.3) representing HGMM as a special case of Gauss-Markov processes. Nevertheless, it is easy to see that by the simple substitution $\mathbf{b}_{t+1} = \mathbf{a}_t$ we convert the system presented in this footnote into (7.4.2) and we obtain equation (7.4.3) with $\mathbf{x}_t = (\boldsymbol{s}_t, \mathbf{a}_{t-1})$

and the states $\boldsymbol{s}_t$ are not observable. To present this process as an HMM, we treat $\mathbf{x}_t$ as not observable and introduce a new process $\mathbf{y}_t = (\mathbf{x}_t, \mathbf{a}_t)$. This new process is described by the equations

$$\mathbf{y}_{t+1} = \mathbf{A}_t^{(1)} \mathbf{x}_t + \mathbf{u}_t^{(1)} \tag{7.4.6}$$

where

$$\mathbf{A}_t^{(1)} = \begin{bmatrix} \mathbf{A}_{ss,t} & \mathbf{A}_{sa,t} \\ \mathbf{A}_{as,t} & \mathbf{A}_{aa,t} \\ \mathbf{A}_{as,t} & \mathbf{A}_{aa,t} \end{bmatrix}$$

which has the form of Eq. (7.4.3). We call this method of constructing the HMM a *trivial augmentation* of the hidden states.

According to Eq. (7.4.3) the one-step transition PDF has the form

$$p(\mathbf{x}_t \mid \boldsymbol{s}_{t-1}) = \mathrm{N}(\mathbf{x}_t - \mathbf{C}_{t-1}\boldsymbol{s}_{t-1}, \mathbf{S}_{t-1}) \tag{7.4.7}$$

This PDF can be factored as $p(\mathbf{x}_t \mid \boldsymbol{s}_{t-1}) = p(\mathbf{a}_t \mid \boldsymbol{s}_{t-1}) p(\boldsymbol{s}_t \mid \mathbf{a}_t, \boldsymbol{s}_{t-1})$ which (according to Eq. (A.7.26) of Appendix 7) can be written as

$$p(\mathbf{x}_t \mid \boldsymbol{s}_{t-1}) = \mathrm{N}(\mathbf{a}_t - \mathbf{H}_{t-1}\boldsymbol{s}_{t-1}, \mathbf{S}_{aa,t-1}) \mathrm{N}(\boldsymbol{s}_t - \boldsymbol{\mu}_{t,t-1}, \boldsymbol{\Sigma}_{t,t-1}) \tag{7.4.8}$$

where the conditional mean $\boldsymbol{\mu}_{t,t-1} = E\{\boldsymbol{s}_t \mid \mathbf{a}_t, \boldsymbol{s}_{t-1}\}$ has the form

$$\boldsymbol{\mu}_{t,t-1} = \mathbf{F}_{t-1}\boldsymbol{s}_{t-1} + \mathbf{S}_{sa,t-1}\mathbf{S}_{aa,t-1}^{-1}(\mathbf{a}_t - \mathbf{H}_{t-1}\boldsymbol{s}_{t-1}) = \mathbf{D}_{t-1}\boldsymbol{s}_{t-1} + \mathbf{K}_{t-1}\mathbf{a}_t$$

and conditional variance matrix

$$\boldsymbol{\Sigma}_{t,t-1} = \mathbf{S}_{ss,t-1} - \mathbf{S}_{sa,t-1}\mathbf{S}_{aa,t-1}^{-1}\mathbf{S}_{as,t-1} = \mathbf{S}_{ss,t-1} - \mathbf{K}_{t-1}\mathbf{S}_{as,t-1}$$

where

$$\mathbf{D}_{t-1} = \mathbf{F}_{t-1} - \mathbf{K}_{t-1}\mathbf{H}_{t-1}, \quad \mathbf{K}_{t-1} = \mathbf{S}_{sa,t-1}\mathbf{S}_{aa,t-1}^{-1}$$

We assume here and in the sequel that the inverse matrices $\mathbf{S}_{aa,t-1}^{-1}$ exist for $t = 1, 2, \ldots$.

Example 7.2. Let

$$s_{t+1} = f s_t + w_t, \quad a_{t+1} = b s_{t+1} + v_t$$

where $w_t \sim N(0, \sigma^2)$ and $v_t \sim N(0, \rho^2)$. After the substitution of s_{t+1}, the system takes the form

$$s_{t+1} = f s_t + w_t, \quad a_{t+1} = h s_t + u_t \tag{7.4.9}$$

where $h = bf$ and $u_t = v_t + b w_t$. Thus, we can write Eq. (7.4.4) with

$$\mathbf{A} = \begin{bmatrix} f & 0 \\ h & 0 \end{bmatrix}, \quad \mathbf{x}_t = \begin{bmatrix} s_t \\ a_t \end{bmatrix}$$

$$\mathbf{S}_t = \begin{bmatrix} \sigma^2 & b\sigma^2 \\ b\sigma^2 & b^2\sigma^2 + \rho^2 \end{bmatrix} = \sigma^2 \begin{bmatrix} 1 & b \\ b & b^2 + \lambda^2 \end{bmatrix}$$

where we denoted as $\lambda = \rho/\sigma$ to simplify notation.

According to Eq. (7.4.8), the one step transition PDF takes the following form:

$$p(\mathbf{x}_t \mid \boldsymbol{s}_{t-1}) = \mathrm{N}[a_t - h s_{t-1}, \sigma^2(b^2 + \lambda^2)] \mathrm{N}(s_t - \mu_{t,t-1}, \Sigma_{t,t-1}) \tag{7.4.10}$$

where

$$\mu_{t,t-1} = fs_{t-1} + (a_t - hs_{t-1})b^2/(b^2+\lambda^2)$$
$$\Sigma_{t,t-1} = \sigma^2 - b^2\sigma^2/(b^2+\lambda^2) = \lambda^2\sigma^2/(b^2+\lambda^2)$$

We will use these equations frequently in the future examples.

7.4.1 Autocovariance Function

Since the complete sequence $\{\mathbf{x}_t\}$ is a Gauss-Markov process its autocovariance function $\mathbf{R}_x(t,T) = \mathbf{E}\{(\mathbf{x}_t-\boldsymbol{\mu}_t)(\mathbf{x}_T-\boldsymbol{\mu}_T)'\}$ has the form [see Appendix 7, Eq. (A.7.48)]:

$$\mathbf{R}_x(t,T) = \begin{cases} \prod_{i=t-1}^{T} \mathbf{A}_i \boldsymbol{\Sigma}_T & \text{for } t>T \\ \boldsymbol{\Sigma}_T & \text{for } t=T \\ \boldsymbol{\Sigma}_t \prod_{i=t}^{T-1} \mathbf{A}_i' & \text{for } t<T \end{cases} \tag{7.4.11}$$

where

$$\boldsymbol{\mu}_T = \mathbf{A}_{T-1}\boldsymbol{\mu}_{T-1} = \prod_{i=T-1}^{0} \mathbf{A}_i \boldsymbol{\mu}_0 \tag{7.4.12}$$

and $\boldsymbol{\Sigma}_t$ are found using the Riccati equations (see Appendix 7):

$$\boldsymbol{\Sigma}_t = \mathbf{A}_{t-1}\boldsymbol{\Sigma}_{t-1}\mathbf{A}_{t-1}' + \mathbf{S}_{t-1} \tag{7.4.13}$$

the initial conditions are $\boldsymbol{\mu}_0 = (\boldsymbol{s}_0, 0)$, $\boldsymbol{\Sigma}_0 = 0$ if the initial state is fixed. If the initial condition is Gaussian $\boldsymbol{s}_0 \sim N(\mathbf{v}_{s,0}, \boldsymbol{\Sigma}_{ss,0})$, then $\boldsymbol{\mu}_0 = (\mathbf{v}_{s,0}, 0)$ and

$$\boldsymbol{\Sigma}_0 = \begin{bmatrix} \boldsymbol{\Sigma}_{ss,0} & 0 \\ 0 & 0 \end{bmatrix}$$

The autocovariance function of the hidden states $\mathbf{R}_{ss}(t,T)$, of observations $\mathbf{R}_{aa}(t,T)$ and their cross-covariance functions $\mathbf{R}_{sa}(t,T)$ represent the subblocks of

$$\mathbf{R}_x(t,T) = \begin{bmatrix} \mathbf{R}_{ss}(t,T) & \mathbf{R}_{sa}(t,T) \\ \mathbf{R}_{as}(t,T) & \mathbf{R}_{aa}(t,T) \end{bmatrix}, \quad \mathbf{R}_x(t,t) = \boldsymbol{\Sigma}_t = \begin{bmatrix} \boldsymbol{\Sigma}_{ss,t} & \boldsymbol{\Sigma}_{sa,t} \\ \boldsymbol{\Sigma}_{as,t} & \boldsymbol{\Sigma}_{aa,t} \end{bmatrix}$$

For $t \neq T$, these subblocks can be evaluated using the recursive Eq. (A.7.47) of Appendix 7 which in our case can be written in the following block form

$$\mathbf{R}_{ss}(t,T) = \mathbf{F}_{t-1}\mathbf{R}_{ss}(t-1,T) = \mathbf{R}_{ss}(t,T-1)\mathbf{F}_{t-1}'$$
$$\mathbf{R}_{as}(t,T) = \mathbf{R}'_{sa}(T,t) = \mathbf{H}_{t-1}\mathbf{R}_{ss}(t-1,T) = \mathbf{R}_{as}(t,T-1)\mathbf{F}_{t-1}'$$
$$\mathbf{R}_{aa}(t,T) = \mathbf{H}_{t-1}\mathbf{R}_{sa}(t-1,T) = \mathbf{H}_{t-1}\mathbf{R}_{ss}(t-1,T-1)\mathbf{H}_{T-1}' \tag{7.4.14}$$

For $t=T$ we need to use the Riccati Eq. (7.4.13) to evaluate the blocks of $\mathbf{R}_x(t,t) = \boldsymbol{\Sigma}_t$ which can be rewritten as

$$\boldsymbol{\Sigma}_{ss,t} = \mathbf{F}_{t-1}\boldsymbol{\Sigma}_{ss,t-1}\mathbf{F}_{t-1}' + \mathbf{S}_{ss,t-1}$$
$$\boldsymbol{\Sigma}_{sa,t} = \mathbf{F}_{t-1}\boldsymbol{\Sigma}_{ss,t-1}\mathbf{H}_{t-1}' + \mathbf{S}_{sa,t-1}, \quad \boldsymbol{\Sigma}_{aa,t} = \mathbf{H}_{t-1}\boldsymbol{\Sigma}_{ss,t-1}\mathbf{H}_{t-1}' + \mathbf{S}_{aa,t-1}$$

The corresponding means are given by

$$\mathbf{v}_{s,T} = \mathbf{F}_{T-1}\mathbf{v}_{s,T-1} = \prod_{i=T-1}^{0} \mathbf{F}_i \mathbf{v}_{s,0}$$
$$\mathbf{v}_{a,T} = \mathbf{H}_{T-1}\mathbf{v}_{s,T-1} = \mathbf{H}_T \prod_{i=T-2}^{0} \mathbf{F}_i \mathbf{v}_{s,0} \tag{7.4.15}$$

Thus, all the block covariance matrices and means can be expressed through the covariance matrices and mean of the hidden states. It follows from Eq. (7.4.11) that

$$\mathbf{R}_{ss}(t,T) = \begin{cases} \prod_{i=t-1}^{T} \mathbf{F}_i \mathbf{\Sigma}_{ss,T} & \text{for } t>T \\ \mathbf{\Sigma}_{ss,t} & \text{for } t=T \\ \mathbf{\Sigma}_{ss,t} \prod_{i=t}^{T-1} \mathbf{F}_i' & \text{for } t<T \end{cases} \tag{7.4.16}$$

The process is homogeneous if the coefficients $\mathbf{C}_t = \mathbf{C}$ and the variance matrices $\mathbf{S}_t = \mathbf{S}$ are constant. If the process is stationary, then $\boldsymbol{\mu}_t = 0$ and $\mathbf{\Sigma}_t = \mathbf{\Sigma}_\infty$ satisfies the Lyapunov equation $\mathbf{\Sigma}_\infty = \mathbf{A}\mathbf{\Sigma}_\infty\mathbf{A}' + \mathbf{S}$ (Eq. (A.7.46) of Appendix 7). This equation is equivalent to the following equations for the matrix blocks

$$\mathbf{\Sigma}_{ss,\infty} = \mathbf{F}\mathbf{\Sigma}_{ss,\infty}\mathbf{F}' + \mathbf{S}_{ss} \tag{7.4.17}$$

$$\mathbf{\Sigma}_{aa,\infty} = \mathbf{H}\mathbf{\Sigma}_{ss,\infty}\mathbf{H}' + \mathbf{S}_{aa}, \quad \mathbf{\Sigma}_{sa,\infty} = \mathbf{F}\mathbf{\Sigma}_{ss,\infty}\mathbf{H}' + \mathbf{S}_{sa} \tag{7.4.18}$$

Thus, we need to solve only the Lyapunov Eq. (7.4.17) corresponding to the hidden states the rest of the subblocks are expressed through this solution.

For the stationary process, $\mathbf{R}_x(t,t+\tau) = \mathbf{R}_x(\tau)$ does not depend on t and Eq. (7.4.11) takes the form

$$\mathbf{R}_x(\tau) = \begin{cases} \mathbf{A}^{-\tau}\mathbf{\Sigma}_\infty & \text{for } \tau \le 0 \\ \mathbf{\Sigma}_\infty(\mathbf{A}')^{\tau} & \text{for } \tau \ge 0 \end{cases}$$

The subblocks of this function can be expressed using the solution of Eq. (7.4.17). The solution is unique if all the eigenvalues of $\mathbf{F}$ lie inside the unit circle $|z|=1$. For example, the autocovariance function of the observation sequence can be written as

$$\mathbf{R}_{aa}(0) = \mathbf{\Sigma}_{aa,\infty}, \quad \mathbf{R}_{aa}(\tau) = \begin{cases} \mathbf{H}\mathbf{F}^{-\tau-1}\mathbf{\Sigma}_{sa,\infty} & \text{for } \tau<0 \\ \mathbf{\Sigma}_{as,\infty}(\mathbf{F}')^{\tau-1}\mathbf{H}' & \text{for } \tau>0 \end{cases} \tag{7.4.19}$$

The power spectral density (PSD) of the complete sequence $\{\mathbf{x}_t\}$ has the form (Eq. (A.7.50) of Appendix 7):

$$\Phi_x(z) = \sum_{\tau=-\infty}^{\infty} \mathbf{R}_x(\tau) z^{\tau} = (\mathbf{I}-\mathbf{A}z^{-1})^{-1}\mathbf{S}(\mathbf{I}-\mathbf{A}'z)^{-1}$$

The PSD of the hidden states and observations are the corresponding subblocks of this matrix. For example, substituting **A** from Eq. (7.4.5) and multiplying the block matrices in the previous equation we obtain the observation sequence PSD as

$$\Phi_{aa}(z) = \mathbf{M}(z)\mathbf{S}_{ss}\mathbf{M}'(z^{-1})+\mathbf{M}(z)\mathbf{S}_{sa}+\mathbf{S}_{as}\mathbf{M}'(z^{-1})+\mathbf{S}_{aa} \qquad (7.4.20)$$

where $\mathbf{M}(z) = \mathbf{H}z^{-1}(\mathbf{I}-\mathbf{F}z^{-1})^{-1}$. $\Sigma_{ss,\infty}$ is the coefficient of z^0 in the expansion of $\Phi_{ss}(z) = (\mathbf{I}-\mathbf{F}z^{-1})^{-1}\mathbf{S}_{ss}(\mathbf{I}-\mathbf{F}'z)^{-1}$ into the power (Laurent's) series and, therefore, can be found using the residue theorem (instead of solving the Lyapunov Eq. (7.4.17)):

$$\Sigma_{ss,\infty} = \frac{1}{2\pi j}\int_{|z|=1}\frac{\Phi_{ss}(z)}{z}dz = \sum Res_{\lambda_i}\frac{\Phi_{ss}(z)}{z}$$

where the sum of residues is taken over all the eigenvalues of **F**.

Example 7.3. For the system in example 7.2, the Lyapunov equation is $\Sigma_{ss,\infty} = f^2\Sigma_{ss,\infty}+\sigma^2$. Its solution has the form $\Sigma_{ss,\infty} = \sigma^2(1-f^2)^{-1}$.
The same result can be obtained using the residue theorem. Indeed,

$$\Phi_{ss}(z) = (1-fz^{-1})^{-1}\sigma^2(1-fz)^{-1} = \frac{\sigma^2 z}{(z-f)(1-fz)}$$

and

$$\Sigma_{ss,\infty} = Res_f\frac{\sigma^2}{(z-f)(1-zf)} = \frac{\sigma^2}{1-f^2}$$

The autocorrelation function of the observation sequence is given by Eq. (7.4.19) which in our case has the form

$$R_{aa}(0) = h^2\Sigma_{ss,\infty}+\sigma^2(b^2+\lambda^2), \quad R_{aa}(\tau) = hf^{|\tau|-1}(fh\Sigma_{ss,\infty}+b\sigma^2)$$

7.4.2 Observation Sequence PDF

Since all the considered sequences are Gaussian, their multidimensional PDFs can be expressed using the corresponding means and autocovariance functions presented in the previous section. For example, $\mathbf{x}_0^T \sim \mathrm{N}$ where $\boldsymbol{\mu}_0^T$ is obtained using Eq. (7.4.12) and the blocks $\mathbf{R}_x(i,j)$ of the covariance matrix

$$D_T = \left[\mathbf{R}_x(i,j)\right]_{T+1,T+1} \qquad (7.4.21)$$

are obtained using Eq. (7.4.11) for $i,j=0,1,...,T$. If the original state s_0 is fixed, we obtain the conditional PDF $p(\mathbf{x}_1^T \mid s_0)i \sim \mathrm{N}(\mathbf{x}_1^T-\boldsymbol{\mu}_1^T,\mathbf{D}_{1,T})$ by removing from $\mathbf{D}_T$ the submatrices $\mathbf{R}(0,j)$ and $\mathbf{R}(i,0)$).

Example 7.4. For the system of Example 7.2, find the PDF $p(\mathbf{x}_1^2 \mid s_0)$.
In this case, the original state s_0 is fixed. Using the matrices **A** and **S** defined in Example 7.2 and Eq. (7.4.12), we can write

$$\boldsymbol{\mu}_1^2 = [fs_0 \;\; hs_0 \;\; f^2s_0 \;\; hfs_0]'$$

Equation (7.4.21) becomes

$$D_2 = \begin{bmatrix} \mathbf{S} & \mathbf{SA}' \\ \mathbf{AS} & \mathbf{ASA}' + \mathbf{S} \end{bmatrix} = \sigma^2 \begin{bmatrix} 1 & b & f & h \\ b & b^2+\lambda^2 & bf & bh \\ f & bf & f^2+1 & fh+b \\ h & bh & fh+b & b^2+h^2+\lambda^2 \end{bmatrix}$$

Thus, $p(\mathbf{x}_1^2 \mid s_0) = N(\mathbf{x}_1^2 - \boldsymbol{\mu}_1^2, \mathbf{D}_2)$.

The PDF of the observation sequence $\mathbf{a}_1^T$ is Gaussian with the mean $\mathbf{v}_{a,1}^T$ and covariance matrix given by

$$\mathbf{v}_{a,1}^T = \begin{bmatrix} \mathbf{v}_{a,i} \end{bmatrix}_{1,T}, \quad \boldsymbol{\Lambda}_T = \begin{bmatrix} \mathbf{R}_{aa}(i,j) \end{bmatrix}_{T,T}$$

where $\mathbf{v}_{a,t}$ is given by Eq. (7.4.15) and the blocks $\mathbf{R}_{aa}(i,j)$ of $\boldsymbol{\Lambda}_T$ are evaluated using Eq. (7.4.14).

Note that if the initial state s_0 is fixed, the observation sequence PDF represent the backward PDF $\boldsymbol{\beta}(\mathbf{a}_1^T) = p(\mathbf{a}_1^T \mid s_0)$ defined by equation (7.2.6).

The forward PDF $\boldsymbol{\alpha}(\mathbf{a}_1^T) = p(\mathbf{a}_1^T, s_T)$ defined by Eq. (7.2.4) is a Gaussian PDF whose mean $\boldsymbol{\xi}_T = E\{\mathbf{a}_1^T, s_T\}$ is obtained using Eq. (7.4.15) as

$$\boldsymbol{\xi}_T = (\mathbf{v}_{a,1}^T, \mathbf{v}_{s,T}) \tag{7.4.22}$$

and the covariance matrix Ξ_T is obtained by adding to $\boldsymbol{\Lambda}_T$ the rows and columns corresponding to s_T:

$$\Xi_T = \begin{bmatrix} \boldsymbol{\Lambda}_T & \mathbf{R}(\mathbf{a}_1^T, s_T) \\ \mathbf{R}(s_T, \mathbf{a}_1^T) & \mathbf{R}_{ss}(T,T) \end{bmatrix} \tag{7.4.23}$$

where

$$\mathbf{R}(s_T, \mathbf{a}_1^T) = [\mathbf{R}_{sa}(T,1) \;\; \mathbf{R}_{sa}(T,2) \;\; \ldots \;\; \mathbf{R}_{sa}(T,T)]$$

and $\mathbf{R}(\mathbf{a}_1^T, s_T) = \mathbf{R}'(s_T, \mathbf{a}_1^T)$.

Example 7.5. Let us find the vector of means and the covariance matrix for the PDF $p(s_2, \mathbf{a}_1, \mathbf{a}_2 \mid s_0)$ using the results of Example 7.4.

According to Eq. (7.4.22) and (7.4.23), we have

$$\boldsymbol{\xi}_2 = \begin{bmatrix} hs_0 \\ hfs_0 \\ f^2 s_0 \end{bmatrix}, \quad \Xi_2 = \sigma^2 \begin{bmatrix} b^2+\lambda^2 & bh & bf \\ bh & b^2+h^2+\lambda^2 & fh+b \\ bf & fh+b & f^2+1 \end{bmatrix}$$

To obtain the mean $\mathbf{v}_{a,1}^2$ and covariance matrix $\boldsymbol{\Lambda}_2$ corresponding to $p(\mathbf{a}_1, \mathbf{a}_2 \mid s_0) \sim N(\mathbf{v}_{a,1}^2, \boldsymbol{\Lambda}_2)$ we need to remove the entries corresponding to s_2 (that is, the last row and column):

$$\mathbf{v}_{a,1}^2 = \begin{bmatrix} hs_0 \\ hfs_0 \end{bmatrix}, \quad \boldsymbol{\Lambda}_2 = \sigma^2 \begin{bmatrix} b^2+\lambda^2 & bh \\ bh & b^2+h^2+\lambda^2 \end{bmatrix}$$

7.4.3 Kalman Filter

Equations (7.4.22) and (7.4.23) allow us to evaluate the forward PDFs $\alpha(s_t, \mathbf{a}_1^t)$. The sizes of the covariance matrices in these equations might be large which makes the direct inversion of Ξ_T a difficult numerical problem. It turns out that the forward algorithm in Eq. (7.2.5) allows us to evaluate this PDF recursively. In general, there is no closed-form expression for the integrals required by this algorithm. However, for the Gauss-Markov processes, we can find the closed-form expression.

Suppose that the initial state PDF is $\mathrm{N}(\boldsymbol{s}_0-\boldsymbol{\mu}_0,\boldsymbol{\Sigma}_0)$. According to Eq. (7.4.7) the transition PDF has the form

$$p(\mathbf{a}_1,\boldsymbol{s}_1 \mid \boldsymbol{s}_0) = \mathrm{N}(\mathbf{x}_1-\mathbf{C}_0\boldsymbol{s}_0,\mathbf{S}_0)$$

Then Eq. (7.1.5) gives us

$$\alpha(\mathbf{a}_1,\boldsymbol{s}_1) = \int_S \mathrm{N}(\mathbf{x}_1-\mathbf{C}_0\boldsymbol{s}_0,\mathbf{S}_0)\mathrm{N}(\boldsymbol{s}_0-\boldsymbol{\mu}_0,\boldsymbol{\Sigma}_0)d\boldsymbol{s}_0 \tag{7.4.24}$$

which, according to Eq. (A.7.22) of Appendix 7, is equal to

$$\alpha(\mathbf{a}_1,\boldsymbol{s}_1) = \mathrm{N}(\mathbf{x}_1-\mathbf{C}_0\boldsymbol{\mu}_0,\mathbf{P}_0) \tag{7.4.25}$$

where $\mathbf{P}_0 = \mathbf{S}_0+\mathbf{C}_0\boldsymbol{\Sigma}_0\mathbf{C}_0'$. The next step of the forward algorithm (7.1.5) gives us

$$\alpha(\mathbf{a}_1,\mathbf{a}_2,\boldsymbol{s}_2) = \int_S \mathrm{N}(\mathbf{x}_2-\mathbf{C}_1\boldsymbol{s}_1,\mathbf{S}_1)\mathrm{N}(\mathbf{x}_1-\mathbf{C}_0\boldsymbol{\mu}_0,\mathbf{P}_0)d\boldsymbol{s}_1 \tag{7.4.26}$$

To evaluate this integral using Eq. (A.7.22) of Appendix 7, we factor the right hand side of the previous equation using the conditional PDF [see Eq. (A.7.26) of Appendix 7]

$$\mathrm{N}(\mathbf{x}_1-\mathbf{C}_0\boldsymbol{\mu}_0,\mathbf{P}_0) = p(\mathbf{a}_1)\mathrm{N}(\boldsymbol{s}_1-\boldsymbol{s}_{1|1},\mathbf{Q}_{1|1})$$

where

$$p(\mathbf{a}_1) = \mathrm{N}(\mathbf{a}_1-\mathbf{H}_0\boldsymbol{\mu}_0,\mathbf{P}_{aa,0}) \tag{7.4.27}$$

is the PDF of the observation $\mathbf{a}_1$

$$\boldsymbol{s}_{1|1} = \mathbf{E}(\boldsymbol{s}_1 \mid \mathbf{a}_1) = \mathbf{F}_0\boldsymbol{\mu}_0+\mathbf{P}_{sa,0}\mathbf{P}_{aa,0}^{-1}(\mathbf{a}_1-\mathbf{H}_0\boldsymbol{\mu}_0)$$

$$\mathbf{Q}_{1|1} = \mathbf{P}_{ss,0}-\mathbf{P}_{sa,0}\mathbf{P}_{aa,0}^{-1}\mathbf{P}_{as,0}$$

In these equations, we use the standard notations $\boldsymbol{s}_{i|j} = \mathbf{E}\{\boldsymbol{s}_i \mid \mathbf{a}_1^j\}$ and for the state $\boldsymbol{s}_i$ conditional mean and $\mathbf{Q}_{i|j} = \mathbf{E}\{(\boldsymbol{s}_i-\boldsymbol{s}_{i|j})(\boldsymbol{s}_i-\boldsymbol{s}_{i|j})' \mid \mathbf{a}_1^j\}$ for the variance matrix given $\mathbf{a}_1^j$.

Now we can rewrite Eq. (7.4.26) as

$$\alpha(\mathbf{a}_1,\mathbf{a}_2,\boldsymbol{s}_2) = p(\mathbf{a}_1)\int_S \mathrm{N}(\mathbf{x}_2-\mathbf{C}_1\boldsymbol{s}_1,\mathbf{S}_1)\mathrm{N}(\boldsymbol{s}_1-\boldsymbol{s}_{1|1},\mathbf{Q}_{1|1})d\boldsymbol{s}_1 \tag{7.4.28}$$

Using this representation and formula (A.7.22) of Appendix 7, we compute the integral and obtain

$$\alpha(\mathbf{a}_1,\mathbf{a}_2,\boldsymbol{s}_2) = \mathrm{N}(\mathbf{x}_2-\mathbf{C}_1\boldsymbol{s}_{1|1},\mathbf{P}_1)p(\mathbf{a}_1) \tag{7.4.29}$$

where $\mathbf{P}_1 = \mathbf{C}_1\mathbf{Q}_{1|1}\mathbf{C}_1' + \mathbf{S}_1$,

Applying Eq. (A.7.26) of Appendix 7, we can present Eq. (7.4.29) as

$$\alpha(\mathbf{a}_1^2,\boldsymbol{s}_2) = \mathrm{N}(\boldsymbol{s}_2-\boldsymbol{s}_{2|2},\mathbf{Q}_{2|2})p(\mathbf{a}_1^2)$$

where

$$p(\mathbf{a}_1^2) = \mathrm{N}(\mathbf{a}_2-\mathbf{H}_1\boldsymbol{s}_{1|1},\mathbf{P}_{aa,1})p(\mathbf{a}_1) \tag{7.4.30}$$

$$\boldsymbol{s}_{2|2} = \mathbf{F}_1 \boldsymbol{s}_{1|1} + \mathbf{P}_{sa,1}\mathbf{P}_{aa,1}^{-1}(\mathbf{a}_2 - \mathbf{H}_1 \boldsymbol{s}_{1|1}) = \mathbf{F}_1 \boldsymbol{s}_{1|1} + \mathbf{K}_1(\mathbf{a}_2 - \mathbf{H}_1 \boldsymbol{s}_{1|1})$$

$$\mathbf{Q}_{2|2} = \mathbf{P}_{ss,1} - \mathbf{P}_{sa,1}\mathbf{P}_{aa,1}^{-1}\mathbf{P}_{as,1} = \mathbf{P}_{ss,1} - \mathbf{K}_1 \mathbf{P}_{as,1}$$

We can assume now that the forward PDF has the form

$$\alpha(\mathbf{a}_1^t, \boldsymbol{s}_t) = \mathsf{N}(\boldsymbol{s}_t - \boldsymbol{s}_{t|t}, \mathbf{Q}_{t|t})p(\mathbf{a}_1^t)$$

and prove it using the method of mathematical induction by computing $\alpha(\mathbf{a}_1^{t+1}, \boldsymbol{s}_{t+1})$.

According to forward algorithm (7.1.5) we have

$$\alpha(\mathbf{a}_1^{t+1}, \boldsymbol{s}_{t+1}) = p(\mathbf{a}_t)\int_S \mathsf{N}(\mathbf{x}_{t+1} - \mathbf{C}_t \boldsymbol{s}_t, \mathbf{S}_t)\mathsf{N}(\boldsymbol{s}_t - \boldsymbol{s}_{t|t}, \mathbf{Q}_{t|t})\, d\boldsymbol{s}_t$$

Similarly to computing the integral in (7.4.28) we obtain

$$\begin{aligned}\alpha(\mathbf{a}_1^{t+1}, \boldsymbol{s}_{t+1}) &= \mathsf{N}(\mathbf{x}_{t+1} - \mathbf{C}_t \boldsymbol{s}_{t|t}, \mathbf{P}_t)p(\mathbf{a}_1^t) \\ &= \mathsf{N}(\boldsymbol{s}_{t+1} - \boldsymbol{s}_{t+1|t+1}, \mathbf{Q}_{t+1|t+1})p(\mathbf{a}_1^{t+1})\end{aligned} \tag{7.4.31}$$

where

$$\boldsymbol{s}_{t+1|t+1} = \mathbf{F}_t \boldsymbol{s}_{t|t} + \mathbf{K}_t(\mathbf{a}_{t+1} - \mathbf{H}_t \boldsymbol{s}_{t|t}), \quad \mathbf{K}_t = \mathbf{P}_{sa,t}\mathbf{P}_{aa,t}^{-1}$$

$$\mathbf{Q}_{t+1|t+1} = \mathbf{P}_{ss,t} - \mathbf{K}_t \mathbf{P}_{as,t}$$

$$\mathbf{P}_t = \mathbf{C}_t \mathbf{Q}_{t|t} \mathbf{C}_t' + \mathbf{S}_t$$

$$p(\mathbf{a}_1^{t+1}) = p(\mathbf{a}_1^t)\mathsf{N}(\mathbf{a}_{t+1} - \mathbf{H}_t \boldsymbol{s}_{t|t}, \mathbf{Q}_{t|t})$$

This algorithm for computing the forward PDFs for the HGMM is called the Kalman filter [10] and the coefficient $\mathbf{K}_t = \mathbf{P}_{sa,t}\mathbf{P}_{aa,t}^{-1}$ is called the Kalman gain. The forward algorithm can be summarized as follows:

Algorithm 7.2.

`Initialize:`

$$\boldsymbol{s}_{0|0} = \boldsymbol{\mu}_0, \quad \mathbf{Q}_{0|0} = \boldsymbol{\Sigma}_0, \quad p(\mathbf{a}_0) = 1 \tag{7.4.32a}$$

`For` $t = 0, 1, ..., T-1$

`Begin`

$$\mathbf{P}_t = \mathbf{C}_t \mathbf{Q}_{t|t} \mathbf{C}_t' + \mathbf{S}_t \tag{7.4.32b}$$

$$p(\mathbf{a}_1^{t+1}) = p(\mathbf{a}_1^t)\mathsf{N}(\mathbf{a}_{t+1} - \mathbf{H}_t \boldsymbol{s}_{t|t}, \mathbf{P}_{aa,t}) \tag{7.4.32c}$$

$$\mathbf{K}_t = \mathbf{P}_{sa,t}\mathbf{P}_{aa,t}^{-1}, \quad \boldsymbol{s}_{t+1|t+1} = \mathbf{F}_t \boldsymbol{s}_{t|t} + \mathbf{K}_t(\mathbf{a}_{t+1} - \mathbf{H}_t \boldsymbol{s}_{t|t}) \tag{7.4.32d}$$

$$\mathbf{Q}_{t+1|t+1} = \mathbf{P}_{ss,t} - \mathbf{K}_t \mathbf{P}_{as,t} \tag{7.4.32e}$$

`End`

A special structure of Eq. (7.4.32c) is usually explored to express matrix blocks of $\mathbf{P}_i$ more explicitly. Indeed, since $\mathbf{C}_i' = [\mathbf{F}_i' \quad \mathbf{H}_i']$, we can express (7.4.32c) as

$$\begin{aligned} \mathbf{P}_{ss,i} &= \mathbf{F}_i \mathbf{Q}_{i|i} \mathbf{F}_i' + \mathbf{S}_{ss,i} & \mathbf{P}_{sa,i} &= \mathbf{F}_i \mathbf{Q}_{i|i} \mathbf{H}_i' + \mathbf{S}_{sa,i} \\ \mathbf{P}_{as,i} &= \mathbf{H}_i \mathbf{Q}_{i|i} \mathbf{F}_i' + \mathbf{S}_{as,i} & \mathbf{P}_{aa,i} &= \mathbf{H}_i \mathbf{Q}_{i|i} \mathbf{H}_i' + \mathbf{S}_{aa,i} \end{aligned} \tag{7.4.33}$$

Since the matrix $\mathbf{P}_i$ is symmetric, $\mathbf{P}_{sa,i} = \mathbf{P}_{as,i}'$ so we need to compute only $\mathbf{P}_{sa,i}$ (or $\mathbf{P}_{as,i}$). Substituting Eq. (7.4.33) into Algorithm 7.2 we can rewrite it more explicitly as

Algorithm 7.3.

`Initialize:`

$$s_{0\,0} = \boldsymbol{\mu}_0, \quad \mathbf{Q}_{0|0} = \boldsymbol{\Sigma}_0, \quad p(\mathbf{a}_0) = 1 \tag{7.4.34a}$$

`For` $t=0,1,\ldots,T-1$

`Begin`

$$p(\mathbf{a}_1^{t+1}) = p(\mathbf{a}_1^t)\mathrm{N}(\mathbf{a}_{t+1} - \mathbf{H}_t s_{t|t}, \mathbf{H}_t \mathbf{Q}_{t|t} \mathbf{H}_t' + \mathbf{S}_{aa,t}) \tag{7.4.34b}$$

$$\mathbf{K}_t = (\mathbf{F}_t \mathbf{Q}_{t|t} \mathbf{H}_t' + \mathbf{S}_{sa,t})(\mathbf{H}_t \mathbf{Q}_{t|t} \mathbf{H}_t' + \mathbf{S}_{aa,t})^{-1} \tag{7.4.34c}$$

$$s_{t+1|t+1} = \mathbf{F}_{t-1} s_{t|t} + \mathbf{K}_t(\mathbf{a}_{t+1} - \mathbf{H}_t s_{t|t}) \tag{7.4.34d}$$

$$\mathbf{Q}_{t+1|t+1} = \mathbf{F}_t \mathbf{Q}_{t|t} \mathbf{F}_t' + \mathbf{S}_{ss,t} - \mathbf{K}_t(\mathbf{F}_t \mathbf{Q}_{t|t} \mathbf{H}_t' + \mathbf{S}_{sa,t})' \tag{7.4.34e}$$

`End`

The forward PDF is obtained at the end of the algorithm using the following formula

$$\alpha(\mathbf{a}_1^T, s_T) = p(\mathbf{a}_1^T, s_T) = \mathrm{N}(s_T - s_{T|T}, \mathbf{Q}_{T|T}) p(\mathbf{a}_1^T) \tag{7.4.35}$$

The normalized forward density

$$\bar{\alpha}(s_T, \mathbf{a}_1^T) = p(s_T \mid \mathbf{a}_1^T) \tag{7.4.36}$$

according to Eq. (7.4.35) can be written as

$$\bar{\alpha}(s_T, \mathbf{a}_1^T) = \mathrm{N}(s_T - s_{T|T}, \mathbf{Q}_{T|T}) \tag{7.4.37}$$

Thus, the Kalman filter can be viewed as a parametric realization of the forward algorithm (7.2.5) for the HGMM. The observation sequence PDF is given by Eq. (7.4.34b).

If the initial state PDF is a mixture of Gaussian PDFs

$$p_0(s_0) = \sum_{i=1}^{M} w_i \mathrm{N}(s_0 - \boldsymbol{\mu}_{i,0}, \boldsymbol{\Sigma}_{i,0}) \tag{7.4.38}$$

then it is easy to see that

$$\alpha(\mathbf{a}_1^T, s_T) = \sum_{i=1}^{M} w_i \alpha_i(\mathbf{a}_1^T, s_T)$$

where $\alpha_i(s_T, \mathbf{a}_1^T)$ is obtained by the Kalman filter with the initial conditions $\boldsymbol{\mu}_{i,0}$ and $\boldsymbol{\Sigma}_{i,0}$.

It follows from Eq. (7.4.32) that the covariance matrices $\mathbf{P}_t$ and $\mathbf{Q}_t$ of the Kalman filter do not depend on the observation sequence. Therefore, if we need to compute the observation PDF values for different observation sequences, these covariance matrices need to be computed and saved only one time.

On the other hand, it also follows from Eq. (7.4.32) that the state conditional means $s_{t|t}$

are linear functions of $\boldsymbol{s}_{t-1|t-1}$ and, therefore, of the initial mean $\boldsymbol{\mu}_0$. To find the coefficients of these linear functions, we need to modify the Kalman filter in the following way.

Equation (7.4.32d) can be written as

$$\boldsymbol{s}_{t|t} = \mathbf{D}_{t-1}\boldsymbol{s}_{t-1|t-1} + \mathbf{d}_{t-1}$$

where

$$\mathbf{D}_{t-1} = \mathbf{F}_{t-1} - \mathbf{K}_{t-1}\mathbf{H}_{t-1}$$

$$\mathbf{d}_{t-1} = \mathbf{K}_{t-1}\mathbf{a}_t$$

Starting with $\boldsymbol{s}_{0|0} = \boldsymbol{\mu}_0$ and $\mathbf{P}_0 = \mathbf{S}_0$, these equations allow us to find $\boldsymbol{s}_{T|T}$ as a function of $\boldsymbol{\mu}_0$ recursively:

$$\boldsymbol{s}_{1|1} = \mathbf{D}_0\boldsymbol{\mu}_0 + \mathbf{d}_0$$

$$\boldsymbol{s}_{2|2} = \mathbf{D}_1\mathbf{D}_0\boldsymbol{\mu}_0 + \mathbf{D}_1\mathbf{d}_0 + \mathbf{d}_1$$

and so on. Finally, we obtain

$$\boldsymbol{s}_{T|T} = \mathbf{U}_T\boldsymbol{\mu}_0 + \mathbf{u}_T \tag{7.4.39}$$

where

$$\mathbf{U}_T = \mathbf{D}_{T-1}\mathbf{D}_{T-2}\cdots\mathbf{D}_0$$

and

$$\mathbf{u}_T = \mathbf{d}_{T-1} + \mathbf{D}_{T-1}\mathbf{d}_{T-2} + \ldots + \mathbf{D}_{T-1}\mathbf{D}_{T-2}\cdots\mathbf{D}_1\mathbf{d}_0$$

Substituting Eq. (7.4.39) into (7.4.35) we obtain the explicit expression for the normalized forward PDF as a function of $\boldsymbol{\mu}_0$:

$$\bar{\alpha}(\mathbf{a}_1^T, \boldsymbol{s}_T; \boldsymbol{\mu}_0) = \mathrm{N}(\boldsymbol{s}_T - \mathbf{U}_T\boldsymbol{\mu}_0 - \mathbf{u}_T, \mathbf{Q}_{T|T}) \tag{7.4.40}$$

It follows from the previous equations that $\mathbf{U}_{t-1}$ and $\mathbf{u}_{t-1}$ can be found recursively as

$$\mathbf{U}_t = \mathbf{D}_{t-1}\mathbf{U}_{t-1} \tag{7.4.41}$$

$$\mathbf{u}_t = \mathbf{d}_{t-1} + \mathbf{D}_{t-1}\mathbf{u}_{t-1}$$

Thus, the coefficients of the linear function can be obtained by the Kalman filter in which Eq. (7.4.32d) is replaced by Eq. (7.4.41):

Algorithm 7.4.

```
Initialize:
```

$$\mathbf{u}_0 = 0, \quad \mathbf{U}_0 = \mathbf{I}, \quad \mathbf{Q}_{0|0} = \boldsymbol{\Sigma}_0 \tag{7.4.42a}$$

For $t = 0, 1, \ldots, T-1$
Begin

$$\mathbf{P}_t = \mathbf{C}_t\mathbf{Q}_{t|t}\mathbf{C}_t' + \mathbf{S}_t \qquad \mathbf{K}_t = \mathbf{P}_{sa,t}\mathbf{P}_{aa,t}^{-1} \tag{7.4.42b}$$

$$\mathbf{D}_t = \mathbf{F}_t - \mathbf{K}_t\mathbf{H}_t \tag{7.4.42c}$$

$$\mathbf{U}_{t+1} = \mathbf{D}_t \mathbf{U}_t, \quad \mathbf{u}_{t+1} = \mathbf{D}_t \mathbf{u}_t + \mathbf{K}_t \mathbf{a}_{t+1} \tag{7.4.42d}$$

$$\mathbf{Q}_{t+1|t+1} = \mathbf{P}_{ss,t} - \mathbf{K}_t \mathbf{P}_{as,t} \tag{7.4.42e}$$

```
End
```

We obtained the Kalman filter using the forward algorithm in its integral form. The same results can be obtained algebraically by Gaussian elimination of the rows and columns corresponding to $\mathbf{a}_1, \mathbf{a}_2, \ldots, \mathbf{a}_T$ from the matrix $[\Xi_T \mid (\mathbf{a}_1^T, s_T)' - \xi_T]$ where ξ_T and Ξ_T are given by Eq. (7.4.22) and (7.4.23), respectively (see Appendix 7). Thus, the k-th step of the Kalman filter can be viewed as a Gaussian elimination of the $\mathbf{a}_k$ from the matrix that has a special structure (see Problem 1).

Example 7.6. Let us illustrate the Kalman filter application to the system of Example 7.2 assuming that the initial state is fixed.
The first step of the Kalman filter gives us

$$s_{1|1} = fs_0 + \frac{b}{b^2+\lambda^2}(a_1 - hs_0)$$

$$Q_{1|1} = \sigma^2 - \frac{b^2\sigma^2}{b^2+\lambda^2} = \frac{\lambda^2\sigma^2}{b^2+\lambda^2}$$

$$\mathbf{P}_1 = \mathbf{CC}'\frac{\lambda^2\sigma^2}{b^2+\lambda^2} + \mathbf{S} = Q_{1|1}\begin{bmatrix} f^2\lambda^2 + b^2 + \lambda^2 & fh\lambda^2 + b^3 + b\lambda^2 \\ fh\lambda^2 + b^3 + b\lambda^2 & h^2\lambda^2 + (b^2+\lambda^2)^2 \end{bmatrix}$$

Next step of the Kalman filter gives us

$$s_{2|2} = fs_{1|1} + \frac{fh\lambda^2 + b^3 + b\lambda^2}{h^2\lambda^2 + (b^2+\lambda^2)^2}(a_2 - hs_{1|1})$$

$$Q_{2|2} = \frac{\sigma^2}{b^2+\lambda^2}\left[f^2\lambda^2 + b^2 + \lambda^2 - \frac{(fh\lambda^2 + b^3 + b\lambda^2)^2}{h^2\lambda^2 + (b^2+\lambda^2)^2}\right]$$

$$\mathbf{P}_2 = \mathbf{CC}' Q_{2|2} + \mathbf{S}$$

Alternatively, we can find the mean and the variance of s_2 by Gaussian elimination of the columns and rows corresponding to a_1 and a_2 from the following matrix (see Example 7.5)

$$[\Xi_2 \mid (\mathbf{a}_1^2, s_2)' - \xi_2] = \begin{bmatrix} \sigma^2(b^2+\lambda^2) & \sigma^2(bh) & \sigma^2(bf) & a_1 - hs_0 \\ \sigma^2(bh) & \sigma^2(b^2+h^2+\lambda^2) & \sigma^2(fh+b) & a_2 - hfs_0 \\ \sigma^2(bf) & \sigma^2(fh+b) & \sigma^2(f^2+1) & s_2 - f^2 s_0 \end{bmatrix} \tag{7.4.43}$$

In the first step of the forward algorithm, we apply the Gaussian elimination of the rows and columns corresponding to a_1 (that is we eliminate the first row and then remove the first column):

$$\begin{bmatrix} \sigma^2(b^2+h^2+\lambda^2 - \frac{b^2h^2}{b^2+\lambda^2}) & \sigma^2(fh + b - \frac{b^2 fh}{b^2+\lambda^2}) & a_2 - hs_{1|1} \\ \sigma^2(fh + b - \frac{b^2 fh}{b^2+\lambda^2}) & \sigma^2(f^2 + 1 - \frac{b^2 f^2}{b^2+\lambda^2}) & s_2 - fs_{1|1} \end{bmatrix}$$

Note that it follows from this expression that the conditional mean $s_{2|1} = \mathbf{E}(s_2 \mid a_1, s_0) = fs_{1|1}$.

In the next step of the forward algorithm we eliminate the rows and the columns corresponding to a_2 (that is the second row an column of the previous matrix) to obtain the desired result

$$\begin{bmatrix} Q_{2|2} & s_2 - s_{2|2} \end{bmatrix}$$

In the backward algorithm, we eliminate the columns and rows from the matrix (7.4.43) corresponding to a_2 first and then the columns and rows corresponding to a_1.

7.4.4 The Innovation Representation

Equation (7.4.32d) has the form

$$s_{t+1|t+1} = \mathbf{F}_k s_{t|t} + \mathbf{K}_t(\mathbf{a}_{t+1} - \mathbf{H}_t s_{t|t})$$

It computes the state conditional (filtered) expectations given the observation sequence. If the observations are random variables, $s_{t|t}$ become random and can be treated as new states. If we denote $\mathbf{e}_t = \mathbf{a}_{t+1} - \mathbf{H}_t s_{t|t}$, then we can write

$$\begin{aligned} s_{t+1|t+1} &= \mathbf{F}_t s_{t|t} + \mathbf{K}_t \mathbf{e}_t \\ \mathbf{a}_{t+1} &= \mathbf{H}_t s_{t|t} + \mathbf{e}_t \end{aligned} \tag{7.4.44}$$

This equation represents an HGMM. To prove it we need to show that $\mathbf{e}_t$ are independent zero mean Gaussian variables.

It follows from Eq. (7.4.27) that $\mathbf{e}_0 \sim \mathsf{N}(\mathbf{e}_0, \mathbf{P}_{aa,0})$. Equation (7.4.30) tells us that $p(\mathbf{e}_0, \mathbf{e}_1) = \mathsf{N}(\mathbf{e}_0, \mathbf{P}_{aa,0})\mathsf{N}(\mathbf{e}_1, \mathbf{P}_{aa,1})$ which means that $\mathbf{e}_0$ and $\mathbf{e}_1$ are independent zero mean Gaussian variables with the variance matrices $\mathbf{P}_{aa,0}$ and $\mathbf{P}_{aa,1}$, respectively. Equation (7.4.32c) allows us to conclude that $\mathbf{e}_t$ are indeed independent zero mean Gaussian variables whose variance matrices are $\mathbf{P}_{aa,t}$. And, therefore, Eq. (7.4.44) represents an HGMM. Equation (7.4.44) is called the *innovation form* of the HGMM since $\mathbf{e}_t$ is indeed the innovation: $\mathbf{e}_t = \mathbf{a}_{t+1} - E\{\mathbf{a}_{t+1} \mid \mathbf{a}_1^t\}$ (see Appendix 7). If $\mathbf{e}_t$ is random, we will denote the model's states as s_t.

The complete state $\mathbf{x}_t = (s_t, \mathbf{a}_t)$ is described by Eq. (7.4.3) which we rewrite here for the convenience of the reader

$$\mathbf{x}_{t+1} = \mathbf{C}_t s_t + \mathbf{u}_t, \quad t = 0, 1, \ldots \tag{7.4.45a}$$

where $\mathbf{C}_t' = [\mathbf{F}_t', \mathbf{H}_t']$ and, according to Eq. (7.4.44),

$$s_0 = \boldsymbol{\mu}_0, \quad \mathbf{u}_t \sim \mathsf{N}(0, \mathbf{S}_t^{(I)}), \quad \mathbf{S}_t^{(I)} = \begin{bmatrix} \mathbf{K}_t \mathbf{P}_{aa,t} \mathbf{K}_t' & \mathbf{K}_t \mathbf{P}_{aa,t} \\ \mathbf{P}_{aa,t} \mathbf{K}_t' & \mathbf{P}_{aa,t} \end{bmatrix} \tag{7.4.45b}$$

The original HGMM and its innovation form are equivalent if they produce the same observation sequence PDF $p(\mathbf{a}_1^T)$, $\forall\ \mathbf{a}_1^T$ and T. This means that the variance matrices $\mathbf{P}_{aa,t}$, the matrices $\mathbf{K}_t$, and initial conditions should be the same for both models. To satisfy this requirement, we use Eq. (7.4.33) and (7.4.34) to have

$$\begin{aligned} \mathbf{P}_{aa,t} &= \mathbf{H}_t \mathbf{Q}_{t|t} \mathbf{H}_t' + \mathbf{S}_{aa,t} \\ \mathbf{K}_t &= (\mathbf{F}_{t-1} \mathbf{Q}_{t-1|t-1} \mathbf{H}_{t-1}' + \mathbf{S}_{sa,t-1})(\mathbf{H}_{t-1} \mathbf{Q}_{t-1|t-1} \mathbf{H}_{t-1}' + \mathbf{S}_{aa,t-1})^{-1} \\ \mathbf{Q}_{t|t} &= \mathbf{F}_{t-1} \mathbf{Q}_{t-1|t-1} \mathbf{F}_{t-1}' + \mathbf{S}_{ss,t-1} - \mathbf{K}_t (\mathbf{F}_{t-1} \mathbf{Q}_{t-1|t-1} \mathbf{H}_{t-1}' + \mathbf{S}_{sa,t-1})' \end{aligned} \tag{7.4.46}$$

Thus, the innovation model is equivalent to the original model if the innovation variance matrix is obtained using Eq. (7.4.46) with the initial condition $\mathbf{Q}_{0|0} = \boldsymbol{\Sigma}_0$ and $s_{0|0} = \boldsymbol{\mu}_0$. Obviously, the Kalman filter for both model is the same and is given by Eq. (7.4.34).

Suppose now that the original HGMM is defined by Eq. (7.4.45) is homogeneous (time invariant): $\mathbf{C}_t' = \mathbf{C} = [\mathbf{F}' \quad \mathbf{H}']$ and $\mathbf{S}_t = \mathbf{S}$ for $t = 0, 1, \ldots$. However, in this case according to Eq. (7.4.46), the corresponding innovation model is not homogeneous since $\mathbf{S}_t^{(l)}$ is time varying. Under certain conditions [2] the innovation model is asymptotically homogeneous in which case there exists a a nonnegative definite matrix $\mathbf{Q} = \lim_{t\to\infty}\mathbf{Q}_{t|t}$. Both models are homogeneous if the process existed for infinite time and reached the steady state. In this case, according to Eq. (7.4.46), $\mathbf{Q}$ is a nonnegative definite solution of the following equation

$$\mathbf{Q} = \mathbf{FQF}' + \mathbf{S}_{ss} - (\mathbf{FQH}' + \mathbf{S}_{sa})(\mathbf{HQH}' + \mathbf{S}_{aa})^{-1}(\mathbf{FQH}' + \mathbf{S}_{sa})'$$

which is called the *steady state Riccati equation*. Using this solution, we find

$$\mathbf{P}_{aa} = \mathbf{HQH}' + \mathbf{S}_{aa}$$
$$\mathbf{K} = (\mathbf{FQH}' + \mathbf{S}_{sa})(\mathbf{HQH}' + \mathbf{S}_{aa})^{-1}$$

Suppose that the original model (7.4.3) has the form of (7.4.45) where $\mathbf{P}_{aa,t}$ and $\mathbf{K}_t$ are the model independent parameters. It is not difficult to prove that this model is also its innovation form if we assume that the variance matrices $\mathbf{P}_{aa,t}$ are invertible and the initial state $s_0 = s_{0|0}$ is fixed. To prove it, we need to derive the Kalman filter for the model structure given by equations (7.4.45) using Algorithm 7.2 and verify that the model parameter $\mathbf{K}_t$ is indeed the Kalman gain. We leave the proof as an exercise for the reader.

If the HGMM is presented by equations of the form (7.4.45), it is called the *direct parametrization* of the HGMM. Usually, this structure has less parameters and is simpler than the general HGMM structure. If the original state is fixed, the Kalman filter is obtained by the trivial substitution, otherwise the model parameter $\mathbf{K}_t$ is not the Kalman gain. If the initial state is fixed, then, according to Eq. (7.4.46), $\mathbf{Q}_{t|t} = 0$ and Algorithm 7.4 is significantly simplified:

Algorithm 7.5.

```
Initialize:
```

$$\mathbf{u}_0 = 0, \quad \mathbf{U}_0 = \mathbf{I}$$

```
For t=0,1,...,T-1
Begin
```

$$\mathbf{D}_t = \mathbf{F}_t - \mathbf{K}_t\mathbf{H}_t$$

$$\mathbf{U}_{t+1} = \mathbf{D}_t\mathbf{U}_t, \quad \mathbf{u}_{t+1} = \mathbf{D}_t\mathbf{u}_t + \mathbf{K}_t\mathbf{a}_{t+1}$$

```
End
```

For the fixed initial state, Eq. (7.4.40) takes the form

$$\bar{\alpha}(\mathbf{a}_1^T, s_T; s_0) = \mathrm{N}(s_T - \mathbf{U}_T s_0 - \mathbf{u}_T, 0) = \delta(s_T - \mathbf{U}_T s_0 - \mathbf{u}_T)$$

For the homogeneous HGMM, $\mathbf{U}_t = \mathbf{D}^t = (\mathbf{F} - \mathbf{KH})^t$. If the eigenvalues of the matrix $\mathbf{F} - \mathbf{KH}$ are inside the unit circle on the complex plane, then $\lim_{t\to\infty}\mathbf{U}_t = 0$ and the stationary Kalman filter exists. Since

$$s_{t|t}(s_0) = \mathbf{U}_t s_0 + \mathbf{u}_t$$

for the stationary Kalman filter, the forward PDF does not depend on the initial state.

The state transition PDF, according to Eq. (7.4.8), has the form

$$p(\mathbf{x}_t \mid \boldsymbol{s}_{t-1}) = \mathsf{N}(\mathbf{a}_t - \mathbf{H}_{t-1}\boldsymbol{s}_{t-1}, \mathbf{S}_{aa,t-1})\mathsf{N}(\boldsymbol{s}_t - \boldsymbol{\mu}_{t,t-1}, 0)$$

where

$$\boldsymbol{\mu}_{t,t-1} = \mathbf{F}_{t-1}\boldsymbol{s}_{t-1} + \mathbf{K}_{t-1}(\mathbf{a}_t - \mathbf{H}_{t-1}\boldsymbol{s}_{t-1}) = \mathbf{D}_{t-1}\boldsymbol{s}_{t-1} + \mathbf{K}_{t-1}\mathbf{a}_t$$

and $\mathsf{N}(\boldsymbol{s}_t - \boldsymbol{\mu}_{t,t-1}, 0) = \delta(\boldsymbol{s}_t - \boldsymbol{\mu}_{t,t-1})$. We can prove formally that $\boldsymbol{\Sigma}_{t,t-1} = \mathbf{Q}_{0|0} = 0$ using equation (7.4.45b), but it also follows from Eq. (7.4.44) that $\boldsymbol{s}_{t+1}$ is uniquely defined, given $\mathbf{a}_{t+1}$ and $\boldsymbol{s}_t$ and, therefore,

$$p(\boldsymbol{s}_{t+1} \mid \boldsymbol{s}_t, \mathbf{a}_{t+1}) = \delta(\boldsymbol{s}_{t+1} - \boldsymbol{\mu}_{t+1,t})$$

The observation PSD is given by Eq. (7.4.20). For the homogeneous innovation model, according to (7.4.45b), this equation takes the form

$$\Phi_{aa}(z) = \mathbf{M}(z)\mathbf{K}\mathbf{P}_{aa}\mathbf{K}'\mathbf{M}'(z^{-1}) + \mathbf{M}(z)\mathbf{K}\mathbf{P}_{aa} + \mathbf{P}_{aa}\mathbf{K}'\mathbf{M}'(z^{-1}) + \mathbf{P}_{aa}$$

where

$$\mathbf{M}(z) = \mathbf{H}z^{-1}(\mathbf{I} - \mathbf{F}z^{-1})^{-1} = \mathbf{H}(\mathbf{I}z - \mathbf{F})^{-1}$$

It is easy to verify by direct multiplication that $\Phi_{aa}(z)$ can be expressed as

$$\Phi_{aa}(z) = \mathbf{W}_{aa}(z)\mathbf{P}_{aa}\mathbf{W}'_{aa}(z^{-1}) \tag{7.4.47}$$

where

$$\mathbf{W}_{aa}^{-1}(z) = [\mathbf{I} + \mathbf{M}(z)\mathbf{K}]^{-1}$$

Using the matrix inversion lemma we can express

$$\mathbf{W}_{aa}^{-1}(z) = \mathbf{I} - \mathbf{H}(\mathbf{I}z - \mathbf{F} + \mathbf{K}\mathbf{H})^{-1}\mathbf{K}$$

Equation (7.4.47) represents the so called spectral factorization.

7.4.5 The Backward Algorithm

Consider now the computation of the backward PDF $\beta(\boldsymbol{s}_{t-1}, \mathbf{a}_t^T) = p(\mathbf{a}_t^T \mid \boldsymbol{s}_{t-1})$. This PDF can be computed using the backward algorithm in Eq. (7.2.7). In this section, we show that there is a closed form expression for the integrals in this algorithm in the case of HGMM.

Indeed, if $t = T$, the closed form expression follows from Eq. (7.4.8):

$$\beta(\boldsymbol{s}_{T-1}, \mathbf{a}_T) = p(\mathbf{a}_T \mid \boldsymbol{s}_{T-1}) = \mathsf{N}(\mathbf{a}_T - \mathbf{H}_{T-1}\boldsymbol{s}_{T-1}, \mathbf{S}_{aa,T-1})$$

Using the notation $\mathsf{G}(\mathbf{x}, \mathbf{B}) = \exp\{-0.5\mathbf{x}'\mathbf{B}\mathbf{x})$ we can rewrite this equation as

$$\beta(\boldsymbol{s}_{T-1}, \mathbf{a}_T) = g_T\mathsf{G}(\mathbf{G}_T\boldsymbol{s}_{T-1} - \mathbf{a}_T, \mathbf{M}_T)$$

with $\mathbf{G}_T = \mathbf{H}_{T-1}$, $\mathbf{b}_T = \mathbf{a}_T$, $\mathbf{M}_T = \mathbf{S}_{aa,T-1}^{-1}$, and $g_T = (2\pi)^{-m/2}|\mathbf{S}_{aa,T-1}|^{-1/2}$ (see Eq. (A.7.10) of Appendix 7). We can assume that the other backward PDFs also have this form:

$$\beta(\boldsymbol{s}_t, \mathbf{a}_{t+1}^T) = g_{t+1}\mathsf{G}(\mathbf{G}_{t+1}\boldsymbol{s}_t - \mathbf{b}_{t+1}, \mathbf{M}_{t+1}) \tag{7.4.48}$$

We use this form because (as we will see later) the matrix $\mathbf{M}_{t+1}$ is might not be positive definite for other backward PDFs. In general, we can only assume that $\mathbf{M}_{t+1}$ is a nonnegative definite matrix, and prove it using the method of mathematical induction by showing that $\beta(\boldsymbol{s}_{t-1},\mathbf{a}_t^T)$ also has this form.

To prove this statement, we use Eq. (7.2.7) which in our case takes the form

$$\beta(\boldsymbol{s}_{t-1},\mathbf{a}_t^T) = g_{t+1}\int_S \mathrm{N}(\mathbf{x}_t-\mathbf{C}_{t-1}\boldsymbol{s}_{t-1},\mathbf{S}_{t-1})\,\mathrm{G}(\mathbf{G}_{t+1}\boldsymbol{s}_t-\mathbf{b}_{t+1},\mathbf{M}_{t+1})\,d\boldsymbol{s}_t \quad (7.4.49)$$

To evaluate this integral, we use Eq. (7.4.8) to factor the integrand:

$$\mathrm{N}(\mathbf{x}_t-\mathbf{C}_{t-1}\boldsymbol{s}_{t-1},\mathbf{S}_{t-1}) = \mathrm{N}(\mathbf{a}_t-\mathbf{H}_{t-1}\boldsymbol{s}_{t-1},\mathbf{S}_{aa,t-1})\,\mathrm{N}(\boldsymbol{s}_t-\boldsymbol{\mu}_{t,t-1},\boldsymbol{\Sigma}_{t,t-1}) \quad (7.4.50)$$

where

$$\boldsymbol{\mu}_{t,t-1} = \mathbf{D}_{t-1}\boldsymbol{s}_{t-1}+\mathbf{K}_{t-1}\mathbf{a}_t \quad (7.4.51)$$

$$\boldsymbol{\Sigma}_{t,t-1} = \mathbf{S}_{ss,t-1}-\mathbf{K}_{t-1}\mathbf{S}_{as,t-1}$$

Substituting (7.4.50) into (7.4.49) we obtain

$$\beta(\boldsymbol{s}_{t-1},\mathbf{a}_t^T) = \mathrm{N}(\mathbf{a}_t-\mathbf{H}_{t-1}\boldsymbol{s}_{t-1},\mathbf{S}_{aa,t-1})\,I(\boldsymbol{s}_{t-1}) \quad (7.4.52)$$

where

$$I(\boldsymbol{s}_{t-1}) = g_{t+1}\int_S \mathrm{N}(\boldsymbol{s}_t-\boldsymbol{\mu}_{t,t-1},\boldsymbol{\Sigma}_{t,t-1})\,\mathrm{G}(\mathbf{G}_{t+1}\boldsymbol{s}_t-\mathbf{b}_{t+1},\mathbf{M}_{t+1})\,d\boldsymbol{s}_t$$

Using Eq. (A.7.19) of Appendix 7, we can write

$$I(\boldsymbol{s}_{t-1}) = g_{t+1} = h_t\mathrm{G}(\mathbf{G}_{t+1}\boldsymbol{\mu}_{t,t-1}-\mathbf{b}_{t+1},\mathbf{V}_t) \quad (7.4.53)$$

where

$$\mathbf{V}_t = \mathbf{M}_{t+1}-\mathbf{M}_{t+1}\mathbf{G}_{t+1}(\mathbf{G}_{t+1}'\mathbf{M}_{t+1}\mathbf{G}_{t+1}+\boldsymbol{\Sigma}_{t,t-1}^{-1})^{-1}\mathbf{G}_{t+1}'\mathbf{M}_{t+1}$$

$$= \mathbf{M}_{t+1}(\mathbf{G}_{t+1}\boldsymbol{\Sigma}_{t,t-1}\mathbf{G}_{t+1}'\mathbf{M}_{t+1} + \mathbf{I})^{-1}$$

$$h_t = g_{t+1}\,|\mathbf{G}_{t+1}'\mathbf{M}_{t+1}\mathbf{G}_{t+1}\boldsymbol{\Sigma}_{t,t-1}^{-1} + \mathbf{I}|^{-1/2}$$

After the substitution of $\boldsymbol{\mu}_{t,t-1}$ from Eq. (7.4.51) into (7.4.53) the latter takes the form

$$I(\boldsymbol{s}_{t-1}) = h_t\mathrm{G}(\mathbf{R}_t\boldsymbol{s}_{t-1}-\mathbf{r}_t,\mathbf{V}_t)$$

where

$$\mathbf{R}_t = \mathbf{G}_{t+1}\mathbf{D}_{t-1}, \quad \mathbf{r}_t = \mathbf{b}_{t+1}-\mathbf{G}_{t+1}\mathbf{S}_{sa,t-1}\mathbf{S}_{aa,t-1}^{-1}\mathbf{a}_t$$

Using this result, we can rewrite Eq. (7.4.52) as

$$\beta(\boldsymbol{s}_{t-1},\mathbf{a}_t^T) = h_t\mathrm{N}(\mathbf{a}_t-\mathbf{H}_{t-1}\boldsymbol{s}_{t-1},\mathbf{S}_{aa,t-1})\,\mathrm{G}(\mathbf{R}_t\boldsymbol{s}_{t-1}-\mathbf{r}_t,\mathbf{V}_t)$$

To complete the proof, we need to show that this function can be expressed in the form of Eq. (7.4.48). This can be achieved by completing the square with respect to $\boldsymbol{s}_{t-1}$ in the exponent of the previous expression:

$$(\mathbf{a}_t-\mathbf{H}_{t-1}\boldsymbol{s}_{t-1})'\mathbf{S}_{aa,t-1}^{-1}(\mathbf{a}_t-\mathbf{H}_{t-1}\boldsymbol{s}_{t-1}) + (\mathbf{R}_t\boldsymbol{s}_{t-1}-\mathbf{r}_t)'\mathbf{V}_t(\mathbf{R}_t\boldsymbol{s}_{t-1}-\mathbf{r}_t) \quad (7.4.54)$$

We use Eq. (A.5.59) of Appendix 5 to complete the square in Eq. (7.4.54) and obtain

$$\beta(\boldsymbol{s}_{t-1},\mathbf{a}_t^T) = g_t\mathrm{G}(\mathbf{G}_t\boldsymbol{s}_{t-1}-\mathbf{b}_t,\mathbf{M}_t) \quad (7.4.55)$$

where $\mathbf{M}_t = \mathbf{G}_t^{\dagger}$

$$\mathbf{G}_t = \mathbf{H}_{t-1}' \mathbf{S}_{aa,t-1}^{-1} \mathbf{H}_{t-1} + \mathbf{R}_t' \mathbf{V}_t \mathbf{R}_{t+1} \tag{7.4.56}$$

$$\mathbf{b}_t = \mathbf{H}_{t-1}' \mathbf{S}_{aa,t-1}^{-1} \mathbf{a}_t + \mathbf{R}_t' \mathbf{V}_t \mathbf{r}_t$$

$$g_t = h_t \exp\{-0.5(\mathbf{a}_t' \mathbf{S}_{aa,t-1}^{-1} \mathbf{a}_t + \mathbf{r}_t' \mathbf{V}_t \mathbf{r}_t - \mathbf{b}_t' \mathbf{G}_t^{\dagger} \mathbf{b}_t)\}$$

As we can see, $\mathbf{G}_t$ defined by Eq. (7.4.56) might not be invertible. Therefore, we use the pseudoinverse matrix $\mathbf{M}_t = \mathbf{G}_t^{\dagger}$ in Eq. (7.4.55).

Putting it all together we can describe the backward algorithm as follows:

Algorithm 7.6.

`Initialize:`

$$\mathbf{G}_T = \mathbf{H}_{T-1}, \quad \mathbf{b}_T = \mathbf{a}_T, \quad \mathbf{M}_T = \mathbf{S}_{aa,T-1}^{-1}, \quad g_T = (2\pi)^{-m/2} |\mathbf{S}_{aa,T-1}|^{-1/2} \tag{7.4.57a}$$

`For` $t = T-1, T-2, \ldots, 1$

`Begin`

$$\mathbf{K}_{t-1} = \mathbf{S}_{sa,t-1} \mathbf{S}_{aa,t-1}^{-1} \tag{7.4.57b}$$

$$\mathbf{\Sigma}_{t,t-1} = \mathbf{S}_{ss,t-1} - \mathbf{K}_{t-1} \mathbf{S}_{as,t-1} \tag{7.4.57b}$$

$$\mathbf{V}_t = \mathbf{M}_{t+1}(\mathbf{G}_{t+1} \mathbf{\Sigma}_{t,t-1} \mathbf{G}_{t+1}' \mathbf{M}_{t+1} + \mathbf{I})^{-1} \tag{7.4.57b}$$

$$h_t = g_{t+1} \, |\mathbf{G}_{t+1}' \mathbf{M}_{t+1} \mathbf{G}_{t+1} \mathbf{\Sigma}_{t,t-1} + \mathbf{I}|^{-1/2} \tag{7.4.57c}$$

$$\mathbf{R}_t = \mathbf{G}_{t+1}(\mathbf{F}_{t-1} - \mathbf{K}_{t-1} \mathbf{H}_{t-1}) \tag{7.4.57d}$$

$$\mathbf{r}_t = \mathbf{b}_{t+1} - \mathbf{G}_{t+1} \mathbf{K}_{t-1} \mathbf{a}_t \tag{7.4.57e}$$

$$\mathbf{b}_t = \mathbf{H}_{t-1}' \mathbf{S}_{aa,t-1}^{-1} \mathbf{a}_t + \mathbf{R}_t' \mathbf{V}_t \mathbf{r}_t \tag{7.4.57f}$$

$$\mathbf{G}_t = \mathbf{H}_{t-1}' \mathbf{S}_{aa,t-1}^{-1} \mathbf{H}_{t-1} + \mathbf{R}_t' \mathbf{V}_t \mathbf{R}_{t+1} \tag{7.4.57g}$$

$$\mathbf{M}_t = \mathbf{G}_t^{\dagger} \tag{7.4.57h}$$

$$g_t = h_t \exp\{-0.5(\mathbf{a}_t' \mathbf{S}_{aa,t-1}^{-1} \mathbf{a}_t + \mathbf{r}_t' \mathbf{V}_t \mathbf{r}_t - \mathbf{b}_t' \mathbf{M}_t)\} \tag{7.4.57i}$$

`End`

This algorithm computes the parameters of the backward PDF; the PDF is presented by Eq. (7.4.48). Note that since $\mathbf{G}_{t+1}' = \mathbf{G}_{t+1}$ and $\mathbf{M}_{t+1} = \mathbf{G}_{t+1}^{\dagger}$ for $t < T-1$, we can simplify these equations by using the identity $\mathbf{G}_{t+1} \mathbf{G}_{t+1}^{\dagger} \mathbf{G}_{t+1} = \mathbf{G}_{t+1}$.

Example 7.7. Let us illustrate the backward algorithm for the system (7.4.9) of Example 7.2. According to Eq. (7.4.10), the one step transition PDF can be written in the following form

$$p(\mathbf{x}_t \mid \mathbf{s}_{t-1}) = \mathsf{N}(a_t - hs_{t-1}, S)\mathsf{N}(s_t - Ds_{t-1} - Ka_t, \Sigma) \tag{7.4.58}$$

where

$$K = b^2/(b^2 + \lambda^2), \quad D = f - hK$$

$$\Sigma = \lambda^2\sigma^2/(b^2+\lambda^2)$$
$$S = \sigma^2(b^2+\lambda^2)$$

Since the transition PDF in Eq. (7.4.58) has a simple form, we rederive the backward algorithm and compare the results with that of the general case.

We have

$$\beta(s_{T-1},a_t) = \mathrm{N}(a_T - hs_{T-1},S)$$

$$\beta(s_{T-2},a_{T-1},a_T) = \int_{-\infty}^{\infty} \mathrm{N}(a_{T-1}-hs_{T-2},S)\mathrm{N}(s_{T-1}-Ds_{T-2}-Ka_{T-1},\Sigma)\beta(s_{T-1},a_t)ds_{T-1}$$
$$= \mathrm{N}(a_{T-1}-hs_{T-2},S)\mathrm{N}(a_T-hDs_{T-2}-hKa_{T-1},S+h^2\Sigma)$$

where we used Eq. (A.7.22) of Appendix 7 to evaluate the integral. In the next step, we complete the square in the $\log\beta(s_{T-2},a_{T-1},a_T)$.

It is easy to verify the following identity (which is a special case of Eq. (A.7.16) of Appendix 7)

$$\frac{(a_1x-b_1)^2}{\Sigma_1}+\frac{(a_2x-b_2)^2}{\Sigma_2} = \frac{(Ax-B)^2}{A\Sigma_1\Sigma_2}+\frac{(a_1b_2-a_2b_1)^2}{A}$$

where

$$A = a_1^2\Sigma_2 + a_2^2\Sigma_1, \quad B = a_1b_1\Sigma_2 + a_2b_2\Sigma_1$$

Using this identity, we can write

$$\mathrm{N}(a_1x-b_1,\Sigma_1)\mathrm{N}(a_2x-b_2,\Sigma_2) = A^2\mathrm{N}(Ax-B,A\Sigma_1\Sigma_2)\mathrm{N}(a_1b_2-a_2b_1,A) \tag{7.4.59}$$

(This identity is a special case of (A.7.20) of Appendix 7.) Applying this identity to $\beta(s_{T-2},a_{T-1},a_T)$ we can write:

$$\beta(s_{T-2},a_{T-1},a_T) = g_{T-1}\mathrm{N}(A_{T-1}s_{T-2}-B_{T-1},\Sigma_{T-1})$$

where

$$g_{T-1} = A_{T-1}^2\mathrm{N}(ha_T-Ua_{T-1},A_{T-1}), \quad \Sigma_{T-1} = A_{T-1}\Delta_{T-1}S$$

where $U = h(hK+D)$

$$\Delta_{T-1} = S+h^2\Sigma, \quad A_{T-1} = h^2(D^2S+\Delta_{T-1}), \quad B_{T-1} = ha_{T-1}\Delta_{T-1}+SDh(a_T-hKa_{T-1})$$

Thus, we can assume that

$$\beta(s_t,a_{t+1}^T) = g_{t+1}\mathrm{N}(A_{t+1}s_t-B_{t+1},\Sigma_{t+1})$$

The backward algorithm is given by Eq. (7.4.49) which in our case takes the form

$$\beta(s_{t-1},a_t^T) = g_{t+1}\mathrm{N}(a_t-hs_{t-1},S)\int_{-\infty}^{\infty}\mathrm{N}(s_t-Ds_{t-1}-Ka_t,\Sigma)\beta(s_t,a_{t+1}^T)ds_t$$
$$= g_{t+1}\mathrm{N}(a_t-hs_{t-1},S)\mathrm{N}(A_{t+1}Ds_{t-1}+A_{t+1}Ka_t-B_{t+1},\Delta_t)$$

where

$$\Delta_t = \Sigma_{t+1}+A_{t+1}^2\Sigma$$

Applying Eq. (7.4.59) we obtain

$$\beta(s_{t-1},a_t^T) = g_t\mathrm{N}(A_ts_{t-1}-B_t,\Sigma_t)$$

where

$$A_t = h^2\Delta_t+D^2SA_{t+1}^2 \; B_t = ha_t\Delta_t+DSA_{t+1}(B_{t+1}-KA_{t+1}a_t), \quad \Sigma_t = SA_{t+1}(\Delta_t+A_{t+1}^2\Sigma)$$

and

$$g_t = g_{t+1}A_t^2\mathrm{N}(UA_{t+1}a_t-hB_{t+1},A_t)$$

We leave it as an exercise to the reader to verify that these equations agree with Eq. (7.4.57).

We would like to note in conclusion that the backward algorithm can be developed using the expression for the kernel of the operator $\mathbf{P}(a_t^T)$ presented in Sec. 7.4.2 and applying the Gaussian elimination to the extended covariance matrix. We have illustrated this approach in Example 7.6 for the forward PDF where we performed the elimination from the "top down."

For the backward PDF, we need to perform the elimination from the "bottom up."

7.4.6 The Forward-Backward Algorithm

Let us now address the problem of evaluating the state smoothing PDF $\gamma_t(\boldsymbol{s}_t, \mathbf{a}_1^T) = p(\boldsymbol{s}_t \mid \mathbf{a}_1^T)$ given by Eq. (7.3.2). One way to present this PDF is by writing the PDF $p(\boldsymbol{s}_t, \mathbf{a}_1^T) = \mathrm{N}[(\boldsymbol{s}_t, \mathbf{a}_1^T) - \boldsymbol{\mu}(t,T), \boldsymbol{\Sigma}(t,T)]$ where $\boldsymbol{\mu}(t,T)$ and $\boldsymbol{\Sigma}(t,T)$ are computed using equations of Sec. 7.3.1. To obtain the smoothing PDF, we need to perform Gaussian elimination of the rows and columns corresponding to $\mathbf{a}_1^T$ from the matrix $[\boldsymbol{\Sigma}(t,T), \boldsymbol{\mu}(t,T)]$. This elimination can be complicated if m is large. Besides, in many applications we need to find the smoothing PDF for all $t = 1,2,...,T$. Therefore, it would be desirable to develop a recursive procedure for computing the PDF.

It follows from Eq. (7.3.2) that the forward-backward algorithm delivers us the recursive procedure. In the forward part, we compute and save in the computer memory $\boldsymbol{s}_{t|t}$, $\mathbf{Q}_{t|t}$, and $p(\mathbf{a}_1^t)$ using the Kalman filter described by Eq. (7.4.32). In the backward, part we compute $\beta(\boldsymbol{s}_t, \mathbf{a}_{t+1}^T)$ using equation (7.4.48) and (7.4.57) and then use equation (7.3.2) which can be written as

$$\gamma_t(\boldsymbol{s}_t, \mathbf{a}_1^T) = \frac{g_{t+1} p(\mathbf{a}_1^t)}{p(\mathbf{a}_1^T)} \mathrm{N}(\boldsymbol{s}_t - \boldsymbol{s}_{t|t}, \mathbf{Q}_{t|t}) \mathrm{G}(\mathbf{G}_{t+1} \boldsymbol{s}_t - \mathbf{b}_{t+1}, \mathbf{M}_{t+1})$$

After completing the square in the exponent of this function, we can rewrite it using Eq. (A.7.18) of Appendix 7 as

$$\gamma_t(\boldsymbol{s}_t, \mathbf{a}_1^T) = \mathrm{N}(\boldsymbol{s}_t - \boldsymbol{s}_{t|T}, \mathbf{Q}_{t|T}) \tag{7.4.60}$$

where

$$\mathbf{Q}_{t|T} = (\mathbf{G}_{t+1}' \mathbf{M}_{t+1} \mathbf{G}_{t+1} + \mathbf{Q}_{t|t}^{-1})^{-1}$$

$$\boldsymbol{s}_{t|T} = \mathbf{Q}_{t|T}(\mathbf{G}_{t+1}' \mathbf{M}_{t+1} \mathbf{b}_{t+1} + \mathbf{Q}_{t|t}^{-1} \boldsymbol{s}_{t|t}) = \boldsymbol{s}_{t|t} + \mathbf{W}_t(b_{t+1} - \mathbf{G}_{t+1} \boldsymbol{s}_{t|t})$$

$$\mathbf{W}_t = \mathbf{Q}_{t|T} \mathbf{G}_{t+1}' \mathbf{M}_{t+1}$$

Equation (7.4.60) shows that we need to compute only the mean $\boldsymbol{s}_{t|T}$ and the variance matrix $\mathbf{Q}_{t|T}$ to evaluate the smoothing PDF $\gamma_t(\boldsymbol{s}_t, \mathbf{a}_1^T)$. Therefore, there is no need to compute $p(\mathbf{a}_1^t)$ in the forward algorithm and the coefficients g_{t+1} in the backward algorithm. Since $\mathbf{G}_{t+1}' = \mathbf{G}_{t+1}$ and $\mathbf{M}_{t+1} = \mathbf{G}_{t+1}^{\dagger}$ for $t < T-1$, we can simplify these equations by using the identity $\mathbf{G}_{t+1} \mathbf{G}_{t+1}^{\dagger} \mathbf{G}_{t+1} = \mathbf{G}_{t+1}$.

7.4.7 The RTS Smoother

In Sec. 7.3, we introduced the alternative forward-backward algorithm for computing the conditional PDF $\gamma_t(\boldsymbol{s}_t, \mathbf{a}_1^T)$ without computing the backward PDFs. The backward part of this algorithm is given by equation (7.3.4d) which can be rewritten as

$$\gamma_t(\boldsymbol{s}_t, \mathbf{a}_1^T) = \int_S \gamma_{t+1}(\boldsymbol{s}_{t+1}, \mathbf{a}_1^T) \phi(\boldsymbol{s}_t, \boldsymbol{s}_{t+1}, \mathbf{a}_1^T) d\boldsymbol{s}_{t+1} \tag{7.4.61}$$

where

$$\phi(s_t, s_{t+1}, \mathbf{a}_1^T) = \alpha(s_t, \mathbf{a}_1^t) p(s_{t+1}, \mathbf{a}_{t+1} \mid s_t) / \alpha(s_{t+1}, \mathbf{a}_1^{t+1}) \tag{7.4.62}$$

To simplify this expression, we use Eq. (7.4.31) to write

$$\alpha(\mathbf{a}_1^t, s_t) p(s_{t+1}, \mathbf{a}_{t+1} \mid s_t) = \mathsf{N}(s_t - s_{t|t}, \mathbf{Q}_{t|t}) p(\mathbf{a}_1^t) \mathsf{N}(\mathbf{x}_{t+1} - \mathbf{C}_t s_t, \mathbf{S}_t)$$

This expression can be factored according to Eq. (A.7.20) of Appendix 7 as

$$\alpha(\mathbf{a}_1^t, s_t) p(s_{t+1}, \mathbf{a}_{t+1} \mid s_t) = \mathsf{N}(s_t - \mathbf{q}_t, \mathbf{R}_t) \mathsf{N}(\mathbf{x}_{t+1} - \mathbf{C}_t s_{t|t}, \mathbf{P}_t) p(\mathbf{a}_1^t) \tag{7.4.63a}$$

where

$$\mathbf{q}_t = s_{t|t} + \mathbf{W}_t(\mathbf{x}_{t+1} - \mathbf{C}_t s_{t|t}) \tag{7.4.63b}$$

$$\mathbf{W}_t = \mathbf{Q}_{t|t} \mathbf{C}_t' \mathbf{P}_t^{-1} \tag{7.4.63c}$$

$$\mathbf{R}_t = \mathbf{Q}_{t|t} - \mathbf{W}_t \mathbf{C}_t \mathbf{Q}_{t|t} \tag{7.4.63d}$$

According to Eq. (7.4.31), $\mathsf{N}(\mathbf{x}_{t+1} - \mathbf{C}_t s_{t|t}, \mathbf{P}_t) p(\mathbf{a}_1^t) = \alpha(s_{t+1}, \mathbf{a}_1^{t+1})$ and Eq. (7.4.63) can be written as

$$\alpha(s_t, \mathbf{a}_1^t) p(s_{t+1}, \mathbf{a}_{t+1} \mid s_t) = \alpha(s_{t+1}, \mathbf{a}_1^{t+1}) \mathsf{N}(s_t - \mathbf{q}_t, \mathbf{R}_t)$$

This expression allows us to simplify Eq. (7.4.62) to

$$\phi(s_t, s_{t+1}, \mathbf{a}_1^T) = \mathsf{N}(s_t - \mathbf{q}_t, \mathbf{R}_t) \tag{7.4.64}$$

With this simplification, the backward recursive Eq. (7.4.61) becomes

$$\gamma_t(s_t, \mathbf{a}_1^T) = \int_S \gamma_{t+1}(s_{t+1}, \mathbf{a}_1^T) \mathsf{N}(s_t - \mathbf{q}_t, \mathbf{R}_t) ds_{t+1} \tag{7.4.65}$$

starting with the initial value

$$\gamma_T(s_T, \mathbf{a}_1^T) = \alpha(s_T, \mathbf{a}_1^T) = \mathsf{N}(s_T - s_{T|T}, \mathbf{Q}_{T|T})$$

obtained at the end of the forward algorithm.

Substituting

$$\gamma_{t+1}(s_{t+1}, \mathbf{a}_1^T) = \mathsf{N}(s_{t+1} - s_{t+1|T}, \mathbf{Q}_{t+1|T})$$

into Eq. (7.4.65) and computing the integral according to Eq. (A.7.22) of Appendix 7 we obtain

$$\gamma_t(s_t, \mathbf{a}_1^T) = \int_S \mathsf{N}(s_{t+1} - s_{t+1|T}, \mathbf{Q}_{t+1|T}) \mathsf{N}(s_t - \mathbf{q}_t, \mathbf{R}_t) ds_{t+1} = \mathsf{N}(s_t - s_{t|T}, \mathbf{Q}_{t|T})$$

where

$$s_{t|T} = s_{t|t} + \mathbf{W}_t(\mathbf{x}_{t+1|T} - \mathbf{C}_t s_{t|t}) \tag{7.4.66}$$

$$\mathbf{x}_{t+1|T} = (s_{t+1|T}, \mathbf{a}_{t+1})$$

$$\mathbf{Q}_{t|T} = \mathbf{R}_t + \mathbf{W}_t \mathbf{Q}_{t+1|T} \mathbf{W}_t' = \mathbf{Q}_{t|t} + \mathbf{W}_t(\mathbf{Q}_{t+1|T} \mathbf{W}_t' - \mathbf{C}_t \mathbf{Q}_{t|t})$$

The forward-backward algorithm that uses the Kalman filter in the forward path and the RTS smoother in the backward part can be summarized as follows.

Algorithm 7.7.

(*Forward Part*)
`Initialize:`

$$s_{0|0} = \mu_0, \quad \mathbf{Q}_{0|0} = \Sigma_0 \tag{7.4.67a}$$

`For` $t=0,1,...,T-1$
`Begin`

$$\mathbf{P}_t = \mathbf{C}_t \mathbf{Q}_{t|t} \mathbf{C}_t' + \mathbf{S}_t \tag{7.4.67b}$$

$$\mathbf{K}_t = \mathbf{P}_{sa,t} \mathbf{P}_{aa,t}^{-1} \tag{7.4.67c}$$

$$s_{t+1|t+1} = \mathbf{F}_t s_{t|t} + \mathbf{K}_t(\mathbf{a}_{t+1} - \mathbf{H}_t s_{t|t}) \tag{7.4.67d}$$

$$\mathbf{Q}_{t+1|t+1} = \mathbf{P}_{ss,t} - \mathbf{K}_t \mathbf{P}_{as,t} \tag{7.4.67e}$$

`Store` $s_{t+1|t+1}$, $\mathbf{Q}_{t+1|t+1}$, `and` $\mathbf{P}_t$
`End`

(*Backward Part*)
`For` $t=T-1,T-2,...,1$
`Begin`

$$\mathbf{W}_t = \mathbf{Q}_{t|t} \mathbf{C}_t' \mathbf{P}_t^{-1} \tag{7.4.67f}$$

$$s_{t|T} = s_{t|t} + \mathbf{W}_t(\mathbf{x}_{t+1|T} - \mathbf{C}_t s_{t|t}) \tag{7.4.67g}$$

$$\mathbf{Q}_{t|T} = \mathbf{Q}_{t|t} + \mathbf{W}_t(\mathbf{Q}_{t+1|T} \mathbf{W}_t' - \mathbf{C}_t \mathbf{Q}_{t|t}) \tag{7.4.67h}$$

`End`

We assumed in our derivations that the initial state PDF is Gaussian. If the initial PDF is a mixture of Gaussian PDFs as in Eq. (7.4.38), then the smoothing PDF is a mixture of smoothing PDFs which are obtained by running M forward-backward algorithms.

$$\gamma_t(s_t, \mathbf{a}_1^T) = \sum_{i=1}^{M} w_i \gamma_{i,t}(s_T, \mathbf{a}_1^T) \tag{7.4.68}$$

where $\gamma_{i,t}(s_t, \mathbf{a}_1^T)$ is obtained by the forward-backward algorithm with the initial conditions $\mu_{i,0}$ and $\Sigma_{i,0}$.

As we can see, the conditional covariance matrices $\mathbf{P}_{t|T}$ and $\mathbf{Q}_{t|T}$ do not depend on the observation sequence and the conditional means are linear functions of the original mean whose coefficients can be found using the following algorithm.

Algorithm 7.8.
(*Forward Part*)
`Initialize:`

$$\mathbf{u}_0 = 0, \quad \mathbf{U}_0 = \mathbf{I}, \quad \mathbf{Q}_{0|0} = \Sigma_0 \tag{7.4.69a}$$

```
For  t=0,1,...,T-1
Begin
```

$$\mathbf{P}_t = \mathbf{C}_t\mathbf{Q}_{t|t}\mathbf{C}_t' + \mathbf{S}_t \tag{7.4.69b}$$

$$\mathbf{K}_t = \mathbf{P}_{sa,t}\mathbf{P}_{aa,t}^{-1} \tag{7.4.69c}$$

$$\mathbf{D}_t = \mathbf{F}_t - \mathbf{K}_t\mathbf{H}_t \tag{7.4.69d}$$

$$\mathbf{d}_t = \mathbf{K}_t\mathbf{a}_{t+1} \tag{7.4.69e}$$

$$\mathbf{U}_{t+1|t+1} = \mathbf{D}_t\mathbf{U}_{t|t} \tag{7.4.69f}$$

$$\mathbf{u}_{t+1|t+1} = \mathbf{D}_t\mathbf{u}_{t|t} + \mathbf{d}_{t|t} \tag{7.4.69g}$$

$$\mathbf{Q}_{t+1|t+1} = \mathbf{P}_{ss,t} - \mathbf{K}_t\mathbf{P}_{as,t} \tag{7.4.69h}$$

Store $\mathbf{u}_{t+1|t+1}$, $\mathbf{U}_{t+1|t+1}$, $\mathbf{Q}_{t+1|t+1}$, and $\mathbf{P}_t$

End

(Backward Part)

```
For  t=T-1,T-2,...,1
Begin
```

$$\mathbf{W}_t = \mathbf{Q}_{t|t}\mathbf{C}_t'\mathbf{P}_t^{-1} \tag{7.4.69i}$$

$$\mathbf{y}_{t|T} = \begin{bmatrix} \mathbf{u}_{t+1|T} \\ \mathbf{a}_{t+1} \end{bmatrix} \tag{7.4.69j}$$

$$\mathbf{u}_{t|T} = \mathbf{u}_{t|t} + \mathbf{W}_t(\mathbf{y}_{t+1|T} - \mathbf{C}_t\mathbf{u}_{t|t}) \tag{7.4.69k}$$

$$\mathbf{U}_{t|T} = (\mathbf{I} - \mathbf{W}_t\mathbf{C}_t)\mathbf{U}_{t|t} + \mathbf{W}_{s,t}\mathbf{U}_{t+1|T} \tag{7.4.69l}$$

$$\mathbf{Q}_{t|T} = \mathbf{Q}_{t|t} + \mathbf{W}_t(\mathbf{Q}_{t+1|T}\mathbf{W}_t' - \mathbf{C}_t\mathbf{Q}_{t|t}) \tag{7.4.69m}$$

End

It is not difficult to verify that

$$\boldsymbol{s}_{t|T} = \mathbf{U}_{t|T}\boldsymbol{s}_0 + \mathbf{u}_{t|T}$$

is the solution of Eq. (7.4.66) and, therefore,

$$\gamma_t(\boldsymbol{s}_t, \mathbf{a}_1^T; \boldsymbol{s}_0) = \mathrm{N}(\boldsymbol{s}_t - \mathbf{U}_{t|T}\boldsymbol{s}_0 - \mathbf{u}_{t|T}, \mathbf{Q}_{t|T})$$

Consider now computing the sliding-window smoothing PDF for the size 2 window. This PDF can be found using equation (7.3.3) which in our case, according to Eq. (7.4.62) and (7.4.64) can be written as

$$\gamma_t(\boldsymbol{s}_t, \boldsymbol{s}_{t+1}, \mathbf{a}_1^T) = \mathrm{N}(\boldsymbol{s}_{t+1} - \boldsymbol{s}_{t+1|T}, \mathbf{Q}_{t|T})\mathrm{N}(\boldsymbol{s}_t - \mathbf{q}_t, \mathbf{R}_t) \tag{7.4.70}$$

Obviously, this is a Gaussian PDF. To present it explicitly, we need to find the covariance matrix for $\boldsymbol{s}_t$ and $\boldsymbol{s}_{t+1}$.

It follows from the right hand side of the previous equation that we can write

$$s_{t+1} = s_{t+1|T} + \varepsilon_1$$
$$s_t = \mathbf{q}_t + \varepsilon_2 = s_{t|t} + \mathbf{W}_t(\mathbf{x}_{t+1} - \mathbf{C}_t s_{t|t}) + \varepsilon_2$$

where $\varepsilon_1 \sim \mathrm{N}(0, \mathbf{Q}_{t|T})$ and $\varepsilon_2 \sim \mathrm{N}(0, \mathbf{R}_t)$. Using this equation, we obtain the mean

$$E\{s_t\} = s_{t|t} + \mathbf{W}_t(\mathbf{x}_{t+1|T} - \mathbf{C}_t s_{t|t}) = s_{t|T}$$

which obviously coincides with (7.4.66). To find the covariance matrix, we use the previous equation to write

$$s_t' - s_{t|t}' = (\mathbf{x}_{t+1}' - \mathbf{x}_{t+1|T}')\mathbf{W}_t' = [(s_{t+1} - s_{t+1|t+1})' \quad 0]\mathbf{W}_t' = \varepsilon_1'\mathbf{W}_{s,t}' \quad (7.4.71)$$

where $\mathbf{W}_{s,t}$ is the subblock of the matrix $\mathbf{W}_t$. Using this expression we obtain the variance matrix

$$E\{(s_{t+1} - s_{t+1|T})(s_t - s_{t|T})'\} = E\{\varepsilon_1 \varepsilon_1'\}\mathbf{W}_{s,t}' = \mathbf{Q}_{t+1|T}\mathbf{W}_{s,t}'$$

Therefore, we can rewrite Eq. (7.4.70) as

$$\gamma_t(s_t, s_{t+1}, \mathbf{a}_1^T) = \mathrm{N}(\xi_t - \xi_{t|T}, \Xi_{t|T})$$

where

$$\xi_t = \begin{bmatrix} s_{t+1} \\ s_t \end{bmatrix}, \quad \xi_{t|T} = \begin{bmatrix} s_{t+1|T} \\ s_{t|T} \end{bmatrix}, \quad \Xi_{t|T} = \begin{bmatrix} \mathbf{Q}_{t+1|T} & \mathbf{Q}_{t+1|T}\mathbf{W}_{s,t}' \\ \mathbf{W}_{s,t}\mathbf{Q}_{t+1|T} & \mathbf{Q}_{t|T} \end{bmatrix}$$

7.4.8 The Viterbi Algorithm

The Viterbi algorithm, given by Eq. (7.3.6) and (7.3.7) or (7.3.8), finds the sequence of conditional means $s_{1|T}, s_{2|T}, \ldots, s_{T|T}$. For the HGMM, the first equation in (7.3.6) is

$$\psi_1(s_1) = \max_{s_0} p(s_1, \mathbf{a}_1 \mid s_0) p_0(s_0) = \max_{s_0} \mathrm{N}(\mathbf{x}_1 - \mathbf{C}_0 s_0, \mathbf{S}_0)\mathrm{N}(s_0 - \boldsymbol{\mu}_0, \boldsymbol{\Sigma}_0)$$

Since the PDF in the right hand side of this equation is Gaussian, its maximum is proportional to the marginal PDF (see Eq. (A.7.30) of Appendix 7). Therefore,

$$\psi_1(s_1) = Cp(\mathbf{x}_1) = C\alpha(s_1, \mathbf{a}_1)$$

We see that $\psi_1(s_1)$ is proportional to the forward PDF given by equation (7.4.25) which in its turn is proportional to the normalized forward PDF given by equation (7.4.37). Therefore, we can write

$$\psi_1(s_1) = C_1 \mathrm{N}(s_1 - s_{1|1}, \mathbf{Q}_{1|1})$$

Repeating the Viterbi algorithm, we find that

$$\psi_k(s_k) = C_k \mathrm{N}(s_k - s_{k|k}, \mathbf{Q}_{k|k})$$

$k = 1, 2, \ldots, T$. This equation shows that, for the HGMM, the forward part of the Viterbi algorithm coincides with the forward algorithm and is performed using the Kalman filter.

At the end of the forward part we have $\psi(s_T) = C_T \mathrm{N}(s_T - s_{T|T}, \mathbf{Q}_{T|T})$. In the first step of the backward part of the Viterbi algorithm we compute

$$s_{T-1|T} = \underset{s_{T-1}}{argmax} \mathrm{N}(\mathbf{x}_{T|T} - \mathbf{C}_{T-1}, \mathbf{S}_{T-1})\mathrm{N}(s_{T-1} - \boldsymbol{\mu}_{T-1}, \boldsymbol{\Sigma}_0)$$

The solution to this type of equation is given by Eq. (A.7.21) of Appendix 7 which in our case has the form

$$s_{T-1|T} = s_{T-1|T-1} + \mathbf{W}_{T-1}(\mathbf{x}_T - \mathbf{C}_{T-1}s_{T-1|T-1})$$

where

$$\mathbf{W}_{T-1} = \mathbf{Q}_{T-1|T-1}\mathbf{C}'_{T-1}(\mathbf{C}_{T-1}\mathbf{Q}_{T-1|T-1}\mathbf{C}'_{T-1} + \mathbf{S}_{T-1})^{-1}$$

which coincides with the first equation of the RTS smoother. We can prove similarly that this is true for the rest of the backward steps of the Viterbi algorithm:

$$s_{k|T} = s_{k|k} + \mathbf{W}_k(\mathbf{x}_{k+1|T} - \mathbf{C}_k s_{k|k}), \quad k = T-1, T-2, ..., 0$$

Therefore, we conclude that the forward part of the Viterbi algorithm coincides with the forward algorithm and is performed using the Kalman filter while the backward part of the Viterbi algorithm is performed using the RTS smoother. The latter, however, computes not only the means $s_{k|T}$, but also the variance matrices $\mathbf{Q}_{k|T}$.

7.5 AUTOREGRESSIVE MOVING AVERAGE PROCESSES

Scalar autoregressive moving average (ARMA) process is defined by the following equation

$$\sum_{i=0}^{p} a_i y_{t-i} = \sum_{i=0}^{q} b_i w_{t-i}, \quad t = 1,2,... \tag{7.5.1}$$

where $a_0 \neq 0$, $b_0 \neq 0$, $w_t \sim \mathrm{N}(w_t, \sigma^2)$ and are independent. If $q = 0$, this equation represents the *autoregressive* (AR) process (see Appendix 7). If $p = 0$, it represents the *moving average* process.

Without loss of generality, we can assume that $a_0 = 1$ (otherwise we divide both sides of the equation by a_0).

Introducing the delay operator z^{-m} defined as $z^{-m}y_t = y_{t-m}$, we can present Eq. (7.5.1) as

$$a(z^{-1})y_t = b(z^{-1})w_t$$

where

$$a(z) = \sum_{i=0}^{p} a_i z^i, \quad b(z) = \sum_{i=0}^{q} b_i z^i$$

Thus, we can write

$$y_t + \sum_{i=1}^{p} a_i y_{t-i} = \sum_{i=0}^{q} b_i w_{t-i}, \quad t = 1,2,... \quad .$$

or

$$y_t = -\mathbf{a}'\mathbf{y}_{t-1} + \mathbf{b}'\mathbf{w}_t + b_0 w_t, \quad t = 1,2,...$$

where $\mathbf{a} = (a_1, a_2, \ldots, a_p)$, $\mathbf{b} = (b_1, b_2, \ldots, b_q)$, $\mathbf{y}_t = (y_t, y_{t-1}, \ldots, y_{t-p+1})$, and $\mathbf{w}_t = (w_t, w_{t-1}, \ldots, w_{t-q+1})$.

7.5.1 State Space Representation

To apply the previously developed theory, we need to represent the ARMA process in the state space form. Obviously, the ARMA process represents a partially observable Gauss-Markov process with states $\mathbf{x}_t = (\mathbf{w}_t, \mathbf{y}_t)$ where $\mathbf{y}_t = (y_t, y_{t-1}, \ldots, y_{t-p+1})$ represents the observable part of the state while $\mathbf{w}_t = (w_t, w_{t-1}, \ldots, w_{t-q+1})$ is not observable. As we know, any partially observable Markov process can be represented as an HGMM using the trivial augmentation of the hidden states described by Eq. (7.4.6). This representation is called the direct type I structure in the signal processing theory. It treats the complete state $\mathbf{x}_t$ as not observable and, therefore, the state dimension is $p+q$ which requires $p+q$ shift registers to store the state variable.

There are many other structures describing ARMA processes which require less shift registers than the direct type I structure. We consider here the most popular structure (which is used by the MATLAB function `filter`) the so-called transposed direct type II structure requiring the minimum number of the shift registers for its implementation. Without loss of generality and to simplify notation we can assume that vectors **a** and **b** have the same length (we can equalize the lengths by setting $a_{p+1} = a_{p+2} = \cdots = a_q = 0$ if $p < q$ or $b_{q+1} = b_{q+2} = \ldots = b_p = 0$ if $p > q$; the MATLAB function `eqtflength` performs this equalization).

To derive the state space equation, we rewrite Eq. (7.5.1) as

$$y_t = b_0 w_t + \sum_{i=1}^{p} (b_i w_{t-i} - a_i y_{t-i})$$

and introduce the state variables recursively as partial sums of the sum in the previous equation:

$$\eta_p^{(t-p)} = b_p w_{t-p} - a_p y_{t-p}, \quad \eta_i^{(t-i)} = \eta_{i+1}^{(t-i-1)} + b_i w_{t-i} - a_i y_{t-i} \tag{7.5.2}$$

$i = p-1, p-2, \ldots, 1$. It follows from these equations that

$$\eta_1^{(t-1)} = \sum_{i=1}^{p} (b_i w_{t-i} - a_i y_{t-i})$$

And, therefore,

$$y_t = b_0 w_t + \eta_1^{(t-1)}$$

The structure implementing these equations is depicted in Fig. 7.1. The state of the system at the moment t is defined by the contents of the shift registers $\boldsymbol{s}_t = (\eta_1^{(t)}, \eta_2^{(t)}, \ldots, \eta_{p-1}^{(t)})$. According to Eq. (7.5.2) and Fig. 7.1 we can write

$$\begin{aligned} \boldsymbol{s}_t &= \boldsymbol{\Gamma}_p \boldsymbol{s}_{t-1} + \mathbf{b} w_t - \mathbf{a} y_t \\ y_t &= \mathbf{H} \boldsymbol{s}_{t-1} + b_0 w_t \end{aligned} \tag{7.5.3}$$

where

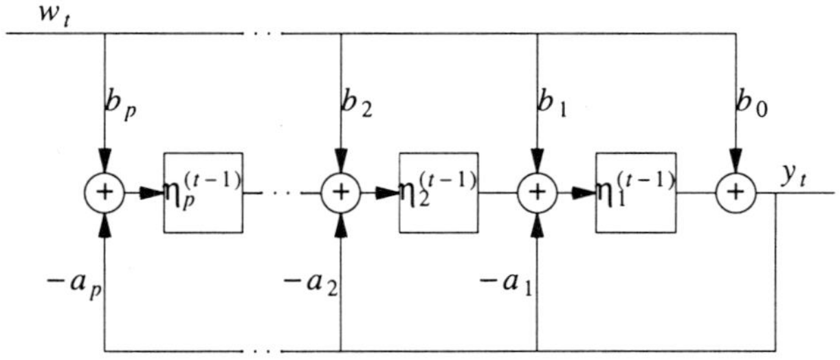

Figure 7.1. Transposed direct form II structure of the ARMA model.

$$\mathbf{\Gamma}_p = \begin{bmatrix} 0 & 1 & \cdots & 0 \\ 0 & 0 & \cdots & 0 \\ \cdots & \cdots & \cdots & \cdots \\ 0 & 0 & \cdots & 1 \\ 0 & 0 & \cdots & 0 \end{bmatrix}, \quad \mathbf{H} = [1 \ \ 0 \ \ 0 \ \ldots \ 0]$$

As we can see, Eq. (7.5.3) describes a partially observable Markov process. We can convert it into the state space form of the HGMM by eliminating y_t from the first equation using the second equation to obtain

$$\begin{aligned} \boldsymbol{s}_t &= \mathbf{F}\boldsymbol{s}_{t-1} + \mathbf{G}w_t \\ y_t &= \mathbf{H}\boldsymbol{s}_{t-1} + b_0 w_t \end{aligned} \tag{7.5.4}$$

where

$$\mathbf{F} = \mathbf{\Gamma}_p - \mathbf{aH} = \begin{bmatrix} -a_1 & 1 & \cdots & 0 \\ -a_2 & 0 & \cdots & 0 \\ \cdots & \cdots & \cdots & \cdots \\ -a_{p-1} & 0 & \cdots & 1 \\ -a_p & 0 & \cdots & 0 \end{bmatrix} \tag{7.5.5}$$

$$\mathbf{G} = \mathbf{b} - b_0\mathbf{a}$$

The MATLAB function `tf2ss` computes the matrices in Eq. (7.5.4).

The initial conditions are defined as the initial contents of the shift registers $\boldsymbol{s}_0 = (\eta_1^{(0)}, \eta_2^{(0)}, \ldots, \eta_p^{(0)})$. If we know first p initial observations $\boldsymbol{y}_1^p$, we can find the initial conditions producing these observations by solving first p equations in (7.5.4) for $\boldsymbol{s}_0$ (see example 7.8). The solution can be found using the MATLAB function `filtic`.

Equation (7.4.3) for the complete state $\mathbf{x}_k = (\boldsymbol{s}_k, y_k)$ has the form

$$\mathbf{x}_k = \mathbf{C}\boldsymbol{s}_{k-1} + \mathbf{u}_k \tag{7.5.6a}$$

where

$$\mathbf{C} = \begin{bmatrix} \mathbf{F} \\ \mathbf{H} \end{bmatrix} = \begin{bmatrix} -a_1 & 1 & 0 & \cdots & 0 \\ -a_2 & 0 & 1 & \cdots & 0 \\ \cdots & \cdots & \cdots & \cdots & \cdots \\ -a_{p-1} & 0 & 0 & \cdots & 1 \\ -a_p & 0 & 0 & \cdots & 0 \\ 1 & 0 & 0 & \cdots & 0 \end{bmatrix} \tag{7.5.6b}$$

$$\mathbf{u}_k = \sigma w_t \begin{bmatrix} \mathbf{G} \\ b_0 \end{bmatrix} \sim \mathrm{N}(\mathbf{u}_k, \mathbf{S}), \quad \mathbf{S} = \begin{bmatrix} \mathbf{S}_{ss} & \mathbf{S}_{sa} \\ \mathbf{S}_{as} & \mathbf{S}_{aa} \end{bmatrix} = \sigma^2 \begin{bmatrix} \mathbf{GG}' & b_0\mathbf{G} \\ b_0\mathbf{G}' & b_0^2 \end{bmatrix} \tag{7.5.6c}$$

As we can see, $\mathbf{u}_t$ is a singular Gaussian variable (see Appendix 7). Since all the columns (and rows) of the variance matrix $\mathbf{S}$ are proportional, $rank\mathbf{S} = 1$.

Example 7.8. Consider the ARMA process

$$y_t + a_1 y_{t-1} + a_2 y_{t-2} = w_t + b_1 w_{t-1}$$

In this case, $p=2$ and $q=1$. Therefore, we assume that $b_2=0$ to equalize the lengths of the vectors $\mathbf{a} = (a_1, a_2)$ and $\mathbf{b} = (b_1, 0)$.

Equation (7.5.4) has the form

$$\boldsymbol{s}_t = \begin{bmatrix} -a_1 & 1 \\ -a_2 & 0 \end{bmatrix} \boldsymbol{s}_{t-1} + \begin{bmatrix} b_1 - a_1 \\ -a_2 \end{bmatrix} w_t, \quad y_t = \begin{bmatrix} 1 & 0 \end{bmatrix} \boldsymbol{s}_{t-1} + w_t$$

If we want the two initial observations to have particular values y_1 and y_2, we can find the corresponding initial conditions $\eta_1^{(0)}, \eta_2^{(0)}$ by solving the system

$$y_1 = \eta_1^{(0)}, \quad y_2 = -a_1 y_1 + \eta_2^{(0)}$$

assuming that $w_1 = w_2 = 0$. We recommend the reader to verify the correctness of this statement using Fig. 7.1 in which shift register contents were originally $\eta_1^{(0)} = y_1$ and $\eta_2^{(0)} = y_2 + a_1 y_1$.

The complete state $\mathbf{x}_t = (\boldsymbol{s}_t, y_t)$ is described by

$$\mathbf{x}_t = \begin{bmatrix} -a_1 & 1 \\ -a_2 & 0 \\ 1 & 0 \end{bmatrix} \boldsymbol{s}_{t-1} + \mathbf{u}_{t-1}$$

where

$$\mathbf{u}_{t-1} \sim \mathrm{N}(\mathbf{u}_{t-1}, \mathbf{S}), \quad \mathbf{S} = \sigma^2 \begin{bmatrix} (b_1 - a_1)^2 & -a_2(b_1 - a_1) & b_1 - a_1 \\ -a_2(b_1 - a_1) & a_2^2 & -a_2 \\ b_1 - a_1 & -a_2 & 1 \end{bmatrix}$$

7.5.2 Autocovariance Function

The autocovariance function for the ARMA process can be obtained using equations developed in Sec. 7.3. In particular, for the stationary process the autocovariance function is given by Eq. (7.4.19). To use this equation, we need to find first the variance matrix $\boldsymbol{\Sigma}_{ss,\infty}$ (which exists only if all the eigenvalues of $\mathbf{F}$ are inside the unit circle of the complex plane). $\boldsymbol{\Sigma}_{ss,\infty}$ can be found by solving the Lyapunov Eq. (7.4.17). The solution can be found using the iterative algorithm A.6.2 or the the Yule-Walker Eq. (A.7.55) of Appendix 7. MATLAB function `dlyap` finds the solution. Then we find $\boldsymbol{\Sigma}_{aa,\infty}$ and $\boldsymbol{\Sigma}_{sa,\infty}$ using equation (7.4.18) and then substitute them into Eq. (7.4.19) to find the covariance function which in our case takes the form

$$R_{aa}(0) = \Sigma_{aa,\infty}, \quad R_{aa}(\tau) = \mathbf{H}\mathbf{F}^{-\tau-1}\Sigma_{sa,\infty} \quad \text{for} \quad \tau<0 \tag{7.5.7}$$

$R_{aa}(\tau) = R_{aa}(-\tau)$ for $\tau>0$.

Example 7.9. Let us find the autocovariance function for the ARMA process

$$y_k - a y_{k-1} = w_k + b w_{k-1}$$

The state space form for this process is given by Eq. (7.5.4) which in our case takes the following form

$$s_t = a s_{t-1} + (a+b) w_t, \quad y_t = s_{t-1} + w_t \tag{7.5.8}$$

In other words,

$$\mathbf{F} = a, \quad \mathbf{H} = 1, \quad \mathbf{G} = a+b, \quad \mathbf{S} = \sigma^2 \begin{bmatrix} (a+b)^2 & a+b \\ a+b & 1 \end{bmatrix}$$

We find $\Sigma_{ss,\infty} = \sigma^2(a+b)^2/(1-a^2)$ (which exists only if $|a|<1$) by solving the Lyapunov Eq. (7.4.17) $\Sigma_{ss,\infty} = a^2\Sigma_{ss,\infty} + \sigma^2(a+b)^2$. Equations (7.4.18) take the form

$$\Sigma_{aa,\infty} = \frac{(a+b)^2\sigma^2}{(1-a^2)} + \sigma^2, \quad \Sigma_{sa,\infty} = \frac{a(a+b)^2\sigma^2}{(1-a^2)} + (a+b)\sigma^2$$

and the autocovariance function is given by (7.5.7):

$$R_{aa}(0) = \Sigma_{aa,\infty}, \quad R_{aa}(\tau) = \Sigma_{sa,\infty} a^{|\tau|-1} \quad \text{for } \tau \neq 0$$

7.5.3 The Forward Algorithm

The forward algorithm for the ARMA process is realized by the Kalman filter. As we can see, Eq. (7.5.6) have the form of Eq. (7.4.45). Therefore, Eq. (7.5.6) represent the direct parametrization of the innovation model. Consequently, if the initial state $\boldsymbol{s}_0 = \boldsymbol{s}_{0|0}$, the Kalman filter is obtained by substituting w_t from the second equation of (7.5.4) into its first equation:

$$\boldsymbol{s}_{t|t} = \mathbf{F}\boldsymbol{s}_{t-1|t-1} + \mathbf{K}(y_t - \mathbf{H}\boldsymbol{s}_{t-1|t-1})$$

where the Kalman gain $\mathbf{K} = b_0^{-1}\mathbf{G} = b_0^{-1}\mathbf{b} - \mathbf{a}$.

Example 7.10. Let us consider the forward algorithm for the ARMA process given in Example 7.11 assuming that the initial state $s_0 \sim \mathrm{N}(s_0 - \mu_0, \sigma_0^2)$.

Using Algorithm 7.3, we obtain the following relations

$$Q_{t+1|t+1} = Q_{t|t}a^2 + \sigma^2(a+b)^2 - K_t[Q_{t|t}a + \sigma^2(a+b)]$$

$$K_t = [Q_{t|t}a + \sigma^2(a+b)](Q_{t|t} + \sigma^2)^{-1}$$

After simplifications, we obtain

$$Q_{t+1|t+1} = \sigma^2 b^2 Q_{t|t}(Q_{t|t} + \sigma^2)^{-1}, \quad K_t = [Q_{t|t}a + \sigma^2(a+b)](Q_{t|t} + \sigma^2)^{-1} \tag{7.5.9}$$

which gives us $K_0 = [\sigma_0^2 a + \sigma^2(a+b)](\sigma_0^2 + \sigma^2)^{-1}$, $s_{1|1} = a\mu_0 + K_0(y_1 - \mu_0)$, $Q_{1|1} = \sigma^2 b^2 \sigma_0^2(\sigma_0^2 + \sigma^2)^{-1}$, $K_1 = [Q_{1|1}a + \sigma^2(a+b)](Q_{1|1} + \sigma^2)^{-1}$, and so on. As we can see, the Kalman filter is time varying even if the HGMM is stationary (in which case $\sigma_0^2 = \Sigma_{ss,\infty}$).

It follows from Eq. (7.5.9) that $\lim_{t\to\infty} Q_{t|t} = 0$ if $|b|<1$. In this case, $\lim_{t\to\infty} K_t = a+b$. Thus, the *stationary Kalman filter* takes the form

$$s_{t+1|t+1} = a s_{t|t} + (a+b)(y_{t+1} - s_{t|t})$$

This equation can be obtained by substituting w_t from the second equation of (7.5.8) into its first equation. If the initial state is fixed ($s_{0|0} = s_0$ and $\sigma_0 = 0$), then Eq. (7.5.8) coincides with the Kalman filter. Using this equation we can write

$$s_{t+1|t+1} = -bs_{t|t} + (a+b)y_{t+1} = b^2 s_{t-1|t-1} - b(a+b)y_t + (a+b)y_{t+1} = \cdots$$
$$= (a+b)\sum_{k=0}^{\infty}(-b)^k y_{t-k+1}$$

Thus, the stationary Kalman filter can be viewed as the weighted sum of all the previous observations.

7.5.4 Power Spectrum

Let us compute the PSD for the ARMA model using Eq. (7.4.47) which in our case has the form $\mathbf{\Phi}_{yy} = W(z)\,W(z^{-1})\,\sigma^2$ where

$$W(z) = 1 + \mathbf{H}z^{-1}(\mathbf{I} - \mathbf{F}z^{-1})^{-1}\mathbf{K} \tag{7.5.10}$$

Without loss of generality and to simplify notation we assume that $b_0 = 1$. (Otherwise, we replace $\mathbf{b}$ with $\mathbf{b}_1 = b_0^{-1}\mathbf{b}$ and replace σ^2 with $\sigma_1^2 = \sigma^2 b_0^2$.) In this case, $\mathbf{K} = \mathbf{b} - \mathbf{a}$ and It is not difficult to show (see Problem 2) that

$$\mathbf{H}(\mathbf{I} - \mathbf{F}z)^{-1} = \frac{1}{a(z)}[1 \;\; z \;\; \ldots \;\; z^{p-1}]$$

where $a(z) = 1 + a_1 z + \ldots + a_p z^p$. Also

$$[1 \;\; z \;\; \ldots \;\; z^{p-1}]\mathbf{K} = [1 \;\; z \;\; \ldots \;\; z^{p-1}](\mathbf{b} - \mathbf{a}) = z^{-1}[b(z) - a(z)]$$

where $b(z) = 1 + b_1 z + \ldots + b_p z^p$. Consequently, Eq. (7.5.10) takes the form

$$W(z) = 1 + [b(z) - a(z)]/a(z) = a(z)/b(z)$$

and Eq. (7.4.47) gives us the PSD

$$\Phi_{yy}(z) = \frac{b(z)b(z^{-1})}{a(z)a(z^{-1})}\sigma^2$$

The PSD on the unit circle represents the Fourier transform of the autocorrelation function and can be written as

$$\Phi_{yy}(e^{j\omega}) = \frac{|b(e^{j\omega})|^2}{|a(e^{j\omega})|^2}\sigma^2$$

7.6 PARAMETER ESTIMATION

7.6.1 HMM Approximation

As in Sec. 3.2.2, consider a problem of approximating a process with multidimensional PDF $f(\mathbf{a}_1^T)$ with an HMM by minimizing the KLD between their PDFs which is equivalent to maximizing the likelihood function (see Sec. 3.1.3):

$$L(\tau) = \int_{A^T} f(\mathbf{a}_1^T)\log p(\mathbf{a}_1^T;\tau)\,d\mathbf{a}_1^T = \mathbf{E}\{\log p(\mathbf{a}_1^T;\tau) \mid f(\mathbf{a}_1^T)\} \tag{7.6.1}$$

where τ is the unknown parameter and $p(\mathbf{a}_1^T;\tau)$ is defined by Eq. (7.1.3) or (7.2.3).

This problem can be solved using the EM algorithm. The EM algorithm for the continuous state HMM can be derived similarly to its discrete counterpart of Sec. 3.2.2:

$$\tau_{p+1} = \underset{\tau}{argmax}\, Q(\tau,\tau_p) \tag{7.6.2}$$

where

$$Q(\tau,\tau_p) = E\{q(\mathbf{a}_1^T,\tau,\tau_p) \mid f(\mathbf{a}_1^T)\}, \quad p=0,1,... \tag{7.6.3}$$

$$q(\mathbf{a}_1^T,\tau,\tau_p) = E\{\log\psi(\mathbf{s}_0^T;\mathbf{a}_1^T,\tau) \mid \kappa(\mathbf{i}_0^T;\mathbf{a}_1^T,\tau_p)\} \tag{7.6.4}$$

As in Sec. 3.2.2, we define the auxiliary function $\kappa(\mathbf{s}_0^T;\mathbf{a}_1^T,\tau) = p(\mathbf{s}_0^T \mid \mathbf{a}_1^T,\tau)$. It follows from this definition that

$$\psi(\mathbf{s}_0^T;\mathbf{a}_1^T,\tau) = \kappa(\mathbf{s}_0^T;\mathbf{a}_1^T,\tau)p(\mathbf{a}_1^t;\tau) = p(\mathbf{s}_0^T,\mathbf{a}_1^T;\tau)$$

is the PDF of the complete state sequence $\mathbf{x}_t = (\mathbf{s}_t,\mathbf{a}_t)$. Since $\mathbf{x}_t$ is a Markov process, then

$$\psi(\mathbf{s}_0^T;\mathbf{a}_1^T,\tau) = p_0(\mathbf{s}_0,\tau)\prod_{t=1}^{T} p(\mathbf{x}_t \mid \mathbf{s}_{t-1};\tau)$$

and, therefore,

$$\log\psi(\mathbf{s}_0^T;\mathbf{a}_1^T,\tau) = \log p_0(\mathbf{s}_0,\tau) + \sum_{t=1}^{T}\log p(\mathbf{x}_t \mid \mathbf{s}_{t-1};\tau)$$

Substituting this expression into Eq. (7.6.4) we obtain

$$q(\mathbf{a}_1^T,\tau,\tau_p) = q_0(\mathbf{a}_1^T,\tau,\tau_p) + q_1(\mathbf{a}_1^T,\tau,\tau_p) \tag{7.6.5}$$

where

$$q_0(\mathbf{a}_1^T,\tau,\tau_p) = E\{\log p_0(\mathbf{s},\tau) \mid \kappa(\mathbf{s}_0^T;\mathbf{a}_1^T,\tau_p)\}$$

$$q_1(\mathbf{a}_1^T,\tau,\tau_p) = \sum_{t=1}^{T} E\{\log p(\mathbf{x}_t \mid \mathbf{s}_{t-1};\tau) \mid \kappa(\mathbf{s}_0^T;\mathbf{a}_1^T,\tau_p)\}$$

Since $\kappa(\mathbf{s}_0^T;\mathbf{a}_1^T,\tau) = p(\mathbf{s}_0^T \mid \mathbf{a}_1^T,\tau)$ we can simplify these expressions:

$$q_0(\mathbf{a}_1^T,\tau,\tau_p) = E\{\log p_0(\mathbf{s},\tau) \mid \gamma_0(\mathbf{s},\mathbf{a}_1^T,\tau_p)\}$$

$$q_1(\mathbf{a}_1^T,\tau,\tau_p) = \sum_{t=1}^{T} E\{\log p(\mathbf{x}_t \mid \mathbf{s}_{t-1};\tau) \mid \gamma_{t-1}(\mathbf{s}_{t-1},\mathbf{s}_t,\mathbf{a}_1^T,\tau_p)\}$$

where $\gamma_{t-1}(\mathbf{s}_{t-1},\mathbf{s}_t,\mathbf{a}_1^T,\tau) = p(\mathbf{s}_{t-1},\mathbf{s}_t \mid \mathbf{a}_1^T,\tau)$ can be evaluated using the forward-backward algorithm described in Sec. 7.2 and 7.3. Using these equations, we can express Eq. (7.6.3) as

$$Q(\tau,\tau_p) = Q_0(\tau,\tau_p) + Q_1(\tau,\tau_p) \tag{7.6.6}$$

where

$$\begin{aligned} Q_0(\tau,\tau_p) &= E\{\log p_0(\mathbf{s},\tau) \mid \Gamma_0(\mathbf{s},\tau,\tau_p)\} \\ \Gamma_0(\mathbf{s},\tau,\tau_p) &= E\{\gamma_0(\mathbf{s},\mathbf{a}_1^T,\tau_p) \mid f(\mathbf{a}_1^T)\} \end{aligned} \tag{7.6.7}$$

$$Q_1(\tau,\tau_p) = \sum_{t=1}^{T} E\{\log p(\mathbf{x}_t \mid \boldsymbol{s}_{t-1};\tau) \mid \rho_{t-1}(\boldsymbol{s}_{t-1},\boldsymbol{s}_t,\mathbf{a}_1^T,\tau_p)\}$$
$$\rho_{t-1}(\boldsymbol{s}_{t-1},\boldsymbol{s}_t,\mathbf{a}_1^T,\tau_p)\} = \gamma_{t-1}(\boldsymbol{s}_{t-1},\boldsymbol{s}_t,\mathbf{a}_1^T,\tau)f(\mathbf{a}_1^T) \qquad (7.6.8)$$

and then apply the EM algorithm given by Eq. (7.6.2).

7.6.1.1 Fitting HMM to Experimental Data

Suppose that we have multiple independent observation sequences $\{\mathbf{a}_1\}_1^{T_1} = (\mathbf{a}_{11},\mathbf{a}_{12},\ldots,\mathbf{a}_{1,T_1})$, $\{\mathbf{a}_2\}_1^{T_2} = (\mathbf{a}_{21},\mathbf{a}_{22},\ldots,\mathbf{a}_{2,T_2})$,..., $\{\mathbf{a}_K\}_1^{T_K} = (\mathbf{a}_{K1},\mathbf{a}_{K2},\ldots,\mathbf{a}_{K,T_2})$ of the same HMM. In this case, the maximum likelihood estimate of the model parameters is

$$\hat{\tau} = \underset{\tau}{arg\max} \sum_{k=1}^{K} \log p(\{\mathbf{a}_k\}_1^{T_k},\tau)$$

As in Sec 3.2.3 we can show that this problem is a special case of the HMM approximation in which the expectation $E\{(\cdot) \mid f(\boldsymbol{x}_1^T)\}$ with respect to $f(\boldsymbol{x}_1^T)$ is replaced by $(1/K)\sum_1^K(\cdot)$. In this case, the EM algorithm described by Eq. (7.6.2) takes the form

$$\tau_{p+1} = \underset{\tau}{arg\max} \sum_{k=1}^{K} q(\{\mathbf{a}_k\}_1^{T_k},\tau,\tau_p), \quad p=0,1,...$$

As we can see, the EM algorithm for the HMM approximation is similar to its discrete counterpart of Sec. 3.2.2 and can be considered as a generalization of the latter (if we use δ-functions). Note that the continuous state algorithm is, generally, more complex than the finite state algorithm, because integration is a more complex operation than multiplication of matrices. However, as we will show in the next sections, it is possible to find closed form expressions for the integrals in the case of HGMM.

7.6.2 HGMM Approximation

We consider a case of approximating a continuous state process with the multidimensional PDF $f(\mathbf{a}_1^T)$ by the homogeneous HGMM which is described by equation (7.4.3) which in our case takes the form

$$\boldsymbol{s}_{t+1} = \mathbf{F}\boldsymbol{s}_t + \mathbf{u}_{s,t}$$
$$\mathbf{a}_{t+1} = \mathbf{H}\boldsymbol{s}_t + \mathbf{u}_{a,t}$$

If we denote as $\mathbf{C}' = [\mathbf{F}',\mathbf{H}']$ and $\mathbf{u}_t = (\mathbf{u}_{s,t},\mathbf{u}_{a,t})$, this equation takes a more compact form:

$$\mathbf{x}_{t+1} = \mathbf{C}\boldsymbol{s}_t + \mathbf{u}_t$$

We assume that $\mathbf{u}_t \sim N(0,\mathbf{S})$ and the initial state PDF is also Gaussian $\boldsymbol{s}_0 \sim N(\boldsymbol{\mu}_0,\boldsymbol{\Sigma}_0)$. Our goal is to find the parameter $\tau = \{\mathbf{C},\mathbf{S},\boldsymbol{\mu}_0,\boldsymbol{\Sigma}_0\}$ maximizing the likelihood function (7.6.1).

To solve this problem using the EM algorithm (7.6.2), we need to find the auxiliary function $q(\mathbf{a}_1^T,\tau,\tau_p)$ given by Eq. (7.6.5). Since $p_0(\boldsymbol{s},\tau) = \mathrm{N}(\boldsymbol{s}-\boldsymbol{\mu}_0,\boldsymbol{\Sigma}_0)$ and

$p(\mathbf{x}_t \mid s_{t-1}) = \mathrm{N}(\mathbf{x}_t - \mathbf{C}_{t-1} s_{t-1}, \mathbf{S})$, Eq. (7.6.7) and (7.6.8) with some abuse of notation can be written as

$$Q_0(\tau,\tau_p) = -0.5E\{[\,(s-\mu_0)'\Sigma_0^{-1}(s-\mu_0)+\log|\Sigma_0|\,] \mid \Gamma_0(s,\tau_p)\}$$

$$Q_1(\tau,\tau_p) = -0.5\sum_{t=1}^{T} E\{[(\mathbf{x}_t-\mathbf{C}s_{t-1})'\mathbf{S}^{-1}(\mathbf{x}_t-\mathbf{C}s_{t-1})+\log|\mathbf{S}|] \mid \rho_{t-1}(s_{t-1},s_t,\mathbf{a}_1^T,\tau_p)\}$$

where we neglected the additive constants that do not depend on τ.

Since the model parameters are independent, we can maximize $Q_0(\tau,\tau_p)$ and $Q(\tau,\tau_p)$ separately. It is not difficult to show (see Example 3.1.2 of Chapter 3) that the maximum of $Q_0(\tau,\tau_p)$ is attained at

$$\mu_{0,p+1} = E\{s \mid \Gamma_0(s,\tau_p)\}$$

and

$$\Sigma_{0,p+1} = E\{(s-\mu_{0,p+1})(s-\mu_{0,p+1})' \mid \Gamma_0(s,\tau_p)\}$$

To find a maximum of $Q_1(\tau,\tau_p)$ we use the identity

$$(\mathbf{x}_t-\mathbf{C}s_{t-1})'\mathbf{S}^{-1}(\mathbf{x}_t-\mathbf{C}s_{t-1}) = tr[\mathbf{S}^{-1}(\mathbf{x}_t-\mathbf{C}s_{t-1})(\mathbf{x}_t-\mathbf{C}s_{t-1})']$$

and the properties of the trace operator (see Appendix 5) to express $Q_1(\tau,\tau_p)$ as

$$Q_1(\tau,\tau_p) = -0.5tr[\mathbf{S}^{-1}(\mathbf{B}_{1,p}-\mathbf{C}\mathbf{B}_{2,p}'-\mathbf{B}_{2,p}\mathbf{C}'+\mathbf{C}\mathbf{B}_{3,p}\mathbf{C}')]-0.5T\log|\mathbf{S}| \qquad (7.6.9)$$

where

$$\mathbf{B}_{1,p} = \sum_{t=1}^{T} E\{\mathbf{x}_t\mathbf{x}_t' \mid \rho_{t-1}(s_{t-1},s_t,\mathbf{a}_1^T,\tau_p)\} \qquad (7.6.10a)$$

$$\mathbf{B}_{2,p} = \sum_{t=1}^{T} \mathrm{E}\{\mathbf{x}_t s_{t-1}' \mid \rho_{t-1}(s_{t-1},s_t,\mathbf{a}_1^T,\tau_p)\} \qquad (7.6.10b)$$

$$\mathbf{B}_{3,p} = \sum_{t=1}^{T} \mathrm{E}\{s_{t-1} s_{t-1}' \mid \rho_{t-1}(s_{t-1},s_t,\mathbf{a}_1^T,\tau_p)\} \qquad (7.6.10c)$$

It is not difficult to show (see Eq. (7.6.9), (A.7.60), and (A.7.62) of Appendix 7) that the maximization step for $Q_1(\tau,\tau_p)$ can be expressed by equations

$$\mathbf{C}_{p+1} = \mathbf{B}_{2,p}\mathbf{B}_{3,p}^{-1} \qquad (7.6.11a)$$

$$\mathbf{S}_{p+1} = \frac{1}{T}(\mathbf{B}_{1,p}-\mathbf{B}_{2,p}\mathbf{B}_{3,p}^{-1}\mathbf{B}_{2,p}') = \frac{1}{T}(\mathbf{B}_{1,p}-\mathbf{C}_{p+1}\mathbf{B}_{2,p}') \qquad (7.6.11b)$$

provided that $\mathbf{B}_{3,p}^{-1}$ exists.

7.6.2.1 Fitting HGMM to Experimental Data

Since fitting the a HGMM to experimental data $\{\mathbf{a}_1\}_1^{T_1} = (\mathbf{a}_{11},\mathbf{a}_{12},\ldots,\mathbf{a}_{1,T_1})$, $\{\mathbf{a}_2\}_1^{T_2} = (\mathbf{a}_{21},\mathbf{a}_{22},\ldots,\mathbf{a}_{2,T_2})$,..., $\{\mathbf{a}_K\}_1^{T_K} = (\mathbf{a}_{K1},\mathbf{a}_{K2},\ldots,\mathbf{a}_{K,T_2})$ is a special case of approximation when $\Gamma_0(s,\tau) = \sum_{k=1}^{K}\gamma_0(s,\{\mathbf{a}_k\}_1^{T^k},\tau)/K$ and

$$\rho_{t-1}(\boldsymbol{s}_{t-1},\boldsymbol{s}_t,\{\mathbf{a}_k\}_1^{T_k},\tau_p) = \gamma_{t-1}(\boldsymbol{s}_{t-1},\boldsymbol{s}_t,\{\mathbf{a}_k\}_1^{T_k},\tau_p)/K$$

we can use the EM algorithm given by Eq. (7.6.11). To simplify notation, we consider only the case of one experimental sequence $\mathbf{a}_1^T$. (As we pointed out previously, the general case is obtained by simple averaging with respect to the multiple sequences.) For the one experimental sequence $(K=1)$, $\Gamma_0(\boldsymbol{s},\tau) = \gamma_0(\boldsymbol{s},\mathbf{a}_1^T,\tau)$, and $\rho_{t-1}(\boldsymbol{s}_{t-1},\boldsymbol{s}_t,\mathbf{a}_1^T,\tau_p) = \gamma_{t-1}(\boldsymbol{s}_{t-1},\boldsymbol{s}_t,\mathbf{a}_1^T,\tau_p)$. In this case we have $\boldsymbol{\mu}_{0,p+1} = \boldsymbol{s}_{0|T,p}$ and $\boldsymbol{\Sigma}_{0,p+1} = \mathbf{Q}_{0|T,p}$ where index p means that the smoothing has been performed using the HGMM parameters obtained at p-th iteration (reestimation) of the EM algorithm.

$$\mathbf{B}_{1,p} = \sum_{t=1}^{T} E\{\mathbf{x}_t\mathbf{x}_t' \mid \gamma_{t-1}(\boldsymbol{s}_{t-1},\boldsymbol{s}_t,\mathbf{a}_1^T,\tau_p)\}, \quad \text{where} \quad \begin{bmatrix} \boldsymbol{s}_t\boldsymbol{s}_t' & \boldsymbol{s}_t\mathbf{a}_t' \\ \mathbf{a}_t\boldsymbol{s}_t' & \mathbf{a}_t\mathbf{a}_t' \end{bmatrix}$$

The variance matrix $E\{(s_t-s_{t|T})(s_t-s_{t|T})'\} = \mathbf{Q}_{t|T}$ can be found using the forward-backward algorithm (the Kalman filtering followed by the RTS smoother). From this equation, we find $E\{s_ts_t'\} = \mathbf{Q}_{t|T} + \boldsymbol{s}_{t|T}\boldsymbol{s}_{t|T}'$. Therefore,

$$\mathbf{B}_{1,p} = \sum_{t=1}^{T} \begin{bmatrix} \mathbf{Q}_{t|T,p}+\boldsymbol{s}_{t|T,p}\boldsymbol{s}_{t|T,p}' & \boldsymbol{s}_{t|T,p}\mathbf{a}_t' \\ \mathbf{a}_t\boldsymbol{s}_{t|T,p}' & \mathbf{a}_t\mathbf{a}_t' \end{bmatrix} \tag{7.6.12a}$$

The rest of the expectations in Eq. (7.6.10) can be found similarly

$$\mathbf{B}_{2,p} = \sum_{t=1}^{T} \begin{bmatrix} \mathbf{Q}_{t|T,p}\mathbf{W}_{s,t}'+\boldsymbol{s}_{t|T,p}\boldsymbol{s}_{t-1|T,p}' \\ \mathbf{a}_t\boldsymbol{s}_{t-1|T,p}' \end{bmatrix} \tag{7.6.12b}$$

where $\mathbf{W}_{s,t}$ is defined by Eq. (7.4.71), and

$$\mathbf{B}_{3,p} = \sum_{t=1}^{T} \mathbf{Q}_{t-1|T,p}+\boldsymbol{s}_{t-1|T,p}\boldsymbol{s}_{t-1|T,p}' \tag{7.6.12c}$$

7.7 ARMA CHANNEL MODELING

7.7.1 Fading Channel

In this section, we describe the ARMA modeling of the wireless fading channels based on the real measured data which was collected at 1900 Mhz along several drive routes in a residential area and on a highway, with vehicle speeds of 30 and 65 mph, and downrange distances between 2 to 5 miles. Further details on the test system and measurements are presented in Ref. 11. We considered two cases. [18] In case R (for "residential"), vehicle moves at 30 mph while in case H (for "highway") it moves at 65 mph. Our results show that ARMA modeling of the fading process is quite adequate while the broadly used theoretical model [3,8] does not agree with the experimental data. [18]

Let $x(t)$ be a low-pass equivalent of the transmitted signal with the inphase component $x_I(t) = \text{Re}\{x(t)\}$ and quadrature component $x_Q(t) = \text{Im}\{x(t)\}$. Consider a frequency-nonselective (flat) fading channel with the additive white Gaussian noise (AWGN) $n(t)$. This channel can be modeled as [12]

$$y(t) = c(t)x(t) + n(t) \tag{7.7.1}$$

where $y(t)$ is the received signal and fading is modeled by the complex random process $c(t)$, channel noise and fading are assumed to be independent. In the case of Rayleigh fading, $c(t)$ is a stationary zero-mean Gaussian processes with independent and identically distributed real and imaginary parts. It is called the Rayleigh fading, because its envelope $\alpha(t) = |c(t)|$ is Rayleigh distributed:

$$Pr\{\alpha(t)<a\} = 1-\exp(-0.5a^2/\mu)$$

If fading mean is non-zero, the fading is called Rician which accounts for the presence of a line-of-sight (LOS) component.

Different models of fading channels are based on different assumptions about the fading PSD $S(f)$ (or autocorrelation function $R(\tau)$). The Clarke's model [3,8] assumes that

$$S(f) = \mu/\pi\sqrt{f_D^2 - f^2}, \qquad R(\tau) = \mu J_0(2\pi f_D|\tau|)$$

where $J_0(\cdot)$ is the Bessel function of the first kind and f_D is the maximal Doppler frequency. The normalized autocorrelation function $J_0(2\pi f_D|\tau|)$ has only one independent parameter and, therefore, fits poorly to the experimental data that are not necessarily agree with the theoretical model. The ARMA model, on the other hand, is more versatile because its PSD is a rational function and, therefore, can approximate reasonably well any measured PSD. It is also more convenient to use the ARMA model because of the rich theory of HGMM and availability of MATLAB tools.

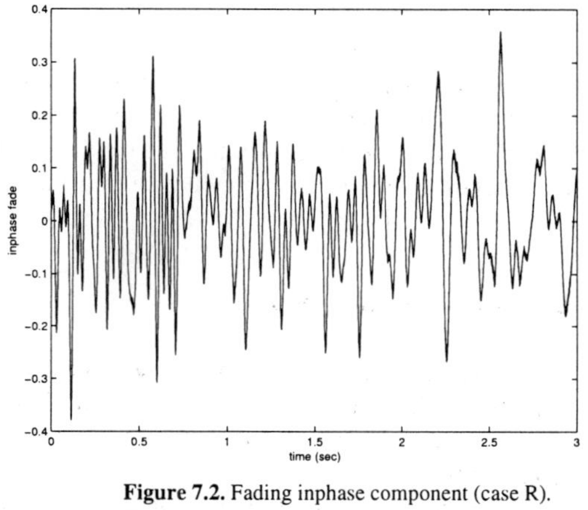

Figure 7.2. Fading inphase component (case R).

Based on the measured data, we found that the inphase and quadrature components of the fading are independent and can be studied independently. A sample of the inphase component (for the case R) is depicted in Figure 7.2. The model's power spectral density of the ARMA model for this component is compared with that of measured data in Figure 7.3. As we can see in this figure, they agree quite well and do not agree with the PSD of the

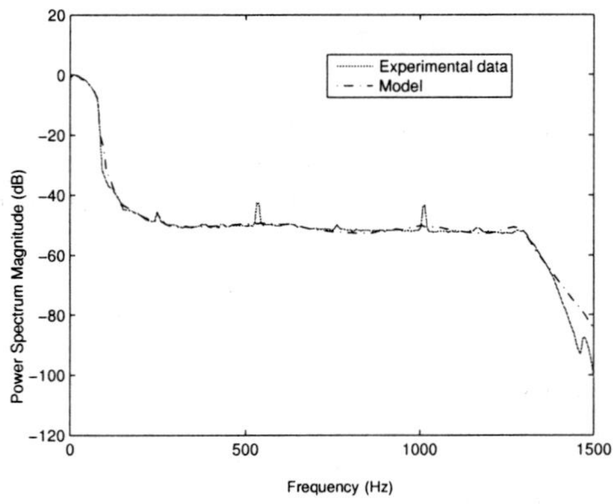

Figure 7.3. Comparison of power spectral densities (case R).

Clarke's model which has a characteristic U-shape. [3,8]

We would like to note in conclusion that, for the mobile applications, a better choice would be to use a switched HGMM model in which the observation sequence is described by the HGMM

$$s_t = \mathbf{F}(r_{t-1})s_{t-1} + \mathbf{G}(r_{t-1})w_t$$

$$y_t = \mathbf{H}(r_{t-1})s_{t-1} + b_0(r_{t-1})w_t$$

whose parameters depend on the switching process r_t. The switching process describes the changing environment (such as velocity, shadowing, etc) in mobile applications while the HGMM models the fading for these various environments. The swithcing process is usually described by a discrete state Markov chain whose parameters are estimated by aggregating the observation process.

7.7.2 Ultra-Wide Bandwidth Radio Channel Modeling

Typically, the ARMA model is applied to describing correlated stochastic processes in the time domain (as we demonstrated in the previous section). In this section, we demonstrate the model application in the frequency domain to describe the correlation between the frequency responses at different frequencies.

Based on the experimental data for the in-home environment [19] we found that the ultra-wide bandwidth (UWB) channel frequency response does not exhibit significant variability in time and can be modeled by the AR model of second order:

$$H(f_i) + a_1 H(f_{i-1}) + a_2 H(f_{i-2}) = w_i$$

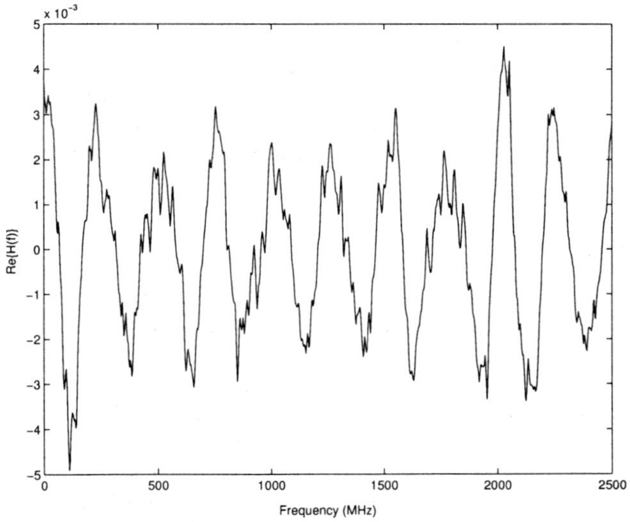

Figure 7.4. Measured real part of frequency response.

where $w_i \sim N(w_i, \sigma^2)$. The coefficients of this equation, the initial conditions, and σ^2 are changing from house to house and also depend on the distance between the transmitter and receiver. A similar approach was used in Ref. 6 and 7.

Typically, for the time domain processes we have one long observation sequence so that we do not have enough statistical data to estimate the PDF of the initial conditions. For the frequency response data, we have many short observation sequences allowing us to estimate this PDF. We applied the MATLAB function `filtic` to obtain the initial conditions $\eta_1^{(0)}$ and $\eta_2^{(0)}$ (see Fig 7.3) from the experimental data and then found they are strongly correlated with each other but independent from the rest of the model parameters. The model coefficients a_1 and a_2 are represented by the poles p_1 and p_2 which are the roots of the polynomial $z^2 + a_1 z + a_2$. We found that the poles are correlated but independent from the rest of the parameters. The white noise standard deviation does not depend on the rest of the model parameters. The results of our study are reported in Ref. 19.

As an illustration, consider one of the models corresponding to 2 ft transmitter-receiver separation. A sample of measured data is depicted in Fig. 7.4 We fitted the AR model to the frequency response and obtained the following parameters

$$a_1 = -1.5025 + j0.8017, \quad a_2 = 0.4427 + j0.7538,$$
$$\eta_1 = (3.168 - j0.003)10^{-3}, \quad \eta_2 = (-2.18 + j2.86)10^{-3}, \quad \sigma = 1.36\ 10^{-4}$$

We found [19] that this model agrees reasonably well with the experimental data. The model based computations of various time domain characteristics also agree with that obtained using the experimental data. For example, we obtained good agreement between the estimated and theoretically predicted delay profiles. [13] The popular time domain models of wireless channels assume that the delay profile samples at different delays are independent

Gaussian variables. We found that these models use more parameters and are less accurate than the frequency domain model for the UWB channel.

Some models assume that the delay profile can be represented by the exponential function with the AWGN. According to our model, however, the mean delay profile is represented by the rational function which agrees better with the experimental data than the exponential function.

7.8 CONCLUSION

Continuous state HMMs are very popular in many applications. Traditionally, their subset presented by the stochastic linear equations was studied thoroughly without referencing the HMMs. The theory of linear stochastic systems was devoted predominantly to estimating of their states and parameters. The theory of discrete state HMMs on the other hand, was developed independently using a more "probabilistic" approach consisting of computing of various probabilities. In this chapter, we applied this approach to studying the continuous state HMMs. This allowed us to consider all the HMMs using the same framework, clearly present the results related general (not necessarily Gaussian) HMMs, and obtain basic equations related to the Gaussian HMMs much simpler than it is done using the traditional approach.

Traditionally, the non-Gaussian HMMs are approximated by the HGMM to be able to apply their theory. By considering them in the framework of the general HMMs it is possible to use the alternative approaches using the quantized state approximation, numerical methods for computing integrals, and mixed discrete-continuous state HMMs.

We show that the continuous state HMMs can be defined as partially observable Markov processes. This definition is more natural and clearer than the traditional one. It covers not only discrete and continuous state HMMs, but also the mixed discrete-continuous state HMMs. The introduction of the operator probabilities allowed us to generalize methods of the matrix probability theory to the operator probability theory. The matrix probability theory is a special case of the operator probability theory.

We do not consider the case of continuous time and space HMM. The theory of these processes is beyond the scope of this book primarily oriented on applications. The theory can be developed as a generalization of the discrete time HMM theory similarly to Chapter 6. We just would like to point out that this theory in some cases is simpler. For example, the backward differential equations are obtained from the forward equations by a simple time variable substitution. Any implementation of a system of differential equations on a *digital* computer is a *discrete time* finite difference system. Therefore, from the practical point of view, it is sufficient to consider only discrete time models.

Problems

1. Derive the Kalman filter by writing the mean and covariance matrix for $(\boldsymbol{s}_t, \mathbf{a}_1^t)$ (see Example 7.6).
2. Let $\mathbf{Y}_t = [y_t \;\; y_{t-1} \;\; \cdots \;\; y_{t-p+1}]$

   — Show that

$$\sum_{i=p-1} \mathbf{Y}_i z^i = [\psi_{p-1}(z) \quad \psi_{p-2}(z) \quad \ldots \quad \psi_0(z)]$$

where

$$\psi_m(z) = z^{-m}(\phi(z) - \sum_{k=0}^{m-1} y_k z^k) \tag{7.P.1}$$

and $\phi(z) = \sum_{i=0}^{\infty} y_i z^i$ is the generating function for the sequence $\mathbf{y}_0^{\infty}$.

— Show that the generating function for the solution of the linear homogeneous equations

$$y_t + \sum_{i=1}^{p} a_i y_{t-i} = 0, \quad t = p, p+1, \ldots \tag{7.P.2}$$

with the initial condition $(y_{p-1}, y_{p-2}, \ldots, y_0)$ has the form

$$\phi(z) = [y_0 a_1(z) + z y_1 a_2(z) + \ldots + z^{p-1} y_{p-1} a_p(z)] / a_0(z) \tag{7.P.3}$$

where $a_m(z) = 1 + a_1 z + a_2 z^2 + \ldots + a_{p-m} z^{p-m}$

— Show that, for the matrix $\mathbf{F}$ given by Eq. (7.5.5),

$$\mathbf{Y}_{p-1}(\mathbf{I} - \mathbf{F}z)^{-1} = [\psi_{p-1}(z) \quad \psi_{p-2}(z) \quad \ldots \quad \psi_0(z)]$$

where $\psi_m(z)$ and $\phi(z)$ are given by Eq. (7.P.1) and (7.P.3), respectively.
(Hint: the left hand side of this equation is the generating function for the solution of equation $\mathbf{Y}_t = \mathbf{Y}_{t-1}\mathbf{F}$ which can be written as (7.P.2).)

3. Show that

$$p(s_{t-1} \mid s_t, \mathbf{a}_1^t) = p(s_{t-1} \mid s_t, \mathbf{a}_1^T)$$

[Hint: use Eq. (7.4.62)]

References

1 P. L. Ainsleigh, N. Kehtarnavaz, and R. L. Streit, "Hidden Gauss-Markov models for signal classification," *IEEE Trans. Signal Proc.*, **50**(6), 1355-1367, June (2002).

2 B. D. O. Anderson and J. B. Moore, *Optimal Filtering*, Prentice-Hall, Inglewood Cliffs, N.J., 1979

3 R. H. Clarke, "A statistical theory of mobile radio reception," *Bell Syst. Tech. J.*, **47** 957-1000, 1968.

4 B. Delyon, "Remarks on linear and nonlinear filtering," *IEEE Trans. Inf. Theory,* **IT-41**(1), 317-322 (1995).

5 V. Digalakis, J. R. Rohlicek, and M. Ostendorf, "ML estimation of a stochastic linear system with the EM algorithm and its application to speech recognition," *IEEE Trans. Speech Audio Processing*, **1**, 431-442, Oct. (1993).

6 S.J. Howard, K. Pahlavan, " Measurement and Analysis of the indoor radio channel in the frequency domain", *IEEE Trans. Instrum. Measure.,*, **39** 751-755, Oct. (1990).

7 S.J. Howard, K. Pahlavan, "Autoregressive Modeling of Wide-Band Indoor Radio Propagation",*IEEE Transaction on Commun.,* **40** 1540-1552, September (1992).

8 W.C. Jakes, *Microwave Mobile Communications*, (Wiley, 1974, New York).

9 M. Morf and T. Kailath, Square-root algorithms for least-squares estimation, *IEEE Trans. AC*, **20(4)**, 487-496, (1975).

10 R. E. Kalman, "A new approach to linear filtering and prediction problems," *Trans. ASME,* **82D** 34-45, Mar. (1960).

11 C. C. Martin, J. H. Winters, and N. R. Sollenberger, "Multiple-input multiple-output (MIMO) radio channel measurements," *Proc. IEEE Veh. Technol. Conf.,* 774-779, Sept. (2000).

12 J.G. Proakis, *Digital Communications,* (McGraw-Hill, New York, 1989).

13 T. S. Rappaport, *Wireless Communications: Principles and Practice.* (Prenice-Hall, New Jersey, 1996).

14 H. E. Rauch, F. Tung, and C. T. Streibel, "Maximum likelihood estimates of linear dynamic systems," *AIAA Journ.*, **3**(8), 1445-1450, Aug. (1965).

15 R. Y. Rubinstein, *Simulation and Monte-Carlo Method,* (Wiley, New York, 1981).

16 R. H. Shumway and D. S. Stoffer, "An approach to time series smoothing and forecasting using the EM algorithm," *J. Time Series Anal.*, **3**(4), 253-264, (1982).

17 T. Soderstrom and P. Stoica, *System Identification*, (Prentice Hall, New York, 1989).

18 W. Turin, R. Jana, S. S. Ghassemzadeh, C. Rice, and V. Tarokh, "Autoregressive modeling of an indoor UWB channel,"*IEEE Conference on Ultra Wideband Systems and Technologies*, 71-74, May (2002).

APPENDIX 1

1.1 MATRIX PROCESSES

The matrix process definition and its application to error source modeling were considered in Sec. 1.3. We provide here proofs of the statements of Sec. 1.3 and also some information that is necessary to understand methods of estimating model parameters. The definitions in the next section play a major role in analyzing matrix processes.

Matrix processes and Markov chains have an interesting common property: There exists a finite set of the so-called basis sequences such that the probability of any sequence can be expressed as a linear combination of the probabilities of these basis sequences.

To prove this statement, we consider the sequences of zeroes and ones of the type $s_i t_j$, where s_i is a sequence preceding the sequence t_j. Let $\varnothing$ be an empty sequence that possesses the properties $\varnothing s_i t_j = s_i \varnothing t_j = s_i t_j \varnothing$ and $Pr(\varnothing) = 1$. Denote

$$\mathbf{P}(s_1,\ldots,s_p;\ t_1,\ldots,t_q) = [\,p(s_i t_j)]_{p,q} \qquad \text{(A.1.1)}$$

as a matrix whose (i,j)-th element is the probability of occurrence of the sequence $s_i t_j$.

Lemma A.1.1: For every matrix process with matrices $\mathbf{M}(e)$ of the order n, there exists an integer $r \le n$ which is equal to the supremum of the ranks of the matrices (A.1.1) (over all possible p and q and sequences $s_i t_j$).

Proof: Indeed, by formula (1.3.5), with $s_i = e_1 e_2 \cdots e_t$, $t_j = e_{t+1} e_{t+2} \cdots e_m$, we have

$$p(s_i t_j) = \mathbf{q}(s_i)\mathbf{r}(t_j) \qquad \text{(A.1.2)}$$

where $\mathbf{q}(s_i) = \mathbf{y}\mathbf{M}(e_1)\cdots\mathbf{M}(e_t)$ and $\mathbf{r}(t_j) = \mathbf{M}(e_{t+1})\cdots\mathbf{M}(e_m)\mathbf{z}$. If we denote

$$\mathbf{Q}(s_1,\ldots,s_p) = block\ col\{\mathbf{q}(s_i)\} \qquad \mathbf{R}(t_1,\ldots,t_q) = block\ row\{\mathbf{r}(t_j)\}$$

then (A.1.1) can be rewritten as

$$\mathbf{P}(s_1,\ldots,s_p;\ t_1,\ldots,t_q) = \mathbf{Q}(s_1,\ldots,s_p)\mathbf{R}(t_1,\ldots,t_q) \qquad \text{(A.1.3)}$$

with $\mathbf{q}(\varnothing) = \mathbf{Q}(\varnothing) = \mathbf{y}$, $\mathbf{r}(\varnothing) = \mathbf{R}(\varnothing) = \mathbf{z}$.

According to the Sylvester inequalities, [2,4] the rank of this matrix product is not greater than ranks of the matrices $\mathbf{Q}(s_1,\ldots,s_p)$ and $\mathbf{R}(t_1,\ldots,t_q)$. But these ranks are less than the sizes of the matrices, so that the rank of their product $\mathbf{P}(s_1,\ldots,s_p;\ t_1,\ldots,t_q)$ cannot exceed n.

The maximal rank of these matrices is called the *rank of the matrix process* and is denoted by r. Sequences for which $\mathbf{P}(s_1,...,s_r;\ t_1,...,t_r) \neq 0$ are called *basis sequences.* They are similar to states of a Markov chain.

Theorem A.1.1: A stationary process is a matrix process if and only if a positive integer r exists such that

$$\det \mathbf{P}(s_1,...,s_r;\ t_1,...,t_r) \neq 0 \tag{A.1.4}$$

and

$$\det \mathbf{P}(s_1,...,s_r,s;\ t_1,...,t_r,t) = 0 \tag{A.1.5}$$

for any sequences s and t.

Proof: The necessity of the conditions follows from the previous lemma. [2] We will prove the conditions' sufficiency if we show that formula (1.3.5) follows from Eq. (A.1.4) and (A.1.5). The determinant in the left-hand side of Eq. (A.1.5) can be expanded by the elements of its last column:

$$p(s\ t) = \sum_{j=1}^{r} a_j(s)p(s_j\ t) \tag{A.1.6}$$

For $s = s_i e$, this equation takes the form

$$p(s_i e\ t) = \sum_{j=1}^{r} a_j(s_i e)p(s_j\ t) \quad i = 1,2,...,r$$

If we denote $\mathbf{A}(e) = [\ a_j(s_i e)\ ]_{r,r}$, then these equations can be presented in the following matrix form

$$\mathbf{P}(s_1,...,s_r;\ e\ t) = \mathbf{A}(e)\mathbf{P}(s_1,...,s_r;\ t) \tag{A.1.7}$$

Denoting $e = e_1$ and $t = e_2 \cdots e_m$, we can rewrite the previous equation as

$$\mathbf{P}(s_1,...,s_r;\ e_1 e_2 \cdots e_m) = \mathbf{A}(e_1)\mathbf{A}(e_2)\mathbf{P}(s_1,...,s_r;\ e_3 \cdots e_m)$$

Repeating this transformation m times, we obtain

$$\mathbf{P}(s_1,...,s_r;\ e_1 e_2 \cdots e_m) = \prod_{i=1}^{m} \mathbf{A}(e_i)\mathbf{P}(s_1,...,s_r;\ \varnothing) \tag{A.1.8}$$

If we choose $s_1 = \varnothing$, then the first element of $\mathbf{P}(s_1,...,s_r;\ e_1 e_2 \cdots e_m)$ becomes $p(\varnothing e_1 e_2 \cdots e_m) = p(e_1 e_2 \cdots e_m)$. This element can be obtained by multiplying $\mathbf{P}(s_1,...,s_r;\ e_1 e_2 \cdots e_m)$ by $\mathbf{y}_l = [\ 1\ 0\ 0 \ldots 0]_{1,r}$ from the left. Therefore, denoting $\mathbf{z}_l = \mathbf{P}(s_1,...,s_r;\ \varnothing)$, we obtain from Eq. (A.1.8)

$$p(e_1 e_2 \cdots e_m) = \mathbf{y}_l \prod_{i=1}^{m} \mathbf{A}(e_i)\mathbf{z}_l \tag{A.1.9}$$

This equation coincides with Eq. (1.3.5), which means that this is a matrix process.

If we expand each column of a square matrix $\mathbf{P}(s_1,...,s_r;\ et_1,...,et_r)$ using Eq. (A.1.7), we obtain

$$\mathbf{P}(s_1,...,s_r;\ et_1,...,et_r) = \mathbf{A}(e)\mathbf{P}(s_1,...,s_r;\ t_1,...,t_r)$$

Since $\mathbf{P}(s_1,...,s_r;\ t_1,...,t_r)$ is nonsingular, it has an inverse matrix and, therefore,

$$\mathbf{A}(e) = \mathbf{P}(s_1,...,s_r;\ et_1,...,et_r)\mathbf{P}^{-1}(s_1,...,s_r;\ t_1,...,t_r) \tag{A.1.10}$$

This means that the matrices $\mathbf{A}(e)$ for a given process and a fixed basis are determined uniquely. They can be found from the probabilities of the basis sequences $p(s_i t_j)$, $p(s_i et_j)$, where $i,j = 1,2,...,r$ and r is the rank of the process.

By expanding the determinant in Eq. (A.1.5) by the elements of its last row we arrive at the formula

$$\mathbf{P}(s\ e;\ t_1,...,t_r) = \mathbf{P}(s;\ t_1,...,t_r)\mathbf{B}(e)$$

which is analogous to Eq. (A.1.7), and then we obtain equations similar to Eq. (A.1.8), (A.1.9), and (A.1.10):

$$\mathbf{P}(e_1 \cdots e_m;\ t_1,...,t_r) = \mathbf{y}_\rho \prod_{i=1}^{m} \mathbf{B}(e_i)$$

$$p(e_1,e_2,...,e_m) = \mathbf{y}_\rho \prod_{i=1}^{m} \mathbf{B}(e_i)\mathbf{z}_\rho \tag{A.1.11}$$

$$\mathbf{P}(s_1 e,...,s_r e;\ t_1,...,t_r) = \mathbf{P}(s_1,...,s_r;\ t_1,...,t_r)\mathbf{B}(e)$$

$$\mathbf{B}(e) = \mathbf{P}^{-1}(s_1 e,...,s_r e;\ t_1,...,t_r)\ \mathbf{P}(s_1,...,s_r;\ t_1,...,t_r) \tag{A.1.12}$$

where $\mathbf{y}_\rho = \mathbf{P}(\varnothing;\ t_1,...,t_r)$ and $\mathbf{z}_\rho = [\ 1\ 0\ 0\ ...\ 0\ ]'$. Combining these two sets of equations, we can write the generalized formula

$$\mathbf{P}(s_1 e,...,s_r e;\ \varepsilon t_1,...,\varepsilon t_r) = \mathbf{A}(\varepsilon)\mathbf{P}(s_1,...,s_r;\ t_1,...,t_r)\mathbf{B}(e)$$

Formulas (A.1.9) and (A.1.11) show that for a process to be a matrix process, it is necessary and sufficient that the probability of any sequence be a linear combination of the probabilities of its basis sequences. Let us now discuss methods of constructing the process basis.

Theorem A.1.1 requires testing of an infinite number of conditions in order to determine the rank of a process. However, it is possible to determine the rank in a finite number of steps. To prove it, we denote $|s|$ as the length of a sequence s; the collection of all sequences having the same length m we call a layer and denote $L_m^{(s)}$. The union of all the layers whose lengths do not exceed m is denoted by $W_m^{(s)} = \bigcup_{i=0}^{m} L_i^{(s)}$.

Theorem A.1.2: Let $Z = \{e_j\!: j \in \mathbf{J}, e_j \in (0,1)\}$ be a matrix process, and let the sequences $s_1,s_2,...,s_p$ and $t_1,t_2,...,t_p$ represent the basis for the sets $s \in W_m^{(s)}$ and $t \in W_m^{(t)}$, that is

$$\det \mathbf{P}(s_1,...,s_p;\ t_1,...,t_p) \neq 0 \tag{A.1.13}$$

and

$$\det \mathbf{P}(s_1,...,s_p,s;\ t_1,...,t_p,t) = 0 \tag{A.1.14}$$

for any $s \in W_m^{(s)}$ and $t \in W_m^{(t)}$. Then, if condition (A.1.14) holds for any sequences $s \in W_{m+1}^{(s)}$ and $t \in W_{m+1}^{(t)}$, the rank of the process is p, and the sequences $s_1,s_2,...,s_p$ and

$t_1, t_2, ..., t_p$ are basis sequences of the process.

Proof: By Eq. (A.1.3), we have

$$\mathbf{P}(s_1, ..., s_p, s; t_1, ..., t_p, t) = \mathbf{Q}(s_1, ..., s_p, s)\mathbf{R}(t_1, ..., t_p, t)$$

Since *rank* $\mathbf{P}(s_1, ..., s_p, s; t_1, ..., t_p, t) = p$, at least one of the matrices $\mathbf{Q}(s_1, ..., s_p, s)$ or $\mathbf{R}(t_1, ..., t_p, t)$ has rank p for any vectors $s \in W_{m+1}^{(s)}$ or $t \in W_{m+1}^{(t)}$. Let *rank* $\mathbf{R}(t_1, ..., t_p, t) = p$ for any $t \in W_{m+1}^{(t)}$. Then the last column of $\mathbf{R}(t_1, ..., t_p, t)$ is a linear combination of its other columns:

$$\mathbf{r}(t) = \sum_{i=1}^{p} \alpha_i \mathbf{r}(t_i)$$

An arbitrary sequence τ from $L_{m+2}^{(t)}$ has the form $\tau = e\,t$, where $e \in (0,1)$, $t \in L_{m+1}^{(t)}$. Equation (1.3.5) gives

$$\mathbf{r}(e\,t) = \mathbf{M}(e)\mathbf{r}(t) = \sum_{i=1}^{p} \alpha_i \mathbf{M}(e)\mathbf{r}(t_i) = \sum_{i=1}^{p} \alpha_i \mathbf{r}(et_i)$$

Since $t_i \in W_m^{(t)}$ and $et_i \in W_{m+1}^{(t)}$, vector $\mathbf{r}(et_i)$ can be expressed as a linear combination of the basis vectors

$$\mathbf{r}(et_i) = \sum_{i=1}^{p} \beta_{ij} \mathbf{r}(t_j)$$

Substituting this expression into the previous equation, we obtain

$$\mathbf{r}(et) = \sum_{i=1}^{p} \alpha_i \sum_{j=1}^{p} \beta_{ij} \mathbf{r}(t_j) = \sum_{j=1}^{p} \left(\sum_{i=1}^{p} \alpha_i \beta_{ij} \right) \mathbf{r}(t_j)$$

This means that the columns of $\mathbf{R}(t_1, t_2, ..., t_p, et)$ are linearly dependent so that *rank* $\mathbf{R}(t_1, t_2, ..., t_p, et) = p$. According to the Sylvester inequalities, [2,4] the rank of $\mathbf{P}(s_1, ..., s_p, s\sigma;\ t_1, ..., t_p, et) = \mathbf{Q}(s_1, ..., s_p, s\sigma)\mathbf{R}(t_1, ..., t_p, et)$. is not greater than p. Thus, $\det \mathbf{P}(s_1, ..., s_p, s\sigma; t_1, ..., t_p, et) = 0$ for all sequences $s\sigma \in W_{m+2}^{(s)}$ and $et \in W_{m+2}^{(t)}$.

Using the method of mathematical induction, we prove that this determinant is equal to zero for any sequences s and t, which, according to Theorem A.1.1, proves that the sequences $s_1, ..., s_p$ and $t_1, ..., t_p$ represent the basis of the process.

This theorem gives us a method of building a matrix process basis. We say that a sequence s_1 precedes a sequence s_2 if the length of s_1 is smaller than the length of s_2; if the lengths of the sequences are equal, then s_1 precedes s_2 if the binary number representing s_1 is smaller than the binary number representing s_2. If s_1 precedes s_2, we write $s_1 \propto s_2$. We say that a pair of sequences $(s_1;\ t_1)$ precedes a pair $(s_2;\ t_2)$ if $s_1 \propto s_2$ or if $s_1 = s_2$, but $t_1 \propto t_2$.

To build the basis, we start with empty sequences $\sigma_1 = \varnothing$, $\tau_1 = \varnothing$. If the process rank is not less than two, then we can find a pair $(\sigma_2;\ \tau_2)$ preceding all the pairs that satisfy the condition $\det \mathbf{P}(\sigma_1, \sigma_2;\ \tau_1, \tau_2) \neq 0$. Similarly, we choose a pair $(\sigma_3;\ \tau_3)$ such that $\det \mathbf{P}(\sigma_1, \sigma_2, \sigma_3;\ \tau_1, \tau_2, \tau_3) \neq 0$, and so on. According to the previous theorem, after a finite number of steps, we will construct the basis of the process. This basis is unique by construction; we call it the *process canonic basis*.

Corollary A.1.1: If a sequence $s_1, s_2, ..., s_r$ (or $t_1, t_2, ..., t_r$) forms a canonic basis of a matrix process and $|s_r| = m$ (or $|t_r| = m$), then this basis includes at least one element of every layer $L_i^{(s)}$ (or $L_i^{(t)}$) for $i < m$.

Corollary A.1.2: If the rank of a process is r, then this process is determined uniquely by the multidimensional distributions of sequences $e_1, e_2, \ldots, e_m$ when $m = |e_1 e_2 \cdots e_m| \leq 2r - 1$.

Indeed, when building the canonic basis, the most unfavorable case is when, starting with $s_1 = \varnothing$ and $t_1 = \varnothing$, we can include in the basis only one sequence s_i and t_j from every layer in which case $\max|s_i e t_j| \leq 2r - 1$.

Corollary A.1.3: For two matrix processes to be equivalent, it is necessary and sufficient that they have the same rank r and the same multidimensional distribution up to the order $2r - 1$.

As we saw in Sec. 1.3, it is important to find equivalent matrix processes to simplify a model description. We showed that if conditions (1.3.6) are satisfied, then processes $\{\mathbf{yT}, \mathbf{M}(e), \mathbf{z}\}$ and $\{\mathbf{y}, \mathbf{M}_1(e), \mathbf{Tz}\}$ are equivalent. We develop here necessary and sufficient conditions for the processes equivalence.

It follows from the results of this Appendix that if the ranks or the canonic bases do not coincide, then the processes are not equivalent. If the ranks and the canonic bases do coincide, it is necessary and sufficient that the matrices $\mathbf{A}(e)$ and $\mathbf{A}_1(e)$ of the canonic representations of these processes coincide which, according to Eq. (A.1.3) and (A.1.10), can be written as:

$$\mathbf{QM}(e)\mathbf{R}[\mathbf{QR}]^{-1} = \mathbf{Q}_1\mathbf{M}_1(e)\mathbf{R}_1[\mathbf{Q}_1\mathbf{R}_1]^{-1} \tag{A.1.15}$$

If the matrices $\mathbf{M}(e)$ and $\mathbf{M}_1(e)$ are order r square matrices, then this relationship implies the following corollary.

Corollary A.1.4: Suppose that $\mathbf{M}(e)$ and $\mathbf{M}_1(e)$ have orders equal to the ranks of the processes. A necessary and sufficient condition for the matrix processes $\{\mathbf{y}, \mathbf{M}(e), \mathbf{z}\}$ and $\{\mathbf{y}_1, \mathbf{M}_1(e), \mathbf{z}_1\}$ to be equivalent is that a matrix $\mathbf{T}$ satisfying the conditions $\mathbf{y}_1 = \mathbf{yT}\quad \mathbf{M}_1(e) = \mathbf{T}^{-1}\mathbf{M}(e)\mathbf{T}\quad e = 0,1\quad \mathbf{z}_1 = \mathbf{T}^{-1}\mathbf{z}$ exists.

Proof: The sufficiency follows from Eq. (1.3.6); let us prove the necessity. It follows from Eq. (A.1.15) that $\mathbf{M}_1(e) = \mathbf{T}^{-1}\mathbf{M}(e)\mathbf{T}$, where $\mathbf{T} = \mathbf{Q}^{-1}\mathbf{Q}_1$. Since $\mathbf{q}(\varnothing) = \mathbf{y}$, $\mathbf{q}_1(\varnothing) = \mathbf{y}_1$ are rows of the matrices $\mathbf{Q}$ and $\mathbf{Q}_1$, then $\mathbf{y} = \mathbf{y}_l\mathbf{Q}\quad \mathbf{y}_1 = \mathbf{y}_l\mathbf{Q}_1$, so that $\mathbf{y}_1 = \mathbf{yQ}^{-1}\mathbf{Q}_1 = \mathbf{yT}$.

Similarly, the identities $\mathbf{r}(\varnothing) = \mathbf{z}, \mathbf{r}_1(\varnothing) = \mathbf{z}_1$ imply, because of the relationships $\mathbf{z}_l = \mathbf{Qr}(\varnothing)$, $\mathbf{z}_l = \mathbf{Q}_1\mathbf{r}_1(\varnothing)$, that $\mathbf{z}_1 = \mathbf{T}^{-1}\mathbf{z}$. Thus, stationary matrix processes with the matrices of the smallest order (equal to the rank of the process) are equivalent only when their defining matrices are similar.

1.1.1 Reduced Canonic Representation

The basis sequences considered in the previous section can also be built using block matrices. To achieve this, we associate basis sequences with symbols e, that is, we consider the sequences of the form $s_i^{(e)} e t_j^{(e)}$, $s_1(e) = \varnothing$, $t_1^{(e)} = \varnothing$. Let $\mathbf{P}_e(s_1^{(e)}, s_2^{(e)}, \ldots, s_p^{(e)};\ t_1^{(e)}, t_2^{(e)}, \ldots, t_q^{(e)})$ denote the matrix whose (i,j)-th element equals $p(s_i^{(e)} e t_j^{(e)})$. Then, in accordance with Eq. (1.3.8), we can represent it in the form

$$\mathbf{P}_e(s_1^{(e)}, \ldots, s_p^{(e)};\ t_1^{(e)}, \ldots, t_q^{(e)}) = \mathbf{Q}_e(s_1^{(e)}, \ldots, s_p^{(e)})\mathbf{R}_e(t_1^{(e)}, \ldots, t_q^{(e)}) \tag{A.1.16}$$

This equation is analogous to Eq. (A.1.3).

Expressions (A.1.16) permit to define ranks r_e corresponding to symbols $e = 0$ or $e = 1$. We can show that the basis sequences exist and that $r_e \le n_e$ (where n_e is the order of the matrix $\mathbf{M}_{ee}$). As in the previous section, we obtain

$$\mathbf{P}_\mu(s_1^{(\mu)},\ldots,s_{r_\mu}^{(\mu)};\ e\ t) = \mathbf{A}_\mu(e)\mathbf{P}_e(s_1^{(e)},\ldots,s_{r_e}^{(e)};\ t) \tag{A.1.17}$$

$$\mathbf{A}_\mu(e) = \mathbf{P}_\mu(s_1^{(\mu)},\ldots,s_{r_\mu}^{(\mu)};\ et_1^{(e)},\ldots,et_{r_e}^{(e)})\mathbf{P}^{-1}(s_1^{(e)},\ldots,s_{r_e}^{(e)};\ t_1^{(e)},\ldots,t_{r_e}^{(e)}) \tag{A.1.18}$$

Note that although these relations are a particular case of those considered in the previous section, the sequences used here in the construction of the basis differ slightly from the sequences used in the previous section. For example, we could have considered previously the sequence $s_1 t_1 = \varnothing$; now the maximal length of a sequence is 1, since we consider sequences of the form $s_i e t_j$, where $e \ne \varnothing$. The matrix $\mathbf{P}(s_1,\ldots,s_p;\ t_1,\ldots,t_q)$ now has the form

$$\mathbf{P}(s_1^{(0)}0,\ldots,s_{p_0}^{(0)}0,s_1^{(1)}1,\ldots,s_{p_1}^{(1)}1;\ t_1^{(0)},\ldots,t_{q_0}^{(0)},t_1^{(1)},\ldots,t_{q_1}^{(1)})$$

$$= \begin{bmatrix} \mathbf{P}_0(s_1^{(0)},\ldots,s_{p_0}^{(0)};\ t_1^{(0)},\ldots,t_{q_0}^{(0)}) & 0 \\ 0 & \mathbf{P}_1(s_1^{(1)},\ldots,s_{p_1}^{(1)};\ t_1^{(1)},\ldots,t_{q_1}^{(1)}) \end{bmatrix} \tag{A.1.19}$$

and

$$\mathbf{P}(s_1^{(0)}0,\ldots,s_{r_0}^{(0)}0,s_1^{(1)}1,\ldots,s_{r_1}^{(1)}1;\ t) = \begin{bmatrix} \mathbf{P}_0(s_1^{(0)},\ldots,s_{r_0}^{(0)};\ t) \\ \mathbf{P}_1(s_1^{(1)},\ldots,s_{r_1}^{(1)};\ t) \end{bmatrix} \tag{A.1.20}$$

$$\mathbf{y}_l = [\mathbf{y}_{l0}\quad \mathbf{y}_{l1}] \qquad \mathbf{z}_l = [\mathbf{z}_{l0}\quad \mathbf{z}_{l1}]$$

$$\mathbf{A}(0) = \begin{bmatrix} \mathbf{A}_{00} & 0 \\ \mathbf{A}_{10} & 0 \end{bmatrix} \qquad \mathbf{A}(1) = \begin{bmatrix} 0 & \mathbf{A}_{01} \\ 0 & \mathbf{A}_{11} \end{bmatrix} \tag{A.1.21}$$

Thus, we see that the use of special basis sequences considered in this section permits us to simplify the process description and, in particular, have matrices $\mathbf{A}(e)$ containing zero blocks. We call this representation $\{\mathbf{y}_l,\mathbf{A}(0),\mathbf{A}(1),\mathbf{z}_l\}$ of a matrix process a *reduced canonic representation.*

1.2 MARKOV LUMPABLE CHAINS

The material in this section is an adaptation of the results of Ref. 1 and 3. It is closely related to that of the previous section and serves as an introduction to the next section.

Let $\mathbf{P} = [\ p_{ij}\ ]_{n,n}$ be a transition matrix and $\mathbf{p} = [\ p_i\ ]_{1,n}$ be a matrix of initial probabilities of a homogeneous Markov chain s_t. Suppose that $\{A_1,A_2,\ldots,A_s\}$ is a partition of the set of all states into nonintersecting subsets: $\bigcup_{i=1}^{s} A_i = \Omega \quad A_i \cap A_j = \varnothing \quad i \ne j$. A process y_t is called a *Markov function* or *lumped process* [3] if $y_t = i$ when $s_t \in A_i$.

To simplify notation, let us assume that $A_1 = \{1,2,...,m_1\}$, $A_2 = \{m_1+1,...,m_2\}$, ..., $A_s = \{m_{s-1}+1,...,m_s\}$. If this assumption is not valid, we can renumber the states to satisfy it (see Appendix 6.1.3). In correspondence with this partition, we can write the matrices **P** and **p** in the block form

$$\mathbf{P} = block\ [\mathbf{P}_{ij}]_{s,s} = \begin{bmatrix} \mathbf{P}_{11} & \mathbf{P}_{12} & \dots & \mathbf{P}_{1s} \\ \mathbf{P}_{21} & \mathbf{P}_{22} & \dots & \mathbf{P}_{2s} \\ . & . & \dots & . \\ \mathbf{P}_{s1} & \mathbf{P}_{s2} & \dots & \mathbf{P}_{ss} \end{bmatrix}$$

$$\mathbf{p} = block\ row\{\mathbf{p}_i\} = [\,\mathbf{p}_1\ \ \mathbf{p}_2\ \ \cdots\ \ \mathbf{p}_s]$$

where matrix block $\mathbf{P}_{ij}$ consists of all transition probabilities $p_{\alpha\beta}$, $\alpha \in A_i$, $\beta \in A_j$.

DEFINITION: A Markov chain is called **p**-lumpable with respect to a partition $\{A_1, A_2, ..., A_s\}$ if there exists a vector **p** such that the lumped process y_t is a Markov chain.

This means that there are scalars $\hat{p}_u$ and $\hat{p}_{uv}$ $(u,v = 1,2,...,s)$ that, in general, depend on **p**, such that $Pr(y_0, y_1, ...\ y_t) = \hat{p}_{y_0}\hat{p}_{y_0y_1} \cdots \hat{p}_{y_{t-1}y_t}$, for any $t \in \mathbf{N}$ and any $y_0, y_1, ..., y_t$.

These conditions can be written in matrix form as follows:

$$\mathbf{p}_{y_0} \prod_{i=1}^{t} \mathbf{P}_{y_{i-1}y_i} \mathbf{1} = \hat{p}_{y_0} \prod_{i=1}^{t} \hat{p}_{y_{i-1}y_i} \tag{A.1.22}$$

Theorem A.1.3: If a Markov chain is **p**-lumpable and $\boldsymbol{\pi}$ is its stationary distribution, then it is $\boldsymbol{\pi}$-lumpable.

Proof: Indeed, the summation of both sides of Eq. (A.1.22), with respect to y_0, gives

$$\mathbf{p}_{y_1} \prod_{i=2}^{t} \mathbf{P}_{y_{i-1}y_i} \mathbf{1} = \hat{p}_{y_1} \prod_{i=2}^{t} \hat{p}_{y_{i-1}y_i}$$

where $\mathbf{p}_1 = \sum_{y_0} \mathbf{p}_{y_0}\mathbf{P}_{y_0y_1}$, $\hat{p}_1 = \sum_{y_0} \hat{p}_{y_0}\hat{p}_{y_0y_1}$. Therefore, if the chain is **p**-lumpable, then it is $\mathbf{p}_1$-lumpable, where $\mathbf{p}_1 = \mathbf{pP}$.

Repeating the same transformations, we observe that the chain is $\mathbf{pP}^2$-lumpable, $\mathbf{pP}^3$-lumpable, and consequently for any positive integer *m* is $\mathbf{pP}^m$-lumpable. Since

$$\boldsymbol{\pi} = \lim_{N \to \infty} \mathbf{p} \sum_{m=0}^{N-1} \mathbf{P}^m / N$$

we conclude that the chain is $\boldsymbol{\pi}$-lumpable. Thus, in order to find out whether a Markov chain is lumpable for any initial distribution, it is sufficient to check that the chain is lumpable for the stationary distribution $\boldsymbol{\pi}$.

The other necessary conditions of lumpability follow directly from Eq. (A.1.22):

$$\hat{p}_i = \mathbf{p}_i\mathbf{1} \tag{A.1.23}$$

$$\hat{p}_{ij} = \mathbf{p}_i\mathbf{P}_{ij}\mathbf{1}/\mathbf{p}_i\mathbf{1}$$

$$\hat{p}_{y_t y_{t+1}} = \mathbf{p}_{y_0} \prod_{i=1}^{t+1} \mathbf{P}_{y_{i-1} y_i} \mathbf{1} / \mathbf{p}_{y_0} \prod_{i=1}^{t} \mathbf{P}_{y_{i-1} y_i} \mathbf{1}$$

Equations (A.1.23) show that if the chain is lumpable, then the initial probabilities of the states of the lumped process are given by $\hat{\mathbf{p}}_0 = [\ \mathbf{p}_1 \mathbf{1}\ \ \mathbf{p}_2 \mathbf{1}\ \ \ldots\ \ \mathbf{p}_s \mathbf{1}\ ]_{1,s}$. The matrix of transition probabilities of the lumped process is also uniquely identified as $\hat{\mathbf{P}} = [\ \mathbf{q}_i \mathbf{P}_{ij} \mathbf{1} / \mathbf{q}_i \mathbf{1}\ ]_{s,s}$, where $\mathbf{q}_i$ is the first different from zero matrix in the series $\mathbf{p}_i$, $\mathbf{p}_1 \mathbf{P}_{1i}$, $\mathbf{p}_2 \mathbf{P}_{2i}$, ..., $\mathbf{p}_s \mathbf{P}_{si}$. If all these matrices are equal to zero, then we delete the set A_i because the process lumpability does not depend on this set. From this reasoning the following theorem is derived.

Theorem A.1.4: For a Markov chain to be lumpable with respect to a partition $\{A_1, A_2, \ldots, A_s\}$, it is necessary and sufficient that, for any $y_0, y_1, \ldots, y_t$ and $t \in \mathbf{N}$,

$$\frac{\mathbf{p}_{y_0} \prod_{i=1}^{t} \mathbf{P}_{y_{i-1} y_i} \mathbf{1}}{\mathbf{p}_{y_0} \prod_{i=1}^{t-1} \mathbf{P}_{y_{i-1} y_i} \mathbf{1}} = \frac{\mathbf{q}_{y_{t-1}} \mathbf{P}_{y_{t-1} y_t} \mathbf{1}}{\mathbf{q}_i \mathbf{1}} = \hat{p}_{y_{t-1} y_t} \tag{A.1.24}$$

in all cases where the denominator is different from zero.

This common value Eq. (A.1.24) gives the transition probability of the lumped chain, while the initial distribution of the lumped chain is given by (A.1.23).

If we introduce the probability vector

$$\mathbf{g}(y_0, y_1, \ldots, y_t) = \mathbf{p}_{y_0} \prod_{i=1}^{t} \mathbf{P}_{y_{i-1} y_i} \Big/ \mathbf{p}_{y_0} \prod_{i=1}^{t} \mathbf{P}_{y_{i-1} y_i} \mathbf{1}$$

then the previous theorem can be formulated in terms of these vectors.

Theorem A.1.5: For a Markov chain to be lumpable with respect to a partition $\{A_1, A_2, \ldots, A_s\}$, it is necessary and sufficient that, for the previously defined probability vectors $\mathbf{g}(y_0, y_1, \ldots, y_{t-1})$ and $t \in \mathbf{N}$, the expression

$$\mathbf{g}(y_0, y_1, \ldots, y_{t-1}) \mathbf{P}_{y_{t-1} y_t} \mathbf{1} = \hat{p}_{y_{t-1} y_t} \tag{A.1.25}$$

assumes fixed values for any given y_{t-1} and y_t.

To test whether the conditions of these theorems are satisfied, one must examine an infinite number of conditions. However, it is sufficient to consider only a finite number of such conditions for the normalized basis of the set

$$\mathbf{Y}_{y_t} = \{\mathbf{p}_{y_t},\ \mathbf{p}_{y_{t-1}} \mathbf{P}_{y_{t-1} y_t},\ \ldots\ , \mathbf{p}_{y_0} \prod_{i=0}^{t} \mathbf{P}_{y_{i-1} y_i}, \ldots\} \tag{A.1.26}$$

for all possible values of $y_0, y_1, \ldots, y_{t-1}$ and fixed y_t. That is, for the vectors

$$\mathbf{g}_{ij} = \mathbf{e}_{ij} / \mathbf{e}_{ij} \mathbf{1} \qquad (i = 1,2,\ldots,r_j;\ j = 1,2,\ldots,s) \tag{A.1.27}$$

where r_j is the rank of the system $\mathbf{Y}_j$, [2] vectors $\mathbf{e}_{1j}, \mathbf{e}_{2j}, \ldots, \mathbf{e}_{r_j j}$ are linearly independent.

Theorem A.1.6: For a Markov chain to be lumpable with respect to a partition $\{A_1, A_2, \ldots, A_s\}$, it is necessary and sufficient that the left-hand side of Eq. (A.1.25) assumes a constant value for the normalized basis vectors (A.1.27) of the sets $\mathbf{Y}_j$, $(j = 1,2,\ldots,s)$.

Proof: The necessity of these conditions is obvious: If condition (A.1.4) is satisfied for all vectors of the set, it is also satisfied for the vectors of their basis.

Let us prove the sufficiency. If we assume that conditions (A.1.25) are satisfied for vectors (A.1.27) $\mathbf{g}_{ij}\mathbf{P}_{jm} = \hat{p}_{jm}$, then we can prove that these conditions are valid for an arbitrary normalized vector $\mathbf{g}_j \in \mathbf{Y}_j$. We decompose this vector with respect to the basis $\mathbf{g}_j = \sum_{i=1}^{r_j} \alpha_i \mathbf{g}_{ij}$ and, since $\mathbf{g}_j\mathbf{1} = \mathbf{g}_{ij}\mathbf{1} = 1$, then $\sum_{i=1}^{r_j} \alpha_i = 1$. But, in this case, $\mathbf{g}_j\mathbf{P}_{jm}\mathbf{1} = \sum_{i=1}^{r_j} \alpha_i \mathbf{g}_{ij}\mathbf{P}_{jm}\mathbf{1} = \hat{p}_{jm}$, which proves the theorem.

Thus, in order to check the process lumpability, we need to build a basis of the set $\mathbf{Y}_j$. To do so it is convenient to consider the layers of the set $\mathbf{Y}_j$:

$$L_j^{(0)} : \mathbf{p}_j$$

$$L_j^{(1)} : \mathbf{p}_1\mathbf{P}_{1j},\ \mathbf{p}_2\mathbf{P}_{2j},\ \ldots, \mathbf{p}_s\mathbf{P}_{sj}$$

$$L_j^{(2)} : \mathbf{p}_1\mathbf{P}_{11}\mathbf{P}_{1j},\ \mathbf{p}_1\mathbf{P}_{12}\mathbf{P}_{2j},\ \ldots, \mathbf{p}_s\mathbf{P}_{ss}\mathbf{P}_{sj}$$

$$\cdots\cdots\cdots\cdots\cdots\cdots$$

$$L_j^{(t)} : \mathbf{p}_{y_0} \prod_{i=1}^{t} \mathbf{P}_{y_{i-1}y_i}, \quad \text{for all } y_0, y_1, \cdots y_{t-1}, \text{ and } y_t = j$$

$$\cdots\cdots\cdots\cdots\cdots\cdots$$

Suppose that we have built the bases of all the sets

$$Y_j^{(t)} = \bigcup_{i=0}^{t} L_j^{(i)} \tag{A.1.28}$$

for $j = 1,2,\ldots,s$, and fixed t. Assume also that by adding the next layer, $L_j^{(t+1)}$, it is possible to keep the basis of $Y_j^{(t+1)}$ unchanged, that is, the ranks of sets $Y_j^{(t)}$ and $Y_j^{(t+1)}$ are the same:

$$rank\ Y_j^{(t)} = rank\ Y_j^{(t+1)} \qquad (j = 1,2,\ldots,s)$$

The following theorem proves that bases of the sets (A.1.28) can serve as bases of the entire sets $\mathbf{Y}_j$.

Theorem A.1.7: If $rank\ Y_j^{(t)} = rank\ Y_j^{(t+1)}$ for all $j = 1,2,\ldots,s$, then $rank\ Y_j^{(t)} = rank\ \mathbf{Y}_j$.

Proof: It is sufficient to show that $rank\ Y_j^{(t+1)} = rank\ Y_j^{(t+2)}$ $(j = 1,2,\ldots,s)$. Let $\mathbf{z}_j \in Y_j^{(t+2)}$ be an arbitrary vector from the set $Y_j^{(t+2)}$. By the set $Y_j^{(t+2)}$ composition vector $\mathbf{z}_j$ either belongs to the set $Y_j^{(t+1)}$ or $\mathbf{z}_j = \mathbf{y}_i\mathbf{P}_{ij}$, where $\mathbf{y}_i \in Y_i^{(t+1)}$. But $rank\ Y_i^{(t+1)} = rank\ Y_i^{(t)}$, so that the vector $\mathbf{y}_i$ can be decomposed with respect to the set $Y_i^{(t)}$ basis: $\mathbf{y}_i = \sum_{k=1}^{r_i} \beta_{ki}\mathbf{e}_{ki}^{(t)}$ and, as a result, $\mathbf{z}_j = \sum_{k=1}^{r_i} \beta_{ki}\mathbf{e}_{ki}^{(t)}\mathbf{P}_{ij}$.

Therefore, any vector from $Y_j^{(t+2)}$ either coincides with a vector from $Y_j^{(t+1)}$, or is a linear combination of vectors from $Y_j^{(t+1)}$, so that $rank\ Y_j^{(t+1)} = rank\ Y_j^{(t+2)}$.

DEFINITION: A Markov chain is called $\mathbf{M}$-lumpable with respect to a partition $\{A_1, A_2, \ldots, A_s\}$ if, for any initial distribution $\mathbf{p}$ from the set $\mathbf{M}$ the lumped process is a Markov chain with transition probabilities independent of $\mathbf{p} \in \mathbf{M}$.

According to Ref. 3, we say that a Markov chain is *strongly lumpable* if it is Ω-lumpable (where Ω is the set of all probability vectors of n-dimensional space).

Let $\mathbf{h}_1, \mathbf{h}_2, \ldots, \mathbf{h}_r$ be the basis of $\mathbf{M}$. If we represent these vectors in the matrix-block form $\mathbf{h}_i = [\ \mathbf{h}_{i1}\ \ \mathbf{h}_{i2}\ \ldots\ \mathbf{h}_{is}\ ]$, corresponding to a partition $\{A_1, A_2, \ldots, A_s\}$, then the previous results lead to the following theorem.

Theorem A.1.8: A necessary and sufficient condition for a Markov chain to be $\mathbf{M}$-lumpable with respect to a partition $\{A_1, A_2, \ldots, A_s\}$ is that expressions (A.1.25) assume constant value for the normalized vectors $\mathbf{h}_{ij}\mathbf{P}_{im}\mathbf{1} = \hat{p}_{im}\mathbf{h}_{ij}\mathbf{1}$ and, in addition, that the chain be $\mathbf{h}_i$-lumpable.

Theorem A.1.9: [3] A necessary and sufficient condition for a Markov chain to be strongly lumpable with respect to a partition $\{A_1, A_2, \ldots, A_s\}$ is that the sums of elements in the rows of the matrices $\mathbf{P}_{ij}$ be equal, that is,

$$\mathbf{P}_{ij}\mathbf{1} = \hat{p}_{ij}\mathbf{1} \qquad (i,j = 1,2,\ldots,s) \tag{A.1.29}$$

Proof: In this case $\mathbf{M} = \Omega$, and the basis can be formed by vectors $\mathbf{h}_1 = [1\ 0\ 0 \cdots 0]$, $\mathbf{h}_2 = [0\ 1\ 0 \cdots 0]$, ..., $\mathbf{h}_s = [0\ 0\ 0 \cdots 1]$. Then, it is necessary and sufficient that $\mathbf{h}_k\mathbf{P}_{ij}\mathbf{1} = \hat{p}_{ij}$, $i,j,k = 1,2,\ldots,s$. These expressions are equivalent to Eq. (A.1.29). They can be written in the following matrix form

$$\hat{\mathbf{P}} = \begin{bmatrix} \mathbf{h}_1 & 0 & \cdots & 0 \\ 0 & \mathbf{h}_2 & \cdots & 0 \\ \cdots & \cdots & \cdots & \cdots \\ 0 & 0 & \cdots & \mathbf{h}_1 \end{bmatrix} \mathbf{P} \begin{bmatrix} \mathbf{1} & 0 & \cdots & 0 \\ 0 & \mathbf{1} & \cdots & 0 \\ \cdots & \cdots & \cdots & \cdots \\ 0 & 0 & \cdots & \mathbf{1} \end{bmatrix}$$

1.3 SEMI-MARKOV LUMPABLE CHAINS

The conditions of semi-Markov lumpability of a Markov chain are largely similar to the conditions for Markov lumpability. We assume that sets $A_1, A_2, \ldots, A_s$ are not absorptive. Relationships (1.4.5), which determine the conditions for a semi-Markov process, are similar to Eq. (A.1.22). As we mentioned in Sec. 1.4, it is convenient to use the notion of the transition process when dealing with semi-Markov processes. Theorem 1.4.1 can now be rephrased as follows.

Theorem A.1.10: For a Markov chain to be $\mathbf{p}$-semi-Markov lumpable according to a partition $\{A_1, A_2, \ldots, A_s\}$, it is necessary and sufficient that, for any $y_0, y_1, \ldots, y_t$; $l_0, l_1, \ldots, l_t$; and $t \in \mathbf{N}$, the following conditions be satisfied:

$$f_{y_t y_{t+1}}(l) = \mathbf{p}_{y_0} \prod_{i=0}^{t} \mathbf{Q}_{y_i}(l_i)\mathbf{Q}_{y_i y_{i+1}} \mathbf{1} / \mathbf{p}_{y_0} \prod_{i=0}^{t-1} \mathbf{Q}_{y_i}(l_i)\mathbf{Q}_{y_i y_{i+1}} \mathbf{1}$$
$$t = 1,2,3,\ldots \tag{A.1.30}$$

These common expressions represent the transition distributions of the semi-Markov process.

Let us consider the sequences of vectors

$$\mathbf{h}_0(y_0) = \mathbf{p}_{y_0}$$
$$\mathbf{h}_t(y_0, y_1, \ldots, y_t; l_0, l_1, \ldots, l_{t-1}) = \mathbf{p}_{y_0} \prod_{i=0}^{t-1} \mathbf{Q}_{y_i}(l_i)\mathbf{Q}_{y_i y_{i+1}} \tag{A.1.31}$$
$$t = 1,2,\ldots$$

and normalized vectors

$$\mathbf{g}_t(y_0,y_1,...,y_t;l_0,l_1,...,l_{t-1}) = \frac{\mathbf{h}_t(y_0,y_1,...,y_t;l_0,l_1,...,l_{t-1})}{\mathbf{h}_t(y_0,y_1,...,y_t;l_0,l_1,...,l_{t-1})\mathbf{1}} \quad \text{(A.1.32)}$$

Using these notations, Theorem A.1.10 can be formulated as follows.

Theorem A.1.11: For a Markov chain to be **p**-semi-Markov lumpable according to a partition $\{A_1,A_2,...,A_s\}$, it is necessary and sufficient that, for all probability vectors of the form of (A.1.32) , the expressions

$$f_{ij}(l) = \mathbf{g}_{t-1}(y_0,y_1,...,y_{t-2},i;l_0,l_1,...,l_{t-2})\mathbf{Q}_i(l)\mathbf{Q}_{ij}\mathbf{1} \quad \text{(A.1.33)}$$

keep a constant value for fixed i, j, l $(i,j = 1,2,...,s; l = 1,2,...)$.

In other words, the right-hand side of Eq. (A.1.33) may depend only on i, j, l for all possible values of $y_0,y_1,...,y_{t-2}$; $l_0,l_1,...,l_{t-2}$; and t.

This theorem requires us to test an infinite number of conditions (A.1.33). In some cases, it is possible to prove that all these conditions are satisfied. Consider as an illustration the following statement, which gives simple sufficient conditions of semi-Markov lumpability.

Theorem 1.4.2: If the sets of states $\{A_1,A_2,...,A_s\}$ of a Markov chain are not absorbing, and the transition matrices have the form

$$\mathbf{P}_{ij} = \mu_{ij}\mathbf{M}_i\mathbf{N}_j \qquad i \neq j \quad (i,j = 1,2,...,s) \quad \text{(A.1.34)}$$

where $\mathbf{N}_j$ is a matrix row, $\mathbf{M}_i$ is a matrix column, and μ_{ij} is a number, then this Markov chain is semi-Markov lumpable.

Proof: Since $\mathbf{M}_i\mathbf{N}_j = [m_{ik}n_{jl}]$ and $\mu_{ij}m_{ik}n_{jl} \geq 0$, without loss of generality one can assume that

$$\mathbf{N}_j\mathbf{1} = 1 \qquad \mathbf{N}_j \geq 0 \quad \text{(A.1.35)}$$

Because **P** is a stochastic matrix, $\mathbf{P1} = \mathbf{1}$, which can be written in the following block form $\sum_j \mathbf{P}_{ij}\mathbf{1} = \mathbf{1}$. Then

$$\mathbf{1} - \mathbf{P}_{ii} = \sum_{j \neq i} \mathbf{P}_{ij}\mathbf{1} = \mu_i\mathbf{M}_i$$

where $\mu_i = \sum_j \mu_{ij}$. Thus, since $\mu_i \neq 0$

$$\mathbf{M}_i = (\mathbf{I} - \mathbf{P}_{ii})\mathbf{1}/\mu_i \quad \text{(A.1.36)}$$

Substituting this matrix into Eq. (A.1.34), we obtain

$$\mathbf{P}_{ij} = \hat{p}_{ij}(\mathbf{I} - \mathbf{P}_{ii})\mathbf{1}\mathbf{N}_j \qquad \hat{p}_{ij} = \mu_{ij}/\mu_i \quad \text{(A.1.37)}$$

But then

$$\mathbf{Q}_{ij} = (\mathbf{I} - \mathbf{P}_{ii})^{-1}\mathbf{P}_{ij} = \hat{p}_{ij}\mathbf{1}\mathbf{N}_j \quad \text{(A.1.38)}$$

To prove the theorem, let us show that the chain is **p**-semi-Markov lumpable if the initial distribution has the form $\mathbf{p} = [\ \hat{p}_1\mathbf{N}_1 \ \ \hat{p}_2\mathbf{N}_2 \ ... \ \hat{p}_s\mathbf{N}_s\ ]$, where $\hat{\mathbf{p}} = [\ \hat{p}_1 \ \ \hat{p}_2 \ ... \ \hat{p}_s\ ]$ is any probability vector.

Indeed, vectors in Eq. (A.1.31) become, after substitutions $\mathbf{p}_{y_0} = \hat{p}_{y_0}\mathbf{N}_{y_0}$ and (A.1.38),

$$\mathbf{h}_0(y_0) = \hat{p}_{y_0}\mathbf{N}_{y_0}$$

$$\mathbf{h}_t(y_0,y_1,...,y_t;l_0,l_1,...,l_{t-1}) = \hat{p}_{y_0}\mathbf{N}_{y_0}\prod_{i=0}^{t-1}\mathbf{Q}_{y_i}(l_i)\hat{p}_{y_iy_{i+1}}\mathbf{1N}_{y_{i+1}} \qquad t = 1,2,... .$$

Taking into account that $\mathbf{N}_{y_i}\mathbf{Q}_{y_i}(l_i)\mathbf{1}$ is a number, we obtain

$$\mathbf{g}_t(y_0,y_1,...,y_t;l_0,l_1,...,l_{t-1}) = \mathbf{N}_{y_t}$$

and, therefore,

$$\mathbf{g}_{t-1}(y_0,y_1,...,y_{t-2},i;l_0,l_1,...,l_{t-2})\mathbf{Q}_i(l)\mathbf{Q}_{ij}\mathbf{1} = \mathbf{N}_i\mathbf{Q}_i(l)\mathbf{Q}_{ij}\mathbf{1}$$

depends only on i, j, l. The conditions of Theorem A.1.11 are satisfied, which proves the theorem.

As for Markov lumpable chains, instead of testing the conditions of lumpability for an infinite number of vectors, it is sufficient to check them for the bases of vector sets (A.1.26).

Theorem A.1.12: For a Markov chain to be **p**-semi-Markov lumpable according to a partition $\{A_1,A_2,...,A_s\}$, it is necessary and sufficient that equations

$$f_{ij}(l) = \mathbf{e}_{ki}\mathbf{Q}_i(l)\mathbf{Q}_{ij}\mathbf{1}/\mathbf{e}_{ki}\mathbf{1} \qquad \text{(A.1.39)}$$

be satisfied for the bases $\{\mathbf{e}_{ki}\}$ of sets $\mathbf{Y}_i$ and for fixed i,j,l $(i,j = 1,2,...,s;\ l = 1,2,...)$.

The proof of this theorem is similar to that of Theorem A.1.6. Again, as we have seen, the bases of the sets $\mathbf{Y}_i$ can be built in a finite number of steps by building bases of the expanding unions of layers (A.1.28).

Even though we have replaced the infinite number of vectors to be tested by a finite number of vectors (the bases), the number of test conditions (A.1.39) is infinite, since l is an arbitrary positive integer. Let us show that it is sufficient to check only a finite number of conditions (A.1.39).

Conditions (A.1.39) can be written in the form $f_{ij}(l) = \mathbf{yP}_{ii}^{l-1}\mathbf{P}_{ij}\mathbf{1}/\mathbf{y1}$ and, therefore, as we have seen in Sec. 1.3.4, the transition distribution $f_{ij}(l)$ satisfies a linear difference equation with constant coefficients: $f_{ij}(l) = \sum_{m=1}^{k} a_{ijm}f_{ij}(l-m)$.

The order k of this equation is not greater than the size of the matrix $\mathbf{P}_{ii}$. It is well known that this equation has a unique solution that satisfies any fixed initial conditions $f_{ij}(1) = y_1, f_{ij}(2) = y_2,..., f_{ij}(k) = y_k$. Thus, if Eq. (A.1.39) is satisfied for all the basis vectors and all l not greater than the order of the matrix $\mathbf{P}_{ii}$, then this equation is valid for all positive l.

DEFINITION: Let **M** be a set of probability vectors. We shall say that a Markov chain is **M**-semi-Markov lumpable according to a partition $\{A_1,A_2,...,A_s\}$ if it is **p**-semi-Markov lumpable for any vector **p** that belongs to the set **M** and if transition distributions do not depend on $\mathbf{p} \in \mathbf{M}$.

Theorem A.1.13: If a regular Markov chain is **M**-semi-Markov lumpable, then the stationary distribution of the transition process belongs to **M**.

Indeed, if a Markov chain is **M**-semi-Markov lumpable, there exists a vector **p** such that

$$\mathbf{p}_{y_0}\prod_{i=0}^{t}\mathbf{Q}_{y_i}(l_i)\mathbf{Q}_{y_iy_{i+1}}\mathbf{1} = \hat{p}_{y_0}\prod_{i=0}^{t}f_{y_iy_{i+1}}(l_i)$$

If we sum these identities with respect to $y_0,y_1,...,y_{k-1}$ and $l_0,l_1,...,l_{k-1}$, we will have

$$\mathbf{p}_{y_k}^{(k)} \prod_{i=k}^{t} \mathbf{Q}_{y_i}(l_i)\mathbf{Q}_{y_i y_{i+1}}\mathbf{1} = \hat{p}_{y_k}^{(k)} \prod_{i=k}^{t} f_{y_i y_{i+1}}(l_i)$$

where $\mathbf{p}_{y_k}^{(k)}$ is the block of the matrix $\mathbf{pQ}^k$, while $\hat{p}_{y_k}^{(k)}$ is an element of the matrix $\hat{\mathbf{p}}\hat{\mathbf{P}}^k$. But, due to matrix multiplication linearity, the previous equality is valid when we replace the block $\mathbf{p}_{y_k}^{(k)}$ with the corresponding block of the matrix $\mathbf{p}\sum_{k=0}^{N-1}\mathbf{Q}^k/N$. For $t = m + N$ and $N\to\infty$, we have

$$\mathbf{q}_{z_0} \prod_{i=0}^{m} \mathbf{Q}_{z_i}(\lambda_i)\mathbf{Q}_{z_i z_{i+1}}\mathbf{1} = \hat{q}_{z_0} \prod_{i=0}^{m} f_{z_i z_{i+1}}(\lambda_i)$$

where $\quad \mathbf{q} = \lim_{N\to\infty} \mathbf{p} \sum_{k=0}^{N-1} \mathbf{Q}^k/N$

The previous equation shows that the process is $\mathbf{q}$-semi-Markov lumpable so that $\mathbf{q} \in \mathbf{M}$.

The invariant vector $\mathbf{q} = [\ \mathbf{q}_1\ \ \mathbf{q}_2\ \dots\ \mathbf{q}_s\ ]$ of the matrix $\mathbf{Q}$ can be found as the solution of the system $\mathbf{q}\,(\,\mathbf{Q} - \mathbf{I}\,) = 0$, $\mathbf{q}\,\mathbf{1} = 1$. This system has a unique solution because $rank(\mathbf{Q}-\mathbf{I}) = rank(\mathbf{P}-\mathbf{I}) = n-1$ as a result of the relationship $(\mathbf{Q}-\mathbf{I}) = (\mathbf{I}-\mathbf{R})^{-1}(\mathbf{P}-\mathbf{I})$, where $\mathbf{R} = blockdiag\{\mathbf{P}_{ii}\}$. It is easy to verify that the unique solution has the form

$$\mathbf{q}_i = \boldsymbol{\pi}_i(\mathbf{1} - \mathbf{P}_{ii})/\sum_{k=1}^{s} \boldsymbol{\pi}_k(\mathbf{1} - \mathbf{P}_{kk})\mathbf{1}$$

where $\boldsymbol{\pi}$ is the stationary distribution of the Markov chain with the transition matrix $\mathbf{P}$.

References

1 C. J. Burke and M. A. Rosenblatt, "Markovian function of a Markov chain," *Ann. Math. Stat.*, **29**(4), 1112-1122, (1958).

2 F. R. Gantmacher, *The Theory of Matrices*, (Chelsea Publishing Co., New York, 1959).

3 J. G. Kemeny and J. L. Snell, *Finite Markov Chains,* (Van Nostrand, Princeton, New Jersey, 1960).

4 M. Marcus and H. Minc, *Survey of Matrix Theory and Matrix Inequalities*, (Allyn & Bacon, Needham Heights, Massachusetts, 1964).

APPENDIX 2

2.1 ASYMPTOTIC EXPANSION OF MATRIX PROBABILITIES

If elements of the z-transform

$$\mathbf{\Phi}(z) = \sum_{m=0}^{\infty} \mathbf{W}_m z^{-m} \tag{A.2.1}$$

are rational fractions $\mathbf{\Phi}(z) = [\, p_{ij}(z) / q_{ij}(z)\,]_{k,k}$, where $p_{ij}(z)$ and $q_{ij}(z)$ are polynomials, then if we denote $q(z)$ a common denominator for all the fractions, we obtain

$$\mathbf{\Phi}(z) = \mathbf{V}(z) / q(z) \tag{A.2.2}$$

where $\mathbf{V}(z)$ is a polynomial with matrix coefficients.

If series (A.2.1) converges outside the disk $|z| < r$, then it is possible to find $\mathbf{W}_m$ by using the inverse z-transform:

$$\mathbf{W}_m = \frac{1}{2\pi j} \oint_{\gamma} \mathbf{\Phi}(z) z^{m-1} dz \qquad \gamma = \{z : |z| = \rho\} \quad \rho > r. \tag{A.2.3}$$

It is possible to calculate this integral using the residue theorem. If we denote by $z_1, z_2, \ldots, z_n$ all the roots of the polynomial $q(z)$, then $q(z) = (z-z_1)^{m_1}(z-z_2)^{m_2} \cdots (z-z_n)^{m_n}$. From the residue theorem, we have

$$\mathbf{W}_m = \sum_{i=1}^{n} res\{\mathbf{\Phi}(z) z^{m-1}; z_i\} \tag{A.2.4}$$

Since all the singularities of the function $\mathbf{\Phi}(z)$ are poles, then

$$res\{\mathbf{\Phi}(z) z^{m-1}; z_i\} = \frac{1}{(m_i-1)!} \frac{d^{m_i-1}}{dz^{m_i-1}} [\mathbf{\Phi}_i(z) z^{m-1}]|_{z=z_i}$$

where $\mathbf{\Phi}_i(z) = (z - z_i)^{m_i} \mathbf{\Phi}(z)$.

After differentiation using the Leibnitz formula, we obtain

$$res\{\mathbf{\Phi}(z) z^{m-1}; z_i\} = \sum_{\nu=0}^{m_i-1} \mathbf{A}_{i\nu} \binom{m-1}{\nu} z_i^{m-\nu-1} \tag{A.2.5}$$

where $\mathbf{A}_{i\nu} = \mathbf{\Phi}_i^{(m_i-\nu-1)}(z_i)/(m_i-\nu-1)!$. Formula (A.2.2) takes the form

$$\mathbf{W}_m = \sum_{i=1}^{n} \sum_{\nu=0}^{m_i-1} \mathbf{A}_{i\nu} \binom{m-1}{\nu} z_i^{m-\nu-1} \tag{A.2.6}$$

In the particular case when all the roots of the polynomial $q(z)$ are simple, the previous formula becomes

$$\mathbf{W}_m = \sum_{i=1}^{n} \mathbf{A}_{i0} z_i^{m-1} \tag{A.2.7}$$

where

$$\mathbf{A}_{i0} = \mathbf{\Phi}_i(z_i) = \mathbf{V}(z_i) / q'(z_i) \tag{A.2.8}$$

Note that the component corresponding to the residue at $z = 0$ is included only in a finite number of the first terms. Therefore, to find asymptotic formulas for matrices $\mathbf{W}_m$ it is possible to drop the components corresponding to the residues at $z = 0$.

Theorem A.2.1: Let the generating function of the sequence $\mathbf{W}_m$ have the form of Eq. (A.2.2) and $z_1, z_2, \ldots, z_a$ be the roots of the polynomial $q(z)$ with the maximal absolute value that have the highest multiplicity factor b, then $\mathbf{W}_m$ has the following asymptotic representation

$$\mathbf{W}_m \sim \frac{m^{b-1}}{(b-1)!} \sum_{i=1}^{a} \mathbf{A}_{i\,b-1} z_i^{m-b} \tag{A.2.9}$$

Proof: Denote $\lambda = \max_i |z_i|$. Since $\lim_{m\to\infty} \binom{m-1}{\nu} z_j^m \lambda^{-m} = 0$ for $|z_j| < \lambda$, we can retain in the asymptotic formula for Eq. (A.2.6) only the components corresponding to roots z_i with modules equal to λ. Furthermore, since for $|z_i| = |z_j|$ and $\alpha > \nu$,

$$\lim_{m\to\infty} \frac{\binom{m-1}{\nu} z_i^m}{\binom{m-1}{\alpha} z_j^m} = 0$$

we can ignore in the remaining sum all the components for which binomial coefficients $\binom{m-1}{\nu} z_i^m$ have ν that is less than the maximum (which is $b-1$). Thus, we obtain the asymptotic representation

$$\mathbf{W}_m \sim \binom{m-1}{b-1} \sum_{i=1}^{a} \mathbf{A}_{i\,b-1}\, z_i^{m-b} \tag{A.2.10}$$

And, finally, since $\lim_{m\to\infty} \binom{m-1}{b-1} (b-1)!/m^{b-1} = 1$, we obtain Eq. (A.2.9).

Note that formula (A.2.9) gives the main term of the asymptotic expansion. It is not difficult to obtain the whole asymptotic expansion using Eq. (A.2.6). To do this it is necessary to write formula (A.2.6) in the form of an asymptotic series for the significance of addends:

$$\mathbf{W}_m \sim \mathbf{W}_m^{(0)} + \mathbf{W}_m^{(1)} + \cdots \tag{A.2.11}$$

where $\mathbf{W}_m^{(0)}$ is defined by formula (A.2.10); $\mathbf{W}_m^{(1)}$ is an expression of the similar form

$$\mathbf{W}_m^{(1)} = \binom{m-1}{b-2} \sum_{i=1}^{a} \mathbf{A}_{i\,b-2} z_i^{m-b+1} \tag{A.2.12}$$

and so on. After we exhaust all components corresponding to the roots with the largest module, we consider the roots with the largest module among the remaining roots, and so on. In this way we obtain asymptotic expansion with respect to sequences of the form $(m-1)...(m-\nu)q^m$. It is possible to obtain an expansion with respect to $m^\nu q^m$ if we use Sterling's numbers. [5]

When we use any approximation formulas, it is important to estimate the error, which is equal to the absolute value of the difference between the exact value and its approximation. As we have seen, the main term of an asymptotic expansion corresponds to the sum of the residues of the z-transform at the poles with the maximal absolute value

$$\mathbf{W}_m \sim \sum_{j=1}^{a} res\{\mathbf{\Phi}(z)z^{m-1}, z_j\} \quad |z_j| = \max_i |z_i| = \lambda$$

The error we calculated by replacing $\mathbf{W}_m$ by the right-hand side of this formula can be written in the integral form

$$\Delta_m = \mathbf{W}_m - \sum_{j=1}^{a} res\{\mathbf{\Phi}(z)z^{m-1}, z_j)\} = \frac{1}{2\pi j} \oint_{|z|=r} \mathbf{\Phi}(z) z^{m-1} dz$$

where $\lambda > r > \lambda_1, \lambda_1 = \max_{i>a} |z_i|$. For elements of Δ_m, we obtain

$$\Delta_{mij} = \frac{1}{2\pi j} \oint_{|z|=r} \phi_{ij}(z) z^{m-1} dz$$

If we denote $M_{ij}(r) = \max_{|z|=r} \phi_{ij}(z)$, then by replacing the integral on the right-hand side of the last formula with its upper bound, we obtain $|\Delta_{mij}| \le M_{ij}(r) r^m$.

Note that to find $M_{ij}(r)$ it is convenient to use the following lemma: If $a_m \ge 0$, then

$$\max_{|z|=r} \left| \sum_m a_m z^m \right| = \sum_m a_m r^m$$

If we consider residues at the poles inside the circle of radius λ_1, then we can replace r by $r_1 < \lambda_1$ in the error upper bound.

2.2 CHERNOFF BOUNDS

The upper bound of probabilities

$$\mathbf{P}(\xi \ge k) = \sum_{x=k}^{\infty} \mathbf{P}_\xi(x) \tag{A.2.13}$$

can be found using the z-transform of the distribution $\mathbf{P}_\xi(x)$:

$$\mathbf{\Phi}(z) = \sum_{x=0}^{\infty} \mathbf{P}_\xi(x) z^{-x} \tag{A.2.14}$$

For real z meeting the conditions $R < z \le 1$, where R is the radius of convergence of the series in Eq. (A.2.14), we have

$$\Phi(z) \geq \sum_{x=k}^{\infty} \mathbf{P}_{\xi}(x) z^{-x} \geq z^{-k} \sum_{x=k}^{\infty} \mathbf{P}_{\xi}(x) = z^{-k} \mathbf{P}(\xi \geq k)$$

This equation gives us the bound

$$\mathbf{P}(\xi \geq k) \leq z^k \Phi(z) \tag{A.2.15}$$

from which we obtain the following inequalities for the elements of $\mathbf{P}(\xi \geq k)$:

$$p_{ij}(\xi \geq k) \leq z^k \phi_{ij}(z) \tag{A.2.16}$$

If, for $z = z_{ij} \in (R, 1]$, the function $z^k \phi_{ij}(z)$ reaches the minimum, then

$$p_{ij}(\xi \geq k) \leq z_{ij}^k \phi_{ij}(z_{ij}) \tag{A.2.17}$$

gives us the desired bound. If we denote z_0 such that $|z_0| = \min_{ij} |z_{ij}|$, then the last formula can be written in the matrix form

$$\mathbf{P}(\xi \geq k) \leq z_0^k \mathbf{F}_k(z_0) \tag{A.2.18}$$

where $\mathbf{F}_k(z_0) = [\ (z_{ij} / z_0)^k \phi_{ij}(z_{ij})\ ]$. Bounds (A.2.16) through (A.2.18) can be generalized to the case of multidimensional distributions.

2.3 BLOCK GRAPHS

A finite graph $G = (X, U)$ [1,4] is called a *block graph* (or *general flow graph*), [4] if it can be partitioned into nonintersecting subgraphs $G_1 = (X_1, U_1), G_2 = (X_2, U_2), \ldots, G_s = (X_s, U_s)$

$$G = \bigcup_{m=1}^{s} G_m \qquad G_i \cap G_j = \varnothing \qquad i \neq j \qquad (i,j = 1,2,\ldots,s) \tag{A.2.19}$$

The partitioning of a signal graph G into subgraphs corresponds to the partition of the matrix of transmission into subblocks

$$\mathbf{T} = block[\mathbf{T}_{ij}]_{s,s} = \begin{bmatrix} \mathbf{T}_{11} & \mathbf{T}_{12} & \cdots & \mathbf{T}_{1s} \\ \mathbf{T}_{21} & \mathbf{T}_{22} & \cdots & \mathbf{T}_{2s} \\ \cdots & \cdots & \cdots & \cdots \\ \mathbf{T}_{s1} & \mathbf{T}_{s2} & \cdots & \mathbf{T}_{ss} \end{bmatrix} \tag{A.2.20}$$

Here the diagonal blocks $\mathbf{T}_{11}, \mathbf{T}_{22}, \ldots, \mathbf{T}_{ss}$ are matrices of internal transmissions of subgraphs $\mathbf{G}_{11}, \mathbf{G}_{22}, \ldots, \mathbf{G}_{ss}$, whereas nondiagonal blocks correspond to transmissions between different subgraphs.

We will depict block graphs in the same way as normal graphs by writing under an arrow the corresponding matrix branch transmission. Matrices of internal transmissions between nodes of subgraphs are represented by loops (Fig. A.1).

Consider now some transformations of signal-flow graphs. The basic transformation of the block graph is the *absorption* [4] of a subgraph, which is frequently used throughout this book when dealing with Markov functions.

Suppose that we are given a graph G with a matrix $\mathbf{T} = [\ \mathbf{T}_{ij}\ ]$ and we want to find the

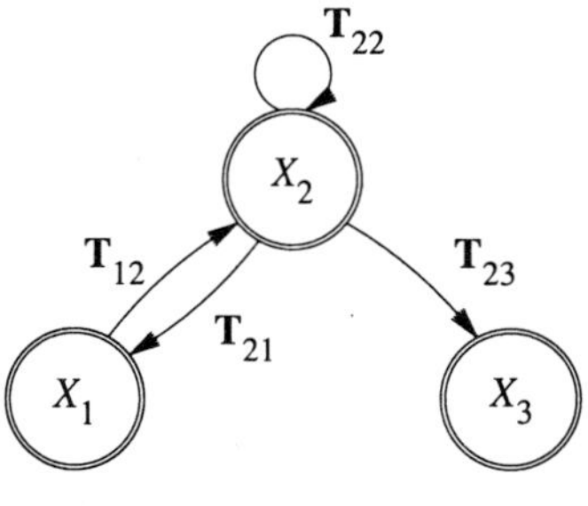

Figure A.1. Block graph.

matrix of $G^{(k)} = G \setminus G_k$ after absorption of the subgraph G_k. Then by the definition of a matrix of a signal-flow graph, [4] we have

$$\mathbf{X}_i = \sum_{j=1}^{s} \mathbf{T}_{ij}\mathbf{X}_j \qquad (i=1,2,\ldots,s) \tag{A.2.21}$$

For $i = k$, we have

$$(\mathbf{I} - \mathbf{T}_{kk})\mathbf{X}_k = \sum_{j \neq k} \mathbf{T}_{kj}\mathbf{X}_j$$

If we assume that the matrix $\mathbf{I} - \mathbf{T}_{kk}$ is nonsingular, then

$$\mathbf{X}_k = \sum_{j \neq k} (\mathbf{I} - \mathbf{T}_{kk})^{-1}\mathbf{T}_{kj}\mathbf{X}_j$$

Substituting the value of $\mathbf{X}_k$ in the remaining equations of system (A.2.21), we obtain

$$\mathbf{X}_i = \sum_{j \neq k} [\mathbf{T}_{ij} + \mathbf{T}_{ik}(\mathbf{I} - \mathbf{T}_{kk})^{-1}\mathbf{T}_{kj}]\mathbf{X}_j. \tag{A.2.22}$$

This leads to the conclusion that if $\det(\mathbf{I} - \mathbf{T}_{kk}) \neq 0$ the matrix branch transmission of the graph $G^{(k)}$ can be found by the formula

$$\mathbf{T}_{ij}^{(k)} = \mathbf{T}_{ij} + \mathbf{T}_{ik}(\mathbf{I} - \mathbf{T}_{kk})^{-1}\mathbf{T}_{kj} \tag{A.2.23}$$

Note that particular cases of transformation (A.2.23) are addition and multiplication of matrices and finding the inverse matrix.

These operations are illustrated in Fig. A.2. Therefore, we can perform elementary operations of matrix algebra using signal-flow graphs.

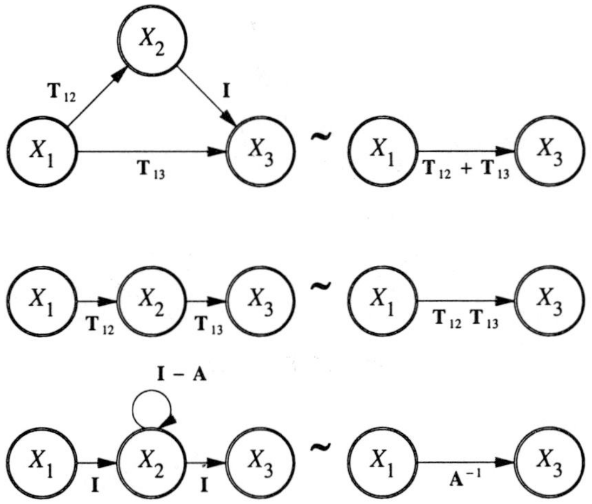

Figure A.2. Elementary matrix operations.

References

1 C. Berge, *Theory of Graphs and Its Applications,* (John Wiley & Sons, New York, 1962).

2 M. A. Evgrafov, *Asymptotic Estimates and Entire Functions*, (Gordon & Breach Science Publishers, New York, 1961).

3 F. R. Gantmacher, *The Theory of Matrices*, (Chelsea Publishing Co., New York, 1959).

4 S. J. Mason and H. J. Zimmermann, *Electronic Circuits, Signals, and Systems,* (John Wiley & Sons, New York, 1965).

5 J. Riordan, *An Introduction to Combinatorial Analysis*, (John Wiley & Sons, New York, 1958).

APPENDIX 3

3.1 STATISTICAL INFERENCE

An error source model development is a problem of mathematical statistics. Our goal is to show that the major steps in model building can be described as particular cases of known statistical methods. We describe here some of the methods that are used at various stages of the statistical analysis.

3.1.1 Distribution Parameter Estimation.

Let $Pr(\xi = x) = f(x,\boldsymbol{\tau})$ be a random variable distribution with the unknown parameter $\boldsymbol{\tau} = (\tau_1, \tau_2, \ldots, \tau_k)$ which should be estimated using a sequence of N independent samples $x_1, x_2, \ldots, x_N$ of this variable. The parameter estimate $\hat{\boldsymbol{\tau}}$ is defined as some function of these samples:

$$\hat{\boldsymbol{\tau}} = \phi(x_1, x_2, \ldots, x_N) \tag{A3.1.1}$$

such that its value $\hat{\boldsymbol{\tau}}$ is, in some sense, close to the actual parameter value.

An estimate is said to be *efficient* if its mean square deviation from the real parameter $\boldsymbol{\tau}$ is the smallest:

$$\mathbf{E}\{[\phi(\xi_1, \xi_2, \ldots, \xi_N) - \boldsymbol{\tau}]^2\} = \min \tag{A3.1.2}$$

It is said to be *unbiased* if its mean is equal to $\boldsymbol{\tau}$:

$$\mathbf{E}\{\phi(\xi_1, \xi_2, \ldots, \xi_N)\} = \boldsymbol{\tau} \tag{A3.1.3}$$

It is also important that an estimate $\hat{\boldsymbol{\tau}}$ converge to $\boldsymbol{\tau}$ in probability:

$$P \lim_{N \to \infty} \phi(\xi_1, \xi_2, \ldots, \xi_N) = \boldsymbol{\tau} \tag{A3.1.4}$$

In this case, the estimate is said to be *consistent*.

Example A.3.1. Consider a channel with independent errors and bit-error probability p. For this model, intervals between consecutive errors are geometrically distributed

$$p(l) = p(1-p)^{l-1} \tag{A3.1.5}$$

This distribution has only one parameter ($\tau = p$). If we have a sequence $x_1, x_2, \ldots, x_N$ of measured intervals between errors, then their mean is inversely proportional to the bit-error rate, which can be used as an estimate of the bit-error probability:

$$\hat{p} = N / \sum_{i=1}^{N} x_i \tag{A3.1.6}$$

This estimate is unbiased, consistent, and efficient.

One of the most powerful methods of parameter estimation, which is called the *maximum likelihood* (ML) method, is based on the assumption that for the best estimate an experiment's result has the largest probability. Suppose that $L(\mathbf{x}, \boldsymbol{\tau})$ is the probability that N samples of a variable ξ are represented by $\mathbf{x} = (x_1, x_2, \ldots, x_N)$. This function is called the *likelihood function*, and the value of the parameter $\hat{\boldsymbol{\tau}}$ that delivers its global maximum is called the ML estimate:

$$\max_{\boldsymbol{\tau}} L(\mathbf{x}, \boldsymbol{\tau}) = L(\mathbf{x}, \hat{\boldsymbol{\tau}}) \tag{A3.1.7}$$

If the set Ω of all possible values of the random variable ξ is partitioned into r disjoint subsets

$$\Omega = \bigcup_i A_i \qquad A_i \cap A_j = \varnothing \quad \text{for } i \neq j$$

then the likelihood function can be expressed as

$$L(\mathbf{x}, \boldsymbol{\tau}) = \prod_{i=1}^{r} P_i^{\nu_i}(\boldsymbol{\tau}) \tag{A3.1.8}$$

where $P_i(\boldsymbol{\tau})$ is the probability that the random variable ξ belongs to the subset A_i, $N = \nu_1 + \nu_2 + \cdots + \nu_r$. The maximization of this function is equivalent to the minimization of the so-called information criterion [12]

$$I = \sum_{i=1}^{r} \nu_i \ln (\nu_i / N) - \ln L(\mathbf{x}, \boldsymbol{\tau}) \tag{A3.1.9}$$

or

$$I = \sum_{i=1}^{r} \nu_i \ln [\nu_i / N P_i(\boldsymbol{\tau})] \tag{A3.1.10}$$

Another measure of the estimate quality, which is very popular in statistical applications, is the χ^2 criterion

$$\chi^2 = \sum_{i=1}^{r} (\nu_i - N P_i)^2 / N P_i \tag{A3.1.11}$$

Using the approximation $\ln x \approx (x - x^{-1})/2$ for $x \approx 1$, it is easy to prove that

$$2 \sum_i a_i \ln (a_i / b_i) \approx \sum_i (a_i - b_i)^2 / b_i \quad \text{if} \quad \sum_i a_i = \sum_i b_i \tag{A3.1.12}$$

It can be shown, applying this formula to (A3.1.9), that the ML estimation method is asymptotically equivalent to the minimum of I or χ^2 criterion, [12] and that under general assumptions [7] it delivers a consistent estimate.

The defined above χ^2 in which $\boldsymbol{\tau}$ is replaced with its ML estimate $\hat{\boldsymbol{\tau}}$ and $N \to \infty$ has a so-called χ^2 distribution with $r-k-1$ degrees of freedom † and the probability density

$$K_{r-k-1}(x) = A_{r-k-1} x^{(r-k-3)} e^{-0.5x} \qquad \text{(A3.1.13)}$$

where

$$A_m = \begin{cases} [2^{i+1} i!]^{-1} & \text{for} \quad m = 2i+2 \\ [1 \cdot 3 \cdots (2i-1) 2^{-i} \sqrt{2\pi}]^{-1} & \text{for} \quad m = 2i+1 \end{cases}$$

Example A.3.2. If $x_1, x_2, ..., x_N$ is a sequence of observed intervals between errors that occur independently with the probability p, then the likelihood function is equal to

$$L(\mathbf{x},p) = \prod_{i=1}^{N} p(1-p)^{x_i - 1}$$

The ML estimate can be found from the necessary condition of this function maximum

$$\frac{\partial L(\mathbf{x},p)}{\partial p} = 0 \qquad \text{(A3.1.14)}$$

This equation has a unique solution, which is given by equation (A3.1.6).

Even though the ML method delivers good parameter estimates, its practical application is rather limited by the complexity of the computations required. In many cases, significantly simpler equations can be obtained using the *method of moments*. According to this method, the distribution characteristics are calculated theoretically and then equated to their experimental estimates, thus creating a system of equations for the unknown parameters.

Example A.3.3. If $x_1, x_2, ..., x_N$ is a sequence of observed intervals between errors that occur independently with probability p, then the mean interval between errors is

$$\bar{l} = \sum_{l=1}^{\infty} l\, p(1-p)^{l-1} = 1/p \qquad \text{(A3.1.15)}$$

By equating $\bar{l}$ to the average observed interval between errors, we obtain the following equation of the method of moments, that is equivalent to (A3.1.6)

$$\frac{1}{N} \sum_{i=1}^{N} x_i = 1/\hat{p}$$

3.1.2 Hypothesis Testing.

Statistical hypothesis testing is one of the basic methods of model building. We make assumptions about the model's structure and parameters and then, on the basis of some experimental data $\mathbf{x} = (x_1, x_2, ..., x_n)$, judge whether this model describes the data accurately. In most cases, any experimental data are compatible with a model. So our

† The distribution parameter $r-k-1$ is called a *number of degrees of freedom* because it is equal to the total number r of independent sets minus the number of connections between them: k equations to estimate parameters $(\tau_1, \tau_2, ..., \tau_k)$ and one identity that states that the total sum of probabilities $P_i(\tau)$ is equal to 1.

inference should be based not on compatibility with the model but rather on the model's likelihood.

To test a hypothesis H_0, we need some criterion (or measure) $w(\mathbf{x})$ of closeness between the experimental data and a theoretical prediction based on the hypothesis. Usually, the criterion grows when the difference between experimental data and theoretical prediction grows. If the criterion value is greater than some critical value w_{cr}, the hypothesis H_0 is rejected and some alternative hypothesis H_1 is therefore accepted. If we define the hypothesis H_0 rejection region as $D_0 = \{\mathbf{x}: w(\mathbf{x}) > w_{cr}\}$, then the hypothesis H_0 is rejected when $\mathbf{x} \in D_0$. The complement $\overline{D}_0$ of the rejection region is usually called the *acceptance region.*

There is a possibility of committing the Type I error by rejecting H_0 when it is true with the probability

$$Pr(\mathbf{x} \in D_0 \mid H_0) = \alpha \tag{A3.1.16}$$

or the Type II error of accepting H_0 when it is false with the probability

$$Pr(\mathbf{x} \notin D_0 \mid H_1) = \beta \tag{A3.1.17}$$

The Type I error probability α is called the criterion *significance level* and the probability $1-\beta$ is called the criterion *power.* Naturally, we want to select a critical region to minimize the probabilities of errors of both types. However, it is not possible to make these probabilities arbitrarily small simultaneously. Usually, the critical region D_0 is selected to minimize the Type II error probability for some small fixed Type I error probability. The region D_0 that maximizes the criterion power subject to $Pr(\mathbf{x} \in D_0 \mid H_0) \le \alpha$ is called the most powerful test of size α.

Example A.3.4. For the geometric distribution (A3.1.5), consider the problem of testing the hypothesis

$$H_0:\ p = p_0 \quad \text{against} \quad H_1:\ p \neq p_0 \tag{A3.1.18}$$

The distribution parameter estimate (A3.1.6) can be rewritten as

$$\hat{p} = N/T \quad \text{where } T = \sum_{i=1}^{N} x_i \tag{A3.1.19}$$

is the total number of bits in all the intervals. Let

$$A(p_0) = (p_0 - \delta_\alpha, p_0 + \delta_\alpha) \tag{A3.1.20}$$

be an acceptance region for the hypothesis H_0. If the hypothesis H_0 is true, then $\hat{p}$ has a binomial distribution

$$Pr(\hat{p} = m/T) = \binom{T}{m} p_0^m q_0^{T-m} \tag{A3.1.21}$$

so that δ_α can be found as a solution of the following equation:

$$Pr(p_0 - \delta_\alpha \le \hat{p} \le p_0 + \delta_\alpha) = \sum_{i=a}^{b} \binom{T}{m} p_0^m q_0^{T-m} = 1-\alpha \tag{A3.1.22}$$

where $a = \max[0, (p_0 - \delta_\alpha)T]$, $b = \min[T, (p_0 + \delta_\alpha)T]$. If this equation does not have a solution, we may find a smaller acceptance region and introduce an uncertainty region, where a decision is made using a random number generator.

For large T the binomial distribution can be approximated by the Gaussian distribution, and Eq. (A3.1.22) becomes

$$Pr(p_0 - \delta_\alpha \le \hat{p} \le p_0 + \delta_\alpha) = erf(\delta_\alpha \sqrt{T/2p_0 q_0}) = 1-\alpha \tag{A3.1.23}$$

when $\delta_\alpha < p_0 < T - \delta_\alpha$. Let Z_α be the root of the equation $erfc(Z_\alpha) = \alpha$. Then the acceptance interval may be expressed as

$$A(p_0) = (p_0 - Z_\alpha\sqrt{2p_0(1-p_0)/T},\ p_0 + Z_\alpha\sqrt{2p_0(1-p_0)/T}) \tag{A3.1.24}$$

3.1.3 Goodness of Fit.

The following problem plays a fundamental role in model building. Suppose that the set Ω of all possible values of the random variable ξ is partitioned into r disjoint subsets:

$$\Omega = \bigcup_i A_i \quad A_i \cap A_j = \varnothing \quad \text{for } i \neq j$$

Let $Pr(\xi \in A_i) = P_i(\tau)$ be the probability that the random variable belongs to the subset A_i and $\hat{P}_i = \nu_i/N$ is the corresponding empirical frequency. Our goal is to estimate the distribution parameters $\tau = (\tau_1, \tau_2, \ldots, \tau_k)$ and decide whether or not the distribution $\{P_i(\tau)\}$ agrees with the experimental data $\hat{P}_i$. We use the χ^2 or the information criterion I to solve this problem.

Denote $\hat{\tau} = (\hat{\tau}_1, \hat{\tau}_2, \ldots, \hat{\tau}_k)$ an estimate of the parameter obtained by the ML (minimum I or χ^2) method. The hypothesis that $\{P_i(\hat{\tau})\}$ is the probability distribution of outcomes can be expressed as

$$H_0:\ Pr(\xi \in A_i) = P_i(\hat{\tau}) \quad (i = 1, 2, \ldots, r) \quad \text{against} \quad H_1 = \overline{H}_0$$

To test this hypothesis, we can use the criterion χ^2 or $2I$ defined by Eq. (A3.1.10) and (A3.1.11), in which τ is replaced by $\hat{\tau}$. Both criteria have the χ^2 distribution (A3.1.13) with $r-k-1$ degrees of freedom. Let $\chi^2_{\gamma,m}$ be a root of the equation

$$Pr(\chi^2 < \chi^2_{\gamma,m}) = \int_0^{\chi^2_{\gamma,m}} K_m(x)\,dx = \gamma$$

Then the significance α rejection region may be defined as

$$D_0 = \{\chi^2:\ \chi^2 > \chi^2_{1-\alpha,r-k-1}\}$$

Thus we say that the distribution agrees with the experimental data if the calculated value χ^2 does not exceed $\chi^2_{1-\alpha,r-k-1}$. The probability of rejecting H_0 when it is true is less than α. Since $\chi^2 \approx 2I$, the same rejection region may be used for the information criterion.

Example A.3.5. Suppose that random vectors $\mathbf{v} = (\xi, \eta)$ are independent. We want to test the null hypothesis that vector components ξ and η are independent.

Let p_{ij} be a vector $\mathbf{v}$ probability distribution and $p_{i.}$, $p_{.j}$ be the distributions of its components

$$p_{i.} = \sum_j^s p_{ij} \quad p_{.j} = \sum_i^r p_{ij}$$

$i = 1, 2, \ldots, r$, $j = 1, 2, \ldots, s$. Then the null hypothesis is equivalent to the hypothesis that the vector probability distribution has the form

$$p_{ij} = p_{i.}p_{.j} \tag{A3.1.25}$$

In other words, we need to test the goodness of fit of the distribution that depends on $r+s-2$ independent parameters.

Suppose that in a sequence of independent trials a vector (i,j) occurs ν_{ij} times. If the hypothesis H_0 is true, then the likelihood function is given by

$$L = \prod_{i=1}^{r}\prod_{j=1}^{s}(p_{i.}p_{.j})^{\nu_{ij}} = \prod_{i=1}^{r} p_{i.}^{\nu_{i.}} \prod_{j=1}^{s} p_{.j}^{\nu_{.j}}$$

where

$$\nu_{i.} = \sum_{j=1}^{s} \nu_{ij} \qquad \nu_{.j} = \sum_{i=1}^{r} \nu_{ij}$$

This function reaches its maximal value when

$$p_{i.} = \hat{p}_{i.} = \nu_{i.}/n \qquad p_{.j} = \hat{p}_{.j} = \nu_{.j}/n \qquad n = \sum_{i=1}^{r} \nu_{i.}$$

The test information criterion (A3.1.10) takes the form

$$I = \sum_{i=1}^{r} \sum_{j=1}^{s} \nu_{ij} \ln (n\nu_{ij}/\nu_{i.}\nu_{.j}) \tag{A3.1.26}$$

It is distributed as $0.5\chi^2$ with $rs-(r+s-2)-1 = (r-1)(s-1)$ degrees of freedom. The expression (A3.1.11) reduces to

$$\chi^2 = \sum_{i=1}^{r} \sum_{j=1}^{s} (n\nu_{ij} - \nu_{i.}\nu_{.j})^2/n\nu_{i.}\nu_{.j} \tag{A3.1.27}$$

This also has a limiting χ^2 distribution with $(r-1)(s-1)$ degrees of freedom. The null hypothesis is rejected if

$$\chi^2 > \chi^2_{(1-\alpha),(r-1)(s-1)}$$

3.1.4 Confidence Limits.

The parameter estimates considered above are called *point estimates* in mathematical statistics. They do not give us an idea of the estimate variations. It is clear that even if the estimate is good, its reliability may be poor for a small number of samples. The quantitative characterization of the estimate reliability is given by the theory of confidence limits. The confidence region $\Theta(\mathbf{x})$ for the parameter τ with the confidence level $1-\alpha$ is defined as a region that contains the parameter with probability $1-\alpha$:

$$Pr(\boldsymbol{\tau} \in \Theta(\mathbf{x})) = 1-\alpha$$

Thus, if α and the region are small, then with high confidence we can say that we have found a reliable estimate. In the case of single-parameter estimation, the confidence region is usually an interval, so that the previous equation becomes

$$Pr(\tau_L < \tau \le \tau_H) = 1-\alpha$$

The interval limits are called the *confidence limits*.

To find a confidence region, let us consider a hypothesis

$$H_0: \quad \boldsymbol{\tau} = \boldsymbol{\tau}_0 \qquad \text{against} \qquad H_1: \quad \boldsymbol{\tau} \ne \boldsymbol{\tau}_0$$

and its acceptance region $A(\boldsymbol{\tau}_0)$ of significance α

$$Pr(\mathbf{x} \in A(\boldsymbol{\tau})) = 1-\alpha$$

A family of acceptance regions $\{A(\boldsymbol{\tau})\}$ for all possible $\boldsymbol{\tau}$ defines a region $U(\alpha)$ in the $(\mathbf{x},\boldsymbol{\tau})$ domain. The confidence region $\Theta(\mathbf{x})$ contains all $\boldsymbol{\tau}$ for which $(\mathbf{x},\boldsymbol{\tau}) \in U(\alpha)$, since

$$Pr(\mathbf{x} \in A(\boldsymbol{\tau})) = Pr\{(\mathbf{x},\boldsymbol{\tau}) \in U(\alpha)\} = 1-\alpha \tag{3.1.28}$$

Example A.3.6. Let us find the confidence limits for the parameter p of the geometric distribution (A3.1.5). Assuming that the number of samples is large and $\delta_\alpha < p_0 < T - \delta_\alpha$, we obtain the family of acceptance regions (A3.1.24)

$$A(p) = (p - Z_\alpha\sqrt{2p(1-p)/T},\ p + Z_\alpha\sqrt{2p(1-p)/T}) \qquad \text{(A3.1.29)}$$

The region $U(\alpha)$ in this case lies inside the ellipse $\hat{p} = p \pm Z_\alpha\sqrt{2p(1-p)/T}$ and square $0 \le p \le 1$, $0 \le \hat{p} \le 1$. Solving these equations with respect to p, we obtain the confidence interval

$$p_{H,L} = \frac{T}{T+2Z_\alpha^2}\left[\hat{p} \pm \frac{Z_\alpha}{T} + Z_\alpha\sqrt{2\hat{p}(1-\hat{p})/T + (Z_\alpha^2/T)^2}\right]$$

Since T is assumed to be large, these equations may be simplified:

$$p_L = \hat{p} - Z_\alpha\sqrt{2\hat{p}(1-\hat{p})/T} \qquad p_H = \hat{p} + Z_\alpha\sqrt{2\hat{p}(1-\hat{p})/T}$$

3.2 MARKOV CHAIN MODEL BUILDING

3.2.1 Statistical Tests for Simple Markov Chains.

There are many publications devoted to statistical inference concerning Markov chains. [1 3 4 5 6 10] We consider here only the methods that are used for HMM building.

Let $x_1, x_2, \ldots, x_n$ be a sequence of samples of some s-ary stochastic process ξ_t ($\xi_t = 1,2,\ldots,s$). We want to test a hypothesis H_0 that this process is a simple (first-order) regular Markov chain with the transition probability matrix $\mathbf{P}(\boldsymbol{\tau}) = [\,p_{ij}(\boldsymbol{\tau})\,]_{s,s}$, with k independent parameters $\boldsymbol{\tau} = (\tau_1, \tau_2, \ldots, \tau_k)$. Within this hypothesis, the likelihood function is given by

$$L(\boldsymbol{\tau}) = p_{x_1}\prod_{i=1}^{n-1} p_{x_i x_{i+1}}(\boldsymbol{\tau}) = p_{x_1}\prod_{ij} p_{ij}^{n_{ij}}(\boldsymbol{\tau}) \qquad \text{(A3.2.1)}$$

where n_{ij} is the observed number of transitions from i to j in the sequence $x_, x_2, \ldots, x_n$. The logarithmic likelihood function is equal to

$$\ln L = J_0 + J_1 \qquad \text{(A3.2.2)}$$

where

$$J_0 = \ln p_{x_1} \qquad J_1(\boldsymbol{\tau}) = \sum_{ij} n_{ij} \ln p_{ij}(\boldsymbol{\tau}) \qquad \text{(A3.2.3)}$$

The ML parameter estimate $\hat{\boldsymbol{\tau}} = (\hat{\tau}_1, \hat{\tau}_2, \ldots, \hat{\tau}_k)$ asymptotically maximizes J_1, since $J_0/J_1 \to 0$ when $n \to \infty$, and can be found as a solution of the system

$$\frac{\partial J_1(\boldsymbol{\tau})}{\partial \tau_m} = \sum_{ij} \frac{n_{ij}}{p_{ij}} \frac{\partial p_{ij}}{\partial \tau_m} = 0 \qquad m = 1,2,\ldots,k \qquad \text{(A3.2.4)}$$

assuming that all the derivatives exist.

It is shown [3,5] that if H_0 is true, the hypothesis information criterion

$$I = \sum_{ij} n_{ij} \ln\left[n_{ij} / n_{i.} p_{ij}(\hat{\boldsymbol{\tau}})\right] \qquad \text{(A3.2.5)}$$

is distributed asymptotically as $0.5\chi^2$ for a family of functions $p_{ij}(\boldsymbol{\tau})$ that satisfy the following conditions. [4] They have third-order continuous partial derivatives in some open subset Θ of the Euclidean space E^k. And the Jacobian matrix

$$\left[\frac{\partial p_u(\tau)}{\partial \tau_v} \right]_{d,k} \qquad u = (i,j) \in D$$

has rank k throughout Θ. Here D is a set of $u = (i,j)$ such that for each $u \in D$ $p_u(\tau) > 0$ throughout Θ, d is the size of D, and the chain is assumed to be regular. The summation in Eq. (A3.2.5) is performed over all $(i,j) \in D$, and the number of degrees of freedom in the limiting distribution is

$$f = N_{nz} - N_{st} - N_{par} \tag{A3.2.6}$$

where $N_{nz} = d$ is the total number of $p_{ij}(\hat{\tau})$ which are not equal to zero, $N_{st} = s$ is the total number of states, and $N_{par} = k$ is the total number of independent parameters that have been estimated.

The hypothesis χ^2 criterion

$$\chi^2 = \sum_{ij} \frac{[n_{ij} - n_{i.} p_{ij}(\hat{\tau})]^2}{n_{i.} p_{ij}(\hat{\tau})} \tag{A3.2.7}$$

is asymptotically equivalent to $2\,I$. [5] For the functions that we use to create an HMM these conditions are satisfied.

Example A.3.7. Consider a particular case of independent and identically distributed variables. The transition probabilities do not depend on i and therefore can be expressed as $p_{ij} = \tau_j$ for all i. The number of independent parameters is equal to $N_{par} = s - 1$, since $\tau_1 + \tau_2 + \cdots + \tau_s = 1$. The likelihood function J_1 reaches its maximal value when $\tau_j = \hat{\tau}_j = n_{.j} / n$ and the information criterion (A3.2.5) coincides with Eq. (A3.1.26). If all the probabilities $\tau_j > 0$, then, according to Eq. (A3.2.6), the number of degrees of freedom is equal to $s^2 - s - (s-1) = (s-1)^2$, which coincides with the results obtained before.

3.2.2 Multiple Markov Chains.

As we have seen in Sec. 1.1.7, the HMM error source model can be built on the basis of order k Markov chains. Different sequences of errors $e_1, e_2, \ldots, e_k$ constitute states of the HMM. The state transition probabilities are given by

$$p_{(d_1 \cdots d_k)(e_1 \cdots e_k)} = \begin{cases} p_{e_1 e_2 \cdots e_k e_{k+1}} & \text{if} \quad e_1 = d_2, \ldots, e_{k-1} = d_k \\ 0 & \text{otherwise} \end{cases} \tag{A3.2.8}$$

where $p_{e_1 e_2 \cdots e_k e_{k+1}}$ is defined as the probability of an error e_{k+1} at the moment t, given that errors e_1, e_2, ..., e_k occurred at the moments $t-k$, $t-k+1$, ..., $t-1$, respectively: $p_{e_1 e_2 \cdots e_{k+1}} = Pr(e_{k+1} \mid e_1, e_2, \ldots, e_k)$. These transition probabilities, along with the chain order (k), are the model parameters, and they must be estimated from the experimental data.

To estimate the chain order, we test the null hypothesis that the chain has order $k-1$ against the alternative hypothesis that its order is k. The hypothesis to be tested may be expressed as

$$H_0: \; p_{e_1 e_2 \cdots e_{k+1}} = p_{e_2 \cdots e_{k+1}} \quad \text{for all} \quad e_1, e_2, \ldots, e_{k+1} \tag{A3.2.9}$$

This is a particular case of the hypothesis considered in the previous section, which can be tested using criterion (A3.2.5) or (A3.2.7). The values on the right-hand side of Eq. (A3.2.9) are the transition probability parameters that must be estimated.

Suppose that $e_1, e_2, \ldots, e_n$ are the outcomes of some experiment. If H_0 is true, the likelihood function is given by

$$L = p(e_1 e_2 \cdots e_{k-1}) \prod_{m=k}^{n-k+1} p_{e_m e_{m+1} \cdots e_{m+k-1}}$$
$$= p(e_1 e_2 \cdots e_{k-1}) \prod_{x_1,\ldots,x_k} p_{x_1 x_2 \cdots x_k}^{n(x_1,\ldots,x_k)}$$

where $n(x_1, x_2, \ldots, x_k)$ is the observed frequency of k-tuples $x_1, x_2, \ldots, x_k$ in the sequence $e_1, e_2, \ldots, e_n$. The logarithmic likelihood function is equal to $\ln L = J_0 + J_1$, where $J_0 = \ln p(e_1 e_2 \cdots e_{k-1})$ and $J_1 = \sum_{x_1,\ldots,x_k} n(x_1, \ldots, x_k) \ln p_{x_1 x_2 \cdots x_k}$.

The maximum likelihood parameter estimates $\hat{p}_{x_1 x_2 \cdots x_k}$ for large n are those that maximize J_1 and can be found as a solution to the system of Eq. (A3.2.4). It is easy to see that

$$\hat{p}_{x_1 x_2 \cdots x_k} = n(x_1, x_2, \ldots, x_k) / n(x_1, x_2, \ldots, x_{k-1}) \tag{A3.2.10}$$

The test information criterion (A3.2.5) takes the following form:

$$I = \sum_{x_1,\ldots,x_{k+1}} n(x_1, x_2, \ldots, x_{k+1}) \ln (\hat{p}_{x_1 x_2 \cdots x_{k+1}} / \hat{p}_{x_2 x_3 \cdots x_{k+1}}) \tag{A3.2.11}$$

and is distributed asymptotically as $0.5\chi^2$. To calculate the number of degrees of freedom, let us assume that all the probabilities Eq. (A3.2.9) are different from zero. This is also sufficient for the chain regularity. Then, according to (A3.2.8), the total number of the chain's nonzero transition probabilities is $N_{nz} = 2^{k+1}$. The number of states is $N_{st} = 2^k$. The total number of independent parameters is $N_{par} = 2^{k-1}$, so that the total number of degrees of freedom is equal to $2^{k+1} - 2^k - 2^{k-1} = 2^{k-1}$. The hypothesis criterion (A3.2.7) takes the form

$$\chi^2 = \sum_{x_1,\ldots,x_{k+1}} n^*(x_1, x_2, \ldots, x_k)(\hat{p}_{x_1 x_2 \cdots x_{k+1}} - \hat{p}_{x_2 \cdots x_{k+1}})^2 / \hat{p}_{x_2 x_3 \cdots x_{k+1}}$$

$$n^*(x_1, x_2, \ldots, x_k) = \sum_l n(x_1, x_2, \ldots, x_k, l)$$

This criterion has a limiting χ^2 distribution with 2^{k-1} degrees of freedom if H_0 is true. Note that $n^*(x_1, x_2, \ldots, x_k)$ may differ from $n(x_1, x_2, \ldots, x_k)$ by 1, so that for a large sample sequence we may replace one by the other.

We can start building an error source model, assuming that it is a classical binary symmetrical channel without memory. In other words, we make an assumption that errors are independent and occur with probability p_1. This model may be considered as a zero-order Markov chain. Equation (A3.2.10) gives us the ML probability of error estimate $\hat{p}_1 = n(1) / n$, where $n(1)$ is the total number of errors in the sequence of n bits. Also, $\hat{p}_0 = 1 - \hat{p}_1 = n(0) / n$, where $n(0)$ is the number of correct bits. Information criterion (A3.2.11) takes the form

$$I = \sum_{x_1 x_2} n(x_1, x_2) \ln (\hat{p}_{x_1 x_2} / \hat{p}_{x_2})$$

where $\hat{p}_{x_1 x_2} = n(x_1, x_2) / n(x_1)$. This equation obviously coincides with Eq. (A3.1.26). We reject the hypothesis that the channel is a binary symmetrical channel without memory if $I > 0.5 \chi^2_{(1-\alpha),1}$.

If the hypothesis that errors are independent is rejected, we try to test the hypothesis that errors form a first-order Markov chain against the alternative hypothesis that they form a second-order Markov chain. The hypothesis information criterion (A3.2.11) is given by

$$I = \sum_{x_1 x_2 x_3} n(x_1, x_2, x_3) \ln (\hat{p}_{x_1 x_2 x_3} / \hat{p}_{x_2 x_3}) \tag{A3.2.12}$$

where $\hat{p}_{x_1 x_2 x_3} = n(x_1, x_2, x_3) / n(x_1, x_2)$. If this hypothesis is rejected, we test the hypothesis that the error source is described by a second-order Markov chain and so on, until an r-th-order Markov chain fits the experimental data reasonably well.

The advantage of this method of model building is that the states of the HMM are well defined (they are different sequences of errors). The disadvantage of this method is the usually tremendous size of matrices that describe the model. If an error source is described by an order r chain, then $\mathbf{P}(0)$ and $\mathbf{P}(1)$ are order 2^r square matrices (see Sec. 1.1.7 and Example 1.1.3). Normally, the size of a channel memory r is measured by hundreds of bits so that it becomes practically impossible to use this model. However, as we will see later, this approach can be successfully used to describe error statistics inside bursts or clusters of short bursts.

3.3 SEMI-MARKOV PROCESS HYPOTHESIS TESTING

3.3.1 ML Parameter Estimation.

Let $x_1, x_2, \ldots, x_n$ be a sequence of samples of some s-ary stochastic process. We want to test the hypothesis H_0 that this process is a semi-Markov process with the interval transition distributions $f_{ij}(l;\boldsymbol{\tau})$ that depends on κ independent parameters $\boldsymbol{\tau} = (\tau_1, \tau_2, \ldots, \tau_\kappa)$. Within this hypothesis, according to the results of Sec. 1.4, the likelihood function is given by

$$L(\boldsymbol{\tau}) = p_{y_0} f^{(0)}_{y_0 y_1}(l_0) \prod_{i=1}^{n} f_{y_i y_{i+1}}(l_i;\boldsymbol{\tau}) \tag{A3.3.1}$$

or

$$L(\boldsymbol{\tau}) = p_{y_0} f^{(0)}_{y_0 y_1}(l_0) \prod_{\xi,\eta,l} [f_{\xi\eta}(l;\boldsymbol{\tau})]^{n(\xi,\eta,l)} \tag{A3.3.2}$$

where $n(\xi,\eta,l)$ is the frequency of transitions from state ξ to state η that took l steps. The logarithmic likelihood function is equal to

$$\ln L(\boldsymbol{\tau}) = J_0 + J_1(\boldsymbol{\tau}) \tag{A3.3.3}$$

where

$$J_0 = \ln p_{y_0} f^{(0)}_{y_0 y_1}(l_0) \qquad \text{(A3.3.4)}$$

$$J_1(\boldsymbol{\tau}) = \sum_{\xi,\eta,l} n(\xi,\eta,l) \ln f_{\xi\eta}(l;\boldsymbol{\tau}) \qquad \text{(A3.3.5)}$$

Since $J_0/J_1 \to 0$ when $n \to \infty$, we can ignore the term J_0 in the expression of the logarithmic likelihood function. The ML parameter estimates are found by minimization of J_1 and represent a solution of the system

$$\frac{\partial J_1(\boldsymbol{\tau})}{\partial \tau_i} = 0 \qquad i = 1,2,\ldots,\kappa$$

If we make no assumptions about the form of the interval transition distributions, this equation has an obvious solution:

$$\hat{f}_{\xi\eta}(l) = n(\xi,\eta,l) \,/\, N(\xi) \qquad \text{(A3.3.6)}$$

where $N(\xi) = \sum_{\eta,l} n(\xi,\eta,l)$.

3.3.2 Grouped Data.

Usually, the interval $[1,\infty)$ is partitioned into a finite number of subintervals:

$$[1,\infty) = \bigcup_{m=1}^{u(\xi,\eta)} \Big[x_m(\xi,\eta), x_{m+1}(\xi,\eta)\Big) \qquad \text{for each type of transition } \xi \to \eta.$$

Let $F_{\xi\eta}(m;\boldsymbol{\tau}) = \sum_{l=x_m}^{x_{m+1}-1} f_{\xi\eta}(l;\boldsymbol{\tau}) = Pr(x_m(\xi,\eta) \le l < x_{m+1}(\xi,\eta),\ \eta \mid \xi)$

be the probability that $l \in [x_m(\xi,\eta), x_{m+1}(\xi,\eta))$, and the process transfers from state ξ to η. Denote $\mathbf{z}_i = (y_{i+1}, m_i)$, where m_i is the index of the subinterval that contains l_i: $x_{m_i}(y_i, y_{i+1}) \le l_i < x_{m_i+1}(y_i, y_{i+1})$. If H_0 is true, the process $\{\mathbf{z}_i\}$ is a Markov chain, because the conditional probability

$$Pr(\mathbf{z}_i \mid \mathbf{z}_0, \mathbf{z}_1, \ldots, \mathbf{z}_{i-1}) = F_{y_i y_{i+1}}(m_i;\boldsymbol{\tau}) = Pr(\mathbf{z}_i \mid \mathbf{z}_{i-1})$$

depends only on $\mathbf{z}_{i-1}$. This means that we may apply the methods developed for the Markov chain hypothesis testing to semi-Markov process hypothesis testing.

If H_0 is true, then

$$Pr(\zeta,k \mid \xi,\eta,m) = F_{\eta\zeta}(k;\boldsymbol{\tau}) \quad \text{for all } \xi,\eta,\zeta,m,k \qquad \text{(A3.3.7)}$$

The ML parameter estimate $\hat{\boldsymbol{\tau}}$ for the grouped data can be found from equation

$$\sum_{\xi,\eta,k} \frac{N(\xi,\eta,k)}{F_{\xi\eta}(k;\boldsymbol{\tau})} \frac{\partial F_{\xi\eta}(k;\boldsymbol{\tau})}{\partial \tau_m} = 0 \qquad \text{(A3.3.8)}$$

where

$$N(\xi,\eta,k) = \sum_{l=x_m}^{x_{m+1}-1} n(\xi,\eta,l)$$

Hypothesis information criterion (A3.3.5) may be written as

$$I = \sum_{\xi,\eta,\zeta,m,k} N(\xi,\eta,\zeta,m,k) \ln \left[N(\xi,\eta,\zeta,m,k) \,/\, N(\xi,\eta,m) F_{\eta\zeta}(k;\hat{\tau}) \right] \tag{A3.3.9}$$

where $N(\xi,\eta,\zeta,m,k)$ is the frequency of transitions $\xi \to \eta \to \zeta$ and where adjacent transition interval lengths x and y satisfy conditions $x_m(\xi,\eta) \le x < x_{m+1}(\xi,\eta)$ and $x_k(\eta,\zeta) \le y < x_{k+1}(\eta,\zeta)$. This criterion is asymptotically distributed as $0.5\chi^2$ with f degrees of freedom defined by Eq. (A3.3.5). The total number of the chain $\{\mathbf{z}_i\}$ states is $N_{st} = \sum_{\xi,\eta} u(\xi,\eta)$. The total number of the nonzero transition probabilities is equal to $N_{nz} = \sum_{\xi,\eta,\zeta} u(\xi,\eta) u(\eta,\zeta)$. Therefore, the number of degrees of freedom of the χ^2 distribution is equal to

$$f = \sum_{\xi,\eta,\zeta} u(\xi,\eta) u(\eta,\zeta) - \sum_{\xi,\eta} u(\xi,\eta) - \kappa \tag{A3.3.10}$$

Criterion (A3.3.6) becomes

$$\chi^2 = \sum_{\xi,\eta,\zeta,m,k} \frac{[\, N(\xi,\eta,\zeta,m,k) - N(\xi,\eta,m) F_{\eta\zeta}(k;\hat{\tau}) \,]^2}{N(\xi,\eta,m) F_{\eta\zeta}(k;\hat{\tau})} \tag{A3.3.11}$$

It is asymptotically equivalent to 2 I.

Consider now a semi-Markov process without making any assumptions about the interval distribution form. This case may be treated as an intermediate stage of HMM building when we just want to test whether the model can be described on the basis of semi-Markov processes. In this case, all the $F_{\xi\eta}(m)$ become parameters and the ML estimates can be found from system (A3.3.8), which has the following solution:

$$\hat{F}_{\xi\eta}(m) = N(\xi,\eta,m) \,/\, N(\xi) \tag{A3.3.12}$$

The total number of independent parameters is equal to $\kappa = \sum_{\xi,\eta} u(\xi,\eta) - s$ since $\sum_{\eta,m} F_{\xi\eta}(m) = 1, \xi = 1,2,\ldots,s$. Hypothesis information criterion (A3.3.9) is given by

$$I = \sum_{\xi,\eta,\zeta,m,k} N(\xi,\eta,\zeta,m,k) \ln \left[\frac{N(\xi,\eta,\zeta,m,k) N(\eta)}{N(\xi,\eta,m) N(\eta,\zeta,k)} \right] \tag{A3.3.13}$$

and criterion (A3.3.11) becomes

$$\chi^2 = \sum_{\xi,\eta,\zeta,m,k} \frac{[\, N(\xi,\eta,\zeta,m,k) - N(\xi,\eta,m) \hat{F}_{\eta\zeta}(k) \,]^2}{N(\xi,\eta,m) \hat{F}_{\eta\zeta}(k)} \tag{A3.3.14}$$

Criteria 2 I and χ^2 are equivalent and have a limiting χ^2 distribution with

$$f = \sum_{\xi,\eta,\zeta} u(\xi,\eta) u(\eta,\zeta) - \sum_{\xi,\eta} u(\xi,\eta) - \sum_{\xi,\eta} u(\xi,\eta) - s \tag{A3.3.15}$$

degrees of freedom.

In the particular case in which $x_1(\xi,\eta) = 1$ and $x_2(\xi,\eta) = \infty$ for all ξ and η, the previous criterion coincides with criterion (A3.3.11) of the first-order Markov chain hypothesis testing. Another particular case, $s = 2$, can be used to test the hypothesis that $x_1, x_2, \ldots, x_n$ are the samples from an alternating renewal process. In this case, the preceding criteria become the well-known criteria of random variable independence testing. [12]

3.3.3 Underlying Markov Chain Parameter Estimation.

As we pointed out in Sec. 1.4, there is an alternative description of semi-Markov processes, which uses the notion of the underlying Markov chain whose transition probabilities are defined by the equation $p_{ij}(\tau) = \sum_{l=1}^{\infty} f_{ij}(l;\tau)$ and state-holding time distributions $w_{ij}(l;\tau) = f_{ij}(l;\tau) / p_{ij}(\tau)$. Let us assume that all the transition probabilities and holding time distributions depend on different sets of parameters. Then, the interval transition distribution takes the form

$$f_{ij}(l;\tau) = p_{ij}(\tau_1) w_{ij}(l;\tau_2) \qquad \text{(A3.3.16)}$$

This assumption, as we will see, may simplify HMM building. Likelihood function (A3.3.5) can be rewritten as

$$J_1(\tau) = J_{11}(\tau_1) + J_{12}(\tau_2) \qquad \text{(A3.3.17)}$$

where

$$J_{11}(\tau_1) = \sum_{\xi,\eta} n(\xi,\eta) \ln p_{\xi\eta}(\tau_1) \qquad n(\xi,\eta) = \sum_l n(\xi,\eta,l) \qquad \text{(A3.3.18)}$$

$$J_{12}(\tau_2) = \sum_{\xi,\eta,l} n(\xi,\eta,l) \ln w_{\xi\eta}(l;\tau_2) \qquad \text{(A3.3.19)}$$

It is clear from Eq. (A3.3.17) that the ML estimate of the parameter τ_1 of the underlying Markov process can be found independently of the parameter τ_2 of the state-holding time distributions. If the individual distributions have independent parameters, the likelihood function $J_{12}(\tau_2)$ can also be decomposed:

$$J_{12}(\tau_2) = \sum_{\xi\eta} G_{\xi\eta}(\tau_{\xi\eta})$$

where

$$G_{\xi\eta}(\tau_{\xi\eta}) = \sum_l n(\xi,\eta,l) \ln w_{\xi\eta}(l;\tau_{\xi\eta})$$

The ML estimates $\hat{\tau}_{\xi\eta}$ can be found independently by maximization of $G_{\xi\eta}(\tau_{\xi\eta})$. In the case in which no assumptions about the distributions' dependency on the parameters have been made, the ML estimates are given by

$$\hat{p}_{\xi\eta} = n(\xi,\eta) / n(\xi) \qquad n(\xi) = \sum_\eta n(\xi,\eta)$$

$$\hat{w}_{\xi\eta}(l) = n(\xi,\eta,l) / n(\xi,\eta)$$

The null hypothesis information criterion can be decomposed into a sum of information criteria of elementary hypotheses: the hypothesis that the transition process is a Markov chain with the transition probabilities $p_{ij}(\tau_1)$, the hypothesis that the state-holding time distribution is $w_{\xi\eta}(l;\tau_{\xi\eta})$, and the hypothesis that the process is semi-Markov with no assumptions about its interval transition distribution form. For the grouped data, we denote

$$W_{\xi\eta}(l;\tau_{\xi\eta}) = \sum_{l=x_m}^{x_{m+1}-1} w_{\xi\eta}(l;\tau_{\xi\eta}) = Pr(x_m(\xi,\eta) \le l < x_{m+1}(\xi,\eta) \mid \xi, \eta)$$

Hypothesis information criterion (A3.3.9), after some elementary transformations, can be rewritten in the form

$$I = I_1 + \sum_{\xi\eta} I_{\xi\eta} + I_3 \tag{A3.3.20}$$

where

$$I_1 = \sum_{\eta,\zeta} N(\eta,\zeta) \ln \left[N(\eta,\zeta) \,/\, N(\eta) p_{\eta\zeta}(\hat{\boldsymbol{\tau}}_1) \right] \tag{A3.3.21}$$

is information criterion (A3.3.5) for the hypothesis that the underlying transition process is a Markov chain, and

$$I_{\xi\eta} = \sum_{\eta,\zeta,k} N(\eta,\zeta,k) \ln \left[N(\eta,\zeta,k) \,/\, N(\eta,\zeta)\, W_{\eta\zeta}(k;\hat{\boldsymbol{\tau}}_{\xi\eta}) \right] \tag{A3.3.22}$$

is the information criterion for the goodness of fit of the state-holding time distribution, and, finally,

$$I_3 = \sum_{\xi,\eta,\zeta,m,k} N(\xi,\eta,\zeta,m,k) \ln \left[\frac{N(\xi,\eta,\zeta,m,k) N(\eta)}{N(\xi,\eta,m) N(\eta,\zeta,k)} \right] \tag{A3.3.23}$$

is information criterion (A3.3.13) for the hypothesis that the process can be described by a semi-Markov process when no assumptions have been made about its transition distributions structure.

Each of the preceding criteria can be used for null hypothesis rejection. First, we can use the criterion I_3 to test whether it is possible to describe the process with a semi-Markov process without making any other assumptions. If the hypothesis passes the first test, then we can test the hypothesis that the transition process is a Markov chain. After that, we can test the goodness of fit of all the state-holding time distributions. And, finally, if all these tests do not reject the null hypothesis, we can apply total information criterion (A3.3.20).

3.3.4 Autonomous Semi-Markov Processes.

If a semi-Markov process is autonomous, then its state-holding time distributions depend only on the transition initial state:

$$w_{\xi\eta}(l;\boldsymbol{\tau}_{\xi\eta}) = w_{\xi}(l;\boldsymbol{\theta}_{\xi}) \tag{A3.3.24}$$

In this case, the likelihood function $J_{12}(\boldsymbol{\tau}_2)$ can be written as

$$J_{12}(\boldsymbol{\tau}_2) = \sum_{\xi} G_{\xi}(\boldsymbol{\theta}_{\xi}) \tag{A3.3.25}$$

where

$$G_{\xi}(\boldsymbol{\theta}_{\xi}) = \sum_{l} n(\xi,l) \ln w_{\xi}(l;\boldsymbol{\theta}_{\xi}) \qquad n(\xi,l) = \sum_{\eta} n(\xi,\eta,l) \tag{A3.3.26}$$

The ML estimates are $\hat{w}_{\xi}(l) = n(\xi,l) \,/\, n(\xi)$ if no assumptions about the distribution form have been made.

The null hypothesis can be tested using information criterion (A3.3.20). For the autonomous semi-Markov processes, this criterion might be rewritten as

$$I = I_1 + \sum_{\eta} I_{a,\eta} + I'_3 \tag{A3.3.27}$$

where I_1 is defined by Eq. (A3.3.21); the remaining terms are given by

$$I_{a,\eta} = \sum_k N(\eta,k) \ln \left[N(\eta,k) / N(\eta) W_\eta(k;\hat{\boldsymbol{\theta}}_\eta) \right] \tag{A3.3.28}$$

and

$$I'_3 = \sum_{\xi,\eta,\zeta,m,k} N(\xi,\eta,\zeta,m,k) \ln \left[\frac{N(\xi,\eta,\zeta,m,k) N(\eta) N(\eta)}{N(\xi,\eta,m) N(\eta,\zeta) N(\eta,k)} \right] \tag{A3.3.29}$$

The autonomous semi-Markov processes are very popular in applications. The number of the state-holding time distributions in this case is equal to the number of states, whereas in general it is equal to the squared number of states.

3.4 MATRIX PROCESS PARAMETER ESTIMATION

In this appendix, we build a matrix process model using its basis sequences introduced in Appendix 1. The method of model parameter estimation consists of finding a set of basis sequences and then estimating the process matrices.

Matrix process rank can be estimated by building different sequences of the form $s_i t_j$ according to the rule of ordering introduced in Appendix 1 (see Theorem A.1.2). If we succeed in constructing a basis from vectors of the sets $W_m^{(s)}$ and $W_m^{(t)}$, and the introduction of vectors of the next layer does not change this basis, then the number of vectors constituting the basis determines the process rank. Therefore, to estimate the rank in accordance with canonical representation (A.1.21), we build sequences of matrices

$$\hat{\mathbf{P}}_e^{(1)} = \hat{p}(e) \qquad \hat{\mathbf{P}}_e^{(2)} = \begin{bmatrix} \hat{p}(e) & \hat{p}(e0) \\ \hat{p}(0e) & \hat{p}(0e0) \end{bmatrix}$$

$$\hat{\mathbf{P}}_e^{(3)} = \begin{bmatrix} \hat{p}(e) & \hat{p}(e1) \\ \hat{p}(1e) & \hat{p}(1e1) \end{bmatrix}, \ldots, \quad \hat{\mathbf{P}}_e^{(k)} = \left[\hat{p}(s_i^{(e)}\, et_j^{(e)}) \right], \ldots$$

where $\hat{p}(s_i^{(e)}\, et_j^{(e)})$ is the relative frequency of the sequence $s_i^{(e)}\, et_j^{(e)}$. According to Theorem A.1.2, the basis building is completed when

$$\det \left[p(s_i^{(e)}\, et_j^{(e)}) \right]_{r,r} \neq 0 \qquad \det \left[p(s_i^{(e)}\, et_j^{(e)}) \right]_{r+1,r+1} = 0$$

for any $s_{r+1}^{(e)}$ and $t_{r+1}^{(e)}$. Therefore, to estimate the process rank we should be able to test a hypothesis that a random matrix determinant is equal to zero.

To test this hypothesis we need to find the distribution of determinants of the previously defined random matrices. Let $\nu(s_i^{(e)}\, et_j^{(e)})$ be a frequency of the sequence $s_i^{(e)}\, et_j^{(e)}$ in the experimental data. We will prove that this frequency is asymptotically normal if the HMM satisfies conditions of regularity. Indeed, consider all sequences of the same length l as

$s_i^{(e)} et_j^{(e)}$.

They can be described as a Markov function whose transition matrix probabilities are given by

$$\mathbf{P}(e_1 e_2 \cdots e_l \mid g_0 g_1 \cdots g_{l-1}) = \begin{cases} \mathbf{P}(e_l) & \text{if } \; e_1 = g_1, \ldots, e_{l-1} = g_{l-1} \\ 0 & \text{otherwise} \end{cases}$$

If we assume that this chain is regular (see Appendix 6), then [2,3] the frequencies of visiting any state of the chain are asymptotically normal random variables. Since the sequence $s_i^{(e)} et_j^{(e)}$ is one of the states of this chain, $\nu(s_i^{(e)} et_j^{(e)})$ is asymptotically normal.

Therefore, to test the hypothesis that the rank of the process is equal to r, we can use the distribution of random determinants whose elements have normal distributions. [8,9,11,13] Denote the determinant distribution density as $f_k(x)$. The hypothesis is rejected if $|\det \hat{\mathbf{P}}_e^{(k)}| > x_\alpha$ where x_α is a root of the equation

$$\int_{-x_\alpha}^{x_\alpha} f_k(x)\, dx = 1 - \alpha$$

After estimating ranks r_0 and r_1, it is necessary to estimate matrices (A.1.20) and (A.1.21), which determine the matrix process. We estimate them by the corresponding relative frequencies

$$\mathbf{A}_\mu(e) = \hat{\mathbf{P}}_\mu(s_1^{(\mu)}, \ldots, s_{r_\mu}^{(\mu)};\, et_1^{(e)}, \ldots, et_{r_e}^{(e)})\, \hat{\mathbf{P}}^{-1}(s_1^{(e)}, \ldots, s_{r_e}^{(e)};\, t_1^{(e)}, \ldots, t_{r_e}^{(e)}) \tag{A3.4.1}$$

where

$$\hat{\mathbf{P}}_\mu(s_1^{(\mu)}, \ldots, s_{r_\mu}^{(\mu)};\, et_1^{(e)}, \ldots, et_{r_e}^{(e)}) = \left[\hat{p}(s_i^{(\mu)} \mu et_j^{(e)})\right]_{r_\mu, r_e}$$

$$\hat{\mathbf{P}}^{-1}(s_1^{(e)}, \ldots, s_{r_e}^{(e)};\, t_1^{(e)}, \ldots, t_{r_e}^{(e)}) = \left[\hat{p}(s_i^{(e)} et_j^{(e)})\right]_{r_e, r_e}$$

and $s_1^{(e)} = \varnothing$, $s_1^{(e)}, \ldots, s_{r_e}^{(e)}$ and $t_1^{(e)} = \varnothing$, $t_1^{(e)}, \ldots, t_{r_e}^{(e)}$ are the corresponding left and right basis sequences. Matrix $\mathbf{Z}_{le}$ is estimated as $\hat{\mathbf{Z}}_{le} = col\ \{\ p(s_i^{(e)})\ \}$.

This method of parameter estimation, arising from results in Sec. 1.3 and Appendix 1, is, in essence, a variant of the method of moments. The most difficult part of this method is estimating ranks r_0 and r_1.

References

1 T. W Anderson and L. A. Goodman, "Statistical inference about Markov chains," *Ann. Math. Statist.*, **28**, 89-110, (1957).

2 M. S. Bartlett, *Introduction to Stochastic Processes with Special Reference to Methods and Applications,* (Cambridge Univ. Press, Cambridge, 1978).

3 M. S. Bartlett, "The frequency goodness of fit test for probability chains," *Proc. Cambridge Philos. Soc.*, **47**, 86-95, (1951).

4 P. Billingsley, "Statistical methods in Markov chains," *Ann. Math. Statist.*, **32**, 12-40, (1961).

5 P. Billingsley, *Statistical Inference for Markov Processes,* (University of Chicago Press, Chicago, 1961).

6 C. Chatfield, "Statistical inference regarding Markov chain models," *Appl. Statist.*, **22**, 7-20, (1973).

7 H. Cramer, *Mathematical Methods of Statistics*, (Princeton Univ. Press, Princeton, NJ, 1946).

8 R. Fortret, "Random determinants," *J. Res. Nat. Bureau Standards*, **47**, 465-470, (1951).

9 V. L. Girco, *Random Matrices* (in Russian), (Vischa Shcola Publishers, Kiev, 1975).

10 P. G. Hoel, "A test for Markoff chains," *Biometrica*, **4a**1, 430-433, (1954).

11 J. Komlós, "On the determinant of random matrices," *Studia Sci. Math. Hungary*, **3**(4), (1968).

12 S. Kullback, *Information Theory and Statistics*, (John Wiley & Sons, New York, 1959).

13 H. Nyquist, S. Rice, and J. Riordan, "The distribution of random determinants," *Quart. Appl. Math.*, **12**(2), (1954).

APPENDIX 4

4.1 SUMS WITH BINOMIAL COEFFICIENTS

It is convenient to use contour integrals to find sums that contain binomial coefficients. From the binomial formula

$$(1+z)^n = \sum_{m=0}^{n} \binom{n}{m} z^m \tag{A.4.1}$$

we obtain

$$\binom{n}{m} = \frac{1}{2\pi j} \oint_{\gamma} (1+z)^n z^{-m-1} dz \tag{A.4.2}$$

where γ is a contour on the complex plane that contains $z = 0$. By applying analytical expansion we may use this equation as a binomial coefficient definition if we assume that the contour γ does not contain $z = -1$. For example, it follows from (A.4.2) that $\binom{n}{m} = 0$ when $m > n$ or $m < 0$. If $n < 0$, Eq. (A.4.2) gives

$$\binom{n}{m} = (-1)^m \binom{m-n-1}{m}$$

which is identical to the definition in Ref. 2.

The utility of the equation for summation is in replacement of binomial coefficients with the powers of $(1+z)$ and z under the sign of the integral. Usually, it is easier to find a sum of a power series than the original sum.

4.2 MAXIMUM-DISTANCE-SEPARABLE CODE WEIGHT GENERATING FUNCTION

As an example, let us find the generating function of the number A_w of code words of the maximum-distance-separable (MDS) codes whose Hamming weight is equal to w. It is well known [1,3] that the Hamming weight of the MDS $(n, n-r)$ code over GF(q) can be expressed as

$$A_w = \binom{n}{w} \sum_{i=0}^{w-r-1} (-1)^i \binom{w}{i} (q^{w-r-i} - 1)$$
$$= \binom{n}{w} (q-1) \sum_{i=0}^{w-r-1} (-1)^i \binom{w-1}{i} q^{w-r-i-1} \qquad \text{for } w > r \tag{A.4.3}$$

$A_0 = 1$ and $A_w = 0$ for $w = 1,2,\ldots,r$. Our goal is to find the generating function

$$A(u) = \sum_{w=0}^{n} A_w u^w \tag{A.4.4}$$

It follows from Eq. (A.4.2) that

$$(-1)^i \binom{w}{i} = \oint_\gamma (1-z)^w z^{-i-1} dz \tag{A.4.5}$$

If we replace the coefficient $(-1)^i \binom{w}{i}$ in Eq. (A.4.3) with contour integral (A.4.5) and sum the geometric progression under the sign of the integral, we obtain, after some elementary transformations,

$$A_w = \frac{(q-1)}{2\pi j} \binom{n}{w} \oint_\gamma \frac{(1-z)^{w-1}}{z^{w-r}(1-qz)} dz \tag{A.4.6}$$

It follows from Cauchy's theorem that this integral is equal to zero for $w \le r$. After substituting (A.4.6) into (A.4.4) and summing under the sign of the integral, using binomial formula (A.4.1), we obtain

$$A(u) = 1 + \frac{(q-1)}{2\pi j} \oint_\gamma \frac{[u+(1-u)z]^n dz}{(1-qz)(1-z)z^{n-r}}$$

This equation represents the generating function in the integral form. We can evaluate this integral by the residue theorem and obtain the following relation:

$$A(u) = 1 + \sum_{i=r+1}^{n} \binom{n}{i} u^i (1-u)^{n-i} (q^{i-r} - 1) \tag{A.4.7}$$

Now that we know the expression of the generating function, we can easily verify its correctness by a direct expansion into power series.

Usually $n-r$ is much greater than r, and we can simplify the expression by replacing the summation from $r+1$ to n by the summation from 0 to r if we make use of the identity

$$\sum_{i=r+1}^{n} \binom{n}{i} a^i b^{n-i} = (a+b)^n - \sum_{i=0}^{r} \binom{n}{i} a^i b^{n-i}$$

The generating function $A(u)$ then takes the form

$$A(u) = q^{-r}(1-u+uq)^n + \sum_{i=0}^{r-1} \binom{n}{i} u^i (1-u)^{n-i} (1-q^{i-r}) \tag{A.4.8}$$

4.3 UNION BOUNDS ON VITERBI ALGORITHM PERFORMANCE

In this appendix, we use transform methods to derive a closed-form expression for the error event probability and performance union bound of the Viterbi algorithm.

Consider a discrete-time system [4]

$$
\begin{aligned}
\mathbf{y}_k &= \mathbf{f}(\mathbf{w}_k) \\
\mathbf{s}_{k+1} &= \mathbf{g}(\mathbf{w}_k) \\
\mathbf{w}_k &= (\mathbf{x}_k, \mathbf{s}_k) \qquad -\infty < k < \infty
\end{aligned}
\tag{A.4.9}
$$

where $\mathbf{x}_k$ is a source symbol, $\mathbf{y}_k$ is the transmitted channel symbol, and $\mathbf{s}_k$ is the corresponding system (transmitter) state. System (4.2.24) is a particular case of this system.

Symbols $\mathbf{y}_k$ are transmitted over a noisy HMM channel, which outputs symbols

$$\mathbf{z}_k = \mathbf{h}(\mathbf{y}_k, \mathbf{n}_k) \tag{A.4.10}$$

where $\mathbf{n}_k$ is the noise.

The receiver outputs symbols $\hat{\mathbf{x}}_k$ if $\hat{\mathbf{w}}_k = (\hat{\mathbf{s}}_k, \hat{\mathbf{x}}_k)$ minimizes the sum

$$M(\mathbf{z}, \mathbf{w}) = \sum_k m(\mathbf{z}_k, \mathbf{W}_k)$$

where $\mathbf{W}_k = (\mathbf{w}_k, \mathbf{w}_{k+1})$ (see Sec. 4.1.6.1). If we interpret $\mathbf{w}_k$ as a node on a trellis diagram (see Fig. 4.5), then $m(\mathbf{z}_k, \mathbf{W}_k)$ can be interpreted as a metric of the branch that connects nodes $\mathbf{w}_k$ and $\mathbf{w}_{k+1}$. Now the problem of minimizing $M(\mathbf{z}, \mathbf{w})$ can be formulated as the problem of finding the shortest path on the trellis diagram. The problem can be solved by the Viterbi algorithm described in Sec. 4.2.6.2:

> Let $M(\mathbf{w}_t)$ be the length of the shortest path (the survivor) that ends at node $\mathbf{w}_t$ at moment t. Then the survivor that ends at node $\mathbf{w}_{t+1}$ is obtained from one of the survivors at moment t and its length is given by
>
> $$M(\mathbf{w}_{t+1}) = \min_{\mathbf{w}_t} \{ M(\mathbf{w}_t) + m(\mathbf{z}_t, \mathbf{W}_t) \}$$

In other words, to find the shortest path at level $t+1$ we need only consider the shortest paths at level t. On the trellis diagram we select a survivor at node $\mathbf{w}_{t+1}$ If several paths have the minimal length, one of them is selected randomly. That is why the shortest path is called a survivor. This algorithm is illustrated in Example 4.2.7.

Depending on the application, we may want to evaluate Viterbi's receiver performance by using some distortion measure $\bar{d} = \mathbf{E}\{ d(\mathbf{W}_k, \hat{\mathbf{W}}_k) \}$, where $d(\mathbf{W}_k, \hat{\mathbf{W}}_k)$ is the distortion characteristic of the symbol $\mathbf{W}_k$, which the receiver identifies as $\hat{\mathbf{W}}_k$.

In order to find the distortion measure union bound, let us consider an error event [5] of length L which is a pair of the correct path $\sigma_L = \{\mathbf{w}_k\}$ and a path $\hat{\sigma}_L = \{\hat{\mathbf{w}}_k\}$, erroneously selected by the Viterbi algorithm, such that $\mathbf{s}_k \neq \hat{\mathbf{s}}_k$ for $k = 1,2,\ldots,L-1$ and $\mathbf{s}_k = \hat{\mathbf{s}}_k$ otherwise.

The distortion measure is upper bounded by the union bound: [4]

$$\bar{d} \leq \mathop{\mathbf{E}}_{\mathbf{x},\, \mathbf{s}_0} \sum_{L=1}^{\infty} \sum_{\hat{\sigma}_L} P_L(\hat{\sigma}_L \mid \sigma_L) d_L(\sigma_L, \hat{\sigma}_L)$$

where the average is taken over all source sequences and system stationary states, $P_L(\hat{\sigma}_L \mid \sigma_L)$ is the conditional probability that the incorrect path $\hat{\sigma}_L$ has a smaller metric than the correct path σ_L, and one-half of the probability of metrics' equality if a tie is resolved randomly:

$$P_L(\hat{\sigma}_L \mid \sigma_L) = Pr\{\sum_{k=0}^{L-1} \Delta m(\mathbf{z}_k,\mathbf{W}_k,\hat{\mathbf{W}}_k) > 0\}$$
$$+ 0.5\, Pr\{\sum_{k=0}^{L-1} \Delta m(\mathbf{z}_k,\mathbf{W}_k,\hat{\mathbf{W}}_k) = 0\} \tag{A.4.11}$$

where $\Delta m(\mathbf{z}_k,\mathbf{W}_k,\hat{\mathbf{W}}_k) = m(\mathbf{z}_k,\mathbf{W}_k) - m(\mathbf{z}_k,\hat{\mathbf{W}}_k)$ and $d_L(\sigma_L,\hat{\sigma}_L)$ is the total distortion along the incorrect path:

$$d_L(\sigma_L,\hat{\sigma}_L) = \sum_{k=0}^{L-1} d(\mathbf{W}_k,\hat{\mathbf{W}}_k)$$

For the HMM channel, the error event probability can be found using matrix probabilities. Define a generating function of a variable $\Delta m(\mathbf{z}_k,\mathbf{W}_k,\hat{\mathbf{W}}_k)$:

$$\mathbf{D}(v;\mathbf{W}_k,\hat{\mathbf{W}}_k) = \sum_{\mathbf{z}_k} \mathbf{P}(\mathbf{z}_k \mid \mathbf{W}_k) v^{\Delta m(\mathbf{z}_k,\mathbf{W}_k,\hat{\mathbf{W}}_k)} \tag{A.4.12}$$

According to the convolution theorem, the generating function of the sum of variables $\Delta m(\mathbf{z}_k,\mathbf{W}_k,\hat{\mathbf{W}}_k)$ is equal to the product of generating functions (A.4.12), and the matrix probability $\mathbf{P}_L(\hat{\sigma}_L \mid \sigma_L)$ is equal to the sum of the coefficients of the positive power terms of the product and one-half of the zero power term coefficient. Using contour integrals, we can express the sum as

$$\mathbf{P}_L(\hat{\sigma}_L \mid \sigma_L) = \frac{1}{2\pi j} \oint_{|v|=\rho} \left(\frac{1}{v-1} - \frac{1}{2v}\right) \prod_{k=0}^{L-1} \mathbf{D}(v;\mathbf{W}_k,\hat{\mathbf{W}}_k)\,dv \tag{A.4.13}$$

where $\rho > 1$. Using the formula of the derivative of the product, we obtain

$$d_L(\sigma_L,\hat{\sigma}_L) = \frac{d}{dv} \prod_{k=0}^{L-1} u^{d(\mathbf{W}_k,\hat{\mathbf{W}}_k)} \Big|_{u=1} \tag{A.4.14}$$

and, therefore, the average distortion is bounded by

$$\bar{d} \le \frac{1}{2\pi j} \frac{d}{du} \{\oint_C \left(\frac{1}{v-1} - \frac{1}{2v}\right) \mathbf{pG}(u,v)\mathbf{1}\,dv\} \Big|_{u=1} \tag{A.4.15}$$

where

$$\mathbf{G}(u,v) = \mathop{\mathbf{E}}_{\mathbf{x},\,\mathbf{s}_0} \sum_{L=1}^{\infty} \sum_{\hat{\sigma}_L} \prod_{k=0}^{L-1} \mathbf{D}(v;\mathbf{W}_k,\hat{\mathbf{W}}_k) u^{d(\mathbf{W}_k,\hat{\mathbf{W}}_k)} \tag{A.4.16}$$

and $C \in \{ v : R_G \cap |v| > 1 \}$ and R_G is the region of convergence of Eq. (A.4.16).

The generating function $\mathbf{G}(u,v)$ can be found from the system state $(\mathbf{W}_k,\hat{\mathbf{W}}_k)$ transition graph with branch weights

$$\mathbf{D}(v;\mathbf{W}_k,\hat{\mathbf{W}}_k) u^{d(\mathbf{W}_k,\hat{\mathbf{W}}_k)} \tag{A.4.17}$$

Indeed, for any path on the graph, the product

$$\prod_{k=0}^{L-1} \mathbf{D}(v;\mathbf{W}_k,\hat{\mathbf{W}}_k) u^{d(\mathbf{W}_k,\hat{\mathbf{W}}_k)}$$

represents the transition weight (which is similar to matrix probabilities and generating functions considered in Secs. 2.1.1, 2.2.1, and 2.4.4). A length L error event may be interpreted as a path that starts at a node with $\mathbf{s}_0 = \hat{\mathbf{s}}_0$ and returns to some node with $\mathbf{s}_L = \hat{\mathbf{s}}_L$ for the first time. The generating function

$$\sum_{L=1}^{\infty} \sum_{\hat{\sigma}_L} \prod_{k=0}^{L-1} \mathbf{D}(v;\mathbf{W}_k,\hat{\mathbf{W}}_k)\, u^{d(\mathbf{W}_k,\hat{\mathbf{W}}_k)}$$

represents the transition weight for all error events that start at $\mathbf{s}_0 = \hat{\mathbf{s}}_0$ and end at $\mathbf{s}_L = \hat{\mathbf{s}}_L$ for $L = 1,2,\ldots$.

To find this generating function let us denote $\mathbf{0}$ as the set of nodes with $\mathbf{s} = \hat{\mathbf{s}}$. As in the derivations of the Markov function interval distributions in Sec. 2.4.3, we obtain the weight generating function for all the paths that start at $\mathbf{0}$ and return to $\mathbf{0}$ for the first time:

$$\mathbf{T}(u,v) = \mathbf{R}_{\mathbf{0},\mathbf{0}} + \mathbf{R}_{\mathbf{0},\mathbf{\Theta}}[\mathbf{I} - \mathbf{R}_{\mathbf{\Theta},\mathbf{\Theta}}]^{-1}\mathbf{R}_{\mathbf{\Theta},\mathbf{0}}$$

where $\mathbf{R}$ is the matrix whose elements are equal to the graph branch weights (A.4.17) that correspond to erroneous transitions and $\mathbf{\Theta}$ is the complement of the set $\mathbf{0}$. In the absence of parallel transitions, $\mathbf{R}_{\mathbf{0},\mathbf{0}} = 0$. This equation coincides with Eq. (A.2.38) and, therefore, the generating function $\mathbf{T}(z)$ can be obtained by absorbing the subgraph $\mathbf{\Theta}$ from the system graph.

Sometimes it is convenient to duplicate the set $\mathbf{0}$ and create a new graph in which all transitions from $\mathbf{\Theta}$ to $\mathbf{0}$ are diverted to the duplicate set $\mathbf{0}'$. In this case, $\mathbf{T}(z)$ can be found as a transfer function from $\mathbf{0}$ to $\mathbf{0}'$ (rather than as a transfer function from $\mathbf{0}$ to itself, as in the original graph).

Generating function (A.4.16) can be found from the generating function $\mathbf{T}(z)$ by averaging. As illustrated in Sec. 4.4.1, the system's symmetry simplifies the construction of the generating function.

References

1 E. R. Berlekamp, *Algebraic Coding Theory*, (McGraw-Hill, New York, 1968).

2 W. Feller, *An Introduction to Probability Theory and Its Applications*, **1**, (John Wiley & Sons, New York, 1962).

3 G. D. Forney, *Concatenated Codes*, Research Monograph 37, (MIT Press, Cambridge, Massachusetts, 1966).

4 J. K. Omura, "Performance bounds for Viterbi algorithms,"ICC'81 Conference Record, Denver, Colorado, 2.2.1-2.2.5, June (1981).

5 A. J. Viterbi, "Error bounds for convolutional codes and asymptotically optimum decoding algorithm," *IEEE Trans. Inform. Theory*, **IT-12**, 260-269, Apr. 1967.

APPENDIX 5

5.1 MATRICES

This appendix provides a review of matrix properties relevant to the material discussed in this book.

5.1.1 Basic Definitions.

A matrix is a rectangular table of numbers:

$$\mathbf{A} = [a_{ij}]_{m,n} = \begin{bmatrix} a_{11} & a_{12} & \cdots & a_{1n} \\ a_{21} & a_{12} & \cdots & a_{2n} \\ \cdots & \cdots & \cdots & \cdots \\ a_{m1} & a_{m2} & \cdots & a_{mn} \end{bmatrix}$$

Some particular cases of matrices and elementary operations are listed on pages xviii and xix and are illustrated by the following simple examples.

Example A.5.1. Suppose that we have two matrices

$$\mathbf{A} = \begin{bmatrix} 1 & 2 \\ 3 & 4 \end{bmatrix} \quad \mathbf{B} = \begin{bmatrix} 5 & 6 \\ 7 & 8 \end{bmatrix}$$

then (see definitions on pages xv and xvi)

$$\mathbf{A} + \mathbf{B} = \begin{bmatrix} 6 & 8 \\ 10 & 12 \end{bmatrix} \quad \mathbf{A}\,\mathbf{B} = \begin{bmatrix} 1\cdot5+2\cdot7 & 1\cdot6+2\cdot8 \\ 3\cdot5+4\cdot7 & 3\cdot6+4\cdot8 \end{bmatrix} = \begin{bmatrix} 19 & 22 \\ 43 & 50 \end{bmatrix}$$

$$3\,\mathbf{A} = \begin{bmatrix} 3 & 6 \\ 9 & 12 \end{bmatrix} \quad \mathbf{A}' = \begin{bmatrix} 1 & 3 \\ 2 & 4 \end{bmatrix} \quad \mathbf{B}\,\mathbf{1} = \begin{bmatrix} 11 \\ 15 \end{bmatrix} \quad \mathbf{B}\,\mathbf{I} = \mathbf{I}\,\mathbf{B} = \mathbf{B}$$

$$\mathbf{A} \otimes B = \begin{bmatrix} 1\cdot\mathbf{B} & 2\cdot\mathbf{B} \\ 3\cdot\mathbf{B} & 4\cdot\mathbf{B} \end{bmatrix} = \begin{bmatrix} 5 & 6 & 10 & 12 \\ 7 & 8 & 14 & 16 \\ 15 & 18 & 20 & 24 \\ 21 & 24 & 28 & 32 \end{bmatrix}$$

The *trace* of a square matrix is the sum of its diagonal elements: $tr(\mathbf{A}) = \sum_{i=1}^{n} a_{ii}$. Obviously, $tr(\mathbf{A}+\mathbf{B}) = tr(\mathbf{A})+tr(\mathbf{B})$. Also if $\mathbf{A} = [a_{ij}]_{m,n}$ and $\mathbf{B} = [b_{ij}]_{n,m}$, then

$$tr(\mathbf{AB}) = \sum_{i=1}^{m}\sum_{j=1}^{n} a_{ij}b_{ji} = tr(\mathbf{BA})$$

The *determinant* of a square $(n \times n)$ matrix can be defined recursively as

$$|\mathbf{A}| = \det \mathbf{A} = \sum_{k=1}^{n} a_{ik}A_{ik} \tag{A.5.1}$$

where A_{ik} is an *adjunct* or *cofactor* of an element a_{ik}:

$$A_{ik} = (-1)^{i+k}M_{ik}$$

M_{ik} is the determinant of the matrix obtained by exclusion of i-th row and k-th column from $\mathbf{A}$.

Example A.5.2. Let us calculate a determinant using an expansion related to its last row.

$$\begin{vmatrix} 1 & 2 & 3 \\ 4 & 5 & 6 \\ 7 & 8 & 9 \end{vmatrix} = 7\cdot\begin{vmatrix} 2 & 3 \\ 5 & 6 \end{vmatrix} - 8\cdot\begin{vmatrix} 1 & 3 \\ 4 & 6 \end{vmatrix} + 9\cdot\begin{vmatrix} 1 & 2 \\ 4 & 5 \end{vmatrix} = 7\cdot(-3) - 8\cdot(-6) + 9\cdot(-3) = 0$$

A matrix whose determinant equals 0 is called *singular*. Every nonsingular matrix $\mathbf{A}$ has a unique *inverse matrix* $\mathbf{A}^{-1}$, which is defined by the relation $\mathbf{A}\,\mathbf{A}^{-1} = \mathbf{A}^{-1}\,\mathbf{A} = \mathbf{I}$. The inverse matrix can be found by the frequently used equation

$$\mathbf{A}^{-1} = \mathbf{A}^{*} / \det \mathbf{A} \tag{A.5.2}$$

where $\mathbf{A}^{*} = [A_{ij}]'_{n,n}$ is called an *adjoint matrix* (a transposed matrix of adjuncts of the elements of $\mathbf{A}$). One can prove the correctness of Eq. (A.5.2) using formula (A.5.1).

Example A.5.3. Let us find the inverse matrix for

$$\mathbf{A} = \begin{bmatrix} 3 & 2 \\ 1 & 1 \end{bmatrix}$$

Since $\det \mathbf{A} = 1 \neq 0$ the inverse matrix exists, we have

$$\mathbf{A}^{*} = \begin{bmatrix} 1 & -2 \\ -1 & 3 \end{bmatrix}$$

In this case $\mathbf{A}^{-1} = \mathbf{A}^{*}$ because $\det \mathbf{A} = 1$. We can test the correctness of this result using the inverse matrix definition

$$\mathbf{A}\,\mathbf{A}^{-1} = \begin{bmatrix} 3 & 2 \\ 1 & 1 \end{bmatrix}\begin{bmatrix} 1 & -2 \\ -1 & 3 \end{bmatrix} = \begin{bmatrix} 1 & 0 \\ 0 & 1 \end{bmatrix} = \mathbf{I}$$

It is often important to find the inverse matrix in the block form

$$\begin{bmatrix} \mathbf{A} & \mathbf{B} \\ \mathbf{C} & \mathbf{D} \end{bmatrix}^{-1} = \begin{bmatrix} \mathbf{X}_{11} & \mathbf{X}_{12} \\ \mathbf{X}_{21} & \mathbf{X}_{22} \end{bmatrix}$$

According to the definition of the inverse matrix, we have

$$\begin{bmatrix} \mathbf{A} & \mathbf{B} \\ \mathbf{C} & \mathbf{D} \end{bmatrix} \begin{bmatrix} \mathbf{X}_{11} & \mathbf{X}_{12} \\ \mathbf{X}_{21} & \mathbf{X}_{22} \end{bmatrix} = \begin{bmatrix} \mathbf{I} & 0 \\ 0 & \mathbf{I} \end{bmatrix}$$

Thus

$$\begin{aligned} \mathbf{AX}_{11} + \mathbf{BX}_{21} &= \mathbf{I} \\ \mathbf{CX}_{11} + \mathbf{DX}_{21} &= 0 \end{aligned}$$

From the second equation we find $\mathbf{X}_{21} = -\mathbf{D}^{-1}\mathbf{CX}_{11}$. Substituting it into the first equation of the previous system, we obtain $\mathbf{X}_{11} = (\mathbf{A} - \mathbf{BD}^{-1}\mathbf{C})^{-1} = \mathbf{M}$ and $\mathbf{X}_{21} = -\mathbf{D}^{-1}\mathbf{CM}$. The rest of the elements of the inverse matrix can be found similarly and the solution is

$$\begin{bmatrix} \mathbf{A} & \mathbf{B} \\ \mathbf{C} & \mathbf{D} \end{bmatrix}^{-1} = \begin{bmatrix} \mathbf{M} & -\mathbf{A}^{-1}\mathbf{BN} \\ -\mathbf{D}^{-1}\mathbf{CM} & \mathbf{N} \end{bmatrix} \tag{A.5.3}$$

where $\mathbf{N} = (\mathbf{D} - \mathbf{CA}^{-1}\mathbf{B})^{-1}$. This formula allows us to obtain the inverse matrix by inverting four matrix blocks of smaller dimensions. It is possible to reduce the number of the block matrix inversions to two if we note that the inverse matrix can be found also by solving the equation

$$\begin{bmatrix} \mathbf{X}_{11} & \mathbf{X}_{12} \\ \mathbf{X}_{21} & \mathbf{X}_{22} \end{bmatrix} \begin{bmatrix} \mathbf{A} & \mathbf{B} \\ \mathbf{C} & \mathbf{D} \end{bmatrix} = \begin{bmatrix} \mathbf{I} & 0 \\ 0 & \mathbf{I} \end{bmatrix} \tag{A.5.4}$$

It follows from this equation that $\mathbf{X}_{11}\mathbf{B} + \mathbf{X}_{12}\mathbf{D} = 0$ and $\mathbf{X}_{21}\mathbf{B} + \mathbf{X}_{22}\mathbf{D} = \mathbf{I}$. Thus, $\mathbf{X}_{12} = -\mathbf{X}_{11}\mathbf{BD}^{-1} = -\mathbf{MBD}^{-1}$ and $\mathbf{X}_{22} = \mathbf{D}^{-1} - \mathbf{X}_{21}\mathbf{BD}^{-1} = \mathbf{D}^{-1} + \mathbf{D}^{-1}\mathbf{CMBD}^{-1}$. Therefore, we can write

$$\begin{bmatrix} \mathbf{A} & \mathbf{B} \\ \mathbf{C} & \mathbf{D} \end{bmatrix}^{-1} = \begin{bmatrix} \mathbf{M} & -\mathbf{MBD}^{-1} \\ -\mathbf{D}^{-1}\mathbf{CM} & \mathbf{D}^{-1} + \mathbf{D}^{-1}\mathbf{CMBD}^{-1} \end{bmatrix} \tag{A.5.5}$$

The next useful identity, which is called the *matrix inversion lemma* can be obtained if we compare the two versions of $\mathbf{X}_{22}$ from Eq. (A.5.3) and (A.5.5):

$$(\mathbf{D} - \mathbf{CA}^{-1}\mathbf{B})^{-1} = \mathbf{D}^{-1} + \mathbf{D}^{-1}\mathbf{CMBD}^{-1} \tag{A.5.6}$$

In particular, if $\mathbf{A} = \mathbf{I}$, $\mathbf{r}$ is a row matrix and $\mathbf{c}$ is a column matrix, this equation has the form

$$(\mathbf{D} + \mathbf{cr})^{-1} = \mathbf{D}^{-1} - \lambda^{-1}\mathbf{D}^{-1}\mathbf{crD}^{-1} \tag{A.5.7}$$

where $\lambda = 1 + \mathbf{rD}^{-1}\mathbf{c}$. Thus, in this case we need to invert only one matrix.

5.1.2 Systems of Linear Equations.

A system of linear equations has the form

$$\begin{aligned} a_{11}x_1 + a_{12}x_2 + \cdots + a_{1n}x_n &= b_1 \\ a_{21}x_1 + a_{22}x_2 + \cdots + a_{2n}x_n &= b_2 \\ \dots\ \dots\ \dots\ \dots\ \dots\ \dots &\quad \dots \\ a_{m1}x_1 + a_{m2}x_2 + \cdots + a_{mn}x_n &= b_m \end{aligned}$$

Using matrix notations this system can be expressed as

$$\mathbf{A}\ \mathbf{x} = \mathbf{b}$$

where $\mathbf{A} = [a_{ij}]_{m,n}$, $\mathbf{x} = [x_1 \quad x_2 \ \dots \ x_n]'$ and $\mathbf{b} = [b_1 \quad b_2 \ \dots \ b_m]'$. (This system can also be expressed as $\mathbf{x}'\ \mathbf{A}' = \mathbf{b}'$.)

Each solution to this system is a matrix column **x** that satisfies it (turns it to a numerical identity). A system of linear equations may have a unique solution, an infinite number of solutions, or none. To analyze solutions of a linear system, we need to introduce the notion of matrix rank.

An order-k minor of the $n \times m$ matrix **A** is a determinant of a $k \times k$ matrix

$$\begin{vmatrix} a_{i_1 j_1} & a_{i_1 j_2} & \cdots & a_{i_1 j_k} \\ a_{i_2 j_1} & a_{i_2 j_2} & \cdots & a_{i_2 j_k} \\ \dots & \dots & \dots & \dots \\ a_{i_k j_1} & a_{i_k j_2} & \cdots & a_{i_k j_k} \end{vmatrix}$$

where $i_1, i_2, \dots, i_k$ and $j_1, j_2, \dots, j_k$ represent the selected rows and columns of **A**, respectively. If all the minors of orders greater than r are equal to 0, but there is an order-r minor different from 0, then r is called matrix **A** *rank*. That is, the rank is the highest order of all the matrix minors that are different from zero.

Example A.5.4. Let us calculate the rank of the matrix

$$\mathbf{A} = \begin{bmatrix} 1 & 2 & 3 \\ 4 & 5 & 6 \\ 7 & 8 & 9 \end{bmatrix}$$

As we saw in Example A.5.2, det $\mathbf{A} = 0$. Therefore, *rank* $\mathbf{A} < 3$. Since the determinant, which is composed out of the first and the second rows and columns

$$\begin{vmatrix} 1 & 2 \\ 4 & 5 \end{vmatrix} = -3$$

is not equal to 0, *rank* $\mathbf{A} = 2$.

It is difficult to calculate matrix rank using its definition, because the number of the minors to be tested might be very large. We can calculate it much faster using the so-called matrix *elementary operations*: multiplying an arbitrary row (column) of the matrix by a number and adding the result to its other row (column). It is easy to prove that these operations do not change matrix rank. (They also do not change the determinant of a square matrix.) To find the rank, we try to make as many zero rows (or columns) as possible by applying the elementary operations. If we were able to transform an $m \times n$ matrix **A** to matrix **B** that has $m-r$ zero rows and also has an order-r minor different from 0, then *rank* **A** = *rank* **B** = r. We say that **A** has *full rank* if *rank* **A** = $\min(m,n)$.

Example A.5.5. Let us calculate the rank of the matrix of Example A.5.4 using elementary operations. After multiplying the first column of **A** by (-2) and adding it to the second column, and then multiplying the first column by (-3) and adding it to the third column, we obtain

$$\mathbf{A} = \begin{bmatrix} 1 & 2 & 3 \\ 4 & 5 & 6 \\ 7 & 8 & 9 \end{bmatrix} \sim \begin{bmatrix} 1 & 0 & 0 \\ 4 & -3 & -6 \\ 7 & -6 & -12 \end{bmatrix} \sim \mathbf{B} = \begin{bmatrix} 1 & 0 & 0 \\ 4 & -3 & 0 \\ 7 & -6 & 0 \end{bmatrix}$$

We obtained matrix **B** after multiplying the second column of the previously transformed matrix by (-3) and adding the result to the third column.

We observe that **B** has two nonzero columns and it has the minor composed out of its first two rows and columns, which is equal to -3. Thus *rank* **A** = *rank* **B** = 2.

Example A.5.6. Vandermonde's determinant, which we often use in this book, has the form

$$\det \mathbf{V} = \begin{vmatrix} 1 & 1 & \cdots & 1 \\ \lambda_1 & \lambda_2 & \cdots & \lambda_n \\ \cdots & \cdots & \cdots & \cdots \\ \lambda_1^{n-1} & \lambda_2^{n-1} & \cdots & \lambda_n^{n-1} \end{vmatrix}$$

Let us calculate third-order Vandermonde's determinant using elementary operations:

$$\begin{vmatrix} 1 & 1 & 1 \\ \lambda_1 & \lambda_2 & \lambda_3 \\ \lambda_1^2 & \lambda_2^2 & \lambda_3^2 \end{vmatrix} = \begin{vmatrix} 1 & 0 & 0 \\ \lambda_1 & \lambda_2-\lambda_1 & \lambda_3-\lambda_1 \\ \lambda_1^2 & \lambda_2^2-\lambda_1^2 & \lambda_3^2-\lambda_1^2 \end{vmatrix} = \begin{vmatrix} \lambda_2-\lambda_1 & \lambda_3-\lambda_1 \\ \lambda_2^2-\lambda_1^2 & \lambda_3^2-\lambda_1^2 \end{vmatrix} = (\lambda_2-\lambda_1)(\lambda_3-\lambda_1)(\lambda_3-\lambda_2)$$

In the general case, this formula can be written as

$$\det \mathbf{V} = \prod_{i \neq j} (\lambda_j - \lambda_i)$$

This means that $\det \mathbf{V} \neq 0$ if and only if all λ_i are different.

The elementary operations that we consider are often called Gauss elimination. This process can be applied also to block matrices. For simplicity, consider second-order matrix

$$\begin{bmatrix} \mathbf{A} & \mathbf{B} \\ \mathbf{C} & \mathbf{D} \end{bmatrix}$$

If we multiply the first row of this matrix by $-\mathbf{CA}^{-1}$ and add the result to the second row, we eliminate **C** in the first column:

$$\begin{bmatrix} \mathbf{A} & \mathbf{B} \\ 0 & \mathbf{D}-\mathbf{CA}^{-1}\mathbf{B} \end{bmatrix} \tag{A.5.8}$$

For the block matrices with more than two rows a similar transformation is applied to eliminate the rest of the matrix blocks in the first column, in the second column, and so on.

Let us denote $\mathbf{B} = [\mathbf{A}\ \mathbf{b}]$ as the matrix consisting of **A** with matrix column **b** attached at the end. System (A.5.9) has a solution if and only if *rank* $\mathbf{A}$ = *rank* $\mathbf{B}$. [1] If r = *rank* $\mathbf{A}$ = *rank* $\mathbf{B}$, then this system has a unique solution if and only if $r = n$. It has an infinite number of solutions if $r < n$.

In particular, if $\mathbf{A}$ is a square matrix ($m = n$), then *rank* $A = n$ if and only if $\det \mathbf{A} \neq 0$, which means that the system has a unique solution if and only if its matrix is nonsingular. This unique solution can be obtained by multiplying both sides of Eq. (A.5.9) by $\mathbf{A}^{-1}$: $\mathbf{A}^{-1}\mathbf{A}\,\mathbf{x} = \mathbf{A}^{-1}\mathbf{b}$, which, because of $\mathbf{A}^{-1}\mathbf{A} = \mathbf{I}$, yields

$$\mathbf{x} = \mathbf{A}^{-1}\mathbf{b}$$

The system is called *homogeneous* when $\mathbf{b} = 0$. Any homogeneous system has a trivial solution $\mathbf{x} = 0$. If r = *rank* $A = n$, this trivial solution is the only solution to the system. Otherwise, the system has $n - r$ linearly independent solutions $\mathbf{x}_1, \mathbf{x}_2, ..., \mathbf{x}_r$, which are called *basis solutions*. Any solution can be expressed as a linear combination of these basis solutions:

$$\mathbf{x} = \alpha_1 \mathbf{x}_1 + \alpha_2 \mathbf{x}_2 + \cdots + \alpha_r \mathbf{x}_r$$

where $\alpha_1, \alpha_2, \ldots, \alpha_r$ are real numbers.

If $\mathbf{A}$ is a square nonsingular matrix ($\det \mathbf{A} \neq 0$), the homogeneous system $\mathbf{A}\,\mathbf{x} = 0$ has only the trivial solution $\mathbf{x} = 0$. Therefore, this system has a nonzero solution if and only if its determinant equals 0.

Example A.5.7. Let us find a solution to the following system of $n = 2$ homogeneous equations

$$0.75\,x_1 - 0.5x_2 = 0$$
$$-0.75\,x_1 + 0.5x_2 = 0$$

In this case

$$\det \mathbf{A} = \begin{vmatrix} 0.75 & -0.5 \\ 0.75 & 0.5 \end{vmatrix} = 0$$

so that r = *rank* $\mathbf{A} = 1$. This system has $n - r = 1$ linearly independent solutions. We see that the second equation is proportional to the first equation. Therefore, we can find all the solutions from the first equation, which can be rewritten as $x_2 = 1.5\,x_1$. We may assign any value to x_1 and determine x_2 from this equation. For $x_1 = 0.4$, we have $x_2 = 0.6$ so that the basis solution is $\mathbf{x} = [0.4 \quad 0.6]$. Every other solution can be expressed as $\alpha\,\mathbf{x}$.

5.1.3 Calculating Powers of a Matrix.

Matrix powers are defined recursively: $\mathbf{A}^2 = \mathbf{A}\,\mathbf{A}$, $\mathbf{A}^3 = \mathbf{A}^2\,\mathbf{A}$, ..., $\mathbf{A}^n = \mathbf{A}^{n-1}\,\mathbf{A}$. Since matrix multiplication is a complex operation, it is difficult to find matrix powers multiplying them directly. Even for a simple matrix

$$\mathbf{A} = \begin{bmatrix} 0 & 1 \\ 1 & 1 \end{bmatrix}$$

it is difficult to compute $\mathbf{A}^{100}$ by direct multiplication.

Powers of matrices play an important role in analyzing Markov chains (see Appendix 6) and error source models. Therefore, we need to develop methods of finding matrix powers.

To find $\mathbf{A}^m$, we use the z-transform

$$\mathbf{S}(z) = \sum_{m=0}^{\infty} \mathbf{A}^m z^{-m-1}$$

which converges for large $|z|$. This sum can be found in a manner similar to the usual geometric progression. After multiplying both sides of this equation by $\mathbf{A}z^{-1}$ and subtracting it from $\mathbf{S}(z)$, all power terms, except for $\mathbf{I}\,z^{-1}$, cancel out and we obtain

$$\mathbf{S}(z)\,(\,\mathbf{I} - \mathbf{A}z^{-1}\,) = \mathbf{I}\,z^{-1}$$

This equation can be rewritten as $\mathbf{S}(z) = (\,\mathbf{I}z - \mathbf{A}\,)^{-1}$. Thus, we obtain the following formula for the sum of the matrix-geometric series

$$\sum_{m=0}^{\infty} \mathbf{A}^m z^{-m-1} = (\,\mathbf{I}z - \mathbf{A})^{-1} \tag{A.5.9}$$

Matrix $\mathbf{A}^m$ is a coefficient of z^{-m-1} in the expansion of this z-transform. To find it, we present the inverse matrix as a ratio of its adjoint matrix $\mathbf{B}(z)$ and its determinant $\Delta(z)$:

$$(\,\mathbf{I}z - \mathbf{A})^{-1} = \mathbf{B}(z)\,/\,\Delta(z)$$

The denominator of this formula $\Delta(z) = \det(\,\mathbf{I}z - \mathbf{A})$ is called a *characteristic polynomial* of $\mathbf{A}$, and its roots $\lambda_1, \lambda_2, \ldots, \lambda_r$ are called matrix *characteristic numbers*:

$$\Delta(z) = (z-\lambda_1)^{m_1}(z-\lambda_2)^{m_2}\cdots(z-\lambda_r)^{m_r}$$

Since $\mathbf{B}(z)\,/\,\Delta(z)$ is a rational function, we can find its series expansion by first decomposing it into partial fractions:

$$(\,\mathbf{I}z - \mathbf{A})^{-1} = \mathbf{B}(z)\,/\,\Delta(z) = \sum_{j=1}^{r}\sum_{i=1}^{m_j} \mathbf{D}_{ij}\,/\,(z-\lambda_j)^i \tag{A.5.10}$$

Matrices $\mathbf{D}_{ij}$ can be determined by multiplying both sides of Eq. (A.5.10) by $\Delta(z)$ and comparing the coefficients of the same powers of z or by using formula (A.2.4), which gives

$$\mathbf{D}_{ij} = \frac{1}{(m_j-i)!}\,\frac{d^{m_j-i}}{dz^{m_j-i}}\left[\frac{\mathbf{B}(z)(z-\lambda_j)^{m_j}}{\Delta(z)}\right]_{z=\lambda_j}$$

To find the power series expansion of Eq. (A.5.10), we can apply a negative binomial formula

$$(z-\lambda)^{-k} = \sum_{m=k-1}^{\infty} \binom{m}{k-1} \lambda^{m-k+1} z^{-m-1}$$

which can be obtained by either expanding $(1 - \lambda x)^{-k}$ into power series using Taylor's formula or by differentiating a geometric progression sum

$$(z - \lambda)^{-1} = \sum_{m=0}^{\infty} \lambda^m z^{-m-1}$$

Expanding Eq. (A.5.10), we find that the coefficient of z^{-m-1} is equal to

$$\mathbf{A}^m = \sum_{j=1}^{r}\sum_{i=1}^{m_j} \mathbf{D}_{ij}\binom{m}{i-1}\lambda_j^{m-i+1} \tag{A.5.11}$$

Here $\binom{m}{i-1} = 1$ when $i = 1$ and $\binom{m}{i-1} = 0$ when $i > m+1$ (see Appendix 4.1).

Example A.5.8. Let us find $\mathbf{A}^m$ for

$$\mathbf{A} = \begin{bmatrix} 0.25 & 0.75 \\ 0.5 & 0.5 \end{bmatrix}$$

We can find $\mathbf{A}^m$ by formula (A.5.11) using decomposition of $(\mathbf{I}z - \mathbf{A})^{-1}$ into partial fractions.

To find the inverse matrix for

$$\mathbf{I}z - \mathbf{A} = \begin{bmatrix} 1 & 0 \\ 0 & 1 \end{bmatrix} z - \begin{bmatrix} 0.25 & 0.75 \\ 0.5 & 0.5 \end{bmatrix} = \begin{bmatrix} z-0.25 & -0.75 \\ -0.5 & z-0.5 \end{bmatrix}$$

we calculate its determinant

$$\Delta(z) = \det(\mathbf{I}z - \mathbf{A}) = \det \begin{bmatrix} z-0.25 & -0.75 \\ -0.5 & z-0.5 \end{bmatrix} = z^2 - 0.75z - 0.25 = (z-1)(z+0.25)$$

and the adjoint matrix

$$\mathbf{B}(z) = \begin{bmatrix} z-0.5 & 0.75 \\ 0.5 & z-0.25 \end{bmatrix} = \begin{bmatrix} 1 & 0 \\ 0 & 1 \end{bmatrix} z + \begin{bmatrix} -0.5 & 0.75 \\ 0.5 & -0.25 \end{bmatrix}$$

(Readers can find computationally efficient methods of simultaneously evaluating $\mathbf{B}(z)$ and $\Delta(z)$ in Ref. 1.) We decompose the inverse matrix into partial fractions

$$(\mathbf{I}z - \mathbf{A})^{-1} = \mathbf{B}(z)/\Delta(z) = \mathbf{D}_{11}/(z-1) + \mathbf{D}_{12}/(z+0.25) \tag{A.5.12}$$

Matrices $\mathbf{D}_{11}$ and $\mathbf{D}_{12}$ can be determined by multiplying both sides of Eq. (A.5.12) by $\Delta(z) = (z-1)(z+0.25)$

$$\mathbf{B}(z) = \mathbf{D}_{11}(z+0.25) + \mathbf{D}_{12}(z-1)$$

and comparing the coefficients of the same powers of z:

$$\mathbf{D}_{11} + \mathbf{D}_{12} = \begin{bmatrix} 1 & 0 \\ 0 & 1 \end{bmatrix} \qquad 0.25\,\mathbf{D}_{11} - \mathbf{D}_{12} = \begin{bmatrix} -0.5 & 0.75 \\ 0.5 & -0.25 \end{bmatrix}$$

Solving this system we obtain

$$\mathbf{D}_{11} = \begin{bmatrix} 0.4 & 0.6 \\ 0.4 & 0.6 \end{bmatrix} \qquad \mathbf{D}_{12} = \begin{bmatrix} 0.6 & -0.6 \\ -0.4 & 0.4 \end{bmatrix}$$

[This solution can also be found by using formula (A.2.13) with $q(z) = \Delta(z)$.]

Thus, for this example, we have according to Eq. (A.5.11)

$$\mathbf{A}^m = \mathbf{D}_{11} + \mathbf{D}_{12}(-0.25)^m \tag{A.5.13}$$

which, after substitutions of $\mathbf{D}_{11}$ and $\mathbf{D}_{12}$, takes the form

$$\mathbf{A}^m = \begin{bmatrix} 0.4 + 0.6(-0.25)^m & 0.6 - 0.6(-0.25)^m \\ 0.4 - 0.4(-0.25)^m & 0.6 + 0.4(-0.25)^m \end{bmatrix}$$

5.1.4 Eigenvectors and Eigenvalues.

An alternative approach to finding $\mathbf{A}^m$ is based on the notion of its eigenvectors and eigenvalues.

We say that matrix column $\mathbf{x} \neq 0$ is a *right eigenvector* of $\mathbf{A}$ if there is a number λ called an *eigenvalue*, such that $\mathbf{A}\,\mathbf{x} = \lambda\,\mathbf{x}$. Similarly, matrix row $\mathbf{y} \neq 0$ is called a *left eigenvector* if $\mathbf{y}\,\mathbf{A} = \mu\,\mathbf{y}$.

It follows from these definitions that $(\mathbf{I}\lambda - \mathbf{A})\mathbf{x} = 0$ and $\mathbf{y}(\mathbf{I}\mu - \mathbf{A}) = 0$. This means that the eigenvectors satisfy linear homogeneous equations. As we know, these equations have nonzero solutions if and only if their determinants are equal to zero:

$$\Delta(\lambda) = \det(\mathbf{I}\lambda - \mathbf{A}) = 0 \qquad \Delta(\mu) = \det(\mathbf{I}\mu - \mathbf{A}) = 0$$

Therefore, both eigenvalues are characteristic numbers of $\mathbf{A}$ and there is no need to make a distinction between them.

Example A.5.9. Let us find eigenvectors and eigenvalues for the matrix of Example A.5.8. Since $\Delta(z) = (z-1)(z+0.25)$, the eigenvalues are $\lambda_1 = 1$ and $\lambda_2 = -0.25$. The left eigenvectors corresponding to $\lambda_1 = 1$ can be found from the system $\mathbf{y}_1(\mathbf{I}\lambda_1 - \mathbf{A}) = 0$, which takes the form

$$[\, y_{11} \quad y_{12} \,] \begin{bmatrix} 0.75 & -0.75 \\ -0.5 & 0.5 \end{bmatrix} = 0 \quad \text{or} \quad \begin{matrix} 0.75\, y_{11} - 0.5 y_{12} = 0 \\ -0.75\, y_{11} + 0.5 y_{12} = 0 \end{matrix}$$

This system has an infinite number of solutions, which can be expressed as $\mathbf{y}_1 = v_1\,[0.4 \quad 0.6]$, where v_1 is an arbitrary real number (see Example A.5.7). Thus, all the eigenvectors corresponding to λ_1 are proportional (colinear). The rest of the eigenvectors can be found similarly. The left eigenvectors corresponding to λ_2 have the form $\mathbf{y}_2 = v_2\,[1 \quad -1]$. The right eigenvectors are $\mathbf{x}_1 = v_3\,[1 \quad 1]'$ and $\mathbf{x}_2 = v_4\,[0.6 \quad -0.4]'$ (these vectors are matrix columns, so we express them as transposed matrix rows to save space).

One can observe that the left eigenvectors are proportional to the rows of the matrices $\mathbf{D}_{11}$ and $\mathbf{D}_{12}$ of Example A.5.8, while the right eigenvectors are proportional to their columns. These properties allow us to construct $\mathbf{D}_{11}$ and $\mathbf{D}_{12}$ using eigenvectors.

We say that matrix $\mathbf{A}$ has a *simple structure* if Eq. (A.5.10) has the following form

$$(\mathbf{I}z - \mathbf{A})^{-1} = \mathbf{D}_{11}/(z-\lambda_1) + \mathbf{D}_{12}/(z-\lambda_2) + \cdots + \mathbf{D}_{1r}/(z-\lambda_r) \quad \text{(A.5.14)}$$

In this case, Eq. (A.5.11) becomes

$$\mathbf{A}^m = \mathbf{D}_{11}\lambda_1^m + \mathbf{D}_{12}\lambda_2^m + \cdots + \mathbf{D}_{1r}\lambda_r^m \quad \text{(A.5.15)}$$

and the coefficients $\mathbf{D}_{1j}$ can be expressed by formula (A.2.8):

$$\mathbf{D}_{1j} = \mathbf{B}(\lambda_j)/\Delta'(\lambda_j)$$

It follows from the results of the previous section that if all the roots of the characteristic polynomial $\Delta(z) = \det(\mathbf{I}z - \mathbf{A})$ are different, then $\mathbf{A}$ has a simple structure. Matrix $\mathbf{A}$ of Example A.5.8 has a simple structure.

Theorem A.5.1: If matrix $\mathbf{A}$ has a simple structure, then rows and columns of the matrices $\mathbf{D}_{1j}$ defined by Eq. (A.5.14) are its eigenvectors.

Proof: To simplify the notation, let us assume that the eigenvalues are sorted according to their absolute values:

$$|\lambda_1| > |\lambda_2| > \cdots > |\lambda_r|$$

Multiplying both sides of Eq. (A.5.15) by λ_1^{-m}, we obtain

$$\lambda_1^{-m}\mathbf{A}^m = \mathbf{D}_{11} + \mathbf{D}_{12}(\lambda_2/\lambda_1)^m + \cdots + \mathbf{D}_{1r}(\lambda_r/\lambda_1)^m \quad \text{(A.5.16)}$$

Since $|\lambda_j/\lambda_1| < 1$

$$\lim_{m\to\infty} \lambda_1^{-m} \mathbf{A}^m = \mathbf{D}_{11} \tag{A.5.17}$$

Passing to the limit in the identity $\lambda_1^{-m}\mathbf{A}^m\ \mathbf{A} = \lambda_1^{-m}\mathbf{A}^{m+1}$, we obtain

$$\mathbf{D}_{11}\ \mathbf{A} = \lambda_1\ \mathbf{D}_{11} \tag{A.5.18}$$

This means that each row of $\mathbf{D}_{11}$ is a left eigenvector corresponding to λ_1. Using the identity $\lambda_1^m \mathbf{A}\ \mathbf{A}^m = \lambda_1^m \mathbf{A}^{m+1}$, we can prove that each column of $\mathbf{D}_{11}$ is a right eigenvector of $\mathbf{A}$: $\mathbf{A}\ \mathbf{D}_{11} = \lambda_1\ \mathbf{D}_{11}$. Thus, we have proven the theorem for the matrix $\mathbf{D}_{11}$ which corresponds to the largest eigenvalue λ_1. To prove it for the rest of the matrices, we need to study some properties of $\mathbf{D}_{11}$.

Property I: $\mathbf{D}_{11}\mathbf{A}^m = \lambda_1^m \mathbf{D}_{11}$ and $\mathbf{A}^m\mathbf{D}_{11} = \lambda_1^m\mathbf{D}_{11}$

We can prove these equations by applying (A.5.18) to $\mathbf{D}_{11}\mathbf{A}^m$:

$$\mathbf{D}_{11}\mathbf{A}^m = \mathbf{D}_{11}\mathbf{A}\ \mathbf{A}^{m-1} = \lambda_1 \mathbf{D}_{11}\mathbf{A}^{m-1} = \lambda_1^2\mathbf{D}_{11}\mathbf{A}^{m-2} = \cdots = \lambda_1^m \mathbf{D}_{11}$$

Property II: $\mathbf{D}_{11}^2 = \mathbf{D}_{11}$

This property can be proven by passing to the limit when $m \to \infty$ in the identity $\mathbf{D}_{11}\lambda_1^{-m}\mathbf{A}^m = \mathbf{D}_{11}$ and using Eq. (A.5.17).

Property III: $\mathbf{D}_{11}\mathbf{D}_{1j} = 0$ and $\mathbf{D}_{1j}\mathbf{D}_{11} = 0$ for $j=2,3,...,r$

Multiplying both sides of Eq. (A.5.15) by $\mathbf{D}_{11}$ and using Property I, we obtain

$$\mathbf{D}_{11}\mathbf{D}_{12}\lambda_2^m + \cdots + \mathbf{D}_{11}\mathbf{D}_{1r}\lambda_r^m = 0 \quad \text{for} \quad m = 1,2,... \tag{A.5.19}$$

This is possible only when $\mathbf{D}_{11}\mathbf{D}_{12} = 0$, ..., $\mathbf{D}_{11}\mathbf{D}_{1r} = 0$ because the determinants of the first $r-1$ systems are Vandermonde's determinants, which are not singular (see Example A.5.6), so that these systems have only trivial solutions. We can prove similarly that $\mathbf{D}_{12}\mathbf{D}_{11} = 0, ..., \mathbf{D}_{1r}\mathbf{D}_{11} = 0$.

To complete the proof of the theorem, consider the matrix $\mathbf{A}_1 = \mathbf{A} - \mathbf{D}_{11}\lambda_1$, which, according to Eq. (A.5.15), for $m = 1$ can be written as

$$\mathbf{A}_1 = \mathbf{A} - \mathbf{D}_{11}\lambda_1 = \mathbf{D}_{12}\lambda_2 + \cdots + \mathbf{D}_{1r}\lambda_r \tag{A.5.20}$$

It follows from the preceding properties of $\mathbf{D}_{11}$ that $\mathbf{A}^m = \mathbf{A}_1^m + \lambda_1^m\mathbf{D}_{11}$. Substituting it into Eq. (A.5.15), we obtain the equation

$$\mathbf{A}_1^m = \mathbf{D}_{12}\lambda_2^m + \cdots + \mathbf{D}_{1r}\lambda_r^m \tag{A.5.21}$$

which is similar to Eq. (A.5.15). Therefore, applying the previous results, we can prove that the rows and columns of $\mathbf{D}_{12}$ are the eigenvectors of $\mathbf{A}_1$ corresponding to λ_2:

$$\mathbf{D}_{12}\ \mathbf{A}_1 = \lambda_2\mathbf{D}_{12} \qquad \mathbf{A}_1\ \mathbf{D}_{12} = \lambda_2\mathbf{D}_{12}$$

But they are also the eigenvectors of $\mathbf{A}$, which is seen from the following transformations

$$\mathbf{D}_{12}\ \mathbf{A} = \mathbf{D}_{12}\ \mathbf{A}_1 + \lambda_1\mathbf{D}_{12}\ \mathbf{D}_{11} = \lambda_2\ \mathbf{D}_{12}$$

Thus, we have proven the theorem for $\mathbf{D}_{11}$ and $\mathbf{D}_{12}$. We can prove it for $\mathbf{D}_{13}$ by considering $\mathbf{A}_2 = \mathbf{A}_1 - \lambda_2\mathbf{D}_{12}$, and so on. Finally, we will prove it for all $\mathbf{D}_{1j}$.

In conclusion, we would like to point out that all $\mathbf{D}_{1j}$ possess properties similar to properties I, II, and III. This can be written as

$$\mathbf{D}_{1j}\,\mathbf{D}_{1j} = \mathbf{D}_{1j} \qquad \mathbf{D}_{1i}\,\mathbf{D}_{1j} = 0 \quad \text{for} \quad i \neq j \;\; \text{and} \;\; i,j = 1,2,\ldots,r \tag{A.5.22}$$

In the particular case when all the roots of the characteristic polynomial are different, all the eigenvectors corresponding to the same eigenvalue are proportional (colinear). Therefore, matrices $\mathbf{D}_{1j}$ can be expressed as

$$\mathbf{D}_{1j} = \mathbf{x}_j\,\mathbf{y}_j \tag{A.5.23}$$

where $\mathbf{x}_j$ is the right eigenvector and $\mathbf{y}_j$ is the left eigenvector of $\mathbf{A}$ corresponding to λ_j. It follows from Eq. (A.5.22) that

$$\mathbf{y}_j\,\mathbf{x}_j = 1 \quad \text{and} \quad \mathbf{y}_i\,\mathbf{x}_j = 0 \quad \text{for} \quad i \neq j \tag{A.5.24}$$

Example A.5.10. It is easy to verify that the matrices $\mathbf{D}_{11}$ and $\mathbf{D}_{12}$ of Example A.5.9 can be expressed as

$$\mathbf{D}_{11} = \begin{bmatrix} 0.4 & 0.6 \\ 0.4 & 0.6 \end{bmatrix} = \begin{bmatrix} 1 \\ 1 \end{bmatrix} [0.4 \quad 0.6] \qquad \mathbf{D}_{12} = \begin{bmatrix} 0.6 & -0.6 \\ -0.4 & 0.4 \end{bmatrix} = \begin{bmatrix} 0.6 \\ -0.4 \end{bmatrix} [1 \quad -1]$$

These equations together with the results of Example A.5.7 illustrate Eq. (A.5.22).

5.1.5 Similar Matrices.

Matrices **A** and **B** are called *similar* if there is a matrix **T** satisfying the following equation

$$\mathbf{T}\,\mathbf{A}\,\mathbf{T}^{-1} = \mathbf{B} \tag{A.5.25}$$

Similar matrices play an important role in simplifying error source models. The following theorem is often used in this book.

Theorem A.5.2: If matrix **A** has a simple structure, then it is similar to a diagonal matrix.

Proof: To simplify the proof, let us assume that all the eigenvalues are different. In this case, matrices $\mathbf{D}_{1j}$ can be expressed by Eq. (A.5.23).

Consider the matrix **T** whose rows are the eigenvectors of **A**:

$$\mathbf{T} = block\ col\{\mathbf{y}_i\} = \begin{bmatrix} \mathbf{y}_1 \\ \mathbf{y}_2 \\ \cdots \\ \mathbf{y}_n \end{bmatrix} \tag{A.5.26}$$

Because of Eq. (A.5.24), its inverse matrix is given by

$$\mathbf{T}^{-1} = block\ row\{\mathbf{x}_i\} = [\mathbf{x}_1 \quad \mathbf{x}_2 \quad \ldots \quad \mathbf{x}_n]$$

To complete the proof of the theorem, let us calculate $\mathbf{T}\,\mathbf{A}\,\mathbf{T}^{-1}$. Since the columns of $\mathbf{T}^{-1}$ are the matrix **A** eigenvectors, we have

$$\mathbf{A}\,\mathbf{T}^{-1} = [\mathbf{A}\,\mathbf{x}_1 \quad \mathbf{A}\,\mathbf{x}_2 \quad \ldots \quad \mathbf{A}\,\mathbf{x}_n] = [\lambda_1\,\mathbf{x}_1 \quad \lambda_2\,\mathbf{x}_2 \quad \ldots \quad \lambda_r\,\mathbf{x}_r]$$

Multiplying this equation from the left by **T** and using Eq. (A.5.26), we obtain

$$\mathbf{T\,A\,T}^{-1} = \begin{bmatrix} \lambda_1 \mathbf{y}_1 \mathbf{x}_1 & \lambda_2 \mathbf{y}_1 \mathbf{x}_2 & \cdots & \lambda_n \mathbf{y}_1 \mathbf{x}_n \\ \lambda_1 \mathbf{y}_2 \mathbf{x}_1 & \lambda_2 \mathbf{y}_2 \mathbf{x}_2 & \cdots & \lambda_n \mathbf{y}_2 \mathbf{x}_n \\ \cdots & \cdots & \cdots & \cdots \\ \lambda_1 \mathbf{y}_n \mathbf{x}_1 & \lambda_2 \mathbf{y}_n \mathbf{x}_2 & \cdots & \lambda_n \mathbf{y}_n \mathbf{x}_n \end{bmatrix}$$

According to Eq. (A.5.24) this equation can be rewritten as

$$\mathbf{T\,A\,T}^{-1} = \begin{bmatrix} \lambda_1 & 0 & \cdots & 0 \\ 0 & \lambda_2 & \cdots & 0 \\ \cdots & \cdots & \cdots & \cdots \\ 0 & 0 & \cdots & \lambda_n \end{bmatrix} = diag\{\lambda_i\}_n = \mathbf{\Lambda} \qquad \text{(A.5.27)}$$

Solving this equation for **A** we obtain

$$\mathbf{A} = \mathbf{T}^{-1}\,\mathbf{\Lambda}\,\mathbf{T} \qquad \text{(A.5.28)}$$

This formula is called matrix **A** *spectral representation.* A matrix **A** is called symmetric if $\mathbf{A} = \mathbf{A}'$. Any symmetric matrix has a simple structure. [1] It follows from the identity $(\mathbf{yA})' = \mathbf{A}'\mathbf{y}'$ that if **y** is a left eigenvector of a symmetric matrix, then $\mathbf{y}'$ is its right eigenvector. Thus, if $\mathbf{y}_i$ are the unit eigenvectors, we can use as $\mathbf{x}_i = \mathbf{y}_i'$ and, therefore, $\mathbf{T}^{-1} = \mathbf{T}'$. We conclude that the symmetric matrix has the following spectral representation

$$\mathbf{A} = \mathbf{T}'\,\mathbf{\Lambda}\,\mathbf{T} \qquad \text{(A.5.29)}$$

Example A.5.11. Let us show that the matrix **A** of Example A.5.8 is similar to a diagonal matrix. Using the results of Example A.5.9 and the previous theorem, we have

$$\mathbf{T} = \begin{bmatrix} 0.4 & 0.6 \\ 1 & -1 \end{bmatrix} \qquad \mathbf{T}^{-1} = \begin{bmatrix} 1 & 0.6 \\ 1 & -0.4 \end{bmatrix}$$

By direct multiplication, we obtain

$$\mathbf{TAT}^{-1} = \begin{bmatrix} 1 & 0 \\ 0 & -0.25 \end{bmatrix}$$

We would like to point out that if **A** does not have simple structure it is similar to the so-called Jordan normal matrix, which is given by

$$\mathbf{J} = \begin{bmatrix} \mathbf{J}_1 & 0 & \cdots & 0 \\ 0 & \mathbf{J}_2 & \cdots & 0 \\ \cdot\cdot & \cdot\cdot & \cdots & \cdot\cdot \\ 0 & 0 & \cdots & \mathbf{J}_s \end{bmatrix} \quad \text{where} \quad \mathbf{J}_i = \begin{bmatrix} \lambda_i & 1 & \cdots & 0 & 0 \\ 0 & \lambda_i & \cdots & 0 & 0 \\ \cdot\cdot & \cdot\cdot & \cdots & \cdot\cdot & \cdot\cdot \\ 0 & 0 & \cdots & \lambda_i & 1 \\ 0 & 0 & \cdots & 0 & \lambda_i \end{bmatrix} \qquad \text{(A.5.30)}$$

Let us illustrate the application of similar matrices to find A^m. Equation (A.5.25) can be rewritten as $\mathbf{A} = \mathbf{T}^{-1}\,\mathbf{B}\,\mathbf{T}$. Therefore

$$\mathbf{A}^2 = \mathbf{T}^{-1}\mathbf{BT}\,\mathbf{T}^{-1}\mathbf{BT} = \mathbf{T}^{-1}\mathbf{B}^2\mathbf{T}$$

Repeating these transformations, we prove that $\mathbf{A}^m = \mathbf{T}^{-1}\mathbf{B}^m\mathbf{T}$. Thus, if we know how to find $\mathbf{B}^m$, we can obtain $\mathbf{A}^m$ using the previous equation. In particular, if $\mathbf{B}$ is a diagonal matrix, we have

$$\mathbf{A}^m = \mathbf{T}^{-1}\begin{bmatrix} \lambda_1^m & 0 & \cdots & 0 \\ 0 & \lambda_2^m & \cdots & 0 \\ \cdots & \cdots & \cdots & \cdots \\ 0 & 0 & \cdots & \lambda_n^m \end{bmatrix}\mathbf{T} \qquad \text{(A.5.31)}$$

Example A.5.12. Let us find $\mathbf{A}^m$ for the matrix $\mathbf{A}$ of Example A.5.8. According to the results of Example A.5.11 and Eq. (A.5.27), we have

$$\mathbf{A}^m = \mathbf{T}^{-1}\begin{bmatrix} 1 & 0 \\ 0 & (-0.25)^m \end{bmatrix}\mathbf{T} = \begin{bmatrix} 0.4 + 0.6\,(-0.25)^m & 0.6 - 0.6\,(-0.25)^m \\ 0.4 - 0.4\,(-0.25)^m & 0.6 + 0.4\,(-0.25)^m \end{bmatrix}$$

5.1.6 Matrix Series Summation.

A matrix

$$\mathbf{S} = \sum_{m_1,\ldots,m_k} \mathbf{A}_{m_1,\ldots,m_k} \qquad \text{(A.5.32)}$$

is called a *matrix series sum* if its elements are the sums of the corresponding elements of the matrices $\mathbf{A}_{m_1,\ldots,m_k}$:

$$s_{ij} = \sum_{m_1,\ldots,m_k} a_{m_1,\ldots,m_k}^{(ij)} \qquad \text{(A.5.33)}$$

For series (A.5.32) to converge, it is necessary and sufficient that all scalar series of type (A.5.33) be convergent. It is convenient to use integral form (2.2.14) to replace a summation of complex expressions by the summation of a geometric progression under the integral sign.

For example, it follows from the matrix-geometric series (A.5.9) that

$$\mathbf{A}^m = \frac{1}{2\pi j}\oint_\gamma (\mathbf{I}z - \mathbf{A})^{-1} z^m dz \qquad \text{(A.5.34)}$$

where $\gamma = \{z : |z| = \rho\}$, $\rho > \max|\lambda_i|$ and λ_i are the matrix $\mathbf{A}$ eigenvalues. This formula is convenient to use for replacing $\mathbf{A}^m$ in sums. It can also be used to find an arbitrary function of the matrix.

A *function of a matrix* is defined [1] as a sum of the series

$$f(\mathbf{A}) = \sum_{m=0}^{\infty} a_m \mathbf{A}^m \qquad \text{(A.5.35)}$$

if eigenvalues of $\mathbf{A}$ lie inside the disk of convergence of the series

$$f(z) = \sum_{m=0}^{\infty} a_m z^m$$

Then, from formula (A.5.34), we obtain

$$f(\mathbf{A}) = \sum_{m=0}^{\infty} a_m \frac{1}{2\pi j} \oint_{\gamma} (\mathbf{I}z - \mathbf{A})^{-1} z^m dz = \frac{1}{2\pi j} \oint_{\gamma} (\mathbf{I}z - \mathbf{A})^{-1} f(z) dz \qquad \text{(A.5.36)}$$

Using formula (A.5.10), we obtain

$$f(\mathbf{A}) = \frac{1}{2\pi j} \sum_{j=1}^{r} \sum_{i=1}^{m_j} \mathbf{D}_{ij} \oint_{\gamma} \frac{f(z)}{(z-\lambda_j)^i} dz$$

which, after applying Cauchy's theorem, gives

$$f(\mathbf{A}) = \sum_{j=1}^{r} \sum_{i=1}^{m_j} \mathbf{D}_{ij} f^{(i-1)}(\lambda_j) / (i-1)! \qquad \text{(A.5.37)}$$

This formula can be used for any function of matrix calculation. Formula (A.5.11) is a particular case of this formula, when $f(z) = z^m$.

Example A.5.13. As an example, let us calculate various functions of

$$\mathbf{A} = \begin{bmatrix} 1 & 1 \\ 0 & 1 \end{bmatrix}$$

For any function of $\mathbf{A}$, we need to find matrices $\mathbf{D}_{ij}$ using decomposition (A.5.10). We have

$$\mathbf{I}z - \mathbf{A} = \begin{bmatrix} 1 & 0 \\ 0 & 1 \end{bmatrix} z - \begin{bmatrix} 1 & 1 \\ 0 & 1 \end{bmatrix} = \begin{bmatrix} z-1 & -1 \\ 0 & z-1 \end{bmatrix}$$

$$\Delta(z) = (z-1)^2 \qquad \mathbf{B}(z) = \begin{bmatrix} z-1 & 1 \\ 0 & z-1 \end{bmatrix}$$

so that

$$(\mathbf{I}z - \mathbf{A})^{-1} = \mathbf{B}(z) / \Delta(z) = \begin{bmatrix} (z-1)^{-1} & (z-1)^{-2} \\ 0 & (z-1)^{-1} \end{bmatrix}$$

The partial fraction decomposition has the form

$$(\mathbf{I}z - \mathbf{A})^{-1} = \mathbf{D}_{11} / (z-1) + \mathbf{D}_{21} / (z-1)^2$$

where

$$\mathbf{D}_{11} = \begin{bmatrix} 1 & 0 \\ 0 & 1 \end{bmatrix} \qquad \mathbf{D}_{21} = \begin{bmatrix} 0 & 1 \\ 0 & 0 \end{bmatrix}$$

and therefore any function of this matrix is given by

$$f(\mathbf{A}) = \mathbf{D}_{11} f(1) + \mathbf{D}_{21} f'(1)$$

For calculating matrix powers, we use $f(z) = z^m$, which gives

$$\mathbf{A}^m = \mathbf{D}_{11} + \mathbf{D}_{21} m = \begin{bmatrix} 1 & m \\ 0 & 1 \end{bmatrix}$$

For calculating $e^{\mathbf{A}t}$, we use $f(z) = e^{zt}$:

$$e^{\mathbf{A}t} = \mathbf{D}_{11} e^t + \mathbf{D}_{21} t e^t = \begin{bmatrix} e^t & te^t \\ 0 & e^t \end{bmatrix}$$

In conclusion, let us prove the Cayley-Hamilton theorem, which we use in Sec 1.3.3. This theorem states that every square matrix is a root of its characteristic polynomial: $\Delta(\mathbf{A}) = 0$.

Indeed, our function is $f(z) = \Delta(z) = \det(\mathbf{I}z - \mathbf{A})$ so that Eq. (A.5.36) becomes

$$\Delta(\mathbf{A}) = \frac{1}{2\pi j}\oint_\gamma (\mathbf{I}z - \mathbf{A})^{-1}\Delta(z)dz = \frac{1}{2\pi j}\oint_\gamma \mathbf{B}(z)dz = 0 \qquad \text{(A.5.38)}$$

This integral is equal to 0 because the adjoint matrix $\mathbf{B}(z)$ is a polynomial and does not have poles inside γ.

5.1.7 Product of Several Matrices.

It is often necessary to use a general formula for the product of several matrices. Let $\mathbf{A}_1, \mathbf{A}_2, \ldots, \mathbf{A}_T$ be a sequence of matrices. The product of two matrices is the matrix $\mathbf{B} = \mathbf{A}_1\mathbf{A}_2$ whose ij-th element is

$$b_{ij} = \sum_m a_{im}^{(1)} a_{mj}^{(2)}$$

where $a_{ij}^{(k)}$ is the ij-th element of $\mathbf{A}_k$. Similarly, the ij-th element of $\mathbf{B} = \mathbf{A}_1\mathbf{A}_2\mathbf{A}_3$ can be written as

$$b_{ij} = \sum_{m_1 m_2} a_{i,m_1}^{(1)} a_{m_1,m_2}^{(2)} a_{m_2,j}^{(3)}$$

and for $\mathbf{B} = \prod_{i=1}^{T} \mathbf{A}_i$, we have

$$b_{ij} = \sum_{m_1, m_2, \ldots, m_{T-1}} a_{i,m_1}^{(1)} a_{m_1,m_2}^{(2)} \cdots a_{m_{T-1},j}^{(T)}$$

which can also be written in the following compact form

$$b_{m_0,m_T} = \sum_{m_1^{T-1}} \prod_{i=1}^{T} a_{m_{i-1}m_i}^{(i)} \qquad \text{(A.5.39)}$$

Suppose that all the matrices depend on some parameter t and we would like to take a derivative of their product

$$\mathbf{B}(t) = \prod_{i=1}^{T} \mathbf{A}_i(t)$$

It follows immediately from Eq. (A.5.39) that the derivative is given by

$$\mathbf{B}' = \mathbf{A}_1' \prod_{i=2}^{T} \mathbf{A}_i + \mathbf{A}_1\mathbf{A}_2' \prod_{i=3}^{T} \mathbf{A}_i + \ldots + \prod_{i=1}^{T-1} \mathbf{A}_i \mathbf{A}_T'$$

which can be written as

$$\mathbf{B}' = \sum_{j=1}^{T} \prod_{i=1}^{j-1} \mathbf{A}_i \mathbf{A}_j' \prod_{i=j+1}^{T} \mathbf{A}_i \qquad \text{(A.5.40)}$$

When $j=1$ in this expression, we have $\prod_{i=1}^{0}$ which does not make sense. If we formally define this product as a unit matrix $\mathbf{I}$ and also define $\prod_{i=T+1}^{T} \mathbf{A}_i = \mathbf{I}$, then we can use the previous compact expression of the derivative.

If we denote the forward and backward matrices

$$\mathbf{C}_j = \prod_{i=1}^{j} \mathbf{A}_i \quad \text{and} \quad \mathbf{D}_j = \prod_{i=j+1}^{T} \mathbf{A}_i$$

respectively, then Eq. (A.5.40) can be written as

$$\mathbf{B}' = \sum_{j=1}^{T} \mathbf{C}_{j-1} \mathbf{A}_j' \mathbf{D}_j$$

which means that we can use the forward-backward algorithm for the derivative evaluation.

In the special case in which all the matrices are identical, we have

$$\mathbf{B}' = \sum_{j=1}^{T} \mathbf{C}_{j-1} \mathbf{A}' \mathbf{D}_j = \sum_{j=1}^{T} \mathbf{A}^{j-1} \mathbf{A}' \mathbf{A}^{T-j} \tag{A.5.41}$$

where

$$\mathbf{C}_j = \mathbf{A}^j \qquad \mathbf{D}_j = \mathbf{A}^{T-j}$$

Equation (A.5.41) is a special case of equation

$$\mathbf{G}_T = \sum_{j=1}^{T} \mathbf{A}^{j-1} \mathbf{U} \mathbf{A}^{T-j} \tag{A.5.42}$$

This sum can be evaluated similarly to the powers of a matrix by using its z-transform:

$$\mathbf{G}(z) = \sum_{T=1}^{\infty} \mathbf{G}_T z^{-T} = (\mathbf{I}z - \mathbf{A})^{-1} \mathbf{U} (\mathbf{I}z - \mathbf{A})^{-1} \tag{A.5.43}$$

According to Eq. (A.5.10), this equation can be written as

$$\mathbf{G}(z) = \mathbf{B}(z) \mathbf{U} \mathbf{B}(z) / \Delta^2(z)$$

which can be expanded into partial fractions as

$$\mathbf{G}(z) = \sum_{j=1}^{r} \sum_{i=1}^{2m_j} \mathbf{E}_{ij} / (z - \lambda_j)^i \tag{A.5.44}$$

Therefore, $\mathbf{G}_T$ has the form of Eq. (A.5.11):

$$\mathbf{G}_T = \sum_{j=1}^{r} \sum_{i=1}^{2m_j} \mathbf{E}_{ij} \binom{m}{i-1} \lambda_j^{m-i+1} \tag{A.5.45}$$

It is usually difficult to find the eigenvalues λ_j and the partial fraction decomposition. Alternatively, the sum can be evaluated recursively using the fast algorithms considered in the next section.

5.1.8 Fast Exponentiation.

Suppose that we would like to evaluate $\mathbf{A}^m$. If $m = 2^k$ is a power of two, we can raise $\mathbf{A}$ to the power of m by sequentially squaring the result

$$\mathbf{A}^2 = \mathbf{A}\mathbf{A}, \ \mathbf{A}^4 = \mathbf{A}^2\mathbf{A}^2, \ldots, \mathbf{A}^{2^k} = \mathbf{A}^{2^{k-1}} \mathbf{A}^{2^{k-1}}$$

For example, $\mathbf{A}^{16}$ can be obtained using 5 matrix multiplications: $\mathbf{A}^2 = \mathbf{A}\mathbf{A}$, $\mathbf{A}^4 = \mathbf{A}^2\mathbf{A}^2$,

$\mathbf{A}^8 = \mathbf{A}^4\mathbf{A}^4, \mathbf{A}^{16} = \mathbf{A}^8\mathbf{A}^8, \mathbf{A}^{32} = \mathbf{A}^{16}\mathbf{A}^{16}$.

If m is not a power of two, we represent it as a binary number, $m = b_0 + b_1 2 + ... + b_k 2^k$. Using this representation, we can calculate m recursively:

$$m_0 = b_0 \quad m_i = m_{i-1} + 2^i b_i \quad i=1,2,...,k \tag{A.5.46}$$

which leads to the following matrix recursion

$$\mathbf{A}^{m_0} = \mathbf{A}^{b_0} \quad \mathbf{A}^{m_i} = \mathbf{A}^{m_{i-1}}\mathbf{A}^{2^i b_i} \quad i=1,2,...,k \tag{A.5.47}$$

In simple terms, this means that if $b_i = 0$, $\mathbf{A}^{m_i} = \mathbf{A}^{m_{i-1}}$; otherwise $\mathbf{A}^{m_i} = \mathbf{A}^{m_{i-1}}\mathbf{A}^{2^i}$ so that the algorithm can be presented as follows **Algorithm A5.1:**

```
Initialize:
```

$$i=0, \; b_0 = REM(m,2) \quad m_0 = m \quad \mathbf{Q}_0 = \mathbf{A}^{b_0} \quad \mathbf{R}_0 = \mathbf{A}$$

```
While m_i>0
Begin
```

$$i+1 \to i, \; m_i = (m_{i-1} - b_{i-1})/2 \qquad b_i = REM(m_i,2) \qquad \mathbf{R}_i = \mathbf{R}_{i-1}^2$$

$$\mathbf{Q}_i = \begin{cases} \mathbf{Q}_{i-1}\mathbf{R}_i, & \text{if } b_i = 1, \\ \mathbf{Q}_{i-1}, & \text{if } b_i = 0 \end{cases}$$

```
End
```

At the end we obtain $\mathbf{A}^m = \mathbf{Q}_i$.

Example A.5.14. To find $\mathbf{A}^{100}$, we present $100 = 2^6+2^5+2^2 = 64+32+4$ (that is, $b_0=b_1=b_3=b_4=0$ while $b_2=b_5=b_6=1$). We initialize

$$\mathbf{Q}_0 = \mathbf{I} \quad \mathbf{R}_0 = \mathbf{A}$$

and compute:

$$\mathbf{R}_1 = \mathbf{A}^2 \quad \mathbf{Q}_1 = \mathbf{I}$$
$$\mathbf{R}_2 = \mathbf{A}^4 \quad \mathbf{Q}_2 = \mathbf{R}_2 = \mathbf{A}^4$$
$$\mathbf{R}_3 = \mathbf{A}^8 \quad \mathbf{Q}_3 = \mathbf{Q}_2$$
$$\mathbf{R}_4 = \mathbf{A}^{16} \quad \mathbf{Q}_4 = \mathbf{Q}_2$$
$$\mathbf{R}_5 = \mathbf{A}^{32} \quad \mathbf{Q}_5 = \mathbf{Q}_2\mathbf{R}_5 = \mathbf{A}^{36}$$

Finally, we obtain $\mathbf{R}_6 = \mathbf{A}^{64}$ and $\mathbf{Q}_6 = \mathbf{Q}_5\mathbf{R}_6 = \mathbf{A}^{100}$.

As we can see, the number of matrix multiplications required by this algorithm is about the logarithm of m.

A similar algorithm can be applied to computing $\mathbf{G}_m$ defined by Eq. (A.5.42). Indeed, it is easy to verify that

$$\mathbf{G}_{m+n} = \mathbf{A}^m\mathbf{G}_n + \mathbf{G}_m\mathbf{A}^n \tag{A.5.48}$$

Thus, if $m = 2^k$ is a power of two, we have

$$\mathbf{G}_1 = \mathbf{U}, \; \mathbf{R}_0 = A$$

$$\mathbf{G}_2 = \mathbf{R}_0\mathbf{G}_1 + \mathbf{G}_1\mathbf{R}_0 = \mathbf{AU} + \mathbf{UA}, \quad \mathbf{R}_1 = \mathbf{R}_0\mathbf{R}_0 = \mathbf{A}^2$$

$$\mathbf{G}_4 = \mathbf{R}_1\mathbf{G}_2 + \mathbf{G}_2\mathbf{R}_1$$

and so on. If m is not a power of two, we represent it as a binary number and obtain, similarly to the fast matrix exponentiation, the following.

Algorithm A5.2:

```
Initialize:
```

$$i=0,\ b_0 = REM(m,2) \quad m_0 = m \quad \mathbf{R}_0 = \mathbf{A} \quad \mathbf{Q}_0 = \mathbf{A}^{b_0} \quad \mathbf{S}_0 = \mathbf{U} \quad \mathbf{V}_0 = b_0\mathbf{U}$$

```
While m_i>0
Begin
```

$$i+1 \rightarrow i$$

$$\mathbf{S}_i = \mathbf{R}_{i-1}\mathbf{S}_{i-1} + \mathbf{S}_{i-1}\mathbf{R}_{i-1} \qquad \mathbf{R}_i = \mathbf{R}_{i-1}\mathbf{R}_{i-1}$$

$$m_i = (m_{i-1} - b_{i-1})/2 \qquad b_i = REM(m_i,2)$$

$$\mathbf{V}_i = \mathbf{V}_{i-1} \qquad \mathbf{Q}_i = \mathbf{Q}_{i-1} \quad \text{if} \quad b_i=0$$

$$\mathbf{V}_i = \mathbf{R}_i\mathbf{V}_{i-1} + \mathbf{S}_i\mathbf{Q}_{i-1} \qquad \mathbf{Q}_i = \mathbf{Q}_{i-1}\mathbf{R}_i \quad \text{if} \quad b_i=1$$

```
End
```

At the end, we obtain $\mathbf{G}_m = \mathbf{V}_k$ and $\mathbf{A}^m = \mathbf{Q}_k$, where k is the number of iterations.

Note that equations for $\mathbf{V}_i$ and $\mathbf{Q}_i$ can be written in a more compact form

$$\mathbf{V}_i = \mathbf{R}_i^{b_i}\mathbf{V}_{i-1} + b_i\mathbf{S}_i\mathbf{Q}_{i-1} \qquad \mathbf{Q}_i = \mathbf{Q}_{i-1}\mathbf{R}_i^{b_i}$$

Example A.5.15. To obtain

$$\mathbf{G}_{100} = \sum_{i=1}^{100}\mathbf{A}^{i-1}\mathbf{U}\mathbf{A}^{100-i}$$

We initialize

$$\mathbf{Q}_0 = \mathbf{I} \quad \mathbf{R}_0 = \mathbf{A} \quad \mathbf{S}_0 = \mathbf{U} \quad \mathbf{V}_0 = 0$$

and compute:

$$\mathbf{R}_1 = \mathbf{A}^2 \quad \mathbf{Q}_1 = \mathbf{I} \quad \mathbf{S}_1 = \mathbf{AU}+\mathbf{UA} \quad \mathbf{V}_1 = 0$$

$$\mathbf{R}_2 = \mathbf{A}^4 \quad \mathbf{Q}_2 = \mathbf{R}_2 \quad \mathbf{S}_2 = \mathbf{R}_1\mathbf{S}_1+\mathbf{S}_1\mathbf{R}_1 \quad \mathbf{V}_2 = \mathbf{S}_2$$

$$\mathbf{R}_3 = \mathbf{A}^8 \quad \mathbf{Q}_3 = \mathbf{Q}_2 \quad \mathbf{S}_3 = \mathbf{R}_2\mathbf{S}_2+\mathbf{S}_2\mathbf{R}_2 \quad \mathbf{V}_3 = \mathbf{V}_2$$

$$\mathbf{R}_4 = \mathbf{A}^{16} \quad \mathbf{Q}_4 = \mathbf{Q}_2 \quad \mathbf{S}_4 = \mathbf{R}_3\mathbf{S}_3+\mathbf{S}_3\mathbf{R}_3 \quad \mathbf{V}_4 = \mathbf{V}_2$$

$$\mathbf{R}_5 = \mathbf{A}^{32} \quad \mathbf{Q}_5 = \mathbf{Q}_2\mathbf{R}_5 = \mathbf{A}^{36} \quad \mathbf{S}_5 = \mathbf{R}_4\mathbf{S}_4+\mathbf{S}_4\mathbf{R}_4 \quad \mathbf{V}_5 = \mathbf{R}_5\mathbf{V}_2+\mathbf{S}_5\mathbf{Q}_2$$

Finally, we obtain

$$\mathbf{R}_6 = \mathbf{A}^{64} \quad \mathbf{S}_6 = \mathbf{R}_5\mathbf{S}_5+\mathbf{S}_5\mathbf{R}_5 \quad \mathbf{Q}_6 = \mathbf{Q}_5\mathbf{R}_6 = \mathbf{A}^{100}$$

and $\mathbf{G}_{100} = \mathbf{V}_6 = \mathbf{R}_6\mathbf{V}_5+\mathbf{S}_6\mathbf{Q}_5$.

5.1.9 Matrix Derivatives.

Previously we considered derivatives of matrices whose elements depend on the same scalar parameter. Sometimes it is convenient to define a derivative of a scalar function

$f(a_{11}, a_{12}, \ldots, a_{mn})$ depending on elements of the matrix $\mathbf{A} = [a_{ij}]_{m,n}$ which are independent variables. For example, a determinant or a trace of the matrix are such functions. A derivative of the scalar function with respect to the matrix **A** is defined as the matrix of the same form as **A** whose elements are the corresponding derivatives:

$$\frac{\partial f}{\partial \mathbf{A}} = \left[\frac{\partial f}{\partial a_{ij}}\right]_{m,n}$$

The reader should always remember that the derivative is defined only with respect to matrices with *independent* elements. Thus, the derivative is not defined if the matrix is, for example, symmetric or stochastic. If the derivative is used to find extreme values of a scalar function of a matrix with dependent elements, one way to circumvent this problem is to use the Lagrange's multipliers as we demonstrate in Appendix 7.

Example A.5.16. Let

$$\mathbf{A} = \begin{bmatrix} a_{11} & a_{12} \\ a_{21} & a_{22} \end{bmatrix}$$

then

$$\frac{\partial |\mathbf{A}|}{\partial \mathbf{A}} = \begin{bmatrix} \frac{\partial |\mathbf{A}|}{\partial a_{11}} & \frac{\partial |\mathbf{A}|}{\partial a_{12}} \\ \frac{\partial |\mathbf{A}|}{\partial a_{21}} & \frac{\partial |\mathbf{A}|}{\partial a_{22}} \end{bmatrix}$$

Since $|\mathbf{A}| = a_{11}a_{22} - a_{12}a_{21}$, we have $\frac{\partial |\mathbf{A}|}{\partial a_{11}} = a_{22}$, $\frac{\partial |\mathbf{A}|}{\partial a_{12}} = -a_{21}$, $\frac{\partial |\mathbf{A}|}{\partial a_{21}} = -a_{12}$, and $\frac{\partial |\mathbf{A}|}{\partial a_{22}} = a_{11}$. Thus

$$\frac{\partial |\mathbf{A}|}{\partial \mathbf{A}} = \begin{bmatrix} a_{22} & -a_{21} \\ -a_{12} & a_{11} \end{bmatrix}$$

As we can see, the determinant's derivative is the matrix of the adjuncts of its elements. Note that if the matrix were symmetric, this result is not valid. Indeed, in this case we have only three independent variables since $a_{12} = a_{21}$. The determinant can be written as $a_{11}a_{22} - a_{12}^2$ and its derivative with respect to a_{12} is $-2a_{21}$. In the case of a more complex dependency, say $a_{11}^5 + \exp(a_{21}) - a_{22}^2 = 0$, the ambiguity is even greater.

Let us study some properties of the derivatives with respect to matrices

1. If $\mathbf{x} = [x_i]_{1,n}$ is a row-vector, then the derivative with respect to this $1 \times n$ matrix is the gradient:

$$\frac{\partial f}{\partial \mathbf{x}} = \left[\frac{\partial f}{\partial x_i}\right]_{1,n} = \nabla f$$

2. The derivative of the determinant of a matrix with respect to the matrix

$$\frac{\partial |\mathbf{A}|}{\partial \mathbf{A}} = \left[\frac{\partial |\mathbf{A}|}{\partial a_{ij}}\right]_{n,n} = \left[\frac{\partial}{\partial a_{ij}} \sum_{k=1}^{n} a_{mk} A_{mk}\right]_{n,n} = [A_{ij}]_{n,n}$$

is obtained from the matrix **A** by replacing its elements with their adjuncts.

Comparing this equation with (A.5.2), we can write

$$\frac{\partial|\mathbf{A}|}{\partial\mathbf{A}} = |\mathbf{A}|(\mathbf{A}^{-1})'$$

This equation can be also rewritten as

$$\frac{\partial\ln(|\mathbf{A}|)}{\partial\mathbf{A}} = (\mathbf{A}^{-1})' \tag{A.5.49}$$

Note that if $\mathbf{A} = x$ is a scalar, this formula becomes $(ln\ x)' = 1/x$.

3. If $\mathbf{A} = [a_{ij}]_{m,n}$ and $\mathbf{B} = [b_{ij}]_{n,m}$, then

$$\frac{\partial tr(\mathbf{AB})}{\partial\mathbf{A}} = \left[\frac{\partial}{\partial a_{ij}}\sum_{k=1}^{m}\sum_{s=1}^{n} a_{ks}b_{sk}\right]_{m,n} = [b_{ji}]_{n,m} = \mathbf{B}' \tag{A.5.50}$$

4. Using the previous equation and the rule of differentiating a product we obtain

$$\frac{\partial tr(\mathbf{DABA}')}{\partial\mathbf{A}} = \frac{\partial tr(\mathbf{D}\overset{v}{\mathbf{A}}\mathbf{BA}')}{\partial\mathbf{A}} + \frac{\partial tr(\mathbf{DAB}\overset{v}{\mathbf{A}}')}{\partial\mathbf{A}} = \mathbf{D}'\mathbf{AB}' + \mathbf{DAB} \tag{A.5.51}$$

Note that we expanded the expression according to the rule of differentiating the product and marked the element to which the derivative is applied. Then we used the properties of the trace and Eq. (A.5.50) to find the derivative.

5. Similarly, we obtain

$$\frac{\partial tr(\mathbf{A}^m)}{\partial\mathbf{A}} = m(\mathbf{A}^{m-1})' = m(\mathbf{A}')^{m-1}$$

6. It follows from the previous equation and Eq. (A.5.35) that

$$\frac{\partial tr[f(\mathbf{A})]}{\partial\mathbf{A}} = f'(\mathbf{A}')$$

where $f'(x)$ is the derivative of the function $f(x)$. In particular, since the derivative of $\exp(x)$ is $\exp(x)$, we have

$$\frac{\partial tr[\exp(\mathbf{A})]}{\partial\mathbf{A}} = \exp(\mathbf{A}')$$

7. Denote as $a_{ij}^{(-1)}$ the ij-th element of the inverse matrix $\mathbf{A}^{-1}$. A derivative of this element with respect to a_{st} can be easily found by differentiating the identity $\mathbf{AA}^{-1} = \mathbf{I}$ to obtain

$$\frac{\partial a_{ij}^{(-1)}}{\partial a_{ts}} = -a_{it}^{(-1)}a_{sj}^{(-1)}$$

Using this formula we find

$$\frac{\partial tr(\mathbf{A}^{-1})}{\partial\mathbf{A}} = = \frac{\partial}{\partial\mathbf{A}}\sum_{k=1}^{n} a_{ii} = -\left[\sum_{i=1}^{n} a_{it}^{-1}a_{si}^{-1}\right]_{n,n} = -(\mathbf{A}^{-2})'$$

5.1.10 Quadratic Polynomials.

A quadratic polynomial is defined by the following expression

$$P(\mathbf{x}) = \mathbf{x}'\mathbf{A}\mathbf{x}+2\mathbf{x}'\mathbf{b}+c \qquad \text{(A.5.52)}$$

where the matrix **A** is symmetric. It is often useful to simplify this expression by completing the square. To solve this problem we perform the substitution $\mathbf{x} = \mathbf{y}+\mathbf{v}$:

$$P(\mathbf{y}+\mathbf{v}) = (\mathbf{y}'+\mathbf{v}')\mathbf{A}(\mathbf{y}+\mathbf{v})+2(\mathbf{y}'+\mathbf{v}')\mathbf{b}+c$$

After some transformations this expression can be written as

$$P(\mathbf{y}+\mathbf{v}) = \mathbf{y}'\mathbf{A}\mathbf{y}+2\mathbf{y}'(\mathbf{A}\mathbf{v}+\mathbf{b})+\mathbf{v}'(\mathbf{A}\mathbf{v}+\mathbf{b})+\mathbf{v}'\mathbf{b}+c$$

To obtain the complete square with respect to **y**, we need to select **v** such that

$$\mathbf{A}\mathbf{v}+\mathbf{b} = 0 \qquad \text{(A.5.53)}$$

As we can see, it is possible to complete the square if and only if this equation has a solution. If **A** is nonsingular, the unique solution is $\mathbf{v} = -\mathbf{A}^{-1}\mathbf{b}$. In this case, we have (after the back substitution)

$$P(\mathbf{x}) = (\mathbf{x}+\mathbf{A}^{-1}\mathbf{b})'\mathbf{A}(\mathbf{x}+\mathbf{A}^{-1}\mathbf{b})-\mathbf{b}'\mathbf{A}^{-1}\mathbf{b}+c$$

which can be also written as

$$P(\mathbf{x}) = (\mathbf{A}\mathbf{x}+\mathbf{b})'\mathbf{A}^{-1}(\mathbf{A}\mathbf{x}+\mathbf{b})-\mathbf{b}'\mathbf{A}^{-1}\mathbf{b}+c$$

If Eq. (A.5.53) has a solution, it can be expressed as $\mathbf{v} = -\mathbf{A}^{\dagger}\mathbf{b}$ where $\mathbf{A}^{\dagger}$ is called the pseudoinverse matrix and can be obtained using the spectral representation $\mathbf{A} = \mathbf{T}'\mathbf{\Lambda}\mathbf{T}$ where $\mathbf{\Lambda} = diag\{\lambda_1\}_n$ [according to Eq. (A.5.29)]. The pseudoinverse matrix is constructed as

$$\mathbf{A}^{\dagger} = \mathbf{T}'\Lambda^{\dagger}\mathbf{T} \qquad \text{(A.5.54)}$$

where the diagonal matrix $\mathbf{\Lambda}^{\dagger} = diag\{\mu_i\}_n$ is obtained from $\mathbf{\Lambda}$ by replacing its nonzero elements with their inverse ($\mu_i = \lambda_i^{-1}$ if $\lambda_i \neq 0$, otherwise $\mu_i = 0$). If matrix **A** is nonsingular, it is clear that $\mathbf{A}^{\dagger} = \mathbf{A}^{-1}$.

If **A** is not symmetric, its psudoinverse can be obtained as

$$\mathbf{A}^{\dagger} = (\mathbf{A}'\mathbf{A})^{\dagger}\mathbf{A}' \qquad \text{(A.5.55)}$$

and, in particular,

$$\mathbf{A}^{\dagger} = (\mathbf{A}'\mathbf{A})^{-1}\mathbf{A}' \qquad \text{(A.5.56)}$$

if the inverse matrix $(\mathbf{A}'\mathbf{A})^{-1}$ exists.

Returning to the problem of completing the square in the quadratic polynomial, we conclude that if Eq. (A.5.53) has a solution, we can complete the square - that is present the polynomial as

$$P(\mathbf{x}) = (\mathbf{x}+\mathbf{A}^{\dagger}\mathbf{b})'\mathbf{A}(\mathbf{x}+\mathbf{A}^{\dagger}\mathbf{b})-\mathbf{b}'\mathbf{A}^{\dagger}\mathbf{b}+c$$

which can be also written as

$$P(\mathbf{x}) = (\mathbf{Ax}+\mathbf{b})'\mathbf{A}^{\dagger}(\mathbf{Ax}+\mathbf{b})-\mathbf{b}'\mathbf{A}^{\dagger}\mathbf{b}+c \qquad \text{(A.5.57)}$$

In this case, the polynomial is called *central* and $-\mathbf{A}^{\dagger}\mathbf{b}$ is called its center. If Eq. (A.5.53) does not have a solution, it is possible to complete the square with respect to the part of the polynomial for which it has a solution, but there will be some linear terms also. (In analytical geometry, the curves or surfaces described by the equation $P(\mathbf{x}) = 0$ are called central, e.g. ellipse and hyperbola, otherwise, they are called noncentral, e.g. parabola.)

Example A.5.17. Complete the square (if possible) in the following polynomial

$$P = x_1^2+2x_1x_2+x_2^2+2x_1+4x_2$$

This equation can be written as (A.5.52) with

$$\mathbf{A} = \begin{bmatrix}1 & 1\\1 & 1\end{bmatrix}, \quad \mathbf{b} = \begin{bmatrix}1\\2\end{bmatrix}, \quad c=0$$

In this case Eq. (A.5.53) does not have a solution since *rank* $\mathbf{A} = 1$ and *rank* $[\mathbf{A}\ \mathbf{b}] = 2$. Thus, it is impossible to compete a square. However, we can complete the square for the part of the polynomial for which this equation has a solution. As we see, if $\mathbf{b} = [1\ \ 1]'$, then *rank* $[\mathbf{A}\ \mathbf{b}] = 1$ and Eq. (A.5.53) would have an infinite number of solutions. Thus, we need to present the polynomial as $P = P_1+2x_2$ where

$$P_1 = x_1^2+2x_1x_2+x_2^2+2x_1+2x_2$$

Eq. (A.5.53) has infinite number of solutions. Any solution allows us to complete the square. Let us find the solution of Eq. (A.5.53) using the pseudoinverse matrix. The characteristic equation

$$\det(\mathbf{I}\lambda-\mathbf{A}) = \begin{vmatrix}1-\lambda & -1\\-1 & 1-\lambda\end{vmatrix} = (1-\lambda)^2-1 = 0$$

has two roots $\lambda_1=2$ and $\lambda_2=0$. The corresponding normalized eigenvectors are $\mathbf{t}_1 = (1/\sqrt{2})[1\ \ 1]'$ and $\mathbf{t}_2 = (1/\sqrt{2})[1\ \ -1]'$ Thus

$$\mathbf{T} = \frac{1}{\sqrt{2}}\begin{bmatrix}1 & 1\\1 & -1\end{bmatrix}$$

and Eq. (A.5.54) takes the form

$$\mathbf{A}^{\dagger} = \mathbf{T}'\begin{bmatrix}2^{-1} & 0\\0 & 0\end{bmatrix}\mathbf{T} = \frac{1}{4}\begin{bmatrix}1 & 1\\1 & 1\end{bmatrix}$$

Substituting this matrix into (A.5.57) we obtain after multiplying the matrices the following

$$P = (x_1+x_2+1)^2-1+2x_2$$

Thus, we have the complete square and some linear term. As we can see, the proposed method is not the most efficient one. Equation (A.5.57) plays an important theoretical role in Chapter 7.

Consider now a polynomial consisting of the several complete squares

$$P(\mathbf{x}) = \sum_{i=1}^{m}(\mathbf{u}_i-\mathbf{V}_i\mathbf{x})'\mathbf{A}_i(\mathbf{u}_i-\mathbf{V}_i\mathbf{x})$$

where all the matrices $\mathbf{A}_i$ are nonnegative definite. This polynomial is central since otherwise we would have a complete square and some linear terms and the polynomial would be negative for some $\mathbf{x}$. To complete the square, we present this polynomial in the standard form of Eq. (A.5.52) with

$$\mathbf{A} = \sum_{i=1}^{m} \mathbf{V}_i' \mathbf{A}_i \mathbf{V}_i, \quad \mathbf{b} = -\sum_{i=1}^{m} \mathbf{V}_i' \mathbf{A}_i \mathbf{u}_i, \quad c = \sum_{i=1}^{m} \mathbf{u}_i' \mathbf{A}_i \mathbf{u}_i \tag{A.5.58}$$

In this case, Eq. (A.5.57) takes the form

$$\sum_{i=1}^{m} (\mathbf{u}_i - \mathbf{V}_i \mathbf{x})' \mathbf{A}_i (\mathbf{u}_i - \mathbf{V}_i \mathbf{x}) = (\mathbf{A}\mathbf{x}+\mathbf{b})' \mathbf{A}^\dagger (\mathbf{A}\mathbf{x}+\mathbf{b}) - \mathbf{b}' \mathbf{A}^\dagger \mathbf{b} + \sum_{i=1}^{m} \mathbf{u}_i' \mathbf{A}_i \mathbf{u}_i \tag{A.5.59}$$

The polynomial acieves its minimum at its center

$$\mathbf{x}_c = -\mathbf{A}^\dagger \mathbf{b} = \left(\sum_{i=1}^{m} \mathbf{V}_i' \mathbf{A}_i \mathbf{V}_i\right)^\dagger \left(\sum_{i=1}^{m} \mathbf{V}_i' \mathbf{A}_i \mathbf{u}_i\right) \tag{A.5.60}$$

and the minimal value is

$$P(\mathbf{x}_c) = -\mathbf{b}' \mathbf{A}^\dagger \mathbf{b} + \sum_{i=1}^{m} \mathbf{u}_i' \mathbf{A}_i \mathbf{u}_i \tag{A.5.61}$$

5.1.11 Quadratic Forms.

If $\mathbf{b} = 0$ and $c = 0$ the quadratic polynomial

$$P(\mathbf{x}) = \mathbf{x}'\mathbf{A}\mathbf{x} = \sum_{i,j=1}^{n} a_{ij} x_i x_j$$

is called a quadratic form. A linear transformation $\mathbf{x} = \mathbf{T}\mathbf{y}$ of variables leads to

$$P(\mathbf{y}) = \mathbf{y}'\mathbf{T}'\mathbf{A}\mathbf{T}\mathbf{y} = \mathbf{y}'\mathbf{B}\mathbf{y}$$

Since $\mathbf{A}$ is symmetric, then, according to Eq. (A.5.29), there exists matrix $\mathbf{T}$ such that $\mathbf{T}'\mathbf{A}\mathbf{T} = \mathbf{\Lambda}$, where $\mathbf{\Lambda} = diag\{\lambda_1^n\}$ is a diagonal matrix of eigenvalues of $\mathbf{A}$. This means that any quadratic form can be transformed to the sum of squares

$$P(\mathbf{y}) = \mathbf{y}'\mathbf{\Lambda}\mathbf{y} = \lambda_1 y_1^2 + \lambda_2 y_2^2 + \ldots + \lambda_n y_n^2 \tag{A.5.62}$$

A symmetric matrix $\mathbf{A}$ is called *positive definite* (*non-negative definite*) if, for any $\mathbf{x} \neq 0$, the corresponding quadratic form $\mathbf{x}'\mathbf{A}\mathbf{x} > 0$ ($\mathbf{x}'\mathbf{A}\mathbf{x} \geq 0$). It follows from Eq. (A.5.62) that it is necessary and sufficient for the matrix to be positive definite (non-negative definite) that all its eigenvalues be positive (non-negative).

It follows from Eq. (A.5.29) that any non-negative definite symmetric matrix can be expressed as

$$\mathbf{A} = \mathbf{B}\mathbf{B}'$$

where $\mathbf{B} = \mathbf{T}'\mathbf{\Lambda}^{1/2}$ and $\mathbf{\Lambda}^{1/2} = diag\{\sqrt{\lambda_i}\}_n$. The matrix $\mathbf{B}$ is sometimes called a square root of $\mathbf{A}$.

The square root is not unique. If $\mathbf{A}$ is positive definite, it is possible to find the square root of it in the form of the lower triangular matrix. We can prove this statement by showing that there exists a lower triangular matrix $\mathbf{T}$ such that $\mathbf{y} = \mathbf{T}\mathbf{x}$ transforms the quadratic form to the sum of squares. The following method of transforming the quadratic form to the sum of squares is called the Lagrange's algorithm.

Since the quadratic form is positive definite, it must be positive for

$\mathbf{x} = \mathbf{e}_1 = [1 \;\; 0 \;\; \dots \;\; 0]'$. Therefore, $\mathbf{e}_1{}'\mathbf{A}\mathbf{e}_1 = a_{11} > 0$ and we can define a new variable

$$y_1 = \frac{1}{\sqrt{a_{11}}} \sum_{j=1}^{n} a_{1j} x_j$$

It is easy to see by inspection that

$$\mathbf{x}'\mathbf{A}\mathbf{x} - y_1^2 = \sum_{i,j=1}^{n} a_{ij} x_i x_j - \frac{1}{a_{11}} \left(\sum_{j=1}^{n} a_{1j} x_j \right)^2$$

does not contain x_1. Thus, we can write

$$\mathbf{x}'\mathbf{A}\mathbf{x} = y_1^2 + \mathbf{x}_1' \mathbf{A}_1 \mathbf{x}_1$$

where $\mathbf{x}_1 = [x_2 \;\; x_3 \;\; \dots \;\; x_n]'$ and the quadratic form $\mathbf{x}_1' \mathbf{A}_1 \mathbf{x}_1$ is positive definite. Defining a new variable

$$y_2 = \frac{1}{\sqrt{a_{22}^{(1)}}} \sum_{j=2}^{n} a_{2j}^{(1)} x_j$$

where $a_{ij}^{(1)}$ with $i,j = 2,3,\dots,n$ denote the elements of $\mathbf{A}_1$, we eliminate x_2. Repeating these steps, we obtain the sum of squares:

$$\mathbf{x}'\mathbf{A}\mathbf{x} = y_1^2 + y_2^2 + \dots + y_n^2 \tag{A.5.63}$$

By construction, the matrix of the transformation $\mathbf{y} = \mathbf{T}\mathbf{x}$ is upper triangular. Therefore, $\mathbf{T}'$ is lower triangular and, according to Eq. (A.5.63) we can write

$$\mathbf{T}'\mathbf{A}\mathbf{T} = \mathbf{I}$$

which leads to

$$\mathbf{A} = \mathbf{L}\mathbf{L}' \tag{A.5.64}$$

where $\mathbf{L} = (\mathbf{T}')^{-1}$ is a lower triangular matrix. Equation (A.5.64) is called the Cholesky decomposition (or factorization) of the positive definite matrix $\mathbf{A}$.

The elements of the matrix $\mathbf{L}$ can be found directly by solving Eq. (A.5.64) since we proved that the solution exists. Denoting

$$\mathbf{L} = \begin{bmatrix} l_{11} & 0 & \cdots & 0 \\ l_{12} & l_{22} & \cdots & 0 \\ \cdots & \cdots & \cdots & \cdots \\ l_{1n} & l_{2n} & \cdots & l_{nn} \end{bmatrix}$$

we can write the part of Eq. (A.5.64) corresponding to the first column (or row) of matrix $\mathbf{A}$ as

$$a_{11} = l_{11}^2, \quad a_{12} = l_{11} l_{12}, \quad \dots, a_{1n} = l_{11} l_{1n}$$

which has the obvious solution

$$l_{11} = \sqrt{a_{11}}, \quad l_{12} = a_{12} l_{11}^{-1}, \quad \dots, \quad l_{1n} = a_{1n} l_{11}^{-1}$$

This solution gives us the first column of $\mathbf{L}$. Substituting it into Eq. (A.5.64) we obtain similarly its second column

$$l_{22} = (a_{22} - l_{12}^2)^{1/2}, \quad l_{23} = (a_{23} - l_{12} l_{13}) l_{22}^{-1}, \quad \ldots, \quad l_{2n} = (a_{2n} - l_{12} l_{1n}) l_{22}^{-1}$$

and so on. The algorithm is summarized by the following recursive equation

$$\begin{aligned} l_{ii} &= \left(a_{ii} - \sum_{k=1}^{i-1} l_{ki}^2\right)^{1/2} \\ l_{ij} &= \left(a_{ij} - \sum_{k=1}^{j-1} l_{ki} l_{kj}\right) l_{ii}^{-1} \end{aligned} \tag{A.5.65}$$

It is implemented in the MATLAB function `chol`.

References

1 F. R. Gantmacher, *The Theory of Matrices*, (Chelsea Publishing Co., New York, 1959).

APPENDIX 6

6.1 MARKOV CHAINS AND GRAPHS

This appendix provides a brief description of the most important properties of Markov chains which are used in this book.

6.1.1 Transition Probabilities

A sequence of random variables $y_0, y_1, y_2, \ldots$ is called a *finite Markov chain* if

$$Pr(\, y_t = i_t \mid y_0 = i_0, y_1 = i_1, \ldots, y_{t-1} = i_{t-1}) = Pr(\, y_t = i_t \mid y_{t-1} = i_{t-1})$$

for any $t \in \mathbf{N} = \{1,2,\ldots\}$ and $i_0, i_1, \ldots, i_t$ from a finite set Ω (usually, $i_k \in \Omega = \{1,2,\ldots,n\}$). In other words, the probability that a Markov chain is in state i_t at the moment t depends only on its state at the moment $t-1$.

If the transition probability

$$p_{ij} = Pr(\, y_t = j \mid y_{t-1} = i)$$

does not depend on t, the chain is called *homogeneous.* Any joint distribution of a homogeneous Markov chain is completely determined by the matrix of its transition probabilities

$$\mathbf{P} = [\, p_{ij} \,]_{n,n}$$

and the initial distribution

$$\mathbf{p} = [\, p_1 \;\; p_2 \;\; \ldots \;\; p_n \,]$$

Indeed, the chain starts from state i_0 with the probability p_{i_0}. Then, it transfers to state i_1 with the probability $p_{i_0 i_1}$. Next, it transfers to state i_2 with the probability $p_{i_1 i_2}$, and so on. Therefore, the joint probability of the sequence of states can be expressed as

$$Pr(y_0 = i_0, y_1 = i_1, \ldots, y_t = i_t) = p_{i_0} \, p_{i_0 i_1} \cdots p_{i_{t-1} i_t} = p_{i_0} \prod_{k=1}^{t} p_{i_{k-1} i_k}$$

Clearly, the elements of the transition matrix are nonnegative and the sum of the elements of each row equals 1:

$$\sum_{j=1}^{n} p_{ij} = 1 \qquad i=1,2,\ldots,n$$

This can be rewritten as $\mathbf{P}\,\mathbf{1} = \mathbf{1}$, where $\mathbf{1}$ is the matrix column of ones. Therefore, $\mathbf{1}$ is an eigenvector of $\mathbf{P}$ with the eigenvalue $\lambda = 1$. This vector is sometimes called a *right invariant vector* of the matrix $\mathbf{P}$, since the multiplication $\mathbf{P}\,\mathbf{1}$ does not change $\mathbf{1}$.

We shall use the notation $p_{ij}^{(m)}$ for the m-step transition probability from state i to state j. This probability can be found recursively by the total probability formula

$$p_{ij}^{(m)} = \sum_{k=1}^{n} p_{ik} p_{kj}^{(m-1)} \tag{A.6.1}$$

which can be written in the matrix form as $\mathbf{P}^{(m)} = \mathbf{P}\mathbf{P}^{(m-1)}$. Therefore, $\mathbf{P}^{(2)} = \mathbf{P}\mathbf{P} = \mathbf{P}^2$, $\mathbf{P}^{(3)} = \mathbf{P}\mathbf{P}^{(2)} = \mathbf{P}^3$, and so on; finally, $\mathbf{P}^{(m)} = [\, p_{ij}^{(m)} \,]_{n,n} = \mathbf{P}^m$. In other words, m-step transition probabilities are the elements of the m-th power of $\mathbf{P}$.

We can find $\mathbf{P}^m$ using the method described in Appendix 5.1.3. In particular, Eq. (A.5.5) gives

$$\mathbf{P}^m = \sum_{j=1}^{r} \sum_{i=1}^{m_j} \mathbf{D}_{ij} \binom{m}{i-1} \lambda_j^{m-i+1} \tag{A.6.2}$$

Example A.6.1. Suppose that we have a Markov chain with two states and the transition matrix

$$\mathbf{P} = \begin{bmatrix} 0.25 & 0.75 \\ 0.5 & 0.5 \end{bmatrix}$$

Let us find the m-step transition probabilities.

The m-step transition matrix is equal to $\mathbf{P}^m$. Fortunately, we have found this matrix in Example A.5.8. It is given by Eq. (A.5.7). Thus,

$$\begin{aligned} p_{11}^{(m)} &= 0.4 + 0.6\,(-0.25)^m & p_{12}^{(m)} &= 0.6 - 0.6\,(-0.25)^m \\ p_{21}^{(m)} &= 0.4 - 0.4\,(-0.25)^m & p_{22}^{(m)} &= 0.6 + 0.4\,(-0.25)^m \end{aligned} \tag{A.6.3}$$

The probability distribution of states after m steps is given by

$$p_i^{(m)} = \sum_{k=1}^{n} p_k p_{ki}^{(m-1)}$$

This can be written using matrix notations as

$$\mathbf{p}^{(m)} = \mathbf{p}\,\mathbf{P}^{(m)} = \mathbf{p}\mathbf{P}^m \tag{A.6.4}$$

Example A.6.2. For the Markov chain considered in Example A.6.1 with the initial distribution $\mathbf{p} = [\,0.2 \quad 0.8\,]$, let us find the probability distribution of its states after m steps.

Using Eq. (A.5.5) and (A.6.4), we obtain the required result:

$$\mathbf{p}^{(m)} = \mathbf{p}\mathbf{D}_{11} + \mathbf{p}\mathbf{D}_{12}\,(-0.25)^m = [\,0.4-0.2(-0.25)^m \quad 0.6+0.2(-0.25)^m\,]$$

6.1.2 Stationary Markov Chains

Matrix $\mathbf{P}$ is called a *regular matrix* if its characteristic polynomial $\Delta(z) = \det(\mathbf{I}z - \mathbf{P})$ has only one root $z = 1$, whose absolute value equals 1. In other words, $\mathbf{P}$ is regular if $\Delta(1) = 0$, $\Delta'(1) \neq 0$, and the absolute values of the remaining roots of the equation $\Delta(z) = 0$ are not equal to 1. If a transition matrix of a Markov chain is regular, then the chain also is called regular.

Theorem A.6.1: If a Markov chain is regular, then its m-step transition probability has a limit, which does not depend on the chain's initial state:

$$\lim_{m \to \infty} p_{ij}^{(m)} = \pi_j \tag{A.6.5}$$

These limits are called *equilibrium* or *stationary* probabilities of states of the Markov chain.

Proof: Since $z = 1$ is a simple root of $\Delta(z)$,

$$\Delta(z) = (z-1)(z-\lambda_2)^{m_2}(z-\lambda_3)^{m_3} \cdots (z-\lambda_r)^{m_r}$$

and formula (A.5.4) becomes

$$(\mathbf{I}z - \mathbf{P})^{-1} = \mathbf{B}(z)/\Delta(z) = \mathbf{D}_{11}/(z-1) + \sum_{j=2}^{r}\sum_{i=1}^{m_j} \mathbf{D}_{ij}/(z-\lambda_j)^i \tag{A.6.6}$$

Expanding the right-hand side of this equation, we obtain from Eq. (A.5.5)

$$\mathbf{P}^m = \mathbf{D}_{11} + \sum_{j=2}^{r}\sum_{i=1}^{m_j} \mathbf{D}_{ij}\binom{m}{i-1}\lambda_j^{m-i+1} \tag{A.6.7}$$

If we assume that $|\lambda_j| > 1$, then $|\lambda_j^m| \to \infty$ when $m \to \infty$, which is impossible since all the elements of $\mathbf{P}^m$, being probabilities, are bounded: $0 \le p_{ij}^{(m)} \le 1$. Thus, $|\lambda_j| < 1$ for $j=2,3,\ldots,r$ because $z = 1$ is the only root whose absolute value is equal to 1. But in this case

$$\lim_{m \to \infty} \binom{m}{i-1}\lambda_j^{m-i+1} = 0 \tag{A.6.8}$$

and we obtain

$$\lim_{m \to \infty} \mathbf{P}^m = \mathbf{D}_{11}$$

This proves that each m-step transition probability has a limit when $m \to \infty$.

Passing to the limit in the identity $\mathbf{P}^m\,\mathbf{P} = \mathbf{P}^{m+1}$, we obtain $\mathbf{D}_{11}\,\mathbf{P} = \mathbf{D}_{11}$. This means that each row of $\mathbf{D}_{11}$ is an eigenvector corresponding to the eigenvalue $\lambda = 1$, so that all the rows of $\mathbf{D}_{11}$ are proportional. Passing to the limit as $m \to \infty$ in the identity $\mathbf{P}^m\,\mathbf{1} = \mathbf{1}$, we obtain $\mathbf{D}_{11}\,\mathbf{1} = \mathbf{1}$, which means that all the rows of $\mathbf{D}_{11}$ are identical and equal to the stationary probability vector $\boldsymbol{\pi} = [\,\pi_1, \pi_2, \ldots, \pi_n\,]$. Therefore, the limits in Eq. (A.6.5) do not depend on the initial state (the number of a row of $\mathbf{D}_{11}$). It is easy to check that [see also Eq. (A.5.17)]

$$\mathbf{D}_{11} = \mathbf{1}\,\boldsymbol{\pi}$$

Example A.6.3. For the Markov chain of Example A.6.1, we had $\Delta(z) = (z-1)(z+0.25)$. This polynomial has a simple root $z = 1$, therefore, the chain is regular. Passing to the limit as $m \to \infty$ in Eq. (A.6.3), we obtain

$$\begin{aligned} p_{11}^{(\infty)} &= 0.4 \quad & p_{12}^{(\infty)} &= 0.6 \\ p_{21}^{(\infty)} &= 0.4 \quad & p_{22}^{(\infty)} &= 0.6 \end{aligned} \tag{A.6.9}$$

So that the stationary distribution has the form

$$\boldsymbol{\pi} = [\, \pi_1 \quad \pi_2 \,] = [\, 0.4 \quad 0.6 \,]$$

It follows from Eq. (A.5.7) that (see also Example A.5.10)

$$\lim_{m\to\infty} \mathbf{P}^m = \mathbf{D}_{11} = \begin{bmatrix} 0.4 & 0.6 \\ 0.4 & 0.6 \end{bmatrix} = \begin{bmatrix} 1 \\ 1 \end{bmatrix} [\, 0.4 \quad 0.6 \,] = \mathbf{1}\,\boldsymbol{\pi}$$

The stationary distribution can be found as a solution of the system:

$$\boldsymbol{\pi}\mathbf{P} = \boldsymbol{\pi} \qquad \boldsymbol{\pi}\mathbf{1} = 1 \tag{A.6.10}$$

since $\boldsymbol{\pi}$ is the left eigenvector of $\mathbf{P}$ corresponding to $\lambda = 1$ (see Appendix 5.1.4). The second equation in Eq. (A.6.10) simply means that the total sum of the stationary probabilities is equal to 1:

$$\sum_{j=1}^{n} \pi_j = \boldsymbol{\pi}\mathbf{1} = 1$$

In its explicit form system (A.6.10) can be written as

$$\pi_j = \sum_{i=1}^{n} \pi_i p_{ij} \qquad \sum_{i=1}^{n} \pi_i = 1 \qquad j = 1,2,\ldots,n$$

It follows from Theorem A.6.1 that this system has a unique solution if the Markov chain is regular.

If the matrix of the initial probabilities $\mathbf{p}$ coincides with the matrix of the stationary probabilities $\boldsymbol{\pi}$, the chain is called *stationary*.

Example A.6.4. System (A.6.10) for the Markov chain of Example A.6.1 has the following form

$$[\, \pi_1 \quad \pi_2 \,] \begin{bmatrix} 0.25 & 0.75 \\ 0.5 & 0.5 \end{bmatrix} = [\, \pi_1 \quad \pi_2 \,] \qquad [\, \pi_1 \quad \pi_2 \,] \begin{bmatrix} 1 \\ 1 \end{bmatrix} = 1$$

In its explicit form this system is

$$\begin{aligned} 0.25\,\pi_1 + 0.5\,\pi_2 &= \pi_1 \\ 0.75\,\pi_1 + 0.5\,\pi_2 &= \pi_2 \\ \pi_1 + \pi_2 &= 1 \end{aligned}$$

The unique solution of this system is

$$\boldsymbol{\pi} = [\, \pi_1 \quad \pi_2 \,] = [\, 0.4 \quad 0.6 \,]$$

6.1.3 Partitioning States of a Markov Chain

It is convenient to partition the set of all states of a Markov chain into subsets of states $\{A_1, A_2, \ldots, A_s\}$, such that

$$\Omega = \bigcup_{i=1}^{s} A_i \qquad A_i \cap A_j = \varnothing \quad \text{for} \quad i \neq j$$

As will be seen shortly, it is convenient to describe the process of transition between the subsets if the numbers of the states belonging to the same subset are consecutive. For example, $A_1 = (1,2,3)$, $A_2 = (4,5)$, and $A_3 = (6,7,8,9)$. If a partition does not satisfy this condition, we can always renumber states that do not satisfy it. Exchanging numbers of two states requires permuting the corresponding rows and columns of the transition matrix. This operation can also be performed by the transformation **UPU**, in which **U** is obtained from the unit matrix by permuting its i-th and j-th rows.

After reordering all the states according to the selected partition, the transition matrix can be rewritten in the following block form

$$\mathbf{P} = \begin{bmatrix} \mathbf{P}_{11} & \mathbf{P}_{12} & \cdots & \mathbf{P}_{1s} \\ \mathbf{P}_{21} & \mathbf{P}_{22} & \cdots & \mathbf{P}_{2s} \\ \cdots & \cdots & \cdots & \cdots \\ \mathbf{P}_{s1} & \mathbf{P}_{s2} & \cdots & \mathbf{P}_{ss} \end{bmatrix}$$

Example A.6.5. Consider a Markov chain with three states an the transition probability matrix

$$\mathbf{P} = \begin{bmatrix} 0.5 & 0 & 0.5 \\ 0.3 & 0.3 & 0.4 \\ 0.2 & 0 & 0.8 \end{bmatrix}$$

Suppose that we want to rewrite this matrix in the block form corresponding to the partition $\{A_1, A_2\}$ in which A_1 is composed of the first and the last states of the chain and A_2 is the second state. In this case, we need to permute the second and third states of the chain. This can be achieved by permuting the second and third columns and rows of the transition matrix. This operation can also be performed by the transformation **UPU**, where **U** is obtained from the unit matrix $\mathbf{I} = diag\,\{1, 1, 1\}$ by permuting its second and third rows:

$$\mathbf{U\,P\,U} = \begin{bmatrix} 0.5 & 0.5 & 0 \\ 0.2 & 0.8 & 0 \\ 0.3 & 0.4 & 0.3 \end{bmatrix} = \begin{bmatrix} \mathbf{P}_{11} & \mathbf{P}_{12} \\ \mathbf{P}_{21} & \mathbf{P}_{22} \end{bmatrix} \tag{A.6.11}$$

where

$$\mathbf{P}_{11} = = \begin{bmatrix} 0.5 & 0.5 \\ 0.2 & 0.8 \end{bmatrix} \qquad \mathbf{P}_{12} = \begin{bmatrix} 0 \\ 0 \end{bmatrix}$$

$$\mathbf{P}_{21} = [\,0.3 \quad 0.4\,] \qquad \mathbf{P}_{22} = 0.3$$

As an illustration of usage of state partitioning, let us consider the problem of exclusion of unobserved states of a Markov chain. [2] Suppose that states of a Markov chain with the transition matrix **P** and initial distribution **p** are partitioned into two subsets A_1 and A_2. We can observe only the states that belong to A_1, and we want to describe the process of their transition.

Clearly, the observed process is a Markov chain. Therefore, to describe it, we need to find its transition matrix $\mathbf{Q} = [q_{ij}]_{s,s}$ and initial distribution **q**. Let x and y be two states that belong to A_1. The transition probability of the visible process can be expressed as

$$q_{xy} = p_{xy} + \sum_{j \in A_2} p_{xj} p_{jy} + \sum_{j,k \in A_2} p_{xj} p_{jk} p_{ky} + \cdots$$

In this equation, p_{xy} corresponds to the transfer $x \rightarrow y$ in one step in the original chain, the first sum over A_2 corresponds to the first transition $x \to y$ in two steps (from the visible state x to an invisible state j and back to the visible state y), the second sum corresponds to the first transition $x \to y$ in three steps , and so on. Obviously, these sums can be expressed as matrix products, so that

$$\mathbf{Q} = \mathbf{P}_{11} + \mathbf{P}_{12}\mathbf{P}_{21} + \mathbf{P}_{12}\mathbf{P}_{22}\mathbf{P}_{21} + \mathbf{P}_{12}\mathbf{P}_{22}^2\mathbf{P}_{21} + \cdots$$

Using Eq. (A.5.3), we can perform summation in the right-hand side of the previous equation

$$\mathbf{Q} = \mathbf{P}_{11} + \mathbf{P}_{12}(\mathbf{I} - \mathbf{P}_{22})^{-1}\mathbf{P}_{21}$$

The initial probability vector $\mathbf{q}$ of the observed process can be obtained similarly to its transition matrix:

$$\mathbf{q} = \mathbf{p}_1 + \mathbf{p}_2(\mathbf{I} - \mathbf{P}_{22})^{-1}\mathbf{P}_{21}$$

where $\mathbf{p} = [\mathbf{p}_1 \quad \mathbf{p}_2]$ is the initial vector of the original process.

6.1.4 Signal Flow Graphs

It is convenient to represent state transitions using signal flow graphs or state transition diagrams. States are represented by the graph vertices and state transitions are represented by the edges.

Example A.6.6.

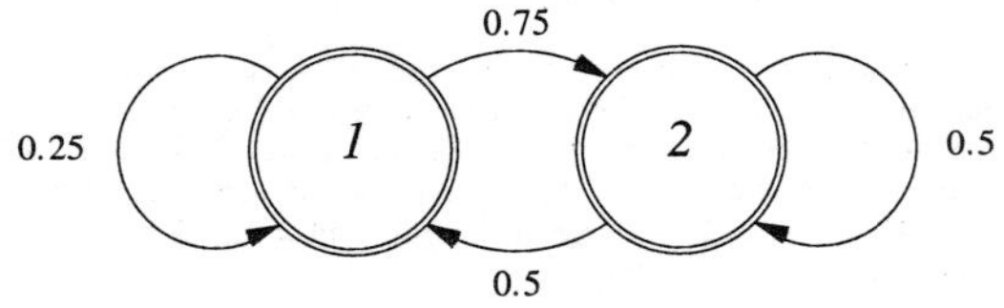

Figure A.3. State diagram of a Markov chain.

The signal flow graph of the Markov chain of Example A.6.1 is shown in Fig. A.3. In Fig. A.3, numbers inside the circles represent states, while numbers near arcs represent the corresponding transition probabilities.

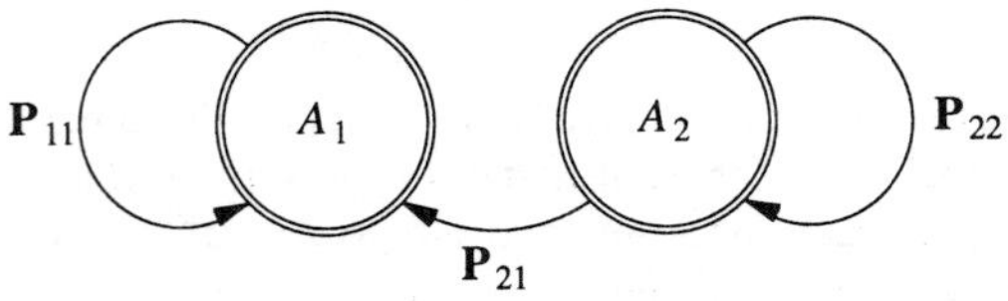

Figure A.4. Block graph of a Markov chain.

If states of a Markov chain are partitioned into subsets of states $\{A_1, A_2, ..., A_s\}$, then the corresponding vertices can be partitioned accordingly to form a block graph whose vertices represent the subsets and the edges represent all the transitions between the subsets. The block graph of the Markov chain of Example A.6.5 is shown in Fig. A.4. In Fig. A.4, symbols inside the circles represent subsets A_1 and A_2, while the symbols near the arcs represent the corresponding subblocks of the transition matrix.

Matrix **P** is called *reducible* if by permutation of its rows and corresponding columns it can be transformed into

$$\begin{bmatrix} \mathbf{P}_{11} & 0 \\ \mathbf{P}_{21} & \mathbf{P}_{22} \end{bmatrix}$$

where $\mathbf{P}_{11}$ and $\mathbf{P}_{22}$ are square matrices. Otherwise, **P** is *irreducible.* The transition matrix of Example A.6.1 is irreducible, while the transition matrix of Example A.6.5 is reducible (see Example A.6.7). Any Markov chain with a reducible transition matrix can be represented with the graph shown in Fig. A.4. There is no path leading from the states of the subset A_1 to the subset A_2, so that a Markov chain with a reducible matrix once entered in the subset A_1 never leaves it.

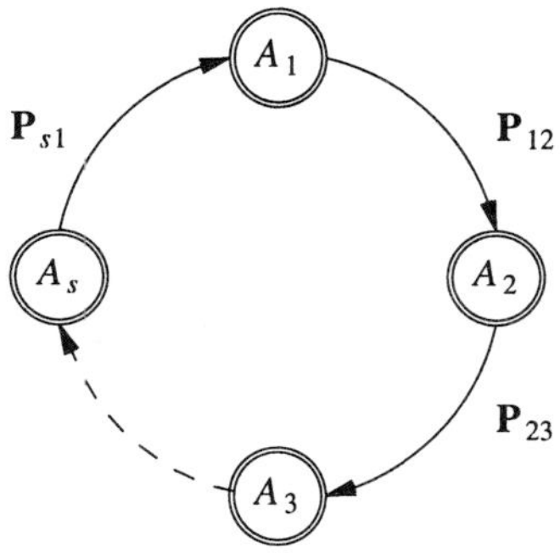

Figure A.5. Block graph of a periodic Markov chain.

If **P** is irreducible, then on the corresponding graph there is a path leading from any vertex to any other vertex.

Matrix **P** is called *cyclic* if by permutation of its rows and corresponding columns it can be transformed into

$$\begin{bmatrix} 0 & \mathbf{P}_{12} & 0 & \cdots & 0 \\ 0 & 0 & \mathbf{P}_{23} & \cdots & 0 \\ \cdots & \cdots & \cdots & \cdots & \cdots \\ 0 & 0 & 0 & \cdots & \mathbf{P}_{s-1,s} \\ \mathbf{P}_{s1} & 0 & 0 & \cdots & 0 \end{bmatrix}$$

where the diagonal blocks are square matrices consisting of zeroes. Otherwise, **P** is *acyclic.* If a transition matrix **P** is cyclic, then the Markov chain is called *periodic*; otherwise it is *aperiodic.* A block graph of a periodic Markov chain is depicted in Fig. A.5.

If a transition matrix **P** is irreducible and acyclic, then it is regular and all the stationary probabilities are positive. This result follows from the classical Perron-Frobenius theorem. [1]

Example A.6.7. Transition matrix (A.6.11) of Example A.6.5 is reducible

$$\mathbf{P} = \begin{bmatrix} 0.5 & 0.5 & 0 \\ 0.2 & 0.8 & 0 \\ 0.3 & 0.4 & 0.3 \end{bmatrix}$$

Its characteristic polynomial $\Delta(z) = (z-1)(z-0.3)^2$ does not have roots, except for $z = 1$ with absolute value 1. Therefore, this chain is regular. But not all the stationary probabilities are positive: $\pi = [\ 2/7 \quad 5/7 \quad 0\ ]$.

References

1 F. R. Gantmacher, *The Theory of Matrices*, (Chelsea Publshing Co., New York, 1959).

2 J. G. Kemeny and J. L. Snell, *Finite Markov Chains,* (Van Nostrand, Princeton, New Jersey, 1960).

APPENDIX 7

7.1 MARKOV PROCESSES

This appendix provides a brief description of the important results related to the Markov processes which are used in Chapter 7. A rigorous definition of Markov processes and derivation of these results can be found, for example, in [1]

7.1.1 Transition Probability Densities

The process $\boldsymbol{x}_t$ is a Markov process, if its future is conditionally independent on the past, given its present. This means that, for any sequence $t_0 < t_1 < ... < t_n$ and all possible sequences $\boldsymbol{x}_1, \boldsymbol{x}_2, \ldots, \boldsymbol{x}_n$ whose elements belong to the set of states $\boldsymbol{X}$, the conditional PDF satisfies the following Markovian property

$$p(\boldsymbol{x}_n; t_n \mid \boldsymbol{x}_0^{n-1}; t_0^{n-1}) = p(\boldsymbol{x}_n \mid \boldsymbol{x}_{n-1}; t_{n-1}) \tag{A.7.1}$$

This equation defines the process transition probability density which we will denote as

$$p_{t_1 t_2}(\boldsymbol{x}_2 \mid \boldsymbol{x}_1) = p(\boldsymbol{x}_2; t_2 \mid \boldsymbol{x}_1; t_1)$$

It is clear that the transition PDF must satisfy the following property

$$p_{t_1 t_3}(\boldsymbol{z} \mid \boldsymbol{x}) = \int_{\boldsymbol{X}} p_{t_1 t_2}(\boldsymbol{y} \mid \boldsymbol{x}) p_{t_2 t_3}(\boldsymbol{z} \mid \boldsymbol{y}) d\boldsymbol{y} \quad \text{where } t_1 < t_2 < t_3 \tag{A.7.2}$$

If X is the m-dimensional vector space, $\int_X$ is a shorthand for the multiple integral and $d\boldsymbol{y} = dy_1 dy_2, \ldots, dy_m$ is a shorthand for the corresponding elemental volume. which generalizes Eq. (A.6.1) and is called the Chapman-Kolmogorov equation. If the process is a discrete homogeneous Markov chain with $t_i = i$ and $\boldsymbol{x}_t \in \{1, 2, ..., m\}$, we have

$$p_{t,t+1}(y \mid i) = \sum_{j=1}^{m} p_{ij} \delta(y - j)$$

and Eq. (A.7.2) takes the form

$$p_{t,t+2}(z \mid x) = \int_X \sum_{j=1}^{m} p_{xj} \delta(y-j) p(z \mid y) dy = \sum_{j=1}^{m} p_{xj} p_{t+1,t+2}(z \mid j)$$

After integrating this equation with respect to z over $(k-0.5, k+0.5)$, we obtain

$$p_{t,t+2}(k \mid x) = \sum_{j=1}^{m} p_{xj} p_{jk}$$

As we can see, this equation is a special case of Eq. (A.6.1) and represents the fact that the transitional probability matrix in two steps is $\mathbf{P}^2$ and the Chapman-Kolmogorov equation for the discrete Markov chain takes the form $\mathbf{P}^{m+n} = \mathbf{P}^m \mathbf{P}^n$.

If the transition probability density depends only on the time difference

$$p_{t_1 t_2}(\boldsymbol{x}_2 \mid \boldsymbol{x}_1) = p_{t_2 - t_1}(\boldsymbol{x}_2 \mid \boldsymbol{x}_1)$$

the process is called *homogeneous*. For the homogeneous discrete-time Markov processes it is usually assumed that $t_i = i$ to simplify notations. The homogeneous processes are completely characterized by the one-step transition probability density

$$p(y \mid x) = p_{t,t+1}(y \mid x)$$

and the initial PDF $p_0(x)$.

Using the Markovian property in Eq. (A.7.1), we can express the PDF of a sequence of states as

$$p(\boldsymbol{x}_0^m; t_0^m) = p_{t_0}(\boldsymbol{x}_0) p_{t_0 t_1}(\boldsymbol{x}_0 \mid \boldsymbol{x}_1) \ldots p_{t_{m-1} t_m}(\boldsymbol{x}_{m-1} \mid \boldsymbol{x}_m) = p_{t_0}(\boldsymbol{x}_0) \prod_{i=1}^{m} p(\boldsymbol{x}_{i-1} \mid \boldsymbol{x}_i)$$

where $p_{t_0}(\boldsymbol{x}_0)$ is the initial state PDF. For the homogeneous Markov processes, the PDF of a sequence of states $\boldsymbol{x}_0^m$ this equation takes the form

$$p(\boldsymbol{x}_0^m) = p_0(\boldsymbol{x}_0) \prod_{i=1}^{m} p(\boldsymbol{x}_i \mid \boldsymbol{x}_{i-1})$$

7.1.2 Transition Probability Operators

As we have seen, it is convenient to use matrix algebra in studying Markov chains. Similarly, it is convenient to use algebra of operators to study Markov processes. We define a transition probability operator $\mathbf{P}_{t_1 t_2}$ that maps an integrable function $\boldsymbol{f} = \{f(x), x \in \boldsymbol{X}\}$ to $\boldsymbol{f}\mathbf{P}_{t_1 t_2}$ as

$$\boldsymbol{f}\mathbf{P}_{t_1 t_2} = \int_{\boldsymbol{X}} f(\boldsymbol{x}_1) p_{t_1 t_2}(\boldsymbol{x}_2 \mid \boldsymbol{x}_1) d\boldsymbol{x}_1$$

Operators of this form are called the Fredholm operators with the *kernel* $p_{t_1 t_2}(y \mid x)$.

Using the same kernel we can define a different operator as mapping

$$\mathbf{P}_{t_1 t_2}\boldsymbol{f} = \int_{\boldsymbol{X}} f(\boldsymbol{x}_2) p_{t_1 t_2}(\boldsymbol{x}_2 \mid \boldsymbol{x}_1) d\boldsymbol{x}_2$$

These two definitions generalize the multiplication of a row-vector by the matrix and the matrix by the column-vector, respectively. The identity operator $\mathbf{I}$, by definition, maps any function to itself. Using this operator, we can define a dot product of two functions as

$f_1 f_2 = f_1 \mathbf{I} f_2$. It is also convenient to define the function $f(x) \equiv 1$ as $\mathbf{1}$. Using this notation, we can write $f\mathbf{1} = \int_X f(x)\,dx$. Any transition probability operator satisfies the condition $\mathbf{P}_{t_1 t_2}\mathbf{1} = \mathbf{1}$.

The right hand side of Eq. (A.7.2) defines the product of operators $\mathbf{P}_{t_1 t_2}\mathbf{P}_{t_2 t_3}$. Thus, the Chapman-Kolmogorov equation can be written in the following operator form

$$\mathbf{P}_{t_1 t_3} = \mathbf{P}_{t_1 t_2}\mathbf{P}_{t_2 t_3}$$

The m-step transition probability operator can be written as

$$\mathbf{P}_{t_0 t_m} = \prod_{i=1}^{m} \mathbf{P}_{t_{i-1} t_i}$$

and the state PDF $\boldsymbol{p}_{t_m} = p(\boldsymbol{x}_m; t_m) = \boldsymbol{p}_0 \mathbf{P}_{t_0 t_m}$ can be evaluated using the following forward algorithm

$$\boldsymbol{p}_{t_k} = \boldsymbol{p}_{t_{k-1}} \mathbf{P}_{t_{k-1} t_k} \quad \text{for} \quad k = 1, 2, \ldots, m \tag{A.7.3}$$

For the homogeneous Markov process, we define $\mathbf{P} = \mathbf{P}_{t,t+1}$. Then the m-step transition probability operator $\mathbf{P}_{t,t+m} = \mathbf{P}^m$. The m-step transition PDF takes the form

$$\boldsymbol{p}_m = \boldsymbol{p}_0 \mathbf{P}^m$$

For a homogeneous process, we define a stationary PDF as a limit of the m-step transition PDF $p_m(\boldsymbol{x}_m \mid \boldsymbol{x}_0)$ as $m \to \infty$ if the limit does not depend on the initial state $\boldsymbol{x}_0$. Since $\boldsymbol{p}_m = \boldsymbol{p}_{m-1}\mathbf{P}$ the stationary PDF $\boldsymbol{\pi}(x)$ must satisfy the following equation

$$\boldsymbol{\pi}\mathbf{P} = \int_X \pi(x) p(y \mid x)\,dx = \pi(y), \quad \boldsymbol{\pi}\mathbf{1} = 1 \tag{A.7.4}$$

This equation looks formally like Eq. (A.6.10). However, (A.6.10) represents a system of linear algebraic equations while (A.7.4) is the Fredholm integral equation of the first type which is much more difficult to solve.

Similarly to eigenvectors, we define left and right *eigenfunctions* of the operator $\mathbf{P}$ corresponding to the eigenvalue λ as solutions of the operator (integral) equations

$$f\mathbf{P} = \lambda f \quad \text{and} \quad \mathbf{P}f = \lambda f$$

respectively. Thus, we can say that the stationary PDF $\boldsymbol{\pi}$ is the left eigenfunction and $\mathbf{1}$ is the right eigenfunction of the transition probability operator $\mathbf{P}$ corresponding to the eigenvalue $\lambda = 1$.

Example A.7.1. For the kernel $p(y \mid x) = (2\pi)^{-1/2}\exp[-0.5(y-ax)^2]$, find the solution of Eq. (A.7.4). We must solve the following equation

$$\pi(y) = \frac{1}{\sqrt{2\pi}} \int_{-\infty}^{\infty} \pi(x)\exp[0.5(y-ax)^2]\,dx, \quad \int_{-\infty}^{\infty} \pi(x)\,dx = 1$$

Let us find a solution in the form $\pi(x) = (2\pi)^{-1/2}\exp(-0.5x^2/\sigma^2)$. For this function, the right hand side of the integral equation is

$$\frac{1}{2\pi\sigma} \int_{-\infty}^{\infty} \exp[-\frac{x^2}{2\sigma^2} - \frac{(y-ax)^2}{2}]\,dx$$

After completing the square with respect to x in the exponent, this integral is evaluated to $(2\pi d)^{-0.5}\exp(-y^2/2d^2)$, where $d = \sqrt{1+a^2\sigma^2}$. This integral is equal to $\pi(y)$ if $d = \sigma$

which is possible only if $|a|<1$ and $\sigma = 1/\sqrt{1-a^2}$. Thus, the solution is $\pi(x) = (2\pi(1-a^2))^{-1/2}\exp[-0.5x^2/(1-a^2)]$.

As we can see, the matrix equations derived in Appendix 6 can be generalized to operator equations for Markov processes. However, the operator equations are more complex, because integrals are rarely found in the closed-form. A class of Markov processes for which there is a closed-form solution is considered in the next section.

7.2 GAUSS-MARKOV PROCESSES

A process is called *Gauss-Markov* if it is Gaussian and Markov. Before we start studying these processes, let us review some facts related to Gaussian random processes.

7.2.1 Gaussian Random Variables and Processes

We begin by reviewing the properties of Gaussian PDF

$$\mathsf{N}(\boldsymbol{x}-\boldsymbol{\mu}_x,\boldsymbol{\Sigma}_{xx}) = (2\pi)^{-m/2}|\boldsymbol{\Sigma}_{xx}|^{-1/2}\exp[-0.5(\boldsymbol{x}-\boldsymbol{\mu}_x)'\boldsymbol{\Sigma}_{xx}^{-1}(\boldsymbol{x}-\boldsymbol{\mu}_x)] \quad \text{(A.7.5)}$$

where $|\boldsymbol{\Sigma}_{xx}|$ denotes the determinant of $\boldsymbol{\Sigma}_{xx}$, $\boldsymbol{x} = (\boldsymbol{x}_1,\boldsymbol{x}_2,\ldots,\boldsymbol{x}_m)'$is the m-dimensional vector, $\boldsymbol{\mu}_x = \mathbf{E}(\boldsymbol{x})$ and $\boldsymbol{\Sigma}_{xx} = \mathbf{E}([\boldsymbol{x}-\boldsymbol{\mu}_x][\boldsymbol{x}'-\boldsymbol{\mu}_x'])$ are the distribution mean and variance matrix:

$$\mathbf{a}_x = \int_X \boldsymbol{x}\mathsf{N}(\boldsymbol{x}-\boldsymbol{\mu}_x,\boldsymbol{\Sigma}_{xx})\,dx$$

$$\boldsymbol{\Sigma}_{xx} = \int_X (\boldsymbol{x}-\boldsymbol{\mu}_x)(\boldsymbol{x}'-\boldsymbol{\mu}_x')\mathsf{N}(\boldsymbol{x}-\boldsymbol{\mu}_x,\boldsymbol{\Sigma}_{xx})\,dx$$

Gaussian variable x with the mean $\boldsymbol{\mu}_x$ and variance matrix $\boldsymbol{\Sigma}_{xx}$ is denoted as $\boldsymbol{x}\sim\mathsf{N}(\boldsymbol{x}-\boldsymbol{\mu}_x,\boldsymbol{\Sigma}_{xx})$. According to Eq. (A.7.5), this definition assumes that the variance matrix $\boldsymbol{\Sigma}$ is invertible. However, we will use this symbol in all the cases as a shorthand of the phrase "$\boldsymbol{x}$ is a Gaussian random variable with the mean $\boldsymbol{\mu}_x$ and the variance matrix $\boldsymbol{\Sigma}_{xx}$."

If $\boldsymbol{x}\sim\mathsf{N}(\boldsymbol{x}-\boldsymbol{\mu}_x,\boldsymbol{\Sigma}_{xx})$ and $\boldsymbol{y} = \mathbf{A}\boldsymbol{x}+\boldsymbol{b}$, then it is easy to see that

$$\boldsymbol{y} \sim \mathsf{N}(\boldsymbol{y}-\mathbf{A}\boldsymbol{\mu}_x+\mathbf{b},\mathbf{A}\boldsymbol{\Sigma}_{xx}\mathbf{A}') \quad \text{(A.7.6)}$$

7.2.1.1 Marginalization

Using Eq. (A.7.6) we can prove that a marginal PDF of a Gaussian PDF is also a Gaussian PDF. Indeed, let $\boldsymbol{z}' = (\boldsymbol{x}',\boldsymbol{y}')\sim\mathsf{N}(\boldsymbol{z}-\boldsymbol{\mu}_z,\boldsymbol{\Sigma}_{zz})$ with $\boldsymbol{\mu}_z' = (\boldsymbol{\mu}_x',\boldsymbol{\mu}_y')$ and

$$\boldsymbol{\Sigma}_{zz} = \begin{bmatrix}\boldsymbol{\Sigma}_{xx} & \boldsymbol{\Sigma}_{xy}\\ \boldsymbol{\Sigma}_{yx} & \boldsymbol{\Sigma}_{yy}\end{bmatrix}$$

We can write $\boldsymbol{y} = \mathbf{A}\boldsymbol{z}$ where $\mathbf{A} = [0 \quad \mathbf{I}]$.

In this case Eq. (A.7.6) takes the form

$$\boldsymbol{y} \sim \mathsf{N}(\boldsymbol{y}-\mathbf{A}\boldsymbol{\mu}_z,\mathbf{A}\boldsymbol{\Sigma}_{zz}\mathbf{A}') = \mathsf{N}(\boldsymbol{y}-\boldsymbol{\mu}_y,\boldsymbol{\Sigma}_{yy})$$

Thus, if $\boldsymbol{z}$ has a Gaussian distribution, its component $\boldsymbol{y}$ has a Gaussian distribution whose mean and variance matrix are obtained from the $\boldsymbol{\mu}_z$ and $\boldsymbol{\Sigma}_{zz}$ by retaining the columns and rows corresponding to $\boldsymbol{y}$.

It is easy to see that the Gaussian PDF attains its global maximum at its mean

$$\boldsymbol{\mu}_z = \underset{z}{argmax} \mathsf{N}(z-\boldsymbol{\mu}_z, \boldsymbol{\Sigma}_z)$$

this property is used sometimes to find the mean.

7.2.1.2 Singular Gaussian distribution

Using Eq. (A.7.6) we can represent any Gaussian variable as a function of independent Gaussian variables $U_i \sim \mathsf{N}(U_i,1)$. Indeed, the vector $\mathbf{u} = (U_1, U_2, \ldots, U_m)' \sim \mathsf{N}(\mathbf{u},\mathbf{I})$ is Gaussian and $\boldsymbol{x} = \mathbf{A}\mathbf{u} + \boldsymbol{\mu}_x \sim \mathsf{N}(\boldsymbol{x}-\boldsymbol{\mu}_x, \mathbf{A}\mathbf{A}')$. Thus, we need to find a matrix $\mathbf{A}$ such that $\mathbf{A}\mathbf{A}' = \boldsymbol{\Sigma}_{xx}$ to obtain $\mathsf{N}(\boldsymbol{x}-\boldsymbol{\mu}_x, \boldsymbol{\Sigma}_{xx})$. Since every variance matrix is nonnegative (positive semidefinite) definite and symmetric, the matrix $\mathbf{A}$ exists (see Appendix 5). This solution can be found using the MATLAB M-file `sprep0`.

If $\boldsymbol{\Sigma}_{xx}$ is singular, then Eq. (A.7.5) is not valid. However, in this case we can define a *singular* Gaussian variable with the mean $\boldsymbol{\mu}_x$ and the variance matrix $\boldsymbol{\Sigma}_{xx}$ as $\boldsymbol{x} = \boldsymbol{\mu}_x + \mathbf{A}_x \boldsymbol{u}$ where $\mathbf{A}_x \mathbf{A}_x' = \boldsymbol{\Sigma}_{xx}$ and $\boldsymbol{u} \sim \mathsf{N}(\boldsymbol{u},\mathbf{I})$.

If the matrix $\boldsymbol{\Sigma}_{xx}$ has rank *r*, then we can use a *r*-dimensional vector *u* in the previous equation. In this case we need to find a matrix $\mathbf{A}_x$ with *r* columns. This matrix can be found using the spectral representation $\boldsymbol{\Sigma}_{xx} = \mathbf{T} diag\{\lambda_i\}\mathbf{T}'$ (see Appendix 5) where the eigenvalues are ordered: $\lambda_1 \geq \lambda_2 \geq \cdots \geq \lambda_r$ and $\lambda_i = 0$ for $i > r$. It is clear from this representation that the matrix $\mathbf{A}_x$ with *r* columns can be presented as

$$\mathbf{A}_x = \mathbf{T} \cdot \begin{bmatrix} \boldsymbol{\Lambda}_r \\ 0 \end{bmatrix}, \quad \boldsymbol{\Lambda}_r = diag\{\sqrt{\lambda_1}, \sqrt{\lambda_2}, \ldots, \sqrt{\lambda_r}\}$$

The MATLAB M-file `sprep` finds this matrix. Note that the pseudoinverse of this matrix has the form

$$\mathbf{A}_x^{\dagger} = (\mathbf{A}_x' \mathbf{A}_x)^{-1} \mathbf{A}_x' \qquad \text{(A.7.7)}$$

according to Eq. (A.5.54) of Appendix 5.

In many applications, it is necessary to evaluate an integral of the form

$$I = \int_X \mathsf{G}(\mathbf{A}\mathbf{x}-\mathbf{b}, \mathbf{C}) \mathsf{N}(\mathbf{x}-\boldsymbol{\mu}, \boldsymbol{\Sigma}) d\mathbf{x} \qquad \text{(A.7.8)}$$

where the matrix $\mathbf{C}$ is nonnegative definite

$$\mathsf{G}(\mathbf{x}, \mathbf{C}) = \exp\{-0.5\mathbf{x}' \mathbf{C}\mathbf{x}\} \qquad \text{(A.7.9)}$$

If matrix $\mathbf{C}$ is positive definite, we can write

$$\mathsf{G}(\mathbf{x}, \mathbf{C}) = (2\pi)^{m/2} |\mathbf{C}|^{1/2} \mathsf{N}(\mathbf{x}, \mathbf{C}^{-1}) \qquad \text{(A.7.10)}$$

We evaluate the integral in Eq. (A.7.8), by completing a square in the exponent of the integrand

$$e(u) = (\mathbf{A}\mathbf{u}-\mathbf{b}_1)' \mathbf{C}(\mathbf{A}\mathbf{u}-\mathbf{b}_1)\} + \mathbf{u}' \boldsymbol{\Sigma}^{-1} \mathbf{u}$$

where we denoted $\mathbf{u} = \mathbf{x}-\boldsymbol{\mu}$ and $\mathbf{b}_1 = \mathbf{b} - \mathbf{A}\boldsymbol{\mu}$ to simplify the derivation. The square is completed using Eq. (A.7.59) of Appendix 5:

$$e(u) = (\mathbf{Pu}-\mathbf{w})'\mathbf{P}^{-1}(\mathbf{Pu}-\mathbf{w}) + d \tag{A.7.11}$$

where

$$d = \mathbf{b}_1'\mathbf{C}\mathbf{b}_1 - \mathbf{w}'\mathbf{P}^{-1}\mathbf{w} \tag{A.7.12}$$

$$\mathbf{P} = \mathbf{A}'\mathbf{C}\mathbf{A} + \mathbf{\Sigma}^{-1} \tag{A.7.13}$$

$$\mathbf{w} = \mathbf{A}'\mathbf{C}\mathbf{b}_1$$

It follows from Eq. (A.7.13) that $\mathbf{P}$ is positive definite and, therefore, $\mathbf{P}^{-1}$ exists. Substituting $\mathbf{w}$, $\mathbf{P}$, and $\mathbf{b}$ into Eq. (A.7.12), we obtain

$$d = \mathbf{b}_1'\mathbf{V}\mathbf{b}_1 = (\mathbf{A}\boldsymbol{\mu}-\mathbf{b})'\mathbf{V}(\mathbf{A}\boldsymbol{\mu}-\mathbf{b}) \tag{A.7.14}$$

where

$$\mathbf{V} = \mathbf{C} - \mathbf{CA}(\mathbf{A}'\mathbf{CA}+\mathbf{\Sigma}^{-1})^{-1}\mathbf{A}'\mathbf{C} = \mathbf{C}(\mathbf{A\Sigma A}'\mathbf{C}+\mathbf{I})^{-1} \tag{A.7.15}$$

We applied the matrix inversion lemma (see Appendix 5)

$$\mathbf{P}^{-1} = (\mathbf{A}'\mathbf{CA}+\mathbf{\Sigma}^{-1})^{-1} = \mathbf{\Sigma}-\mathbf{\Sigma A}'\mathbf{C}(\mathbf{A\Sigma A}'\mathbf{C}+\mathbf{I})^{-1}\mathbf{A\Sigma}$$

to have the alternative expression for $\mathbf{P}^{-1}$. Which expression to use must be decided by considering the sizes of the matrices that need to be inverted. It is easy to prove that $\mathbf{V}$ in Eq. (A.7.15) is a nonnegative definite matrix.

Substituting d from Eq. (A.7.14) into (A.7.11), we can write

$$\begin{aligned}&(\mathbf{Ax}-\mathbf{b})'\mathbf{C}(\mathbf{Ax}-\mathbf{b}) + (\mathbf{x}-\boldsymbol{\mu})'\mathbf{\Sigma}^{-1}(\mathbf{x}-\boldsymbol{\mu})\\ &= (\mathbf{Px}-\mathbf{v})'\mathbf{P}^{-1}(\mathbf{Px}-\mathbf{v}) + (\mathbf{A}\boldsymbol{\mu}-\mathbf{b})'\mathbf{V}(\mathbf{A}\boldsymbol{\mu}-\mathbf{b})\end{aligned} \tag{A.7.16}$$

where

$$\mathbf{v} = \mathbf{A}'\mathbf{Cb} + \mathbf{\Sigma}^{-1}\boldsymbol{\mu}$$

This identity leads to

$$\mathrm{G}(\mathbf{Ax}-\mathbf{b},\mathbf{C})\,\mathrm{N}(\mathbf{x}-\boldsymbol{\mu},\mathbf{\Sigma}) = |\mathbf{\Sigma}|^{-1/2}|\mathbf{P}|^{1/2}\mathrm{N}(\mathbf{Px}-\mathbf{v},\mathbf{P})\,\mathrm{G}(\mathbf{A}\boldsymbol{\mu}-\mathbf{b},\mathbf{V}) \tag{A.7.17}$$

which can also be written as

$$\mathrm{G}(\mathbf{Ax}-\mathbf{b},\mathbf{C})\,\mathrm{N}(\mathbf{x}-\boldsymbol{\mu},\mathbf{\Sigma}) = |\mathbf{\Sigma}|^{-1/2}|\mathbf{P}|^{-1/2}\mathrm{N}(\mathbf{x}-\mathbf{q},\mathbf{Q})\,\mathrm{G}(\mathbf{A}\boldsymbol{\mu}-\mathbf{b},\mathbf{V}) \tag{A.7.18}$$

where

$$\mathbf{q} = \mathbf{P}^{-1}\mathbf{v} = (\mathbf{A}'\mathbf{CA}+\mathbf{\Sigma}^{-1})^{-1}(\mathbf{A}'\mathbf{Cb} + \mathbf{\Sigma}^{-1}\boldsymbol{\mu}) = \boldsymbol{\mu}+\mathbf{K}(\mathbf{b}-\mathbf{A}\boldsymbol{\mu})$$

$$\mathbf{K} = (\mathbf{A}'\mathbf{CA}+\mathbf{\Sigma}^{-1})^{-1}\mathbf{A}'\mathbf{C} = \mathbf{\Sigma A}'\mathbf{C}(\mathbf{A\Sigma A}'\mathbf{C} + \mathbf{I})^{-1}$$

$$\mathbf{Q} = (\mathbf{A}'\mathbf{CA}+\mathbf{\Sigma}^{-1})^{-1} = \mathbf{\Sigma}-\mathbf{KA\Sigma}$$

Integrating both sides of Eq. (A.7.18) and using the fact that the integral of the Gaussian PDF is equal to one, we obtain the desired result

$$\int_X \mathsf{G}(\mathbf{Ax}-\mathbf{b},\mathbf{C})\,\mathsf{N}(\mathbf{x}-\boldsymbol{\mu},\boldsymbol{\Sigma})\,d\mathbf{x} = |\boldsymbol{\Sigma}|^{-1/2}|\mathbf{P}|^{-1/2}\mathsf{G}(\mathbf{A}\boldsymbol{\mu}-\mathbf{b},\mathbf{V}) \tag{A.7.19}$$

If the matrix $\mathbf{C}$ is positive definite and $\boldsymbol{\Sigma}_1 = \mathbf{C}^{-1}$ is its inverse, then $\mathbf{V} = (\mathbf{A}\boldsymbol{\Sigma}\mathbf{A}' + \boldsymbol{\Sigma}_1)^{-1}$ and Eq. (A.7.18) takes the form

$$\mathsf{N}(\mathbf{Ax}-\mathbf{b},\boldsymbol{\Sigma}_1)\,\mathsf{N}(\mathbf{x}-\boldsymbol{\mu},\boldsymbol{\Sigma}) = \mathsf{N}(\mathbf{x}-\mathbf{q},\mathbf{Q})\,\mathsf{N}(\mathbf{A}\boldsymbol{\mu}-\mathbf{b},\mathbf{A}\boldsymbol{\Sigma}\mathbf{A}'+\boldsymbol{\Sigma}_1) \tag{A.7.20}$$

where

$$\mathbf{q} = (\mathbf{A}'\boldsymbol{\Sigma}_1^{-1}\mathbf{A}+\boldsymbol{\Sigma}^{-1})^{-1}(\mathbf{A}'\boldsymbol{\Sigma}_1\mathbf{b} + \boldsymbol{\Sigma}^{-1}\boldsymbol{\mu}) = \boldsymbol{\mu}+\mathbf{K}(\mathbf{b}-\mathbf{A}\boldsymbol{\mu}) \tag{A.7.21}$$

$$\mathbf{K} = (\mathbf{A}'\boldsymbol{\Sigma}_1^{-1}\mathbf{A}+\boldsymbol{\Sigma}^{-1})^{-1}\mathbf{A}'\boldsymbol{\Sigma}_1^{-1} = \boldsymbol{\Sigma}\mathbf{A}'(\mathbf{A}\boldsymbol{\Sigma}\mathbf{A}'+\boldsymbol{\Sigma}_1)^{-1}$$

$$\mathbf{Q} = \mathbf{P}^{-1} = (\mathbf{A}'\boldsymbol{\Sigma}_1^{-1}\mathbf{A}+\boldsymbol{\Sigma}^{-1})^{-1} = \boldsymbol{\Sigma}-\mathbf{K}\mathbf{A}\boldsymbol{\Sigma}$$

Thus, the integral in Eq. (A.7.19) takes the form

$$\int_X \mathsf{N}(\mathbf{Ax}-\mathbf{b},\boldsymbol{\Sigma}_1)\,\mathsf{N}(\mathbf{x}-\boldsymbol{\mu},\boldsymbol{\Sigma})\,d\mathbf{x} = \mathsf{N}(\mathbf{A}\boldsymbol{\mu}-\mathbf{b},\mathbf{A}\boldsymbol{\Sigma}\mathbf{A}'+\boldsymbol{\Sigma}_1) \tag{A.7.22}$$

7.2.1.3 Conditioning

If the variance matrix is positive definite (all its eigenvalues are positive), then we can find a lower triangular matrix $\mathbf{L}$ to obtain the following Cholesky decomposition $\mathbf{LL}' = \boldsymbol{\Sigma}_{xx}$. The MATLAB function `chol` performs this decomposition. Thus we have the following representation for a Gaussian variable $x \sim \mathsf{N}(\boldsymbol{x}-\boldsymbol{\mu}_x,\boldsymbol{\Sigma}_{xx})$

$$\mathbf{x} = \mathbf{Lu} + \boldsymbol{\mu}_x \tag{A.7.23}$$

This representation is often used for simulating Gaussian variables $\mathbf{x} \sim \mathsf{N}(\boldsymbol{x}-\boldsymbol{\mu}_x,\boldsymbol{\Sigma}_{xx})$ using $u \sim \mathsf{N}(\boldsymbol{u},1)$.

Let us find now an expression for the conditional PDF $p(\boldsymbol{y} \mid \boldsymbol{x}) = p(\boldsymbol{x},\boldsymbol{y})/p(\boldsymbol{x})$ provided that $z' = (\boldsymbol{x}',\boldsymbol{y}')$ is Gaussian $z \sim \mathsf{N}(z-\boldsymbol{\mu}_z,\boldsymbol{\Sigma}_{zz})$. The Cholesky decomposition can be written in the following block-matrix form

$$\boldsymbol{\Sigma}_{zz} = \begin{bmatrix}\boldsymbol{\Sigma}_{xx} & \boldsymbol{\Sigma}_{xy}\\ \boldsymbol{\Sigma}_{yx} & \boldsymbol{\Sigma}_{yy}\end{bmatrix} = \begin{bmatrix}\mathbf{L}_{11} & 0\\ \mathbf{L}_{21} & \mathbf{L}_{22}\end{bmatrix}\begin{bmatrix}\mathbf{L}_{11}' & \mathbf{L}_{21}'\\ 0 & \mathbf{L}_{22}'\end{bmatrix} \tag{A.7.24}$$

Equation (A.7.23) for the variable z takes the form

$$\boldsymbol{x} = \mathbf{L}_{11}\boldsymbol{u}_1 + \boldsymbol{\mu}_x$$

$$\boldsymbol{y} = \mathbf{L}_{21}\boldsymbol{u}_1 + \mathbf{L}_{22}\boldsymbol{u}_2 + \boldsymbol{\mu}_y$$

where $\boldsymbol{u}_1$ and $\boldsymbol{u}_2$ are the Gaussian vectors with independent $\mathsf{N}(U_i,1)$ components. Solving the first equation for u_1 and substituting it into the second equation, we obtain

$$\boldsymbol{y} - \boldsymbol{\mu}_y = \mathbf{L}_{21}\mathbf{L}_{11}^{-1}(\boldsymbol{x} - \boldsymbol{\mu}_x) + \mathbf{L}_{22}\boldsymbol{u}_2$$

which, according to Eq. (A.7.6), means that the conditional PDF is Gaussian:

$$p(\boldsymbol{y} \mid \boldsymbol{x}) = \mathsf{N}(y - \boldsymbol{\mu}_{y/x},\boldsymbol{\Sigma}_{y/x})$$

with

$$\begin{aligned} \boldsymbol{\mu}_{y/x} &= \boldsymbol{\mu}_y + \mathbf{L}_{21}\mathbf{L}_{11}^{-1}(\boldsymbol{x} - \boldsymbol{\mu}_x) \\ \boldsymbol{\Sigma}_{y/x} &= \mathbf{L}_{22}\mathbf{L}_{22}' \end{aligned} \tag{A.7.25}$$

Thus, we have the following formula

$$\mathsf{N}(\boldsymbol{z}-\boldsymbol{\mu}_z,\boldsymbol{\Sigma}_{zz}) = \mathsf{N}(\boldsymbol{x}-\boldsymbol{\mu}_x)\mathsf{N}(\boldsymbol{y}-\boldsymbol{\mu}_{y/x},\boldsymbol{\Sigma}_{y/x}) \tag{A.7.26}$$

Using Eq. (A.7.24) it is easy to show that Eq. (A.7.25) can be also written in as

$$\begin{aligned} \boldsymbol{\mu}_{y/x} &= \boldsymbol{\mu}_y + \boldsymbol{\Sigma}_{yx}\boldsymbol{\Sigma}_{xx}^{-1}(\boldsymbol{x}-\boldsymbol{\mu}_x) \\ \boldsymbol{\Sigma}_{y/x} &= \boldsymbol{\Sigma}_{yy} - \boldsymbol{\Sigma}_{yx}\boldsymbol{\Sigma}_{xx}^{-1}\boldsymbol{\Sigma}_{xy} \end{aligned} \tag{A.7.27}$$

Note that, if we perform the Gaussian elimination (see Eq. (A.5.8) of Appendix 5) of the block matrix $\boldsymbol{\Sigma}_{yx}$ from the matrix

$$[\boldsymbol{\Sigma}_{zz} \mid \boldsymbol{z}-\boldsymbol{\mu}_z] = \begin{bmatrix} \boldsymbol{\Sigma}_{xx} & \boldsymbol{\Sigma}_{xy} & \boldsymbol{x}-\boldsymbol{\mu}_x \\ \boldsymbol{\Sigma}_{yx} & \boldsymbol{\Sigma}_{yy} & \boldsymbol{y}-\boldsymbol{\mu}_y \end{bmatrix}$$

we obtain

$$\begin{bmatrix} \boldsymbol{\Sigma}_{xx} & \boldsymbol{\Sigma}_{xy} & \boldsymbol{x}-\boldsymbol{\mu}_x \\ 0 & \boldsymbol{\Sigma}_{yy}-\boldsymbol{\Sigma}_{yx}\boldsymbol{\Sigma}_{xx}^{-1}\boldsymbol{\Sigma}_{xy} & \boldsymbol{y}-\boldsymbol{\mu}_y-\boldsymbol{\Sigma}_{yx}\boldsymbol{\Sigma}_{xx}^{-1}(\boldsymbol{x}-\boldsymbol{\mu}_x) \end{bmatrix} \tag{A.7.28}$$

As we can see, the second row of this matrix gives us $\boldsymbol{\Sigma}_{y/x}$ and $\boldsymbol{y}-\boldsymbol{\mu}_{y/x}$. Thus, the conditional PDF mean and covariance matrices can be found by the Gaussian elimination of the entries corresponding to the condition. Note that if the inverse matrix $\boldsymbol{\Sigma}_{xx}^{-1}$ does not exist, a pseudoinverse matrix $\boldsymbol{\Sigma}_{xx}^{\dagger}$ can be used.

It follows from (A.7.26) that

$$\boldsymbol{\mu}_{y/x} = \underset{y}{argmax}\ p(\boldsymbol{x},\boldsymbol{y}) = \underset{y}{argmax}\ \mathsf{N}(\boldsymbol{z}-\boldsymbol{\mu}_z,\boldsymbol{\Sigma}_z) \tag{A.7.29}$$

and

$$\max_{y}\ p(\boldsymbol{x},\boldsymbol{y}) = C\cdot p(\boldsymbol{x}) = C\cdot\mathsf{N}(\boldsymbol{x}-\boldsymbol{\mu}_x,\boldsymbol{\Sigma}_{xx}) \tag{A.7.30}$$

where the constant $C = p(\boldsymbol{\mu}_{y/x} \mid \boldsymbol{x}) = \mathsf{N}(0,\boldsymbol{\Sigma}_{y/x})$. Thus, for the Gaussian distribution, the conditional mean can be found by maximizing the joint PDF $p(\boldsymbol{x},\boldsymbol{y})$ and the maximum is proportional to the marginal PDF $p(\boldsymbol{x})$. Also, it follows from these equations and formula (A.7.5) that $\boldsymbol{\mu}_{y/x}$ can be found by solving for $\boldsymbol{y}$ the following equation

$$\frac{\partial \log\mathsf{N}(\boldsymbol{z}-\boldsymbol{\mu}_z,\boldsymbol{\Sigma}_z)}{\partial \boldsymbol{y}} = 0 \tag{A.7.31}$$

and

$$\boldsymbol{\Sigma}_{y/x}^{-1} = -\frac{\partial^2 \log\mathsf{N}(\boldsymbol{z}-\boldsymbol{\mu}_z,\boldsymbol{\Sigma}_z)}{\partial \boldsymbol{y}^2} \tag{A.7.32}$$

Note also that the unconditional means can be found by solving for $\boldsymbol{x}$ and $\boldsymbol{y}$ the system

$$\frac{\partial \log\mathsf{N}(\boldsymbol{z}-\boldsymbol{\mu}_z,\boldsymbol{\Sigma}_z)}{\partial \boldsymbol{x}} = 0, \quad \frac{\partial \log\mathsf{N}(\boldsymbol{z}-\boldsymbol{\mu}_z,\boldsymbol{\Sigma}_z)}{\partial \boldsymbol{y}} = 0 \tag{A.7.33}$$

and

$$\Sigma_{zz}^{-1} = -\begin{bmatrix} \dfrac{\partial^2 \log N(z-\mu_z,\Sigma_z)}{\partial x^2} & \dfrac{\partial^2 \log N(z-\mu_z,\Sigma_z)}{\partial x \partial y} \\ \dfrac{\partial^2 \log N(z-\mu_z,\Sigma_z)}{\partial y \partial x} & \dfrac{\partial^2 \log N(z-\mu_z,\Sigma_z)}{\partial x^2} \end{bmatrix} \quad \text{(A.7.34)}$$

These equations are the alternative to completing the square with respect to z in the $\log N(z-\mu_z,\Sigma_z)$ if the PDF is represented as a product of several functions.

Let us consider now Gaussian processes. A process x_t is called Gaussian if for any sequence $t_1, t_2, \ldots, t_m$ and any m the vector $x_1^m \sim N(x_1^m - \mu_m, \Sigma_m)$.

Note that, according to Eq. (A.7.27), a Markov process with the Gaussian transitional PDF $p_{t_1 t_2}(x_2 \mid x_1) = N(x_2 - \mu_{2/1}(x_1), \Sigma_{2/1}(x_1))$ can be at the same time Gaussian only if its conditional mean $\mu_{2/1}(x_1)$ is a linear function of x_1 and $\Sigma_{2/1}(x_1)$ is independent of x_1. An important class of such processes is considered in the next section.

7.2.1.4 The Orthogonality Principle

The following equation is called the *orthogonality principle*

$$E\{[\mathbf{y}-\mu_y-\Sigma_{yx}\Sigma_{xx}^{-1}(x-\mu_x)]\mathbf{x}'\} = 0 \quad \text{(A.7.35)}$$

To prove it, we can replace $\mathbf{x}'$ with $(\mathbf{x}-\mu_x)'$ since the expectation of the expression inside the square brackets is equal to zero. We have

$$E\{[(\mathbf{y}-\mu_y)-\Sigma_{yx}\Sigma_{xx}^{-1}(x-\mu_x)](\mathbf{x}-\mu_x)\} = \Sigma_{yx}-\Sigma_{yx}\Sigma_{xx}^{-1}\Sigma_{xx} = 0$$

For the Gaussian variables, according to Eq. (A.7.27) the orthogonality principle can be written as

$$E\{(\mathbf{y} - E\{\mathbf{y} \mid \mathbf{x}\})\mathbf{x}'\} = 0 \quad \text{(A.7.36)}$$

The variable $\mathbf{e} = \mathbf{y}-E\{\mathbf{y} \mid \mathbf{x}\}$ is called the *innovation* and the orthogonality principle states that the innovation is the part of **y** which is independent of **x** (or orthogonal to **x**).

7.2.2 Linear Systems

Consider a system described by linear stochastic equations

$$x_{k+1} = \mathbf{A}_k x_k + \mathbf{B}_k w_k, \quad k=0,1,... \quad \text{(A.7.37)}$$

where $\mathbf{A}_k$ and $\mathbf{B}_k$ are matrices and $w_k \sim N(w_k, \mathbf{Q}_k)$ is a sequence of independent zero-mean Gaussian variables. The initial state $x_0 \sim N(x_0 - \mu_0, \Sigma_0)$ is also Gaussian.

With these assumptions, it is easy to see that x_k is a Gauss-Markov process. Indeed, if x_k is fixed, then x_{k+1} depends only on w_k and, therefore, is a Markov process whose transition PDF [according to Eq. (A.7.6)] is $N(x_{k+1} - \mathbf{A}_k x_k, \mathbf{B}_k \mathbf{Q}_k \mathbf{B}_k')$. On the other hand, the process is Gaussian since, according to Eq. (A.7.37), it represents a linear transformation of Gaussian variables.

For the Gauss-Markov process, the m-step transition probability operator $\mathbf{P}_{0,m} = \mathbf{P}_1 \mathbf{P}_2 \cdots \mathbf{P}_m$ can be found using the forward algorithm (A.7.3). The kernel of this operator is the m-step transition PDF $p(x_m \mid x_0)$ of transferring from state x_0 to state x_m

after m steps. Since the process is Gaussian, $p(\boldsymbol{x}_m \mid \boldsymbol{x}_0) = \mathsf{N}(\boldsymbol{x}_m - \boldsymbol{\mu}_m, \boldsymbol{\Sigma}_m)$, thus, we need to find only the mean $\boldsymbol{\mu}_m$ and variance matrix $\boldsymbol{\Sigma}_m$. Using Eq. (A.7.6) and (A.7.37), we can find them recursively:

$$\boldsymbol{\mu}_k = \mathbf{A}_{k-1}\boldsymbol{\mu}_{k-1}, \quad \boldsymbol{\mu}_0 = \boldsymbol{x}_0 \tag{A.7.38}$$

$$\boldsymbol{\Sigma}_k = \mathbf{A}_{k-1}\boldsymbol{\Sigma}_{k-1}\mathbf{A}_{k-1}^{'} + \mathbf{R}_{k-1}, \quad \text{where } \mathbf{R}_{k-1} = \mathbf{B}_{k-1}\mathbf{Q}_{k-1}\mathbf{B}_{k-1}^{'} \tag{A.7.39}$$

with $\boldsymbol{\Sigma}_0 = \mathbf{R}_0$ and $k = 1,2,\ldots,m$. These equations are called the algebraic Riccati equations. In other words, we found a closed form expression for the Chapman-Kolmogorov equation $p(\boldsymbol{x}_2 \mid \boldsymbol{x}_0) = \int_X p(\boldsymbol{x}_2 \mid \boldsymbol{x}_1) p(\boldsymbol{x}_1 \mid \boldsymbol{x}_0) d\boldsymbol{x}_1$:

$$\mathsf{N}(\boldsymbol{x}_2 - \mathbf{A}_1\mathbf{A}_0\boldsymbol{x}_0, \mathbf{R}_1 + \mathbf{A}_1\mathbf{R}_0\mathbf{A}_1^{'}) = \int_X \mathsf{N}(\boldsymbol{x}_2 - \mathbf{A}_1\boldsymbol{x}_1, \mathbf{R}_1)\mathsf{N}(\boldsymbol{x}_1 - \mathbf{A}_0\boldsymbol{x}_0, \mathbf{R}_0)d\boldsymbol{x}_1$$

which represents a special case of Eq. (A.7.22). This formula represents the kernel of the product of operators $\mathbf{P}_1\mathbf{P}_2$. Thus, for the Gauss-Markov processes, the multiplication of transition operators is performed using the Riccati equations.

It follows from the previous equation that we can evaluate the product of operators using the backward algorithm

$$\mathbf{D}_{m-1} = \mathbf{A}_{m-1}, \quad \boldsymbol{\Xi}_{m-1} = \mathbf{R}_{m-1}$$

$$\mathbf{D}_k = \mathbf{D}_{k+1}\mathbf{A}_k, \quad \boldsymbol{\Xi}_k = \boldsymbol{\Xi}_{k+1} + \mathbf{D}_{k+1}\mathbf{R}_k\mathbf{D}_{k+1}^{'} \tag{A.7.40}$$

for $k = m-2, m-3, \ldots, 0$. At the end, we obtain $\boldsymbol{\mu}_m = \mathbf{D}_0\boldsymbol{x}_0$ and $\boldsymbol{\Sigma}_m = \boldsymbol{\Xi}_0$.

If the initial state is selected randomly: $\boldsymbol{x}_0 \sim \mathsf{N}(\boldsymbol{x}_0 - \boldsymbol{\mu}, \boldsymbol{\Sigma})$, then the state $\boldsymbol{x}_m$ PDF (A.7.3) can be evaluated using the same Riccati equations, only the initial conditions change to $\boldsymbol{\mu}_0 = \boldsymbol{\mu}$ and $\boldsymbol{\Sigma}_0 = \boldsymbol{\Sigma}$. Note that if the initial condition is not Gaussian, the process is not Gaussian. However, we can still find the PDF using Eq. (A.7.3) and the Riccati equations to evaluate the kernel of the transition operator.

If $\mathbf{A}_k = \mathbf{A}$, $\mathbf{B}_k = \mathbf{B}$, and $\mathbf{Q}_k = \mathbf{Q}$ are constant, the process is homogeneous. For this process, the Riccati equations take the form

$$\boldsymbol{\mu}_k = \mathbf{A}\boldsymbol{\mu}_{k-1} = \mathbf{A}^k\boldsymbol{x}_0 \tag{A.7.41}$$

$$\boldsymbol{\Sigma}_k = \mathbf{A}\boldsymbol{\Sigma}_{k-1}\mathbf{A}^{'} + \mathbf{R}, \quad \mathbf{R} = \mathbf{B}\mathbf{Q}\mathbf{B}^{'} \tag{A.7.42}$$

7.2.3 Operator $\mathbf{P}^m$

Matrix $\mathbf{P}^m$ plays an important role in the theory of Markov chains. In Appendix 6 we considered fast methods for computing this matrix and analyzed its asymptotic behavior. Operator $\mathbf{P}^m$ plays a similar role in the theory of the homogeneous Gauss-Markov processes. Expanding the Riccati equations of the previous section, we can write

$$\boldsymbol{\Sigma}_2 = \mathbf{A}\boldsymbol{\Sigma}_1\mathbf{A}^{'} + \mathbf{R} = \mathbf{A}\mathbf{R}\mathbf{A}^{'} + \mathbf{R}$$

$$\boldsymbol{\Sigma}_3 = \mathbf{A}\boldsymbol{\Sigma}_2\mathbf{A}^{'} + \mathbf{R} = \mathbf{A}^2\mathbf{R}(\mathbf{A}^{'})^2 + \mathbf{A}\mathbf{R}\mathbf{A}^{'} + \mathbf{R}$$

and so on. Thus,

$$\boldsymbol{\Sigma}_m = \sum_{k=0}^{m-1} \mathbf{A}^k \mathbf{R} (\mathbf{A}')^k \tag{A.7.43}$$

According to the Chapman-Kolmogorov equations, we can write $\mathbf{P}^m = \mathbf{P}^k \mathbf{P}^{m-k}$. Therefore, we can apply algorithm (A5.1) of Appendix 5 for the fast operator exponentiation:

Algorithm A7.1:
`Initialize:`

$$i=0,\ b_0 = REM(m,2),\quad m_0 = m,\quad \mathbf{M}_0 = \mathbf{P}^{b_0},\quad \mathbf{N}_0 = \mathbf{P}$$

`While` $m_i > 0$
`Begin`

$$i+1 \rightarrow i,\ m_i = (m_{i-1} - b_{i-1})/2,\quad b_i = REM(m_i,2),\quad \mathbf{N}_i = \mathbf{N}_{i-1}^2$$

$$\mathbf{M}_i = \begin{cases} \mathbf{M}_{i-1}\mathbf{N}_i, & \text{if } b_i = 1, \\ \mathbf{M}_{i-1}, & \text{if } b_i = 0 \end{cases}$$

`End`

At the end we obtain $\mathbf{P}^m = \mathbf{M}_i$.

Using the Riccati equations, we can rewrite this algorithm explicitly in terms of the mean and variance matrix of the operator $\mathbf{P}^m$ kernel (see M-file `Ricc`)

Algorithm A7.2:
`Initialize:`

$$i=0,\ b_0 = REM(m,2),\quad m_0 = m,\quad \mathbf{N}_0 = \mathbf{R},\quad \mathbf{M}_0 = \mathbf{A},\ \mathbf{V}_0 = b_0\mathbf{R},\quad \mathbf{Q}_0 = \mathbf{A}^{b_0}$$

`While` $m_i > 0$
`Begin`

$$i+1 \rightarrow i$$

$$\mathbf{N}_i = \mathbf{M}_{i-1}\mathbf{N}_{i-1}\mathbf{M}_{i-1}' + \mathbf{N}_{i-1},\quad \mathbf{M}_i = \mathbf{M}_{i-1}^2$$

$$m_i = (m_{i-1} - b_{i-1})/2,\quad b_i = REM(m_i,2)$$

$$\mathbf{V}_i = \mathbf{V}_{i-1},\quad \mathbf{Q}_i = \mathbf{Q}_{i-1},\quad \text{if}\quad b_i = 0$$

$$\mathbf{V}_i = \mathbf{Q}_{i-1}\mathbf{N}_i\mathbf{Q}_{i-1}' + \mathbf{V}_{i-1},\quad \mathbf{Q}_i = \mathbf{Q}_{i-1}\mathbf{M}_i,\ \text{if}\ b_i = 1$$

`End`
At the end, we obtain $\boldsymbol{\Sigma}_m = \mathbf{V}_i$ and $\mathbf{A}^m = \mathbf{Q}_i$.

7.2.4 Stationary Distribution

To find the stationary distribution $\pi(x)$ for the Gauss-Markov process, we need to find the limit of its mean μ_m and variance matrix $\boldsymbol{\Sigma}_m$ as $m \rightarrow \infty$.

Let us show that if all the eigenvalues of the matrix $\mathbf{A}$ are inside the unit circle: $|\lambda_i| < 1$, then the limits of μ_m and $\boldsymbol{\Sigma}_m$ exist. Indeed, it follows from Eq. (A.5.5) that $\mu_\infty = \lim_{m\rightarrow\infty} \mathbf{A}^m \boldsymbol{x}_0 = 0$. From Eq. (A.7.43) we obtain

$$\Sigma_\infty = \sum_{k=0}^{\infty} \mathbf{A}^k \mathbf{R}(\mathbf{A}')^k \tag{A.7.44}$$

if this series converges. To prove the convergence, we use equation (A.5.5) for computing the matrix powers:

$$\mathbf{A}^k \mathbf{R}(\mathbf{A}')^k = \sum_{j=1}^{r} \sum_{i=1}^{m_j} \sum_{s=1}^{r} \sum_{m=1}^{m_s} \Sigma_{ij} \mathbf{R} \Sigma'_{ms} \binom{k}{i-1} \lambda_j^{k-i+1} \binom{k}{m-1} \lambda_s^{k-m+1}$$

It follows from this expression that the matrix series in Eq. (A.7.44) converges. The limit can be found if we let $k \to \infty$ in Eq. (A.7.42):

$$\Sigma_\infty = \mathbf{A}\Sigma_\infty\mathbf{A}' + \mathbf{R} \tag{A.7.45}$$

This equation is called the Lyapunov equation. It has a unique solution if all the eigenvalues of $\mathbf{A}$ lie inside the unit circle. Indeed, suppose that Σ is another solution: $\Sigma = \mathbf{A}\Sigma\mathbf{A}' + \mathbf{R}$. Then, subtracting Eq. (A.7.45) from this equation, we obtain $(\Sigma - \Sigma_\infty) = \mathbf{A}(\Sigma - \Sigma_\infty)\mathbf{A}'$. Applying this recursion to its right hand side, we obtain

$$(\Sigma - \Sigma_\infty) = \mathbf{A}(\Sigma - \Sigma_\infty)\mathbf{A}' = \mathbf{A}^2(\Sigma - \Sigma_\infty)(\mathbf{A}')^2 = \ldots = \mathbf{A}^m(\Sigma - \Sigma_\infty)(\mathbf{A}')^m$$

Thus $(\Sigma - \Sigma_\infty) = \lim_{m\to\infty} \mathbf{A}^m(\Sigma - \Sigma_\infty)(\mathbf{A}')^m = 0$ hence $\Sigma = \Sigma_\infty$ which proves the uniqueness of the solution. If this solution is positive definite, then it can be used as a variance matrix. In this case, the Markov process has a stationary Gaussian distribution $\mathsf{N}(x, \Sigma_\infty)$.

To find Σ_∞ and we need solve Eq. (A.7.45) which is a linear equation. Alternatively, we can use iterative method (A.7.42) or the accelerated algorithm A.7.2. MATLAB function `dlyap` solves this equation (M-file `liap.m` solves it using Algorithm A7.2).

If the initial state PDF is $\pi(x)$ the homogeneous Gauss-Markov process is stationary which follows from the fact that Σ_∞ is the fixed point of the Riccati equations: If $\Sigma_0 = \Sigma_\infty$, then $\Sigma_1 = \mathbf{A}\Sigma_\infty\mathbf{A}' + \mathbf{R} = \Sigma_\infty$; by induction we prove that $\Sigma_m = \Sigma_\infty$. The stationary process can be viewed as a limiting case of a homogeneous process that started at $t_0 \to -\infty$.

Example A.7.2. Let $x_{k+1} = ax_k + w_k$ with $w_k \sim \mathsf{N}(w_k, \sigma)$ and $x_0 = 0$.
Here $\mathbf{A} = a$ and a is also the only eigenvalue of $\mathbf{A}$. The Riccati equation takes the form $\Sigma_m = a^2\Sigma_{m-1} + \sigma^2$ and its solution is

$$\Sigma_m = \sum_{i=0}^{m} a^{2(m-1)}\sigma^2 = \frac{\sigma^2(1-a^{2m})}{1-a^2}$$

The m-step transition PDF is $p_{t,t+m}(x \mid 0) = \mathsf{N}(x, \Sigma_m)$. We see that $\lim_{m\to\infty} \Sigma_m = \sigma^2/(1-a^2)$ only if $|a| < 1$. Thus the stationary distribution is $\pi(x) = \mathsf{N}(x, \sigma^2/(1-a^2))$. Compare this result with that of example A.7.1.

7.2.5 Autocovariance Function

The autocovariance function $\mathbf{R}(k,m) = \mathbf{E}((x_k - \mu_k)(x_m - \mu_m)')$ can be found using Eq. (A.7.37).

For $m = k$, $\mathbf{R}(m,m) = \Sigma_m$ which is evaluated by the Riccati Eq. (A.7.39). Since $x_k - \mu_k$ satisfy Eq. (A.7.37), for $k \neq m$ we have

$$\mathbf{R}(k,m) = \mathbf{A}_{k-1}\mathbf{R}(k-1,m) = \mathbf{R}(k,m-1)\mathbf{A}'_{m-1} \tag{A.7.46}$$

Thus, for $k>m$ $\mathbf{R}(k,m) = \mathbf{A}_{k-1}\mathbf{A}_{k-2}\cdots\mathbf{A}_m\mathbf{\Sigma}_m$. For $k<m$ we have $\mathbf{R}(k,m) = \mathbf{R}(m,k)'$ $= \mathbf{\Sigma}_k\mathbf{A}'_k\mathbf{A}'_{k+1}\cdots\mathbf{A}'_{m-1}$.

In summary,

$$\mathbf{R}_x(k,m) = \begin{cases} \prod_{i=k-1}^{m} \mathbf{A}_i\mathbf{\Sigma}_m & \text{for } k>m \\ \mathbf{\Sigma}_m & \text{for } k=m \\ \mathbf{\Sigma}_k \prod_{i=k}^{m-1} \mathbf{A}'_i & \text{for } k<m \end{cases} \tag{A.7.47}$$

For the stationary process, $\mu_0 = 0$ and $\mathbf{\Sigma}_k = \mathbf{\Sigma}_\infty$. In this case $\mathbf{R}(k,k+\tau) = \mathbf{R}(\tau)$ does not depend on k and Eq. (A.7.47) can be written as

$$\mathbf{R}(\tau) = \begin{cases} \mathbf{A}^{-\tau}\mathbf{\Sigma}_\infty & \text{for } \tau\le 0 \\ \mathbf{\Sigma}_\infty(\mathbf{A}')^{\tau} & \text{for } \tau\ge 0 \end{cases} \tag{A.7.48}$$

The power spectrum of a discrete-time stationary process is defined as

$$\Phi(z) = \sum_{\tau=-\infty}^{\infty} \mathbf{R}(\tau)z^{\tau}$$

According to Eq. (A.7.48), for the stationary Gauss-Markov process, we have

$$\Phi(z) = \sum_{-\infty}^{0} \mathbf{A}^{-\tau}\mathbf{\Sigma}_\infty z^{\tau} + \sum_{\tau=1}^{\infty} \mathbf{\Sigma}_\infty(\mathbf{A}')^{\tau}z^{\tau}$$

Using Eq. (A.5.3) for summation of the matrix geometric series, we obtain

$$\Phi(z) = (\mathbf{I}-\mathbf{A}z^{-1})^{-1}\mathbf{\Sigma}_\infty + \mathbf{\Sigma}_\infty(\mathbf{A}'z)(\mathbf{I}-\mathbf{A}'z)^{-1}$$

By factoring out the inverse matrices, we obtain

$$\Phi(z) = (\mathbf{I}-\mathbf{A}z^{-1})^{-1}[\mathbf{\Sigma}_\infty - \mathbf{A}\mathbf{\Sigma}_\infty(\mathbf{A}'z)](\mathbf{I}-\mathbf{A}'z)^{-1}$$

which, according to Eq. (A.7.45), takes the form

$$\Phi(z) = (\mathbf{I}-\mathbf{A}z^{-1})^{-1}\mathbf{R}(\mathbf{I}-\mathbf{A}'z)^{-1} \tag{A.7.49}$$

Thus, we can express $\Phi(z)$ in terms of the system parameters without computing the stationary variance matrix $\mathbf{\Sigma}_\infty$. Using this fact, we can find $\mathbf{\Sigma}_\infty$ as a coefficient of z^0 in the expansion of $\Phi(z)$:

$$\mathbf{\Sigma}_\infty = \frac{1}{2\pi j}\int_{|z|=1} \frac{\Phi(z)}{z}\,dz = \sum Res_{\lambda_i}\frac{\Phi(z)}{z} \tag{A.7.50}$$

where the sum of residues is taken over all the eigenvalues of **A**.

Example A.7.3. For the system in example A.7.2, we have

$$\Phi(z) = (1-az^{-1})^{-1}\sigma^2(1-az)^{-1} = \frac{\sigma^2 z}{(z-a)(1-az)}$$

$$\Sigma_\infty = Res_a \frac{\sigma^2}{(z-a)(1-za)} = \frac{\sigma^2}{1-a^2}$$

Since that the power (Laurent's) series expansion of $\Phi(z)$ in the ring $a<|z|<1/a$ has the form

$$\Phi(z) = \Sigma_\infty \sum_{\tau=-\infty}^{\infty} a^{|\tau|} z^\tau \quad \text{so that} \quad r(\tau) = \Sigma_\infty a^{|\tau|}$$

7.2.6 Autoregressive Processes

Scalar autoregressive (AR) process is defined by the following finite difference equation

$$y_k + \sum_{i=1}^{s} a_i y_{k-i} = w_k \tag{A.7.51}$$

where $\boldsymbol{w}_k \sim \mathrm{N}(\boldsymbol{w}_k, \sigma^2)$ are i.i.d. random variables. It follows from this equation that y_k is s-th order Markov chain since y_k depends on its s previous s values. Obviously, $\boldsymbol{x}_k = [y_k, y_{k-1}, \ldots, y_{k-s+1}]'$ is a simple (first-order) Markov chain (see Sec. 1.1.7 for a similar argument) and is described by the following equation:

$$\boldsymbol{x}_k = \mathbf{A}\boldsymbol{x}_{k-1} + \mathbf{B}w_k \tag{A.7.52}$$

where $\mathbf{B} = [1 \;\; 0 \;\; \cdots \;\; 0]'$ and

$$\mathbf{A} = \begin{bmatrix} -a_1 & -a_2 & \cdots & -a_{s-1} & -a_s \\ 1 & 0 & \cdots & 0 & 0 \\ 0 & 1 & \cdots & 0 & 0 \\ \cdots & \cdots & \cdots & \cdots & \cdots \\ 0 & 0 & \cdots & 1 & 0 \end{bmatrix}$$

We denote the initial state as $\boldsymbol{x}_0 = [y_0, y_{-1}, \ldots, y_{-s+1}]'$ to comply with the state x_k definition. Equation (A.7.52) is a special case of Eq. (A.7.37). Therefore, we can apply previously developed theory to AR processes.

The stationary distribution exists if magnitudes of all the eigenvalues of $\mathbf{A}$ are less than 1. These eigenvalues are the roots of the characteristic polynomial

$$\det(\mathbf{A} - \lambda\mathbf{I}) = \lambda^s + \sum_{i=1}^{s} a_i \lambda^{s-i} = 0 \tag{A.7.53}$$

The stationary variance matrix

$$\boldsymbol{\Sigma}_\infty = \mathbf{E}(\boldsymbol{x}_k \boldsymbol{x}_k') = \begin{bmatrix} r(0) & r(1) & \cdots & r(s-1) \\ r(1) & r(0) & \cdots & r(s-2) \\ \cdots & \cdots & \cdots & \cdots \\ r(s-1) & r(s-2) & \cdots & r(0) \end{bmatrix}$$

where $r(\tau) = \mathbf{E}\{y_t y_{t+\tau}\}$ satisfies the Lyapunov Eq. (A.7.45) if all the roots of the characteristic polynomial lie inside the unit cirle. It is not difficult to show that this Lyapunov equation can be simplified to

$$\begin{bmatrix} r(0) & r(1) & \cdots & r(s) \\ r(1) & r(0) & \cdots & r(s-1) \\ \cdots & \cdots & \cdots & \cdots \\ r(s) & r(s-1) & \cdots & r(0) \end{bmatrix} \begin{bmatrix} 1 \\ a_1 \\ \cdots \\ a_s \end{bmatrix} = \begin{bmatrix} \sigma^2 \\ 0 \\ \cdots \\ 0 \end{bmatrix} \tag{A.7.54}$$

which is called the Yule-Walker equation. Usually this equation is applied to the system identification: finding the unknown coefficients of the system given the autocorrelation function (see Example A.7.5). As we can see, this equation can be applied to finding $r(0), r(1), \ldots, r(s-1)$ if we take into account that, according to Eq. (A.7.46), $r(\tau)$ satisfies Eq. (A.7.51) with $w_k = 0$. The process is stationary, if its initial state $x_0 = [y_0, y_{-1}, \ldots, y_{-s+1}]' \sim \mathrm{N}(x_0, \Sigma_\infty)$.

Example A.7.4. Let us find the the correlation matrix for the system of the previous example using the Yule-Walker equation. This equation takes the form

$$\begin{bmatrix} r(0) & r(1) \\ r(1) & r(0) \end{bmatrix} \begin{bmatrix} 1 \\ -a \end{bmatrix} = \begin{bmatrix} \sigma^2 \\ 0 \end{bmatrix}$$

and its solution is $\Sigma_\infty = r(0) = \sigma^2/(1-a^2)$ which is the same as in Example A.7.3.

7.2.7 Parameter Estimation

Let $f(\boldsymbol{x}_1^T)$ be a PDF of some stochastic process which we would like to approximate by a Markov process by minimizing the KLD between their PDFs which is equivalent to maximizing the likelihood function (see Sec. 3.1.3):

$$L(\tau, \boldsymbol{x}_0) = \int_{\boldsymbol{x}_1^T} f(\boldsymbol{x}_1^t) \log p(\boldsymbol{x}_1^T; \tau)\, d\boldsymbol{x}_1^T$$

where τ is an unknown parameter and

$$p(\boldsymbol{x}_1^T; \tau) = \prod_{k=1}^{T} p(\boldsymbol{x}_k \mid \boldsymbol{x}_{k-1}; \tau)$$

After the substitution, we obtain

$$L(\tau, \boldsymbol{x}_0) = \sum_{k=1}^{T} \int_{\boldsymbol{x}_1^T} f(\boldsymbol{x}_1^t) \log p(\boldsymbol{x}_k | \boldsymbol{x}_{k-1}; \tau)\, d\boldsymbol{x}_1^T \tag{A.7.55}$$

The maximum likelihood approximation is obtained by maximizing $L(\tau, \boldsymbol{x}_0)$ with respect to τ and $\boldsymbol{x}_0$. If $\boldsymbol{x}_0$ is random, the likelihood function takes the form

$$L(\tau) = \sum_{k=0}^{T} \int_{\boldsymbol{x}_0^T} f(\boldsymbol{x}_1^t) \log p(\boldsymbol{x}_k | \boldsymbol{x}_{k-1}; \tau)\, d\boldsymbol{x}_0^T \tag{A.7.56}$$

where we denoted as $p(\boldsymbol{x}_0 | \boldsymbol{x}_{-1}; \tau) = p_0(\boldsymbol{x}_0; \tau)$ the initial state PDF.

Maximum likelihood parameter estimation on the basis of experimental data $\boldsymbol{x}_0^T$ is achieved by maximizing the log-likelihood function which in our case takes the form

$$L(\tau) = \sum_{k=0}^{T} \log p(\boldsymbol{x}_k | \boldsymbol{x}_{k-1}; \tau) \tag{A.7.57}$$

which can be treated as a special case of the maximum likelihood approximation.

For the homogeneous Gauss-Markov process, Eq. (A.7.56) takes the form

$$L(\mathbf{A},\mathbf{D},\boldsymbol{x}_0) = \int_{\boldsymbol{x}_1^T} f(\boldsymbol{x}_1^T)[-0.5\sum_{k=1}^{T}(\boldsymbol{x}_k-\mathbf{A}\boldsymbol{x}_{k-1})'\mathbf{D}(\boldsymbol{x}_k-\mathbf{A}\boldsymbol{x}_{k-1})+0.5T\log|\mathbf{D}|]\,d\boldsymbol{x}_1^T + const$$

where we denoted as $\mathbf{D} = \mathbf{R}^{-1}$ and $\mathbf{R}$ defined by Eq. (A.7.42). The constant in the previous equation does depend on the model parameters and, therefore, can be neglected. Using the identity

$$(\boldsymbol{x}_k-\mathbf{A}\boldsymbol{x}_{k-1})'\mathbf{D}(\boldsymbol{x}_k-\mathbf{A}\boldsymbol{x}_{k-1}) = tr[\mathbf{D}(\boldsymbol{x}_k-\mathbf{A}\boldsymbol{x}_{k-1})(\boldsymbol{x}_k-\mathbf{A}\boldsymbol{x}_{k-1})']$$

we can write

$$L(\mathbf{A},\mathbf{D},\boldsymbol{x}_0) = = -0.5tr[\mathbf{D}(\mathbf{C}_1-\mathbf{A}\mathbf{C}_2'-\mathbf{C}_2\mathbf{A}'+\mathbf{A}\mathbf{C}_3\mathbf{A}')]+0.5T\log|\mathbf{D}| \quad \text{(A.7.58)}$$

where

$$\mathbf{C}_1 = \sum_{k=1}^{T}\mathbf{E}\{\boldsymbol{x}_k\boldsymbol{x}_k' \mid f(\boldsymbol{x}_1^T)\},\quad \mathbf{C}_2 = \sum_{k=1}^{T}\mathbf{E}\{\boldsymbol{x}_k\boldsymbol{x}_{k-1}' \mid f(\boldsymbol{x}_1^T)\}$$

$$\mathbf{C}_3 = \sum_{k=1}^{T}\mathbf{E}\{\boldsymbol{x}_{k-1}\boldsymbol{x}_{k-1}' \mid f(\boldsymbol{x}_1^T)\}\}$$

where $\mathbf{E}\{\cdot \mid f(\boldsymbol{x}_1^T)\}$ denotes the expectation with respect to $f(\boldsymbol{x}_1^T)$.

Let us assume that $\boldsymbol{x}_0$ is fixed. To find the maximum likelihood approximation we need to take derivatives with respect to $\mathbf{A}$ and $\mathbf{D}$ of $L(\mathbf{A},\mathbf{D},\boldsymbol{x}_0)$ and equate them to zero. Using Eq. (A.5.49) and (A.5.51), we obtain $\mathbf{D}[-\mathbf{C}_2+\mathbf{A}\mathbf{C}_3] = 0$ whose solution is

$$\mathbf{A} = \mathbf{C}_2\mathbf{C}_3^{-1} \quad \text{(A.7.59)}$$

provided that the inverse matrix exists.

Differentiating with respect to $\mathbf{D}$ is not that straightforward because it is symmetric and, therefore, its elements are not independent while we defined derivatives with respect to matrices whose elements are independent. In order to circumvent this problem, we use the theory of conditional extrema: The conditional maximum of $L(\mathbf{A},\mathbf{D},\boldsymbol{x}_0)$ can be found as an unconditional maximum of the auxiliary function

$$L(\mathbf{A},\mathbf{D},\boldsymbol{x}_0)+\sum_{i=1}^{n}\sum_{j=i+1}^{n}\lambda_{ij}(d_{ij}-d_{ji})$$

where λ_{ij} are the auxiliary variables that are called the Lagrange multipliers. Now we can use Eq. (A.5.49) and (A.5.50) of Appendix 5 to find the derivative of the auxiliary function with respect to $\mathbf{D}$:

$$0.5T(\mathbf{D}^{-1})' - 0.5\mathbf{G}' + \boldsymbol{\Lambda} = 0$$

where

$$\mathbf{G} = \mathbf{C}_1-\mathbf{A}\mathbf{C}_2'-\mathbf{C}_2\mathbf{A}'+\mathbf{A}\mathbf{C}_3\mathbf{A}' = \mathbf{C}_1-\mathbf{C}_2\mathbf{C}_3^{-1}\mathbf{C}_2' \quad \text{(A.7.60)}$$

and

$$\mathbf{\Lambda} = \begin{bmatrix} 0 & \lambda_{12} & \cdots & \lambda_{1n} \\ -\lambda_{12} & 0 & \cdots & \lambda_{2n} \\ \cdots & \cdots & \cdots & \cdots \\ -\lambda_{1n} & -\lambda_{2n} & \cdots & 0 \end{bmatrix}$$

The derivatives with respect to λ_{ij} give us $d_{ij} - d_{ji} = 0$ which means that we are looking for a symmetric matrix $\mathbf{D}$.

It is clear from Eq. (A.7.60) that $\mathbf{G}$ is symmetric. Therefore, $\mathbf{D}$ can be symmetric only if $\mathbf{\Lambda} = 0$. Since $\mathbf{D} = \mathbf{R}^{-1}$, we have

$$\mathbf{R} = \frac{1}{T}(\mathbf{C}_1 - \mathbf{C}_2\mathbf{C}_3^{-1}\mathbf{C}_2') = \frac{1}{T}(\mathbf{C}_1 - \mathbf{A}\mathbf{C}_2') \tag{A.7.61}$$

In the previous derivations we assumed that $\boldsymbol{x}_0$ is fixed. If it is not random, we can maximize the likelihood function with respect to it quite similarly to maximizing with respect to $\mathbf{A}$ and we leave it as an exercise to the reader. If $\boldsymbol{x}_0 \sim \mathrm{N}(\boldsymbol{x}_0 - \boldsymbol{\mu}_0, \boldsymbol{\Sigma}_0)$, than we add

$$-0.5(\boldsymbol{x}_0 - \boldsymbol{\mu}_0)'\boldsymbol{\Sigma}_0^{-1}(\boldsymbol{x}_0 - \boldsymbol{\mu}_0) - 0.5\log|\boldsymbol{\Sigma}_0| + const$$

to $L(\mathbf{A}, \mathbf{R}, \boldsymbol{x}_0)$ and replace $f(\boldsymbol{x}_1^T)$ with $f(\boldsymbol{x}_0^T)$ in the previous equations. The maximization of the likelihood function with respect to $\boldsymbol{\mu}_0$ and $\boldsymbol{\Sigma}_0$ gives us

$$\boldsymbol{\mu}_0 = \mathbf{E}\{\boldsymbol{x}_0 \mid f(\boldsymbol{x}_1^T)\}, \quad \boldsymbol{\Sigma}_0 = \mathbf{E}\{(\boldsymbol{x}_0 - \boldsymbol{\mu}_0)(\boldsymbol{x}_0 - \boldsymbol{\mu}_0)' \mid f(\boldsymbol{x}_1^T)\}$$

In particular, if $\boldsymbol{\mu}_0 = 0$, $\mathbf{E}\{\boldsymbol{x}_k\boldsymbol{x}_k' \mid f(\boldsymbol{x}_1^T\} = \boldsymbol{\Sigma}$, and $\mathbf{E}\{\boldsymbol{x}_k\boldsymbol{x}_{k-1}' \mid f(\boldsymbol{x}_1^T\} = \mathbf{R}(-1) = \mathbf{R}(\tau)|_{\tau=-1}$ are constant, equations (A.7.59) and (A.7.61) take the form

$$\mathbf{R}(-1) = \mathbf{A}\boldsymbol{\Sigma} \tag{A.7.62}$$

$$\mathbf{R} = \boldsymbol{\Sigma} - \mathbf{A}\boldsymbol{\Sigma}\mathbf{A}' \tag{A.7.63}$$

Equation (A.7.62) coincides with (A.7.47) and Eq. (A.7.63) is the Lyapunov equation (A.7.45). However, they are used in the opposite way: $\boldsymbol{\Sigma}$ and $\mathbf{R}(-1)$ are evaluated with respect to the given PDF $f(\boldsymbol{x}_0^T)$ and $\mathbf{A}$ and $\mathbf{R}$ are found from Eq. (A.7.62) and (A.7.63). This means that the maximum likelihood approximation of a wide-sense stationary process with the Gauss-Markov process can be achieved using the method of moments.

Example A.7.5. Consider a zero-mean wide-sense stationary process whose autocorrelation function is $r(\tau)$. Suppose that we want to approximate this process with the autoregressive process (A.7.51). For this process, equation of (A.7.62) takes the form

$$\begin{bmatrix} r(1) & r(2) & \cdots & r(s) \\ r(0) & r(1) & \cdots & r(0) \\ \cdots & \cdots & \cdots & \cdots \\ r(s-2) & r(s-3) & \cdots & r(1) \end{bmatrix} = \mathbf{A} \begin{bmatrix} r(0) & r(1) & \cdots & r(s-1) \\ r(1) & r(0) & \cdots & r(s-2) \\ \cdots & \cdots & \cdots & \cdots \\ r(s-1) & r(s-2) & \cdots & r(0) \end{bmatrix}$$

where the matrix $\mathbf{A}$ is defined by Eq. (A.7.54). After simplifications this equation can be rewritten as

$$\begin{bmatrix} r(1) & r(0) & \cdots & r(s-1) \\ r(2) & r(1) & \cdots & r(s-2) \\ \cdots & \cdots & \cdots & \cdots \\ r(s) & r(s-1) & \cdots & r(0) \end{bmatrix} \begin{bmatrix} 1 \\ a_1 \\ \cdots \\ a_s \end{bmatrix} = 0 \tag{A.7.64}$$

The variance matrix $\mathbf{R} = \mathbf{B}\sigma^2\mathbf{B}'$, according to Eq. (A.7.52), has only one nonzero element which can be found using Eq. (A.7.63) as

$$\sigma^2 = r(0)+r(1)a_1+...+r(s)a_s$$

Combining this equation with (A.7.64) we conclude that we can obtain the unknown coefficients and σ^2 from the Yule-Walker Eq. (A.7.54)

Consider now the problem of estimating parameters **A** and **R** of the homogeneous Gauss-Markov process on the basis of the observation sequence x_0^T. The log-likelihood function can be written as

$$L(\mathbf{A},\mathbf{R}) = -0.5\sum_{k=1}^{T}(x_k-\mathbf{A}x_{k-1})'\mathbf{R}^{-1}(x_k-\mathbf{A}x_{k-1})-0.5T\log|\mathbf{R}|+const$$

Comparing this equation with (A.7.58), we can immediately obtain the maximum likelihood estimates from Eq. (A.7.59) and (A.7.61) as

$$\hat{\mathbf{A}} = \mathbf{C}_2\mathbf{C}_3^{-1} \tag{A.7.65}$$

$$\hat{\mathbf{R}} = \frac{1}{T}(\mathbf{C}_1-\mathbf{C}_2\mathbf{C}_3^{-1}\mathbf{C}_2') = \frac{1}{T}(\mathbf{C}_1-\mathbf{A}\mathbf{C}_2') \tag{A.7.66}$$

with

$$\mathbf{C}_1 = \sum_{k=1}^{T}x_kx_k',\ \mathbf{C}_2 = \sum_{k=1}^{T}x_kx_{k-1}',\ \mathbf{C}_3 = \sum_{k=1}^{T}x_{k-1}x_{k-1}' \tag{A.7.67}$$

Equation (A.7.66) can be also written as

$$\hat{\mathbf{R}} = \frac{1}{T}\sum_{k=1}^{T}(x_k-\hat{\mathbf{A}}x_{k-1})(x_k-\hat{\mathbf{A}}x_{k-1})' \tag{A.7.68}$$

Example A.7.6. If we apply Eq. (A.7.65) and (A.7.66) to estimate parameters of the autoregressive process, we obtain, similarly to Example A.7.5, the Yule-Walker Eq. (A.7.54) in which the autocorrelation function is replaced, according to Eq. (A.7.67) by its estimate

$$\hat{r}(\tau) = \frac{1}{T}\sum_{k=1}^{T}y_ky_{k-\tau}$$

where y_t with $t \le 0$ represents the initial condition.

References

1. W. Feller, *An Introduction to Probability Theory and Its Applications,* (Wiley, New York, 1966).

INDEX